# Nuclease Methods and Protocols

# METHODS IN MOLECULAR BIOLOGY™

## *John M. Walker,* Series Editor

METHODS IN MOLECULAR BIOLOGY™

# Nuclease Methods and Protocols

Edited by

## Catherine H. Schein

*University of Texas Medical Branch, Galveston, TX*

**Humana Press** ☀ Totowa, New Jersey

Library of Congress Cataloging in Publication Data

Main entry under title:

Methods in molecular biology™.

Nuclease methods and protocols/edited by Catherine H. Schein.
    p.    cm.—(Methods in molecular biology; v. 160)
  ISBN 0-89603-679-0 (alk. paper)
  1. Nucleases—Laboratory manuals.    I.  Schein, Catherine H.    II.  Methods in molecular biology (Totowa, NJ); v. 160.

# Preface

Nucleases, enzymes that restructure or degrade nucleic acid polymers, are vital to the control of every area of metabolism. They range from "housekeeping" enzymes with broad substrate ranges to extremely specific tools *(1)*. Many types of nucleases are used in lab protocols, and their commercial and clinical uses are expanding. The purpose of *Nuclease Methods and Protocols* is to introduce the reader to some well-characterized protein nucleases, and the methods used to determine their activity, structure, interaction with other molecules, and physiological role. Each chapter begins with a mini-review on a specific nuclease or a nuclease-related theme. Although many chapters cover several topics, they were arbitrarily divided into five parts:

Part I, "Characterizing Nuclease Activity," includes protocols and assays to determine general (processive, distributive) or specific mechanisms. Methods to assay nuclease products, identify cloned nucleases, and determine their physiological role are also included here.

Part II, "Inhibitors and Activators of Nucleases," summarizes assays for measuring the effects of other proteins and small molecules. Many of these inhibitors have clinical relevance.

Part III, "Relating Nuclease Structure and Function," provides an overview of methods to determine or model the 3-D structure of nucleases and their complexes with substrates and inhibitors. A 3-D structure can greatly aid the rational design of nucleases and inhibitors for specific purposes.

Part IV, "Nucleases in the Clinic," summarizes assays and protocols suitable for use with tissues and for nuclease based therapeutics.

Part V, "Nucleases in the Lab," includes protocols for the use of nucleases in cloning and in determining the activity of other proteins.

The experienced reader will immediately recognize several of the nucleases used as examples throughout this book, especially RNase A and restriction enzymes. However, new nucleases with novel specificity, often performing unexpected functions, are constantly being discovered. For example, a regulator of the unfolded protein response, identified initially as a kinase, is also a sequence-specific ribonuclease (Chapter 3). A human analog of a plant ribonuclease was discovered in the search for a tumor suppressor protein (Chapter 7), whereas angiogenin (Chapter 25) was cloned as a factor stimulating blood vessel formation. RNase L is one of the mediators of Interferon activity (Chapter 12).

Researchers who unmask a nuclease disguised as a cytokine, mating factor, toxin, and so forth should find the methods for characterizing their protein described in the first section of this book particularly useful. These chapters suggest questions to ask about the nuclease's activity or primary structure. Is the amino acid sequence novel or similar to one of the major families of nucleases (Chapters 7 and 18)? Is the cleavage processive or distributive, i.e., does the nuclease scan the nucleic acid polymer and cleave repeatedly before separating, as has been shown for restriction endonu-cleases and glycosylases (Chapter 1) or, does it, in the fashion of RNase A (Chapter 2), cleave and simultaneously release the polymer, generating products that are at the same time novel substrates? Does the back (synthesis) reaction affect the kinetics of the cleavage process? Is the nuclease activity essential for metabolic activity, as McClure and coworkers (Chapter 5) have elegantly demonstrated for the stylar RNases? Finally, is the activity sensitive to known inhibitors or activators (Chapters 8, 9, and 12)?

These questions all pave the way for characterization of the 3-D structures of nucleases and their complexes with substrates and activity modulators. The chapters at the center of this book were selected to introduce the reader to methods that can be used to define the tertiary structure of nucleases. Of course, a complete tertiary struc-ture determination by X-ray crystallography (Chapters 13, 14, 17, 19, and 20) or NMR (Chapter 16) requires a good deal of time and specialized techniques too complicated to be summarized here. However, if the sequence has significant identity to a protein for which a structure has been determined, tools now available on the Internet allow one to model the probable 3-D structure (Chapter 18). The methods described aid in the design of nucleases with new properties (Chapters 15 and 20) and improved inhibitors (Chapters 13 and 14).

Nuclease-based therapies and diagnostics are slowly coming into the clinic. DNase I therapy (Chapters 20 and 21) has improved the lives of thousands of cystic fibrosis victims. Nucleases with demonstrated antitumor activity (Chapter 24) stimu-lated clinical trials of other members of the RNase A family and modified forms (Chapters 23 and 26). Better understanding of nucleases that repair damaged DNA (Chapters 1 and 18), mediate retroviral integration and replication (Chapters 10 and 22), or play a role in cytokine and growth factor mechanisms (Chapters 9, 12, and 25) is important both in understanding disease progression and developing better thera-peutic modalities. Antisense therapies, for example, depend on directing the activity of intracellular RNase H (Chapter 11).

As more medical professionals become aware of the importance of nucleases in metabolism and the improved assays for their activity, there will almost certainly be an increase in their use in diagnostics. A recent report, for example *(2),* correlated increases in the levels of eosinophil cationic protein in asthmatics allergic to grass pollen with the risk of onset of prolonged asthmatic symptoms.

Finally, nucleases are a major research tool in molecular biology. The exquis-ite specificity of restriction endonucleases (Chapters 19, 27, and 28) is routinely used in gene cloning. Exploiting the special qualities of a subclass, hapaxoterminers, can make subcloning and gene modification easier (Chapter 29). There are many uses for nonspecific nucleases as well. DNase I can be used to locate the binding sites of proteins on DNA (Chapter 30) and S1-nuclease (Chapter 31) or ribonu-clease (Chapter 32) mapping to quantitate specific mRNAs. Degradation of nucleic acid polymers with nonspecific nucleases, including DNase I, RNase A, and the endonuclease from *Serratia* (Chapter 17), can be used to clarify lysates and ease protein purification.

All the chapters describe why and when to use the assays, and the thinking that went into the development of the protocols. These comments, and the Notes on the method, provide guidance and insight when things go wrong (i.e., not as planned), and for how to go about correcting them. Protocols change in their details constantly. A Northern blot to check the expression in a tissue from various organs of mRNA for a specific protein, which 10 years ago required weeks to prepare, can be done today in a few hours starting with a commercially prepared membrane. The reader is welcome to simplify these protocols further as new developments allow.

We can anticipate that the model proteins used to develop new biophysical methods and clinical therapies, which have changed little in the past 30 years, will show more variety in the future. Most scientists will claim that they use RNase A since the protein is small, soluble, and easy to assay and to refold from a completely denatured state (Chapter 15). However, ready availability and force of habit contribute to the attention paid this "ideal" protein. RNase A was first crystallized *(3,4)*, Bill Wyckoff has suggested only partially in jest, because a meat packing company made it available free of charge in a highly purified form. The easy purification of pancreatic RNase from the cadavers of zoo animals made this family a Rosetta stone for comparative biology as well. However, many new nucleases are commercially available, and the genome projects are revealing copious sequence information about nuclease families that may be more important metabolically. The examples in *Nuclease Methods and Protocols*, as varied as they are, are only starting points for exploration in the wide world of nucleases.

*Catherine H. Schein*

## References

1. Schein, C. H. (1997) From housekeeper to microsurgeon: the diagnostic and therapeutic potential of ribonucleases. *Nature Biotechnol.* **15,** 529–536.
2. Di Gioacchino, M., Cavallucci, E., Di Stefano, F., Verna, N., Ramondo, S., Ciuffreda, S., Riccioni, G., and Boscolo, P. (2000) Influence of total IgE and seasonal increase of eosinophil cationic protein on bronchial hyperreactivity in asthmatic grass-sensitized farmers. *Allergy* **55,** 1030–1034.
3. Wyckoff, H. W., Hardman, K. D., Allewell, N. M., Inagami, T., Johnson, L. N., and Richards, F. M. (1967) The structure of ribonuclease-S at 3.5 A resolution. *J. Biol. Chem.* **242(17),** 3984–3988.
4. Kim, E. E., Varadarajan, R., Wyckoff, H. W., and Richards, F. M. (1992) Refinement of the crystal structure of ribonuclease S. Comparison with and between the various ribonuclease A structures. *Biochemistry* **31(49),** 12,304–12,314.

# List of Abbreviations

AMBER, assisted model building with energy refinement

AMV, avian myeloblastosis virus

Ang, angiogenin

APS, ammonium persulfate

ATP, adenosine triphosphate

β-ME, 2-mercaptoethanol

BAP, bacterial alkaline phosphatase

bp, base pairs

BPB, bromophenol blue

BSA, bovine serum albumin

CHAPS, 3-((3-cholamidopropyl) dimethylammonio)-1-propane-sulfonate

CHARMM, Chemistry at Harvard Macro-molecular Mechanics

CIP, calf intestinal alkaline phosphatase

Cot, value representing the speed of reasso-ciation of complementary nucleic acids; the product of Co (the starting DNA concentration) and t (time, expressed in seconds)

C>p, cytidine 2',3'-cyclic phosphate

(Cp) C>p, an oligocytidylic acid of $n + 1$ residues ending in a 2',3'-cyclic phos-phate

CPD, citrate-phosphate-dextrose

CPK, Corey, Pauling, and Kultun

cpm, counts per minute

DDS, dimethyldichlorsilane

DEAE, diethylaminoethyl

DEPC, diethyl pyrocarbonate

DMF, dimethyl formamide

DNA, deoxyribonuclease acid

DQF-COSY, double-quantum-filtered corre-lation 2D spectroscopy

DS, dextran sulfate

dsRBD, double-stranded RNA binding domain

dsRNA, double-stranded RNA

DTNB, 5,5'-dithiobis(2-nitrobenzoic acid)

DTT, dithiothreitol

e, fundamental charge of an electron

ECP, eosinophil cationic protein

EDN, eosinophil-derived neurotoxin

EDTA, ethylenediamine tetra acetic acid

EGTA, ethyleneglycol-bis-(2-aminoethyl ether)-$N,N,N',N'$-tetraacetic acid

ER, endoplasmic reticulum

ESMS, electrospray mass spectrometry

FABMS, fast atom bombardment mass spectrometry

FCS, fetal calf serum

FDPB, finite difference Poisson-Boltzmann

F-Met, formyl-methionine

GRASP, graphical representation and analy-sis of protein structures

GRASS, graphical representation and analy-sis of structures server

Grx, *E. coli* glutaredoxin

GSH, reduced glutathione

GS-RNase A, ribonuclease A with the eight cysteine residues involved in mixed disulfides with glutathione

GSSG, oxidized glutathione

GST, glutathione-*S*-transferase

HBS, HEPES-buffered saline

HEPES, $N$-(2-hydroxyethyl)piperazine-$N'$-(2-ethane sulfonic acid)

HIV-1, human immunodeficiency virus type 1

HPLC, high performance liquid chromatog-raphy

HSC, hepatic stellate cell

HTML, hypertext markup language

IAM, iodoacetic acid

IFN, interferon

IFNAR, cell surface receptors

IN, integrase

IPTG, isopropyl thiogalactose

Ire1p(k+t), Ire1p kinase and nuclease domain recombinant protein

JAK, Janus kinases

k, Boltzmann's constant

LOH, loss of heterozygosity

LPA, linear polyacrylamide
LRR, leucine-rich repeats
LTR, long terminal repeat
MAb, monoclonal antibody
MALDIMS, matrix assisted laser desorption ionization mass spectrometry
MALDI-TOF, matrix assisted laser desorption ionization-time of flight
MDL, Molecular Design Limited
MMLV, Moloney murine leukemia virus
MOI, multiplicity of infection
MOPS, 3-($N$-morpholino)propanesulfonic acid
$Na_2MoO_4$, sodium molybdate
NaTCA, sodium trichloracetate
NDB, nuclear dialysis buffer
$Ni^{2+}$-NTA, nickel-nitrilotriacetic acid
NLB, nuclear lysis buffer
NMR, nuclear magnetic resonance
NOE, nuclear Overhauser effect
NOESY, nuclear Overhauser effect 2D spectroscopy
NP40, Nonidet P-40
nt, nucleotide
ON, oligodeoxynucleotide
OPLS, optimized potentials for liquid simulations
PAC, P1-based artificial chromosome
PAGE, polyacrylamide gel electrophoresis
PARSE, parameters for solvation energy
PB, Poisson-Boltzmann
PBS, phosphate-buffered saline
PC, methylphosphonodiester
PCR, polymerase chain reaction
PCV, packed cell volume
PDB, protein data bank
PDI, protein disulfide isomerase
PEG, polyethylene glycol
PFU, plaque-forming unit
PI, propidium iodide
PIC, preintegration complex
PKR, protein kinase
PMSF, phenylmethyl sulfonic fluoride
PNK, polynucleotide kinase
PNPP, paranitrophenylphosphate
PO, phosphodiester
Poly(A), polyadenylic acid
Poly(C), polycytidylic acid
Poly(U), polyuridylic acid
PPT, polypurine tract
PS, phosphorodiester

$(Pur)_nPyr>p$, a purine core oligonucleotide ending in a pyrimidine 2',3'-cyclic phosphate nucleotide
py/TEAA, pyridine/triethylammonium acetate
RAM, random access memory
rd-RNase A, reduced and denatured ribonuclease A
rd-RNase T1, reduced and denatured ribonuclease T1
RI, ribonuclease inhibitor
RLI, RNase L inhibitor
RL-PCR, reverse ligation mediated, reverse transcription dependent polymerase chain reaction
RM, restriction modification
rmsd, root mean square deviation
RNA, ribonuclease acid
RNase, ribonuclease
RT, reverse transcriptase
SAX, strong anion exchange
SDS, sodium dodecyl sulfate
SLO, streptolysin O
SRB, storage reconstitution buffer
SSC, saline sodium citrate
ssRNA, single-stranded RNA
STAT, signal-transducing proteins
T4-PNK, T4 polynucleotide kinase
T7-RNAP, T7 RNA polymerase
TBE, Tris-borate-EDTA
TEAA, triethylammonium acetate
TEMED, $N,N,N',N'$-tetramethylethylenediamine
TEN, Tris-EDTA-sodium chloride
TFA, trifluoroacetic acid
TLC, thin layer chromatography
TOCSY, total correlation 2D spectroscopy
Tris, Tris(hydroxymethyl)aminomethane
TSOPB, tris-sucrose-orthophenanthroline-benzamidine
tsp, transcription start point
$(Up)_nU>p$, an oligouridylic acid of $n + 1$ residues ending in a 2',3'-cyclic phosphate
USB, urea sample buffer
UV, ultraviolet
VRML, virtual reality modeling language
VvE3, Vaccinia virus E3
XC, xylene cyanol
XRBP, *Xenopus laevis* RNA binding protein
YAC, yeast artificial chromosome

# Contents

# Contents

# Contributors

FRANCESCO ACQUATI • *Unité di Genetica Umana, Dipartimento di Biologia Strutturale e Funzionale, Université degli Studi dell'Insubria, Varese, Italy*

BRIAN BEECHER • *Department of Biochemistry, University of Missouri, Columbia, MO*

SHELBY L. BERGER • *Section on Genes and Gene Products, National Cancer Institute, National Institutes of Health, Bethesda, MD*

MARCO G. BIANCHI • *Unité di Genetica Umana, Dipartimento di Biologia Strutturale e Funzionale, Université degli Studi dell'Insubria, Varese, Italy*

CATHERINE BISBAL • *Institut Genetique Moleculaire, Montpellier, France*

WERNER BRAUN • *Human Biological Chemistry and Genetics, Sealy Center for Structural Biology, University of Texas Medical Branch, Galveston, TX*

DAVID A. BRENNER • *Department of Medicine, University of North Carolina, Chapel Hill, NC*

PETER CHEREPANOV • *Rega Institute for Medical Research, Katholieke Universiteit, Leuven, Belgium*

CLAUDI M. CUCHILLO • *Departament de Bioquimica et Biologia Molecular, Universitat Autonoma de Barcelona, Bellaterra, Spain*

ZEGER DEBYSER • *Rega Institute for Medical Research, Katholieke Universiteit, Leuven, Belgium*

ERIK DE CLERCQ • *Rega Institute for Medical Research, Katholieke Universiteit, Leuven, Belgium*

JOSEPH B. DOMACHOWSKE • *Department of Pediatrics, State University of New York Health Science Center, Syracuse, NY*

PETER FRIEDHOFF • *Fachbereich Biologie, Institut für Biochemie, Justus-Liebig-Universität, Giessen, Germany*

RICHARD V. GILES • *Department of Haematology, University of Liverpool, UK*

OLEG GIMADUTDINOW • *Department of Genetics, University of Kazan, Russian Federation*

TANIA N. GONZALEZ • *Department of Biochemistry and Biophysics, Howard Hughes Medical Institute, University of California, San Francisco, CA*

TATIANA GORLETTA • *Unité di Genetica Umana, Dipartimento di Biologia Strutturale e Funzionale, Universita degli Studi dell'Insubria, Varese, Italy*

ALBERT JELTSCH • *Fachbereich Biologie, Institut für Biochemie, Justus-Liebig-Universität, Giessen, Germany*

BOSTJAN KOBE • *Structural Biology Laboratory, St. Vincent's Institute of Medical Research, Fitzroy, Victoria, Australia*

JAN KORMANEC • *Institute of Molecular Biology, Slovak Academy of Sciences, Bratislava, Slovak Republic*

KURT L. KRAUSE • *Department of Biology and Biochemistry, University of Houston, TX*

ROBERT A. LAZARUS • *Department of Protein Engineering, Genentech, Inc., South San Francisco, CA*

BERNARD LEBLEU • *Institut de Genetique Moleculaire, Montpellier, France*

STUART F. J. LE GRICE • *Reverse Transcriptase Biochemistry Section, HIV Drug Resistance Program, Division of Basic Sciences, Frederick Cancer Research and Development Center, National Cancer Institute, Frederick, MD*

FLORENCE LE ROY • *Institut Genetique Moleculaire, Montpellier, France*

R. STEPHEN LLOYD • *Sealy Center for Molecular Science, University of Texas Medical Branch, Galveston, TX*

GENNARO MARINO • *Dipartimento di Chimica Organica e Biologica, Universita degli studi di Napoli Federico II, Naples, Italy*

CAMILLE MARTINAND • *Institut Genetique Moleculaire, Montpellier, France*

BRUCE A. McCLURE • *Department of Biochemistry, University of Missouri, Columbia, MO*

GREGOR MEISS • *Fachbereich Biologie, Institut für Biochemie, Justus-Liebig-Universität, Giessen, Germany*

JENNIFER T. MILLER • *Reverse Transcriptase Biochemistry Section, HIV Drug Resistance Program, Division of Basic Sciences, Frederick Cancer Research and Development Center, National Cancer Institute, Frederick MD*

MITCHELL D. MILLER • *Department of Biology and Biochemistry, University of Houston, TX*

DIANNE L. NEWTON • *SAIC Frederick, Frederick Cancer Research and Development Center, National Cancer Institute, Frederick, MD*

KAZUO NITTA • *Cancer Research Institute, Tohoku Pharmaceutical University, Sendai, Japan*

M. VICTÒRIA NOGUÉS • *Departament de Bioquimica et Biologia Molecular, Facultat de Ciencies, Universitat Autonoma de Bacelona, Bellaterra, Spain*

CINZIA NUCCI • *Laboratorio di Genetica Umana, IRCCS Ospedale San Raffaele, Milan, Italy*

CLARK Q. PAN • *Department of Protein Engineering, Genentech, Inc., South San Francisco, CA. Current address: Bayer Corp., Berkeley, CA*

BRITTAN L. PASLOSKE • *Ambion, Inc., The RNA Company, Austin, TX*

ANNALISA PASTORE • *National Institute of Medical Research—The Ridgeway, London, UK*

ALFRED M. PINGOUD • *Fachbereich Biologie, Institut für Biochemie, Justus-Liebig-Universität, Giessen, Germany*

WIM PLUYMERS • *Rega Institute for Medical Research, Katholieke Universiteit, Leuven, Belgium*

ELLEN A. PREDIGER • *Ambion, Inc., Austin, TX*

PIERO PUCCI • *Universita di Napoli—CNR, International Mass Spectrometry Facilities Center, Naples, Italy*

ANDRES RAMOS • *National Institute of Medical Research—The Ridgeway, London, UK*

JASON W. RAUSCH • *Reverse Transcriptase Biochemistry Section, HIV Drug Resistance Program, Division of Basic Sciences, Frederic Cancer Research and Development Center, National Cancer Institute, Frederick, MD*

JAMES F. RIORDAN • *Center for Biochemical and Biophysical Sciences and Medicine, Harvard Medical School, Boston, MA*

RICHARD A. RIPPE • *Division of Digestive Diseases, Department of Medicine, University of North Carolina, Chapel Hill, NC*

HELENE F. ROSENBERG • *Laboratory of Host Defenses, National Institute of Allergy and Infectious Diseases, National Institutes of Health, Bethesda, MD*

MARGHERITA RUOPPOLO • *Dipartimento di Chimica, Universita degli studi di Salerno, Salerno, Italy*

SUSANNA M. RYBAK • *Frederick Cancer Research and Development Center, National Cancer Institute, Frederick, MD*

TAMIM SALEHZADA • *Institut Genetique Moleculaire, Montpellier, France*

CATHERINE H. SCHEIN • *Human Biological Chemistry and Genetics, Sealy Center for Structural Biology, University of Texas Medical Branch, Galveston, TX*

GIDEON SCHREIBER • *Department of Biological Chemistry, Weizmann Institute of Science, Rehovot, Israel*

ROBERT SHAPIRO • *Center for Biochemical and Biophysical Sciences and Medicine, Harvard Medical School, Boston, MA*

MICHELLE SILHOL • *Institut Genetique Moleculaire, Montpellier, France*

DOMINICK V. SINICROPI • *Department of BioAnalytical Technology, Genentech, Inc., South San Francisco, CA*

KIZHAKE V. SOMAN • *Human Biological Chemistry and Genetics, Sealy Center for Structural Biology, University of Texas Medical Branch, Galveston, TX*

ROBERTO TARAMELLI • *Unité di Genetica Umana, Dipartimento di Biologia Strutterale e Funzionale, Universitá degli Studi dell'Insubria, Varese, Italy*

DAVID M. TIDD • *School of Biological Sciences, University of Liverpool, UK*

ANTONIO TUGORES • *Axys Pharmaceuticals, Inc., La Jolla, CA*

PETER WALTER • *Department of Biochemistry and Biophysics, Howard Hughes Medical Institute, University of California, San Francisco, CA*

RAYMOND J. WILLIAMS • *Promega Corp., Madison, WI*

HONGYAO ZHU • *College of Pharmacy, University of Texas, Austin, TX*

# I

# CHARACTERIZING NUCLEASE ACTIVITY

# 1

# Processivity of DNA Repair Enzymes

## R. Stephen Lloyd

## 1. Introduction
### 1.1. General Considerations of Processivity

DNA repair enzymes monitor a host's genome for structural aberrations caused by exogenous damage (ionizing or UV radiation, chemical exposure) or spontaneous damage (deamination, oxidation, or base loss). Lack of repair at such sites can result in replication and transcription blockage, and potentially error-prone replication bypass that could eventually lead to cell transformation. Several human diseases, such as xeroderma pigmentosum, Cockayne's syndrome, trichothiodystrophy, and human hereditary nonpolyposis colon cancer, are associated with inefficient DNA repair mechanisms (reviewed in **ref. *1***). As these lesions occur infrequently in DNA, enzymes must discriminate between normal, intact DNA and damaged DNA in circumstances where the former predominates.

One of the mechanisms used by DNA repair enzymes to locate damaged sites has been described as a processive search mechanism (reviewed in **refs. *2* and *3***). In this context, "processive" does not relate to the activity of enzymes (DNA or RNA polymerases or exonucleases) that create their substrate through a processive activity. Here, "processive" means that multiple catalytic events occur on a defined piece of DNA (whether a plasmid or a chromatin domain) prior to the enzyme's macroscopic diffusion from the DNA to which it was initially bound. After it has been released from that DNA, the enzyme may become bound to the same or another DNA molecule, upon which multiple catalytic events can again take place.

The factors that effect processivity have been extensively reviewed for several proteins *(2–5)*. These studies indicate that the major attractive force between the proteins and DNA are electrostatic, but the energy of net movement between the DNA and protein comes from simple Brownian motion. Some enzymes hydrolyze ATP to track unidirectionally along DNA, but these are not discussed in this chapter.

From: *Methods in Molecular Biology, vol. 160: Nuclease Methods and Protocols*
Edited by: C. H. Schein © Humana Press Inc., Totowa, NJ

## 1.2. Discovery of In Vitro Processive Nicking Activity of DNA Glycosylase/Abasic (AP)-Site Lyase on DNA Containing Cyclobutane Pyrimidine Dimers

T4 endonuclease V (now referred to as T4-pdg for pyrimidine dimer glycosylase) is an enzyme that initiates repair at UV light-induced cyclobutane pyrimidine dimers in double-stranded DNA (reviewed in **refs. *3*** and ***6***). The mechanism of action involves:

1. Recognition and binding to the dimer;
2. Flipping the nucleotide that is opposite the 5' pyrimidine of the dimer to an extrahelical position on the enzyme;
3. Scission at the N-C-1 glycosyl bond; and
4. Phosphodiester-bond scission via a β-elimination reaction (AP lyase activity).

The processive nicking activity of T4-pdg was originally observed by Dr. M. L. Dodson: when limiting concentrations of T4-pdg were added to heavily UV irradiated supercoiled (form I) plasmids, double-strand breaks appeared in the plasmid population (linear form III DNA) prior to all form I DNA being converted to either nicked circles (form II DNA) or form III DNA. The early appearance of plasmids containing double-stranded breaks indicated that T4-pdg incised most, if not all, dimers on one DNA molecule prior to enzyme dissociation *(7)*. Form III linear DNA is formed by the incision at two dimers in complementary strands in close proximity. Processive nicking activity can be most easily measured by damaging a plasmid at multiple sites and monitoring for the accumulation of form III DNA as a function of the remaining form I DNA (**Fig. 1**) *(7,8)*. This assay is described in **Subheading 3.1.1.** As the interactions are primarily electrostatic, increases in salt concentrations decrease processive nicking, i.e., encourage the enzyme to release its substrate before completing the reaction (**Fig. 2**).

Processive nicking reactions can also be readily distinguished from a random accumulation of breaks within a population of plasmids by measuring the number average molecular weight of denatured DNAs by either denaturing agarose gel electrophoresis or velocity sedimentation in alkaline sucrose gradients following enzymatic treatment. These techniques will be described in **Subheadings 3.1.2.** and **3.1.3.**, respectively.

## 1.3. Discovery of Processive Plasmid Repair Within Intact Cells

The processivity of enzymes, such as T4-pdg, as described above, is not merely an in vitro artifact of well-controlled biochemical enzymology, but rather is operative in intact, living cells *(9,10)*. Dr. Elliott Gruskin pioneered the concepts and methodologies to obtain evidence for the processivity of T4-pdg and the *Escherichia coli* nucleotide-excision repair system. His experimental procedure (described in Method 4) was to irradiate cells harboring plasmids in order to introduce 5 or 10 pyrimidine dimers per plasmid within *E. coli*, that contained limiting concentrations of T4-pdg, and then measure the kinetics of the accumulation of fully repaired plasmid molecules. This experimental design assumed that the rate-limiting step in the complete repair of a dimer site was the recognition of the lesion and incision at that site. The subsequent steps of AP endonucleolytic cleavage, polymerization, and ligation were presumed to be rapid. If T4-pdg initiated repair at all dimer sites within a subset of plasmids, then that subset of plasmids should be completely free of dimers prior to any repair in the remaining plasmids. Thus, if repair occurs processively, the kinetics of the accumulation of damage-

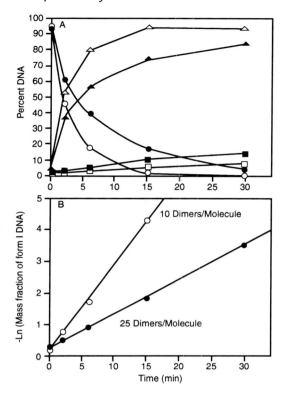

Fig. 1. (**A**) Time course analysis of T4 endonuclease V-nicking of form I DNA containing 10 or 25 dimers per molecule. Endonuclease V at 0.5 ng/μL was reacted in 25 m*M* NaCl with form I DNA containing 10 and 25 dimers per molecule. The three topological forms of DNA are as follows: form I DNA, supercoiled covalently closed circular DNA (○, 10 dimers per molecule; ●, 25 dimers per molecule); form II DNA, nicked or open circular DNA (△, 10 dimers per molecule; ▲, 25 dimers per molecule); and form III DNA, monomer length linear DNA (□, 10 dimers per molecule; ■, 25 dimers per molecule). (**B**) Change in the rate of disappearance of form I DNA as a function of the average number of dimers per molecule. The negative natural logarithm of the mass fraction of form I DNA was calculated for each time point taken in the time course endonuclease V-nicking assay (A) on DNA containing 10 (○), or 25 (●) dimers per molecule.

free plasmids will be linear with repair time, and the rate of accumulation of fully repaired DNAs will be inversely proportional to the UV dose. If these assumptions were valid, the repair rate of plasmids containing five dimers should be twice as fast as for DNAs containing 10 dimers on average. In contrast, if repair occurred randomly within a cell, then the kinetics of "distributive" repair would be characterized by a significant time lag to the accumulation of fully repaired plasmid molecules (*see* **Fig. 3**).

Intracellular plasmid repair kinetics initiated by T4-pdg displayed all the predicted characteristics of processive nicking at dimers and a processive completion of the base-excision repair pathway (*9*). Additionally, repair initiated by photolyase, an enzyme known to have low affinity for nontarget DNA, and thus predicted not to repair dimers processively, appeared to be distributive (*10*). For photolyase, fully repaired plasmid

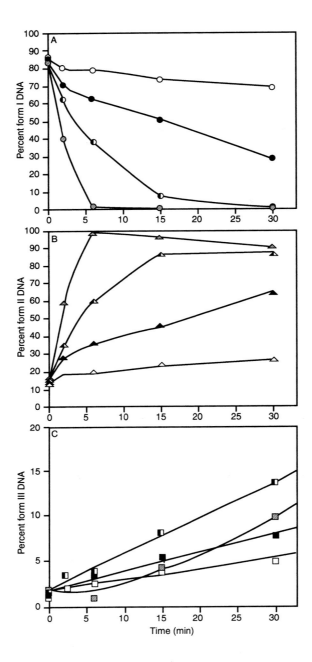

Fig. 2. Modulation of the processive nicking activity of T4 endonuclease V on dimer-containing form I DNA through changes in the NaCl concentration of the reaction. Time course reactions were analyzed in which form I DNA containing 25 dimers per molecule were incubated with endonuclease V at 0.17 ng/μL in various NaCl concentrations. **(A)** Loss of form I DNA at the following NaCl concentrations: ○, 0 m$M$; ●, 25 m$M$; ◐, 50 m$M$; and ◓, 100 m$M$. **(B)** Accumulation of form II DNA at the following NaCl concentrations: △, 0 m$M$; ▲, 25 m$M$; ◭, 50 m$M$; and ◮, 100 m$M$. **(C)** Accumulation of form III DNA at the following NaCl concentrations: □, 0 m$M$; ■, 25 m$M$; ◪, 50 m$M$; and ▨, 100 m$M$.

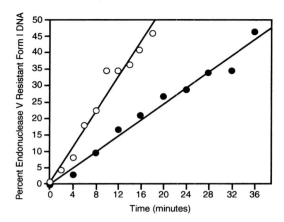

Fig. 3. In vivo quantitation of accumulation of endonuclease V-resistant form I DNA. In vivo kinetic analyses of endonuclease V-initiated excision repair were performed on *uvr*A⁻ *rec*A⁻*den*V⁺ *E. coli* grown at 40°C in Luria broth containing 100 µg/mL ampicillin, 0.8% glucose, and 5 mCi/mL [³H]thymidine. *E. coli* was UV-irradiated and incubated at 37°C. Plasmid DNA was isolated at the specified time points and treated in vitro with purified endonuclease V. Reaction products were resolved by electrophoresis through 1.2% agarose gels. Accumulation of endonuclease V-resistant form I DNA was determined by liquid scintillation spectroscopy of dissolved agarose slices containing form I, II, and III DNAs at each time point. ○, *uvr*A⁻*rec*A⁻*den*V⁺, UV dose of 450 J/m²; ●, *uvr*A⁻*rec*A⁻*den*V⁺, UV dose of 900 J/m².

molecules accumulated only after several hours time lag, but then accumulated exponentially. Repair initiated by the nucleotide excision-repair system, UvrABC, showed characteristics of limited processivity *(10)*.

## 1.4. Biological Significance of Processive Repair

Since electrostatic interactions are the primary attractive forces between the repair enzymes and DNA, it was hypothesized that neutralization of basic residues on the surface of T4-pdg that binds DNA could decrease in vitro and in vivo processivity of the mutant enzyme, without necessarily diminishing catalytic activity. Dr. Diane Dowd was first in testing this hypothesis by creating a series of site-directed mutants of T4-pdg that retained full catalytic activity, but were compromised with respect to their processivity *(11–13)*. Since no crystal structure was known for T4-pdg at the time of these studies, the choice of which sites to mutate was based on molecular modeling analyses. These predictions utilized circular dichroism data and turn prediction computer programs of all α-helical proteins to identify probable helix nucleation sites. Helix wheel diagrams suggested α-helices with a series of positively charged residue along one face of the helix, and these sites were mutated. These studies revealed a good correlation between in vitro and in vivo processivity assays, and showed that those mutants with diminished processivity had significantly reduced survival following UV challenge as measured by colony-forming ability. Specifically, some mutant enzymes were created that incised heavily irradiated plasmids by a distributive mechanism, even at low salt concentrations (**Fig. 4**). Mutant enzymes that displayed this loss in processive nicking activity (**Fig. 5**) were also unable to enhance UV survival in DNA

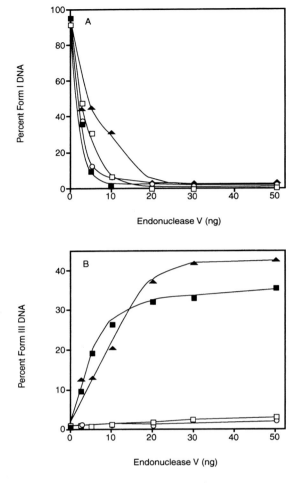

Fig. 4. Analysis of T4 endonuclease V-nicking of form I DNA containing 20–25 dimers/molecule. Cellular lysates containing endonuclease V were added to 1.0 μg of UV-irradiated [³H] pBR322 in 20 μL 10 m*M* Tris-HCl (pH 8.0), 1 m*M* EDTA, and 10 m*M* KCl. Solutions were incubated at 37°C for 30 min. Shown is a representative graph; measurements were reproducible to within 10%. ▲, endonuclease V (wild-type); ○, Gln-26; □, Gln-33; ■, Gln-26,33.

repair-deficient cells (**Fig. 6**). These data were the first that demonstrated the biological significance of nontarget DNA binding.

Attempts to enhance the nontarget DNA binding of T4-pdg by converting neutral amino-acid side chains to basic ones were successful in increasing nontarget DNA binding, but these mutants were unable to enhance UV survival above that of the wild type enzyme and in some cases could depress survivals *(14–16)*.

## 2. Materials
### *2.1. Buffers and Solutions*

1. 10 m*M* Tris-HCl, pH 7.5, 1 m*M* EDTA.
2. 1 *M* NaCl.

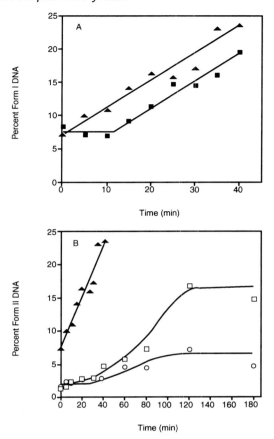

Fig. 5. Analysis of in vivo plasmid repair. Cells expressing endonuclease V wild type, Gln-26, Gln-33, or Gln-26,33 were harvested, resuspended in 43 m$M$ Na$_2$HPO$_4$, 20 m$M$ KH$_2$PO$_4$, 9 m$M$ NaCl, and 19 m$M$ NH$_4$Cl. They were irradiated at 900 J/m$^2$ at 0°C, mixed with LB media, and incubated at 37°C for varying times in order to initiate repair. Plasmid DNA was extracted, incubated with endonuclease V for 1 h at 37°C, and analyzed for endonuclease V-resistant DNA. These data were the average of two experiments. ▲, endonuclease V (wild-type); ○, Gln-26; ■, Gln-33; □, Gln-26,33.

3. 10 m$M$ Tris-HCl (pH 7.5), 1 m$M$ EDTA, 100 mg/mL bovine serum albumin (BSA), varying [NaCl].
4. Reaction stop buffers:
    a. For neutral agarose gels: 10 mM Tris-HCl (pH 7.5), 1 m$M$ EDTA, 20% (w/v) sucrose, 2% (w/v) SDS, 0.2% (w/v) Bromophenol blue.
    b. For denaturing agarose gels: same as **item 4a,** except also contains 30 mM NaOH.
    c. For velocity sedimentation through alkaline sucrose gradients: same as **item 4a** except also contains 100 mM NaOH.
5. 1X TAE.
6. 100 μg/mL ethidium bromide; store in brown bottle.
7. 30 m$M$ NaOH, 1 m$M$ EDTA.
8. Luria broth: 10 g bacto-tryptone, 5 g bacto-yeast extract, 10 g NaCl/L deionized H$_2$O pH to 7.0 with 5 g NaOH sterilized by autoclaving.

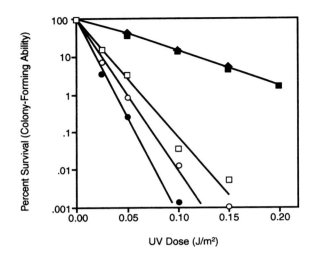

Fig. 6. Colony-forming ability of UV-irradiated repair-deficient *E. coli* containing *den*V⁺, *den*V⁻, or mutant *den*V constructs. Shown is the average of three measurements. ●, AB2480 with pGX2608-*den*V⁻; ▲, *den*V⁺; ○, *den*V Gln-26; ■, *den*V Gln-33; □, *den*V Gln-26,33.

9. 2X SSC = 17.53 g NaCl, 8.82 g sodium citrate, per 1 L deionized $H_2O$; pH to 7.0 with NaOH.
10. 5% alkaline sucrose = 5% (w/v) sucrose.
11. 20% alkaline sucrose = 20% (w/v) sucrose.
12. Buffered salt solution for irradiation of intact *E. coli* cells: 43 m$M$ $Na_2$ $HPO_4$, 22 m$M$ $KH_2PO_4$, 9 m$M$ NaCl, 19 m$M$ $NH_4Cl$.
13. Cell lysis suspension buffer: 50 m$M$ Tris-HCl (pH 8.0), 50 m$M$ EDTA, 8% (w/v) sucrose, 5% (v/v) Triton X-100.
14. 10 mg/mL lysozyme in $H_2O$ (Sigma Chemical).
15. 2 propanol.
16. Buffer-saturated phenol.
17. Chloroform/isoamyl alcohol (24:1).
18. 2.5 $M$ sodium acetate.
19. 95% ethanol.
20. 10 m$M$ Tris-HCl (pH 8.0), 0.2 m$M$ EDTA, 100 m$M$ NaCl, 5% (v/v) glycerol, 2 mg/mL BSA.
21. [14]C or [3]H-labeled thymidine.
22. Scintillation fluid.
23. *E. coli uvrA–recA–* cells (i.e., AB2480).

## 2.2. Equipment

1. Short-wave (254 nm) UV lamp (Spectroline [Westbury, NY] model EF-16 spectronics).
2. Short-wave (254 nm) UV monitor (International Light, Newbury Port, MA) radiometer/photometer.

## 2.3. Enzyme

1. T4-pdg is commercially available through Pharmingen (San Diego, CA).

## 3. Methods

### 3.1. In Vitro Plasmid-Nicking Assay

#### 3.1.1. Analyses by Native Agarose Gel Electrophoresis

##### 3.1.1.1. PREPARATION OF UV-IRRADIATED DNA

1. Dialyze covalently closed circular DNA (form I DNA) into 10 m$M$ Tris-HCl (pH 7.5), 1 m$M$ EDTA.
2. Dilute DNA to 0.1 μg/μL in 10 m$M$ Tris-HCl (pH 7.5), 1 m$M$ EDTA; aliquot-sufficient DNA for the complete experiment, plus an additional 10% to serve as a master mixture.
3. Prewarm short-wave UV lamps at least 15 min prior to irradiation of DNA.
4. Adjust UV lamp height so that at the height where the DNA is to be exposed, the UV meter reads 100 μW/cm$^2$.
5. Using continuous stirring, expose DNA to UV light (100 μW/cm$^2$) for varying times in which a 4 kb plasmid will accumulate 1 cyclobutane pyrimidine dimer per plasmid DNA mol in 10 s *(7)*.
6. Remove 50-μL aliquots of the irradiated DNA since for each experimental variable condition (different salt concentrations; enzyme concentrations, and so on), at least five time points will be needed, each containing 1 μg DNA.
7. Add an equal volume of 10 m$M$ Tris-HCl (pH 7.5), 1 m$M$ EDTA, 100 μg/mL BSA; this buffer will also contain varying NaCl concentrations to adjust the final salt between 0.01 and 0.2 $M$.

##### 3.1.1.2. INCISION KINETICS

1. Prewarm the 100-μL master mixture to the desired temperature for at least 5 min.
2. Dilute pure T4-pdg in 10 m$M$ Tris-HCl (pH 7.5), 1 m$M$ EDTA, 100 μg/mL BSA (nanogram quantities of T4-pdg are generally sufficient to incise microgram quantities of DNA for 15 min at 37°C).
3. Remove a 20-μL DNA aliquot and add it to 20 μL of the reaction stop buffer (10 m$M$ Tris-HCl, pH 7.5, 1 m$M$ EDTA, 20% [w/v] sucrose, 2% [w/v] SDS, 0.2% [w/v] bromophenol blue) for a no enzyme control.
4. Add 5 μL T4-pdg to the remaining 80-μL master reaction mixture, mix, and return to appropriate temperature.
5. Remove 21-μL aliquots at selected times and add to an equal volume of reaction stop buffer.

##### 3.1.1.3. DATA ANALYSIS

1. Separate the form I, II, and III DNAs by electrophoresis (4 h at 100 V constant) through a 1% horizontal agarose gel, and run in 1X TAE buffer.
2. Stain agarose gel for at least 2 h in 1X TAE supplemented with 0.5 μg/mL ethidium bromide; avoid prolonged exposure to visible or UV light by covering staining container with aluminum foil.
3. Place fully stained gel on a short-wave UV light box and capture the image, using a camera attached to an imaging system, such as Appligene Imager, in which the data can be analyzed and quantitated as a TIFF file.
4. Prior to quantitating the percentiles of forms I, II, and III DNAs, the raw data values obtained for form I DNA should be multiplied by 1.42, a correction factor that normalizes for the reduced binding of ethidium bromide by covalently closed circular DNA.

### 3.1.2. Analyses by Denaturing Alkaline Agarose Gel Electrophoresis

   1–9. Same as described in **Subheading 3.1.1.**

 10–12. Same as described in **Subheading 3.1.1.**, except the stop buffer also contains 30 m$M$ NaOH.

    13. Separate the denatured DNAs (including denatured mol-length markers) for 4 h at 100 V through a 1.2% agarose gel that was originally prepared as a native gel, but after solidifying, was soaked overnight in excess 30 m$M$ NaOH, 1 m$M$ EDTA.

    14. Neutralize the gel to pH 7.0–8.0 by repeated buffer exchanges in 2X SSC.

    15. Transfer DNAs to nitrocellulose paper using Southern blot technique.

    16. Probe filter with $^{32}$P-labeled plasmid DNA and analyze by autoradiography.

### 3.1.3. Analyses by Alkaline Sucrose Velocity Sedimentation

   1–9. Same as described in **Subheading 3.1.1.**, except $^3$H- or $^{14}$C-labeled DNAs should be used.

 10–12. Same as described in **Subheading 3.1.1.**, except the reactions are terminated by the addition of NaOH to 100 m$M$.

    13. Layer the denatured DNA solution on top of a 5–20% continuous 4.2-mL alkaline sucrose gradient.

    14. Separate the DNAs by size by centrifugation in a SW50.1 rotor (or equivalent) for 5 h at 45,000 rpm (~270,000$g$) at 20°C.

    15. Dispense the gradient in 200-µL aliquots from the top onto strips of 3MM paper.

    16. Immerse Whatman #17 paper successively in 4°C 20% trichloroacetic acid (TCA), 5% TCA, and 70% ethanol for 5 min each.

    17. Dry strips completely under a heat lamp.

    18. Cut each strip between the DNA aliquots and place into a scintillation tube containing 20 mL of scintillant.

    19. Count the radioactivity on each strip for at least 5 min and plot the data.

## 3.2. In Vivo Plasmid Repair Assay

   1. Overnight, grow *E. coli* AB2480 cells (*uvrA–, recA–*) containing a plasmid that encodes T4-pdg at 40°C in Luria broth (LB) supplemented with 0.8% glucose and $^3$H-thymidine (5 mCi/mL).

   2. Pellet cells by centrifugation at 5000 rpm (~4500$g$) using a fixed angle rotor.

   3. Decant supernatant and resuspend the cells in one-half the original volume in a buffer containing 43 m$M$ Na$_2$HPO$_4$, 22 m$M$ KH$_2$PO$_4$, 9 m$M$ NaCl, and 19 m$M$ NH$_4$ Cl.

   4. Perform the following steps under yellow light or in the dark to minimize photoreactivation of UV damage.

   5. UV irradiate cells at 0°C for 5 min at 150 or 300 mW/cm$^2$ (or 450 or 900 J/m$^2$, respectively); to generate on average 5 or 10 dimers per 4.5 kb plasmid.

   6. Mix equal volumes of LB containing 0.8% glucose at 75°C with the 0°C UV irradiated cells and incubate at 37°C.

   7. Remove 10-mL aliquots at various times and immediately freeze in an ethanol-dry-ice bath to terminate repair.

   8. After all samples have been collected and frozen, centrifuge cells for 2 h at 4°C. The vast majority of this time is necessary to thaw cells from –80°C to 4°C.

   9. Decant supernatant and resuspend cell pellet in 50 m$M$ Tris-HCl (pH 8.0), 50 m$M$ EDTA, 8% (w/v) sucrose, and 5% (v/v) Triton X-100.

 10. Add lysozyme (in water) to a final concentration of 1 mg/mL for 10 min at 22°C.

 11. Boil samples for 5 min and remove cell debris by centrifugation at 20,000$g$ for 20 min at 4°C.

12. Collect the plasmid containing supernatant; add an equal volume of 2-propanol, mix, and store at –20°C for 2 h.
13. Pellet DNA by centrifugation at 10,000*g* for 10 min and resuspend the DNA pellet in 1 mL of 10 m*M* Tris-HCl (pH 8.0), 1 m*M* EDTA.
14. Extract the DNA solution twice with equal volumes of a 1:1 mixture of buffer-saturated (pH 7.0) phenol:chloroform-isoamyl alcohol (24:1).
15. Collect the aqueous (top) phase and precipitate the DNA at –20°C for 16 h by the addition of 0.1 vol 2.5 *M* sodium acetate (pH 5.2) and 2.5 vol 95% ethanol.
16. Pellet DNA and resuspend in 300 µL of 10 m*M* Tris-HCl (pH 8.0), 0.2 m*M* EDTA; 100 m*M* NaCl, 5% (v/v) glycerol, and 2 mg/mL BSA.
17. Aliquots of these DNAs can be treated or not with T4-pdg and analyzed by native or denaturing agarose gel electrophoresis, as described in **Subheadings 3.1.1.** or **3.1.2.**

## 4. Notes

1. The form I DNA to be used in these assays should be free of protein contaminants prior to UV-irradiation to avoid DNA-protein crosslinks. This is best achieved by banding the DNA in ethidium bromide CsCl equilibrium gradients. The DNA should be at least 90% form I at time zero, and concatenated DNAs should be avoided.
2. Protective clothing and eyewear is essential while irradiating DNAs with short-wavelength UV light. Severe sunburns and retinal damage will occur even with brief exposure.
3. During the UV irradiation, the DNAs should be continuously stirred. UV irradiation of cells should be at 0°C to prevent the initiation of repair. Cells and a sterile stir bar should be placed in a Petri dish without the lid; the Petri dish is placed in an ice bath on a magnetic stirrer.
4. Dilute T4-pdg into a BSA-containing solution, to prevent adsorption onto plasticware.
5. When separating forms I, II, and III DNAs through native agarose gels, a 1% (w/v) agarose concentration should fully separate all three DNA forms. However, different suppliers and grades of agarose may not achieve full separation at 1%, and thus different percentages may be used. Generally, form I DNA has the fastest migration, followed by forms III and II.
6. Ethidium bromide acts as a frame-shift mutagen, and must be handled with caution. Following the staining of agarose gels, ethidium bromide can be removed from the solution by absorption onto Dowex beads. However, individual state or university rules may dictate how disposal of ethidium bromide waste should be handled; thus, consult with the university chemical safety officer for specific recommendations.
7. The length of time for staining agarose gels with ethidium bromide varies with the thickness of the gel. It is essential that the gel be fully stained prior to quantitation.

## References

1. Sarasin, A. and Stary, A. (1997) Human cancer and DNA repair-deficient diseases. *Cancer Detect. Prev.* **21,** 406–411.
2. Lloyd, R. S. and Van Houten, B. (1995) DNA damage recognition, in *DNA Repair Mechanisms: Impact on Human Diseases and Cancer* (Voss, J. M. H., ed.), R. G. Landes, Austin, TX, pp. 25–66.
3. Lloyd, R. S. (1998) Base excision repair of cyclobutane pyrimidine dimers. *Mutat. Res.* **408,** 159–170.
4. Lohman, T. M. (1986) Kinetics of protein-nucleic acid interactions: use of salt effects to probe mechanisms of interaction. *CRC Crit. Rev. Biochem.* **19,** 191–245.

5. von Hippel, P. H. and Berg, O. G. (1989) Facilitated target location in biological systems. *J. Biol. Chem.* **264,** 675–678.
6. Lloyd, R. S. (1998) The initiation of DNA base excision repair of dipyrimidine photo-products. *Prog. Nucleic Acid Res.* **62,** 155–175.
7. Lloyd, R. S., Hanawalt, P. C., and Dodson, M. L. (1980) Processive action of T4 endonu-clease V on ultraviolet-irradiated DNA. *Nucleic Acids Res.* **8,** 5113–5127.
8. Gruskin, E. A. and Lloyd, R. S. (1986) The DNA scanning mechanism of T4 endonu-clease V. Effect of NaCl concentration on processive nicking activity. *J. Biol. Chem.* **261,** 9607–9613.
9. Gruskin, E. A. and Lloyd, R. S. (1988) Molecular analysis of plasmid DNA repair within ultraviolet-irradiated Escherichia coli. I. T4 endonuclease V-initiated excision repair. *J. Biol. Chem.* **263,** 12,728–12,737.
10. Gruskin, E. A. and Lloyd, R. S. (1988) Molecular analysis of plasmid DNA repair within ultraviolet-irradiated Escherichia coli. II. UvrABC-initiated excision repair and photo-lyase- catalyzed dimer monomerization. *J. Biol. Chem.* **263,** 12,738–12,743.
11. Dowd, D. R. and Lloyd, R. S. (1989) Biological consequences of a reduction in the non-target DNA scanning capacity of a DNA repair enzyme. *J. Mol. Biol.* **208,** 701–707.
12. Dowd, D. R. and Lloyd, R. S. (1989) Site-directed mutagenesis of the T4 endonuclease V gene: The role of arginine-3 in the target search. *Biochemistry* **28,** 8699–8705.
13. Dowd, D. R. and Lloyd, R. S. (1990) Biological significance of facilitated diffusion in protein-DNA interactions. Applications to T4 endonuclease V-initiated DNA repair. *J. Biol. Chem.* **265,** 3424–3431.
14. Nickell, C. and Lloyd, R. S. (1991) Mutations in endonuclease V that affect both protein-protein association and target site location. *Biochemistry* **30,** 8638–8648.
15. Nickell, C., Anderson, W. F., and Lloyd, R. S. (1991) Substitution of basic amino acids within endonuclease V enhances nontarget DNA binding. *J. Biol. Chem.* **266,** 5634–5642.
16. Nickell, C., Prince, M. A., and Lloyd, R. S. (1992) Consequences of molecular engineer-ing enhanced DNA binding in a DNA repair enzyme. *Biochemistry* **31,** 4189–4198.

# 2

## Analysis by HPLC of Distributive Activities and the Synthetic (Back) Reaction of Pancreatic-Type Ribonucleases

M. Victòria Nogués and Claudi M. Cuchillo

## 1. Introduction

### 1.1. Distributive vs Processive Cleavage of Polymeric Substrates

Nucleic-acid cleavage can be processive, with the enzyme moving from one site to the next in the polymer before dissociating from the substrate, or distributive, with partially cleaved substrates released to the medium after the initial reaction. Kinetics indicate that either mechanism can be seen in reactions with polymerases, helicases, or nucleases (1). However, many factors affect these kinetics, including the strength of the enzyme–substrate binding, which can be altered by the salt concentration of the buffer, and intermediates in the formation of the complex or the release of the products that may be poorly characterized. The distinction between processivity and distributivity is based primarily on the ratio of rate constants for cleavage and dissociation. Unlike the distinction between endo- and exo-nucleases (i.e., those cleaving within or from one end of the nucleic acid), which will not change with reaction conditions, although in some cases enzymes show only a preference from one or the other activity, assay conditions can blur the distinction between processivity and distributivity.

An example of the analysis of a processive reaction, that of DNA repair by bacterial glycosylases at low salt concentrations, is presented in Chapter 1. Here we present HPLC-based methods to analyze the distributive cleavage of RNA by bovine pancreatic ribonuclease (RNase A), and other enzymes of this family.

### 1.2. The Enzymatic Mechanism of Pancreatic Ribonucleases

Bovine pancreatic ribonuclease A (RNase A) and related enzymes catalyze the breakdown of their natural substrate, RNA, in two steps. The first step in the cleavage of the 3'-5' phosphodiester bonds of RNA is a transphosphorylation reaction from the 5' position of one nucleotide to the 2' position of an adjacent pyrimidine nucleotide to form two polynucleotides, one ending in a 2',3'-cyclic phosphate and another with a free 5'-OH end (**Fig. 1A**). These RNases have some degree of base specificity, cleaving polyadenylic acid (poly(A)) poorly in comparison to polycytidylic acid (poly(C)) or

From: *Methods in Molecular Biology, vol. 160: Nuclease Methods and Protocols*
Edited by: C. H. Schein © Humana Press Inc., Totowa, NJ

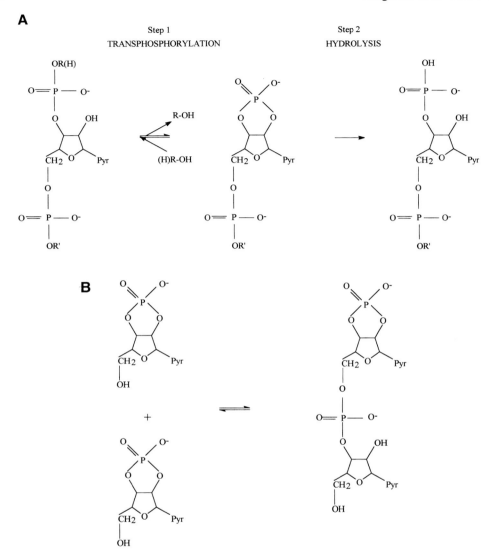

Fig. 1. (**A**) The reaction catalyzed by RNase A. The reaction for the complete breakdown of RNA takes place in two steps: transphosphorylation and hydrolysis. The pyrimidine 2',3'-cyclic phosphate intermediates are released to the medium as true products of the reaction. The hydrolysis of these cyclic compounds only begins when the transphosphorylation reaction is nearly finished. (**B**) The RNase A synthetic (back) reaction from cytidine 2',3'-cyclic phosphate (C>p) to cytidylyl 3',5'-cytidine 2',3'-cyclic phosphate (CpC>p).

polyuridylic acid (poly(U)) *(2)*. Thus the RNA cleavage will yield purine-core oligonucleotides ending with a pyrimidine 2',3'-cyclic phosphate nucleotide ((Pur)$_n$Pyr>p). In the second step of the enzyme reaction, the terminal 2',3'-cyclic phosphodiester of the core oligonucleotides is hydrolyzed to a 3'-nucleotide *(3)* (**Fig. 1A**). The transphosphorylation reaction needs a dinucleoside monophosphate as the minimum size for the substrate, but the hydrolysis reaction can take place with a 2',3'-cyclic pyrimidine mononucleotide.

### 1.3. Cleavage of Polymeric Substrates

Characterizing the RNA breakdown kinetically is complicated by the structural complexity of the RNA molecule, the difficulty of monitoring a very fast reaction, and the fact that most products of the reaction are also substrates, although with different specificity constants *(4)*. In **Subheadings 3.1.** and **3.2.** we describe assays to study different aspects of the processes catalyzed by RNase A *(4–6)* and the eosinophil cationic protein (ECP) *(7)*, using methods based on the separation of the reaction products by HPLC. The pattern of product formation from poly(C) or poly(U) to oligonucleotides of different size, $(Cp)_nC>p$ or $(Up)_nU>p$ respectively, indicates the endo- or exonuclease preference of these "nonspecific" enzymes. The pattern of poly(C) cleavage, for example, indicates that RNase A does not act in a random fashion. Rather, it preferentially binds longer-substrate molecules and cleaves phosphodiester bonds 6–7 residues from the end of the chain *(5)*. In contrast, ECP cleaves predominantly in an exonuclease-like manner *(7)*.

### 1.4. Characterizing the Synthetic (Back) Reaction

At sufficiently high concentrations of cytidine 2',3'-cyclic phosphate (C>p) as substrate, RNase A generates the hydrolysis product 3'-CMP, or the synthesis product cytidylyl 3',5'-cytidine 2',3'-cyclic phosphate (CpC >p) (*see* **Fig. 1B**). The hydrolysis reaction must be considered formally as a special case of the transphosphorylation back reaction in which the R group of the R-OH substrate is H (*see* **Fig. 1A**). We show in **Subheading 3.3.** how the products of the RNase A reaction of C>p at concentrations above 10 m$M$ (pH 5.5), 3'-CMP and CpC>p, can be separated and quantified by means of anion-exchange HPLC. This method was used to characterize the kinetics of both reactions that take place in the same assay mixture *(4,8,9)*.

The methods described here, developed for RNase A, can be applied to different RNases by adjusting the enzyme and substrate concentrations and the assay conditions (pH, ionic strength, temperature, and reaction time).

## 2. Materials

### 2.1. Equipment

1. Apparatus: An HPLC system with two pumps (or one pump with a gradient mixer system), a liquid chromatography injector, a UV absorbance detector, and a computer for both control of the system and data processing. We have used the following systems: Amersham Pharmacia Biotech, Waters Corp. (Mildford, MA), and Varian Associates (Sunnyvale, CA).
2. Chromatography columns: reversed-phase HPLC column, Nova Pak $C_{18}$, 4 µm, 3.9 × 150 mm (Waters Corp.), anion-exchange column, Nucleosil 10 SB, 300 mm × 4 mm I.D., and Vydac-310 SB precolumn stationary phase (Macherey, Nagel and Co., Düren, Germany).
3. Use distilled water treated with a MilliQ water purification system (Millipore Corp.) for the preparation of all solutions.
4. HPLC solvents: acetonitrile 240/farUV HPLC grade (Scharlau Chemie [Barcelona, Spain] or Carlo Erba [Milano, Italy]), ammonium acetate GR (Merck) and acetic acid (analytical grade) (Carlo Erba).
5. Substrates and enzymes: poly(C), C>p ($\varepsilon_{254}$ = 7214 $M^{-1}$/cm), 3'-CMP, 2'-CMP, pepsin from porcine stomach mucosa and bovine pancreatic RNase A (Type XII-A) ($\varepsilon_{277.5}$ = 9800 $M^{-1}$/cm *[10]*) from Sigma. All other reagents are of analytical grade.

## 2.2. Solutions

1. Poly(C) solution: 5 mg/mL in 10 m$M$ HEPES, pH 7.5, with KOH (or 10 m$M$ Tris-HCl, pH 7.5).
2. C>p solution: between 10 and 40 m$M$ in 0.2 $M$ sodium acetate, pH 5.5.
3. Dissolve RNase A (or the enzyme that is analyzed) in the same buffer as the substrate, and adjust the enzyme concentration according to its activity.
4. Pepsin solution: 3.6 mg/mL in 0.2 $M$ $H_3PO_4$ in 0.1 $M$ HCl.
5. Solvent A: 10% ammonium acetate (w/v) in water, degassed before use, to which 1% acetonitrile (v/v) is added.
6. Solvent B: 10% ammonium acetate (w/v) in water, degassed before use, to which 11% acetonitrile (v/v) is added.
7. 0.1 $M$ ammonium acetate, pH 5.5, and 0.6 $M$ ammonium acetate, pH 5.5, solutions are degassed before use.

## 3. Methods

## 3.1. Analysis of the Digestion Products of Poly(C) by Ribonuclease A (Fig. 2)

1. Combine 50 μL of poly(C) solution with 10 μL of 30 n$M$ RNase A at 25°C. Stop the reaction by applying samples at different time intervals depending on the enzyme activity (between 0 and 45 min in this case) directly to the HPLC.
2. HPLC procedure:
   a. Nova Pak $C_{18}$ column conditions: flow-rate: 1 mL/min, pressure: 1000–2000 psi (68–136 atm). Fix the maximum pressure limit to 6000 psi (408 atm) to avoid damaging the column. Store the column in 100% acetonitrile, and wash for 20 min with 1% acetonitrile in HPLC-grade water before use.
   b. Separation procedure: Wash the column for 15 min with solvent A. Inject 20 μL of the reaction mixture onto the column. Wash for 10 min, and then elute with a 50-min linear gradient from 100% solvent A to 10% solvent A plus 90% solvent B.
   c. After each run, wash the system for 5 min with water containing 1% acetonitrile and for 10 min with 100% acetonitrile.
   d. Before reequilibrating with solvent A, wash 5 min with water containing 1% acetonitrile. Slight differences in the retention times of oligonucleotides can be produced, depending on the equilibration time of the column. The intermediate washes with water are important to avoid contact between a highly concentrated saline solution and a concentrated organic solvent, which can produce some precipitates with the subsequent clogging of the column and a sharp increase in the pressure of the system.
   e. Monitor the absorbance at 260 nm of the eluate to detect and quantify product. Previously, obtain the number of integration counts per absorbance unit using a standard solution of nucleotide.
   f. Identification of the products: at the initial conditions (**Fig. 2**, $t = 0$), although poly(C) is not electrophoretically homogeneous (according to the information provided by Sigma), all high-molecular-mass components elute as a single peak, and no oligonucleotides or other small-molecular-mass contaminants are present in the sample. The elution position of the small oligonucleotides (**Fig. 3**) is deduced from the pattern of the poly(C) digestion by RNase A after a long incubation time (100 min) when no high molecular mass poly(C) is left, the pattern found by McFarland and Borer *(11)* for the chemical hydrolysis of polynucleotides, and the MALDI-TOF (matrix-assisted laser desorption ionization-time of flight) mass-spectrometry analysis of the individual peaks, except for the two mononucleotide species (3'-CMP and C>p) which yield

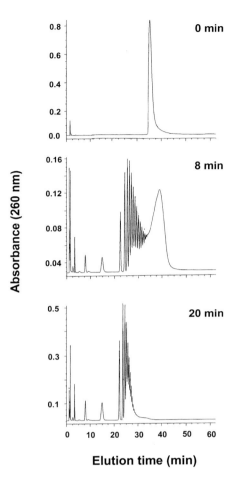

Fig. 2. Analysis by reversed-phase HPLC column (Nova-Pak $C_{18}$ column) of products obtained from poly(C) digestion by RNase A at different time intervals. In each case, 50 µL of a 5 mg/mL poly(C) solution in 10 m$M$ HEPES-KOH, pH 7.5, is digested with 10 µL of 30 n$M$ RNase A at 25°C. At the initial conditions ($t = 0$ min) poly(C) elute as a single peak and small molecular mass contaminants are not observed. Poly(C) fraction diminished with the increase of the reaction time ($t = 8$ and 20 min) with the subsequent formation of oligonucleotides.

unclear spectra. The two mononucleotide products 3'-CMP and C>p elute sequentially, but very close in the first peak, and oligomers of increasing size elute with increasing retention times (**Fig. 3**). Oligonucleotides up to nine residues can be separated with a good resolution.

3. Quantification of results:

   a. Determine the relative formation of mononucleotides and low-mol-mass oligonucleotides (up to nine nucleotides) by integration of the 260-nm peak area and, to normalize the relation between the peak area and the quantity of oligonucleotides, divide the area of each peak by the number of nucleotide residues. As the system cannot resolve nucleotides >10 bases, for the calculation of the relative distribution of each oligonucleotide, the area corresponding to all oligonucleotides with $n > 9$ can be considered as a high-mol-mass product together with the undigested substrate fraction (*9*).

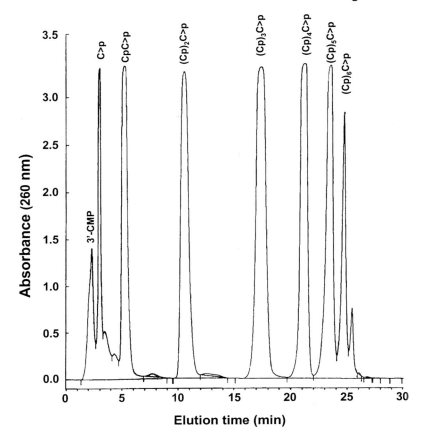

Fig. 3. Elution profile on a reversed-phase HPLC column (Nova-Pak $C_{18}$ column) of oligocytidylic acids $(Cp)_nC{>}p$ ($n = 0–6$) from poly(C) digestion. 200 μL of 10 mg/mL poly(C), in 10 m$M$ HEPES-KOH, pH 7.5, are digested with 15 μL of 0.5 μ$M$ RNase A at 25°C for 16 min.

b. To quantify the relative amount of oligonucleotides up to seven residues, divide the 260-nm peak area by the extinction coefficient at 260 nm ($\varepsilon_{260}$): 15,175 $M^{-1}$/cm for CpC>p, 20,745 $M^{-1}$/cm for $(Cp)_2C{>}p$, 24,282 $M^{-1}$/cm for $(Cp)_3C{>}p$, 28,683 $M^{-1}$/cm for $(Cp)_4C{>}p$, 37,711 $M^{-1}$/cm for $(Cp)_5C{>}p$, and 42,428 $M^{-1}$/cm for $(Cp)_6C{>}p$.

## 3.2. Analysis of Product Formation from Oligocytidylic Acids (Cp)ₙC>p by Ribonuclease A (Fig. 4)

The method described in **Subheading 3.1.** can also be used for the large-scale preparation of oligonucleotides of the general structure $(Cp)_nC{>}p$ or $(Up)_nU{>}p$, in which $n$ ranges from 0–6. These oligonucleotides are good substrates for assessing the role of the noncatalytic binding subsites, adjacent to the active site, in the catalytic process.

1. Preparation of oligocytidylic acids: Digest 500 μL of poly(C) solution with 50 μL of 7 μ$M$ RNase A at 25°C for 5 min. Separate oligonucleotides of different length according to the method described in **Subheading 3.1.** Collect the fractions that correspond to tetra, penta, hexa, and heptacytidylic acids, pool with the corresponding fractions of several chromatographic runs, freeze-dry, and keep at –20°C until use.

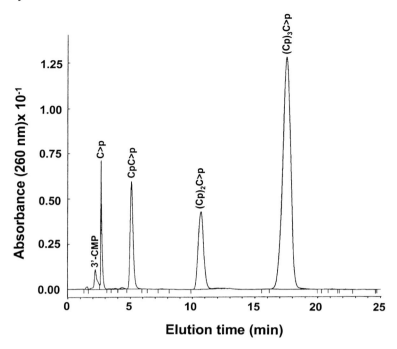

Fig. 4. HPLC separation by reversed-phase HPLC (Nova-Pak $C_{18}$ column) of the products obtained from (Cp)$_3$C>p digestion by RNase A. 100 μL of (Cp)$_3$C>p solution (0.3 U of *A* at 260 nm) in 10 m*M* HEPES-KOH, pH 7.5, are digested with 10 μL of 0.1 n*M* RNase A at 25°C for 15 min.

2. Digestion of oligonucleotides: Use the general reaction conditions described for poly(C) digestion. Specific characteristics are related to the substrate and enzyme concentrations. Use substrate solutions with an approximate $A_{260}$ of 0.3; the enzyme concentration depends both on the substrate and the enzyme species used. **Figure 4** is an example, which shows the specific conditions applied to the digestion of (Cp)$_3$C>p by RNase A.
3. Separate the reaction products by the HPLC procedure described in **Subheading 3.1.**
4. Calculate the amount of each oligonucleotide product by integration of the area of the peaks corresponding to the absorbance at 260 nm divided by the corresponding extinction coefficient ($\varepsilon_{260}$). The number of integration counts per *A*U, using a standard solution of nucleotide, should be obtained beforehand.

### 3.3. Analysis of the Hydrolysis (3'-CMP Formation) and Synthesis (CpC>p Formation) Reactions by RNase A Using C>p as Substrate (Fig. 5)

The synthetic reaction is only measurable at substrate concentrations above 10 m*M*. At lower substrate concentrations, only the hydrolysis reaction is observed.

1. In a typical assay the reaction mixture contains 30 μL substrate solution and 5 μL RNase A.
2. Quenching procedure: Incubate the assay mixture (35 μL) at 25°C for the desired time. Quench the reaction by the addition of 25 μL 0.2 *M* $H_3PO_4$ in 0.1 *M* HCl; the final pH of the reaction is 2.0. Immediately add 5 μL of the pepsin solution and incubate at 25°C for 15 min. Pepsin cleaves the Phe120–Asp121 bond in RNase A, and irreversible inactivation takes place. The mixture can be analyzed immediately or stored at –20°C. Storage

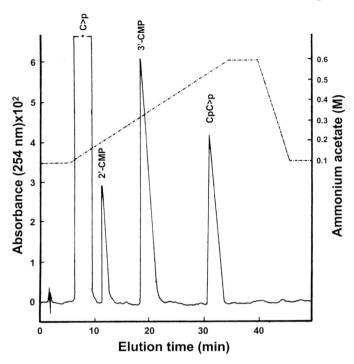

Fig. 5. HPLC separation by anion-exchange chromatography (Nucleosil 10 SB) of the products of the simultaneous hydrolysis and synthesis activities of RNase A. In this case, a reaction mixture containing 40 m*M* C>p and 0.6 μ*M* RNase A in 0.2 *M* sodium acetate, pH 5.5, is incubated at 25°C for 5 min, and the reaction is stopped by the quenching procedure described in the text. Product separation is carried out with a gradient from 0.1–0.6 *M* ammonium acetate, pH 5.5.

of the samples for several months at –20°C does not affect the reproducibility of the results if the correct controls are included.

3. Increase the pH of the sample to 5.5 by the addition of 4 μL 2 *M* NaOH before injecting the sample onto the HPLC column.

4. HPLC procedure:

   a. Nucleosil 10 SB anion-exchange column and Vydac-310 SB precolumn stationary phase conditions: flow-rate: 1 mL/min, pressure: 500–600 psi (34–40 atm). Fix the maximum pressure limit to 3000 psi (204 atm) to avoid problems in the event of collapse of the column. The precolumn prevents the clogging of the HPLC column with protein (pepsin or digested RNase). Change the precolumn after approx 20 runs. Store the column in 20% methanol and wash the column for 20 min with MilliQ-water before use.

   b. Separation procedure: wash the column with 0.1 *M* ammonium acetate, pH 5.5 for 30 min. Inject 20 μL of the sample onto the column. Elute with a linear-salt gradient from 0.1 to 0.6 ammonium acetate, pH 5.5.

   c. Monitor and quantify product elution from the absorbance at 254 nm.

   d. Identification of the products: **Fig. 5** shows a typical chromatogram. Although the hydrolytic action of RNase A on C>p produces only 3'-CMP, a small amount of 2'-CMP appeared in the chromatogram. Both nucleotide monophosphates are produced by direct hydrolysis of C>p in the acid quench. The peaks corresponding to C>p, 2'-CMP, and 3'-CMP can be confirmed by the use of the respective standards. C>p, the substrate of the reaction, elutes with a retention time of 10 min, the 2'-CMP a product of the acid treatment elutes at 12 min; the 3'-CMP (the product of the hydrolysis reaction) at 16 min, and CpCp (the product of the synthesis reaction) at 34 min. The effect of the acid in the formation of 2'-CMP and 3'-CMP is evaluated by a blank containing all the components of the assay except RNase A. It must be considered that the chemical hydrolysis yields equal amounts of 2'-CMP and 3'-CMP; thus, in order to quantify the amount of 3'-CMP formed by RNase A, an amount of 3'-CMP equal to the amount of 2'-CMP is subtracted from the total amount of 3'-CMP.

e. Quantification of the results: Determine the amount of each nucleotide from the integration of the peak areas divided by the corresponding extinction coefficient, $\varepsilon_{254}$, which are 7214 $M^{-1}$/cm for C>p, 6686 $M^{-1}$/cm for 3'-CMP, and 14,400 $M^{-1}$/cm for Cp >p. The number of integration counts per $AU$ from standard solutions of C>p or 3'-CMP should be determined beforehand.

## 4. Notes

1. The method described for poly(C) digestion analysis (**Subheading 3.1.**) has been applied to poly(U) digestion with similar results, although the digestion pattern depends on the specificity of the enzyme *(7)*.

2. In the digestion of oligocytidylic acids $(Cp)_nC>p$ by RNase A (**Subheading 3.2.**) the enzyme concentrations must be extremely low (in the nanomolar range). To avoid denaturation, prepare a concentrated solution of the enzyme and dilute further immediately before the assay. It is also advisable to distribute the substrate in Eppendorf tubes, each one with the exact volume for an individual assay, and keep them frozen. They are thawed immediately before use, equilibrated to 25°C, and then the enzyme solution is added.

3. In the quenching procedure described in **Subheading 3.3.** the final pH (2.0) is critical. Below this pH, too much acid hydrolysis to 2'-CMP and 3'-CMP occurs. At higher pH levels, residual RNase activity can falsify the results.

## Acknowledgments

We thank M. Moussaoui (Departament de Bioquímica i Biologia Molecular, UAB) for technical support, S. Bartolomé (Laboratori d'Anàlisi i Fotodocumentació, Departament de Bioquímica i Biologia Molecular, UAB) for the processing of the figures, and F. Canals (Servei de Seqüenciació de Proteïnes, IBF, UAB) for assessment in MALDI-TOF measurements. This work was supported by Grant PB96-1172-C02-01 from the Dirección General de Enseñanza Superior of the Ministerio de Educación y Cultura, Spain, and 1996SGR-00082 from Comissió Interdepartamental de Recerca i Tecnologia of the Generalitat de Catalunya. We thank the Fundació M. F. de Roviralta (Barcelona, Spain) for grants for the purchase equipment.

## References

1. Kornberg, A. and Baker, T. A. (1992) in *DNA Replication*, W. H. Freeman and Company, New York, NY.
2. DelCardayré, S. B. and Raines, R. T. (1994) Structural determinants of enzymatic processivity. *Biochemistry* **33,** 6031–6037.
3. Cuchillo, C. M., Parés, X., Guasch, A., Barman, T., Travers, F., and Nogués, M. V. (1993) The role of 2',3'-cyclic phosphodiesters in the bovine pancreatic ribonuclease A catalysed cleavage of RNA: intermediates or products? *FEBS Lett.* **333,** 207–210.

4. Moussaoui, M., Nogués, M. V., Guasch, A., Barman, T., Travers, F., and Cuchillo, C. M. (1998) The subsites structure of bovine pancreatic ribonuclease A accounts for the abnormal kinetic behavior with cytidine 2',3'-cyclic phosphate. *J. Biol. Chem.* **273,** 25,565–25,572.

5. Moussaoui, M., Guasch, A., Boix, E., Cuchillo, C. M., and Nogués, M. V. (1996) The role of non-catalytic binding subsites in the endonuclease activity of bovine pancreatic ribonuclease A. *J. Biol. Chem.* **271,** 4687–4692.

6. Nogués, M. V., Moussaoui, M., Boix, E., Vilanova, M., Ribó, M., and Cuchillo, C. M. (1998) The contribution of noncatalytic phosphate-binding subsites to the mechanism of bovine pancreatic ribonuclease A. *Cell Mol. Life Sci.* **54,** 766–774.

7. Boix, E., Nikolovski, Z., Moiseyev, G. P., Rosenberg, H. F., Cuchillo, C. M., and Nogués, M. V. (1999) Kinetic characterization of human eosinophil cationic protein. The subsites structure favors a predominantly exonuclease type activity. *J. Biol. Chem.* **274,** 15,605–15,615.

8. Guasch, A., Barman, T., Travers, F., and Cuchillo, C. M. (1989) Coupling of proteolytic quenching and high-performance liquid chromatography to enzyme reactions. Application to bovine pancreatic ribonuclease. *J. Chromatogr.* **473,** 281–286.

9. Boix, E., Nogués, M. V., Schein, C. H., Benner, S. A., and Cuchillo, C. M. (1994) Reverse transphosphorylation by ribonuclease A needs an intact p2-binding site. *J. Biol. Chem.* **269,** 2529–2534.

10. Sela, M. and Anfinsen, C. B. (1957) Some spectrophotometric and polarimetric experiments with ribonucleases. *Biochem. Biophys. Acta* **24,** 229–235.

11. McFarland, G. D. and Borer, P. (1979) Separation of oligo-RNA by reverse-phase HPLC. *Nucleic Acids Res.* **7,** 1067–1079.

# 3

# Ire1p: A Kinase and Site-Specific Endoribonuclease

## Tania N. Gonzalez and Peter Walter

## 1. Introduction
### 1.1. Ire1p in the Unfolded Protein Response

The lumen of the endoplasmic reticulum (ER) is a highly specialized compartment in eukaryotic cells. Here, secretory and most membrane proteins are folded, covalently modified, and oligomerized with the assistance of specialized ER resident proteins *(1)*. Perturbation of the ER lumen interferes with the production of many essential cellular components and can thus be highly deleterious. Indeed, in humans, defects in protein folding in the ER can lead to devastating diseases, such as cystic fibrosis, alpha$_1$-antitrypsin deficiency, and osteogenesis imperfecta *(2)*. One way in which cells cope with the accumulation of unfolded proteins in the ER is by activating the unfolded protein response (UPR), an ER-to-nucleus signal transduction pathway *(3–5)*. In the yeast *Saccharomyces cerevisiae*, Ire1p is an essential component of this pathway.

Ire1p is a transmembrane protein localized to ER membranes. When Ire1p was initially identified as a component of the UPR, it was thought to function primarily as a serine/threonine kinase because of its homology to other kinases and its demonstrated kinase activity *(6–8)*. Subsequent studies revealed that Ire1p functions as a site-specific endoribonuclease as well *(9)*. Both the kinase and endoribonuclease activities map to the carboxy-terminal half of Ire1p. The amino-terminal portion of Ire1p lies in the ER lumen *(6,7)*, where it senses increases in the concentration of misfolded proteins. Since the ER and nuclear membranes are continuous, the carboxy-terminal portion of Ire1p is predicted to be located in either the cytoplasm and/or nucleus, where it induces downstream events in the UPR pathway. Immunoprecipitation of Ire1p from yeast-cell extracts has demonstrated that oligomerization of Ire1p coincides with its phosphorylation, suggesting that, as is the case for transmembrane receptor kinases, Ire1p can trans-autophosphorylate *(10)*. Oligomerization of Ire1p coincides with induction of its endoribonuclease activity in vivo *(9,10)*. The only cellular substrate known to be cleaved by Ire1p is the mRNA encoding the UPR-specific transcription factor, Hac1p.

From: *Methods in Molecular Biology, vol. 160: Nuclease Methods and Protocols*
Edited by: C. H. Schein © Humana Press Inc., Totowa, NJ

### 1.2. Splicing of HAC1 mRNA by Ire1p and tRNA Ligase

Hac1p upregulates transcription of genes encoding ER resident proteins by binding to a common regulatory sequence in their promoters *(11–14)*. Cellular levels of Hac1p are controlled by the regulated splicing of its mRNA *(13,15)*. In the absence of splicing, *HAC1* mRNA translation is inhibited by a 252 nucleotide intron located at the 3'-end of the Hac1p coding region *(15,16)*. Unlike the spliceosomal catalyzed splicing of all other pre-mRNAs known to date, removal of the *HAC1* intron is catalyzed by the combined actions of Ire1p and tRNA ligase *(9)*. When unfolded proteins accumulate in the ER, Ire1p initiates splicing by cleaving *HAC1ᵘ* mRNA (*u* for *uninduced* or *unspliced*) to liberate the intron and 5' and 3' exons. The exons are then joined by tRNA ligase, an enzyme previously thought to function exclusively in the splicing of pre-tRNAs. The ligated product, *HAC1ⁱ* mRNA (*i* for *induced*), goes on to be efficiently translated to produce Hac1p. Thus the regulated splicing of *HAC1* mRNA is a key step in the UPR signaling pathway.

The splicing of *HAC1* mRNA by Ire1p and tRNA ligase is unprecedented. Significantly, Ire1p and tRNA ligase are sufficient to splice *HAC1ᵘ* mRNA in vitro *(9,17)*. This is in striking contrast to the splicing of all other studied pre-mRNAs, which require the more than 100 proteins and small nuclear RNAs which constitute the spliceosome and its associated components *(18)*. The splicing of *HAC1ᵘ* mRNA mechanistically resembles the splicing of pre-tRNAs *(19,20)*. Ire1p and tRNA endonuclease both cleave to produce 5'-OH and 2',3'-cyclic phosphate RNA termini. Exon ligation by tRNA ligase in both splicing reactions follows the same chemical steps *(17)*. Aside from these similarities, intriguingly, Ire1p and tRNA endonuclease are quite different. The two endoribonucleases lack any significant similarity in amino acid sequence or subunit composition, and recognize different structural features in their RNA substrates. Stem-loop structures predicted to form at the *HAC1ᵘ* mRNA splice junctions are required and sufficient to direct Ire1p cleavage. The stem-loop substrates that define the minimal Ire1p cleavage site are described elsewhere *(17)*.

### 1.3. Ire1p Is Similar to RNase L

The kinase and nuclease domains of Ire1p share significant homology with RNase L *(8)*. RNase L cleaves RNA nonspecifically in cells infected with double-stranded RNA viruses, and has no known role in pre-mRNA processing *(20,21)*. In vitro dimerization of RNase L induces its nuclease activity *(23)*. Analogously, oligomerization of Ire1p correlates with its nuclease activity in vivo *(9,10)*. In contrast to Ire1p, the kinase domain of RNase L appears to be catalytically inactive and has been shown to be involved in mediating dimerization of the protein *(23)*. The relationship between the kinase domain and nuclease activity of Ire1p has yet to be determined.

We can follow the Ire1p kinase *(9)* and *HAC1ᵘ* RNA-splicing activities *(9,17)* in vitro by using a recombinant fragment carrying the Ire1p kinase and putative nuclease tail domains, Ire1p(k+t). The reconstituted splicing reaction has been used to define the minimal substrate requirements for Ire1p cleavage, determine the chemical nature of the RNA termini produced by Ire1p cleavage, and map Ire1p cleavage sites. Overall, we found that small stem-loop minisubstrate RNAs are sufficient to direct cleavage by Ire1p; that, like tRNA endonuclease, Ire1p endonucleolytic cleavage produces RNA fragments with 2',3'-cyclic phosphate and 5'-OH termini; and that Ire1p cleaves 3' of the guanosine residue found at each splice site *(17)*. This chapter focuses on the methods we have used during our studies of the kinase and nuclease activities of yeast Ire1p.

## 2. Materials

### 2.1. Equipment, Chemicals, and Molecular Biology Reagents

1. Equipment: Microfluidizer (Microfluidics, Newton, MA), scintillation counter.
2. Chemicals: Glycogen (Promega, Madison, WI), *N*-butanol, phenylmethylsulfonyl flouride (PMSF), tRNA, diethyl pyrocarbonate (DEPC), 10% Triton X-100, carbenicillin, isopropyl-1-thio-β-D-galactoside (IPTG), glutathione, LiCl, phenol, chloroform, formamide.
3. Nucleotides: ADP, 5'pN, 5'pN2', 5'pN 3'p, and pN2'3'(cyclic)p (Sigma); [α-$^{32}$P]NTP and [γ-$^{32}$P]ATP (3000 Ci/m$M$; Amersham, Arlington Heights, IL); NTPs; $N$ = A, C, G, or U.
4. Enzymes: T7 RNA polymerase, nuclease P1, AMV reverse transcriptase, and mutant T4 polynucleotide kinase lacking 3' phosphatase activity (Roche Molecular Biochemicals); T4 polynucleotide kinase and *Sac*I restriction enzyme (New England Biolabs, MA); PreScission Protease and ribonuclease U2 (Pharmacia, Uppsala, Sweden).
5. Miscellaneous: glutathione-Sepharose (Pharmacia, Uppsala, Sweden); RNasin ribonuclease inhibitor (Promega, Madison, WI); Centricon-50 ultrafiltration units (Amicon, Beverly, MA); PEI cellulose thin-layer chromatography plates (EM Science, Gibbstown, NJ); X-ray film; Glogos II Autorad Markers (Stratagene, La Jolla, CA); plastic wrap (i.e., Saran Wrap), bovine serum albumin (BSA).

### 2.2. Solutions (see Note 1)

1. Crude extract buffer: 20 m$M$ HEPES, pH 7.6, NaCl (0.35 or 1 $M$), 2 m$M$ EDTA, 1 m$M$ DTT, 10% glycerol, 1% Triton X-100.
2. 100 m$M$ PMSF in ethanol.
3. Ire1p elution buffer: 20 m$M$ HEPES, pH 7.6, 0.35 m$M$ NaCl, 2 m$M$ EDTA, 20 m$M$ glutathione, pH 7.6.
4. Ire1p kinase buffer: 20 m$M$ HEPES, pH 7.6, 250 m$M$ K-acetate, 10 m$M$ Mg(acetate)$_2$, 1 m$M$ DTT, 100 m$M$ ATP.
5. Ire1p cleavage buffer: 20 m$M$ HEPES, pH 7.6, 50 m$M$ K-acetate, 1 m$M$ Mg(acetate)$_2$, 1 m$M$ DTT, 2 m$M$ ADP, 40 U RNasin.
6. PreScission protease buffer: 50 m$M$ Tris-HCl, pH 7.0, 150 m$M$ NaCl, 1 m$M$ EDTA, 1 m$M$ DTT, 10% glycerol.
7. Transcript elution buffer: 0.3 $M$ Na-acetate, pH 5.2, 10 m$M$ Mg(acetate)$_2$.
8. Phenol/chloroform: mix equal parts phenol and chloroform.
9. TE: 10 m$M$ Tris-HCl, pH 7.5, 1 m$M$ EDTA, pH 8.0.
10. Stop solution: 50 m$M$ Na-acetate, pH 5.2, 1 m$M$ EDTA, 0.1% SDS.
11. RNA gel-loading buffer: 99% formamide, 1 m$M$ Tris-HCl, pH 7.8, 0.1 m$M$ EDTA, 0.1% SDS, xylene cyanol FF, Bromophenol blue.
12. Ribonuclease U2 buffer: 20 m$M$ Na-citrate, pH 3.5, 1 m$M$ EDTA, 7 $M$ urea, 0.5 μg/μL tRNA.
13. Alkaline hydrolysis buffer: 50 m$M$ NaHCO$_3$, pH 9.0, 1 m$M$ EDTA, 0.5 μg/μL tRNA.
14. Nuclease P1 buffer: 20 m$M$ Na-acetate, pH 5.2, 50 ng/μL tRNA.
15. 1 $M$ LiCl.
16. Primer extension buffer: 20 m$M$ NaCl, 15 m$M$ HEPES, pH 7.6.

### *2.3. Plasmids and Oligonucleotides*

1. Vector for expression of recombinant *S. cerevisiae* Ire1p(k+t) in *Escherichia coli*: We expressed and purified a fragment of Ire1p fused at its amino-terminal end to glutathione-*S*-transferase (GST). This fragment (Ire1p amino acids 556–1115) carries the kinase (k) and putative nuclease tail (t) domains of Ire1p, and is termed GST-Ire1p(k+t). The expression vector pCF210 *(9)* was made by subcloning the Ire1p(k+t) coding region into pGEX-6p-2 (Pharmacia, Uppsala, Sweden). In this construct, a PreScission Protease cleavage site is encoded in the region that links GST and Ire1p(k+t) in the fusion protein.

2. *HAC1^U* 508 RNA in vitro transcription vector: The vector pCF187 *(9)* was constructed by subcloning a PCR fragment carrying *HAC1^U* mRNA sequences into pBluescript IISK(–) (Stratagene). Using linearized pCF187 as a template for T7 RNA polymerase transcription produces a *HAC1^U* RNA carrying 181 nucleotides of the 5' exon, the 252 nucleotide intron, and 75 nucleotides of the 3' exon, for a total of 508 nucleotides. Full-length *HAC1^U* mRNA is 1560 nucleotides.

3. Oligonucleotides for *HAC1^U* stem-loop minisubstrate in vitro transcriptions: Make *HAC1^U* stem-loop minisubstrates by in vitro transcription using annealed oligonucleotides as templates. Use the following oligonucleotides: T7 promoter, 5'TAATACGACT CACTATAG; hactng-38 (encoding the wild-type 5' splice-site stem-loop RNA), 5'TGAGCCGGTC ATCGTAATCA CGGCTGGATT ACGCCAACCG GCTATAGTGA GTCGTATTA; and hactng-10 (encoding the wild-type 3' splice-site stem-loop RNA), 5'TGAGGTCAAA CCTGACTGCG CTTCGGACAG TACAAGCTTG ACCTATAGTG AGTCGTATTA *(17)*.

4. Oligonucleotides for primer extension reactions: Use the sequencing primers TGSP-3 (5'GAAGAAATCA TTCAATTCAA ATGAATTC) and TGSP-1 (5'GCTAGTGTTC TTGTTCACTG) to map the 5' and 3' Ire1p(k+t) cleavage sites in *HAC1^u* 508 RNA, respectively *(17)*.

## 3. Methods
### *3.1. Expression and Purification of Ire1p(k+t)*

The portion of *S. cerevisiae* Ire1p containing its kinase and C-terminal tail domains fused to glutathione-*S*-transferase (GST-Ire1p(k+t)) is expressed in and purified from *E. coli*. A 1-L culture yields about 0.5 mg GST-Ire1p(k+t). Ire1p(k+t) is produced by proteolytic cleavage of GST-Ire1p(k+t) with PreScission Protease (*see* **Note 2**).

1. Grow *E. coli* strain DH5α carrying plasmid pCF210 at 37°C with shaking (~200 rpm) in 1 L LB media plus 100 µg/mL carbenicillin. When the culture reaches an $OD_{600}$ of 0.5, induce expression of GST-Ire1p(k+t) by adding IPTG to a final concentration of 0.1 m$M$ and grow cells for another 5 h.

2. Harvest cells by centrifugation and quick-freeze the pellet on dry ice. At this point, the cell pellet can be stored at –80°C until ready to continue. Thaw and resuspend the cell pellet on ice in 40 mL Ire1p crude extract buffer (0.35 $M$ NaCl) with 1 m$M$ PMSF (*see* **Note 3**). Break cells by passing three times through the microfluidizer. Pellet cell debris by centrifugation at 30,000*g* for 20 min.

3. Capture GST-Ire1p(k+t): In a 50-mL conical tube, incubate the supernatant at 4°C for 1 h with 0.5 mL glutathione-Sepharose pre-equilibrated in Ire1p crude-extract buffer (0.35 $M$ NaCl). Rotate the tube to gently mix its contents.

4. Wash resin: Spin down the resin (5 min at 1500*g*), discard the supernatant, and add 50 mL cold Ire1p crude-extract buffer (1 $M$ NaCl). Spin down the resin and discard the supernatant. Add 50 mL of the same buffer, and rotate gently for 30 min at 4°C. Spin down the resin and discard the supernatant.

5. Elute GST-Ire1p(k+t): Add 3 mL cold Ire1p elution buffer to the resin and rotate at 4°C for 15 min. Spin and remove eluate to a tube on ice. Repeat elution 2–3 more times. Pool the eluates and dialyze in PreScission Protease buffer to remove the glutathione. Dialyze at 4°C in an amount of buffer that is 1000 times the volume of the eluate you are dialyzing. After 3–12 h, replace the dialysis buffer and dialyze for an additional 3–12 h.

6. Cleave GST-Ire1p(k+t) to produce Ire1p(k+t): Add 20 U PreScission Protease (~4 U/100 μg fusion protein) and rotate for 4 h at 4°C. Remove PreScission Protease (*see* **Note 4**), GST, and residual GST-Ire1p(k+t) by addition of 0.5 mL glutathione-Sepharose pre-equilibrated in PreScission Protease buffer. Rotate for 1 h at 4°C.

7. Spin down the resin and remove the supernatant containing Ire1p(k+t). Wash the resin with Ire1p crude-extract buffer (0.35 *M* NaCl), and pool the supernatants. Dialyze into Ire1p cleavage buffer *lacking* ADP, and concentrate the protein by ultrafiltration in Centricon-50 concentrators following the manufacturer's recommendations (*see* **Note 5**). Determine the purity of your Ire1p(k+t) preparation by electrophoresis of the protein through a polyacrylamide sodium dodecyl sulfate (SDS) gel followed by Coomassie blue staining. Ire1p(k+t) resolves to a position between the 50 and 60 kDa protein markers on a 12% polyacrylamide gel. Estimate the Ire1p(k+t) concentration by comparing the intensity of the Ire1p(k+t) band to that of BSA of known concentration separated on the same gel. Aliquot Ire1p(k+t) and store frozen at –80°C (*see* **Note 6**).

### 3.2. Ire1p(k+t) RNA Substrates (see Note 7)

### 3.2.1. In Vitro Transcription of HAC1$^U$ 508 RNA

1. Linearize plasmid pCF187 by digestion with *Sac*I restriction enzyme.
2. Set up transcription reactions (20 μL) containing 1 m*M* each of ATP, CTP, GTP, 0.1 m*M* UTP, 25 mCi of [α-$^{32}$P]UTP, 1 μg linearized pCF187 DNA, 40 U RNasin, and 20 U T7 RNA polymerase. Incubate at 37°C for 1.5 h. To generate unlabeled RNA, omit [α-$^{32}$P]UTP and add 1 m*M* UTP to your reactions. Add 180 μL water and extract with phenol-chloroform. Ethanol precipitate the RNA using 40 μg glycogen as a carrier.
3. Prepare transcripts for electrophoresis by addition of 25 μL RNA gel-loading buffer followed by heating at 95–100°C for 3 min. Separate the transcription reaction products by electrophoresis through denaturing 5% polyacrylamide gels. Visualize the transcripts by exposing the gel to X-ray film for 3–5 min. When exposing the gel, use glow-in-the-dark stickers (Glogos II Autorad Markers) to help align the gel with the film later. On the developed X-ray film, cut out the most prominent band. Use this X-ray film mask to guide you as you cut out a slice of gel containing your transcript. Elute the RNA from the gel slice by shaking vigorously overnight at 4°C in 400 μL transcript elution buffer plus 400 μL phenol-chloroform. Ethanol precipitate and resuspend the RNA in 10 μL water. Determine the counts per minute (cpm) for 1 μL of the radiolabeled RNA.

### 3.2.2. In Vitro Transcription of HAC1$^U$ Minisubstrates

*HAC1$^U$* minisubstrates are transcribed using single-stranded DNA oligonucleotide templates to which the "T7 promoter" oligonucleotide is annealed to create a double-stranded T7 RNA polymerase promoter *(24)*. These minisubstrates should form stable stem-loop structures that mimic the structures predicted to form at the 5' and 3' splice sites of *HAC1$^U$* mRNA *(17)*. Mutant minisubstrate RNAs can be made simply by using DNA oligonucleotide templates that carry the mutations of interest.

1. Gel-purify each oligonucleotide by electrophoretically fractionating each through a denaturing 12% polyacrylamide gel. Cover the gel with plastic wrap and place it on top of a

PEI cellulose thin-layer chromatography (TLC) plate. Visualize the oligonucleotide containing band by illuminating the gel from above with long-wavelength UV light. The oligonucleotide containing band will cast a shadow on the TLC plate. On the plastic wrap, mark the location of the most prominent band for each oligonucleotide and cut the bands out of the gel. Elute the oligonucleotides from the gel slices by shaking vigorously overnight in TE. Ethanol precipitate the oligonucleotides. Resuspend them in water or TE and determine the oligonucleotide concentration by measuring the absorbance at 260 nm. At 260 nm, an absorbance of 1 is equal to 40 µg/mL of single stranded DNA.

2. Anneal the oligonucleotides by heating 15 p$M$ T7 promoter oligonucleotide with 0.25 p$M$ template oligonucleotide (i.e., hactng-10) at 95–100°C for 3 min. Immediately place on ice. The amount of [α-$^{32}$P]NTP and NTPs to use in each 20 µL transcription reaction is given in **Table 1**. Add NTPs, [α-$^{32}$P]NTP, 40 U RNasin, and 20 U T7 RNA polymerase to the annealed oligonucleotides. Incubate for 1.5 h at 37°C. To make unlabeled stem-loop RNAs, omit [α-$^{32}$P]NTP and include 1 m$M$ of each NTP. Phenol-chloroform extract and ethanol precipitate the RNA. Add 20 mL RNA gel-loading buffer and heat at 95–100°C for 3 min.

3. Separate the transcription products on a 15% denaturing polyacrylamide gel. Purify and quantitate the transcripts as in **Subheading 3.2.1.**

### 3.3. Kinase Assay

Kinase reactions containing 0.5 µg Ire1p(k+t) should produce a band visible by overnight autoradiography.

1. In 20 µL, incubate Ire1p(k+t), and 50 mCi [γ-$^{32}$P]ATP in kinase buffer at 30°C for 30 min.
2. Fractionate the protein by electrophoresis through a 12% polyacrylamide gel.
3. Visualize the phosphorylated Ire1p(k+t) by autoradiography of the gel.

### 3.4. RNA Cleavage Assay

#### 3.4.1. Cleavage Assay

A 30-min cleavage reaction containing 0.5 µg Ire1p(k+t) should result in 100% cleavage of *HAC1$^U$* 508 RNA. A 2-h cleavage reaction containing 1 µg Ire1p(k+t) should result in about 50% cleavage of a wild-type stem-loop minisubstrate.

1. In 20 µL, combine Ire1p(k+t) and Ire1p cleavage buffer. For cleavage of *HAC1$^U$* 508 RNA, use 20,000 cpm of labeled *HAC1$^U$* 508 RNA per reaction. For cleavage of *HAC1$^U$* stem-loop minisubstrates, use 2000 cpm of labeled *HAC1$^U$* stem-loop minisubstrate per reaction.
2. Incubate at 30°C for 0.5–2 h. Stop the reaction with 400 µL stop solution, and extract with 400 µL vol of phenol-chloroform. Ethanol precipitate the RNA using 40 µg glycogen as carrier. Add 8 µL of RNA gel-loading gel buffer and heat at 95–100°C for 3 min.
3. Separate the cleavage products of the *HAC1$^U$* 508 RNA reactions and the *HAC1$^U$* stem-loop minisubstrate reactions on 5 and 15% gels, respectively.
4. Visualize RNA by autoradiography of the gels.

#### 3.4.2. Marker Preparation

Alkaline hydrolysis of a 5'-end-labeled RNA will produce a ladder of bands differing by one nucleotide each. A partial ribonuclease U2 RNA digest of the same end-labeled RNA will help you align your alkaline hydrolysis ladder with your Ire1p(k+t) cleavage reaction products. This approach works well with small RNAs, such as the *HAC1$^U$* stem-loop minisubstrates (*see* **Note 8**).

**Table 1**
**Transcription Reaction Mixtures for Internally Radiolabeled *HAC1ᵘ* Stem-Loop Minisubstrates**

| Radiolabeled nucleotide | µCi, [α-³²P]NTP | mM ATP | mM CTP | mM GTP | mM UTP |
|---|---|---|---|---|---|
| A | 300 | 0.1 | 1 | 1 | 1 |
| C | 300 | 1 | 0.1 | 1 | 1 |
| G | 300 | 1 | 1 | 0.1 | 1 |
| U | 50 | 1 | 1 | 1 | 0.1 |

1. Make unlabeled minisubstrate RNA as in **Subheading 3.2.2.**, and 5'-end label it using [γ-³²P]ATP and T4 polynucleotide kinase.
2. Alkaline hydrolysis RNA ladder: Incubate 30,000 cpm of 5'-end labeled minisubstrate RNA with 10 µL freshly made alkaline hydrolysis buffer at 90°C for 15 min. Immediately place on ice and add 5 µL of RNA gel-loading buffer. Load one-half of the reaction per lane on your gel.
3. Partial RNA digestion with Ribonuclease U2: Incubate 20,000 cpm 5'-end-labeled minisubstrate in 10–20 µL freshly made Ribonuclease U2 buffer for 1 min at 80°C. Immediately place on ice, and add ribonuclease U2 freshly diluted into water. Incubate at 55°C for 15 min. Add 400 µL stop solution and 400 µL phenol-chloroform. Extract and ethanol precipitate the RNA. Load the entire reaction in one lane on your gel.

### 3.5. Nearest Neighbor Analysis of the Cleavage Termini (see Fig. 1)

This analysis is made possible for the following reasons. First, the dinucleotide at the Ire1p(k+t) cleavage site occurs only once in the loop of the *HAC1ᵁ* stem-loop minisubstrate. Second, nuclease P1 cleaves to the 3' side of nucleotides to produce 5'pN products. However, nuclease P1 *cannot* cleave 2',3'-cyclic phosphate groups.

### 3.5.1. Nuclease P1 Digests

1. Make *HAC1ᵁ* stem-loop minisubstrate RNAs internally radiolabeled with either A, C, or G.
2. Cleave these RNAs with Ire1p(k+t).
3. Separate the RNA fragments by electrophoresis through a 12% denaturing polyacrylamide gel. Visualize the bands, and cut out the gel slices containing the 5' cleavage fragments as in **Subheading 3.2.1.**
4. Elute the 5' cleavage fragments from the gel slices by shaking vigorously overnight at 4°C in 400 µL water and 400 µL phenol-chloroform.
5. Precipitate the eluted RNA by extracting with *N*-butanol to dryness (*see* **Note 9**). Wash the pelleted RNA with ethanol and dry.
6. Digest the pelleted RNA at 37°C for 30 min by addition of 10 µL nuclease P1 buffer and 0.2 U of nuclease P1. Place on ice until ready to spot onto your TLC plate.

### 3.5.2. Marker Preparation for Thin-Layer Chromatography

1. Make radiolabeled marker 5'pN (where *N* = A, C, G, or U) by phosphorylating N3'p nucleotide, in the presence of [γ-³²P]-ATP and wild-type T4 polynucleotide kinase.
2. Make radiolabeled markers 5'pN2'p, 5'pN3'p, and pN2',3'(cyclic)p by phosphorylating N2'p, N3'p, and N2',3'(cyclic)p nucleotides in the presence of [γ-³²P]-ATP, and mutant T4 polynucleotide kinase lacking the 3' phosphatase activity of the wild-type enzyme.

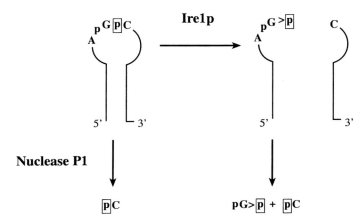

Fig. 1. Experimental scheme for the nearest neighbor analysis. In this scheme, the *HAC1*[U] stem-loop minisubstrate was transcribed in the presence of [$\alpha$-$^{32}$P]CTP. Thus, the phosphate 5' of every C residue is labeled and is here highlighted with a box. When the *HAC1*[U] stem-loop minisubstrate is digested with nuclease P1, the only radiolabeled nucleotide produced is [$^{32}$P]C. However, if the stem-loop minisubstrate is first cleaved by Ire1p(k+t) and then digested with nuclease P1, [$^{32}$P]C as well as pG>[$^{32}$P] are produced. Here, ">p" represents a 2',3'-cyclic phosphate group.

### 3.5.3. Thin-Layer Chromatography

1. Lightly draw a line in pencil across a PEI cellulose thin-layer chromatography (TLC) plate about 3 cm from the bottom of the plate.
2. Spot nuclease P1 digests and radiolabeled markers onto the TLC plate evenly along this line. Spot 1–2 µL of each sample and let dry. To spot more of each sample, spot another 2 µL at the same location, and let dry. Repeat until the entire sample is loaded.
3. Develop the plates with 1 *M* LiCl (*see* **Note 10**).
4. Visualize the separated nuclease P1 digestion products by autoradiography of the TLC plates.

### 3.6. Band Mobility Shift Analysis of the Cleavage Termini (see *Fig. 2*)

This assay is quick relative to the nearest neighbor analysis. Also, it allows one to determine that Ire1p(k+t) cleavage produces 2',3'-cyclic phosphate groups as well as 5'-OH groups. This approach takes advantage of the fact that the presence or absence of a terminal phosphate group will increase (because of negative charge) or decrease (because of loss of negative charge) the mobility of a *small* RNA electrophoresed through a polyacrylamide gel. Calf intestinal phosphatase (CIP) is used to remove noncyclic terminal phosphates. T4 polynucleotide kinase (PNK) is used to remove 2',3'-cyclic terminal phosphates in the absence of ATP. In the presence of ATP, it is used to phosphorylate 5'-OH groups.

1. Set up a 80-µL Ire1p(k+t) cleavage reaction containing Ire1p cleavage buffer, 10,000 cpm *HAC1*[U] stem-loop minisubstrate RNA, and Ire1p(k+t).
2. Incubate at 30°C for 2 h. Add stop solution and phenol-chloroform extract. Set one-quarter of the reaction aside. This is your untreated sample.

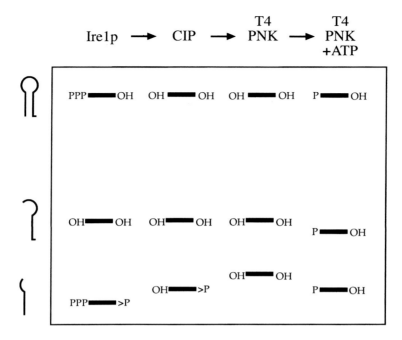

Fig. 2. Experimental scheme for the band mobility shift analysis of the RNA termini produced following cleavage of a *HAC1$^U$* stem-loop minisubstrate by Ire1p(k+t). The ">p" represents a 2',3'-cyclic phosphate termini.

3. Ethanol-precipitate the remaining 3/4 of the reaction. Treat the precipitated RNA with 30 U of CIP for 1.5 h at 37°C.
4. Phenol-chloroform extract the CIP reaction and set one-third of it aside. This is your CIP-treated sample.
5. Ethanol precipitate the remaining 2/3 of the CIP reaction. Treat the precipitated RNA with 30 U T4 PNK in the absence of ATP for 1.5 h at 37°C.
6. Phenol-chloroform extract the T4 PNK reaction and set one-half of it aside. This is your T4 PNK (minus ATP)-treated sample.
7. Ethanol precipitate the remaining one-half of the T4 PNK reaction. Treat the precipitated RNA with 30 U T4 polynucleotide kinase plus 2 m*M* ATP for 1.5 h at 37°C.
8. Phenol-chloroform extract and ethanol precipitate the RNA. This is your T4 PNK (plus ATP)-treated sample.
9. Ethanol precipitate all samples that you have set aside.
10. Add 8 μL RNA gel-loading buffer to the precipitated RNA samples and heat them at 95–100°C for 3 min.
11. Separate the RNA fragments on a 15% denaturing polyacrylamide gel.
12. Visualize the RNA fragments by autoradiography of the gel.

### *3.7. Mapping Ire1p(k+t) Cleavage Sites*

### *3.7.1. Primer Extension of Cleaved* HAC1$^u$ *508 RNA*

1. Use the sequencing primers TGSP-3 and TGSP-1 to map the 5' and 3' Ire1p(k+t) cleavage sites in *HAC1$^U$* 508 RNA, respectively. Each of these primers lies within 20–30 nucleotides of the 5' or 3' Ire1p(k+t) cleavage site.

2. Set up primer extension reactions containing 1 p*M* 5' end-labeled sequencing primer and 10 ng Ire1p(k+t) cleaved or uncleaved *HAC1*$^U$ 508 RNA in primer extension buffer. Primer extension reactions on uncleaved *HAC1*$^U$ 508 RNA provide a control for the pausing of AMV reverse transcriptase that naturally occurs during primer extension on the template.
3. Heat at 100°C for 3 min in a metal block. Anneal the primer to the *HAC1*$^U$ 508 RNA by slowly cooling the metal block to 40°C on your benchtop.
4. Start the extension reaction by adding dNTPs to a final concentration of 0.1 m*M* and 3 U AMV reverse transcriptase.
5. Incubate at 40°C for 30 min.
6. Add RNA gel-loading buffer.
7. Analyze samples by electrophoresis through a 10% denaturing polyacrylamide gel followed by autoradiography.
8. Generate sequencing ladders using the primer extension reaction protocol, but include 0.1 m*M* of ddATP, ddCTP, ddGTP, or ddTTP in each reaction.

### 3.7.2. Nearest Neighbor Analysis of Cleaved Stem-Loop HAC1ᵘ Minisubstrates

Mapping of the Ire1p(k+t) cleavage site on each stem-loop minisubstrate can be unambiguously achieved by the nearest neighbor analysis outlined in **Subheading 3.5.** as long as the dinucleotide at the cleavage site only occurs once in the loop of the stem-loop minisubstrate.

## 4. Notes

1. Your solutions need to be RNase-free. RNases can be inactivated by addition of diethyl pyrocarbonate (DEPC), which is carcinogenic. Wear gloves and avoid its fumes. Add 1 mL DEPC/L of solution and stir the mixture for 15 min in a fume hood. Autoclave the solution for 20 min to inactivate the DEPC. Buffers containing Tris-HCl cannot be treated directly with DEPC. Instead, use DEPC-treated water when making them.
2. You can use Ire1p(k+t) or GST-Ire1p(k+t) in all of the assays outlined here. However, the Ire1p(k+t) preparations tend to be of higher purity when compared to the GST-Ire1p(k+t) preparations.
3. PMSF is highly unstable in water, and it is very toxic. Wear gloves and avoid its fumes.
4. PreScission Protease is also a GST fusion protein. This allows for removal of the protease from solution by incubating it with glutathione-sepharose resin.
5. Occasionally Ire1p(k+t) will form a white, flaky precipitant during the concentration step. To avoid this, concentrate the protein as the last step before aliquoting it for storage at –70 to –80°C. Also, to reduce the amount of time it takes to concentrate the protein, use an ultrafiltration unit with the highest molecular weight cut-off possible. You should be able to concentrate Ire1p(k+t) down to 0.5 µg/mL without it precipitating out of solution.
6. To retain Ire1p(k+t) activity, flash-freeze aliquots in liquid nitrogen and store at –70 to –80°C. Ire1p(k+t) loses activity over time when stored at 4°C.
7. Gel purification of RNA transcripts, especially the stem-loop minisubstrates, leads ultimately to cleaner RNA cleavage assays. T7 RNA polymerase tends to add nontemplate nucleotides to the 3' end of transcripts.
8. Ribonuclease U2 cleaves after adenosine residues to produce 3'-phosphate termini. You will need to determine the optimal enzyme to RNA ratio for your partial digestion reactions. Alkaline hydrolysis produces 2'- and 3'-phosphate termini.
9. *N*-butanol extraction allows you to precipitate small quantities of RNA in the absence of carriers such as glycogen or tRNA. Glycogen interferes with sample migration during thin-layer chromatography.

10. To reduce smudging and smearing of samples during thin-layer chromatography, prerun the PEI cellulose TLC plates in distilled water and allow them to dry. Store the plates at 4°C until ready for use. Further improvements in sample separation can sometimes be achieved by soaking the TLC plate in methanol for 10 min following spotting of your samples onto the plate. Allow the plate to dry, and develop it as usual.

## Acknowledgments

This work was supported by a UCSF Biomedical Science Research Career Enhancement Fellowship from the National Institute of General Medical Science to T.N.G., and by grants from the National Institute of Health to P. W. P. W. is an investigator of the Howard Hughes Medical Institute.

## References

1. Gething, M.-J. and Sambrook, J. (1992) Protein folding in the cell. *Nature* **355,** 33–45.
2. Kuznetsov, G. and Nigam, S. K. (1998) Folding of secretory and membrane proteins. *N. Engl. J. Med.* **339,** 1688–1695.
3. Shamu, C. E., Cox, J. S., and Walter, P. (1994) The unfolded protein response pathway in yeast. *Trends Cell Biol.* **4,** 56–60.
4. Chapman, R., Sidrauski, C., and Walter, P. (1998) Intracellular signaling from the endoplasmic reticulum to the nucleus. *Annu. Rev. Cell Dev. Biol.* **14,** 459–485.
5. Sidrauski, C., Chapman, R., and Walter, P. (1998) The unfolded protein response: an intracellular signalling pathway with many surprising features. *Trends Cell Biol.* **8,** 245–249.
6. Cox, J. S., Shamu, C. E., and Walter, P. (1993) Transcriptional induction of genes encoding endoplasmic reticulum resident proteins requires a transmembrane protein kinase. *Cell* **73,** 1197–1206.
7. Mori, K., Ma, W., Gething, M. J., and Sambrook, J. (1993) A transmembrane protein with a cdc2+/CDC28-related kinase activity is required for signaling from the ER to the nucleus. *Cell* **74,** 743–756.
8. Bork, P. and Sander, C. (1993) A hybrid protein kinase-RNase in an interferon-induced pathway? *FEBS Lett.* **334,** 149–152.
9. Sidrauski, C. and Walter, P. (1997) The transmembrane kinase Ire1p is a site-specific endonuclease that initiates mRNA splicing in the unfolded protein response. *Cell* **90,** 1–20.
10. Shamu, C. E. and Walter, P. (1996) Oligomerization and phosphorylation of the Ire1p kinase during intracellular signaling from the endoplasmic reticulum to the nucleus. *EMBO J.* **15,** 3028–3039.
11. Mori, K., Sant, A., Kohno, K., Normington, K., Gething, M. J., and Sambrook, J. F. (1992) A 22 bp cis-acting element is necessary and sufficient for the induction of the yeast KAR2 (BiP) gene by unfolded proteins. *EMBO J.* **11,** 2583–2593.
12. Kohno, K., Normington, K., Sambrook, J., Gething, M. J., and Mori, K. (1993) The promoter region of the yeast KAR2 (BiP) gene contains a regulatory domain that responds to the presence of unfolded proteins in the endoplasmic reticulum. *Mol. Cell Biol.* **13,** 877–890.
13. Cox, J. S. and Walter, P. (1996) A novel mechanism for regulating activity of a transcription factor that controls the unfolded protein response. *Cell* **87,** 391–404.
14. Nikawa, J., Akiyoshi, M., Hirata, S., and Fukuda, T. (1996) Saccharomyces cerevisiae IRE2/HAC1 is involved in IRE1-mediated KAR2 expression. *Nucleic Acids Res.* **24,** 4222–4226.
15. Kawahara, T., Yanagi, H., Yura, T., and Mori, K. (1997) Endoplasmic reticulum stress-induced mRNA splicing permits synthesis of transcription factor Hac1p/Ern4p that activates the unfolded protein response. *Mol. Biol. Cell* **8,** 1845–1862.

16. Chapman, R. E. and Walter, P. (1997) Translational attenuation mediated by an mRNA intron. *Curr. Biol.* **7,** 850–859.
17. Gonzalez, T. N., Sidrauski, C., Dörfler, S., and Walter, P. (1999) Mechanism of non-spliceosomal mRNA splicing in the unfolded protein response pathway. *EMBO J.* **18,** 3119–3132.
18. Moore, M. J., Query, C. C., and Sharp, P. A. (1993) Splicing of precursors to mRNAs by the spliceosome, in *The RNA World* (Gesteland, R. F. and Atkins, J. F., eds.), Cold Spring Harbor Laboratory, Cold Spring Harbor, NY, pp. 303–357.
19. Westaway, S. K. and Abelson, J. (1995) Splicing of tRNA precursors, in *tRNA: Structure, Biosynthesis, and Function* (Söll, D. and RajBhandary, U., eds.), ASM Press, Washington, DC, pp. 79–92.
20. Abelson, J., Trotta, C. R., and Li, H. (1998) tRNA splicing. *J. Biol. Chem.* **273,** 12,685–12,688.
21. Dong, B. and Silverman, R. H. (1997) A bipartite model of 2-5A-dependent RNase L. *J. Biol. Chem.* **272,** 22,236–22,242.
22. Dong, B., Xu, L., Zhou, A., Hassel, B. A., Lee, X., Torrence, P. F., and Silverman, R. H. (1994) Intrinsic molecular activities of the interferon-induced 2-5A-dependent RNase. *J. Biol. Chem.* **269,** 14,153–14,158.
23. Dong, B. and Silverman, R. H. (1999) Alternative function of a protein kinase homology domain in 2', 5'-oligoadenylate dependent RNase L. *Nucleic Acids Res.* **27,** 439–445.
24. Milligan, J. F., Groebe, D. R., Witherell, G. W., and Uhlenbeck, O. C. (1987) Oligoribonucleotide synthesis using T7 RNA polymerase and synthetic DNA templates. *Nucleic Acids Res.* **15,** 8783–8798.

# 4

## Microtiter-Plate Assay and Related Assays for Nonspecific Endonucleases

**Gregor Meiss, Oleg Gimadutdinow, Peter Friedhoff, and Alfred M. Pingoud**

## 1. Introduction
### 1.1. Nucleases

Nucleases catalyze the cleavage of phosphodiester bonds in nucleic acids. They range in size from divalent cations, such as $Pb^{2+}$, which specifically cleaves $tRNA^{Phe}$ from yeast (*1*), and $Mg^{2+}$, which specifically acts on squid $tRNA^{Lys}$ (*2*) to divalent cations complexed to chelating agents, for example $Fe^{2+}$-bleomycin (*3*) and $Cu^{2+}$-phenanthroline (*4*) or imidazole (*5*), to complex nucleic acids and proteins, such as ribozymes (*6,7*) or the ribonucleoprotein RNaseP (*8,9*). Protein nucleases are involved in a variety of important cellular functions, such as DNA restriction, DNA repair and recombination, proofreading of DNA replication, DNA cleavage during programmed cell death and RNA processing, maturation and editing (*10,11*). Nucleases may be extremely specific; for example, homing endonucleases, or nonspecific. Examples of the latter include nucleases found in the gastrointestinal tract of higher vertebrates and those secreted into the medium by various microorganisms.

### 1.2. Nonspecific Nucleases

Nonspecific endonucleases are enzymes with the capability to attack DNA and/or RNA, in their single- and/or double-stranded form, catalyzing the hydrolytic cleavage of phosphodiester bonds without pronounced base specificity. These enzymes are widely used as biochemical tools in molecular biology, for example, to digest DNA and/or RNA in cell extracts or for footprinting studies, or probing nucleic-acid structure. It must be emphasized that certain sequence or structural preferences have been described for these so-called nonspecific nucleases, e.g., for DNase I (*12*) or the extracellular *Serratia marcescens* endonuclease (*13*). In our discussion of the mechanism of action of nonspecific nucleases, we will focus on bovine pancreatic DNase I (E.C.3.1.21.1) (*14,15*), staphylococcal nuclease (E.C.3.1.31.1) (*16,17*), and *Serratia* nuclease (E.C.3.1.30.2) (Benzonase®) that is used in industrial biotechnology for the downstream processing of various pharmaceutical and biotechnological products (*18–27*; Chapter 17).

From: *Methods in Molecular Biology, vol. 160: Nuclease Methods and Protocols*
Edited by: C. H. Schein © Humana Press Inc., Totowa, NJ

### 1.3. General Acid-Base Catalysis

The mechanism of hydrolysis of phosphodiester bonds by nonspecific endonucleases which act as phosphodiesterases normally involves a general base that generates the attacking nucleophile by deprotonating a water molecule, an electrophile that stabilizes the additional negative charge of the pentacovalent phosphorane transiently formed during catalysis, and a general acid that protonates the leaving group *(28–30)*. These catalytic functions might be directly taken over by certain amino-acid residues or by a divalent metal ion(s) located at the enzyme's active site. A comprehensive presentation of the mechanisms of action of various phosphodiesterases is provided in two books edited by Linn et al. *(10)* and by D'Alessio and Riordan *(11)*, respectively. Useful information regarding their enzymology may be found in the book by Eun *(31)*.

### 1.4. Catalytic Mechanisms of Nonspecific Nucleases

The mechanisms of action of several nonspecific nucleases have been analyzed in detail by biochemical as well as crystallographic studies. For DNase I, a nonspecific endonuclease that hydrolyzes double-stranded DNA, His134, Asp168, Asn170, and His252 are key residues in the mechanism of phosphodiester-bond hydrolysis *(32)*, but their individual role in catalysis is controversially discussed *(33–37)*. Residues His134 and His252 could act as the general base and general acid, respectively, whereas Asp168 and Asn170 could bind the essential cofactor $Mg^{2+}$ or $Mn^{2+}$, stabilizing the transition state and/or protonating the leaving group. DNase I binds to the minor groove of the DNA, as indicated by its sensitivity to variations in minor groove width *(12,38–41)*. Depending on the divalent metal ion acting as a cofactor, DNase I either nicks DNA in the presence of $Mg^{2+}$ ("haplotomic mechanism") or introduces double-strand breaks into DNA with $Mn^{2+}$ ("diplotomic mechanism") *(42)*.

The preferred substrate of staphylococcal nuclease is single-stranded DNA. The essential cofactor $Ca^{2+}$ is bound by Asp21, Asp40, Glu43, and Thr41. The attacking nucleophile is generated from a water molecule in the hydration spehere of the $Ca^{2+}$ ion. Arg35 and Arg87 act as electrophiles stabilizing the additional negative charge occuring during the transition state *(17,43)*.

*Serratia* nuclease, which belongs to a family of ubiquitous endonucleases with diverse functions, is one of the most active nucleases known. A $Mg^{2+}$ ion in its catalytic center is bound to Asn119. A histidine residue, His89, seems to be the general base, and an arginine residue, Arg57, might be involved in transition-state stabilization. A water molecule in the hydration sphere of the $Mg^{2+}$ may be responsible for protonation of the leaving group *(23,44*; Chapter 17). As *Serratia* nuclease is not specific for the sugar in the backbone, it degrades single- and double-stranded RNA and DNA. It preferentially cleaves double-stranded nucleic acids in the A-conformation *(45)*. A *Serratia* nuclease homodimer forms substrate-mediated complexes with high-mol-wt DNA, thereby increasing the rate of cleavage at low enzyme and substrate concentrations *(46)*.

### 1.5. Nuclease Activity Assays

Many different assays exist for nonspecific nucleases. In the most accurate, but time-consuming, assays for determining steady-state kinetic parameters, oligonucleotides are used as substrates, and the products are analyzed by electrophoresis or chromatography. Simpler and more rapid assays use macromolecular nucleic acids (usually tRNA, plasmids, or fragments of these) as substrates and measure the nuclease activity through a change in UV-*A* (hyperchromicity, caused by the conversion of double strands into small single strands), change in fluorescence (decrease in ethidium fluorescence quantum yield caused by the liberation of the intercalated drug), or change in color, e.g., of the pH-indicator dyes methyl green or toluidine blue (caused by the release of protons during nucleic acid cleavage). Generally these assays are semiquantitative, but some can be used for a quantitative determination of nuclease activity. They are suitable for simultaneous measurements of many samples in parallel. Here we present three convenient, rapid assays that can be used to determine the activity of nonspecific nucleases (**Table 1**).

## 2. Materials

### 2.1. In-Gel Assay

#### 2.1.1. Equipment

1. UV-transilluminator.

#### 2.1.2. Solutions

1. Standard solutions, reagents, and materials for SDS-PAGE *(47)*.
2. High molecular weight herring sperm DNA (Sigma), calf thymus DNA (Sigma), or total yeast RNA (Merck) or 16S + 23S rRNA (Roche Molecular Diagnostics). For the preparation of 10X nucleic-acid stock solutions, dissolve 1.5 mg high-molecular-weight herring sperm or calf thymus DNA, 6.5 mg total yeast RNA, or 16S + 23S rRNA in 10 mL 10 m$M$ Tris-HCl, pH 7.5, by gentle agitation at 50°C overnight.
3. Renaturation and cleavage buffer: 10 m$M$ Tris-HCl, pH 7.5, 5 m$M$ MgCl$_2$ (or other appropriate buffer).
4. Ethidium bromide stock solution (10 mg/mL water).

### 2.2. LB-Agar and Agar-Plate Assay

#### 2.2.1. Equipment

1. Petri dishes.

#### 2.2.2. Solutions

1. Agar–agar and LB-broth. Autoclave 200-mL LB-broth containing 1.5% (w/v) agar–agar or 200 mL 2.5% agar–agar in 10 m$M$ Tris-HCl, pH 7.5. Leave solutions at 50°C until pouring the plates.
2. Methyl green or toluidine blue pH-indicator solution. Dissolve 12.5 mg toluidine blue or methyl green in 50 mL in 10 m$M$ Tris-HCl, pH 7.5 (final concentration 0.25 mg/mL).
3. High-molecular-weight herring sperm DNA (Sigma), calf thymus DNA (Sigma), total yeast RNA (Merck), or 16S + 23S RNA (Roche Molecular Diagnostics). Dissolve 30 mg high-molecular-weight herring sperm DNA in 150 mL 10 m$M$ Tris-HCl, pH 7.5, by gentle agitation at 50°C overnight.
4. Nuclease reaction buffer (1X): 10 m$M$ Tris-HCl, pH 7.5, 5 m$M$ MgSO$_4$ or MnSO$_4$.
5. To prepare 50 mL of the final solution for the LB-agar-plate assay mix: 20 mL LB-broth/ 1.5% agar agar + 5 mL toluidine blue solution + 25 mL DNA + 250 μL of a 1 $M$ solution of MgSO$_4$ or MnSO$_4$. To prepare 50 mL of the final solution for the agar plate assay mix:

**Table 1**
**Nuclease Activity Assays**

| Assay | Substrate | Detection method | Principle | Ref. |
|---|---|---|---|---|
| In-gel assay | High mol-wt DNA or RNA | Ethidium bromide/DNA or RNA fluorescence | The enzyme containing sample is separated on an SDS-PAGE gel containing the nucleic-acid substrate. The enzyme is then renatured in the gel and cleavage of substrate in its vicinity is detected by reduced fluorescence. | (48) |
| LB-agar plate assay (I) | High mol-wt DNA or RNA | Color change of pH-Indicator dye methyl green or toluidine blue | E. coli colonies secreting the nuclease are grown on LB-broth agar plates containing substrate and pH-indicator dye. | (49) |
| LB-agar plate assay (II) | High mol-wt DNA or RNA | Ethidium bromide/DNA or RNA fluorescence | E. coli colonies secreting the nuclease are grown on LB-broth agar plates containing substrates, which are subsequently stained in an ethidium bromide solution. | (49) |
| Agar plate assay | High mol-wt DNA or RNA | Color change of pH indicator dye | Cellular extracts or purified nuclease are spotted on agar plates containing substrate and pH-indicator dye. | (43) |
| Microtiter plate assay | High mol-wt DNA or RNA, ds oligonucleotides | Ethidium bromide/DNA or RNA fluorescence | Cellular extracts or purified nuclease are incubated in reaction buffer containing substrate and ethidium bromide. | (21) |

20 mL 2.5% agar-agar + 5 mL toluidine blue solution + 25 mL DNA + 250 μL of a 1 *M* solution of $MgSO_4$ or $MnSO_4$

Incubate the mix for 30 min at 50°C with agitation and pour approx 15 mL into a standard petri dish (60 mm diameter × 15 mm height).
6. Ethidium bromide stock solution (10 mg/mL water).

## 2.3. Microtiter-Plate Assay

### 2.3.1. Equipment

1. Microtiter plates.
2. UV-transilluminator.
3. Video documentation system.

### 2.3.2. Solutions

1. Ethidium bromide stock solution (10X): Dissolve 0.2 mg ethidium bromide solution in 10 mL 10 m*M* Tris-HCl, pH 7.5 (50 μ*M*).
2. High-molecular-weight herring sperm DNA (Sigma), calf thymus DNA (Sigma), total yeast RNA (Merck), or 16S + 23S RNA (Roche Molecular Diagnostics) (10X): Dissolve 25 mg of high-molecular-weight herring sperm DNA in 100 mL 10 *M* Tris-HCl, pH 7.5, by gentle agitation at 50°C overnight.
3. Buffer: prepare a suitable 10X reaction buffer for the nuclease to be tested. A good 1X starting buffer is 10 m*M* Tris-HCl, pH 7.5, 5 m*M* $MgCl_2$ or $MnCl_2$ (or any other divalent metal ion that acts as a cofactor).

# 3. Methods

## 3.1. In-Gel Assay

This method is well-suited to follow nuclease activity during purification of nucleases from cellular extracts, provided that the nuclease is a monomer or not composed of different subunits and can be renatured following electrophoresis (*see* **Fig. 1**). The easiest way to detect the activity of a nonspecific endonuclease in cellular extracts or in a protein preparation is to use PAGE to monitor the reaction (*48,49*). Proteins are separated by electrophoresis on an SDS-polyacrylamide gel that contains high-molecular-weight herring sperm DNA or RNA, which is too large to elute from the gel during electrophoresis. After the run, the SDS-gel is incubated in a buffer which allows renaturation of the enzyme and nucleolytic activity. The gel is stained with ethidium bromide, and after illumination by UV-light nucleolytic activity is detectable as a non- or weakly fluorescent band against a highly fluorescent background at a position characteristic for the subunit molecular weight of the nuclease in the SDS-gel.

### 3.1.1. In-Gel Assay Protocol

1. SDS-PAGE: Prepare a standard SDS-polyacrylamide gel containing 15 μg/mL high-molecular-weight herring sperm, calf thymus DNA, 65 μg/mL total yeast RNA, or 16S + 23S rRNA added as solution to the separating SDS-gel solution (for nucleic-acid stock solution *see* **Subheading 2.1.**).
2. Renaturation: After electrophoresis incubate the SDS-polyacrylamide gel with an equal vol of renaturation buffer at 4°C for approx 2 h with 2–3 changes of the buffer.
3. Substrate cleavage: Incubate the SDS-polyacrylamide gel containing the renatured nuclease in reaction buffer at 37°C for several hours to allow for the digestion of the embedded nucleic acid.

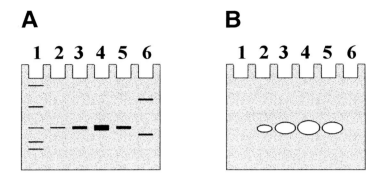

Fig. 1. In-gel assay. **(A)** Nuclease samples are separated by SDS-PAGE using a gel containing high molecular weight nucleic acids. **(B)** After renaturation of the enzymes *in situ* and subsequent incubation in cleavage buffer, digestion of the embedded nucleic acids is visualized by soaking the gel in ethidium bromide solution under UV light. 1. Protein size marker. 2, 3, 4. Increasing amounts of the same monomeric nuclease; the size of the halos produced by the nuclease corresponds to its concentration in the sample. 5. Homodimeric form of a nuclease; nucleases composed of identical subunits, like *Serratia* nuclease, can be renatured *in situ* after SDS-PAGE to give an active enzyme. 6. Nuclease composed of different subunits that are required for activity; nucleases composed of different subunits that are separated by SDS-PAGE cannot be renatured *in situ* to give an active enzyme.

4.  Staining: Soak the SDS-polyacrylamide gel in ethidium bromide solution and visualize the nuclease activity as a non- or weakly fluorescent band against the highly fluorescent background of the gel by using a UV-light source.

### 3.2. LB-Agar and Agar-Plate Assay

#### 3.2.1. LB-Agar-Plate Assay

If a nuclease is secreted by its host organism or genetically engineered to be secreted from *Escherichia coli*, the activity of individual clones can be detected by an LB-agar plate assay *(49)*.

1.  Add pH-indicator dyes such as toluidine blue (25 µg/mL final concentration) or methyl green, and high-molecular-weight DNA or RNA to the agar solution cooled to 50°C before pouring the plates.
2.  Detect secreted nuclease activity as a visible ring of altered color around the colonies. This pH-based change in color is caused by the release of one proton per phosphodiester bond hydrolyzed (**Fig. 2A**).
3.  Alternatively, nucleic acids plates may be soaked with 10 mL/plate of a 1 µg/mL ethidium bromide solution in 10 m$M$ Tris-HCl, pH 7.5, for 5 min after colony growth (**Fig. 2B**). Activity is then detected via non- or weakly fluorescent areas around the colonies against

Fig. 2. LB-agar and agar-plate assay. **(A)** Bacterial colonies were grown on LB-broth agar containing DNA and toluidine blue. Colonies secreting different amounts of active nuclease or differently active mutant forms of a nuclease produce an area of color change, corresponding to the concentration and activity of the secreted enzyme. **(B)** LB-agar plate containing DNA. Colonies secreting different amounts of active nuclease or mutant forms of different specific activity are identified after ethidium-bromide staining under UV-light. **(C) (1)** Agar plate with-

## A

**toluidine blue LB-broth DNA agar**

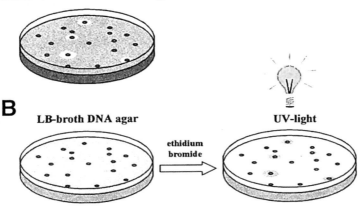

## B

**LB-broth DNA agar**     **UV-light**

ethidium
bromide

## C1

**toluidine blue DNA agar without nutrients**

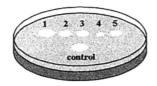

1   2   3   4   5

control

wild type NucA
Concentration: 1/1000;
Volume: 3μl

NucA variants
Concentration: 1;
Volume: 3μl

## C2

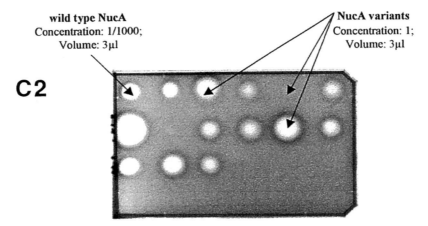

out nutrients containing toluidine blue and DNA. Aliquots of nuclease preparations of different concentration or activity when spotted onto the agar surface produce an area of color change corresponding to the concentration and activity of the nuclease. Spots 1, 2, 3, 4, and 5, for example, represent different mutant forms of the same nuclease, and for control of the wild-type enzyme. (2) Example of the above using the sugar-nonspecific nuclease NucA from *Anabaena* sp. PCC 7120 *(49)*. Three-microliter aliquots of each variant were spotted onto the plates. The concentration of the wild-type enzyme was 1000-fold less than that of the nuclease active-site variants (Meiss, unpublished).

a highly fluorescent background of the agar-plate *(49)*. If ethidium bromide is used as the indicator dye, the pH-sensitive dye may be omitted.

### 3.2.2. Agar-Plate Assay

This variation of the assay is sensitive enough to measure and quantify the relative activity of even weakly active nucleases *(44,50)*.

1. Spot defined amounts (up to several microliters) of solutions containing the nuclease, e.g., cellular extracts or chromatographic fractions onto agar plates containing nucleic acids and pH-sensitive dyes but no nutrients (**Fig. 2C1**; **Subheading 3.2., step 1**). If only low concentrated nuclease is available, one should spot only up to 5 μL of the preparation, repeatedly allowing each aliquot to be soaked into the agar. The agar should have a thickness of approx 0.5 cm.
2. The areas of color change at the spots are a direct measurement of relative nuclease activities, if equal amounts of nucleases are compared. Quantitation can be best performed using a video documentation system as well as a computer program that allows to determine the areas of the spots and to compare these with the areas produced by different dilutions of a nuclease whose specific activity is known, e.g., a wild-type nuclease compared to mutant variants (*see* **Fig. 2C2**).
3. Incubation times up to several days can be used to detect low nuclease activities, such as active-site variants of a nuclease with little residual activity.

### 3.3. Microtiter-Plate Assay

While detection and semiquantitative measurements of nuclease activity are best performed using the in-gel or the agar-plate assay, the microtiter-plate assay allows quantitative measurement of nuclease activity *(21)*. The assay uses double-stranded high-molecular-weight DNA, partially double-stranded RNA, or double-stranded oligonucleotides as substrates, which are visualized with a UV-light source after ethidium bromide staining (**Fig. 3**). Ethidium ions intercalate between the bases of double-stranded nucleic acids, which when hydrolyzed by a nonspecific nuclease are converted into smaller fragments. The dissociation of these into single-stranded nucleic acids is accompanied by liberation of ethidium ions with a concomitant decrease in fluorescence intensity. For quantification, the fluorescence intensities are converted to substrate concentrations using a calibration curve from a serial dilution of nucleic acids of known concentration. The nuclease activity is calculated from the maximum slope (to be preferred over the initial slope when macromolecular substrates are used *[22]*) of a substrate concentration vs time plot.

### 3.3.1. Microtiter-Plate Assay Protocol

1. Reaction mixture: mix 10 μL of the DNA solution (final concentration 0.025 μg/mL), 10 μL of the ethidium bromide solution (final concentration 5 μ*M* (2 μg/mL) and 10 μL of the 10X reaction buffer, and 50 μL sterile water in a well of a microtiter plate.
2. Start the reaction by adding 20 μL of a suitable nuclease preparation in reaction buffer and mix the reactants by pipetting in and out once or twice.
3. Follow the decrease in *A* of the DNA/ethidium bromide solution using a video gel documentation system.

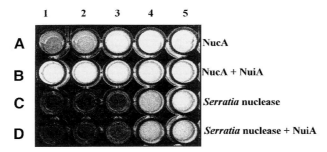

Fig. 3. Microtiter-plate assay. The microtiter plate assay was performed using the nonspecific nucleases from *Anabaena* sp. (NucA) and *Serratia marcescens* (*Serratia* nuclease) in the absence or presence of the NucA-specific inhibitor NuiA. Ethidium-DNA fluorescence decreases in wells where nucleolytic activity is present. The reactions were performed in the presence of 5 µ$M$ ethidium bromide in 50 m$M$ Tris-HCl, pH 8.2, 5 m$M$ MgCl$_2$ using a DNA concentration of 25 µg/mL. (**A**) NucA; (**B**) NucA + NuiA; (**C**) *Serratia* nuclease; (**D**) *Serratia* nuclease + NuiA. From left to right a 1:1-serial dilution of the respective nuclease is shown: 1, 10 n$M$; 2, 5 n$M$; 3, 2.5 n$M$; 4, 1.25 n$M$; 5, 0.63 n$M$.

## 4. Notes

1. In-gel assay: At **step 2**, the gel can be soaked in any buffer where the nuclease renatures well. One may also need to first renature the enzyme in one buffer, and then change to a buffer that is optimized for the cleavage reaction, for example with respect to salt or divalent metal ion concentration. At **step 3**, incubation can be prolonged up to 12 h, depending on the specific activity of the nuclease under investigation.
2. LB-agar and agar-plate assay: Depending on the number of variants to be tested, different sizes of Petri dishes can be used, but for most purposes, a standard Petri dish will be sufficient.
3. Microtiter-plate assay: The decrease in ethidium fluorescence upon cleavage of the substrate is best monitored on-line using a video documentation system that records the data in defined time intervals. We have also used double-stranded (11 bp long) oligodeoxyribonucleotides and oligoribonucleotides in this assay.

## Acknowledgments

This work was supported by the Deutsche Forschungsgemeinschaft (Pi 122/9-2), INTAS, the BMBWFT (Bundesministerium für Bildung, Wissenschaft, Forschung und Technologie), and the Fonds der Chemischen Industrie.

## References

1. Brown, R. S., Dewan, J. C., and Klug, A. (1985) Crystallographic and biochemical investigation of the lead (II)-catalyzed hydrolysis of yeast phenylalanine tRNA. *Biochemistry* **24,** 4785–4801.
2. Matsuo, M., Yokogawa, T., Nishikawa, K., and Watanabe, K. (1995) Highly specific and efficient cleavage of squid tRNALys catalyzed by magnesium ions. *J. Biol. Chem.* **270,** 10,097–10,104.
3. Umezawa, H., Takita, T., Sugiura, Y., Otsuka, M., Kobayashi, S., and Ohno, M. (1984) DNA-bleomycin interaction: nucleotide sequence specific binding and cleavage of DNA by bleomycin. *Tetrahedron Lett.* **40,** 501–509.

4. Hertzberg, R. P. and Dervan, P. B. (1982) Cleavage of double helical DNA by (methidiumpropyl-EDTA) iron (II), *J. Am. Chem. Soc.* **104,** 313–315.

5. Vlassov, V. V., Zuber, G., Felden, B., Behr, J. P., and Giege, R. (1995) Cleavage of tRNA with imidazole and imidazole constructs: a new approach for probing RNA structure. *Nucleic Acids Res.* **23,** 3161–3167.

6. James, H. A. and Turner, P. C. (1997) Ribozymes. *Methods Mol. Biol.* **74,** 1–9.

7. Tanner, N. K. (1999) Ribozymes: the characteristics and properties of catalytic RNAs. *FEMS Microbiol. Rev.* **23,** 257–275.

8. Frank, D. N. and Pace, N. R. (1998) Ribonuclease P: unity and diversity in a tRNA processing ribozyme. *Annu. Rev. Biochem.* **67,** 153–180.

9. Schoen, A. (1999) Ribonuclease P: the diversity of a ubiquitous RNA processing enzyme. *FEMS Microbiol. Rev.* **23,** 391–406.

10. Linn, S. M., Lloyd, R. S., and Roberts, R. J. (eds.) *Nucleases*, 2nd ed., Cold Spring Harbor Laboratory, Cold Spring Harbor, NY.

11. D'Alessio, G. and Riordan, J. F., eds. (1997) *Ribonucleases: Structures and Functions.* Academic, San Diego, CA.

12. Drew, H. R. and Travers, A. A. (1984) DNA structural variations in the *E. coli* tyrT promoter. *Cell* **37,** 491–502.

13. Meiss, G., Friedhoff, P., Hahn, M., Gimadutdinow, O., and Pingoud, A. (1995) Sequence preferences in cleavage of dsDNA and ssDNA by the extracellular *Serratia marcescens* endonuclease. *Biochemistry* **34,** 11,979–11,988.

14. Kunitz, M. (1950) Crystalline deoxyribonuclease I. Isolation and general properties. Spectrophotometric method for the measurement of deoxyribonuclease activity. *J. Gen. Physiol.* **33,** 349–362.

15. Laskowski, S. M. (1959) Enzymes hydrolyzing DNA. *Ann. NY Acad. Sci.* **81,** 776–781.

16. Moravek, L., Anfinsen, C. B., Cone, J. L., and Taniuchi, H. (1969) The large scale preparation of an extracellular nuclease of *Staphylococcus aureus. J. Biol. Chem.* **244,** 497–499.

17. Hale, S. P., Poole, L. B., and Gerlt, J. A. (1993) Mechanism of the reaction catalyzed by staphylococcal nuclease: identification of the rate-determining step. *Biochemistry* **32,** 7479–7487.

18. Nestle, M. and Roberts, W. K. (1969) An extracellular nuclease from *Serratia marcescens* I. Purification and some properties of the enzyme. *J. Biol. Chem.* **244,** 5213–5218.

19. Nestle, M. and Roberts, W. K. (1969) An extracellular nuclease from *Serratia marcescens* II. Specificity of the enzyme. *J. Biol. Chem.* **244,** 5219–5225.

20. Friedhoff, P., Gimadutdinow, O., and Pingoud, A. (1994) Identification of catalytically relevant amino acids of the extracellular *Serratia marcescens* endonuclease by alignment-guided mutagenesis. *Nucleic Acids Res.* **22,** 3280–3287.

21. Friedhoff, P., Matzen, S. E., Meiss, G., and Pingoud, A. (1996) A quantitative microtiter plate nuclease assay based on ethidium/DNA fluorescence. *Anal. Biochem.* **240,** 283–288.

22. Friedhoff, P., Meiss, G., Kolmes, B., Pieper, U., Gimadutdinow, O., Urbanke, C., and Pingoud, A. (1996) Kinetic analysis of the cleavage of natural and synthetic substrates by the *Serratia* nuclease. *Eur. J. Biochem.* **241,** 572–580.

23. Friedhoff, P., Franke, I., Krause, K. L., and Pingoud, A. (1999) Cleavage experiments with deoxythymidine-3'-5'-bis-(*p*-nitrophenylphosphate) suggest that the homing endonuclease I-*Ppo*I follows the same mechanism of phosphodiester bond hydrolysis as the non-specific *Serratia* nuclease. *FEBS Lett.* **443,** 209–214.

24. Friedhoff, P., Franke, I., Meiss, G., Wende, W., Krause, K., and Pingoud, A. (1999) A similar active site for non-specific and specific endonucleases. *Nat. Struct. Biol.* **6,** 11–113.

25. Miller, M. D., Tanner, J., Alpaugh, M., Benedik, M. J., and Krause, K. L. (1994) 2.1 Å structure of *Serratia* endonuclease suggests a mechanism for binding to double-stranded DNA. *Nature Struct. Biol.* **1,** 461–468.

26. Miller, M. D. and Krause, K. L. (1996) Identification of the *Serratia* endonuclease dimer: Structural basis and implications for catalysis. *Protein Sci.* **5,** 24–33.

27. Benedik, M. J. and Strych, U. (1998) *Serratia marcescens* and its extracellular nuclease. *FEMS Microbiol. Lett.* **165,** 1–13.

28. Saenger, W. (1991) Structure and catalytic function of nucleases. *Curr. Opin. Struc. Biol.* **1,** 130–138.

29. Suck, D. (1992) Nuclease structure and catalytic function. *Curr. Opin. Struc. Biol.* **2,** 84–92.

30. Gerlt, J. A. (1993) Mechanistic principles of enzyme-catalyzed cleavage of phosphodiester bonds, in *Nucleases* (Linn, S. M., Lloyd, R. S., and Roberts, R. J., eds.), 2nd ed., Cold Spring Harbor Laboratory, Cold Spring Harbor, NY, pp. 1–34.

31. Eun, H.-M. (1996) *Enzymology Primer for Recombinant DNA Technology.* Academic Press, San Diego, CA.

32. Suck, D. and Oefner, C. (1986) Structure of DNase I at 2.0 Å resolution suggests a mechanism for binding to and cutting DNA. *Nature* **321,** 621–625.

33. Worrall, A. F. and Connolly, B. A. (1990) The chemical synthesis of a gene coding for bovine pancreatic DNase I and its cloning and expression in *Escherichia coli. J. Biol. Chem.* **265,** 21,889–21,895.

34. Doherty, A. J., Worrall, A. F., and Connolly, B. A. (1991) Mutagenesis of the DNA binding residues in bovine pancreatic DNase I: an investigation into the mechanism of sequence discrimination by a sequence selective nuclease. *Nucleic Acids Res.* **22,** 6129–6132.

35. Doherty, A. J., Connolly, B. A., and Worrall, A. F. (1993) Overproduction of the toxic protein, bovine pancreatic DNase I, in *Escherichia coli* using a tightly controlled T7-promoter-based vector. *Gene* **136,** 337–340.

36. Doherty, A. J., Worrall, A. F., and Connolly, B. A. (1995) The roles of arginine 41 and tyrosine 76 in the coupling of DNA recognition to phosphodiester bond cleavage by DNase I: a study using site directed mutagenesis. *J. Mol. Biol* **251,** 366–377.

37. Warren, M. A., Steven, J. E., and Connolly, A. (1997) Effects of non-conservative changes to tyrosine 76, a key DNA binding residue of DNase I, on phosphodieseter bond cleavage and DNA hydrolysis selectivity. *Protein Eng.* **10,** 279–283.

38. Hogan, M. E., Roberson, M. W., and Austin, R. H. (1989) DNA flexibility variation may dominate DNase I cleavage. *Proc. Natl. Acad. Sci. USA* **86,** 9273–9277.

39. Brukner, I., Jurokovski, V. and Savic, A. (1990) Sequence dependent structural variations of DNA revealed by DNase I, *Nucleic Acids Res.* **18,** 891–894.

40. Brukner, I., Sanchez, R., Suck, D., and Pongor, S. (1995) Sequence-dependent bending propensity of DNA as revealed by DNase I: parameters for trinucleotides. *EMBO J.* **14,** 1812–1818.

41. Travers, A. (1993) *DNA-Protein Interactions*, Chapman and Hall, London.

42. Laskowski, S. M. (1971) Deoxyribonuclease I, in *The Enzymes, vol. IV: Hydrolysis*, 3rd ed. (Boyer, P. D., ed.), Academic Press, New York.

43. Serpersu, E. H., Shortle, D., and Mildvan, A. S. (1987) Kinetic and magnetic resonance studies of active-site mutants of staphylococcal nuclease: factors contributing to catalysis. *Biochemistry* **26,** 1289–300.

44. Miller, M. D., Cai, J., and Krause, K. L. (1999) The active site of *Serratia* endonuclease contains a conserved magnesium-water cluster. *J. Mol. Biol.* **288,** 975–987.

45. Meiss, G., Gast, F. U., and Pingoud, A. (1999) The DNA/RNA non-specific *Serratia* nuclease prefers double stranded *A*-form nucleic acids as substrates, *J. Mol. Biol.* **288,** 377–390.

46. Franke, I., Meiss, G., and Pingoud, A. (1999) On the advantage of being a dimer: a case study using the dimeric *Serratia* nuclease and the monomeric nuclease from *Anabaena* sp. strain PCC 7120. *J. Biol. Chem.* **274,** 825–832.

47. Laemmli, U. K. (1970) Cleavage of structural proteins during the assembly of bacterio-phage T4. *Nature* **227,** 680–685.
48. Rosenthal, A. L. and Lacks, S. A. (1977) Nuclease detection in SDS-polyacrylamide gel electrophoresis. *Anal. Biochem.* **80,** 76–90.
49. Muro-Pastor, A. M., Flores, E., Herrero, A., and Wolk, C. P. (1992) Identification, genetic analysis and characterization of a sugar-non-specific nuclease from the cyanobacterium *Anabaena* sp. strain PCC 7120. *Mol. Microbiol.* **6,** 3021–3030.
50. Weber, D. J., Meeker, A. K., and Mildvan, A. S. (1991) Interactions of the acid and base catalysts on Staphylococcal nuclease as studied in a double mutant. *Biochemistry* **30,** 6103–6114.

# 5

# Quantitating mRNAs with Relative and Competitive RT-PCR

## Ellen A. Prediger

## 1. Introduction
### 1.1. Why Use RT-PCR for mRNA Quantitation?

Reverse transcription coupled with the polymerase chain reaction (RT-PCR), permits amplification of cellular RNA. Gene expression can be measured even when the message copy number is very low (1–10 copies per cell) *(1)* or the sample size is very small (1–1000 cells) *(2,3)*. In RT-PCR, an RNA sample is primed with gene-specific primers, oligo dT, or random primers, and copied into a complementary DNA sequence, or cDNA, using a retroviral reverse transcriptase. The first-strand cDNA from the RT reaction is subsequently amplified using gene-specific primers and a thermostable DNA polymerase (e.g., Taq polymerase).

To make RT-PCR experiments quantitative, the amount of the amplified product is measured in reference to an internal control (relative RT-PCR) or competing synthetic target sequence (competitive RT-PCR) that is coamplified. However, accurate quantitation requires overcoming several hurdles which are intrinsic to PCR. With relative RT-PCR, small differences between samples in RNA sample quantitation, sample loading, and reaction component concentrations become magnified during in vitro amplification. Coamplifying an internal standard provides a way to relate product yield to the initial amount of transcript *(4,5)*.

Competitive RT-PCR requires a synthetic target that can be reverse transcribed and amplified with the same efficiency as the mRNA of interest *(6)*. As other techniques to quantify RNA (Northern analysis *(7)*, nuclease protection assay *(8; see* Chapters 31 and 32) are easier to perform and analyze, quantitative RT-PCR should only be used when extreme sensitivity is required.

### 1.2. Relative RT-PCR

Results for relative quantitation are commonly expressed as "percent (or *x*-fold) difference" in the level of a specific RNA; e.g., 2.5× more IFN-γ receptor mRNA in sample two than in sample one. An internal control with invariant expression is included to normalize samples for errors in sample quantitation and pipeting, making sample comparisons possible. Several commonly used internal controls include β-actin, glyceral-

From: *Methods in Molecular Biology, vol. 160: Nuclease Methods and Protocols*
Edited by: C. H. Schein © Humana Press Inc., Totowa, NJ

dehyde phosphate dehydrogenase (GAPDH), cyclophilin, L32, 18S rRNA, and 28S rRNA *(9–12)*. rRNAs are the most invariant internal controls. This is because rRNAs make up 80% of total RNA samples, and mass measurements of total RNA are thus normalized to rRNA mass.

For relative RT-PCR results to be meaningful, the PCR products must be measured while they are accumulating at a constant rate; i.e., while amplification falls within "linear range." Pilot experiments are performed to determine the minimal number of cycles necessary for product detection, and the maximum number of PCR cycles that meet linear-range conditions for the messages being studied. However, if the internal control is too abundant, it may be overamplified in a few PCR cycles, while a rare message will require still more cycles just to be detected.

One solution to this problem is to use nonfunctional primers ("competimers") *(13; see* **Note 1**), identical to the actual primers used to amplify the internal control, but blocked at the 3'-end so they cannot be extended. Altering the ratio of competimers to primers attenuates amplification of the internal control to any level, such that the linear-range conditions of an abundant internal control can be adjusted to mimic that of a rare message (**Fig. 1**). The relative RT-PCR protocol provided here includes optional experimental details on the use of competimers. Relative RT-PCR, however, only provides an estimate of the relative changes in gene expression between samples.

### 1.3. Competitive Quantitative RT-PCR

Absolute quantitation by RT-PCR, or "competitive RT-PCR," is currently the most sensitive and rigorous way to quantitate specific messages in RNA samples *(14–18)*. Competitive RT-PCR determines exactly how much of a specific RNA transcript is present. Results can be expressed as transcript mass or copies per microgram of total RNA. Competitive RT-PCR requires the addition of known amounts of a synthetic transcript, or "competitor RNA." The competitor is titrated into replicate RNA samples to create a standard curve for quantitation of the endogenous target. Both the competitor and the endogenous target are then reverse-transcribed and coamplified in the same RT-PCR reaction. If the amplification rates of the competitor and endogenous target are the same, the amount of PCR product for each can be directly compared (**Fig. 2**). Although both RNA and DNA competitors and "mimics" have been used, only a synthetic RNA transcript derived from the endogenous target will provide accurate results *(6*; Dr. David Brown, personal communication). The PCR product generated from the synthetic RNA competitor can be distinguished that of the endogenous target if, for example, it contains a small internal deletion of the endogenous product sequence (*see* **Fig. 3**).

Competitive RT-PCR requires less optimization than relative RT-PCR. Since both the endogenous and synthetic target are amplified with equal efficiency, the product will accumulate with the same kinetics, even after PCR components become limiting. Therefore, it is unnecessary to be in linear range of amplification when measuring the PCR products. Several pilot reactions must be conducted over a narrowing range of competitor RNA transcript dilutions for each sample, which makes handling multiple samples more difficult.

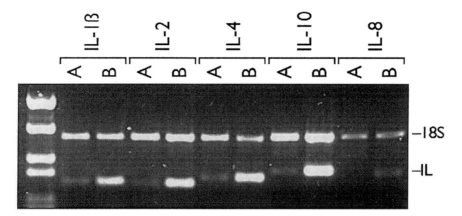

Fig. 1. Quantitation of several mRNAs using Relative RT-PCR. Mouse-liver total RNA was spiked with 1X (**A**) or 10X (**B**) amounts of sense strand RNA for five human cytokines. Following the protocols in **Subheading 3.**, cDNA was synthesized using random decamers. Each PCR reaction contained a pair of gene-specific primers (Ambion) and 18S rRNA primers and competimers (Ambion). The 18S rRNA amplification efficiency was attenuated by the non-functional primers (competimers) so that it was amplified within linear range under the same cycling parameters used for the much less abundant interleukin targets. The interleukin-specific signals reflect the 10X difference in input between sample A and sample B.

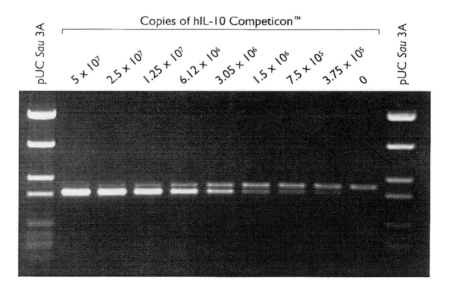

Fig. 2. Quantitation of human IL-10 mRNA using competitive RT-PCR. Two micrograms of sample RNA was reverse-transcribed in a 20 µL RT reaction, using the protocol described here. A separate RT was performed on the human IL-10 competitive RNA. One-microgram aliquots of dilutions of this RT, corresponding to the copy number specified in the figure, were added to the sample RT. The cDNA mixture was subjected to 30 rounds of PCR, and 5 µL of the reaction was run in an ethidium bromide-stained agarose gel. The "no template" reaction contained PCR primers and reagents only. The top band represents the endogenous IL-10 PCR product. The bottom band represents PCR product amplified from the IL-10 competitive RNA.

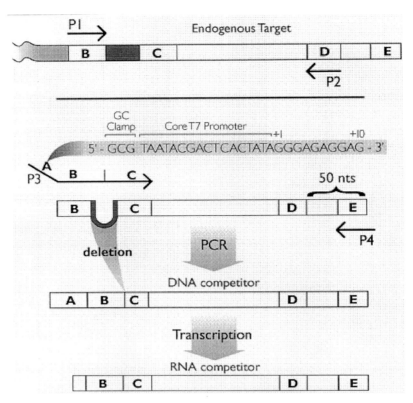

Fig. 3. Design and primer binding sites for synthetic competitive RNA construct. Where two primers (P1 and P2) are being used for PCR amplification of a given target, two additional primers are used to generate a deletion construct for competitive RT-PCR. Primer P3 includes a T7 promoter primer at its 5'-end, the P1 sequence in its middle, and a downstream target-specific sequence at its 3'-end. P3 creates the deletion and incorporates a transcription promoter for subsequent synthesis of an RNA competitor. P4 primes synthesis from the target cDNA at a site approx 50 nucleotides downstream of the P2 binding site. The extra 50 nucleotides ensure that the competitor is reverse-transcribed as efficiently as the endogenous target.

## 1.4. RNA Sample

Either total or poly(A) RNA can be used for relative and competitive quantitative RT-PCR experiments, but it is typically unnecessary to use the polyadenylated fraction. A recommended protocol for total RNA isolation is outlined in Chapter 32, Subheading 3.1., and uses acidified phenol:chloroform extraction of a guanidinium cell lysate to separate RNA from proteinacious cellular components and DNA. The reagents are standard in most laboratories, and the procedure can be scaled up or down to accommodate different sample sizes. However, the method of RNA isolation used for RT-PCR is not critical, and there are a variety of other protocols that will work equally well (*19–21*).

The RNA sample must be DNA-free to avoid false-positive products during the PCR step. There is no method of RNA isolation that can consistently eliminate genomic DNA from an RNA preparation (*see* **Note 2**). Therefore, it is important to treat the

RNA sample with DNase I and run appropriate controls to detect false positives amplified from contaminating genomic DNA. Primers should be designed to span an intron in the genomic sequence when possible (*see* **Subheading 1.5.**). Products generated from a "No RNA Control" point to contamination of reagents or equipment, while those from a "No RT Control" point to contaminating genomic DNA in the RNA sample.

## 1.5. Primer Choice and Design

The RT reaction can be primed with gene-specific primers, oligo dT, or random primers (hexamers or decamers). Whereas the primer annealing step of PCR takes place at temperatures above 50°C, primer annealing in the reverse transcription reaction occurs at 42°C, where secondary structure can form and interfere. Gene-specific primers bind to only one site, and thus are more likely to suffer from inefficient priming. Oligo dT primers require highly processive reverse transcription through potentially long poly(A) sequence and possible secondary structure. Random sequence primers bind along the length of the mRNAs, thus overcoming the issues of priming efficiency and RT processivity (*1,22,23*). If, in relative RT-PCR, one of the rRNA species is used as an internal control, random primers must be used in the RT step.

Gene-specific primers are used in the PCR step. These primers should be between 17–22 bases in length and with an average GC content. Select primers to amplify fragments of approx 200–500 bp and avoid amplifying regions high in secondary structure. For eukaryotic mRNAs where the genomic structure is known, choose primers that span an intron. If the genomic structure is not known, design primers to amplify a 300–400 bp fragment in the upstream end of the coding sequence. Hawkins et al. (*24*) have shown that exons longer than 300 bp are statistically rare in this part of the message. Therefore, primers 300–400 bp apart in the cDNA sequence should span an intron in the genomic sequence. If contaminating genomic DNA is present, the primers will either be too far apart to amplify product, or the amplified product will be distinctly larger than that expected from cDNA (*16*). Primers with GC, GG, or CC as the last two bases at their 3'-end will help prevent melting at this end, increasing extension efficiency. Also, in relative RT-PCR experiments, primers for the internal control and the gene(s) under study must be compatible to avoid amplification of nondesired targets. Software programs are recommended for design of primers free of features that contribute to the production of unwanted side products (e.g., primer dimers), and to determine annealing temperature (*1*; *see* **Note 3**).

## 1.6. Enzyme Choice

The reverse transcriptases, Molony Murine Leukemia Virus Reverse Transcriptase (M-MLV RT) and Avian Myeloblastosis Virus Reverse Transcriptase (AMV RT), are most commonly used in RT-PCR. They differ in stability (M-MLV > AMV), processivity (AMV > M-MLV), and lack of contaminating RNase H activity (M-MLV > AMV) (*1*). The source of Taq polymerase for RT-PCR applications where the amplified fragment size is <1 kb, is not critical.

## 1.7. Quantitation

Relative RT-PCR requires a method for detecting PCR products that is sensitive, has a wide dynamic range, and that is quantifiable. Unfortunately, EtBr staining is relatively insensitive, and has a narrow dynamic range (1–2 log). Alternative staining agents that will provide a greater range of sensitivity include Syber Green (Molecular Probes) and silver. Radiolabeled PCR products currently provide the greatest sensitivity and dynamic range of detection (approx 5 logs; *see* **Note 4**). Target and internal standard PCR products are distinguished by denaturing polyacrylamide gel electrophoresis. The products are then quantitated by one of two methods: phosphoimaging ($6.5 \times 10^4$ log dynamic range), or cutting out and counting the PCR products. In the following protocol, a trace radiolabel for quantitation of relative RT-PCR products, as well as the synthetic competitor RNA used in competitive RT-PCR, are included. Competitive RT-PCR does not require the sensitivity and wide dynamic range needed for relative RT-PCR, since there is no amplification cycle limit for the PCR. Therefore, either EtBr or radiolabeling can be used.

## 2. Materials

### 2.1. Buffers and Solutions

1. Use RNase-free water for all buffers and solutions.
2. Proteinase K/SDS solution: 10% SDS, 10 mg/mL Proteinase K, 250 µg/mL yeast RNA. Make up fresh.
3. 0.5 $M$ EDTA.
4. Phenol:$CHCl_3$ (1:1).
5. 5 $M$ $NH_4OAc$.
6. 100% EtOH.
7. 0.1 m$M$ EDTA.
8. RNase-free $H_2O$.
9. 10X RT-PCR buffer: 100 m$M$ Tris-HCl, pH 8.3, 500 m$M$ KCl, and 15 m$M$ $MgCl_2$.
10. 6% denaturing polyacrylamide urea gel: For 15 mL, enough for a 13 cm × 15 cm × 0.75 mm gel, mix 7.2 g urea, 1.5 mL 10X TBE, 2.25 mL 40% acrylamide (acrylamide:bis acrylamide = 19:1) and d$H_2O$ to 15 mL. Add 120 µL 10% ammonium persulfate, made up fresh in d$H_2O$. Finally, add 16 mL TEMED, mix briefly, and pour gel immediately (*see* **Note 5**).
11. 1X TBE: 89 m$M$ Tris, 89 m$M$ boric acid, 2 m$M$ EDTA.
12. 2X Denaturing gel loading buffer: 95% formamide, 18 m$M$ EDTA, 0.025% SDS, 0.025% each of Bromophenol blue and xylene cyanol.
13. Scintillation fluid.
14. 10X transcription buffer (use buffer provided with the polymerase; *see* **Note 6**).
15. 70% EtOH.
16. 10X DNase I reaction buffer (provided with DNase I).
17. TE: 10 m$M$ Tris-HCl, pH 8.0, 1 m$M$ EDTA.
18. 6X Agarose gel-loading buffer: 37% glycerol, 20 m$M$ Tris-HCl, pH 8.0, 5 m$M$ EDTA, 0.025% xylene cyanol, 0.025% bromophenol blue.
19. Agarose; molecular biology grade.
20. EtBr Stock; 1 mg/mL.

### 2.2. Proteins and Nucleic Acids

1. RNase-free DNase I (5 U/µL; supplied with 10X reaction buffer; Ambion).
2. Ribonuclease Inhibitor Protein (25 U/µL) or antiRNase™ (25 U/µL; Ambion).
3. M-MuLV RT (100–200 U/µL).
4. Thermostable DNA polymerase (4–5 U/µL).

5. RNA polymerase; 20 U/μL (*see* **Note 7**).
6. Sample RNA: total or poly(A) RNA, DNase I-treated; up to 2.5 μg in 16 μL or less for relative RT-PCR and 1–5 mg in up to 11 μL for competitive RT-PCR. Less sample can be used, depending on the target concentration. Sample RNA is isolated using the protocol in Chapter 32 and is DNase I treated as described in **Subheading 3.1.**
7. dNTP Mix; 2.5 m*M* each.
8. Random primers; 500 ng/μL; hexamers or decamers.
9. Gene-specific PCR primers; 10 μ*M* each and 5 μ*M* each.
10. [α-$^{32}$P]dCTP or dATP (10 mCi/μL); [α-$^{32}$P]CTP or ATP (*see* **Note 8**).
11. *(optional)* 18S rRNA primers and competimers (Ambion).
12. 5X rNTPs (2.5 m*M* each).
13. DNA PCR template (or 1 μg plasmid DNA template); 0.2 μg in 12 μL or less.
14. Competitive transcript; 0.1 μ*M* or 6 × 10$^{10}$ copies/μL.

## 2.3. Equipment and Supplies

1. Thermal cycler.
2. Filter paper, the size of the gel.
3. Plastic wrap to cover the gel.
4. Phosphoimager or XAR or BlueMax film (Kodak) and scintillation counter.
5. Logarithmic graph paper.
6. Scintillation counter.
7. Siliconized microfuge tubes.

# 3. Methods
## 3.1. DNase I-Treating the RNA Sample

1. Resuspend the RNA pellet in 50–200 μL of RNase-free water.
2. Add 1/10 vol of 10X DNase I reaction buffer to the RNA sample.
3. Add 1 U DNase I /g of starting tissue (1 × 10$^8$ cells).
4. Incubate 15–30 min at 37°C.
5. Inactivate by either:
   a. Add EDTA to 4 m*M* and then heat to 70°C for 5 min (*25*; see **Note 9**).
   b. Proteinase K/SDS treatment and phenol:CHCl$_3$ extraction.
6. Reprecipitate the RNA. Add 5 *M* NH$_4$OAc to a final concentration of 0.5 *M* and 2.5 vol of 100% EtOH. Mix and store at –20°C for at least 30 min.
7. Centrifuge at 10,000–12,000*g* (e.g., full speed in a microfuge) for 15 min to collect the precipitated RNA.
8. Resuspend the RNA pellet in 50–100 μL of 0.1 m*M* EDTA/100 mg tissue (1 × 10$^7$ cells).
9. The RNA concentration can be determined from sample absorbance at 260 nm.
10. Check the RNA sample integrity by running a 1-μg aliquot on an agarose gel and staining with EtBr. The rRNA bands should be sharp, and the 28S rRNA band should be about twice the intensity of the smaller 18S rRNA band.
11. Aliquot the RNA and store at –20°C short-term, or –80°C long-term.

## 3.2. Relative RT-PCR

First, the PCR cycle number is determined such that all PCR products will be quantitated while still in the linear range of amplification. The transcript of greatest abundance in your sample will exceed linear range at the earliest, i.e., at the lowest number of cycles. At this number of cycles, transcripts of lesser abundance will still be amplified within linear range. The transcript in the sample predicted to have the highest expression level, therefore, should be used in the first pilot experiment. The internal control transcript is usually chosen unless competimer technology will be used. If competimer technology will be used, select the experimental sample where expression of the transcript under study is believed to be highest. A second pilot experiment is then performed to determine the 18S rRNA competimer primer ratio needed to attenuate 18S rRNA amplification. Random decamers are used to prime cDNA synthesis in the RT reaction. For the PCR step, gene-specific primers are used to amplify the transcripts under study, and primers and optional competimers are used to amplify the internal control. A trace amount of radiolabeled dNTP and scintillation or Cherinkov counting are used to make measurements in the PCR steps.

### 3.2.1. Pilot Experiment: Determine Cycle Number for Final Experiment

#### 3.2.1.1. RT REACTION

1. Assemble the reaction on ice in thin wall PCR tubes: sample RNA, 4 μL dNTP mix (2.5 μM each), 2 μL random primers (500 ng/μL); bring volume to 16 μL with RNase-free $H_2O$.
2. Heat for 3 min at 85°C to remove RNA secondary structure (this should be skipped if there is $Mg^{2+}$ present in the RNA sample; *see* **Note 9**).
3. Add the remaining components: 2 μL 10X RT-PCR buffer, 1 μL ribonuclease inhibitor protein or AntiRNase™ (25 U), and 1 μL M-MuLV RT (100–200 U).
4. Flick the tube to mix, and spin briefly.
5. Incubate at 42°C for 1 h.
6. If you do not plan to proceed directly to amplification, or do not use up all of the RT reaction, store at –20°C.

#### 3.2.1.2. PCR

1. On ice, set up a PCR "Master Mix" for 10 reactions: 10 μL RT reaction (from above), 50 μL 10X RT-PCR buffer, 40 μL dNTP mix (2.5 m*M* each), 40 μL gene-specific primers (5 μM each), 2.5 μL Taq polymerase (4–5 U/μL), 344 μL $H_2O$, and 5 μL [α-$^{32}$P]dCTP or dATP (10 mCi/μL).
2. Mix by flicking tube, spin down briefly, and split the Master Mix into 10 tubes of 50 μL each.
3. Program the PCR machine for 33 cycles; 94°C for 30 s; annealing temperature of gene-specific primers for 30 s; 72°C for 30 s.
4. Preheat the thermal cycler before adding the samples to the machine.
5. Remove one of the samples after each odd-numbered cycle starting with 15 and ending with 33: 15, 17, 19, and so on.
6. Add denaturing gel-loading buffer to 10 μL of each reaction, heat at 95°C for 3 min, load all or half of samples onto a 6–8% denaturing acrylamide gel, and electrophorese in 1X TBE. While native gels can be used, the bands may appear as doublets, because PCRs often contain both double- and single-stranded products.
7. Transfer gel to filter paper, cover with plastic wrap, expose to film. Use the developed film to localize bands, excise them, and count in a scintillation counter. Alternatively, the gel products can be quantitated using a phosphoimager.
8. Plot the log of the PCR product signal vs the number of cycles. Find the middle of the linear part of the curve to determine the ideal cycle number since, as you add other primer sets (e.g., for amplification of the internal control), the linear range curve will be shifted to a lesser number of cycles.

## 3.2.2. Pilot Experiment to Determine 18S rRNA Primer:Competimer Ratio

If competimer technology will be used to attenuate amplification of the internal control, 18S rRNA, then an additional pilot experiment is run to determine the appropriate ratio of 18S rRNA primers to competimers. The goal is to select a primer:competimer ratio that result in similar yields of product for both the 18S rRNA internal control and the target RNA under study. A 3:7 ratio is appropriate for most target transcripts, although a 2:8 or 1:9 ratio may be needed for extremely rare messages.

1. Prepare the primer:competimer mixes as follows:

   18S rRNA primer:competimer ratio

   |  | 1:9 | 2:8 | 3:7 |
   |---|---|---|---|
   | 18S rRNA competimers | 9 μL | 8 μL | 7 μL |
   | 18S rRNA primers | 1 μL | 2 μL | 3 μL |

2. On ice mix: 5 μL RT reaction (from **Subheading 3.2.1.**), 27.5 μL 10X RT-PCR buffer, 22 μL dNTP mix (2.5 m$M$ each), 1.38 μL Taq polymerase (4–5 U/μL), 175 μL H$_2$O, and 2.5 μL [α-$^{32}$P]dCTP or [α-$^{32}$P]dATP.
3. Aliquot 42 μL of the PCR cocktail into five PCR tubes labeled 1–5.
4. Add 4 μL of gene specific primers to tubes 1–4.
5. Add 4 μL of 18S rRNA primer and competimer mixtures as follows: 4 μL of 1:9 to tube 2, 4 μL of 2:8 to tube 3, and 4 μL of 3:7 to tubes 4 and 5.
6. Perform PCR as in **Subheading 3.2.1.** using the cycle number determined there.
7. Assess the results by denaturing polyacrylamide urea gel electrophoresis (**Subheading 3.2.1., steps 6** and **7**).
8. Identify the lane in which the level of 18S rRNA product is most similar to the level of gene-specific product. The primer:competimer ratio used in this sample is selected for the final experiment.

## 3.2.3. Final Experiment

Once cycling parameters for linear range, and 18S rRNA primer:competimer ratio have been established, set up the actual relative quantitative RT-PCR experiment for all of the RNA samples.

### 3.2.3.1. RT REACTION

1. Use the same amount of input RNA for all samples as used in the pilot experiments. Set up Master Mixes of the dNTPs, random primers, and RNase-free H$_2$O minus template; and 10X RT-PCR buffer, ribonuclease inhibitor protein or antiRNase, and M-MuLV RT as before (**Subheading 3.2.1.**).
2. Add template (and RNase-free H$_2$O to adjust for different template input vols) to the sample-specific reaction tubes.
3. Then add the RT master mix.

### 3.2.3.2. PCR

1. Again set up a PCR master mix (as in **Subheading 3.2.1.**), this time with 10% overage for use with individual samples. Use the same amount of RT reaction in the PCR as that used for the pilot experiment.
2. Set the thermal cycler profile for the number of cycles determined in the pilot experiment in **Subheading 3.2.1.**
3. Analyze RT-PCR products as in **Subheading 3.2.1., steps 6** and **7**.

4. Quantitate PCR products by for each sample, dividing the signal obtained for the endogenous target by the signal obtained for the internal control. This will provide a corrected relative value for the gene under study for each sample. Compare these values across samples for an estimate of the relative expression of endogenous target RNA in the samples.

## 3.3. Competitive RT-PCR

### 3.3.1. Competitor RNA Construct Design and Transcription

Synthesis of an RNA competitor involves using PCR to create a deletion in the target cDNA sequence, and to add a T7 RNA phage promoter. The resultant PCR product is then in vitro transcribed, and the RNA competitor is quantitated. A schematic representation of competitor template design and location of the primer binding sites is shown in **Fig. 3**. The most convenient way to add a polymerase promoter is to include it in the 5' PCR primer used to construct the deletion (**Fig. 3**, Primer P3). In addition, at least 50 base pairs of extra sequence 3' to the downstream (3') binding site (**Fig. 3**, Primer P4) is required for optimal primer binding during reverse transcription.

1. Assemble the following components at room temperature in the following order: Up to 20 μL final volume RNase-free H$_2$O, 2 μL 10X Transcription buffer, 4 μL 5X rNTPs (2.5 mM each), 0.2 μg DNA PCR template (or 1 μg plasmid DNA template), 0.4 μL [α-$^{32}$P]CTP or ATP, and 2 μL RNA polymerase.
2. Flick tube to mix, and spin down briefly.
3. Incubate 2–4 h at 37°C.
4. Remove an aliquot from the reaction for quantitation (total input counts).
5. Precipitate the transcription reaction with 0.5 *M* NH$_4$OAc and 2.5 vol of 100% EtOH. Incubate at –20°C for 30 min, then collect the pellet by microfuging at 4°C at maximum speed for 30 min. Aspirate off the supernatant and wash the pellet with 70% EtOH.
6. Remove template DNA by dissolving the pellet in 20 μL of 1X transcription buffer or 1X DNase I reaction buffer. Add 2 μL DNase I (5 U/μL) to the reaction, flick to mix, and incubate at 37°C for 1 h.
7. Gel-purify the competitive RNA transcript on a denaturing polyacrylamide urea gel, typically a 5–6% gel will sufficiently separate transcripts of 151–500 nt in length. Use 1X TBE as the running buffer. Load the samples in gel-loading buffer.
8. Leave the gel on the bottom glass plate; cover with plastic wrap, and expose to film (30 s–2 min). Align the film under the gel and cut out the gel-purified transcript. Cut out the smallest gel fragment possible (no more than 1 mm in thickness); the gel can always be re-exposed to determine whether or not the transcript has been successfully removed.
9. Chop up the gel fragment into small pieces and transfer into a microfuge tube. Cover with 350 μL of a buffer, such as 0.1 m*M* EDTA or 0.1% SDS, and incubate the tube overnight at 37°C.
10. Transfer the solution containing the eluted RNA away from the gel fragment, into a new microfuge tube.
11. Remove an aliquot from the purified RNA competitor for quantitation (purified competitor RNA counts).
12. Quantitate the competitive transcript. One microliter of gel-purified, full-length competitor and 1 μL of the transcription reaction are placed in scintillant and counted. The resulting numbers can be entered into the equations below to determine the competitor concentration in micromolarity (m*M*) or in copies/μL (*see* **Note 10**).

$$\mu M \text{ competitor RNA} = \frac{\text{cpm/\mu L purified competitor RNA}}{\text{cpm/\mu L of transcription reaction}} \times \frac{500 \text{ m}M \text{ CTP or ATP in reaction}}{\text{\# Gs or As in the competitor RNA}}$$

$$\begin{array}{l}\text{Copies of competitor}\\ \text{RNA/\mu L}\end{array} = \frac{\text{cpm/\mu L purified competitor RNA}}{\text{cpm/\mu L of transcription reaction}} \times \frac{3 \times 10^{14} \text{ molecules CTP or ATP/\mu L rxn}}{\text{\# Cs or As in the competitor RNA}}$$

13. The competitive RNA transcript should be stored at concentrations >0.1 $\mu M$ or $6 \times 10^{10}$ copies/$\mu$L. Use siliconized tubes to prevent copy number change resulting from adherence to the tube surface.

### 3.3.2. Competitive RT-PCR

Perform pilot experiments to estimate how much synthetic competitor RNA transcript to use for the final experiment by reverse-transcribing $10^9$ copies of synthetic RNA transcript and separately reverse-transcribing each of the sample RNAs. Set up replicate PCRs for each sample RT, and add a dilution series of the synthetic RNA transcript RT to the replicates. This is faster and requires less reagents than doing separate RT reactions with different amounts of competitive RNA transcript mixed with the experimental RNA. Note that variations in input and RT efficiency are not accounted for until the final experiment, since, in the pilot experiments, the competitive RNA transcript and sample are reverse-transcribed separately. The pilot experiment is only meant to provide an estimate of copy number that will form a basis for the more precise final experiment(s). The PCR reactions are analyzed on a native agarose gel; the dilution of competitive RNA transcript that yields a similar amount of PCR product as the experimental samples becomes the basis for subsequent experiments.

#### 3.3.2.1. RT REACTION

1. The proper amount of input-sample RNA will depend upon the relative abundance of the target message in a given sample. A reasonable starting amount is 1–5 µg of total RNA.
2. Assemble RT reactions on ice. The template will be the competitive RNA transcript in one reaction, and the RNA samples in the other reactions.

    | | |
    |---|---|
    | 11 µL | RNA + RNase-free $H_2O$ |
    | 4 µL | dNTP mix (2.5 m$M$ each) |
    | 1 µL | Random primers (500 ng/µL) |
    | 2 µL | 10X RT-PCR buffer |
    | 18 µL | Final vol |

3. Flick tube to mix and spin down briefly.
4. Heat at 75°C for 3 min to denature the RNA, then return to ice to cool down.
5. Add the remaining reaction components: 1 µL ribonuclease inhibitor protein or anti-RNase (25 U) and 1 µL M-MuLV RT (100–200 U).
6. Flick tube to mix, and spin down briefly.
7. Incubate at 42°C for 1 h.
8. Inactivate the RT by heating at 95°C for 5 min.
9. Cool the reactions on ice.

#### 3.3.2.2. PCR

1. Dilute the RT reaction containing the competitive RNA transcript as follows:
    Undiluted: 1 µL of RT = $5 \times 10^7$ copies/µL;
    $10^{-2}$ dilution: dilute 1 µL of RT into 99 µL TE = $5 \times 10^5$ copies/µL;
    $10^{-4}$ dilution: dilute 1 µL of RT into 99 µL TE = $5 \times 10^3$ copies/µL;
    $10^{-6}$ dilution: dilute 1 µL of RT into 99 µL TE = $5 \times 10^1$ copies/µL.

2. For each sample RNA, set up six reactions; #1–4 will contain the sample RNA + one of the competitive RNA-transcript dilutions, #5 will contain no competitive transcript, and #6 will not contain sample RNA or competitive transcript (negative control):

Sample A

| Tube # | 1 | 2 | 3 | 4 | 5 | 6 |
|---|---|---|---|---|---|---|
| Sample RNA RT (see **Note 11**) | $x$ µL | $x$ µL | $x$ µL | $x$ µL | $x$ µL | $x$ µL $H_2O$ |
| Competitive RNA transcript RT | 1 µL | 1 µL | 1 µL | 1 µL | 1 µL $H_2O$ | 1 µL $H_2O$ |
| (copies of competitive sequence) | $5 \times 10^7$ | $5 \times 10^5$ | $5 \times 10^3$ | $5 \times 10^1$ | 0 | 0 |
| $H_2O$ to final volume | 36 µL | 36 µL | 36 µL | 36 µL | 36 µL | 36 µL |

3. For each sample (set of six reaction tubes), prepare the following master mix: 35 µL 10X RT-PCR buffer, 35 µL dNTPs (2.5 m$M$ each), 28 µL gene-specific PCR primers (10 µ$M$ each), and 1.75 µL thermostable DNA polymerase (4–5 U/µL).
4. Perform PCR using the following cycling profile: Soak at 94°C for 3 min; cycle 30X: 94°C, 20 s; 55–60°C, 30 s; 72°C, 30 s; soak at 72°C for 5 min; and soak at 4°C indefinitely.
5. Assess the results of the pilot experiment by running of the samples on a 2–2.5% agarose gel, containing 1 µg/mL EtBr, or a 5% nondenaturing polyacrylamide gel, poststained with EtBr.
6. Do a second pilot experiment to further narrow the range of competitive transcript copies equivalent to the endogenous target. Use the number of copies of competitive RNA transcript that yields a similar amount of PCR product as the experimental sample in the first pilot to narrow the range of competitive RNA transcript dilutions for the second pilot experiment.
7. The RT reactions do not need to be repeated in the second pilot experiment; the same RT reactions from the first pilot will be used in this PCR. Set up the PCR as in **steps 1–5** above, but make smaller dilutions of competitive RNA transcript covering only 1–2 logs across the five tubes.
8. Again select the number of copies of competitive RNA transcript that yields PCR product of approximately equal intensity as that from the endogenous target. This number becomes the focus for competitive RNA transcript dilutions in the final experiment.
9. To control for variations in RT efficiency, the final experiment requires that the diluted competitive RNA transcript be added directly to the sample RNA prior to reverse transcription. Use dilutions within one-half log of the number of competitive RNA transcript copies indicated by the second pilot experiment results. Typically, 3–5 dilutions are sufficient.
10. The number of competitive RNA transcript copies needed to produce an RT-PCR product, the intensity of which matches that produced by the endogenous message, represents the amount of endogenous message present in the RNA sample.

## 4. Notes

1. US Patent pending #06057134.
2. We have found that total RNA isolated by acid phenol chloroform extraction, including single-reagent extraction methods (26), glass binding (27,28) and CsCl gradient centrifugation (29), and mRNA purification using oligo dT (30,31), still contained amplifiable DNA contamination (Ambion's TechNotes 4(3) p. 6, 1997).
3. We have primarily used Oligo 7 (Medprobe, Oslo, Norway); however, other available programs may be equally suitable. Be aware that some primer-design programs force both primers to be the same length, and therefore limit the number of possible primer pairs over a given region.
4. Incorporate $^{32}P$ (or another radiolabel, such as $^{35}S$ or $^{33}P$) into the PCR products either by end-labeling the PCR primers with [γ-$^{32}P$]ATP using T4 polynucleotide kinase or by

including a trace label (e.g., 1 μL of 10 mCi/mL, 3000 Ci/mmol [$\alpha$-$^{32}$P]dCTP) in the PCR reaction.

5. Adding the last reagent will start the polymerization reaction. By modulating the temperature of the gel solution, the polymerization rate can be slowed (cold gel solution) or hastened (warm gel solution, or heat lamp applied to poured gel).

6. Each of the three different RNA phage polymerases requires a different salt concentration for optimal activity. Commercial sources of these polymerases usually provide a 10X transcription buffer with the enzyme, but vary regarding whether they add the salt to the transcription buffer or the enzyme. Therefore, it is important to use the 10X transcription buffer provided with the enzyme.

7. T7, T3, or SP6 RNA phage polymerase; be sure to use the transcription buffer provided with the enzyme to ensure the correct final concentration of salt in the reaction, *see* **Note 6**.

8. The specific activity of the radiolabel tracer (Ci/mmol) is not important.

9. Although **ref. 25** recommends heat inactivation of the DNase I reaction directly, there is also evidence that heating RNA in the presence of $Mg^{2+}$ cations found in the DNase I reaction buffer can cause some enzyme-independent strand scission (Ambion, personal communication). Therefore, this protocol calls for chelating $Mg^{2+}$ cations with EDTA prior to heating or using an alternative method of DNase I inactivation.

10. If [$\alpha$-$^{32}$P]CTP was used to trace-label the competitor RNA, use the number of Cs in the competitor sequence; if [$\alpha$-$^{32}$P]ATP was used, count the number of As.

11. The input volume of sample RT used will depend on target abundance and primer efficiency, and may need to be adjusted. Try starting with 2 μL of the RT reaction. A band should be visible after 30 cycles of PCR on an EtBr-stained agarose gel.

## Acknowledgments

I thank Dr. Eric Lader and Dr. David Brown (Ambion) for protocol development and technical advice, and Lader, Brown, and Lori Martin for critical editing of this manuscript.

## References

1. Kawasaki, E. S. (1990) Amplification of RNA, in *PCR Protocols: A Guide to Methods and Applications* (Innis, M. A., Gelfand, D. H., Sninsky, J. J., and White, T. J., eds.), Academic Press, London and New York, pp. 21–27.

2. Kelso, A., Groves, P., Ramm, L., and Doyle A. G. (1999) Single-cell analysis by RT-PCR reveals differential expression of multiple type 1 and 2 cytokine genes among cells within polarized CD4+ T cell populations. *Int. Immunol.* **11(4),** 617–621.

3. Malnic, B., Hirono, J., Sato, T., and Buck, L. B. (1999) Combinatorial receptor codes for odors. *Cell* **96(5),** 713–723.

4. Uchide, T., Masuda, H., Mitsui, Y., and Saida, K. (1999) Gene expression of vasoactive intestinal contractor/endothelin-2 in ovary, uterus, and embryo: comprehensive gene expression profiles of the endothelin ligand-receptor system revealed by semi-quantitative reverse transcription-polymerase chain reaction analysis in adult mouse tissues and during late embryonic development. *J. Mol. Endocrinol.* **22(2),** 161–171.

5. Wong, H., Anderson, W. D., Cheng, T., and Riabowol, K. T. (1994) Monitoring mRNA expression by polymerase chain reaction: the "primer dropping" method. *Anal. Biochem.* **223,** 251–258.

6. Freeman, W. M., Walker, S. J., and Vrana, K. E. (1999) Quantitative RT-PCR: pitfalls and potential. *BioTechniques* **26,** 112–125.

7. Wang, S., Murtagh, J. J., Jr., Luo, C., and Martinez-Maldonado, M. (1993) Internal cRNA standards for quantitative northern analysis. *Biotechniques* **14(6),** 935–942.

8. Davis, M. J., Bailey, C. S., and Smith, C. K., II (1997) Use of internal controls to increase quantitative capabilities of the ribonuclease protection assay. *Biotechniques* **23(2),** 280–285.

9. DeLeeuw, W., Siaboom, P., and Vijg, J. (1989) Quantitative comparison of mRNA levels in mammalian tissues: 28S ribosomal RNA level as an accurate internal control. *Nucleic Acids Res.* **17,** 10,137–10,138.

10. Mansur, N., Meyer-Siegler, K., Wurzer, J., and Sirover, M. (1993) Cell cycle regulation of the glyceraldhyde-3-phosphate dehydrogenase/uracil DNA glycosylase gene in normal human cells. *Nucleic Acids Res.* **4,** 993–998.

11. Spanakis, E. (1993) Problems related to the interpretation of autoradiographic data on gene expression using common constitutive transcripts as controls. *Nucleic Acids Res.* **16,** 3809–3819.

12. Bhatia, P., Taylor, W., Greenberg, A., and Wright, J. (1994) Comparison of glyceraldehyde phosphate dehydrogenase and 28S ribosomal RNA gene expression as RNA loading controls for northern blot analysis of cell lines of varying malignant potential. *Analyt. Biochem.* **216,** 223–226.

13. Vogt, T., Stolz, W., Welsh, J., Jung, B., Kerbel, R. S., Kobayashi, H., Landthaler, M., and McClelland, M. (1998) Over expression of Lerk-5/Eplg5 messenger RNA: a novel marker for increased tumorigenicity and metastatic potential in human malignant melanomas. *Clin. Cancer Res.* **4,** 791–797.

14. Borson, N. D., Strausbauch, M. A., Wettstein, P. J., Oda, R. P., Johnston, S. L., and Landers J. P. (1998) Direct quantitation of RNA transcripts by competitive single-tube RT-PCR and capillary electrophoresis. *Biotechniques* **25(1),** 130–137.

15. Liu, Z. F. and Burt, D. R. (1998) A synthetic standard for competitive RT-PCR quantitation of 13 GABA receptor type A subunit mRNAs in rat and mice. *J. Neurosci. Meth.* **85,** 89–98.

16. El-Osta, A., Kantharidis, P., and Salcberg, J. (1999) Absolute quantitation of MDR1 transcripts using heterologous DNA standards–validation of the competitive RT-PCR (CRT-PCR) approach. *Biotechniques* **26,** 1114–1116.

17. Klein, S. A., Ottmann, O. G., Ballas, K., Dobmeyer, T. S., Pape, M., Weidmann, E., Hoelzer, D., and Kalina, U. (1999) Quantification of human interleukin 18 mRNA expression by competitive reverse transcriptase polymerase chain reaction. *Cytokine* **11,** 451–458.

18. Riedy, M. C., Timm, E. A., Jr., and Stewart, C. C. (1995) Quantitative RT-PCR for measuring gene expression. *BioTechniques* **18,** 70–76.

19. Ausubel, F. M. Brent, R., Kingston, R. F., Moore, D. D., Seidman, J. G., Smith, J. A., and Struhl, K., eds. (1987) *Current Protocols in Molecular Biology,* Greene Publishing Associates and Wiley-Interscience, New York, NY.

20. Rapley, R. and Manning, D. L., eds. (1998) *Methods in Molecular Biology, vol. 86: RNA Isolation and Characterization Protocols,* Humana Press, Totowa, NJ.

21. Maniatis, T., Fritsch, E. F., and Sambrook, J., eds. (1989) *Molecular Cloning. A Laboratory Manual,* 2nd ed., Cold Spring Harbor Laboratory, Cold Spring Harbor, NY.

22. Veres, G., Gibbs, R. A., Scherer, S. E., and Caskey, C. T. (1987) The molecular basis of the sparse fur mouse mutation. *Science* **237,** 415–417.

23. Noonan, K. E. and Roninson, I. B. (1988) mRNA phenotyping by enzymatic amplification of randomly primed cDNA. *Nucleic Acids Res.* **16,** 10,366.

24. Hawkins, J. D. (1988) A survey of intron and exon lengths. *Nucleic Acid Res.* **16,** 9893–9908.

25. Huang, Z., Fasco, M. J., and Kaminsky, L. S. (1996) Optimization of DNase I removal of contaminating DNA from RNA for use in quantitative RNA-PCR. *BioTechniques* **20(6),** 1012–1020.

26. Chomzynski, P. and Sacchi, N. (1987) Single-step method of RNA isolation by acid guanidinium thiocyanate-phenol-chloroform extraction. *Analyt. Biochem.* **162,** 156–159.

27. Boom, R., Sol, C. J. A., Salimans, M. M. M., Jansen, C. L., Wertheim-Van Dillen, P. M. E., and Van Der Noordaa, J. (1990) Rapid and simple method for purification of nucleic acids. *J. Clin. Microbiol.* **283,** 495–503.

28. Marko, M., Chipperfield, R., and Birnboim, H. C. (1982) A procedure for the large scale isolation of highly purified plasmid DNA using alkaline extraction and binding to glass powder. *Analyt. Biochem.* **121,** 382–387.

29. Glisin, V., Czkvenjakov, R., and Byus, C. (1973) Ribonucleic acid isolated by cesium chloride centrifugation. *Biochemistry* **13,** 2633.

30. Chirgwin, J. M., Przybyla, A. E., MacDonald, R. J., and Rutter, W. J. (1979) Isolation of biologically active ribonucleic acid from sources enriched in ribonuclease. *Biochemistry* **24,** 5294–5299.

31. Thompson, J. and Gillespie, D. (1987) Molecular hybridization with RNA probes in concentrated solution of guanidine thiocynate. *Anal. Biochem.* **163,** 281–291.

# 6

# Expressing Self-Incompatibility RNases (S-RNases) in Transgenic Plants

## Brian Beecher and Bruce A. McClure

## 1. Introduction

The most obvious functions of nucleases are nucleic-acid turnover and phosphate mobilization, but they also have more subtle roles. By its nature, nuclease activity may disrupt cellular information flow. Familiar examples include degradation of specific mRNAs, and bacterial restriction endonucleases. In some plant families, RNases have been recruited for highly specific pollen rejection systems. This chapter reviews methods and protocols used to manipulate plant breeding behavior using these RNases. Here, the "assay" for RNase activity is different from the others described in this volume. In our system, RNases determine mating behavior; thus, the ultimate "assay" is whether or not a fruit is formed following pollination.

### 1.1. Reproduction in Plants

In angiosperm pollination, pollen grains germinate on the stigma to form a pollen tube that carries the sperm cells through the extracellular matrix to the ovule (*see* **Note 1**). Mating can be controlled throught interference with any stage of this process. There is a window of genetic relatedness for optimal mating, so mechanisms for determining relationships are essential. Inbreeding depression may result from crossing closely related plants, whereas crosses that are "too wide" may lead to inviable offspring. In some families, RNases directly mediate pollen–pistil interactions that help to define this "window of relatedness" for compatible mating.

### 1.2. Self-Incompatibility

Self-incompatibility (SI) systems are genetically programmed to prevent self-fertilization *(1)* thereby maintaining hybrid vigor and genetic diversity. The underlying genetic and biochemical mechanisms of SI may be totally unrelated in different plant families, but in most systems a multiallelic S-locus provides for specific recognition and rejection of self-pollen or pollen from closely related plants. SI systems are classified as sporophytic or gametophytic (*see* **Note 2**), depending on whether the pollen's behavior is determined by the plant that produced it (i.e., the sporophyte) or by its own haploid genome (i.e., the gametophyte).

From: *Methods in Molecular Biology, vol. 160: Nuclease Methods and Protocols*
Edited by: C. H. Schein © Humana Press Inc., Totowa, NJ

In gametophytic SI, pollen is rejected if its single S-allele is the same as an S-allele in the pistil (*see* **Note 3**). Therefore, compatible matings can occur between plants that share one S-allele; complete rejection only occurs if both S-alleles in both parents are the same. In the Solanaceae, Scrophulariaceae, and Rosaceae (*see* **Note 4**), the S-locus encodes RNases that control the specificity of SI. In *Nicotiana*, these RNases also control interspecific pollination.

### 1.3. RNase Based Pollen Rejection

The most extensive studies of RNase-based systems have been in the Solanaceae. In *Nicotiana alata*, it was noted that certain abundant style proteins cosegregate with the pollen rejection phenotype *(2)*. Anderson et al. (1986) showed that in *N. alata*, a 32 kDa stylar glycoprotein was present in plants that rejected $S_2$-pollen *(3)*. S-glycoprotein cDNAs were quickly cloned from *N. alata* and *Petunia* species *(4–6)*. The S-glycoproteins accumulate in the extracellular matrix, forming a pathway from the stigma surface to the ovary *(5,7)*. They are typically basic glycoproteins of 28–35 kDa, and are extraordinarily abundant, reaching approximately millimolar concentrations (10–50 mg/mL) *(8)*. Allelic sequences are only about 50% identical, a startling level of divergence for alleles of a single locus *(9)*. This sequence diversity is probably related to the unique role the S-locus plays in SI species. Other RNases would not be expected to show such divergence. S-alleles are under balancing selection. As a consequence, they may be much older than the species in which they reside. Thus, gene trees constructed with allelic cDNA sequences often cluster across species or even genus boundaries *(10,11)*. However, recent results suggest that balancing selection alone may not be sufficient to explain the diversity of S-alleles *(12)*. S-RNase sequence diversity has thus been useful in studies of plant population genetics and demography.

The discovery that the S-glycoproteins are RNases was initiated by noting that the *N. alata* $S_2$-cDNA contained the sequences FTIHGLWP and KHGTC, which are similar to active-site sequences in *Aspergillus oryzae* RNase T2 *(13,14)*. Direct enzymatic assays confirmed that five S-glycoproteins from *N. alata* were indeed active RNases *(14)*. Since they are products of the S-locus, they are now referred to as S-RNases Experiments showing that S-RNases can enter pollen tubes, and that pollen RNA is degraded after incompatible pollinations, suggested that RNase activity was directly responsible for pollen rejection *(15,16)*. Thus, S-RNases would be both the carriers of S-allelic-specificity information and agents that directly inhibit incompatible pollen-tube growth.

Less is known about interspecific pollination, but systems where inter- and intraspecific compatibility are related are experimentally the most tractable. Lewis and Crow (1958) observed that crosses between SI species and their self-compatible (SC) relatives often follow the SI × SC rule: the SI pistil rejects pollen from the SC species, but the reciprocal cross is compatible *(17)*. Although genetic studies implicate the S-locus *(15)*, this is controversial because there are important exceptions to the SI × SC rule *(18)*.

As RNase-based pollen rejection occurs in at least three families, the mechanism may have evolved early in the angiosperm lineage, perhaps from a pathogen rejection system. S-RNase homologs, members of the Rh/T2/S-glycoprotein family of RNases, occur widely in nature, although there is no evidence for extensive polymorphisms in non-S-RNase homologs. Family members have been identified in human and mouse *(19)*, fruit fly *(20)*, *Escherichia coli* *(21)*, yeast *(22)*, and *C. elegans* *(23)*, as well as *A.*

*oryzae.* These genes probably have functions as diverse as the organisms that express them; some may play roles in host response to pathogens. There is no evidence for extensive allelic polymorphism similar to the S-RNases among these nonplant genes. It is noteworthy that RNase A superfamily members are present in human eosinophil granules where they may play a role in defense *(24,25)*.

## 1.4. Transgenic Plant Results

### 1.4.1. Dual Roles for S-RNase in Pollen Rejection

The cytotoxic model predicts that S-RNases have both specificity and cytotoxic functions. These predictions have been directly tested using transgenic plant methods *(26–28)*. For example, expressing $S_{A2}$-RNase in transgenic (*N. langsdorffii* × SC *N. alata*) hybrids caused specific rejection of $S_{A2}$-pollen *(28)*, providing definitive evidence that S-RNases are the carriers of S-allele-specificity information. Kao's group showed that RNase activity is required for pollen rejection by expressing an inactive $S_3$-RNase mutant in *P. inflata* *(26)*. These results provided critical support for the cytotoxic hypothesis.

### 1.4.2. Allelic Specificity of S-RNases

To determine which sequence domains are required for recognition, chimeric S-RNases containing sequences from $S_{A2}$- and $S_{C10}$-RNase were generated and expressed in (*N. langsdorffii* × SC *N. alata*) hybrids. In four constructs, single domains were exchanged; and in another five constructs, larger contiguous domains were exchanged. In control transgenic hybrids, recombinant $S_{A2}$- and $S_{C10}$-RNase genes behaved as expected. $S_{A2}$-RNase caused rejection of $S_{A2}$-pollen, but not $S_{C10}$-pollen. $S_{C10}$-RNase caused rejection of $S_{C10}$-pollen, but not $S_{A2}$-pollen. The chimeric S-RNases were expressed at similar levels and were enzymatically active. However, none of the chimeras caused rejection of either $S_{A2}$- or $S_{C10}$-pollen *(29)*. This suggests that allelic specificity information is distributed throughout the sequence. However, experiments in *Solanum* showed that it is possible to alter the allelic specificity of S-RNases by changing a small number of amino acids *(30)*. Thus, the determinants of allelic specificity are still controversial.

### 1.4.3. RNases in Interspecific Pollination

Similar experiments showed that rejection of pollen from SC *N. plumbaginifolia* by SI *N. alata* (i.e., a system that follows the SI × SC rule), is mechanistically similar to SI *(31)*. S-RNase is also implicated in rejecting pollen from *N. tabacum* (i.e., a system that does not follow the SI × SC rule), but the mechanism is distinct from *N. plumbaginifolia* pollen rejection and SI *(31)*. Therefore, at least in *Nicotiana*, it is clear that RNases are involved in pollen rejection at both the inter- and intraspecific levels, but the genetic mechanisms are very complex.

## 1.5. Strategy for Directing Expression of Stylar S-RNases in Plants

Experimentally, the transformation procedure involves the following steps:

1. Creating an appropriate gene construct.
2. Culturing a suitable explant, often a leaf or an embryonic organ.

3. Introducing a cloned gene, often by *Agrobacterium*-mediated transformation (*see* **Note 5**).
4. Selection for transformed cells.
5. Regeneration of a mature plant.
6. Analysis.

The three most important considerations are the choice of plant materials, the construct design, and the analytical methods.

### 1.5.1. Plant Materials

Whenever possible, transgenic plant experiments involve analysis of multiple transformants. Transgene insertion occurs randomly, and position effects cause wide variation in expression levels. It is also possible that insertion could inactivate an important gene. Most studies also employ second-generation plants, because primary transformants may show aberrant behavior from passage through culture. Because of these considerations, we generate 26 primary transformed lines for each construct. Four to eight second-generation plants are then grown from each of the five or six best primary lines. It is important to remember that there will be a delay of many months between starting a transformation experiment and testing pollination phenotypes. It is therefore beneficial to plan for as much flexibility as possible from the outset.

It is important to carefully consider the genetic background of the transgene recipient. The controls must be thoroughly tested. Although S-RNases determine allelic specificity in SI, other genes are also required *(31,32)*. Preliminary genetic experiments will be required to show that a full complement of these "modifier" genes is present. We have observed that genetic background profoundly effects expression levels, which in turn are critical for determining effects of a gene on pollination behavior. For example, **Fig. 1** shows that an $S_{A2}$-RNase transgene was expressed at threefold higher levels in a (*N. plumbaginifolia* × SC *N. alata*) hybrid than in *N. plumbaginifolia*.

### 1.5.2. Construct Design

For expression studies, we recommend using the tomato *Chi2;1* gene promoter *(28,33)*. With this promoter, one-half to two-thirds of transformants express sufficient S-RNase to cause a change in pollination phenotype. For uniformity, we use $S_{A2}$-RNase gene 3' sequences, but there is no evidence that this sequence is unique. Others have successfully used 3' sequences from different genes in combination with the *Chi2;1* promoter *(30)*.

For antisense studies, we recommend the CaMV 35S promoter with duplicated enhancer *(34)*. This recommendation is contrary to conventional wisdom, which suggests that a gene's native promoter is best for antisense experiments. However, at least three labs have reported success using the CaMV 35S promoter to suppress expression of style-specific genes *(35–37)*.

### 1.5.3. Analysis

The critical aspects of the analysis phase are to establish for the recombinant RNase:

1. Enzymatic activity.
2. Expression level.
3. Effect on pollination behavior.

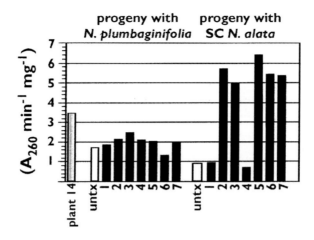

Fig. 1. Effect of genetic background on $S_{A2}$-RNase expression. A *N. plumbaginifolia* (Plant 14) with a single transgene locus expressing $S_{A2}$-RNase was crossed with untransformed *N. plumbaginifolia* (**left**) or untransformed SC *N. alata* (**right**). Style extracts were prepared and RNase specific activities ($A_{260}$/min/mg) were determined for the primary transformant (plant 14, gray), for individual progeny plants (1–7, solid), and for untransformed control progeny (untx, white). The variation in transgene expression level among the progeny plants is typical of transgenic plant experiments. However, expression is consistently higher in the (*N. plumbaginifolia* × SC *N. alata*) hybrids than in the *N. plumbaginifolia* genetic background (left). Reprinted with kind permission of Kluwer Academic Publishers, from *Plant Molecular Biology 37* (1998), pp. 561–569, J. Murfett and B. A. McClure "Expressing foreign genes in the pistil: A comparison of S-RNase constructs in different *Nicotiana* backgrounds."

The behavior of the recombinant RNase should be compared with a control gene.

We use a crude acid solubilization assay to determine enzymatic activity *(14,38)*. This assay is not as sensitive as the modern methods described in this volume, but it is inexpensive and does not require special equipment. The specific activity of S-RNases vary over at least a 40-fold range, which complicates analysis. First, increased stylar RNase activity may be difficult to discern when a low-specific activity S-RNase is introduced into a background expressing a high-specific activity protein. In such cases, the recombinant protein can be purified or zymograms can be used to demonstrate activity *(30)*. These difficulties can be avoided by using a transgene recipient with low stylar RNase activity (e.g., a SC background or the [*N. langsdorffii* × SC *N. alata*] hybrid).

Because enzyme specific activity varies between different RNases, measuring the amount of protein expressed is critical. Immunoblotting is a convenient way to measure accumulation, but requires generating an antibody. Other methods in the literature include quantitative FPLC and RNA blot analysis *(17,28,36)*. **Figures 2** and **3** illustrate the importance of determining both RNase activity and expression levels. The study compared the pollen rejection activity of $S_{A2}$-RNase and RNaseI from *E. coli* *(39)*. The activity of $S_{A2}$-RNase in inter- and intraspecific pollen rejection is well-established *(28,31,36)*, and was therefore used as a positive control. The experiment was designed to test whether a non-S-RNase could also cause pollen rejection. Plants expressing RNaseI had approximately a fourfold greater RNase specific activity than plants

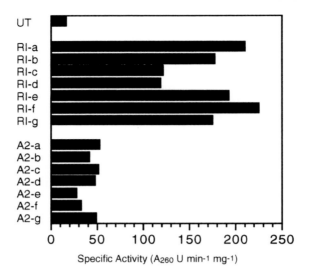

Fig. 2. RNase-specific activity in style extracts. (**Top**) Untransformed plant (UT). (**Middle**) RI-a to RI-g, transgenic (*N. plumbaginifolia* × SC *N. alata*) hybrids expressing *E. coli* RNaseI. (**Bottom**) A2-a to A2-g, transgenic hybrids expressing *N. alata* $S_{A2}$-RNase. Reprinted with kind permission of Kluwer Academic Publishers, from *Plant Molecular Biology 35* (1998), pp. 553–563, B. Beecher, J. Murfett, and B. A. McClure "RNaseI from *E. coli* cannot substitute for S-RNase in rejection *Nicotiana plumbaginifolia* pollen."

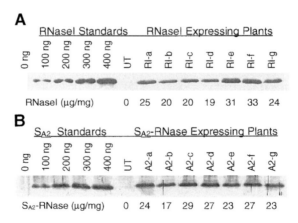

Fig. 3. Quantitating RNase levels by immunoblot analysis. Style extracts equivalent to 1 mg fresh weight were separated by SDS-PAGE and blotted to nitrocellulose. In both panels, purified protein standards are shown at left and style extracts from transgenic hybrids are at right. UT, untransformed hybrid. (**A**) Blot immunostained with anti-RNaseI showing RNaseI-expressing hybrids RI-a to RI-g. (**B**) Blot stained with anti-$S_{A2}$-RNase showing $S_{A2}$-RNase-expressing hybrids A2-a to A2-g. Estimated RNase concentrations are shown below (i.e., mg RNaseI or $S_{A2}$-RNase/mg style protein). Band intensities were compared with the standards on the left and corrected for the amount of total protein loaded in each lane. Reprinted with kind permission of Kluwer Academic Publishers, from *Plant Molecular Biology 35* (1998), pp. 553–563, B. Beecher, J. Murfett, and B. A. McClure "RNaseI from *E. coli* cannot substitute for S-RNase in rejection *Nicotiana plumbaginifolia* pollen."

**Table 1**
**Pollination Phenotypes**

| | Pollen donor (compatible/total pollinations) | |
|---|---|---|
| | *N. plumbaginifolia* | *N. alata* $S_{C10}$ |
| Untransformed control | | |
| UT | 5/5 | 5/5 |
| $S_{A2}$-RNase-expressing hybrids | | |
| A2-a | 0/5 | 5/5 |
| A2-b | 0/5 | 4/5 |
| A2-c | 0/5 | 5/5 |
| A2-d | 0/5 | 5/5 |
| A2-e | 0/5 | 4/5 |
| A2-f | 0/5 | 5/5 |
| A2-g | 0/5 | 5/5 |
| RNaseI-expressing hybrids | | |
| RI-a | 5/5 | 5/5 |
| RI-b | 5/5 | 5/5 |
| RI-c | 5/5 | 5/5 |
| RI-d | 5/5 | 5/5 |
| RI-e | 5/5 | 5/5 |
| RI-f | 5/5 | 5/5 |
| RI-g | 5/5 | 5/5 |

expressing $S_{A2}$-RNase (**Fig. 2**), but immunoblot analysis showed that the proteins were expressed at very similar levels (**Fig. 3**). For the purposes of this study, these results thus established the validity of comparing the pollination behavior of plants expressing $S_{A2}$-RNase and RNase I.

Pollen tube staining and fruit set are used to determine pollination behavior. For example, **Table 1** shows pollination data for the plants expressing RNaseI and $S_{A2}$-RNase. In this experiment, fruit set was determined by ovary swelling rather than seed formation, because the hybrid is infertile. All plants were capable of supporting pollen-tube growth (when *N. alata* $S_{C10}$-pollen was used as a control). **Table 1** shows that plants expressing $S_{A2}$-RNase accepted $S_{C10}$-pollen and rejected *N. plumbaginifolia* pollen (i.e., pollen rejection is specific). In contrast, hybrids expressing RNaseI from *E. coli* were compatible with both types of pollen. Style-squashes show that differences in fruit set are due to pollen-tube growth. In this procedure, styles are fixed and pollen tubes are visualized with decolorized aniline blue *(40)*. As shown in **Fig. 4**, hybrids expressing RNaseI resemble untransformed controls, and a large number of pollen tubes are seen at the base of the style. No pollen tubes are visible in the plants expressing $S_{A2}$-RNase, thus confirming the fruit set results.

In this example, both $S_{A2}$-RNase from *N. alata* and RNaseI from *E. coli* were shown to be active and expressed at similar levels. Yet $S_{A2}$-RNase caused rejection of pollen from *N. plumbaginifolia*, but RNaseI did not. This suggests that S-RNases are specially adapted to function in pollen rejection, and their role in determining mating behavior is not simply a function of their enzymatic activity.

**Untransformed Style**

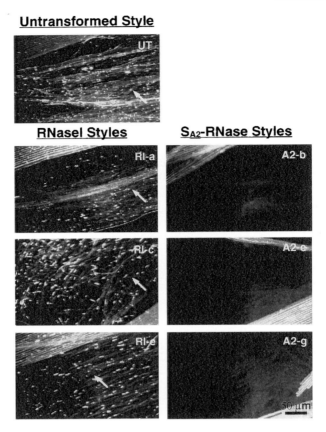

Fig. 4. Style-squash results. Transgenic hybrids were pollinated by *N. plumbaginifolia*, then fixed and stained 48 h later. In the untransformed control and in hybrids expressing RNaseI, many pollen tubes (arrows) are visible in the lower style. No pollen tubes are visible at the base of the style in hybrids expressing $S_{A2}$-RNase. Scale bar, 50 mm. Reprinted with kind permission of Kluwer Academic Publishers, from *Plant Molecular Biology 35* (1998), pp. 553–563, B. Beecher, J. Murfett, and B. A. McClure "RNaseI from *E. coli* cannot substitute for S-RNase in rejection *Nicotiana plumbaginifolia* pollen."

## 2. Materials

### 2.1. Gene Constructs

1. pHK7-X: contains the tomato *Chi2;1* promoter *(33)* flanked by a *Hin*dIII site on the 5' end, and a *Nco*I site containing the ATG initiator.
2. pSA23': contains 2.7 kb of downstream sequence from the *N. alata* $S_{A2}$-RNase gene. It has a *Xba*I site engineered into the 3' UTR, and an *Eco*RI site at the extreme 3' end. Plasmids pHK7-X and pSA23' are available for research purposes from the authors.
3. pAGUS1: contains the CaMV35S promoter with duplicated enhancer and no 3' sequences recommended for antisense constructs *(41)*. These two elements are available from many sources.
4. pBIN19 binary vector *(42)*.
5. pPZP series of binary vectors *(43)*.
6. XL1-Blu: *E. coli* strain for routine manipulations.

7. DH5-α: *E. coli* strain for routine manipulations.
8. LBA4404: *Agrobacterium tumefaciens* strain.
9. GV3101 *Agrobacterium tumefaciens* (grows faster than LBA4404).

## 2.2. Plant Materials

1. Self-incompatible *N. alata* lines: derived from stocks obtained from Digger's Seeds (105 LaTrobe Parade, Dromona, Victoria 3936, Australia). Similar materials may be obtained from any well-stocked nursery.
2. Self-compatible *N. alata* accession: inbred derivative of the cultivar Breakthrough (Thompson and Morgan, Jackson, NJ) and is available from the authors.
3. *N. tabacum* cv. Praecox (inventory #TI1347), *N. langsdorffii* (inventory number TW74 accession 28A), and *N. plumbaginifolia* (inventory number TW107, accession 43B): obtained from the US Tobacco Germplasm Collection, Crops Research Laboratory, Oxford, NC.
4. (*N. langsdorffii* × SC *N. alata* cv. Breakthrough) hybrid: made by emasculating *N. langsdorffii* before anthesis, and pollinating with SC *N. alata* cv. Breakthrough at maturity. Gram quantities of seed are required for hypocotyl transformation.

## 2.3. Bacterial Media and Stocks

1. YEP: 10 g/L peptone, 5 g/L yeast extract, 5 g/L NaCl.
2. Kanamycin sulfate (Fisher BP906-5) stored as a frozen filter-sterilized stock at 50 mg/mL.
3. Rifampicin (Sigma R-3501) stored as a 50 mg/mL stock in DMSO at 4°C.
4. Acetosyringone (Aldrich D13,440-6) is kept as a 20 mg/mL filter-sterilized stock in ethanol at 4°C.

## 2.4. Plant Tissue Culture Media and Stocks

A list of the stocks needed for plant tissue culture is given in **Table 2**. Standard medium contains Murashige and Skoog macro- and micronutrients, and Gamborg's vitamins (i.e., all at 1X concentration), 1 g/L MES, 30 g/L of the appropriate sugar, adjusted to pH 5.7 with KOH (*see* **Note 6**). Dispense in 250 mL aliquots containing either 0.44 g Phytagel or 3.75 g agar (*see* **Note 7**). One 250-mL aliquot is sufficient for ten 15 × 100 mm Petri plates. Media bottles are autoclaved 35 min, hormones are added, and the plates are poured in a laminar flow cabinet (*see* **Note 8**). The hormone ratio effective for each species is notoriously variable, and can only be determined by trial and error. **Table 3** provides a summary of effective media for our materials.

Rooted explants are transitioned (*see* **Note 9**) to soil by sterile transfer to an autoclaved humid chamber (Sigma P-4928) containing a 4-in. azalea pot (e.g., filled with a medium such as Metromix 350, Scotts-Sierra, Marysville, OH). Standard greenhouse materials are suitable for propagation of mature plants.

## 2.5. Analysis of Transgenic Plants Expressing Foreign RNase Genes

### 2.5.1. RNase Protein and Activity in Crude Style Extracts

1. Protein extraction buffer: 0.15 *M* Tris-HCl, 5 m*M* EDTA, 1% 2-mercaptoethanol, pH 8.3.
2. RNase-assay buffer: 25% glycerol, 0.1 *M* potassium phosphate, pH 7.0, 0.05 *M* KCl, 4 mg/mL purified Torula yeast RNA (*see* **Subheading 3.5.2.1.**).
3. RNase stop solution: 25% (v/v) perchloric acid containing 0.75% (w/v) uranylacetate.
4. SDS-loading buffer: 2% (w/v) SDS, 0.025% (w/v) Bromophenol blue, 5% (v/v) 2-mercaptoethanol, 10% v/v) glycerol, 62.5 m*M* Tris-HCl, pH 6.8.

**Table 2**
**Plant Tissue Culture Stock Solutions**

| Stock | Supplier cat. no. | Stock conc. | Working conc. |
|---|---|---|---|
| Basic media components | | | |
|    Murashige and Skoog macronutrient | Sigma M-0654 | $10X^a$ | 1X |
|    Murashige and Skoog micronutrients | Sigma M-0529 | $10X^a$ | 1X |
|    Gamborg's vitamins | Sigma G-1019 | $1000X^a$ | 1X |
|    MES buffer | Sigma M-8652 | Solid | 1 g/L |
| Sugars | | | |
|    Glucose | Fisher BP350-1 | Solid | 30 g/L |
|    Sucrose | Sigma S-5391 | Solid | 30 g/L |
| Hormones | | | |
|    BAP (6-benzylaminopurine) | Sigma B3274 | $1 mg/mL^a$ | 2 µg/mL |
|    IAA (indoleacetic acid) | Sigma I-2886 | $2 mg/mL^a$ | 0.5 µg/mL |
|    IBA (indolebutyric acid) | Sigma I-7512 | $2 mM^a$ | 1 µM |
|    kinetin (6-furfurylaminopurine) | Sigma K0753 | $2 mM^a$ | 1 µM |
|    NAA (naphthaleneacetic acid) | Sigma N-1641 | $1 mg/mL^a$ | 0.1 µg/mL |
| Antibiotics | | | |
|    Ampicillin | Sigma A-9518 | $100 mg/mL^a$ | 100 µg/mL |
|    Kanamycin | Fisher BP906-5 | $50 mg/mL^a$ | 100 µg/mL |
|    Timentin | Pharmacy[b] | $100 mg/mL^a$ | 100 µg/mL |
| Solidifying agents | | | |
|    Agar | Sigma A-1296 | Solid | 15 g/L |
|    Phytagel | Sigma P-8169 | Solid | 1.75 g/L |

[a]Purchased as a sterile solution or prepared and filter sterilized.

[b]Timentin is a prescription antibiotic that can be obtained for research purposes through a hospital pharmacy.

### 2.5.2. Style Squash Assay

1. Fixing and storage solution: ethanol:acetic acid (3:1). Clearing and softening solution: 10% (w/v) sodium sulfite.
2. Decolorized aniline blue: Dissolve 1 g/L water-soluble aniline blue (Aldrich 86,102-2) in 0.1 $M$ $K_3PO_4$. Allow to stand at room temperature in the dark for at least 3 d, or until the blue color disappears (*see* **Note 10**). Store at 4°C.

## 3. Methods
### 3.1. Gene Constructs

No general method can be described to prepare an RNase gene for transformation. For expression constructs, we use recombinant PCR to create genes with an engineered *Nco*I site at the initiator and a *Xba*I site downstream of the stop codon (*see* **Note 11**). A high-fidelity DNA polymerase should be used in PCR to avoid mutations in expression constructs. The final structure must be confirmed by sequencing. Expression constructs are assembled as follows:

1. Digest pHK7-X with *Nco*I and *Eco*RI, and isolate the 4.5-kb band containing the vector and the *Chi2;1* promoter.

**Table 3**
**Plant Tissue Culture Media**

| Plant material | Medium | Sugar | Cytokinin | Auxin | Other |
|---|---|---|---|---|---|
| *N. tabacum* | | | | | |
| | Cocultivation[a] | Glucose | BAP | IAA | Acetosyringone |
| | Shoot[a] | Sucrose | BAP | IAA | Kanamycin, timentin |
| | Root[b] | Sucrose | | | Kanamycin, timentin |
| *N. plumbaginifolia* | | | | | |
| | Cocultivation[a] | Glucose | BAP | NAA | Acetosyringone |
| | Shoot[a] | Sucrose | BAP | NAA | Kanamycin, timentin |
| | Root[b] | Sucrose | | | Kanamycin, timentin |
| (*N. langsdorffii* × SC *N. alata*) hybrid | Cocultivation[a] | Glucose | Kinetin | IBA | Acetosyringone |
| | Shoot[a] | Glucose | Kinetin | IBA | Kanamycin, timentin |
| | Root[b] | Glucose | | | Kanamycin, timentin |

[a]Cocultivation and shooting media are solidified with 1.75 g/L phytagel.
[b]Rooting medium is solidified with 15 g/L agar.

2. Digest pSA23' with *Xba*I, *Eco*RI, and *Bgl*I, and isolate the 2.7-kb band containing the 3' UTR.
3. Finally, ligate these two fragments with your engineered RNase gene (i.e., containing *Nco*I and *Xba*I ends).

Confirm the structure of the final construct using restriction digests or DNA sequencing. Excise the expression construct as a *Hin*dIII *Eco*RI fragment and reclone it in a binary vector, such as pBIN19 *(42)* or a pPZP *(43)* series plasmid.

### 3.2. Agrobacterium *Transformation*

We recommend direct transformation by the freeze–thaw method as an inexpensive and reliable method for moving binary vectors into *Agrobacterium*. The method rarely results in rearrangements and does not require an electroporation apparatus, but it is extremely inefficient compared to *E. coli* transformation methods. It is important to again confirm the structure of the construct once it is in *Agrobacterium*. LBA4404 is the most commonly used *Agrobacterium* strain. However, GV3101 grows faster and is better for preparing DNA.

### 3.2.1. Preparation of Competent Agrobacterium *Cells*

1. Inoculate a colony of *Agrobacterium tumefaciens* LBA4404 into 3 mL YEP plus 50 µg/mL rifampicin and grow overnight at 28°C with gentle shaking.
2. Add 1 mL of culture to 75 mL YEP in a 500-mL flask. Shake at 250 rpm, 28°C until the $OD_{600}$ reaches 0.5–0.75.

3. Place culture on ice for 5 min.
4. Spin down the culture at 3000*g*; 10 min; 4°C.
5. Resuspend in 1.5 mL of prechilled 20 m*M* CaCl2, 15% glycerol.
6. Flash freeze 0.1-mL aliquots in 1.5-mL tubes.
7. Store at –70°C.

### 3.2.2. Transforming Competent Agrobacterium Cells

1. Add 5 µg of plasmid DNA (e.g., from a sequencing prep) to a frozen aliquot of competent *Agrobacterium* cells.
2. Thaw with mixing in a 37°C water bath. Mixing can be accomplished by hand using a sterile pipet tip.
3. Flash-freeze cell aliquot.
4. Repeat the thawing and mixing steps.
5. Add 1 mL YEP and incubate at 28°C for 4 h.
6. Spin down cells 1 min at 9000*g* and resuspend in 100 µL YEP.
7. Spread the cells on YEP plates containing 50 µg/mL rifampicin, and appropriate selection for the binary plasmid (i.e., 100 µg/mL kanamycin). Spread all the cells on two plates. Place 10 µL on one plate, and 90 µL on another.
8. Incubate the plates at 22–28°C. Colonies will appear in 2–4 d.

## 3.3. Preparation of Plant Material

For materials such as *N. tabacum* and *N. plumbaginifolia* that will be transformed from leaf explants, we recommend maintaining "clean" stocks that are free of pests (*see* **Note 12**). Plants are seeded 6–8 wk prior to transformation on artificial soil mix and bottom-watered to minimize splash contamination of the leaves. For materials such as (*N. langsdorffii* × SC *N. alata*) hybrids that transform better from hypocotyls, the seeds are surface sterilized (*see* **Subheading 3.3.1.**) and seeded at very high density on standard media. The object is to produce a "lawn" of sterile seedlings at such a density that the hypocotyls extend to approx 1 cm.

### 3.3.1. Surface Sterilization

Young leaves that are free of soil particles are picked when approx 3 cm × 2 cm. For a standard transformation with about 10 plates of primary explants, sterilize 10–15 leaves. Some leaves may be lost during the sterilization procedure.

The procedure is the same for surface sterilizing seeds, but they are first washed with 70% ethanol, and the procedure is performed in 1.5-mL tubes. *Nicotiana* seeds are small, so it is more convenient to discuss a vol than a number. Plan on sterilizing approx 0.25 mL of seeds for each transformation. When sterilizing seeds, it is best to decant using a sterile 1-mL pipet tip. After sterilization, spread the seeds evenly on a plate of standard media. Incubate 1–3 wk at 28°C under a 16-h light/8-h dark cycle. A Petri plate filled with sterile wet sand can also be used to germinate the seeds.

1. Place two or three leaves in a 5-mL sterile disposable screw-cap tube. Do not pack the leaves, and try to minimize bruising and contact between leaves. Cut the leaves in half if necessary.
2. Dilute 1 vol of household bleach with 4 vol of sterile water.
3. Completely fill each tube with diluted bleach. Care must be taken to avoid air bubbles on the leaves. Bubbles can be removed by gentle tapping. If necessary, empty the tube and

gently refill it avoiding bubbles. Tighten the cap and invert the tube. Make sure the leaves are completely submerged in dilute bleach. Allow sterilization to proceed for 8 min, occasionally inverting the tubes.

4. Decant the bleach, and rinse five times with autoclaved deionized water.

### 3.3.2. Preparing Explants

Sterile leaves are cut into pieces 0.5–0.7 cm on a side using a #11 scalpel. Discard large veins. Arrange the explants on cocultivation plates. The explants are allowed to recover at room temperature overnight (i.e., leave them in the laminar flow hood).

When they are ready (i.e., about 1 cm long), hypocotyls are prepared by holding a group of seedlings with forceps and removing the roots and cotyledons with a scalpel. Hypocotyls can be arranged on cocultivation plates as described here. Alternatively, they can be suspended in about 10 mL of liquid cocultivation medium in a Petri plate.

## 3.4. Transformation and Regeneration

### 3.4.1. Cocultivation

Transformed *Agrobacterium* is grown in rich medium under appropriate selection supplemented with 12 µg/mL acetosyringone.

1. Grow a 50-mL culture in a 250-mL flask for each transformation.
2. The culture is ready when it is slightly turbid, usually 2–3 d when shaken at 250 rpm, 28°C.
3. Pellet the cells (4000*g*, 10 min) in a sterile 50-mL tube.
4. Wash with 50 mL of liquid standard plant-tissue culture media supplemented with glucose (30 g/L) and acetosyringone (12 µg/mL).
5. Resuspend in another 50 mL of the same media (*see* **Note 13**).
6. Add 5 mL of cell suspension to each plate of explants. Take care not to wash the explants on top of each other.
7. After 5–10 min, remove as much of the *Agrobacterium* culture as possible. Close the lid and seal each plate with a few layers of Handiwrap to prevent contamination and reduce evaporation. For hypocotyl transformation in liquid medium, add 1 mL of *Agrobacterium* to the hypocotyls, cover, and seal with Handiwrap.
8. Cocultivate the explants and the *Agrobacterium* for 3–5 d at 22°C to allow transformation to occur (*see* **Note 14**).

### 3.4.2. Shoot Regeneration

After cocultivation, excess *Agrobacterium* are removed by vigorously shaking the explants in several changes of sterile cocultivation media (approx 40 mL medium in a 50-mL sterile tube). The explants are then placed approx 1 cm apart on appropriate shoot regeneration medium and maintained at 28°C with a 16-h light/8-h dark cycle. Transfer them to fresh medium every 10–14 d. During this period, the explants will expand and it will be necessary to increase the number of plates to maintain the proper spacing. Transformed cells will generate callus and shoots on the edges of the explants. Untransformed cells turn yellow and necrotic over the course of several months. The time required for regeneration varies depending on the plant material. Shoots may form on *N. tabacum* explants with 1 mo. Other materials may require as long as 1 yr. As soon as they are large enough to handle, shoots may be cut away from the callus and transferred to rooting medium (*see* **Note 15**).

### 3.4.3. Rooting

The cut ends of the shoots are placed into agar rooting medium and maintained at 28°C under a 16-h light/8-h dark cycle. Unrooted shoots are recut and transferred to fresh plates at 4-wk intervals (*see* **Note 16**). We continue transferring shoots to rooting medium until at least 26 vigorous rooted plantlets are obtained, each from a separate callus. Saving only one rooted plantlet from each callus ensures that each of the 26 rooted plantlets arose from an independent transformation event. Rooted plantlets are transitioned to soil in sterile 4-in. pots inside humid chambers (Sigma P-4928). The lid is kept on the chamber for 5–15 d, after which it is left slightly opened for another week. The new plants are then acclimated to ambient humidity and can be treated as ordinary (albeit valuable) seedlings, and grown to maturity in the greenhouse.

### 3.5. Analysis

The transgenic plants can be screened for intact copies of the construct by Southern blotting or PCR. However, as long as the transforming *Agrobacterium* culture had the construct intact, and a full complement of 26 independent lines are available for analysis, it is unlikely that rearrangements will be prevalent enough to cause problems. The basic goals in the analysis phase of the experiment are to determine expression levels, enzymatic activity, and the effects on pollination behavior.

### 3.5.1. Determining Expression Levels

Only plants expressing very high levels of RNase protein in the style are likely to cause a change in pollination behavior. A simple preliminary screen for high-level expressors uses Coomassie-stained SDS gels (*see* **Note 17**):

1. Weigh one to three mature styles on a milligram balance.
2. Add 10 µL of SDS loading buffer per mg fresh weight.
3. Homogenize thoroughly using a microfuge tube pestle attached to a variable speed power drill (e.g., Sears and Roebuck Co. Craftsman 3/8 in. variable-speed drill model 312.101421).
4. Boil 3 min.
5. Pellet the debris and pull it out with a toothpick.
6. Run extract equivalent to 1.0–2.5-mg fresh weight (i.e., 10–25 µL) for each transgenic plant along with markers and untransformed controls.

In most cases, this procedure will quickly identify the most interesting plants. Usually these are the highest expressors.

We use quantitative immunoblotting to estimate the absolute concentration of recombinant protein expressed in transgenic plants. To prepare concentration standards, the protein of interest is purified and known amounts are run alongside extracts from the transgenic plants (*see* **Note 18**). As shown in **Fig. 3**, this allows the amount of recombinant protein in the transgenic plants to be determined.

### 3.5.2. Determining RNase Activity

As noted previously, the specific activity of S-RNases varies widely. Thus, the absolute level of enzyme activity is not usually a major factor in analyzing the transgenic plants (*see* **Note 19**). Therefore, the most important point is simply to establish whether the recombinant protein is active or not. The acid solubilization assay is used to determine RNase activity.

### 3.5.2.1. RNase Activity in Crude-Style Extracts

The RNA substrate for the acid solubilization assay must first be purified. This procedure neutralizes the free acid and reduces background RNase activity.

1. Dissolve 5 g Torula yeast RNA (Sigma R-6625) in 20 mL deionized water using dropwise addition of $NH_4OH$.
2. Precipitate RNA by adding 5 mL 10 $M$ ammonium acetate and 25 mL isopropanol and centrifuge.
3. Decant supernatant, resuspend RNA in 15 mL deionized water, and repeat **steps 2** and **3** twice.
4. Slowly add ice-cold 8 $M$ LiCl just until a precipitate forms and persists upon shaking (15–20 mL LiCl).
5. Spin immediately for 10 min at 1000$g$. Resuspend pellet in 20 mL deionized water.
6. Add 15 mL of a phenol:chloroform:isoamyl alcohol (25:24:1) mixture (BP1752I-100). Close cap tightly and shake by hand 5 min.
7. Spin 10 min at 2500$g$, and repeat extraction step twice more.
8. Precipitate RNA by adding 5 mL 10 $M$ ammonium acetate and 25 mL isopropanol. Mix and spin at 2500$g$ for 10 min.
9. Resuspend pellet in 20 mL deionized water. Repeat **step 8**.
10. Wash pellet sequentially in 95% ethanol then 100% methanol.
11. Dry pellet under vacuum, record its weight, and store at $-20°C$.
12. Test purity of RNA substrate preparation by running a blank RNase activity assay (**Subheading 3.4.2.**). $A_{260}$ of a 1/20 dilution should be $\leq 0.1$.

### 3.5.2.2. Extract Preparation

1. Weigh three to four mature styles on a milligram balance.
2. Place them in a 1.5-mL tube and freeze in liquid nitrogen. Store at $-70°C$ until needed.
3. Add 4 µL of protein extraction buffer/mg fresh weight to the frozen tissue and place on ice.
4. As the buffer thaws, grind the styles briefly but thoroughly using a microfuge pestle driven by a variable-speed power drill. Keep the tube cold at all times. Grind samples in lots of three to five to minimize thawing of the sample prior to homogenization. Between each tube, clean the pestle with Windex and pat dry.
5. Microfuge the extracts 13,000$g$ for 20 min at 4°C.
6. Remove approx 90% of the extract and transfer to a clean tube, leaving the debris pellet and the lipid pellicle behind.
7. Repeat **steps 5** and **6** twice to remove all cell debris.
8. Flash-freeze the supernatant in small aliquots so that multiple freeze-thaw cycles can be avoided.

### 3.5.2.3. RNase Assay

1. Label 1.5-mL tubes and chill on ice 5 min.
2. Add 0.2 mL of ice-cold 1X RNase activity assay buffer to each tube.
3. Add 0.01 mL of diluted extract. Vortex briefly at low setting, and immediately place each tube in a 37°C circulating water bath.
4. Remove each tube after precisely 30 min and immediately add 0.04 mL stop solution.
5. Vortex well and place on ice at least 15 min to precipitate.
6. Spin tubes for 20 min at high speed in a refrigerated microfuge. Keep tubes on ice.
7. Dilute 0.05 mL reaction mix in 0.950 mL water (i.e., 1:20 dilution) and read $A_{260}$.

Assays are performed in duplicate along with negative (i.e., blank) and positive controls. Activity is expressed as $A_{260}$ U/min/mL/mg.

$$(A_{260}\text{sample-blank}) \times (1/30 \text{ min}) \times (1/0.25 \text{ mL reaction volume}) \times$$
$$(1/\text{mg protein present in } 0.01 \text{ mL sample})$$

This assay measures the amount of substrate RNA rendered acid-soluble by the enzyme. The assay is not linear over a wide range, so it is best to assay several dilutions of each sample. Only use dilutions that give $A_{260}$ readings in the range 0.05–0.3 (i.e., $A_{260}$-blank, diluted 1:20).

### 3.5.3. Determining Pollination Phenotypes

To test for pollen rejection, flower buds are emasculated 1–2 d before anthesis. Emasculation is accomplished by making a small slit in the side of the bud with jeweler's forceps and carefully removing the anthers. It is important not to damage the pistil. When the flower is mature, it may be pollinated by gently brushing it with a mature pollen-bearing anther. The pollination phenotype is then scored by fruit set or style squash.

1. Clean the forceps with ethanol after every flower. This prevents contamination by stray pollen.
2. It is extremely important to make sure that all pollinations are performed at a comparable developmental stage. Transgene expression levels are constantly changing because the *Chi2;1* promoter continues to be expressed after anthesis.
3. Be sure to check the viability of each batch of pollen by performing compatible control pollinations.

#### 3.5.3.1. FRUIT SET ASSAY

If the transgenic plants are fertile, a compatible pollination results in fruit set. In this case, the flower will senesce and ovary swelling will begin within 3 d. Mature fruits require 3–4 wk to develop. However, flowers will stay fresh for 5–7 d after an incompatible pollination. After that, they will senesce and fall off. Sterile hybrids respond similarly, but the fruit never matures. Ovary swelling can be scored 4 d after pollination, but it is advisable to make observations every day. The ovary will swell to three or four times its prepollination size after a compatible pollination, and the corolla will wilt. When scoring phenotypes by ovary swelling, it is advisable to perform compatible and incompatible controls at the same time as the test pollinations.

#### 3.5.3.2. STYLE SQUASH ASSAY

1. Forty-eight hours after pollination, the styles are fixed in ethanol:acetic acid (3:1). They can be stored in fixative indefinitely at 4°C.
2. Decant the fixative and add sufficient 10% sodium sulfite to cover the styles. Loosely cover the tubes and autoclave for 5 min.
3. Transfer the softened styles to a microscope slide.
4. Add a few drops of decolorized aniline blue (*see* **Note 20**).
5. Smash the style under a large coverslip. A slight rolling motion splits the style, exposing the pollen tubes.
6. Seal the edges of the coverslip with nail polish.
7. Observe the pollen tubes with an epifluorescence microscope. Pollen tubes will appear to be bright blue-green in color.

## 4. Notes

1. A simplistic view is that the pistil imposes barriers to fertilization that the pollen must overcome. In the angiosperms, the carpels are fused to enclose the ovules and present a

barrier to fertilization. The name "angiosperm" (sperms carried in tubes or vessels) emphasizes that pollen tubes must carry the sperms through this barrier to the ovule.

2. Plants alternate between sporophytic and gametophytic generations. Cells in the sporophyte (usually 2n) undergo meiosis, forming spores that develop into the (usually 1n) gametophyte. Gametophytic cells then develop into gametes. In higher plants, the sporophyte is prominent and the gametophytic generations are small and rarely seen. On the female side, the megaspore divides three times in the ovule, producing an eight-celled megagametophyte. One of these cells is the egg cell. On the male side, the microspore divides once, forming the vegetative and generative cells. The generative cell divides again, forming the two sperm cells. The vegetative cell forms the pollen grain and pollen tube. The sperm cells are held within the pollen-tube cytoplasm and carried to the ovule. One sperm fuses with the egg cell to form the zygote (i.e., the new sporophytic generation), and the other participates in forming the endosperm. The critical distinction between sporophytic and gametophytic SI is that in the former pollen behaves as if it were 2n, and in the latter it behaves as if it is expressing its own haploid genome. This is explained by noting that much pollen development occurs within sporophytic tissue, and the sporophyte contributes material to the pollen coat. In sporophytic SI, the determinants of specificity (on the pollen side) are believed to reside in the pollen coat. In gametophytic SI, they are believed to be expressed by the growing pollen tube.

3. Gametophytic SI is formally an interaction between distinct organisms: the gametophyte and the sporophyte. Thus, it is not true self-recognition.

4. Familiar solanaceous plants include tomato, potato, eggplant, chilies, petunia, and tobacco. Snapdragon is a representative of the *Scrophulariaceae*. Rosaceous plants include cherry, apple, and peach.

5. *Agrobacterium tumefaciens* is a plant pathogen that causes crown gall. Pathogenesis involves transfer of DNA from the Ti plasmid (i.e., T-DNA) into the host plant nucleus, where it integrates randomly. Wild-type T-DNA includes genes for synthesis of plant hormones and specific compounds for bacterial growth. "Disarmed" laboratory strains have T-DNA lacking these genes, but retain their transfer functions (i.e., *vir* genes). Binary transformation vectors (e.g., pBIN and pPZP series [42,43]) can replicate in *Agrobacterium* and *E. coli*. They contain one or more selectable markers and a multiple cloning site flanked by the T-DNA border sequences necessary for transfer to the plant cell nucleus.

6. Be careful when adjusting the pH. **Do not** overshoot and then correct the pH by adding acid. This changes the salt concentration in the medium.

7. To save time and avoid variability, prepare large amounts of medium and store at –20°C. To prevent contamination from laboratory reagents, store the medium in 250-mL polypropylene bottles designated for tissue culture media only. When needed, the media bottles are thawed, autoclaved for 35 min, and appropriate hormones and vitamins are added. Each bottle is used to make 10 plates.

8. The success of your project is critically dependent on good, sterile technique! In this regard, obsessive-compulsive behavior should be encouraged. Carefully review every aspect of your media preparation, transfer procedures, and culture storage to insure that sterility will be maintained. Some materials will be in culture for months, or even years, so even low rates of attrition because of contamination are unacceptable.

9. This procedure minimizes loss of plants at this critical stage.

10. The active flourochrome is a minor contaminant in aniline blue. Often, an old batch of aniline blue works better than a new batch. Alternatively, the pure flourochrome can be purchased from Biosupplies Australia (Parkville, Vic 3052, Australia), although it is very expensive.

11. S-RNase genes contain a single intron in a conserved location. We have prepared *E. coli* RNaseI gene constructs with and without an engineered intron. Both constructs were expressed efficiently, so the intron is not required for efficient expression (B. Beecher, unpublished).

12. Many laboratories expend considerable effort to maintain axenic plant materials for transformation. This has the advantage of eliminating contamination. However, such materials desiccate very rapidly and make manipulations difficult. Plants grown under normal conditions are more robust and less expensive to maintain. With proper care, contamination is seldom a problem.

13. Be sure to save aliquots of the culture used for transformation. *Agrobacterium* is notorious for DNA rearrangements. Therefore, it is essential to perform one last quick prep to confirm the structure of the construct in the culture actually used for transformation.

14. The *Agrobacterium* should grow to form a translucent lawn. Longer cocultivation times will yield more transformed cells. However, if the bacteria grow to a very high density, they can damage the explants, resulting in poor regeneration.

15. A numbering or grid scheme should be used to keep track of which shoots were derived from which calli. It is fine to move several shoots from a given callus onto rooting medium. However, to ensure that the same transformant is not reanalyzed, it is best to keep only one rooted explant from each callus.

16. Sometimes untransformed shoots form on the explants. This is more likely to occur early in the procedure. Such "escapes" will not root. Presumably, this is because roots must grow in direct contact with selective medium. Transformed shoots either will root vigorously or die. Do not be discouraged if your first shoots do not form roots.

17. The staining properties of proteins may vary considerably, so this does not work in all cases. RNase A, for example, does not appear to stain well with Coomassie blue.

18. It is best to mix known amounts of purified recombinant protein with extracts from untransformed plants so the total amount of protein loaded in each lane is the same. This produces better-looking figures, because the standard protein will have the same mobility as the protein in the transgenic plants.

19. This is not always true. In some circumstances, we have used RNase activity itself as a measure of expression level as in **Fig. 1** and elsewhere *(28,31,36)*. In general however, immunoblotting is the preferred method.

20. Some people prefer to prepare the squashes in stain containing 50% glycerol to minimize evaporation.

## Acknowledgments

We thank Dr. Charles Gasser, University of California-Davis, and the members of the McClure lab who contributed to developing this system, including Drs. Jane Murfett, Daniel Zurek, and Tim Strabala. This work was supported by the University of Missouri Research Board, the Food for the 21st Century Program, and National Science Foundation Grants 93-16152, 96-04645, and 99-82686.

## References

1. de Nettancourt, D. (1977) *Incompatibility in Angiosperms*. Monographs on Theoretical and Applied Genetics 3, Springer-Verlag, Berlin, Germany.

2. Bredemeijer, G. M. M. and Blaas, J. (1981) S-Specific proteins in styles of self-incompatible *Nicotiana alata*. *Theor. Appl. Genet.* **59,** 185–190.

3. Anderson, M. A., et al. (1986) Cloning of cDNA for a stylar glycoprotein associated with expression of self-incompatibility in *Nicotiana alata*. *Nature* **321,** 38–44.

4. Ai, Y., Singh, A., Coleman, C. E., Ioerger, T. R., Kheyr-Pour, A., and Kao, T.-H. (1990) Self-incompatibility in *Petunia inflata*: isolation and characterization of cDNAs encoding three S-allele-associated proteins. *Sex. Plant Reprod.* **3,** 130–138.

5. Anderson, M. A., McFadden, G. I., Bernatzky, R., Atkinson, A., Orpin, T., Dedman, H., Tregear, G., Fernley, R., and Clarke, A. E. (1989) Sequence variability of three alleles of the self-incompatibility gene of *Nicotiana alata*. *Plant Cell* **1,** 483–491.

6. Clark, K. R., Okuley, J. J., Collins, P. D., and Sims, T. L. (1990) Sequence variability and developmental expression of S-alleles in self-incompatible and pseudo-self-compatible *Petunia*. *Plant Cell* **2,** 815–826.

7. Cornish, E. C., Pettitt, J. M., Bönig, I., and Clarke, A. E. (1987) Developmentally controlled expression of a gene associated with self-incompatibility in *Nicotiana alata*. *Nature* **326,** 99–102.

8. Jahnen, W., Batterham, M. P., Clarke, A. E., Moritz, R. L., and Simpson, R. J. (1989) Identification, isolation, and N-terminal sequencing of style glycoproteins associated with self-incompatibility in *Nicotiana alata*. *Plant Cell* **1,** 493–499.

9. Tsai, D.-S., Lee, H.-S., Post, L. C., Kreiling, K. M., and Kao, T.-H. (1992) Sequence of an S-protein of *Lycopersicon peruvianum* and comparison with other solanaceous S-proteins. *Sex. Plant Reprod.* **5,** 256–263.

10. Ioerger, T. R., Clark, A. G., and Kao, T.-H. (1990) Polymorphism at the self-incompatibility locus in Solanaceae predates speciation. *Proc. Natl. Acad. Sci. USA* **87,** 9732–9735.

11. Richman, A. D., Uyenoyama, M. K., and Kohn, J. R. (1996) Allelic diversity and gene geneology at the self-incompatibility locus in the Solanaceae. *Science* **273,** 1212–1216.

12. Richman, A. D. and Kohn, J. R. (1999) Self-incompatibility alleles from *Physalis*: implications for historical inference from balanced genetic polymorphisms. *Proc. Natl. Acad. Sci. USA* **96,** 168–172.

13. Kawata, Y., Sakiyama, F., and Tamaoki, H. (1988) Amino acid sequence of ribonuclease T2 from *Aspergillus oryzae*. *Eur. J. Biochem.* **176,** 683–697.

14. McClure, B. A., Haring, V., Ebert, P. R., Anderson, M. A., Simpson, R. J., Sakiyama, F., and Clarke, A. (1989) Style self-incompatibility gene products of *Nicotiana alata* are ribonucleases. *Nature* **342,** 955–957.

15. Gray, J. E., McClure, B. A., Bönig, I., Anderson, M. A., and Clarke, A. E. (1991) Action of the style product of the self-incompatibility gene of *Nicotiana alata* (S-RNase) on in vitro-grown pollen tubes. *Plant Cell* **3,** 271–283.

16. McClure, B. A., Gray, J. E., Anderson, M. A., and Clarke, A. E. (1990) Self-incompatibility in Nicotiana alata involves degradation of pollen rRNA. *Nature* **347,** 757–760.

17. Lewis, D. and Crowe, L. K. (1958) Unilateral interspecific incompatibility in flowering plants. *Heredity* **12,** 233–256.

18. Mutschler, M. A. and Liedl, B. E. (1994) Interspecific crossing barriers in Lycopersicon and their relationship to self-incompatibility, in *Genetic Control of Self-Incompatibility and Reproductive Development in Flowering Plants* (Williams, E. G., Clarke, A. E., and Knox, R. B., eds.), Kluwer, Dordrecht, Germany, pp. 164–188.

19. Trubia, M., Sessa, L., and Taramelli, R. (1997) Mammalian Rh/T2/S-glycoprotein ribonuclease family genes: cloning of a human member located in a region of chromosome 6 (6q27) frequently deleted in human malignancies. *Genomics* **42,** 342–344.

20. Hime, G., Prior, L., and Saint, R. (1995) The *Drosophila melanogaster* genome contains a member of the Rh/T2/S-glycoprotein family of ribonuclease-encoding genes. *Gene* **158,** 203–207.

21. Meador, J. I. and Kennell, D. (1990) Cloning and sequencing the gene encoding *Escherichia coli* ribonuclease I: exact physical mapping using the genome library. *Gene* **95,** 1–7.

22. Bussey, H., et al. (1997) The nucleotide sequence of *Saccharomyces cerevisiae* chromosome XVI. *Nature* **387,** 103–105.

23. Wilson, R., et al. (1994) 2.2 Mb of contiguous nucleotide sequence from chromosome III of *C. elegans*. *Nature* **368,** 32–38.

24. Rosenberg, H. F., Ackerman, S. J., and Tenen, D. G. (1989) Human eosinophil cationic protein. Molecular cloning of a cytotoxin and helminthotoxin with ribonuclease activity. *J. Exp. Med.* **170,** 163–176.

25. Rosenberg, H. F., Tenen, D. G., and Ackerman, S. J. (1989) Molecular cloning of the human eosinophil-derived neurotoxin: a member of the ribonuclease gene family. *Proc. Natl. Acad. Sci. USA* **86,** 4460–4464.

26. Huang, S., Lee, H.-S., Karunanandaa, B., and Kao, T.-H. (1994) Ribonuclease activity of Petunia inflata S proteins is essential for rejection of self-pollen. *Plant Cell* **6,** 1021–1028.

27. Lee, H.-S., Huang, S., and Kao, T.-H. (1994) S proteins control rejection of incompatible pollen in *Petunia inflata*. *Nature* **367,** 560–563.

28. Murfett, J., Atherton, T. L., Mou, B., Gasser, C. S., and McClure, B. A. (1994) S-RNase expressed in transgenic *Nicotiana* causes S-allele-specific pollen rejection. *Nature* **367,** 563–566.

29. Zurek, D. M., Mou, B., Beecher, B., and McClure, B. (1996) Exchanging sequence domains between S-RNases from *Nicotiana alata* disrupts pollen recognition. *Plant J.* **11,** 797–808.

30. Matton, D. P., Maes, O., Laublin, G., Xike, Q., Bertrand, C., Morse, D., and Cappadocia, M. (1997) Hypervariable domains of self-incompatibility RNases mediate allele-specific pollen recognition. *Plant Cell* **9,** 1757–1766.

31. Murfett, J. M., Strabala, T. J., Zurek, D. M., Mou, B., Beecher, B., and McClure, B. A. (1996) S RNase and interspecific pollen rejection in the genus *Nicotiana*: multiple pollen rejection pathways contribute to unilateral incompatibility between self-incompatible and self-compatible species. *Plant Cell* **8,** 943–958.

32. Ai, Y., Kron, E., and Kao, T.-H. (1991) S-alleles are retained and expressed in a self-compatible cultivar of *Petunia hybrida*. *Mol. Gen. Genet.* **230,** 353–358.

33. Harikrishna, K., Jampates-Beale, R., Milligan, S. B., and Gasser, C. S. (1996) An endochitinase gene expressed at high levels in the stylar transmitting tissue of tomatoes. *Plant. Mol. Biol.* **30,** 899–911.

34. Kay, R., Chan, A., Daly, M., and McPherson, J. (1987) Duplication of CaMV 35S promoter sequences creates a strong enhancer for plant genes. *Science* **236,** 1299–1302.

35. Cheung, A. Y., Wang, H., and Wu, H.-M. (1995) A floral transmitting tissue-specific glycoprotein attracts pollen tubes and stimulates their growth. *Cell* **82,** 383–393.

36. Murfett, J., Bourque, J. E., and McClure, B. A. (1995) Antisense suppression of S-RNase expression in *Nicotiana* using RNA polymerase II- and III-transcribed gene constructs. *Plant. Mol. Biol.* **29,** 210–212.

37. Sessa, G. and Fluhr, R. (1995) The expression of an abundant transmitting tract-specific endoglucanase (Sp41) is promoter-dependent and not essential for the reproductive physiology of tobacco. *Plant. Mol. Biol.* **29,** 969–982.

38. Brown, P. H. and Ho, T.-H. D. (1986) Barley aleurone layers secrete a nuclease in response to giberellic acid. *Plant Physiol.* **82,** 801–806.

39. Beecher, B., Murfett, J., and McClure, B. A. (1998) RNaseI from *E. coli* cannot substitute for S-RNase in rejection *Nicotiana plumbaginifolia* pollen. *Plant. Mol. Biol.* **36,** 553–563.

40. Kho, Y. and Baer, J. (1968) Observing pollen tubes by means of fluorescence. *Euphytica* **17,** 299–302.

41. Skuzeski, J. M., Nichols, L. M., and Gesteland, R. F. (1990) Analysis of leaky viral translation termination codons *in vivo* by transient expression of improved β-glucuronidase vectors. *Plant. Mol. Biol.* **15,** 65–69.

42. Bevan, M. (1984) Binary *Agrobacterium* vectors for plant transformation. *Nucleic Acids Res.* **12,** 8711–8721.

43. Hajdukiewicz, P., Svab, Z., and Maliga, P. (1994) The small, versatile pPZP family of *Agrobacterium* binary vectors for plant transformation. *Plant. Mol. Biol.* **25,** 989–994.

# 7

## Molecular Cloning, Tissue Distribution, and Chromosomal Localization of the Human Homolog of the R2/Th/Stylar Ribonuclease Gene Family

**Francesco Acquati, Cinzia Nucci, Marco G. Bianchi, Tatiana Gorletta, and Roberto Taramelli**

## 1. Introduction

### 1.1. A Chromosomal Region Is Often Deleted in Human Malignancies

Allelic losses in the peritelomeric region of the long arm of human chromosome 6 (6q26-27) are associated with several human malignancies. These range from solid neoplasms such as ovarian carcinomas *(1–8)* to hematological cancers, among which B-cell non-Hodgkin's lymphomas and acute lymphoblastic leukemias are the most representative examples *(9,10)*.

Several cytogenetic, and more recently, molecular studies, assaying for loss of heterozygosity (LOH), have highlighted a specific genomic interval within the 6q27 region that is frequently lost or rearranged in ovarian carcinomas, other neoplasms, and SV-40 immortalized human fibroblasts *(1,2,6,8,11,12)*. The latter findings suggest that one or more tumor-suppressor genes and a senescence gene are localized in this subchromosomal region. This interval is flanked by markers D6S193 and D6S149, which are 2 c$M$ apart *(2,6)*. We constructed a physical long-range restriction map of the genomic region bracketed by the above markers by YAC cloning and YAC, PAC, and cosmid contig assembling *(11)*. According to the genomic mapping data, the physical distance between those two markers is no more than 1000 kb *(11)*. The physical map around the D6S149-D6S193 region prompted us to start the construction of a transcriptional map. Thus, we attempted to identify by means of cDNA selection methodologies most of the genes localized in this genomic region to provide candidate genes to be individually tested for structural and/or biological evidence suggesting a role in the pathogenesis of cancer.

### 1.2. Cloning of a Human Homolog of the Rh/T2/S-Glycoprotein RNase Family from This Region

We screened a YAC recombinant (clone 4Hhe8 from the ICI library) for cDNAs from fetal brain. A cDNA clone (p321) was isolated and sequenced, and BLASTX and

From: *Methods in Molecular Biology, vol. 160: Nuclease Methods and Protocols*
Edited by: C. H. Schein © Humana Press Inc., Totowa, NJ

BLASTN algorithms were used to search for sequence similarities *(13)*. Clone p321 matched three ESTs perfectly: H91956, derived from retina; R71946, obtained from breast tissue, and T39686 from fetal spleen. Sequencing of the largest of these clones (H91956) allowed us to expand the original sequences of clone p321 both in the 3' and 5' directions for a total of 1182 bp. Simultaneously a human-heart cDNA library was screened, and clone p321-A (1204 bp long) was isolated. The sequence indicated the presence of a 257-aa ORF with a hydrophobic, putative signal peptide at its N terminus (**Fig. 1**) *(13)*.

When the p321-A 257-aa ORF was compared to the sequences of other known proteins using the FastA and BLAST programs, two short stretches of amino acids were found to be conserved between the human ORF of p321-A and the RNase gene from the tomato plant *Lycopersicon esculentum* *(14)*, which is a member of the Rh/T2/S-glycoprotein class of RNase proteins. Although the sequence similarities were limited overall, two pentamers HGLWP and KHGTC, known to be the catalytic sites of the RNase enzyme *(15)*, were perfectly conserved (10/10 identity) between these two sequences (**Fig. 2**). The genes belonging to this class are well known in plants (e.g., S-glycoproteins from *Nicotiana alata*), where they control the gametophytic self-incompatibility *(16)* and fungal species (e.g., RNases T2 and Rh from *Aspergillus oryzae* and *Rhizopus niveus* *[16]*). Recently a homologous gene from *Drosophila melanogaster* has also been isolated and characterized *(17)*. Further searching in the database allowed us to identify the mouse homologue (EST:W82885) of this class of ribonucleases. Southern blotting analysis using the human full-length cDNA as a probe indicated that the human RNase gene is present in single copy in the human genome, and related sequences in the EST data banks indicate the expressed gene is not multiallelic. We thus discovered, serendipitously, a gene product specifying a new ribonuclease.

### 1.3. Does This Protein Play a Role in Tumor Suppression?

Northern blotting analysis of RNA from several human tissues revealed that the transcript corresponding to clone p321-A (of approx 1200 bp) is present in all the tissues surveyed, although at different levels. The transcript is most abundant in heart, skeletal muscle, liver, and pancreas (**Fig. 3**). Recent (unpublished) data indicate that the expression of the RNA for this gene is reduced in about 50% of cancer-cell lines derived from ovary and breast tissues compared with normal ovary and breast tissues. We are now trying to determine whether this gene can play a causal role in cancer development by checking for the presence of inactivating mutations in human malignancies or the reversion of tumor phenotype after expression of the gene into immortal or neoplastic cell lines by means of cell culture transfection experiments. For the latter, one can compare the growth kinetics of the matched transfected/parental cell lines during anchorage independent growth on soft agar *(18)*, or in Matri-gel invasion assays *(19)*. Of course, the most reliable indication of suppression would be provided by the injection of the transfected cell lines into nude mice, and the observation of an inhibition of tumor formation. Experiments of these types are being done by us and related groups with the human RNase gene.

RNases are now the focus of a great deal of attention because of newly discovered activities and applications, particularly in cancer biology *(20)*. It would be interesting to demonstrate whether the new mouse and human RNases that we have described have any cytotoxic activity that could be instrumental in the treatment of cancer.

```
              10          20          30          40          50          60
ggc gac tga ccg tgg tcg tgg gcg gac ggc ggc ttg cag cgt gga gga gct ggg gtc gct

              70          80          90         100         110         120
gtg ggt cgc gaa gca gag ccc ggg acg tgc gcg ctt ggt gca cga tcc tga agg gga gct

             130         140         150         160         170         180
ccg agg ggc ccg ggt cgc cag ggc tgc tgc ggc cat tcc cgg agc ccg gcg cgg ggc ccg

             190         200         210         220         230         240
cga gat act ggt tta ggc cgt ccc agg gct ccg ggc gca ccc ggt ggc cgc tgc tgc agc

             250         260         270         280         290         300
gga ggg agc gcg gcg gcg cgg ggg ctc gga gac agc gtt tct ccc gga agt ctt cct cgg

             310         320         330         340         350         360
gca gca ggt ggg aag tgg gag ccg gag cgg cag ctg gca gcg ttc tct ccg cag gtc ggc

             370         380         390         400         410         420
acc atg cgc cct gca gcc ctg cgc ggg gcc ctg ctg ggc tgc ctc tgc ctg gcg ttg ctt
     M   R   P   A   A   L   R   G   A   L   L   G   C   L   C   L   A   L   L

             430         440         450         460         470         480
tgc ctg ggc ggt gcg gac aag cgc ctg cgt gac aac cat gag tgg aaa aaa cta att atg
 C   L   G   G   A   D   K   R   L   R   D   N   H   E   W   K   K   L   I   M

             490         500         510         520         530         540
gtt cag cac tgg cct gag aca gta tgc gag aaa att caa aac gac tgt aga gac cct ccg
 V   Q   H   W   P   E   T   V   C   E   K   I   Q   N   D   C   R   D   P   P

             550         560         570         580         590         600
gat tac tgg aca ata cat gga cta tgg ccc gat aaa agt gaa gga tgt aat aga tcg tgg
 D   Y   W   T   I   H   G   L   W   P   D   K   S   E   G   C   N   R   S   W

             610         620         630         640         650         660
ccc ttc aat tta gaa gag att aag gat ctt ttg cca gaa atg agg gca tac tgg cct gac
 P   F   N   L   E   E   I   K   D   L   L   P   E   M   R   A   Y   W   P   D

             670         680         690         700         710         720
gta att cac tcg ttt ccc aat cgc agc cgc ttc tgg aag cat gag tgg gaa aag cat ggg
 V   I   H   S   F   P   N   R   S   R   F   W   K   H   E   W   E   K   H   G

             730         740         750         760         770         780
acc tgc gcc gcc cag gtg gat gcg ctc aac tcc cag aag aag tac ttt ggc aga agc ctg
 T   C   A   A   Q   V   D   A   L   N   S   Q   K   K   Y   F   G   R   S   L

             790         800         810         820         830         840
gaa ctc tac agg gag ctg gac ctc aac agt gtg ctt cta aaa ttg ggg ata aaa cca tcc
 E   L   Y   R   E   L   D   L   N   S   V   L   L   K   L   G   I   K   P   S

             850         860         870         880         890         900
atc aat tac tac caa gtt gca gat ttt aaa gat gcc ctt gcc aga gta tat gga gtg ata
 I   N   Y   Y   Q   V   A   D   F   K   D   A   L   A   R   V   Y   G   V   I

             910         920         930         940         950         960
ccc aaa atc cag tgc ctt cca cca agc cag gat gag gaa gta cag aca att ggt cag ata
 P   K   I   Q   C   L   P   P   S   Q   D   E   E   V   Q   T   I   G   Q   I

             970         980         990        1000        1010        1020
gaa ctg tgc ctc act aag caa gac cag cag ctg caa aac tgc acc gag ccg ggg gag cag
 E   L   C   L   T   K   Q   D   Q   Q   L   Q   N   C   T   E   P   G   E   Q

            1030        1040        1050        1060        1070        1080
ccg tcc ccc aag cag gaa gtc tgg ctg gca aat ggg gcc gcc gag agc cgg ggt ctg aga
 P   S   P   K   Q   E   V   W   L   A   N   G   A   A   E   S   R   G   L   R

            1090        1100        1110        1120        1130        1140
gtc tgt gaa gat ggc cca gtc ttc tat ccc cca cct aaa aag acc aag cat tga tgc cca
 V   C   E   D   G   P   V   F   Y   P   P   P   K   K   T   K   H  STOP

            1150        1160        1170        1180        1190        1200
agt ttt gga aat att ctg ttt taa aaa gca aga gaa att cac aaa ctg cag ccc gga att

 1204
cct g
```

Fig. 1. Nucleotide and deduced amino-acid sequences of the human RNase cDNA.

```
CLUSTAL W (1.74) multiple sequence alignment


Rnase-Hs      ------------------------------MR-PAALRGALLG-----------CLCLAL  18
RNASE-Mm      ------------------------------LRSPGQCDGAGGGSRPLPGWISVLGWGLAL  30
RNASE-Rh      ------------------------------MKAVLALA-TLIGST-----LASSCSSTAL  24
RNASE-Ao-T2   ------------------------------MKAVLALA-TLIGST-----LASSCSSTAL  24
RNASE-Le      ------------------------------MASNSAFS--------------LFLIL  13
RNASE-Dm      WQSQRFLFVAILACFLVTIKSTPLSDISDSDESTSVDEKTQKRPDDDSEFDGFPFREDDD  60


Rnase-Hs      LCLGGADKRLRDNHEWKKLIMVQHWPETVCEKIQNDCRD---PP----DYWTIHGLWPDK  71
RNASE-Mm      CSLCGAGPLWSGSHEWKKLILTQHWPPTVCKEVN-SCQD---SL----DYWTIHGLWPDR  82
RNASE-Rh      SCSNSANSDTCCSPEYGLVVLNMQWAPGYGPDNA---------------FTLHGLWPDK  68
RNASE-Ao-T2   SCSNSANSDTCCSPEYGLVVLNMQWAPGYGPNAA---------------FTLHGLWPDK  68
RNASE-Le      LIITQCLSVLNAAKDFDFFYFVQQWPGSYCDTKQSCCYP---TTGKPAADFGIHGLWPNN  70
RNASE-Dm      SLQDSSREMSVQDHNWDVLIFTQQWPVTTCYHWREENPDQECSLPQKKEFWTIHGIWPTK 120
                 .             ::  . :  :*.               :  :**:**   .


Rnase-Hs      SE-------GC--NRSWPFNLEEIKD----LLPEMRAYWPDVIHSFPNRSRFWKHEWEKH 118
RNASE-Mm      AE-------DC--NQSWHFNLDEIKD----LLRDMKIYWPDVIHRSSNRSQFWKHEWVKH 129
RNASE-Rh      CSGAYAPSGGCDSNRASSSIASVIKSKDSSLYNSMLTYWP---SNQGNNNVFWSHEWSKH 125
RNASE-Ao-T2   CSGAYAPSGGCDSNRASSSIASVIKSKDSSLYNSMLTYWP---SNQGNNNVFWSHEWSKH 125
RNASE-Le      NDGTYP--SNC--DPNSPYDQSQISD----LISSMQQNWPTLACPSGSGSTFWSHEWEKH 122
RNASE-Dm      LHQMGP--NFC--NNSANFDPSKLNP----IEDRLETFWPD-LKGMDSTEWLWKHEWQKH 171
                    *   :     . :.            **  :      :     .:*.*** **


Rnase-Hs      GTCAAQVD---ALNSQK-----KYFGRSLELYRELDLNSVLLKLGIKPSINYYQVADFKD 170
RNASE-Mm      GTCAAQVD---ALNSEK-----KYFGKSLDLYKQIDLNSVLQKCGIKPSINYYQLADFKD 181
RNASE-Rh      GTCVSTYDPDCYDNYEEGEDIVDYFQKAMDLRSQYNVYKAFSSNGITPGG-TYTATEMQS 184
RNASE-Ao-T2   GTCVSTYDPDCYDNYEEGEDIVDYFQKAMDLRSQYNVYKAFSSNGITPGG-TYTATEMQS 184
RNASE-Le      GTCAESVL----TNQHA------YFKKALDLKNQIDLLSILQGADIHPDGESYDLVNIRN 172
RNASE-Dm      GTCAMLVEE--LDNELK------YFEQGLTWREEYIMSRILDASDIHPDS-NNTVAAINN 222
              ***         *         ** :.:   :   :   :  . * *.      . :..


Rnase-Hs      ALARVYGVIPKIQCLP-PSQDEEVQTIGQIEL----CLTKQDQQLQN-CTEPGEQPSPKQ 224
RNASE-Mm      ALTRIYGVVPKIQCLM-PEQGENVRPFAR------------------------------ 209
RNASE-Rh      AIESYFGAKAKIDCSSGTLSDVALYFYVRGRD----TYVITDALSTGSCSGDVEYPTK-- 238
RNASE-Ao-T2   AIESYFGAKAKIDCSSGTLSDVALYFYWRGRD----TYVITDALSTGSCSGDVEYPTK-- 238
RNASE-Le      AIKSAIGYTPWIQCNVDQSGNSQLYQVYICVDGSGSSLIECPIFPGGKCGTSIEFPTF-- 230
RNASE-Dm      AIVKALGKNPSIHCLYDGKHGISYLSEIRICFSKSLELIDCDGIKQG-DAVPVGVPGGTI 281
              *:      *   . *.*             .


Rnase-Hs      EVWLANGAAESRGLRVCEDGPVFYPPPKKTKHZ----------- 257
RNASE-Mm      --------------------------------------------
RNASE-Rh      --------------------------------------------
RNASE-Ao-T2   --------------------------------------------
RNASE-Le      --------------------------------------------
RNASE-Dm      ITNCHIGSLVHYPSLVPPLQRKSHWKLPLVNVYKLLQFLMWFTL 325
```

Fig. 2. Sequence comparison of the region containing the bipartite RNase motif from different proteins. Multiple sequence comparisons were obtained using the MACAW program. Numbers at the right refer to the amino-acid positions encoded by the RNases from the different species. RNase-Hs: human, RNase Mm: mouse, RNase Rh: *Rhizopus niveus*, RNase Ao-T2: *Aspergillus oryzae*, RNase Le: *Lycopersicon esculentum*, RNase Dm: *Drosophila Melanogaster*.

## 2. Materials

### 2.1. Preparation of cDNA

All solutions listed are prepared using bidistilled, deionized water.

1. Eppendorf tubes, 1.5 mL and 0.5 mL.
2. Bidistilled sterile water (dH$_2$O).
3. PCR apparatus.
4. 10X PCR buffer with MgCl$_2$ (Promega).
5. cDNA library from the tissue or cell-line of interest, cloned in Lambda GT 10 vector and having a titer >10$^6$ PFU/μL.

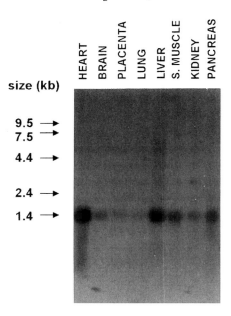

Fig. 3. Northern analysis of human RNase from adult tissues. Northern blots were hybridized with the RNase full-length cDNA.

6. 2 m$M$ dNTP mix, pH 7.0.
7. Primer 10F: 5'-GCAAGTTCAGCCTGGTTAAG-3'; Primer 10R: 5'-GAGTATTTCTTC CAGGGTA-3'. Dissolve both primers in sterile water at a 100 m$M$ concentration.
8. Taq DNA Polymerase (Promega).
9. ART (Aerosol Resistant Tips) for P20 and P200 micropipeters.
10. Apparatus for gel electrophoresis.
11. 10X loading dye: 50% glycerol, 0.2 $M$ EDTA, 0.05% (w/v) bromophenol blue.
12. Ethidium bromide 10 mg/mL. Dissolve 1 mg in 100 mL of dH$_2$O by means of a magnetic rod, leaving the solution on a magnetic stirrer for 2 h in the dark, then remove the rod and store at 4°C in the dark.
13. TAE electrophoresis buffer, 50X stock solution: add 242 g Tris-HCl base, 57 mL glacial acetic acid, and 37.2 g Na$_2$EDTA·2H$_2$O to 1 L dH$_2$O.
14. Agarose MP (Boehringer).
15. 1-kb Ladder mol-wt marker (Promega).
16. Qiaquick PCR purification kit (Qiagen).
17. Sodium acetate 3 $M$: dissolve 408 g sodium acetate·3 H$_2$O in H$_2$O. Adjust pH to 5.2 with 3 $M$ acetic acid. Add H$_2$O to 1 L.
18. Ethanol, 99.7–100% (DBH).
19. Ethanol, 70% in water.
20. Microcentrifuge.
21. Hybridization solution: prepare fresh by mixing the following reagents in a sterile glass beaker: 50 mL formamide, 4.38 g NaCl, 5 mL NAPI, pH 7.2 (*see below* for recipe), 1 mL 0.5 $M$ EDTA, pH 8.0 (*see below* for recipe), 5 mL 5X Denhart's solution (*see below* for recipe), 0.5 mL 10% SDS. Add dH$_2$O to 100 mL.
22. NAPI 1 $M$: dissolve 14.19 g Na$_2$HPO$_4$ and 9.79 g H$_3$PO$_4$ in 80 mL of dH$_2$O, then bring to a final volume of 100 mL and filter sterilize.

23. Denhart's solution, 100X: Dissolve 10 g Ficoll 400, 10 g polyvinylpyrrolidone, and 10 g BSA (fraction V) in 500 mL dH$_2$O. Filter-sterilize and store 10-mL aliquots at –20°C.

## 2.2. Preparation of Genomic Target DNA

1. Plasmid Midi kit (Qiagen).
2. 10X *Mbo*I restriction buffer.
3. *Mbo*I (Promega).
4. Water bath at 37°C.
5. EDTA 0.5 *M*: dissolve 186.1 g Na$_2$EDTA·2 H$_2$O in 700 mL H$_2$O. Adjust pH to 8.0 with 10 *M* NaOH and add H$_2$O to 1 L.
6. Thermoblock at 65°C.
7. 1:1 Phenol-chloroform pH 6.6–8.0 (BDH).
8. Chloroform (DBH)
9. Linker oligonucleotides: T-Mbo: 5'-GATCCACGAATTCAAGCTCTGG-3'; T-i: 5'-CCAGAGCTTGAATTCGTG-3'. The T-Mbo primer must be phosphorylated at the 5' terminus. Half of the T-i primer must be biotynilated.
10. 10X Ligation buffer (Boehringer Mannheim).
11. 100 m*M* ATP, pH 7.0 (Boehringer Mannheim).
12. T4 DNA ligase (Boehringer Mannheim).
13. Water bath at 16°C.
14. 10 *M* ammonium acetate: Dissolve 385.4 g ammonium acetate in 150 mL dH$_2$O. Add dH$_2$O to 500 mL, filter sterilize and store at room temperature.
15. Biotynilated T-i primer.
16. Cot-1 DNA (Gibco-BRL).
17. Cosmid vector DNA (approx 50 µg).

## 2.3. Direct cDNA Selection

1. Thermoblock at 95°/42°C.
2. Mineral oil (Sigma).
3. Dynabeads (Dynal, Oslo).
4. TE, pH 7.5: Dissolve 1.21 g Tris base and 0.358 g Na$_3$-EDTA in 800 mL dH$_2$O, adjust the pH to 7.5 with HCl, and bring the volume to 1 L.
5. NaCl 5 *M*: Dissolve 292 g NaCl in dH$_2$O to a final volume of 1 L. Filter-sterilize and store at room temperature.
6. SSC 20X: Dissolve 175.2 g NaCl and 88.2 g Na$_3$-citrate in 800 mL of dH$_2$O and bring to a final volume of 1 L.
7. 10% SDS (w/v) in water.
8. Magnetic particle concentrator.

## 2.4. Cloning of the Selected cDNAs

1. *Eco*RI 10X restriction buffer (Promega).
2. *Eco*RI (Promega).
3. pBluescript SK cloning vector (Stratagene).
4. 10X dephosphorilation buffer (Boehringer Mannheim).
5. Calf intestinal alkaline phosphatase (Boehringer Mannheim).
6. LB plates: dissolve in a 2-L flask 10 g bacto tryptone, 5 g yeast extract, 5 g NaCl, and 15 g Agar in 1000 mL H$_2$O, add 1 mL 1 *N* NaOH, and autoclave 15 min at 15 lb/in.[2]. Cool the medium to 50°C before pouring 20-mL aliquots into 100-mm Petri dishes.
7. Ampicillin, 50 mg/mL in dH$_2$O, stored at –20°C.

8. IPTG 100 m$M$ in dH$_2$O, stored at –20°C.
9. X-Gal stock solution: 40 mg/mL in *N-N*-dimethylformamide. Store at –20°C in the dark.
10. Plastic Petri dishes 100 mm (ICI).
11. Hybond N+ hybridization filters (Amersham).
12. Whatman 3-MM filter paper.
13. Denaturing solution: prepare a 10 $M$ NaOH stock solution by dissolving 400 g NaOH in 450 mL dH$_2$O, then add dH$_2$O to 1 L. Add 50 mL of this solution to 300 mL of 5 $M$ NaCl and 650 mL dH$_2$O.
14. Neutralizing solution: Dissolve 87.5 g NaCl and 60.55 g Tris-HCl base to 800 mL dH$_2$O. Add 2 mL 0.5 $M$ EDTA and adjust pH to 7.5 with concentrated HCl. Bring the final volume to 1 L with dH$_2$O.
15. 2X SSC. Add 50 mL of 20X SSC to 450 mL dH$_2$O.
16. Oven (80°C).
17. Cot-1 DNA (Gibco-BRL).
18. Cosmid vector DNA (50 μg).
19. Rediprime labeling kit (Amersham).
20. [α-$^{32}$P]-dCTP (Amersham).
21. Hybridization oven.
22. Thermoblock at 95°/37°C.
23. Quick-spin G-50 column (Roche).
24. Premix: Prepare fresh by mixing the following reagents: 15 mL 20X SSC, 2 mL 100X Denhart's solution, 5 mL 10% SDS, dH$_2$O to 100 mL. Just before using the solution, add 100 μg/mL denatured salmon sperm DNA (Sigma).
25. To prepare washing solutions: 20X SSC and 10% SDS.
26. Kodak X-AR5 autoradiography films.
27. Autoradiography cassettes with intensifying screens.

## 3. Methods

The following section describes several experimental procedures that have been used in our lab and have led us to the identification of a new member of the mammalian ribonuclease family. All these procedures have a broad application to projects whose final aim is the identification of expressed sequences from a cloned genomic region of interest. The first method (**Subheading 3.1.**) describes the procedure for the preparation of a PCR-amplified cDNA population to be used as the starting DNA from which expressed sequences representing genes localized in the genomic region under investigation will be enriched and isolated.

Thereafter, a method for the preparation of "cDNA selection-grade" DNA from the genomic region itself is described (**Subheading 3.2.**). Once the two DNA sources are properly prepared, the cDNA selection protocol is eventually carried out, and the relative experimental procedure is reported in the third method (**Subheading 3.3.**). Finally, the fourth method (**Subheading 3.4.**) deals with the procedures to be followed for the isolation by cloning of the enriched cDNAs.

### 3.1. PCR Method to Isolate cDNA from Human Tissues

The most important consideration for a successful cDNA selection procedure is that the transcripts derived from the genomic region of interest must be represented in the starting mRNA population (*see* **Note 1**). In many instances, the biological problem to be addressed allows the investigator to choose the mRNA source to be used, but there are often cases where candidate RNA sources are not immediately evident *a priori*. In such cases, complex mRNA sources such as fetal brain or different pooled tissues are adequate. The direct cDNA selection approach can be applied either to an uncloned population of cDNA that has been obtained by reverse transcription, or alternatively, to a cloned cDNA library in the form of hundreds of thousands of recombinant phage particles. In the former case, a total cDNA sample from the selected tissue must be manipulated by attaching proper oligonucleotide linkers to the cDNA in order to permit PCR amplification of the entire cDNA population.

To avoid such handlings (which sometimes prove not to be straightforward), it is preferable to use a premade, cloned cDNA library as the starting material such as the fetal brain cDNA library cloned into the λ-gt 10 phage vector that we have used.

1. Amplify the inserts from the λ-GT 10 cDNA library by PCR using 1, 3, and 5 μL of a starting library having a titer of approx $10^6$ PFU/μLl (*see* **Note 1**). Assemble the PCR reaction as follows: 1, 3, or 5 μL aliquots from the cDNA library, 5 μL 10X PCR reaction buffer with $MgCl_2$, 5 μL 2 m*M* dNTPs, 4 μL 3 μ*M* primer 10F/10R mix, $H_2O$ to 49.5 μL and 0.5 μL Taq polymerase. As a negative control, assemble a PCR reaction without cDNA (*see* **Notes 2** and **3**).
2. Perform PCR under the following conditions: 94°C for 2 min, then 35 cycles at 94°C for 30 s, 55°C for 30 s, 72°C for 2 min, plus a single step at 72°C for 3 min.
3. Check the PCR by loading 5 μL of each reaction on a 1% agarose gel containing ethidium bromide at a concentration of 5 μg/mL. Ideally, a smear ranging from <100 bp up to few kilobases from the cDNA library-amplified reactions should be observed.
4. Determine which of the three reactions gave the best results, then assemble a scaled-up reaction by repeating the PCR in multiple tubes (up to 20) exactly as in **Subheading 3.1.2.** (*see* **Note 4**).
5. Check again a 5-μL aliquot from each reaction on a 1% agarose gel; if the reactions have run as expected, pool them and purify the amplified DNA from primers, salts, dNTPs, and Taq polymerase using Qiaquick PCR purification columns following the manufacturer's instructions. Elute the DNA in 100 μL.
6. Precipitate the DNA by adding 10 μL of 3 *M* sodium acetate and 220 μL of ice-cold ethanol. Keep the tube at –80°C for 15 min, then centrifuge at full speed for 20 min at 4°C. Remove the supernatant and wash the pellet briefly with 1 mL of 70% ethanol, centrifuge for 3 min at full speed at 4°C, again remove the supernatant and dry the DNA pellet briefly by leaving the tube open on the bench for 10 min. Resuspend the purified cDNA in hybridization mix at a concentration of approx 200 ng/μL.

### 3.2. Subcloning and Biotin Labeling of a Recombinant YAC

The genomic region of interest to be used for cDNA selection usually consists of a rather large contiguous stretch of genomic DNA available as several overlapping recombinant genomic clones in such vectors as lamba phages, cosmids, PACs, or YACs. In most cases, a set of overlapping YAC clones spanning the region of interest is used. This is because YACs are thus far the cloning vectors with the highest insert-size capacity *(21)*, allowing the coverage of several megabases of genomic DNA with just few overlapping clones.

Although single YAC clones from such contigs could be directly used for cDNA selection, the very large size of some YACs poses some problems with the sensitivity of cDNA selection methods when applied to such huge stretches of genomic DNA; an efficient alternative is therefore to use pools of cosmid clones derived from the subcloning of such large YACs (*see* **Note 5**). Among the methods that can be applied to isolate cosmid clones representing the genomic insert of a YAC clone, probably the most effective is the direct screening of cosmid libraries with YAC clones *(23)*.

Accordingly, the following protocol describes the preparation of biotin-labeled DNA from pools of cosmids representing the genomic insert of the YAC clone of interest. The number of cosmid clones covering the genomic insert of the YAC clone of interest is obviously related to the insert size of the YAC itself, and can vary from less than 10 from a few hundred for very large YACs.

Whatever the number of cosmid clones to use, each pool usually consists of five different cosmid clones, since this number will maximize the sensitivity of the cDNA selection method. The DNA from each cosmid clone is prepared with the Qiagen midi kit, and labeling is achieved by partial digestion of DNA with *Mbo*I followed by ligation with an *Mbo*I-compatible adaptor and PCR amplification using an adaptor-specific biotynilated primer.

1. Pools are prepared by mixing 1 µg of DNA from each cosmid in a final volume of 50 µL. Assemble five partial *Mbo*I digestion reactions in Eppendorf tubes as follows: 20 µL cosmid pool DNA (5 µg DNA), 5 µL *Mbo*I 10X digestion buffer, H$_2$O to 50 µL, and 1 U *Mbo*I. Incubate the five tubes at 37°C for 0, 10, 20, 30, and 45 min. At each time point, add 5 mL 0.5 *M* EDTA pH 8.0 (final concentration: 50 m*M*) and put the Eppendorf tubes at 65°C for 20 min, then spin for 10 s and transfer them on ice.
2. Load the digested DNA into a 1% agarose gel containing ethidium bromide at a concentration of 5 µg/mL, together with a molecular weight marker; at least one of the reaction conditions used should yield a smear with most DNA migrating in the range of 0.4–2 kb.
3. Reassemble three partial *Mbo*I digestion reactions exactly as above, incubating the samples at 37°C for *X* – 5, *X*, and *X* + 5, where *X* indicates the incubation time (in min) who gave the desired DNA migration pattern in the previous experiment.
4. Pool the three samples together and load a 5-µL aliquot on a 1% agarose gel to check the quality of the digestion reaction.
5. Purify the digested DNA: Add to the sample 1 vol of phenol-chloroform, vortex the solution vigorously for 10 s, and centrifuge at room temperature for 3 min at 12,000 rpm. Recover the upper aqueous phase from the tube and transfer it in a clean Eppendorf tube. Add to the sample 1 vol of chloroform, vortex the solution vigorously for 10 s, and centrifuge at room temperature for 3 min at 12,000 rpm. Recover the upper aqueous phase from the tube and again transfer it in a new clean Eppendorf tube.

   Precipitate the DNA by adding 1/10 vol of 3 *M* sodium acetate and 2 vol of ice-cold absolute ethanol. Keep the tubes at –80°C for 15 min, then centrifuge at full speed for 20 min at 4°C and discard the supernatant. Wash the pellet briefly with 1 mL of 70% ethanol and centrifuge for 3 min at full speed at 4°C. Again remove the supernatant and briefly dry the DNA pellet at 37°C. Finally, resuspend the DNA pellet in 10 µL of water.
6. Prepare the linker adaptor to be ligated to the partially digested cosmid DNA: put together 100 pmol of phosphorylated T-*Mbo* oligonucleotide and 100 pmol of T-i oligonucleotide in a final volume of 10 µL of sterile water; heat denature the sample at 100°C for 5 min in a thermoblock, then switch it off and let the sample cool to room temperature (about 2 h) to let the two oligonucleotide anneal together.
7. Assemble the ligation reaction as follows: 10 µL *Mbo*I digested target genomic DNA, 2 µL adaptor linker (annealed oligonucleotides T-*Mbo* and T-i), 2 µL 10X ligation buffer, 2 µL 2 m*M* ATP, H$_2$O to 19 mL, and 1 µL T4 DNA ligase. Mix by gentle pipeting and incubate at 16°C for 2 h.

8. Add sterile water to a final volume of 200 μL, extract the sample once with phenol-chloroform, once with chloroform, and precipitate the ligated DNA by adding ammonium acetate at a final 2 *M* concentration and 2 vol of cold absolute ethanol, as described in **Subheading 3.2.5.** Resuspend the DNA in 10 μL of sterile water.

9. Assemble the biotin-labeling PCR reaction as follows: 1 μL linker-ligated genomic DNA, 5 μL 10X PCR reaction buffer with MgCl$_2$, 5 μL 2 m*M* dNTPs, 4 μL 3 μ*M* biotinylated T-i primer, H$_2$O to 49.5 μL, and 0,5 μL Taq polymerase. As controls, include a reaction with no DNA and a reaction without the primer.

10. Run the PCR under the following conditions: 94°C for 2 min, then 35 cycles 94°C for 30 s, 52°C for 30 s, 72°C for 3 min. Finally, make a single step at 72°C for 5 min.

11. Check the PCR by loading a 5-μL aliquot on a 1% agarose gel containing ethidium bromide at a concentration of 5 μg/mL; a DNA smear resembling the starting DNA from the control reaction without primer, but with higher intensity, should be visible.

12. Purify the amplified DNA using Qiaquick PCR purification columns following the manufacturer's instructions. Elute the DNA in 100 μL and precipitate it by adding 10 μL of 3 *M* sodium acetate and 220 μL of cold ethanol.

    Keep the tube at −80°C for 15 min, then centrifuge at full speed for 20 min at 4°C. Wash the pellet briefly with 1 mL of 70% ethanol, centrifuge for 3 min at full speed at 4°C. Briefly dry the DNA pellet at 37°C.

13. Block the repetitive and vector sequences in the target genomic DNA by resuspending the DNA in 25 μL of hybridization mix containing 20 μg of Cot-1 DNA (*see* **Note 6**) and 1 mg of cosmid vector DNA (which have been previously ethanol-precipitated and resuspended in hybridization mix), and denaturing for 5 min at 95°C. Let the blocking reaction proceed for 4 h at 42°C.

## *3.3. Direct cDNA Selection*

Once both the amplified cDNA population and the biotin-labeled blocked target genomic DNA have been prepared, the cDNA selection method is carried out. The biotin-labeled genomic DNA works as a hook to pick up cDNAs which specifically hybridize with it.

In practice, the two DNA sources are denatured and hybridized in solution. During this step, only the cDNAs representing genes localized in the genomic region of interest will find complementary sequences in the region itself and hybridize to it, while all unrelated cDNAs will not hybridize and will be washed away in the subsequent steps. The hybridized cDNAs are then cocaptured with the genomic DNA by means of streptavidin-coated Dynabeads which selectively bind the biotin-labeled target genomic DNA. After several washes, the hybridized cDNAs are eluted and passed through a second round of selection before being cloned and analyzed.

1. Cover 5 μL (1 μg) of amplified cDNA with a small layer of mineral oil and denature the sample by incubating at 95°C for 5 min.

2. Add 25 μL of prehybridized biotin-labeled genomic DNA to the denatured cDNA and incubate at 42°C for 48 h.

3. Wash 100 μL of dynabeads three times with 500 μL TE + 1 *M* NaCl. Remove the beads after each wash by means of a magnetic concentrator. Finally, resuspend the beads in 100 μL of TE + 1 *M* NaCl.

4. Recover the hybridized DNA from **step 2** (taking care not to include the mineral oil above). Add it to the beads and incubate for 10 min at room temperature, briefly shaking the tube every 2 min.

5. Concentrate the beads on one side of the tube using the magnetic concentrator and perform the following washing steps in a final volume of 500 µL: once for 15 min at room temperature in 3X SSC-0.1% SDS, once for 15 min at 50°C in 0.1X SSC–0.1% SDS and three times for 15 min at 65°C in 0.1X SSC-0.1% SDS. During each washing step, the beads are resuspended homogeneously and repelletted by means of the magnetic concentrator before the next wash.

6. Following the last wash, elute the bound cDNA by resuspending the beads in 50 µL of sterile water and incubating at 80°C for 5 min. Again remove the beads using the magnetic concentrator and transfer the supernatant into a separate tube.

7. Assemble PCR reactions with 1, 3, and 5 µL of the eluted cDNAs as follows: 1-, 3-, or 5-µL aliquots from the eluted cDNA, 5 µL 10X PCR reaction buffer with MgCl$_2$, 5 µL 2 m*M* dNTPs, 4 µL 3 µ*M* primer 10F/10R mix, H$_2$O to 49.5 µL, and 0.5 µL Taq polymerase. As controls, assemble PCR reactions with primers alone (no cDNA) and 1 µL of cDNA without primers.

8. Perform the PCR under the following conditions: 94°C for 2 min, then 35 cycles at 94°C for 30 s, 55°C for 30 s, 72°C for 2 min, and a single cycle at 72°C for 3 min.

9. Check the PCR by loading a 5-µL aliquot on a 1% agarose gel containing ethidium bromide at a concentration of 5 µg/mL; at least one of the three reactions with the eluted cDNA should yield a smear ranging from few hundred to few thousand kilobases; discrete bands can also be observed at this stage (*see* **Note 7**).

## 3.4. Cloning of the Enriched cDNAs

Following the elution and amplification steps described in **Subheading 3.3.**, a population of enriched cDNAs is obtained and all members of this population must be individually identified and analyzed. A cDNA "minilibrary" is made by cloning the eluted cDNA in a plasmid vector in order to isolate single cDNA clones. After background clones are identified by Southern hybridization and discarded, the remaining cDNA clones are characterized in detail by DNA sequencing, and now represent candidate genes (such as the ribonuclease gene that we have identified) that can be tested for their putative role in the biological process or disease of interest.

1. Repeat in triplicate the PCR reaction with primers 10F and 10R, using the amount of eluted cDNA that performed best in the previous experiment; perform the reaction exactly as in **Subheading 3.3., step 8.**

2. Pool the three reactions and check a 5-µL aliquot on a 1% agarose gel, then purify the amplified DNA from primers, salts, dNTPs, and Taq polymerase using Qiaquick PCR purification columns, following the manufacturer's instructions. Elute the DNA in 100 µL and precipitate it by adding 10 µL of 3 *M* sodium acetate and 220 µL ice-cold Ethanol. Keep the tube at –80°C for 15 min, then centrifuge at full speed for 20 min at 4°C. Wash the pellet briefly with 1 mL of 70% ethanol, centrifuge for 3 min at full speed at 4°C, and resuspend the purified DNA in 180 µL water.

3. Digest the cDNAs with *Eco*RI as follows in one Eppendorf tube: 180 µL amplified cDNA, 20 µL *Eco*RI 10X restriction buffer, and 10 U (1 µL) *Eco*RI. Incubate the reaction at 37°C for 2 h, then immediately load the sample on a 0.7% agarose preparative gel containing ethidium bromide at a concentration of 5 µg/mL. Load a molecular weight marker on each side of the *Eco*RI-digested DNA.

4. Under the light of a UV lamp, excise from the gel the DNA which migrated in the range of 0.2–3 kb, taking care to minimize the DNA's exposure to UV radiation; extract the DNA from the gel by means of a Qiaquick gel extraction column, following the manufacturer's instructions. Elute the DNA in a final volume of 20 µL.

5. Digest the pBluescript SK vector with *Eco*RI into one Eppendorf tube as follows: 10 μL vector DNA (1 μg), 3 μL 10X *Eco*RI restriction buffer, $H_2O$ to 30 μL, and 0.5 U *Eco*RI. Incubate for 90 min at 37°C.

6. Check a 1-μL aliquot on a 1% agarose gel together with 50 ng undigested vector as a control. If the reaction is complete, add water to the sample to a final volume of 180 μL, then add 20 μL 10X dephosphorilation buffer and 1 U calf intestinal alkaline phosphatase. Incubate 1 h at 37°C, then extract the sample once with phenol-chloroform, and once with chloroform, and precipitate it with 1/10 vol of 3 *M* sodium acetate and 2 vol ice-cold ethanol as described in **Subheading 3.2., step 5**. Keep the tube at –80°C for 15 min, then centrifuge at full speed for 20 min at 4°C. Wash the pellet briefly with 1 mL 70% ethanol, centrifuge for 3 min at full speed at 4°C, and and resuspend the purified DNA in 20 μL water.

7. Load a 2-μL aliquot from both the *Eco*RI digested vector and cDNAs on a 1% agarose gel to check for quality of the samples and to roughly quantitate them.

8. Assemble on ice the ligation reaction as follows: 1 μL *Eco*RI-digested, dephosphorylated vector, 10 μL *Eco*RI-digested cDNA, 1.5 μL 10X ligation buffer, 1.5 μL 10 m*M* ATP, and 1 μL (10 U) T4 DNA ligase. Incubate the reaction 2 h at 16°C, then use three 5-μL aliquots to transform competent DH5-α cells according to the manufacturer's instructions. Spread the transformed cells on the surface of LB plates containing 50 μg/mL ampicillin, 0.1 m*M* IPTG, and 20 μg/mL X-Gal. Incubate the plates overnight at 37°C.

9. The next morning, pick all colorless together with five blue-stained colonies with sterile toothpicks and make a double spot on an LB-Amp plate and on several grilled circular Hybond N+ hybridization filters (Amersham) laid on the surface of LB-Amp plates.

10. Incubate the plates at 37°C until 1-mm-sized colonies have formed, then store the membraneless master plate at 4°C and process the nylon filters for hybridization as follows. Remove the membrane and place it for 5 min (colony side up) on a Whatman 3-MM paper sheet soaked in denaturing solution, then place the membrane for 3 min on a second 3-MM paper sheet soaked in neutralizing solution. Wash the filter by submerging it in 2X SSC and rub away the cellular debris with a paper towel soaked in the same solution. Briefly dry the filter on a new 3-MM sheet and fix the DNA to the filter by baking for 2 h at 80°C. Store the filters wrapped in aluminum foil at 4°C until needed.

11. To set apart background clones containing either vector sequences or repetitive elements, the filters are hybridized separately with 100 ng of [32]P-labeled COT-1 or 10 ng of cosmid vector DNA. The radiolabeled probes are prepared using the Rediprime labeling kit (Amersham) as follows: dilute the DNA with water to a final volume of 45 μL and denature for 5 min at 95°C. Briefly spin the tube, then add the DNA to the labeling mix tube, mix by gentle flicking, and add 5 μL of [α-[32]P]-dCTP, mix by gentle pipeting, and incubate for 10 min at 37°C. Stop the reaction by adding 2 μL of 0.5 *M* EDTA, pH 8.0, and remove unincorporated nucleotides with a Quick-spin column (Roche) following the manufacturer's instructions.

12. Prehybridize the filters in a minimal volume of premix at 65°C for 2 h in a hybridization bag. Denature the radiolabeled probe for 5 min at 95°C, then add it to the bag containing the prehybridized filters. Incubate overnight at 65°C.

13. Wash filters once with an excess of 3X SSC–0.1% SDS, once with 0.3X SSC–0.1% SDS, and twice with 0.1X SSC–0.1% SDS. All washes are made at 65°C for 15 min.

14. Briefly dry the filters on 3-MM sheets. Then wrap them in Saran wrap and carry out autoradiography. Colonies giving strong hybridization signals with either probes are discarded for further analysis. The remaining clones are recovered from the master plate and grown individually, and their inserts are sequenced to completion (*see* **Note 8**).

15. A computer-assisted comparative and structural analysis is carried out for all sequenced clones by means of the BLAST-N, CLUSTALW, EST ASSEMBLY MACHINE, and ORF FINDER programs, all of which were available at the TIGEM web site (http://www.tigem.it).

## 4. Notes

1. Low-abundance mRNAs may be present at less than one copy per $10^6$ transcripts, especially when using complex mRNA sources, such as total brain, for preparing the cDNA library.
2. Use of cDNA inserts amplified by PCR using vector-derived primers can introduce a bias against very large cDNAs. One way to minimize this problem is to start from cDNA libraries constructed by random priming instead of poly-dT priming. However, this procedure will often select partial cDNAs whose full-length counterpart must be isolated by screening a second poly-dT-primed cDNA library.
3. The final concentration of $MgCl_2$ in the PCR reaction is usually 2 m$M$, but this value can be modified from 0.5–3 m$M$.
4. Scaling-up is done assembling many tubes containing the same 50-µL reaction volume as described above, rather than one single reaction with an increased final volume.
5. cDNA selection protocols from cosmid pools rather than purified YAC DNA give better results because of the high background of ribosomal sequences containing cDNAs that are enriched using the entire YAC clone as the source of genomic DNA. This is caused by copurification of yeast chromosome fragments (containing ribosomal DNA) with the YAC DNA during the PFGE preparative electrophoresis that usually precedes the biotin labeling step.
6. Approximately 10% of human cDNAs are known to contain repetitive elements, so it is crucial to effectively block these repeats by adding COT-1 DNA to the biotynilated genomic DNA before the hybridization step.
7. The level of enrichment obtained after cDNA selection can be assessed if a cloned gene has already been mapped to the genomic region of interest and the corresponding cDNA is known to be represented in the cDNA library to be used. In this case, equal amounts (by weight) of amplified inserts from both the starting and the enriched cDNA population are spotted onto nylon membranes and hybridized to [$\alpha$-$^{32}$P]dCTP-labeled cDNA. Following autoradiography, the intensity ratio between the two samples provides a rough estimation of the level of enrichment attained.
8. It is also crucial to map back each of the selected cDNA clones to the genomic source from which the cDNA selection was carried out, because sometimes partial local homology between the genomic DNA and an unrelated cDNA (not mapping to the cognate genomic fragment) is enough to rescue undesired cDNA clones. This is usually accomplished through of Southern hybridization of the cDNA clones to a filter containing total human genomic DNA, a somatic-cell hybrid DNA carrying the chromosome of interest, the DNA from the rodent parental-cell line, and the cosmid clones used to assemble the genomic DNA pool.

## Acknowledgments

RT was supported by a grant from AIRC (Associazione Italiana Ricerca Cancro) and from Telethon (grant E209).

## References

1. Cooke, I., Shelling, A., Le Meuth, V., Charnock, M., and Ganesan, T. (1996) Allele loss on chromosome arm 6q and fine mapping of the region at 6q27 in epithelial ovarian cancer. *Genes. Chromosom. Cancer* **15,** 223–233.
2. Foulkes, W., Ragoussis, J., Stamp, G., Allan, G., and Trowsdale, J. (1993) Frequent loss of heterozygosity on chromosome 6 in human ovarian carcinoma. *Br. J. Cancer* **67,** 551–559.
3. Devilee, P., Van Vliet, M., Van Sloun, P., Dijkshoorn, N., Hermans, J., Pearson, C. C. (1990) Allelotype of human breast carcinoma: a second major site for loss of heterozygosity is on chromosome 6q. *Oncogene* **6,** 1705–1711.

4. Millikin, D., Meese, E., Vogelstein, B., Witkowsky, C., and Trent, J. (1991) Loss of heterozygosity for loci on the long arm of chromosome 6 in human malignant melanoma. *Cancer. Res.* **51,** 5449–5443.

5. Morita, R., Saito, S., Ishikawa, J., Ogawa, O., Yoshida, O., Yamakawa, K., and Nakamura, Y. (1991) Common region of deletion on chromosomes 5q, 6q, and 10q in renal cell cancer. *Cancer Res.* **51,** 5817–5820.

6. Saito, S., Saito, H., Koi, S., Kudo, R., Noda, K., and Nakamura, Y. (1992) Fine-scale deletion mapping of the distal long arm of chromosome 6 in 70 human ovarian cancers. *Cancer. Res.* **52,** 5815–5817.

7. Thomas, G. and Raffel, C. (1991) Loss of heterozygosity on 6q, 16q and 17p in human central nervous system primitive neuroectodermal tumours. *Cancer. Res.* **51,** 639–643.

8. Tibiletti, M., Benasconi, B., Furlan, D., Riva, C., Trubia, M., Buraggi, G., et al. (1996) Early involvement of 6q in surface epithelial ovarian tumours. *Cancer Res.* **56,** 4493–4498.

9. Gaidano, G., Hauptschein, R., Parsa, N., Offit, K., Rao, P., Lenoir, G., et al. (1992) Deletions involving two distinct regions of 6q in B-cell non-Hodgkin's lynphoma. *Blood* **80,** 1781–1787.

10. Kitchingman, P., Rivera, G., and Williams, D. (1990) Abnormalities of the long arm of chromosome 6 in childhood acute lymphoblastic leukemia. *Blood* **76,** 1626–1630.

11. Tibiletti, M. G., Trubia, M., Ponti, E., Sessa, L., Acquati, F., Furlan, D., et al. (1998) Physical map of the D6S149-D6S193 region on chromosome 6q27 and its involvement in benign surface epithelial ovarian tumors. *Oncogene* **16,** 1639–1642.

12. Banga, S. S., Kim, S., Hubbard, K., Dasgupta, T., Jha, K. K., Patsalis, P., et al. (1997) SEN6, a locus for SV40-mediated immortalization of human cells, maps to 6q26-27. *Oncogene* **14,** 313–321.

13. Trubia, M., Sessa, L., and Taramelli, R. (1997) Mammalian Rh/T2/S-glycoprotein ribonuclease family genes: cloning of a human member located in a region of chromosome 6 (6q27) frequently deleted in human malignancies genomics. **42,** 342–344.

14. Jost, W., Bak, H., Glund, K., Terpstra, P., and Beintema, J. (1991) Amino acid sequence of an extracellular phosphate-starvation induced ribonuclease from cultured tomato (*Lycopersicon esculentum*) cells. *Eur. J. Biochem.* **198,** 1–6.

15. Kurihara, H., Mitsui, Y., Ohgi, K., Irie, M., Mizuno, H., and Nakamura, K. (1992) Crystal and molecular structure of RNAse Rh, a new class of microbial ribonucleases from *Rhizopous niveus. FEBS Lett.* **306,** 189–192.

16. McClure, B., Haring, V., Ebert, P., Anderson, M., Simpson, R., Sakiyama, F., and Clarke, A. (1989) Style self-incompatibility gene products of *Nicotiana alata* are ribonucleases. *Nature* **342,** 955–957.

17. Hime. G., Prior, L., and Saint, R. (1995) The *Drosophila melanogaster* genome contains a member of the Rh/T2/S-glycoprotein family of ribonuclease-encoding gene. *Gene* **158,** 203–207.

18. Senger, D., Perruzzi, C., and Ali, I. (1998) T24 human bladder carcinomas cells with activated ha-ras proto-oncogene: non tumorigenic cells susceptible to malignant transformation with carcinogen. *Proc. Natl. Acad. Sci. USA* **85,** 5105–5111.

19. Kundu, G., Mantile, G., Miele, L., Cordella-Miele, E., and Mukherjee, A. (1996) Recombinant human uteroglobin suppresses cellular invasiveness via a novel class of high-affinity cell surface binding. *Proc. Natl. Acad. Sci. USA.*

20. Schein, C. H. (1997) From housekeeper to microsurgeon: the diagnostic and therapeutic potential of ribonucleases. *Nat. Biotechnol.* **15,** 529–536.

21. Schlessinger, D. (1990) Yeast artificial chromosomes: tools for mapping and analysis of complex genomes. *Trends Genetics* **6,** 248–258.
22. Chumakov, I., Rigault, P., Le Gall, I., et al. (1995) A YAC contig map of the human genome. *Nature* **377(suppl.),** 175–297.
23. Baxendale, S., Bates, G. P., MacDonald, M. E., Gusella, J. F., and Lehrach, H. (1991) The direct screening of cosmid libraries with YAC clones. *Nucleic Acid Res.* **19(23),** 6651.

# II

# INHIBITORS AND ACTIVATORS OF NUCLEASES

# 8

## Ribonuclease Inhibitors

### Brittan L. Pasloske

### 1. Introduction

RNA analysis and quantification require completely intact, nondegraded RNA samples to produce optimal results. Although nonenzymatic hydrolysis of phosphodiester bonds is favored by high temperature or pH and the presence of divalent cations ($Mg^{2+}$, $Mn^{2+}$), an RNA sample is most likely to be rapidly degraded by a contaminating ribonuclease (RNase). RNases are difficult to completely remove or inactivate during RNA isolation procedures, and they may be introduced into the sample inadvertently during its handling. There are several possible sources for RNases in the laboratory. RNases are ubiquitous in the environment, and are found on pollen, dust, and fingerpaint grease. Routine lab procedures, such as ribonuclease protection assays or degrading RNA in plasmid preparations, introduce highly purified, concentrated RNases. RNase may be in the powdered reagents used to make the stock solutions or in the tips and tubes used for handling the RNA.

RNases from many different sources have been isolated and characterized. RNase A isolated from bovine pancreas is probably the best studied, and has been used as a model system for structure–function studies (*see* Chapter 13). Once introduced into the lab, it is nearly impossible to remove. Many researchers boil RNase A preparations to inactivate any DNase contamination. Other commercially available RNases, RNase 1 from *Escherichia coli* and RNase T1 from *Aspergillus oryzae*, are also very heat tolerant. Each of these three enzymes has a different substrate specificity and a different amino-acid sequence. In *E. coli*, at least 20 different RNases have been identified, and more will be named *(1)*. Based on this data, it is easy to imagine that the numbers are much greater for mammalian cells.

RNase inhibitors have been developed to minimize RNA degradation during RNA isolation and RNA handling, or to study their mechanism for hydrolysis. Some inhibitors are active on a broad range of different RNases, while others are very specific. Some inhibitors are to be used for RNA isolation, and others are most effective as prophylactics during enzymatic reactions. The type of RNase inhibitor used is dependent on the application.

From: *Methods in Molecular Biology, vol. 160: Nuclease Methods and Protocols*
Edited by: C. H. Schein © Humana Press Inc., Totowa, NJ

## 1.1. Proteinaceous Ribonuclease Inhibitors

Natural proteinaceous RNase inhibitors have been detected or isolated in a variety of tissues, such as placenta, erythrocytes, eye lens, spleen kidney, thymus, heart, lung, and bone marrow *(2,3)*. A number of RNases have been identified as having important biological functions behaving as angiogenins, neurotoxins, immunosuppressants, and antiviral agents. Molecules that inhibit the ribonuclease activities of these proteins may also reduce their other corresponding biological effects. Therefore, the appropriate application of RNase inhibitors may have a wide range of therapeutic benefits.

RNase L is activated by the unusual oligonucleotide 2-5A that is generated by the 2-5A synthetases from ATP. 2-5A synthetase is induced by double-stranded RNA and interferon, which are produced during viral infection. The human RNase L is a homodimer, involved in antiviral activities. It inhibits protein synthesis by the cleavage of mRNAs at the 3' end of UpNp sequences. RNase L is inhibited by the RNase L inhibitor (RLI; *4*; *see* Chapter 13). The overexpression of RLI in HeLa cells inhibits the interferon-activated 2-5A pathway. Thus, RLI may be a regulator of RNA stability in cells.

Of all the natural inhibitors, human placental RNase inhibitor (PRI) has been the best characterized *(5,6)*. PRI's natural target is angiogenin, a member of the RNase A superfamily that stimulates the formation of new blood vessels and is important in the growth of several types of human tumors. PRI is a potent inhibitor of RNase A, B, and C (different glycosylated versions of RNase A), noncovalently binding to the RNases in a 1:1 ratio. It does not inhibit RNase T1, RNase 1, or S1 nuclease. It has a molecular weight of ~50,000, and requires a reducing environment for activity. PRI is mainly used for the synthesis of cDNA by reverse transcription, for in vitro transcription, and for in vitro translation. PRI binds very tightly to RNase A and to angiogenin with $K_i$ values of $\sim 5 \times 10^{-14}$ and $8 \times 10^{-16}$ M, respectively *(7,8, see* Chapter 13 for the structural basis of its mode of inhibition). PRI is very sensitive to oxidation and requires a reducing environment. Therefore, for practical purposes, PRI would serve as a poor cancer therapeutic.

As an alternative to PRI, a humanized mouse monoclonal antibody (cAb 26-2F) was designed to inhibit human angiogenin. This antibody inhibited the ribonuclease activity of angiogenin and was able to suppress the growth of human breast tumors in athymic mice *(9)*. Polyclonal antibodies have been generated against other RNases by immunizing animals with purified preparations of the RNase *(10,11)*. Commercial preparations of anti-RNase A antibodies are now available that have some distinct advantages over PRI *(12)*. These reagents are useful for inhibiting potential RNase A contaminations in enzymatic reactions.

## 1.2. Small Molecule Ribonuclease Inhibitors

There are a number of small molecules that bind to the active sites of RNases and inhibit their activity. The disadvantage of these inhibitors is that large quantities may be required, as their affinities are in the low micromolar range.

Vanadyl ribonucleoside complexes (VRC) are added to cell lysis solutions to inhibit the endogenous RNase activity of the cells. These complexes consist of the oxovanadium ion and one or more of the ribonucleosides *(13)*. VRC act as transition-state analogs that bind and irreversibly inactivate RNase. VRC bind to a broad spectrum of RNases such as RNase A and T1, but not RNase H. The reported $K_i$ for VRC is 10 μ$M$ for RNase A at pH 7.0 *(14)*. The main disadvantage of VRC is that trace amounts can inhibit in vitro translation and reverse transcriptase activity. Thus, it may not be the most appropriate choice of inhibitor to isolate RNA to be used for RT-PCR.

Another small inhibitor of RNase A, 5'-diphosphoadenosine 3'-phosphate *(15)*, has a $K_i$ of 1.3 μ$M$ at pH 7.0, which is eightfold stronger than uridine-vanadate. It is the most potent nucleotide inhibitor of the previously reported nucleotides including 2'-CMP, 2'-UMP, 4-thiouridine 3'-phosphate, thymidine 3',5'-diphosphate, and pAp, which have $K_i$ values ranging from 1 to 10 μ$M$, depending on assay conditions *(16,17)*.

Other RNase A inhibitors are diribonucleoside 2'-5' monophosphates (G2'-5'G, C2'-5'G, and U2-5'G) and diribonucleoside 3'-5' monophosphates (ApU, ApC, GpU) *(18)*. The most potent nucleotide inhibitors of RNase T1 are guanylyl-2',5'-guanosine (2',5'-GpG), and 2'-guanylic acid (2'-GMP) *(19)*.

## 1.3. Ribonuclease Inactivators

Diethyl pyrocarbonate (DEPC) reacts with the amino acids within proteins, including RNases, thereby inactivating them. More precisely, DEPC forms amide bonds with the side chains of lysine, aspartic acid, and glutamic acid *(20)*. It can also split the adenine major ring of RNA, thereby destroying the biological activity of an RNA preparation *(21)*. DEPC is primarily used as a universal inactivator of RNases in solutions, especially for water.

Proteinase K in the presence of sodium dodecyl sulfate (SDS) will digest contaminating RNases in solution *(22)*. SDS inhibits RNase activity to some extent on its own.

Solutions can also be treated with bentonite, a clay chiefly composed of the clay mineral montmorillonite, to remove RNase activity. RNases are absorbed through a combination of an ion exchange and by capturing the nucleases within the intermicellar spaces *(23)*. It has been used effectively in the isolation of sea urchin RNA *(24)*.

## 1.4. RNA Storage

Reagents are available that inhibit or inactivate RNase and are useful for the storage of RNA. By dissolving RNA in concentrated formamide, RNase A activity can be inhibited at concentrations as high as 50 μg/mL *(25)*. Guanidine isothiocyanate at 4 $M$ has also been used to store RNA up to 18 mo at –20°C with no degradation *(26)*.

Another storage solution that can be used as a universal inactivator of RNases is RNA*secure*™ Resuspension Solution, Ambion, Inc. RNA preparations treated with this solution become free of any RNase activity. Unlike formamide or guanidinium, it offers the advantage of compatibility with the enzymes used in conjunction with RNA, such as AMV-RT, MMLV-RT, the T7, T3, and SP6 RNA polymerases, DNase I, and Taq DNA polymerase.

Until recently, tissue specimens had to be stored at –80°C or in liquid nitrogen to preserve the integrity of the RNA in the tissue. Now, recently harvested tissues and cells can be immersed in a solution (RNA*later*™, patent pending, Ambion, Inc.) and stored at 37°C for 1 d, 21°C for 1 wk, at 4°C for 1 mo and –20 or –80°C for an indefinite period, and the RNA in the tissue will remain undegraded. Material frozen in this solution can be thawed without causing RNA degradation. The tissue may be removed from the preservation solution and placed directly into a lysis solution such as guanidinium for RNA isolation.

## 1.5. Summary

Some RNase inhibitors are specific for certain RNases, and certain RNase inactivators react with a broad range of different RNases. The choice of appropriate inhibitor will depend on the application or experiments of the researcher.

## 2. Materials
### 2.1. Proteinaceous RNase Inhibitors

1. RNase Inhibitor (recombinant PRI) (40 U/μL; Ambion, Inc., Austin, TX) (*see* **Note 1**).
2. Anti-RNase-RNase Inhibitor (30 U/μL; Ambion, Inc.) (*see* **Note 1**).

### 2.2. Vanadyl Complex

1. Vanadium sulfate (Fisher, F.W. 163).
2. Adenosine (Sigma, F.W. 267.2).
3. Lysis buffer: 140 m$M$ NaCl, 10 m$M$ Tris-HCl, pH 8.0, 1.5 m$M$ MgCl$_2$.

### 2.3. Other Ribonuclease Inhibitors and Inactivators

1. Proteinase K (20 mg/mL; Ambion, Inc.).
2. Proteinase K digestion buffer (10X): 100 m$M$ Tris-HCl, pH 7.5, and 50 m$M$ EDTA.
3. 10% SDS.
4. Phenol.
5. DEPC (Sigma).
6. Bentonite. Grind 100 g of bentonite (Fisher Scientific) in a power mortar (Torsion Balance Co.) and sieve through a 60-mesh screen. Suspend the ground bentonite in 500 mL of water, stir at ~21°C for 1 h. Centrifuge at 2500$g$ for 10 min. Discard pellet. Centrifuge the supernatant at 8500$g$ for 10 min. Resuspend the pellet in 2 vol of water. Centrifuge at 8500$g$ for 10 min. Remove the supernatant and then dry the pellet under vacuum at 40°C. Sieve the dried material through a 60-mesh screen.
7. Cell lysis solution: 1 m$M$ MgCl$_2$, 100 m$M$ NaCl, and 10 m$M$ potassium acetate (pH 5.2).
8. RNA*secure*™ (Ambion, Inc.).

## 3. Methods
### 3.1. Natural RNase Inhibitors

The RNase inhibitors are commonly added to in vitro transcription and reverse transcription reactions to minimize RNA degradation.

1. Add RNase inhibitor or the anti-RNase antibodies to a final concentration of 1 U/μL to the enzymatic reaction (*see* **Note 2**).

### 3.2. Use of Small Chemical Inhibitors

### 3.2.1. Vanadyl Complex

1. 2 $M$ VOSO$_4$: Dissolve 2.6 g vanadium sulfate in 8 mL of RNase-free, distilled water. Mix well by vortexing. **Caution: Vanadium sulfate is very poisonous.**
2. 0.25 $M$ adenosine: Add 3.34 g adenosine to 48 mL of RNase-free, distilled water in an RNase-free, 100 mL beaker with a stir bar. Autoclave the beaker with the stir bar in it. Place the beaker on a stirring, hot plate, heat until the adenosine is dissolved, and then turn off the heat.

3. Continue stirring the 0.25 *M* adenosine. Spray a pH probe with ElectroZap™ (Ambion, Inc.) and rinse with RNase-free water to remove contaminating RNases from the pH probe. Place the pH probe in the 0.25 *M* adenosine and monitor the pH during the subsequent steps.

4. Add 6 mL of the 2 *M* $VOSO_4$ dropwise to the 0.25 *M* adenosine. The formation of complex is signaled by a color change (opaque to gray to dark green) and a drop to pH 2.5. Even if the color change does not occur, move quickly to the next step.

5. Immediately raise the pH to between 6.0 and 7.0 by squirting in 1–2 Pasteur pipets full of 10 *N* NaOH. Quickly add 10 *N* NaOH dropwise until the pH is >6.5. Add 1 *M* NaOH dropwise to pH 7.5. Add RNase-free water to bring the final volume to 60 mL. If the volume is >60 mL, do not be concerned.

6. Freeze the complex in 0.5–1.0 mL volumes in RNase-free microfuge tubes (Ambion, Inc.). Freeze quickly in dry ice or liquid nitrogen. The final concentration is 200 m*M*. Store at –70° or –80°C.

7. Warm a tube of the complex at room temperature or at 37°C. The solution should be clear or a dark green. If a white, insoluble precipitate appears in the thawed complex, then centrifuge at high speed, and then use the supernatant for the RNA isolation.

8. Dilute the 0.2 *M* vanadyl complex 1:10 in a lysis buffer prior lysing cells for RNA isolation. The final working concentration is 0.02 *M* (*see* **Note 3**).

## 3.3. Inactivation of RNase

### 3.3.1. Proteinase K

Proteinase K treatment is often used to remove RNase and other RNA-binding proteins from an RNA sample. High temperatures and adding SDS and EDTA are important for inactivating the RNase until the RNases are digested.

1. Add a 1:20 vol of 10% SDS to the RNA sample.
2. Add a 1:10 vol of the Proteinase K digestion buffer.
3. Add Proteinase K to a final concentration of 50 µg/mL.
4. Incubate 50°C for at least 60 min.
5. Remove Proteinase K from the RNA by performing repeated phenol extractions until the interface is clean. Precipitate the RNA with ethanol and resuspend in buffer.

### 3.3.2. DEPC

To inactivate contaminating RNases in solutions that need to be RNase-free, it is common practice to treat them with 0.1% DEPC (*see* **Note 4**). DEPC is most commonly used to prepare RNase-free water (*see* **Note 5**). *See* Chapter 13 for protocol.

### 3.3.3. Bentonite

Bentonite is most commonly used during RNA isolation procedures. It is added to cell lysates to inhibit RNase activity *(24)*. It has also been added to different enzymatic reactions that involve RNA at 0.04% to inhibit RNase activity *(27)*.

1. Add bentonite to 3 mg/mL in the cell lysis solution. Chill on ice.
2. Add 15 vol of the bentonite-cell lysis solution to sea-urchin eggs and homogenize rapidly.
3. Centrifuge at 8500*g* for 10 min to pellet the bentonite and cell debris. The supernatant contains the RNA (*see* **Note 6**).

### 3.3.4. RNAsecure ™

A reaction mixture that contains RNA as a substrate or will contain RNA as a product may be purged of RNase activity through a simple heat treatment in the presence of

RNA*secure*™ (patent pending). This treatment is commonly used for reverse transcription reaction mixtures and for in vitro transcription mixtures. An RNA pellet may also be dissolved in RNA*secure* Resuspension solution to inactivate contaminating RNases in the RNA preparation.

The following is a method to inactivate a potential RNase contamination from all the components in a reverse transcription, except the reverse transcriptase. The heating step also mediates the hybridization of the random primers.

1. 10X reverse-transcription reaction buffer: 2.5 μL.
2. Water: 20.5 μL.
3. RNA (1.0 mg/mL): 1 μL.
4. RNA*secure*™ reagent: 1 μL.
5. Random primers: 1 μL.
6. Incubate 10 min at 65°C.
7. Add reverse transcriptase: 1 μL. Continue the reverse transcription reaction as usual.

## 4. Notes

1. One unit is the amount of inhibitor needed to inhibit by 50% the activity of 5 ng of RNase A.
2. Placental RNase inhibitor requires 5–10 m*M* DTT (Calbiochem, La Jolla, CA) for activity, whereas the anti-RNase antibodies are active without reducing reagent. In fact, the anti-RNase antibodies are slightly inhibited in the presence of 20 m*M* DTT.
3. Vanadyl complex can be removed from the RNA by organic extraction. Typically, several phenol extractions are performed until the green color of the vanadyl complex is no longer visible.
4. Adding 0.1% DEPC to water and autoclaving will inactivate RNase A up to a concentration of 0.5 mg/mL (Ambion Tech Bulletin 178).
5. Do not use DEPC with Tris-HCl or other solutions with free amino groups. DEPC will react with the high-concentration amino group leaving no DEPC remaining to inactivate RNase. Instead, use DEPC-treated water to make such a solution and then autoclave this solution.
6. The RNA prepared by this method is compatible for use with *E. coli* in vitro translation systems *(24)*. Polyvinylsulfate, a compound that has also been used to inhibit RNase, greatly reduces protein synthesis, presumably by interacting with the ribosomes.

## References

1. Nicholson, A. W. (1997) *Escherichia coli* ribonucleases: paradigms for understanding cellular RNA metabolism and regulation, in *Ribonuclease: Structures and Functions* (D'Alessio, G. and Riordan, J. F., eds.), Academic Press, New York, pp. 1–49.
2. Bloemendal, H. and Jansen, K. (1988) Detection of RNAase inhibitor from different species and organs. *Biochim. Biophys. Acta* **966,** 117–121.
3. Nadano, D., Yasuda, T., Takeshita, H., and Kishi, K. (1995) Activity staining of mammalian ribonuclease inhibitors after electrophoresis in sealed vertical slab polyacrylamide gels. *Anal. Biochem.* **227,** 210–215.
4. Bisbal, C., Martinand, C., Silhol, M., Lebleu, B., and Salehzada, T. (1995) Cloning and characterization of RNase L inhibitor. *J. Biol. Chem.* **270,** 13,308–13,317.
5. Blackburn, P., Wilson, G., and Moore, S. (1977) Ribonuclease inhibitor from human placenta. Purification and properties. *J. Biol. Chem.* **252,** 5904–5910.
6. Blackburn, P. and Moore, S. (1982) Pancreatic ribonuclease, in *The Enzymes* (Boyer, P. D., ed.), Academic Press, New York, pp. 416–424.

7. Lee, F. S., Shapiro, R. S., and Vallee, B. L. (1989) Tight-binding inhibition of angiogenin and ribonuclease A by placental ribonuclease inhibitor. *Biochemistry* **28**, 225–230.

8. Shapiro, R. and Vallee, B. L. (1991) Interaction of human placental ribonuclease with placental ribonuclease inhibitor. *Biochemistry* **31**, 2246–2255.

9. Piccoli, R., Olson, K. A., Vallee, B. L., and Fett, J. W. (1998) Chimeric anti-angiogenin antibody cAb 26-2F inhibits the formation of human breast cancer xenografts in athymic mice. *Proc. Natl. Acad. Sci. USA* **95**, 4579–4583.

10. Hastie, A. T. (1981) Monospecific antibodies to rabbit lung ribonucleases. *J. Biol. Chem.* **256**, 12,553–12,560.

11. Weickmann, J. L. and Glitz, D. G. (1982) Human ribonucleases. Quantitation of pancreatic-like enzymes in serum, urine, and organ preparations. *J. Biol. Chem.* **257**, 8705–8710.

12. Murphy, N. R., Leinbach, S. S., and Hellwig, J. H. (1995) A potent, cost-effective RNase inhibitor. *BioTechniques* **18**, 1068–1073.

13. Berger, S. L. and Birkenmeier, C. S. (1979) Inhibition of intractable nucleases with ribonucleoside-vanadyl complexes: isolation of messenger ribonucleic acid from resting lymphocytes. *Biochemistry* **18**, 5143–5149.

14. Lindquist, R. N., Lynn, J. L., Jr., and Lienhard, G. E. (1973) Possible transition-state analogs for ribonuclease. The complexes of uridine with oxovanadium (IV) ion and vanadium (V) ion. *J. Am. Chem. Soc.* **95**, 8762–8768.

15. Russo, N, Shapiro, R., and Vallee, B. L. (1997) 5'-diphosphoadenosine 3'-phosphate is a potent inhibitor of bovine pancreatic ribonuclease A. *Biochem. Biophys. Res. Comm.* **231**, 671–674.

16. Anderson, D. G., Hammes, G. G., and Walz, F. G., Jr. (1968) Binding of phosphate ligands to ribonuclease A. *Biochemistry* **7**, 1637–1645.

17. Irie, M., Watanabe, H., Ohgi, K., Tobe, M., Matsumura, G., Arata, Y., Hirose, T., and Inayama, S. (1984) Some evidence suggesting the existence of P2 and B3 sites in the active site of bovine pancreatic ribonuclease A. *J. Biochem. (Tokyo)* **95**, 751–759.

18. White, M. D., Bauer, S., and Lapidot, Y. (1977) Inhibition of pancreatic ribonuclease by 2'-5' and 3'-5' oligonucleotides. *Nucleic Acids Res.* **4**, 3029–3038.

19. Koepke, J., Maslowska, M., Heinemann, U., and Saenger, W. (1989) Three-dimensional structure of ribonuclease T1 complexed with guanylyl-2',5'-guanosine at 1.8 Å resolution. *J. Mol. Biol.* **206**, 475–488.

20. Wolf, B., Lesnaw, J. A., and Reichmann, M. E. (1970) A mechanism of the irreversible inactivation of bovine pancreatic ribonuclease by diethylpyrocarbonate. *Eur. J. Biochem.* **13**, 519–525.

21. Wiener, S. L., Wiener, R., Urivetzky, M., and Meilman, E. (1972) Inhibition of ribonuclease by diethyl pyrocarbonate and other methods. *Biochim. Biophys. Acta* **259**, 378–385.

22. Mendelsohn, S. L. and Young D. A. (1978) Inhibition of ribonuclease: efficacy of sodium dodecyl sulfate, diethyl pyrocarbonate, proteinase K and heparin using a sensitive ribonuclease assay. *Biochim. Biophys. Acta* **519**, 461–473.

23. Daigneault, R., Bellemare G., and Cousineau, G. H. (1971) Effect of various inhibitors on the activity of sea urchin ribonucleases. *Lab Pract.* **20**, 487–488.

24. Jacoli, G. G., Ronald, W. P., and Lavkulich, L. (1973) Inhibition of ribonuclease activity by bentonite. *Can. J. Biochem.* **51**, 1558–1565.

25. Chomczynski, P. (1992) Solubilization in formamide protects RNA from degradation. *Nucleic Acids Res.* **20**, 3791,3792.

26. Gilleland, R. C. and Hockett, R. D., Jr. (1998) Stability of RNA molecules stored in GITC. *BioTechniques* **25**, 944–948.

27. Tyulkina, L. G. and Mankin, A. S. (1984) Inhibition of ribonuclease contamination in preparations of T4 RNA ligase, polynucleotide kinase, and bacterial alkaline phosphatase with bentonite. *Anal. Biochem.* **138**, 285–290.

# 9

## Producing Soluble Recombinant RNases and Assays to Measure Their Interaction with Interferon-γ In Vitro

### Catherine H. Schein

## 1. Introduction
### 1.1. Cytokine Regulation of mRNA Half-Lives

Many cytokines alter the half-lives and stabilities of mRNAs for other cytokines *(1)* and enzymes that may be related to disease progression. For example, IL-1β stabilizes the mRNA for human interstitial collagenase-1 in normal fibroblasts, but not in breast cancer cells *(2)*. Altering mRNA stability through direct control of specific RNase activity is a rapid way for cells to respond to environmental changes. Small differences in half-life can have dramatic effects on expressed products; a sixfold increase in μ heavy-chain mRNA half-life in differentiated B-cells largely accounted for a 100-fold increase in its messenger concentration compared to the undifferentiated cells *(3)*. The high specific activity of RNases makes them ideal cellular targets for interaction with cytokines *(4)*.

However, relating the half-lives of mRNAs to the activity of RNases is difficult in the complex environment within cytokine-treated cells. Cytokines may also simultaneously affect transcription of the target mRNA and induce other cytokines with opposing effects *(5–7)*. For this reason, assays of purified proteins, done in vitro, provide a useful way to explore interactions that may play a physiological role. This chapter describes sample protocols for isolating soluble, recombinant RNases and cytokines and assays to study their interactions.

### 1.2. Direct vs Indirect RNase Activation by Interferons

Most of the assays described in **Subheading 3.2.** were designed to study the effect of interferon-γ (IFN-γ) on the activity of ribonucleases *(8,9)*. One of the earliest defined effects of treating cells with interferons was an increase in RNase activity *(10)*. A mechanism for indirect activation of 2'5'A-dependent RNase (RNase L) by IFN-α/β and -γ has been extensively studied *(11; see* other chapters in this book). Interferons -α and -β indirectly activate a latent RNase (RNase L) by inducing an enzyme that synthesizes an RNase L activator, 2'5'-linked poly A. RNase L forms a maximally active dimer in the presence of 2'5'-oligoadenylate *(12)*. A specific inhibitor of RNase L, a

From: *Methods in Molecular Biology, vol. 160: Nuclease Methods and Protocols*
Edited by: C. H. Schein © Humana Press Inc., Totowa, NJ

68-kDa polypeptide whose expression is not regulated by IFN, does not cause 2-5A degradation or irreversibly modify RNase L (*see* Chapter 12). Overexpression of RLI in stably transfected HeLa cells inhibits the antiviral activity of IFN on encephalo-myocarditis virus but not on vesicular stomatitis virus *(13)*.

IFN-γ, a protein with only minimal sequence similarity to IFN-α/β, does not induce the synthesis of 2'5'-A-synthetase *(14)*. However, IFN-γ can directly modulate the activity of mammalian ribonucleases in vitro in a nuclease specific fashion and in the presence of serum molecules. Specific modulation of intracellular nucleases could explain how IFN-γ stabilizes some mRNAs, e.g., for IFN-β *(15)* and the complement components C1 inhibitor and factor B *(16)* in monocytes, while destabilizing others–including c*fos* *(17)* in macrophages, IL-8 in neutrophils *(18)* and the cystic fibrosis transmembrane conductance regulator in epithelial cells *(19)*. Although bacterial LPS, an important coregulator of the cellular effects of IFN-γ, increases ribonuclease expression in rat lung *(20)* and total RNase increased in the sera of patients treated with IFN-α, little is known about the changes in cellular RNase after treatment with IFN-γ.

In vitro, recombinant IFN-γ from three different species activates the cleavage of ds-RNA by bovine seminal ribonuclease (BS-RNase) but not by the closely related RNases of the pancreatic family *(8)*. Only biologically active IFN-γ modulates the activity of ribonucleases; its effects are reversed by IFN-γ specific-antibodies or by heating IFN-γ *(9)*. Both bovine and murine pancreatic RNases are inhibited, but their close relative, BS-RNase, is stimulated. The increased activity of BS-RNase is particularly interesting because of this nuclease's ability to inhibit human HIV-1 replication *(21)* and its cytotoxicity *(22)*, properties which distinguish it from RNase A. The activation of BS-RNase can be inhibited by heparin (Schein, unpublished data), a compound that also abrogates the biological activities of IFN-γ *(23,24)*. Time course assays, using a 300-bp-long ds-RNA substrate made by annealing two complementary strands from a section of the T7 phage genome, as described in **Subheading 3.2.**, show that stimulation depends on fine differences in the RNA binding *(25)* of these ancient enzymes *(26)*.

## 1.3. Biological Significance of Direct RNase Activation

The ability of IFN-γ to slow the growth of cells and induce an antiviral state make it a potential treatment for certain diseases. IFN-γ inhibits the growth of tumor cells and may be useful alone or as an adjuvant to treat certain cancers. However, basal production of IFN-γ has been implicated in worsening some disorders *(27,28)*, and clinical use is hindered by side effects and eventual resistance to the cytokine. For example, although IFN-γ did reduce swelling and morning stiffness in rheumatoid arthritis (RA) patients, perhaps by altering the production of cytokines, such as IL-8, involved in the pathogenicity of RA *(29)*, it was not useful for long-term treatment. Better knowledge of the mechanism of IFN action could help in distinguishing patients most suited for treatment and appropriate adjuvants *(30)*.

IFN-γ's modulation of ribonucleases is important at two levels. First, high concentrations of IFN-γ in the serum or loosely associated with cells and basement membranes (perhaps bound to heparan sulfate or heparin-like molecules, as has been shown for another heparin-binding growth factor, bFGF *[31,32]*) would activate serum ribonucleases, which may be active in destroying nonencapsulated parasitic RNAs. As IFN-γ is rapidly cleared from the serum, subsequent effects would be caused by IFN transported into the cell with its receptor. The species specificity of IFN-γ, which is due to the structural requirements for IFN-γ to interact with its receptor *(33)*, would seem to indicate that the cytokine activity is exerted only indirectly *(34)*. In the IFN antiviral assay (**Subheading 3.4.3.**), human IFN-α subtypes, but not human IFN-γ, protect murine cells from virus. However, if high intracellular concentrations of IFN-γ are attained by the use of liposomes *(35)* or through transient transfection with an expression vector lacking the signal sequence needed to transport IFN out of the cell *(36)*, species specificity is abrogated. Because the specific activity of IFN-γ is high, visualization of the few molecules of protein that would be needed within cells to exert activity* might be difficult. Still, significant amounts of IFN-γ protein were seen in the nucleus of murine cells *(37)*. These data, in addition to evidence that only a small area of the cytokine actually binds to its receptor, all suggest a more direct role for the protein.

### 1.4. Identifying RNase Targets in Cells

The effect of IFN-γ is very specific for the RNase. Similar in vitro assays using isolated RNases, identified by gene expression studies, will be necessary to determine possible physiological targets. Judging from *Escherichia coli*, there must be many yet undiscovered ribonucleases with distinct cleavage specificities in mammalian cells *(38,38a)*. The "alkaline ribonucleases" in human sera fractionate into at least five distinct peaks on ion-exchange chromatography, all of which increased in patients responding to treatment with IFN-α *(39)*. Although activity-based cloning is difficult, because RNases are present at low levels in cells or are bound up with specific inhibitors *(40)*, novel mammalian RNases are constantly being discovered.

RNase A homologs isolated to date were found serendipitously while looking for other activities (angiogenin induces angiogenesis [blood vessel formation] in the rabbit-eye model system [*41*; *see also* Chapter 25] a neurotoxin and cationic protein that are released by eosinophils at the site of injury *[42]*; an "antitumor protein" isolated from frog tissue *[43]*). These proteins, although clearly related to RNases by sequence, have amino-acid identity <30%, and would thus be difficult to detect by nucleic-acid hybridization techniques using probes based on the pancreatic gene sequence (*4*; *see also* Chapter 8). Serendipitous isolation of an RNase of a type previously only known in plants and microorganisms is described in Chapter 7. New details about viral RNases are also being discovered. For example, the recombinant glycoprotein E2 of classic swine fever virus has intrinsic RNase activity and two small stretches of homology to the active site of fungal and plant RNases *(45)*.

The best way to demonstrate that a RNase plays a role in cell protection by IFN-γ would be to delete the gene in cells (or in viruses used to infect the cells). Assuming that the RNase was not required for viability, cells in which it was deleted should become wholly or partially resistant to the effects of the cytokine. Alternatively, expression of the RNase in nonresponsive cell lines could convert them to reponsive. The traditional IFN assay (**Subheading 3.4.3.**) or one based on fluorescent cell-sorting could be used to characterize the response of the altered cell lines to the cytokine.

---

*A "back of the envelope" calculation, based on a specific activity of $5 \times 10^7$ interferon units/mg protein, a molecular weight of 28,000 for the dimeric protein, $10^6$ cells/mL in the IFN assay and a 10% binding efficiency to the cell receptors, would suggest that 100–400 molecules/cell are necessary to exert an antiviral effect.

## 1.5. Using These Assays for Drug Selection and Design

In addition to their usefulness in characterizing the mechanism of action of IFN-γ, there are many practical applications for these assays. They could be used to determine the specific activity of batches of IFN-γ for treatment purposes. The activity assay for IFN is long (~2 d if run routinely), and requires culturing mammalian cells and titrating virus (*see* **Subheading 3.4.3.**). Results are only valid to a factor of 2–3. In contrast, in vitro assays quickly yield quantitative results that one can compare directly to an enzymatic standard, and require no mammalian cell culture or virus.

The most promising use is to rapidly screen for substances that specifically inhibit IFN-γ. Such inhibitors could be used help to distinguish which alterations within the cell contribute to beneficial effects from this cytokine. They might also supply lead compounds for use in treating diseases, such as psoriasis and multiple sclerosis, which are exacerbated by IFN-γ.

## 1.6. Producing Soluble Recombinant RNases

To demonstrate that the differential effects of IFN-γ on RNase activity were genuine, a variety of different assays were needed. All the enzymatic components were produced in *E. coli* (**Subheading 3.1.**) to eliminate the chance that components from different cellular backgrounds affected the results, and several different preparations of IFN-γ, obtained from different research groups, were used. An *E. coli* secretion system based on the signal sequence from a mammalian protein was developed specifically for this project *(52)*. While three different mammalian RNases were expressed with this system, the highest yield was obtained using a synthetic gene for RNase A that had been altered to reflect the bacterial codon preferences. The different purification protocols developed for the proteins are included here (**Subheading 3.1.**). Various ribonuclease assays were used to eliminate the possibility of an artifact and to show that IFN-γ has no intrinsic nuclease activity (assays for RNase activity, **Subheading 3.2.**).

Several assays use a $^{32}P$-labeled, defined RNA, synthesized with T7 RNA polymerase, as the substrate. This highly sensitive assay permitted the study of the interaction at low, pg/mL concentrations of IFN and nuclease such as one would expect in vivo. The specificity of IFN-γ for the interaction was demonstrated by comparing its effects with those of a protein with related cellular activity but little chemical similarity (interferon-α2) and a protein with similar high positive surface charge (lysozyme) *(8)*. The specificity was further demonstrated by neutralizing the effect of the cytokine by adding monoclonal antibodies (MAbs) that neutralized its biological activity (*see* neutralization with MAbs, **Subheading 3.3.**). Finally, the area of the protein responsible for the interaction was demonstrated to be, at least in part, at the C-terminus of the protein by systematically removing this section of the protein through proteolysis *(25)*. Several useful related methods, for preparing proteins used in the assays, are described in **Subheading 3.4.**

## 2. Materials

### 2.1. Viral and Mammalian-Based Assays and Cell Culture

Cells are cultured in Dulbecco's modified Eagle's medium (DMEM) plus glutamine and 10% heat inactivated horse serum (L-cells) or fetal bovine serum (FBS) (WISH cells) in a 10% $CO_2$ atmosphere. I do not routinely add antibiotics to the culture medium except, occasionally, when using cell lines from other labs.

1. L929 cells (ATCC(R) 1-CCL) (American Type Culture Collection [ATCC], Rockville, MD).
2. WISH Cells ((ATCC(R) 25-CCL) (ATCC).
3. Murine EMC virus (ATCC(R) 129B-VR) (ATCC).
4. DMEM with horse serum or FBS (Gibco or equivalent).
5. Crystal violet dye: combine 5 mL formaldehyde (35%), 50 mL ethanol (96%), 100 mL $H_2O$, 0.25 g NaCl, and 0.75 g crystal violet dye. Store, tightly closed, at 4°C.
6. Trypsin solution (Gibco) diluted 1:10 in sterile PBS buffer.
7. Phosphate-buffered saline (PBS), a 10× solution contains per liter: 80 g NaCl, 2 g KCl, 14.4 g $Na_2HPO_4 \cdot 2H_2O$, 2 g $KH_2PO_4$. Dissolve in water, check pH (should be 7.2), set vol to 1 L, and autoclave. Dilute 1:10 for PBS buffer in sterile water.
8. Incubator, sterile hood (preferably with UV light sterilization), and normal facilities for mammalian cell cultivation.
9. 96-well sterile coated microtiter plates with lids.
10. Multichannel pipet (or a microtiter plate dilution system).

## 2.2. Bacterial Cultivation

1. Sterile 40% glycerol/water, aliquoted into culture tubes or snap lock Eppendorf cups (0.5 mL/tube).
2. Fermentor medium: 5 g/L casein enzymatic hydrolysate (Sigma), 0.75 g/L yeast extract (Difco), 4 g/L $K_2HPO_4$, 1 g/L $KH_2PO_4$, 1 g/L $NH_4Cl$, 2.6 g/L $K_2SO_4$, 0.01 g/L $CaCl_2$, 20 g/L glycerol (80%, Fluka), 5 mL/L 200X trace salt solution, and about 500 µL Dow-Corning 1510 silicon antifoam were sterilized in the fermentor for 20 min after heat up to 120°C. $MgSO_4$ (to 2 m$M$) and Na ampicillin (dissolved in water and filter sterilized before use, to 100 mg/L) were added after cooling to below 50°C.
3. Fermentor: 10–20 L vessel with running vol of 10 L; should have both pH control and dissolved oxygen monitoring (Chemap, Bioengineering, or equivalent).
4. Sorvall centrifuge with 1 L buckets and various rotors.
5. Eppendorf centrifuge with cooling.

### 2.2.1. Bacterial Strains

The following *E. coli* strains were used:

1. JM101 (for initial cloning after ligation of plasmids).
2. *DS410* (minicell phenotype *(46)*; and *lon⁻ hptR⁻ (47)* for expression after trp promotor induction.

### 2.2.2. Plasmids

The following plasmids were all based on Plasmid *pHR148*, which contains the trp promotor and ribosome binding site *(48)*, was cleaved with *Nco*I and *Bam*HI.

1. pTRPmuRN+ss, which contains the whole coding sequence for the murine RNase cDNA, including the signal sequence.
2. pTRPmuRN, which codes for mature murine RNase with one additional N-terminal methionine.
3. pTrpmuSSboRN, which contains a *Nco*I/*Fok*I fragment with most of the murine signal sequence preceding an *Nde*I/*Bam*HI synthetic gene-fragment coding for most of the mature sequence of bovine pancreatic ribonuclease *(49)*, joined in a four-way ligation to *Nco*I/*Bam*HI cleaved pHR148 with a synthetic linker.

4. pTrpbsRN, which contains the cDNA for mature BS-RNase *(50)* cloned after the signal sequence of murine pancreatic RNase and the trp promotor of pHR 148, as described previously for the expression of RNase A and mutants thereof *(51)*.

## *2.3. Protein Purification*

1. TTG buffer: Tris-HCl, pH 7.9, 10% glycerol, 0.02% Tween-20 (TTG buffer). Sterilize and store at room temperature. Add 1 mL/L 20 m$M$ PMSF in isopropanol before using. 20% TTG contains 20% glycerol.
2. DB: dialysis buffer, TTG set to pH 7.0 plus 0.02 m$M$ PMSF.
3. Tris-HCl buffer: maintain sterile 1 $M$ Tris-HCl buffer stocks of pH 7.0, 7.5, and 8.0 to dilute to the concentrations indicated in the protocols with RNase-free water.
4. NaCl: 5 $M$ solution in RNase-free water, autoclave, and store at room temperature.
5. 2-Mercaptoethanol: store in the cold room, add to buffers immediately before use.
6. PMSF: 20 m$M$ phenyl-methyl sulfonyl fluoride in 1-isopropanol (store at room temperature, dilute into buffer immediately before use.) This inhibitor of serine proteases is very toxic! Wear gloves and face mask when preparing the stock solution and weigh out the powder in an enclosed balance or in a fume hood.
7. Na-EDTA: 0.5 $M$ ethylene diamine tetraacetic acid, pH 8.0. Autoclave and store at room temperature.
8. Ammonium bicarbonate, 25 m$M$ in RNase-free water.
9. Protein purification resins: SP-Trisacryl (IBF Biotechnics), Matrex Blue Gel (Amicon), CM-Sephadex, Phenyl-sepharose, and DEAE Sephacel (Pharmacia).
10. Centricon 10 microconcentrators (Amicon).
11. pH meter.
12. Conductivity meter: preferably hand-held.
13. Protein purification columns and disposable 10-mL columns (Bio-Rad)
14. Polyacrylamide-gel electrophoresis (PAGE): Pour gels from stock acrylamide solutions (Sigma A9666) and buffers, as directed, or purchased premade from Biorad.
15. Hoeffer minigel apparatus (or equivalent).
16. Protein sequencer: e.g., model 810 from Knauer (Dr. Herbert Knauer, Wissenschaftliche Geräte, Henchelheimer Str. 9, D-6380 Bad Homburg, Germany), modified for isocratic reversed phase HPLC determination of the PTH-amino acids.

## *2.4. Synthesis of Labeled and Unlabeled RNA*

1. 80% FAHB: contains per mL 0.91 g recrystallized formamide, 200 µL of 2 $M$ NaCl, 0.2 $M$ PIPES, pH 6.4, 5 m$M$ EDTA.
2. FAHB + dyes: Add 0.05% each bromophenol blue and xylene cyanol FF to 80% FAHB.
3. AGUC: contains 2.8 m$M$ of each nucleotide triphosphate, approx 1.5 mg/mL of ATP, CTP, UTP, and GTP (which can be obtained as a kit from Promega).
4. 10X T7 buffer: 400 m$M$ Tris-HCl, pH 7.9, 60 m$M$ MgCl$_2$, 20 m$M$ spermidine (29 mg/10 mL).
5. 1 $M$ DTT: store frozen in aliquots.
6. Linearized (i.e., cleaved with a restriction enzyme at a unique site) plasmid DNA (1–20 ng/reaction).
7. α-$^{32}$P-GTP. 400 mCi/mmol, 10 mCi/mL (Amersham or NEN).
8. 1 $M$ TBE: per L, 121.1 g Trisma base, 61.83 g boric acid, add Na-EDTA solution (0.5 $M$, pH 8.0) to a final concentration of 20 m$M$ EDTA. Autoclave and store at room temperature.
9. RNasin: RNase inhibitor, 40 u/mL, from Promega, Ambion, or other source.
10. 20% TCA: Dissolve tricholoracetic acid in water and freeze in aliquots. The TCA must be of very high quality (crystals should not be even slightly yellow, and the final solution should not smell like vinegar) and the solution should not be stored in the cold room, even overnight.

11. Ethanol: undiluted or 80% in RNase-free water, keep at –20°C.
12. Phenol: recrystallized and equilibrated with TE buffer to pH above 7.0.
13. TE: 10 mL Tris-HCl, pH 7.6, 0.1 mL EDTA.
14. CIA: 98 mL chloroform stabilized by the addition of 2 mL isoamyl alcohol.
15. Scintillation vials and counter.
16. Microtiter-plate reader (Dynatech or equivalent).

## 2.5. Assays

1. $^{32}$P- labeled-ss- and ds-RNA: produced from T7 phage DNA with T7 RNA polymerase (*see* **Subheading 3.2.1.** *[16]*).
2. 3H-poly A/poly U combine 50 µL (~40 pmol) $^{3}$H-polyadenylic acid (n~50) with 0.8 mg polyuridylic acid (Pharmacia) in 200 µL $H_2O$. Heat to 80°C for 15 min and cool slowly to room temperature.
3. 10 or 20% TCA; *see* precautions above.
4. Carrier bovine serum albumin (BSA)/RNA: 20 mg/mL of macaloid-treated BSA and 40 µg/mL of yeast tRNA in TE. The yeast tRNA can be dissolved to 2–10 mg/mL in TE, extracted with an equal volume of Phenol/CIA, precipitated with ethanol, and then redissolved in TE. Check the $OD_{259}$ of the solution to determine the mg/mL of RNA before combining with the BSA. Highly purified BSA, tested to be nuclease-free, can also be obtained from various sources.
5. TLC plates: Silica gel thin-layer chromatography plates, glass, with fluorescence indicator (Merck).
6. Macaloid powder (from Walter Schaffner, Mol. Biology, University of Zurich, ex Georg Langer Co., Ritterhude, Bremen, Germany). Macaloid may also be obtained from other sources, including NL company (New Jersey), and National Lead Company (Houston, TX).
7. Total protein assay, such as the Bradford Coomassie brilliant blue assay (Bradford, 1976). Because RNase A reacts with the reagent in this assay with about 1/15 the intensity of BSA and is detected only slightly better with the Lowry protein reagent, final yields of RNases are best estimated from Coomassie blue-stained minigels using Boehringer RNase A as a standard.
8. Alkaline Cu reagent: 50 mL 1.2% $Na_2CO_3$ in 0.15 $M$ NaOH combined just before use with 0.5 mL each of 1% $CuSO_4$ in water and 1% $Na_2$ Tartrate in water (for Lowry assay).
9. Folin Ciocalteu phenol reagent: purchase from Merck, usually as a 2X solution (for Lowry).
10. 2' and 3'-mononucleotides (Sigma, kit N-8) for standards in the TLC assay.

## 2.6. Enzymes, Cytokines, and Other Proteins

1. r-RNase A was purified from *E. coli* *(52)*, **Subheading 3.1.**
2. BS-RNase. Bovine seminal RNase isolated from bovine seminal fluid was from Stephen Benner (Dept. Chemistry, E.T.H. Zürich). rBS-RNase was isolated from *E. coli* *(25)* (**Subheading 3.1.3.**).
3. Recombinant human IFN-γ, which, like that isolated from human cells, lacks the initial N-terminal Cys-Tyr-Cys residues encoded in the gene sequence, was obtained from C. Weissmann (Molecular Biology I, University of Zurich) and had a specific activity of 2–3 × 10$^7$ IFU/mg on WISH cells, but no activity on L-cells. IFN-γΔ14 was prepared proteolytically from recombinant ΔCYC-IFN-γ and purified and characterized by mass spectroscopy as described previously *(9)*.
4. Recombinant human IFN-α2 from Schering-Plough (in a solution containing human serum albumin, 0.25 mg/mL IFN-α2, 5 × 10$^7$ IFU/mL); used as a standard for the WISH cell (human) IFN-assay.

5. Recombinant murine IFN-γ was from G. Garotta (Hoffmann-LaRoche Research, Basel); specific activity of $3 \times 10^6$ IFU/mg on L-cells.

6. Recombinant murine IFN-α (prepared by B. Alberti in the group of Prof. C. Weissmann U. Zürich Mol. Biol. I) with a specific activity of $10^7$ IFU/mg on L-cells and $10^6$ IFU/mg on WISH cells was used as a standard for the L-cell assay.

7. Antibodies to IFN-γ (in PBS) were obtained from G. Garotta of Roche Research, Basel; neutralization of IFN-γ activity was determined by inhibition of IFN-γ activity in the antiviral assay (in Zürich) (*see* **Subheading 3.4.3.**) and its binding to cells (in Basel). None of the antibody solutions alone contained appreciable nuclease activity or activated the nuclease.

8. Macaloid-treated BSA: 1 mg/mL final concentration in PBS (*see* **Subheading 3.4.1.**).

## 3. Methods

### 3.1. *Purification of Recombinant RNases from an* E. coli *Secretion System*

The following protocols were used to isolate RNase proteins expressed in a secretion system in *E. coli*, based on the trp promoter and the signal sequence from murine pancreatic type RNase (the gene for which was isolated from a murine spleen cDNA library) *(52)*. While the isolation of recombinant proteins from the in bacterial cells grown at 37°C or higher is described in several chapters in this book, isolation of proteins from the soluble fraction of cells grown at lower temperature *(53)* or from the culture supernatant has significant advantages *(54,55)*. The protocols resemble one another in the columns used, but the starting material and the purification path are different. The different protocols are given to suggest ways to optimize purification of a novel RNase (*see also* Chapter 28 for protocols for unknown nucleases).

### 3.1.1. *Bacterial Cultivation and Expression of the RNase*

1. Use shaker flask culture to determine expression levels from differing clones (*see* **ref. 53** for some tips and suggested media; **Note 2**). To ensure that the medium pH in shaker flask cultures does not drop below pH 6.5, use a high-buffer concentration in the medium (up to 100 m*M* PIPES buffer, for example, is tolerated well by *E. coli*); start cultivation at pH 7.8–8.0, and/or add small additions (carefully!) of 0.1 *M* NaOH to the culture during growth.

2. To express RNases in a larger volume, inoculate 10 L sterile fermentor medium with 25 mL of an L-broth culture in late growth or early stationary phase. Control culture pH to 7.0 by addition of 10% aqueous ammonia or 5 *M* NaOH, and maintain dissolved oxygen to above 20% of saturation using aeration of not more than 0.9 vol air/liquid vol/min (VVM), increasing vessel backpressure as needed to maintain at least 20% of saturation. Growth at T <30°C, besides generally resulting in more protein in the soluble fraction of cells *(53)*, will also allow a higher concentration of dissolved oxygen at the same aeration rate.

3. If using a T7 promotor plasmid (pET series), induce the culture with IPTG at up to 1 m*M* final concentration. Many groups have found that lower concentrations of inducers may be preferable to obtain more soluble protein *(55)*. For the tryptophan promotor, induction occurs when the medium is exhausted for tryptophan.

4. To measure RNase activity during the cultivation, centrifuge 1-mL samples in Eppendorf cups in an Eppendorf centrifuge at 10,000*g* for 5 min. This supernatant is referred to as "culture supernatant" or simply "culture medium." Freeze the cell pellets at –20°C or take up in 100 µL $H_2O$/0.2 m*M* PMSF/10 m*M* EDTA and lyse the cells by freezing and thawing three times (cycle between liquid $N_2$ and cold water; **Note 3**). Centrifuge 10 min in an

Eppendorf centrifuge (as before). This supernatant is the "freeze-thaw supernatant" (FTS). For very dense samples, one may need to shear the DNA by drawing the sample up and down several times in a 1-mL disposable syringe fitted with a narrow gage needle (brown tip). Any of the assays in **Subheading 3.2.** can be used to measure RNase activity as a function of time.

5. At the end of the induction period (**Note 2**), centrifuge the culture at 10,000$g$ for 10 min in the cold room. Add 1 μL/mL PMSF solution (final [PMSF] 0.02 m$M$) to the supernatant to inhibit proteases and process immediately or freeze. Resuspend the cell pellet in buffer containing PMSF (approx 1 mL/g wet pellet) and freeze in a disposable tube. Do not freeze the pellet in the centrifuge bottle; besides taking too much space in the freezer, surface drying of the pellet reduces yields.

6. The RNase can be isolated from the culture supernatant (**Subheadings 3.1.2.** and **3.1.3.**) or from the supernatant of lysed cells (**Subheadings 3.1.4.** and **3.1.5.**). Purification schemes should capitalize on the unique solubility characteristics of the desired protein *(56)*. In the protocols in **Subheadings 3.1.2.** onward, the nucleotide-binding ability, high pK, and resistance to precipitation with high salt of typical RNases is used to separate them from other proteins in the *E. coli* lysate.

## 3.1.2. Isolation of Bovine RNase from the Culture Supernatant (**Note 3**)

The first protocol starts by diluting the supernatant to lower the salt concentration enough that the RNase will bind to a column of Matrex blue gel. Nucleic-acid-binding proteins often have a very high affinity for the blue dye, which is supposed to be an analog of NADH. The resin itself is very stable, and can be reused repeatedly if washed in high salt and reequilibrated. The column can also be cleaned with 0.1 $M$ NaOH and solvents such as ethanol, but these lead to dye leakage from the column. Of course, when purifying many mutants of the same protein, small columns with fresh packing material should be used every time to avoid cross-contamination. In the next step, the salt is removed from the (now quite concentrated) eluates of the column by dialysis, and the RNase is further purified by passing it over a DEAE-column at a pH where it will not bind, and then binding it to a CM-column.

The protocols are all designed to run in 1–2 lab days, and should not be interrupted. Avoid stopping and freezing your sample between steps. If you must stop, check that freezing the sample will not denature the activity.

1. Dilute 400 mL fermentor culture supernatant at least threefold with 10 m$M$ Tris-HCl, pH 7.5 (CB; *see* **Note 4**) and apply to a 20-mL column of Matrex blue gel, washed with 2 $M$ salt in CB and equilibrated in CB without salt (check that the salt wash has been completely removed from the column by measuring conductivity). Elute the column with a step gradient from 0.05–1 $M$ NaCl in 10 m$M$ Tris-HCl, pH 7.5. RNase A should elute around 200 m$M$ NaCl (*see* **Note 6**). Measure RNase activity (**Subheading 3.2.**) and protein in the flow-through and all eluates.

2. Place the RNase-containing eluates (<10% of the initial volume of culture) in dialysis sacks (10,000 Daltons molecular weight cutoff, rinse with water and EDTA, but do not boil), seal, and dialyze against 1–2 changes of 10 m$M$ Tris-HCl, pH 7.5, until the conductivity indicates <20 m$M$ residual NaCl.

3. Pass the dialyzed eluates through a 5-mL DEAE column equilibrated with 10 m$M$ Tris-HCl, pH 7.5. The RNase should remain in the flow-through.

4. Pass the DEAE flow-through and buffer washes through a 2-mL column of CM-Sephadex G-25. The RNase should bind to this negatively charged column and can be eluted using a

step-gradient elution like that described for Matrex blue. The small column size allows one to concentrate the RNase in this step (elute with one CV at each salt concentration; the salt gradient can be prepared in labeled tubes from a stock solution of sterile 5 *M* NaCl and the CB. Apply all of the appropriate buffer to the top of the column and collect the eluate into the same tube).

5. Run a PAGE (e.g., 10–20% gradient; **Note 7**) to check for purity of the RNase. It may be necessary to add a gel-sizing column or an HPLC step to obtain completely pure RNase. Stain the gel first with Coomassie blue and completely destain before using silver stain to detect trace impurities.

6. To determine the N-terminal sequence: Concentrate ~0.1 mg protein to about 40 µL in a Centricon, remove into a siliconized Eppendorf cup, add 10 µL SDS-sample buffer, heat 2 min in a boiling water bath, and apply to a 17% SDS-polyacrylamide minigel (*see* **Note 7**).

7. To locate the position of the protein on the gel without staining, the gel order on a 10-slot minigel would be: empty wells on both sides, 2 µL sample, prestained standard, then three lanes of 10-µL sample, prestained standard, another 2 µL of the sample, and blank lanes. Cut the gel longitudinally through the middle of the prestained standard lanes and stain the outer lanes with Coomassie blue protein dye for 15–30 min. Destain with the typical acetic acid/methanol/water mixture, which causes the gel to shrink. After destaining, put the gel into water for a few minutes until it is approximately the same length as the original gel. Align the three gel fragments on Saran wrap on a light box and cut out very thin strips, lengthwise, in the sample portion of the gel around the position of the band in the stained portions. Place the strips in siliconized Eppendorf cups (*see* **Note 8**), labeled to reflect the position of the gel strip (e.g., 3.1–3.25 cm).

8. Elute each strip overnight into 25 m*M* ammonium bicarbonate (approx 500 µL/strip).

9. Check 10 µL of each eluate for protein on a silver-stained minigel; lyophilize protein containing samples.

10. The protein is now ready for N-terminal amino-acid sequence determination in a suitably equipped lab, or to be sent to a protein-sequencing service.

11. Some services will run the gel for you (usually in one lane on a gel with other samples) and transfer the proteins from the gel to a protein-blotting membrane. The band is then identified by staining with a dye that does not interfere with subsequent operations. The membrane piece containing the band can be placed directly in the sequencing chamber.

### 3.1.3. Isolation of Murine Spleen RNase from Culture Supernatants

This cytokine was expressed at lower levels than the bovine RNase, perhaps because the mammalian gene sequence was used. In this protocol, the culture was desalted and concentrated during the first step. High salt inhibits many proteases and breaks up ionic interactions with cell components. If the protein binds to hydrophobic interaction chromatography, one can concentrate/purify/desalt in one step. For very hydrophobic proteins, such as RNases, which do not bind, a high degree of purification is obtained in one step, but the sample must be extensively dialyzed to remove salt before proceeding.

1. Combine the supernatant of the culture (40 mL) with 8 mL of 5 *M* NaCl.

2. Pass the sample through a 1-mL column of Phenylsepharose. Wash the column with 2–3 mL of 1 *M* NaCl in 20 m*M* Tris-HCl, pH 7.5.

3. Combine the flowthrough and buffer wash and dialyze against 2 × 1 L of buffer (20 m*M* Tris-HCl, pH 7.5). Change the buffer after 4 h and then dialyze overnight to equilibrium.

4. Pass the dialyzed sample through a 1-mL column of DEAE-Sephacel equilibrated with the dialysis buffer and then washed with 2 CV TTG buffer.

5. Combine the flowthrough and buffer washes and apply directly to a 1-mL column of SP-Trisacryl, equilibrated with the dialysis buffer.
6. Wash the SP-Trisacryl column with the dialysis buffer and elute with 1.5-mL aliquots of the TTG containing progressively more NaCl (*see* **Note 9**). RNase activity elutes from the column between 120 and 180 m$M$ NaCl.
7. Combine the peak active fractions and concentrate to about 0.3 mL in a Centricon 10. Several milliliters can be concentrated to achieve a high protein concentration. Be careful not to centrifuge at speeds higher than recommended by the manufacturer.
8. Determine purity and N-terminal sequence as described in the last steps of **Subheading 3.1.2.**

### 3.1.4. Purification from the Cell Pellet

1. Make a slurry of the frozen cell pellet from approx 0.3 L culture in 100 mL 10 m$M$ Tris-HCl, pH 7.5, 10 m$M$ EDTA plus 0.02 m$M$ PMSF, and lyse the cells using a French Press at 500–1000 psi.
2. Centrifuge the lysate for 30 min at 16,000$g$ at 4°C. Discard the pellet. For the supernatant, measure total protein as directed with the Bradford protein assay and determine RNase activity with one of the assays in **Subheading 3.2.**
3. Dilute the supernatant to 400 mL and pass through an ~70-mL column of DEAE Sephacryl (use approx 1 mL gel for 5–10 mg total protein applied).
4. Pass the flowthrough and buffer wash from the DEAE over a 2-mL column of CM-Sephadex G-25 (Pharmacia).
5. Elute the column with a step gradient from 50–750 m$M$ NaCl in 20% TTG (*see* **Note 9**).
6. Combine the RNase-containing eluates, dilute with 20% TTG to a conductivity indicating <20 m$M$ NaCl, and apply to a small Matrex blue column
7. Elute the RNase with a step salt gradient.
8. Follow the last steps in the protocol for **Subheading 3.1.2.** to determine the purity and N-terminal sequence of the RNase.

### 3.1.5. Purification of Recombinant Bovine Seminal Ribonuclease (BS-RNase) from Lysed Cells

This is another purification protocol, using soluble protein obtained after cell lysis with a freeze/thaw technique (which is only useful for small volumes, but is especially practical when many different samples should be analyzed simultaneously). If the equipment is available, try always to use a larger volume and break the cells with a French Press, as in **Subheading 3.1.4.** BS-RNase was expressed at a lower level in bacteria than RNase A *(25)*. BS-RNase from the supernatant was purified with a contaminant (identified by N-terminal sequencing as oligopeptide-binding protein of *E. coli*) that also inhibited its activity. Thus the protein used for kinetic studies was purified from supernatant of lysed cells, and the first step was ammonium sulfate fractionation, where the high salt was useful for breaking up any complexes that might have formed. The small column size during the purification allows the protein to be concentrated.

1. Suspend 7.5 g frozen cell pellet (approx 0.5 L original fermentor culture) in 20 mL TTG buffer, add 500 μL 0.5 $M$ Na EDTA, pH 8.0, and vortex the sample. Freeze (in liquid $N_2$) and thaw in room temperature water for two cycles.
2. Combine the sample (26 mL) with 20 mL of 4 $M$ $NH_4SO_4$ (set to pH 8.0 with $NH_4OH$) and vortex well to mix; set pH to 8.0 with a few drops of 25% $NH_4OH$. Centrifuge at 11,000$g$ in a Beckman table-top centrifuge for 75 min.

3. Apply the clear supernatant to a 2-mL column of Phenyl-sepharose (Pharmacia) equilibrated with 1.3 $M$ NH$_4$SO$_4$ in TTG buffer. The flow-through and buffer wash should be pooled and dialyzed 3 h against 500 mL 20 m$M$ Tris-HCl, pH 7.0, 0.2 m$M$ PMSF, and then overnight against two changes of "DB." The sample will expand as salt is dialyzed away, then contract as the concentration of glycerol increases.

4. Centrifuge the dialyzed sample 30 min at 11,000$g$ and discard the small pellet.

5. Apply the supernatant to a 1-mL column of DEAE-Sephacryl equilibrated in DB.

6. Collect the flow-through and buffer wash and apply to a 0.8-mL column of SP-trisacryl. Elute with a step gradient of NaCl in DB. BS-RNase should elute around 0.3 $M$ NaCl.

7. Dilute the RNase-containing salt eluates 40-fold with DB, apply to a 0.2-mL column of Matrex Blue (Amicon) equilibrated in DB, and elute with a step gradient of NaCl in DB. BS-RNase activity should elute from 0.2 to 0.5 $M$ NaCl.

8. Check eluates for purity on PAGE.

## 3.2. Assays for RNase Activity

### 3.2.1. Preparation of Defined ss- and ds-RNA Substrates for Assays

Complementary ss-RNA transcripts of a section of the T7 phage genome (approx 300 bp long, cloned in opposite orientations downstream of the promotor for T7 RNA polymerase at the *Bam*HI site of plasmid pET-1 *(57)* are prepared from linearized, purified plasmids using T7-RNA polymerase (the polymerase reaction works more efficiently on linearized than circular DNA). To make ds-RNA, the complementary strands are combined, heated to remove secondary structure, and allowed to hybridize.

1. Cleave the plasmid containing the T7 recognition site, followed by the RNA sequence of interest, with *Eco*RV or another enzyme with a unique site in the plasmid following the sequence to be amplified. Check for linearization on a 1% agarose gel (the linearized plasmid should run above the intact plasmid) and ethanol-precipitate the plasmid. Resuspend in TE buffer.

2. For the reaction, in siliconized cups with a lock lid (*see* **Note 8**), combine 25 μL AGUC cocktail, 10 μL 10X buffer, 1 μL 1 $M$ DTT, 1 μL RNasin, 1–20 ng linearized plasmid DNA, 2 μL α-$^{32}$P-GTP (for labeled substrate) and RNase-free water (*see* **Note 1**) to 100 μL. Remove 1 μL into a cup to use to quantitate the reaction (*see* **Note 10**). Start the reaction by adding 1 μL T7 RNA polymerase and incubate for 1 h at 37°C.

3. Stop the reaction by adding 100 μL TE buffer and putting on ice. Add 100 μL phenol, vortex, and add 100 μL CIA. Vortex again, and centrifuge 10 min, 10,000$g$ in the cold. Remove the top phase into a fresh siliconized cup and add 500 μL ice-cold ethanol.

4. Leave for about 30 min at –70°C. Centrifuge, discard the supernatant (into radioactive waste), and wash the (probably barely visible) pellet with 500 μL 80% ethanol. Remove the wash ethanol into radioactive waste.

5. Dissolve the pellet in 20–30 μL FAHB + dyes by pipeting up and down. Check both the cup and the pipet to make the sure the RNA has dissolved.

6. Purify the ss-RNA by applying the material in the pipet to a polyacrylamide gel run in 50 m$M$ TBE. For the 300-bp fragment used in these studies, a 4% gel was sufficient. Identify the RNA band by transferring the gel, on plastic wrap, onto a TLC plate containing fluorescent marker. Illuminate the gel with a mineral lamp from above to see the bands. Cut out the band with a razor blade and put in a siliconized cup. Count the band in a scintillation counter. Cover with approx 0.5 mL sterile TE buffer and elute a few hours or overnight. Remove the TE, add fresh TE, and allow to elute a second time.

7. To quantitate the RNA, measure the $OD_{260}$ of the eluted RNA, or run standards on the gel of a control RNA preparation of known concentration. For labeled RNA, the amount can be calculated from the specific activity of the GTP in the starting reaction, which contains 0.7 m$M$ GTP, and determining the initial CPM/1 μL of the starting reaction mix (*see* **Note 10**).

8. To make ds-RNA, combine equal amounts of the labeled ss-RNAs for each strand (be sure to correct for the different G base content of the two strands, if using CPM-based quantitation), heat for 5 min to 55°C, and allow to cool slowly to room temperature.

9. The double stranded substrate is extremely resistant to degradation by RNase A (1 μL of a $10^{-6}$ mg/mL dilution of RNase A in TTG will degrade 10 ng of the ss-RNA to >80% in 10 min at 37°C, whereas 1 μL of a $10^{-4}$ to $10^{-3}$ dilution degrades a comparable amount of the annealed substrate in 1 h at 37°C to the same degree) and migrates slower on a PAG. The substrate, when incubated for 1 h at 37°C in assay buffer, should not increase appreciably in its degree of acid solubility (<2%) or show any signs of strand separation on PAGE.

10. Store RNA in buffered formamide or TE buffer at liquid nitrogen temperature.

## 3.2.2. Basic Assay for Solubilization of a Labeled RNA Substrate

This assay measures RNase activity by determining the generation of acid-soluble radioactivity from an initially acid-insoluble substrate. To precipitate the partially degraded RNA completely, cold RNA and protein are added as carriers. The assay uses labeled ss- or ds-RNA substrate, is easy to run, and can be used to quantitate the nuclease activity of 20–30 samples simultaneously. Similar assays can also be used to test the effect on the rate of reaction of affector molecules, such as IFN-γ. To calculate specific activity, protein concentration should be determined with any commercially available protein assay (e.g., Coomassie blue or Lowry methods) or by estimating the amount of protein from Coomassie blue-stained minigels. An RNase A equivalent can be defined by comparing the percent of total RNA solubilized by the sample to that degraded in the same period by a known amount of RNase A (1 μL of a $1 \times 10^{-4}$ mg/mL solution of RNase A would degrade 60% of the labeled ssRNA in the assay within 10 min). The specific activity of the purified murine spleen RNase in this assay was approx 30% that of RNase A.

1. Prepare a master reaction mix (approx 200,000 cpm/mL ss- or ds-$^{32}$P-labeled RNA in 25 m$M$ Tris-HCl, pH 7.9; the mix should have at least 1000 cpm of labeled RNA substrate per 5 μL). Prepare more mix than you need; the mix can be frozen (put the Eppendorf cup in a leaded container) and used for several days if not left overnight in the ice bucket.

2. Prepare dilutions of the RNase to be tested in TTG buffer. The minimum detectable amount of RNase A is in the range of 0.02 ng/assay. Prepare dilutions of human and bovine IFN-γ, or other cytokine or affector, prepared in the same buffer. The glycerol and Tween-20 in the buffer stabilize the RNase and increase its specific activity. This buffer may need to be modified for RNases that are sensitive to viscosity. Keep dilutions on ice.

3. In prelabeled Eppendorf cups, combine 2 μL of enzyme dilutions with 2 μL of TTG buffer or dilutions of the affector. Add Tris-HCl buffer to 10 μL. Prepare a no RNase cup (4 μL TTG buffer, 6 μL Tris-HCl) to check for RNase contamination of the buffer or substrate degradation.

4. Start the reaction by, at timed intervals (20 s should suffice), adding 5 μL of the nuclease master reaction mix (as in **step 1**) and placing the samples in a 37°C water bath.

5. Incubate the reactions for 5–10 min for ss-RNA, 30–60 min for ds-RNA substrates.

6. Stop the reactions in the same time series by adding 15 μL of cold 10% TCA, and 5 μL BSA/RNA carrier. Vortex and leave 10–30 min on ice. Do not leave overnight.

7. Centrifuge the samples 10 min at 4°C in an Eppendorf centrifuge, with the cup hinge facing outward. Pipet the supernatant into a labeled cup and count it and the pellet separately. The added carrier RNA and protein yield a visible pellet on the hinge side of the cup.
8. Express the data as percent of total RNA solubilized. This can be related to ng RNA solubilized by knowing the specific activity of the GTP in the initial synthesis reaction.

### 3.2.3. Thin-Layer Chromatography of RNA Degradation Products.

This assay (**Fig. 1**) was useful for showing that the RNA fragment length after degradation, an indication of turnover rate with a distributive enzyme (*see* Chapters 1 and 2) was affected by the presence of IFN-γ.

1. Set up assays as in **steps 1–4**, using up to 10X the volume. Stop the samples by moving onto an ice bath and adding 1 vol phenol/chloroform (stabilized with 2% [v/v] iso-amylalcohol) 1:1.
2. Vortex, centrifuge the samples 10,000*g* for 10 min. Remove the top phase into a fresh cup (discard the bottom organic phase) and extract twice with diethyl ether to remove residual phenol, dry on a Rotovac, and redissolve in <10 μL water.
3. Apply the resuspended samples to a PEI-Cellulose F plate (Merck 5725; 0.1 m*M* thick, prewashed in water and vacuum dried). Dissolve 2' and 3'-mononucleotides standards to 0.1 *M* in 25 m*M* Tris-HCl, pH 7.9 (add ~20 μL 1 *M* NaOH to *A* and *C* to adjust pH to approx 7.0 by pH paper; no adjustment should be necessary for *U* and *G*) and let run as standards in the outer lanes of the plate.
4. Develop the chromatogram in 1 *M* acetic acid/3 *M* LiCl 9:1. Mark the plate with glow-in-the-dark or radioactive ink to orient the film image. Dry the plate under vacuum.
5. Detect unlabeled standards as fluorescent spots using surface illumination with a mineral lamp.
6. Expose the plate to X-ray film. Indicate the position of the standard spots either by circling these with radioactive ink before exposure or by indicating the spots on the film after developing.

### 3.2.4. PAGE Assay of Products

While the above assay distinguishes very low molecular weight reaction products, PAGE shows the initial stages of cleavage within a defined substrate (*see* **Fig. 2** for examples using both ss- and ds-RNA substrates).

1. Perform **steps 1** and **2** as above, but redissolve the dried samples in 5–10 μL of FAHB + dyes instead of water. Do not phenol extract if protein binding is to be assayed for simultaneously.
2. Apply the samples to a PAG (4–10%, depending on the size of the original substrate), 25 m*M* in TBE, and use 25 m*M* TBE as the electrophoresis buffer.
3. Electrophorese just until the Bromophenol blue line is about 75% down the gel.
4. Wearing gloves, open the glass plates (carefully) and lay a piece of filter paper cut slightly larger than the gel over it. Invert the bottom glass plate so the gel is on top of the filter paper and slide the gel/filter paper out from underneath (let the gel fall from the glass plate onto the filter paper). Press gently (pour a bit of buffer on top to help) to remove any bubbles between the gel and paper. Make a mark with radioactive ink on the upper left-hand corner of filter paper (or write a message that indicates the start of the gel).
5. Cover the gel with plastic wrap and expose to film (the gel can also be sealed between layers of heavier plastic). Drying the gel can make the bands sharper, but may cause cracking.
6. The RNA substrate in the control samples should be intact and move as a distinct band of the appropriate size. A band shift to lower mobility at higher concentrations of IFN-γ can

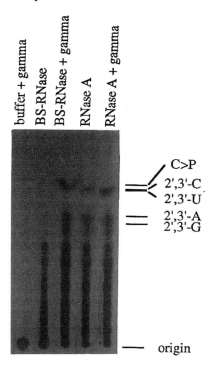

Fig. 1. Thin-layer chromatography assay of the products of degradation by bovine seminal RNase (BS-RNase) and RNase A in the presence and absence of human IFN-γ. Although the activity of BS-RNase is activated by the cytokine, as seen by more low molecular weight (higher mobility) material, RNase A is inhibited. Reprinted with permission from **ref. 25**.

be seen, even in samples containing RNase. The lower mol-wt products of the RNA usually form a product ladder.

### 3.2.5. Kinetic Assay

Kinetic experiments should start with freshly synthesized and annealed substrate to maximize the specific activity.

1. Prepare a master reaction mix (**Subheading 3.2.2., step 1**). Prepare a series of labeled Eppendorf cups for each assay, containing 20–50 μL ice-cold 10% TCA (the volume of TCA should be at least one-half the volume of the sample to be taken). Larger assay volumes should be used for the initial time points to reduce the counting error when relatively little of the substrate has been degraded. Keep the sample cups on ice or an ice/salt mixture.
2. Put assay mixture containing the labeled RNA, sufficient for 5–8 time points, in a siliconized cup. Then add TTG-buffer alone, or an IFN dilution in TTG-buffer. Remove an aliquot for the "zero time point," which is held at 37°C for the duration of the assay (20–30 min) to check for any RNase contamination of the sample or the affector.
3. Prewarm the rest of the mixture for 5 min at 37°C, and start the assay by adding prewarmed BS-RNase diluted in buffer (*see* **Note 11**).
4. At intervals (10–20 s is practical), remove aliquots into the labeled Eppendorf cups containing cold 10% TCA and mix by gentle vortexing.

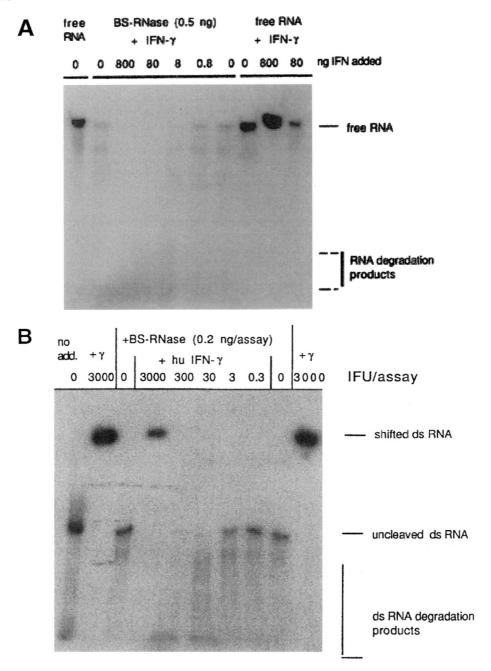

Fig. 2. Assay by PAGE for RNA binding and activation of BS-RNase by IFN-γ. In A, the substrate was ss-RNA, and bovine IFN-γ was added in the amounts per assay shown. In B, the substrate was ds-RNA, and human IFN-γ was used. Note that gel density (10 and 6%, respectively) must be carefully chosen for such assays, to visualize the maximum number of products while still seeing the protein gel shift. In other experiments (not shown here), the IFN was first incubated with the RNA, and the sample was then treated with SDS and protease K to remove the cytokine. The resulting RNA was still intact, thus showing that IFN-γ has no intrinsic RNase activity.

5. At the end of the assay, add cold TCA to the remaining reaction mixture to stop the reaction.
6. Add protein/RNA carrier to all cups, and, after leaving 10–30 min on ice, centrifuge, remove the supernatants into a separate labeled cups, and count as usual.

### 3.2.6. Ribonuclease Assays for Cultured Mammalian Cells

This assay can be run with many samples simultaneously.

1. Remove supernatant samples from cultured cells and store in 96-well microtiter plates. Cover the plates with parafilm (use your fingertips to indent over each well) and then the lid to avoid mixing from one well to the other.
2. To measure intracellular nuclease activity, cover the cell layers with 100 μL PBS buffer and freeze at –20°C overnight or until assay. Thaw the frozen cell layers quickly and homogenize by drawing the well contents up and down in a pipet.
3. Remove aliquots (e.g., 25 μL of supernatant or lysed cell suspension) to labeled wells in another microtiter plate. Add 4 μL (approx 40 ng) of $^{32}$P-labeled ds-RNase assay mix and incubate for 30–60 min at 37°C.
4. Add 25 μL cold 10% TCA and 5 μL carrier BSA/RNA per well using a multichannel pipet, cover, and shake gently for a few min on a rotating platform shaker in the cold room.
5. Centrifuge the plates and count the supernatant and pellet separately for each sample.
6. Human serum samples should be diluted 1:500-fold with TTG buffer before assay, as they contain much more RNase activity than the medium from cultured cells.

### 3.2.7. Lowry Assay for Total Cell Protein

This is a useful assay to give results as specific activity (RNase activity/mg cell protein) when correlating the effects of cytokine treatment on cellular RNase activity if the cytokine also inhibits growth. To run the assay on cells used for nuclease assays as above, one can add 20 μL of the cell suspension (frozen cells in buffer after resuspension with pipeting, **step 2**) to the NaOH.

1. Suspend cell layers (or frozen cells in buffer) in 100–300 μL 0.1 *M* NaOH by drawing up and down several times in an eppendorf yellow tip. Transfer 30 μL sample to a well of a microtiter plate.
2. Add 200 μL alkaline Cu reagent and shake the plate for 10 min at room temperature.
3. Add 20 μL Folin Ciocalteu phenol reagent (dilute to 1 *N* [1:2] in water), cover the plate with Parafilm and incubate with shaking another 30 min at room temperature. Mix samples again thoroughly by drawing up and down several times in an eppendorf yellow tip and read the OD$_{750}$ with a microtiter plate reader.
4. Run a bovine serum albumin (Sigma Fraction V) (dissolved in 0.1 *M* NaOH) standard curve in a row of the microtiter plate. The plate reader can be programmed to analyze the data using this curve and express the samples in mg/mL total protein.

### 3.3. Neutralization of IFN-γ Effect Using MAbs

Antibodies to human IFN-γ neutralize its activation effect on bovine seminal ribonuclease in the percent of RNA solublized assay using double stranded RNA (3H-poly A/poly U). Run all samples in duplicate or triplicate.

1. Add IFN-γ (250 ng, diluted with 1 mg/mL macaloid treated (to remove ribonucleases [58]) BSA in PBS) to PBS/BSA buffer (control) or to 10 mL of antibody in PBS.
2. Incubate the samples for 5 h at 4°C and centrifuge in an Eppendorf centrifuge for 15 min.

3. For the assay, in triplicate, combine 2 µL of a $5 \times 10^{-3}$ mg/mL solution of BS-RNase in PBS/BSA with 2 µL of the preincubated IFN-γ containing solution, and start the reaction by adding the ds-RNA nuclease assay mix (*see* **Note 11**).
4. Incubate the samples 1 h at 37°C, and stop by adding 10 µL TCA and 5 µL carrier protein and RNA. Note that the buffer control must be subtracted from all numbers to determine the percent of RNA solubilized by the enzyme. "Percent control activity" is the activity of the sample containing IFN-γ relative to that of the nuclease alone.

### *3.4. Additional Useful Techniques Related to These Assays*

### *3.4.1. Concentrating Dilute Protein Solutions for PAGE with TCA Precipitation*

One should ideally have a functional assay to test for expression of protein. In the event that gels are used to follow expression, or to look for leakage through a filter concentration device, the following method, which removes all the salt and most of the short peptides, can be used to prepare dilute samples for PAGE.

1. Combine 100 µL of culture supernatant or other dilute protein sample with 100 µL ice-cold 20% trichloroacetic acid (w/v) in water, in a siliconized Eppendorf cup. Leave samples for 10–30 min on ice (do not store overnight in TCA).
2. Centrifuge 10 min in an Eppendorf at 4°C. You will know where your pellet should be if you always centrifuge with the hinge of the cup facing outward.
3. Discard all the supernatant (go down into the cup with a pipet to get the last drops and then wash the pellet several times with ether to get rid of the TCA (you can check with pH paper that the ether is neutralized and no longer acidic). You may need to centrifuge the pellet back out of the ether each time.
4. Then dissolve the pellet in gel-loading buffer. The sample should be blue, if not increase the ether washes when you do the technique again. Then heat and apply to the gel.
5. Although silver staining should be the most sensitive detection method for proteins in gels, always stain your gel first with Coomassie blue and destain, to fix the proteins in the gel and to detect the occasional protein that does not react with the silver stain. If you are using the Biorad silver-stain kit, plan 25 mL of each reagent for a minigel.

### *3.4.2. Removing RNase Activity from Proteins with Macaloid Powder*

Macaloid powder can be used to remove ribonuclease activity from proteins *(58)*, such as the RNase activity that contaminates preparations of DNase. The following protocol was used to remove detectable RNase activity from BSA to be used as a carrier protein in assays.

1. Boil 0.5 g macaloid powder in 50 mL of 50 m$M$ Tris-HCl, pH 7.6, for 5 min with constant agitation; and then centrifuge the suspension 5 min, 2500$g$ at room temperature.
2. Discard the supernatant and resuspend the sticky macaloid pellet in 40 mL of the same buffer. Recentrifuge. After one more similar wash step, resuspend the pellet in 30 mL of buffer to a final macaloid concentration of ~16 mg/mL. This suspension may be stored for several years if refrigerated.
3. Dissolve 1 g BSA (Cohn fraction V from Sigma) in 40 mL 25 m$M$ Tris-HCl, pH 7.6.
4. Add 10 mL of the washed macaloid suspension (**step 2**) and agitate vigorously for 10 min. Centrifuge 10 min, 2500$g$, at 4–10°C.
5. Decant the supernatant into a fresh tube and add another 10 mL of the macaloid slurry. Centrifuge as in **step 4**, then pass through a 0.2-µ sterile filter into sterile tubes.

6. Measure the $OD_{280}$ of the filtered solution at a 1:30 dilution in water to determine the final protein concentration, using $OD_{280}$~0.66 for a 1 mg/mL solution of BSA. The final protein concentration should be about 10 mg/mL.

7. Check for residual nuclease activity by incubating labeled RNA substrate with 10 μL of the concentrated protein solution for 30 min at 37°C. The labeled substrate should not be solubilized to more than 2% in the acid solubility assay, and should still migrate as a single band on PAGE (**Subheading 3.2.1.**).

8. Macaloid-treated BSA solutions can be stored frozen for several years. Tubes stored in the cold room should be checked for nuclease activity before use.

### 3.4.3. Cell-Based Antiviral Assay

This assay can be used to check IFN preparations for antiviral activity against EMC virus before use in in vitro assays. The following protocol assumes familiarity with sterile cell-culture technique, with precautions to be taken when dealing with replication-competent mammalian viruses. Variations of these techniques can be used to prepare cells treated by virus; for example, to measure increases in the amounts of RNase produced or for measuring mRNA stability after treating cells with cytokines or viruses.

1. Propagate the assay cell lines and check for contamination with yeast or fungi. Murine L929 cells are used to measure the activity of murine interferon and to cultivate the EMC virus, which lyses L cells, and the human WISH cells used to assay for human IFN activity.

2. Propagate the EMC-virus. Grow L929 cells in a T25-culture flask to about half confluence. Remove the medium, add 10 mL fresh medium, and 0.5 mL of EMC virus (from the ATCC sample). Incubate for 24 h or until lysis of the cell layer is complete.

3. Remove the lysate and centrifuge for 30 min at 10,000*g*. Discard the cell debris pellet. Mix the supernatant 1:1 with 40% glycerol in PBS (heat-sterilized), and store 1-mL aliquots in cell-culture tubes in liquid N2.

4. The virus can be titered in several ways. If prepared as directed here, adding 5 μL of the thawed virus solution, diluted $10^6$-fold with medium, to 100 μL cell medium should lyse a confluent layer of L929 cells in a well of a 96-well microtiter plate within 20–48 h incubation at 37°C in a cell incubator.

5. For assay, remove WISH or L-cells from the cell-culture bottles by trypsinization, and dilute the cell suspension in DMEM/10% FBS (WISH) or DMEM/10% HS (L-cells) to $10^5$ cells/mL. Add 100 μL of this suspension to each well of a 96-well microtiter plate, and incubate the plates for up to 3 d until a confluent cell layer is established.

6. Remove the spent medium and add fresh medium (100 μL/microtiter well).

7. Then add 50 μL of the interferon-containing sample (diluted in FMX-8 or DMEM/serum medium) or a buffer control to the first row. Dilute the samples 1:3 serially down the plate by mixing (i.e., by drawing the medium up and down in the multichannel pipet tips) and removing 50 μL of the mixed medium into the next row of the plate. Change tips every third row. Typically, the control lanes are on the outside rows of each plate. Here, add only medium without sample. Add virus to one lane, and reserve the first for a cell-growth control. If there are many samples, you can cut the controls to one lane and add virus to only a few wells instead of a whole row.

8. Incubate the plates 8–24 h in the incubator.

9. Add EMC virus (20 μL of a 1:5000 dilution for L-cells, or 1:1000 dilution for WISH cells) and incubate the cultures a further 15–18 h for L-cells or 24–28 h for WISH cells. Do not add virus to the cell control lane or wells.

10. Inspect the plates for cell lysis using a cell-culture microscope. When sufficient viral lysis is seen in the virus-added control lane, remove the culture supernatants with a multichannel rake into an autoclavable container.

11. Stain the cell layers by filling the wells with the crystal violet solution. Remove the stain (we autoclave this as well) and wash the plate with warm water several times. Confluent blue layers in the wells indicate where cells were protected from the virus. Typically, wells treated with high concentrations of IFN are less blue than the control cell lanes where no virus was added, because of cell-growth inhibition by the cytokine. The plates may be evaluated visually by a trained operator or by using a microtiter plate reader.

# 4. Notes

1. Water used for assays, preparing RNA (including buffers used for gels), or in the final stages of protein purification is completely deionized (to >18 MegaO resistivity) in a typical apparatus and then bidistilled in a glass unit. The distillate is collected directly into bottles reserved for RNase-free water (adding buffer components to these bottles is strictly forbidden!), and the bottles are closed and autoclaved as soon as possible. RNase-free water must always be checked to ensure that there is no detectable nuclease activity, and partially used bottles should be reautoclaved or reserved for less sensitive purposes. We use similar water to prepare cell-culture media. Deionized water is sufficient for preparation of bacterial growth media. Always autoclave buffers and stock solutions used for protein purification, except those dissolved in organic solvents.

2. Optimizing a large-scale cultivation based on shaker flask data is difficult. Shaker-flask cultures may be used to select inoculum cultures with relatively high production capacity or detect intrinsic proteolysis problems. However, growth densities in a fermentor are typically at least 10–20× higher than that attainable in shaker-flask cultures, which means that the optimal cell/inducer ratio changes. Several fermentor runs may be necessary to determine the optimal concentration of inducer or induction time. If only one run is possible (when one is using a borrowed fermentor, or when using a service, for example), remove at least 1 L of culture per hour after induction and centrifuge immediately, saving the cell paste in a labeled tube in the freezer. This will ensure that if the protein increases and then decreases (because of proteolysis) with longer induction, there will be enough cells from the run to try a small purification. Microidentification methods allow one to run small purifications, such as that in **Subheading 3.1.5.**

3. To obtain more complete lysis, one can add lysozyme HCl in Tris-HCl buffer, pH 8.0 (5 mg/mL solution, add 20 μL/sample) to the cell pellet with the EDTA and incubate briefly (~5–10 min) at 30°C before freezing and thawing the cells. This procedure works well for larger proteins or those clearly different in physical properties from the hydrophilic, positively-charged lysoyzme. Lysozyme is close in size and charge to RNases, and may complicate the purification.

4. For the purification of RNases, 0.02 m$M$ PMSF (a protease inhibitor) and 1 m$M$ β-mercaptoethanol (to prevent intermolecular crosslinking of free disulfides) were added to all buffers.

5. Short, squat columns such as this can be run using a pump (or simply by gravity elution if the amount of material applied does not exceed 500 mL). Small columns of <1 mL, prepared, for example, in 10 mL total volume disposable columns from Biorad, can be run to determine elution conditions initially, and the column scaled up for the preparative run. Clean and autoclave the columns before use. Loop the tubing up so that the end is slightly above the top of the gel; **never** allow your column to run dry after the sample has been applied.

6. There are two efforts during the purification: to concentrate the protein from the supernatant and to remove protein impurities and residual small molecular weight contamination

from the culture. A cell lysate contains a higher concentration of the desired protein initially, but there are usually more contaminants than when starting with the culture supernatant. Depending on the starting volumes, it is usually a good idea to bind the protein to the first column and elute into a relatively small volume. Although the DEAE flowthrough step used here removes a good deal of impurities, ideal purification steps will completely change the buffer around the protein. A good purification scheme should move the protein from one buffer to another at least three times. In these schemes, binding to CM-Sepharose and then to Matrex blue should be followed by a third step that again changes the buffer, such as dialysis, before concentration. Using small columns can help to concentrate a protein binding to them, as long as the rule of not more than 5 mg protein/mL column volume is followed. If a protein at neutral pH does not bind to either negatively or positively charged ion-exchange resins, it may, if the salt concentration is high enough, bind to hydrophobic interaction columns, such as Phenylsepharose. If it does not bind to any of these columns, it is probably aggregated and will be difficult to purify using any technique.

7. One can choose from a variety of PAGE conditions to achieve the desired separation. To check the purity of the protein, wide-range gradient gels are appropriate (e.g., commercially available 10–20% tricine gels allow one to visualize both high- and low-molecular weight contaminants) However, to separate different forms of a protein, select the gel and buffer concentration that provides a good separation around the molecular weight of your protein, even if this means pouring your own gels. For example, the small, very positively charged histone proteins in nuclear extracts separate well on an 18% gel containing three times the normal Tris-HCl buffer concentration (end concentration 0.75 $M$), while a 20% gel with 0.25 $M$ Tris is ideal for visualizing C-terminal proteolyzed forms of IFN-$\gamma$ and separating the dimer from the monomer. Optimize your staining techniques to visualize the minimum amount of protein, and do not apply too much sample to a gel. A wide band on a gel, especially when the standards are running as thin bands, indicates that the protein is heterogeneous. Ascertain that your protein does not leach out of the gel into the stain/destain solutions. IFN-$\alpha$, for example, is highly soluble in acidic alcohol solutions.

8. Siliconized Eppendorf cups can now be purchased. To siliconize your own cups, wash them with methanol, and then fill with a solution of 10% dichlor-dimethyl silane in methanol (the solution can be tipped from one cup to the next and reused 10–20 times). Turn the cups upside down on lint free tissue and allow to air dry. Rinse the cups with methanol and dry autoclave. Wash the cups with 0.1% SDS in water and rinse three times with sterile filtered water, turning the cups upside down to dry between rinses. Finally, wash twice with ethanol, let dry, and autoclave in jars labeled "siliconized." Students who use such cups for plasmid preps should be encouraged to make up for their error by replacing the stock. The procedure can also be used to siliconize glassware (e.g., cover-slips for crystal trials).

9. The high concentration of glycerol or other osmolytes in the latter steps of the protocols increases the yield of protein. A discussion of protein osmolytes is given elsewhere *(56)*.

10. To determine the specific activity of the GTP, count the CPM in 1 μL of the original reaction mixture, which is 0.7 m$M$ in GTP (or 0.7 nmol GTP in 1 μL). One can retain this cup to count every time assay samples are run, to directly calculate the total n$M$ GTP in the samples as the $^{32}$P decays.

11. Usually, enzymatic reactions are started by adding the enzyme and then beginning the incubation. However, when many different enzyme samples are being tested with a single reaction mix, adding the substrate to start the reaction makes more sense.

## Acknowledgments

Most of this work was supported by grants from the Zürcher Krebsliga. I especially thank Monika Haugg for refining many of these techniques and David Volk for carefully reading the manuscript.

## References

1. Rai, R. M., Loffreda, S., Karp, C. L., Yang, S. Q., Lin, H. Z., and Diehl, A. M. (1997) Kupffer cell depletion abolishes induction of interleukin-10 and permits sustained overexpression of tumor necrosis factor alpha messenger RNA in the regenerating rat liver. *Hepatology* **25,** 889–895.
2. Rutter, J. L., Benbow, U., Coon, C. I., and Brinckerhoff, C. E. (1997) Cell-type specific regulation of human interstitial collagenase-1 gene expression by interleukin-1 beta (IL-1 beta) in human fibroblasts and BC-8701 breast cancer cells. *J. Cell. Biochem.* **66,** 322–336.
3. Cox, A. and Emtage, J. S. (1989) A 6-fold difference in the half-life of immunoglobulin μ heavy chain mRNA in cell lines representing two stages of B-cell differentiation. *Nucleic Acids Res.* **17,** 10,439–10,454.
4. Schein, C. H. (1997) From housekeeper to microsurgeon: the diagnostic and therapeutic potential of ribonucleases. *Nat. Biotechnol.* **15,** 529–536.
5. Schluns, K. S., Cook, J. E., and Le, P. T. (1997) TGF-beta differentially modulates epidermal growth factor-mediated increases in leukemia-inhibitory factor, IL-6, IL-1 alpha, and IL-1 beta in human thymic epithelial cells. *J. Immunol.* **158,** 2704–2712.
6. Calleja, C., Eeckhoutte, C., Larrieu, G., Dupuy, J., Pineau, T., and Galtier, P. (1997) Differential effects of interleukin-1 beta, interleukin-2, and interferon-gamma on the inducible expression of CYP 1A1 and CYP 1A2 in cultured rabbit hepatocytes. *Biochem. Biophys. Res. Comm.* **239,** 273–278.
7. Zhou, X., Polgar, P., and Taylor, L. (1998) Roles for interleukin-1beta, phorbol ester and a post-transcriptional regulator in the control of bradykinin B1 receptor gene expression. *Biochem. J.* **330,** 361–366.
8. Schein, C. H., Haugg, M., and Benner, S. A. (1990) Interferon-γ activates the cleavage of double stranded RNA by bovine seminal ribonuclease. *FEBS Lett.* **270,** 229–232.
9. Schein, C. H. and Haugg, M. (1995) The role of the C-terminus of human interferon-γ in RNA binding and activation of the cleavage of ds-RNA by bovine seminal ribonuclease. *Biochem. J.* **307,** 123–127.
10. Lengyel, P. (1993) Tumor-suppressor genes: news about the interferon connection. *Proc. Natl. Acad. Sci. USA* **90,** 5893–5895.
11. Dong, B., Xu, L., Zhou, A., Hassel, B. A., Lee, X., Torrence, P. F., and Silverman, R. H. (1994) Intrinsic molecular activities of the interferon-induced 2-5A-dependent RNase. *J. Biol. Chem.* **269,** 14,153–14,158.
12. Carroll, S. S., Cole, J. L., Viscount, T., Geib, J., Gehman, J., and Kuo, L. C. (1997) Activation of RNase L by 2',5'-oligoadenylates. Kinetic characterization. *J. Biol. Chem.* **272,** 19,193–19,198.
13. Bisbal, C., Martinand, C., Silhol, M., Lebleu, B., and Salehzada, T. (1995) Cloning and characterization of a RNase L inhibitor. A new component of the interferon-regulated 2-5A pathway. *J. Biol. Chem.* **270,** 13,308–13,317.
14. Der, S. D., Zhou, A., Williams, B. R. G., and Silverman, R. H. (1998) Identification of genes differentially regulated by interferon α, β or γ using oligonucleotide arrays. *Proc. Natl. Acad. Sci. USA* **95,** 623–628.
15. Gessani, S., DiMarzio, P., Rizza, P., Belardelli, F., and Baglioni, C. (1991) Posttranscriptional regulation of interferon mRNA levels in peritoneal macrophages. *J. Virol.* **65,** 989–991.

16. Lappin, D. F., Guc, D., Hill, A., McShane, T., and Whaley, K. (1992) Effect of interferon-γ on complement gene expression in different cell types. *Biochem. J.* **281,** 437–442.

17. Radzich, D. and Varesio, L. (1991) c-fos mRNA expression in macrophages is downregulated by interferon-γ at the post-transcriptional level. *Mol. Cell. Biol.* **11,** 2718–2722.

18. Kasama, T., Strieter, R. M., Lukacs, N. W., Lincoln, P. M., Burdick, M. D., and Kunkel, S. L. (1995) Interferon gamma modulates the expression of neutrophil-derived chemokines. *J. Invest. Med.* **43,** 58–67.

19. Besançon, F., Przewlocki, G., Baró, I., Hongre, A., Escande, D., and Edelman, A. (1994) Interferon-γ downregulates CFTR expression in epithelial cells. *Am. J. Physiol. (Cell Physiol.* **36,** C1398–C1404.

20. Clerch, L. B., Wright, A., and Massaro, D. (1993) Endotoxin induces rat lung ribonuclease activity. *FEBS Lett.* **328,** 250–252.

21. Youle, R. J., Wu, Y., Mikulski, S. M., Shogen, K., Hamilton, R. S., Newton, D., D'Alessio, G., and Gravell, M. (1994) RNase inhibition of human immunodeficiency virus infection of H9 cells. *Proc. Natl. Acad. Sci. USA* **91,** 6012–6016.

22. Wu, Y., Saxena, S. K., Ardelt, W., Gadina, M., Mikulski, S. M., De Lorenzo, C., D'Alessio, G., and Youle, R. J. (1995) A study of the intracellular routing of cytotoxic ribonucleases. *J. Biol. Chem.* **270,** 17,476–17,481.

23. Däubener, W., Nockemann, S., Gutsche, M., and Hadding, U. (1995) Heparin inhibits the antiparasitic and immune modulatory effects of human recombinant interferon-γ. *Eur. J. Immunol.* **25,** 688–692.

24. Korlipara, L. V. P., Pleass, H. C. C., Taylor, R. M. R., Proud, G., Forsythe, J. L. R., and Kirby, J. A. (1994) Heparin antagonizes the induction of class II MHC molecules by interferon-γ. *Transplantation* **58,** 1426–1429.

25. Schein, C. H. (1997) An enzymatic cell-free assay to detect direct inhibitors of interferon-γ. *In Vitro Toxicol.* **10,** 275–285.

26. Jermann, T. M., Opitz, J. G., Stackhouse, J., and Benner, S. A. (1995) Reconstructing the evolutionary history of the artiodactyl ribonuclease superfamily. *Nature* **374,** 57–59.

27. Buschle, M., Campana, D., Carding, S. R., Richard, C., Hoffbrand, A. V., and Brenner, M. K. (1993) Interferon-γ inhibits apoptotic cell death in B cell chronic lymphocyte leukemia. *J. Exp. Med.* **177,** 213–218.

28. Gu, D., Wogensen, L., Calcutt, N. A., Xia, C., Zhu, S., Merlie, J. P., Fox, H. S., Lindstrom, J., Powell, H. C., and Sarvetnick, N. (1995) Myasthenia gravis-like syndrome induced by expression of interferon-γ in the neuromuscular junction. *J. Exp. Med.* **181,** 547–557.

29. Koch, A. E., Kunkel, S. L., and Streiter, R. M. (1995) Cytokines in rheumatoid arthritis. *J. Invest. Med.* **43,** 28–38.

30. Gutterman, J. U. (1994) Cytokine therapeutics: lessons from interferon α. *Proc. Natl. Acad. Sci. USA* **91,** 1198–1205.

31. Yayon, A., Klagsbrun, M., Esko, J. D., Leder, P., and Ornitz, D. M. (1991) Cell surface, heparin-like molecules are required for binding of basic fibroblast growth factor to its high affinity receptor. *Cell* **64,** 841–848.

32. Ornitz, D. M., Yayon, A., Flanagan, J. G., Svahn, C. M., Levi, E., and Leder, P. (1992) Heparin is required for cell-free binding of basic fibroblast growth factor to a soluble receptor and for mitogenesis in whole cells. *Mol. Cell. Biol.* **12,** 240–247.

33. Hemmi, S., Merlin, G., and Aguet, M. (1992) Functional characterization of a hybrid human-mouse interferon γ receptor: evidence for species-specific interaction of the extracellular receptor domain with a putative signal transducer. *Proc. Natl. Acad. Sci. USA* **89,** 2737–2741.

34. Huang, S., Hendriks, W., Althage A., Hemmi, S., Bluethmann, H., Kamijo, R., Vilček, J., Zinkernagel, R. M., and Aguet, M. (1993) Immune response in mice that lack the interferon-γ receptor. *Science* **259,** 1742–1745.

35. Fidler, I. J., Fogler, W. E., Kleinerman, E. S., and Saiki, I. (1985) Abrogation of species specificity for activation of tumoricidal properties in macrophages by recombinant mouse or human interferon-γ encapsulated in liposomes. *J. Immunol.* **135,** 4289–4296.

36. Sancéau, J., Sondermeyer, P., Béranger, F., Falcoff, R., and Vaquero, C. (1987) Intracellular human γ-interferon triggers an antiviral state in transformed murine cells. *Proc. Natl. Acad. Sci. USA* **84,** 2906–2910.

37. McDonald, H. S., Kushnayov, V. N., Sedmak, J. J., and Grossberg, S. E. (1986) Transport of gamma interferon into the cell nucleus may be mediated by nuclear membrane receptors. *Biochem. Biophys. Res. Comm.* **138,** 254–260.

38. Deutscher, M. P. (1988) The metabolic role of RNases. *TIBS* **13,** 136–139.

38a. Deutscher, M. P. (1990) Introduction to an issue devoted to E. coli ribonucleases. *Biochemie* **72,** 769.

39. Tsuji, H., Murai, K., Akagi, K., and Fujishima, M. (1990) Effects of recombinant leukocyte interferon on ribonuclease activities in serum in chronic hepatitis B. *Clin. Chem.* **36,** 913–916.

40. Lee, F. S. and Vallee, B. L. (1993) Structure and action of mammalian ribonuclease (angiogenin) inhibitor. *Prog. Nucleic Acid Res. Mol. Biol.* **44,** 1–30.

41. Shapiro, R., Weremocicz, S., Riordan, J. F., and Vallee, B. L. (1987) Ribonucleolytic activity of angiogenin: essential histidine, lysine, and arginine residues. *Proc. Natl. Acad. Sci. USA* **84,** 8783–8787.

42. Rosenberg, H. F., Tenen, D. G., and Ackerman, S. J. (1989) Molecular cloning of the human eosinophil-derived neurotoxin: a member of the ribonuclease gene family. *Proc. Natl. Acad. Sci. USA* **86,** 4460–4464.

43. Ardelt, W., Mikulski, S. M., and Shogen, K. (1991) Amino acid sequence of an anti-tumor protein from *Rana pipiens* oocytes and early embryos. *J. Biol. Chem.* **266,** 245–251.

44. Mosimann, S. C., Ardelt, W., and James M. N. G. (1994) Refined 1.7 Å X-ray crystallographic structure of P-30 protein, an amphibian ribonuclease with anti-tumor activity. *J. Mol. Biol.* **236,** 1141–1153.

45. Hulst, M. M., Himes G., Newbign, E., and Moormann, R. J. M. (1994). Glycoprotein E2 of classical swine fever virus; expression in insect cells and identification as a ribonuclease. *Virology* **200,** 558–565.

46. Roozen, K. J., Fenwick, R. G., Jr., and Curtiss, R. (1971) Synthesis of ribonucleic acid and protein in plasmid-containing minicells of *Escherichia coli* K-12. *J. Bacteriol.* **107,** 21–33.

47. Goff, S. A., Casson, L. P., and Goldberg, A. L. (1984) Heat shock regulatory gene htpR influences rates of protein degradation and expression of the lon gene in *Escherichia coli. Proc. Natl. Acad. Sci. USA* **81,** 6647–6651.

48. Rink, H., Liersch, M., Sieber, P., and Meyer, F. (1984) A large fragment approach to DNA synthesis: total synthesis of a gene for the protease inhibitor eglin c from the leech Hirudo medicinalis and its expression in E. coli. *Nucleic Acids Res.* **12,** 6369–6387.

49. Nambiar, K. P., Stackhouse, J., Presnell, S. R., and Benner, S. A. (1987). Expression of bovine pancreatic ribonuclease A in *Escherichia coli. Eur. J. Biochem.* **163,** 67–71.

50. Preuss, K.-D., Wagner, S., Freudenstein, J., and Scheit, K. H. (1990) Cloning of cDNA encoding the complete precursor for bovine seminal ribonuclease. *Nucleic Acids Res.* **18,** 1057.

51. Boix, E., Nogués, M. V., Schein, C. H., Benner, S. A., and Cuchillo, C. M. (1994) Reverse transphosphorylation by ribonuclease A needs an intact p2 binding site. Point mutations at Lys-7 and Arg-10 alter the catalytic properties of the enzyme. *J. Biol. Chem.* **269,** 2529–2534.

52. Schein, C. H., Boix, E., Haugg, M., Holliger, K. P., Hemmi, S., Frank, G., and Schwalbe, H. (1992) Secretion of mammalian ribonucleases from *E. coli* using the signal sequence of murine spleen ribonuclease. *Biochem. J.* **283,** 137–144.

53. Schein, C. H. and Noteborn, M. H. M. (1988) Production of soluble recombinant mammalian proteins in Escherichia coli is favored by lower growth temperature. *Bio/Technology* **6,** 291–294.

54. Schein, C. H. (1989) Production of soluble recombinant proteins in bacteria. *Bio/Technology* **7,** 1141–1149.

55. Schein, C. H. (1999) Protein expression, soluble, in *The Encyclopedia of Bioprocess Technology: Fermentation, Biocatalysis and Bioseparation* (Flickinger, M. C. and Drew, S. W., eds.), Wiley, New York, pp. 2156–2169.

56. Schein, C. H. (1990) Solubility as a function of protein structure and solvent components. *Bio/Technology* **8,** 308–317.

57. Rosenberg, A. H., Lade, B. N. L., Chui, D. S., Dunn, J. J., and Studier, F. W. (1987) Vectors for selective expression of cloned DNAs by T7 RNA polymerase. *Gene* **56,** 125–135.

58. Picard, D. and Schaffner, W. (1983) Correct transcription of a cloned mouse immunoglobulin gene in vivo. *Proc. Natl. Acad. Sci. USA* **80,** 417–421.

# 10

# Assays for the Evaluation of HIV-1 Integrase Inhibitors

## Zeger Debyser, Peter Cherepanov, Wim Pluymers, and Erik De Clercq

## 1. Introduction

Integration is an essential step in the replication cycle of a retrovirus, such as the human immunodeficiency virus type 1 (HIV-1) (for a review, *see* **ref.** *1*). After reverse transcription of the RNA genome, the DNA copy is transported into the nucleus and integrated in the host chromosome. The only viral enzyme required for HIV-1 integration is integrase (IN), a protein of 32 kDa encoded by the 3'-end of the *pol* gene. The enzyme is produced by protease-mediated cleavage of the gag-pol precursor during virion maturation. Integrase can be considered a site-specific endonuclease; the removal of two nucleotides creates a hydroxyl residue at each 3'-end of the viral DNA that will carry out a nucleophilic attack on the phosphodiester backbone of the host DNA.

Integrase recognizes specific sequences in the long-terminal repeat (LTR) elements that flank the viral cDNA. The terminal 15 bp of the LTR are necessary and sufficient for site-specific cleavage and integration. Although the recognition sites are virus-specific, a CA dinucleotide immediately upstream of the cleavage site is conserved in all retroviral LTRs. In the first step of the integration reaction, termed 3'-end processing, two nucleotides are removed from each 3'-end to produce new 3'-hydroxyl ends (CA-3'OH). This reaction occurs in the cytoplasm, within a nucleoprotein complex, referred to as the preintegration complex (PIC). After entering the nucleus, the processed viral dsDNA is joined to host-target DNA. The joining reaction involves a coordinated 4–6 bp staggered cleavage of the host DNA and the ligation of the processed CA-3'OH viral DNA ends to the 5' phosphate ends of the target DNA. Repair of the remaining gaps, although not understood at this time, is probably accomplished by host–cell DNA repair enzymes. Staggered strand transfer and gap repair result in the duplication of host-cell sequences immediately flanking the inserted provirus. The duplication size is virus-specific, and in the case of HIV-1, consists of a 5 bp direct repeat.

Purified IN also catalyzes the reverse-strand transfer reaction (termed disintegration) when provided with specific gapped and branched DNA substrates *(2)*. Concerted integration of both HIV LTR ends has not been reproduced in vitro with recombinant integrase. However, using cytoplasmatic extracts of acutely HIV-1 infected cells that contain PICs in vitro concerted integration has been accomplished *(3)*.

From: *Methods in Molecular Biology, vol. 160: Nuclease Methods and Protocols*
Edited by: C. H. Schein © Humana Press Inc., Totowa, NJ

The HIV IN protein is composed of three functional domains. The N-terminal region (residues 1–50) is characterized by an "HHCC" zinc finger-like motif. The central region (residues 60–160) is characterized by three highly conserved amino-acid residues D, D (35) E, and represents the catalytic domain. The central core domain is also responsible for the disintegration reaction. The carboxyl-terminal domain (200–270) is involved in nonspecific DNA binding. In vitro complementation studies have established the existence of functional subdomains in an oligomeric complex *(4,5)*.

This chapter focuses on the experimental procedures used to analyze inhibitors of HIV-1 integrase. Various molecules have been reported to interfere with HIV-1 integrase activity in vitro (For a review, *see* **ref. 6**). With some modifications (in the sequence of the specific oligonucleotides used) the assays can be performed with other retroviral integrases, as has been described for avian myeloblastosis virus (AMV) *(7)*, avian sarcoma virus (ASV) *(8)*, Moloney murine leukemia virus (MoMLV) *(9)*, human immunodeficiency virus type 2 (HIV-2) *(10)*, feline immunodeficiency virus (FIV) *(11,12)*, caprine arthritis encephalitis virus (CAEV), and visna virus *(13)*.

## 2. Materials

### 2.1. Equipment

1. The chromatography media used are from Pharmacia Biotech (HT-heparin columns and Sephadex G-25), Qiagen (Ni-NTA resin and columns) (Hilden, Germany) and Bio-Rad (Bio-Gel A5m) (Hercules, CA).
2. The X-press was from LKB.
3. The ultrafiltration media (Centricon-10) are obtained from Amicon (Beverly, MA).
4. Polyethylene glycol-8000 (PEG-8000) is from Sigma (St. Louis, MO), 3-{[3-cholamido-propyl]dimethylammonio}-1-propane-sulfonate (CHAPS) is from Aldrich (Bornem, Belgium), imidazole from Sigma.
5. Binding analysis is done using a BIAcore2000 (Biacore, Uppsala, Sweden) and sensor chips SA™ that contain preimmobilized streptavidin.
6. For microtiter-plate integration assays, use NucleoLink polystyrene microtiter plates (NUNC, Roskilde, Denmark), EDC {0.2 *M* 1-ethyl-3-[3-dimethylaminopropyl]-carbodiimide} (Acros, New Jersey) and 1-methylimidazole (Sigma), alkaline phosphatase-conjugated avidin (Dako, Glostrup, Denmark), and alkaline phosphatase substrate (Sigma FAST para-nitrophenylphosphate (pNPP) substrate tablet setN-2770).

### 2.2. Reagents

1. Bacterial lysis buffer: 10 m*M* Tris-HCl, pH 7.5, 150 m*M* NaCl, 1 m*M* β-mercaptoethanol, and 0.05 m*M* EDTA
2. Integrase-solubilization buffer: 10 m*M* Tris-HCl, pH 7.5, 1 *M* NaCl, 1 m*M* β-mercapto-ethanol, 0.05 m*M* EDTA, and 10 m*M* CHAPS (*see* **Note 1**).
3. CMTG buffer: 7.5 m*M* CHAPS, 1 m*M* β-mercaptoethanol, 10 m*M* Tris-HCl, pH 7.6, and 10% glycerol.
4. Integrase-storage buffer: CMTG buffer + 700 m*M* NaCl.
5. Integrase-binding buffer B: 20 m*M* HEPES, pH 7.5, 10 m*M* MgCl$_2$, and 5 m*M* DTT.
6. Formamide dye solution: 80% formamide, 0.2% SDS, 10 m*M* EDTA (pH 8.0), 1 mg/mL xylene cyanol FF, 1 mg/mL bromophenol blue (*see* **Note 2**).
7. Integrase-dilution buffer: 10 m*M* Tris-HCl, pH 7.5, 750 m*M* NaCl, 10% glycerol, 1 m*M* β-mercaptoethanol.

8. Microtiter plate assay blocking buffer: phosphate-buffered saline (PBS) (made from dissolved tablets (Oxoid, Basingstoke, UK), 1% bovine serum abumin (BSA), fraction, 98% pure (Sigma), and 0.2% sodium azide.
9. Alkaline phosphatase buffer: 1 $M$ Tris-HCl, pH 9.5, 1 $M$ NaCl, 50 m$M$ MgCl$_2$.
10. PIC buffer K: 20 m$M$ HEPES, pH 7.4, 150 m$M$ KCl, 5 m$M$ MgCl$_2$, 1 m$M$ dithiothreitol (DTT), 25 m$M$ aprotinin (Sigma).

## 2.3. Integrase

### 2.3.1. Sources

#### 2.3.1.1. EXPRESSION VECTORS FOR ESCHERICHIA COLI

1. The following expression constructs are available through the AIDS Research and Reference Reagent Program (http://www.aidsreagent.org) (*see* **Note 3**):
    Plasmids contributed by Dr. R. Craigie (NIH):
        pINSD: HIV-1 integrase
        pINSD.His: integrase with amino-terminal polyhistidine tag
        pINSD.His.Sol: *idem*, but with two amino-acid substitutions (F185K and C280S) that improve solubility of the protein.
    Plasmids contributed by Drs. J. M. Groarke, J. V. Hughes, and F. J. Dutko:
        pLJS10: HIV-1 integrase
2. In our laboratory, we use the plasmid pRP1012 in PC1 cells for IN expression. The plasmid pRP1012 (from Dr. R. H. A. Plasterk, Netherlands Cancer Institute, Amsterdam), encodes HIV-1 IN under control of the T7 promoter. The plasmid (a derivative of pET15b) also codes for a stretch of six histidine residues ("His tag") linked to the N-terminus of the native integrase polypeptide. The *E. coli* strain PC1 lacks EndoI, a nonspecific endonuclease for duplex DNA *(14)*. This strain was constructed in our laboratory by P1-phage-mediated transduction of the $\Delta endA$::Tc$^R$ mutation from BT333 *(15)* into the *E. coli* B strain BL21(DE3)(pLysS). Integrase preparations purified from extracts derived from this strain contain less nonspecific nuclease activity.

#### 2.3.1.2. PURIFIED RECOMBINANT INTEGRASE

Recombinant integrase expressed from pINSD.His.Sol, HIV-1$_{NL4-3}$ (cat. no. 2959) can be obtained through the AIDS Research and Reference Reagent Program.

#### 2.3.1.3. VIRION-DERIVED INTEGRASE

Nonionic detergent lysates of purified and concentrated HIV-1 virions can be used as an alternative enzyme source *(16)*.

### 2.3.2. Expression and Purification of Recombinant HIV-1 Integrase

Integrase is produced in the *E. coli* strain BL21(DE3)(pLysS) or PC1 from pRP1012 after induction with isopropyl-β-D-thiogalactopyranoside (IPTG).

1. Grow the cells in 1.5 L of LB medium to an A (OD$_{600}$) of 0.85 in a shaker flask at 300 rpm. Add IPTG to a final concentration of 0.4 m$M$ and grow the cells for an additional 3 h.
2. Resuspend the bacteria in 25 mL of cold lysis buffer (10 m$M$ Tris-HCl, pH 7.5, 150 m$M$ NaCl, 1 m$M$ β-mercaptoethanol, and 0.05 m$M$ EDTA) and rupture cells by five passages through an X-press.
3. Wash the insoluble fraction, which contains integrase, twice with 25 mL of lysis buffer, and extract with 25 mL of solubilization buffer supplemented with 10 m$M$ CHAPS using a Dounce homogenizer. Stir the suspension for 30 min at 4°C.

4. Centrifuge the extract at 30,000*g* for 25 min and add supernatant to 4 mL of Ni-NTA (nitrilotriacetic acid) resin preequilibrated with the solubilization buffer in the presence of 30 m*M* imidazole. Stir the slurry for 30 min at 4°C and pour into an empty disposable polypropylene column. Wash the column extensively (60–100 mL) with CMTG buffer plus 60 m*M* imidazole and 1 *M* NaCl, and elute the protein with 200 m*M* imidazole.

5. Dilute the eluate immediately with salt-free CMTG buffer to a NaCl concentration of 300 m*M* and load onto a 5 mL HT-heparin column. Wash the column with 5 vol of CMTG containing 300 m*M* NaCl, and elute integrase with a linear gradient of NaCl. The protein elutes at approx 700 m*M* NaCl.

6. Collect the fractions containing integrase, concentrate by ultrafiltration using Centricon-10, and store frozen at –70°C. The protein concentration should not exceed 1 mg/mL to avoid precipitation. The protein obtained this way is >95% pure, as determined by SDS-PAGE and silver-staining. Activity is tested in oligonucleotide- based assays (*see* **Subheading 3.1.2.**).

## 2.4. Oligonucleotides/DNA Templates

### 2.4.1. Substrate and Target DNA for Oligonucleotide Assay

The following HPLC-purified oligodeoxynucleotides (Pharmacia Biotech) are used (*see* **Note 4**):

INT1: 5'-TGTGGAAAATCTCTAGCAGT
INT2: 5'-ACTGCTAGAGATTTTCCACA
T35: 5'-ACTATACCAGACAATAATTGTCTG GCCTGTACCGT
SK70: 5'-ACGGTACAGGCCAGACAATTATTGTCTGGTATAGT

For the DNA-binding assay, first couple the 3'-biotinylated 39-mer 5'-ACTGCT AGAGATTTTCCACACTGACTAAAAGGGTCAAAA-3' to the sensor chip. Then, the complementary 35-mer 5'-GACCCTTTTAGTCAGTGTGGAAAATCTCTA GCAGT-3' can be annealed to the first oligonucleotide.

For the disintegration assay, the following oligonucleotides are required:

T1 (16-mer): 5'-CAGCAACGCAAGCTTG
T3 (30-mer): 5'-GTCGACCTGCAGCCCAAGCTTGCGTTGCTG
V2 (21-mer): 5'-ACTGCTAGAGATTTTCCACAT

V1/T2 (33-mer): 5'-ATGTGGAAAATCTCTAGCAGGCTGCAGGTCGAC

For the microtiter-plate assay, INT3/INT4 substrate is used. INT3 is 5'-phosphorylated and contains a three-base overhang facilitating carbodiimide-mediated condensation to the plates. INT5 and INT6 (target DNA) are biotinylated at the 3'-end (-b).

INT3 (38-mer): 5'-P-TCAGACCCTTTTAGTCAGTGTGGAAAATCTCTAGCAGT
INT4 (35-mer): ACTGCTAGAGATTTTCCACACTGACTAAAAGGGTC
INT5 (20-mer): TGACCAAGGGCTAATTCACT-b
INT6 (20-mer): AGTGAATTAGCCCTTGGTCA-b

### 2.4.2. Mini-HIV DNA Substrate

This DNA substrate is referred to as "mini-HIV" in analogy with previously reported integrase DNA substrates flanked by the enzyme-recognition sequences from both LTR ends *(17)*. Our mini-HIV DNA is prepared by linearization of the plasmid pU3U5 with *Sca*I. This plasmid contains the HIV-1 LTR U3 and RU5 terminal sequences joined to produce a unique *Sca*I site and cloned between *Eco*RI and *Kpn*I sites of the multiple cloning site of pBKRSV *(18)*. The plasmid pU3U5 can be obtained from us upon request.

## 2.5. Cells and Virus

Cells persistently infected with HIV-1 are cocultured with susceptible cells prior to the isolation of preintegration complexes from the cytoplasm. Molt/IIIB (which can be obtained from our Laboratory) is used as a persistently infected cell line, whereas Sup-T1 cells (AIDS Research and Reference Reagent Program no. 100) are used as target cells.

## 3. Methods

### 3.1. Oligonucleotide-Based Assays (Fig. 1)

Simple assays have been designed to study the biochemical characteristics of integrase and to search for inhibitors of the reactions *(17,19–21)*. These assays are based on oligonucleotide substrates that correspond to the LTR termini, and are used to monitor binding of integrase to the DNA substrate, 3'-processing, and DNA strand transfer. Modified substrates are used to detect the disintegration activity. The enzymatic test systems have been adapted for high throughput screening of integrase inhibitors *(22,23)*.

### 3.1.1. Substrate and Target DNA

The substrate DNA is a double-stranded DNA oligonucleotide that mimics the 3'- or 5'-HIV LTR terminus. Any double-stranded DNA molecule can be used as target DNA. If target DNA is not in excess, integration will occur into the substrate DNA. The oligonucleotides INT1 and INT2 correspond to the U5 end of the HIV-1 long-terminal repeat (LTR), since integrase activity in vitro is higher with 5'-LTR substrates.

1. Purify the oligonucleotide INT1 by electrophoresis through a 20% polyacrylamide denaturing gel (*see* **Note 4**). Extract the gel slice using the "crush and soak" method *(24)*.
2. Label the 5'-end of 50 pmol INT1 using polynucleotide T4 kinase and $\{\gamma\text{-}^{32}P\}$ATP (50 μL). Heat-inactivate the enzyme for 10 min at 80°C. Add NaCl to 100 m*M* and 1 μL containing 50 pmol INT2. Anneal the oligonucleotides: heat an equimolar mixture (5 μ*M*) of both oligonucleotides in the presence of 100 m*M* NaCl shortly at 95°C, and allow to cool slowly to room temperature. Dilute fivefold in water (final concentration 1 μ*M*; use 0.3 μL per assay).
3. Annealing of SK70 and T35 results in a 35-bp double-stranded (ds) DNA molecule that can be used as a target DNA molecule (T35/SK70).

### 3.1.2. Integration Assay (*Fig. 1*)

There are two distinct catalytic steps during retroviral integration. The viral dsDNA ends are processed to remove two nucleotides from the 3'-ends. Subsequent DNA strand-transfer leads to the integration of the processed viral DNA ends into target DNA. In this assay, integrase reaction products are analyzed using denaturing polyacrylamide gels. The first step results in the formation of a –2 band, whereas strand-transfer produces a DNA ladder of variable length. The effect of the addition of various concentrations of an integrase inhibitor to the reaction is shown in **Fig. 1**. A concentration-dependent inhibition of both the 3'-processing and the formation of strand-transfer products is observed.

## A   Principle

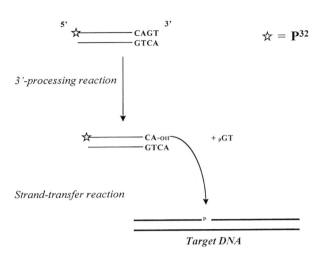

*3'-processing reaction*

☆ = P³²

*Strand-transfer reaction*

+ ₚGT

*Target DNA*

*Product of integration reaction*

*Separate on a denaturing polyacrylamide gel*

## B   Example

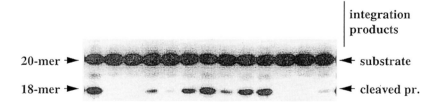

inhibitor

1. Make the master mix (8 μL per sample). The final reaction mixture for the 3'-processing assay contains 20 m$M$ HEPES, pH 7.5, 5 m$M$ DTT, 10 m$M$ MgCl$_2$, 75 m$M$ NaCl, 15% (v/v) dimethylsulfoxide (DMSO), 5% (v/v) PEG-8000, and 0.3 pmole (30 n$M$) of the oligonucleotide substrate INT1/INT2.
2. Add 1 μL of inhibitor dissolved in water or DMSO.
3. Start the reactions by the addition of 1 μL of His-tag integrase (final concentration 330 n$M$).
4. Allow the reactions to proceed at 37°C for 40 min and stop by the addition of formamide dye solution. Separate the reaction products on a 17% polyacrylamide/urea gel. Perform autoradiography by exposing the wet gel between the glass plate and plastic bag to X-ray film (CURIX RP1 from Agfa).
5. For quantification of the enzymatic activity, the PhosphoImager from Molecular Dynamics is used. In the presence of an active integrase inhibitor, a concentration-dependent decrease of the –2 band (cleavage reaction) and the upper bands (strand transfer reaction) is observed.

### 3.1.3. Kinetic 3'-Processing Assay (*Fig. 2*)

In the integration reaction (**Subheading 3.1.2.**) both compounds that interfere with DNA binding or 3'-processing will result in a decrease of the –2 band. To distinguish between both mode of actions, we established a kinetic cleavage assay *(25)*. Reaction conditions are as in **Subheading 3.1.2.**, except for the modifications listed below.

1. Incubate integrase with the specific oligonucleotide substrate (INT1/INT2) for 3 min at 37°C before adding dextran sulfate (DS 5000) (5 μ$M$) to a final concentration of 0.3 μ$M$. As a polyanion, dextran sulfate prevents the formation of new DNA-integrase complexes by trapping free integrase. When present before integrase, 0.3 μ$M$ of DS5000 inhibits the reaction completely (**Fig. 2B**, lane C2).
2. To measure the cleavage reaction directly, take 5-μL aliquots after different time intervals (e.g., 0.5, 2, 4, 6, 8, ... min) and denature in an equal volume of formamide loading buffer. Separation and quantification of the –2 band is as described in **Subheading 3.1.2.** The cleavage process is slow and takes minutes to proceed (**Fig. 2B**).
3. The curve fitting and the calculation of k$_{cat}$ values (slope) can be performed using the SigmaPlot 5.0 curve fitter and the equation: $A - B \times e^{-kt}$.
4. In order to examine the inhibitory effect of compounds on the k$_{cat}$ of 3'-processing, the same experiment can be carried out in the presence of inhibitor.

    We determined the rate constant for the formation of cleaved product by HIV-1 integrase to be 0.08 ± 0.01 min$^{-1}$. This is in a good agreement with the reported k$_{cat}$ values for the HIV-1 (0.069 min$^{-1}$) and RSV (0.18 min$^{-1}$) integrases.

### 3.1.4. DNA-Strand Transfer Assay (*Fig. 3*)

The composition of the reaction mixture is identical to that used for the integration assay (**Subheading 3.1.2.**).

1. Preincubate 20 n$M$-labeled DNA substrate INT1/INT2 with 330 n$M$ HIV-1 integrase at 37°C for 3 min, to allow the cleavage reaction to occur.

---

Fig. 1. *(opposite page)* Integration assay. Autoradiogram of denaturing PAGE separating integrase reaction products (strand transfer products and cleaved product (–2 band). Lane 1 is a control reaction. In the other lanes, decreasing concentrations of investigational integrase inhibitors were added.

## A  Principle

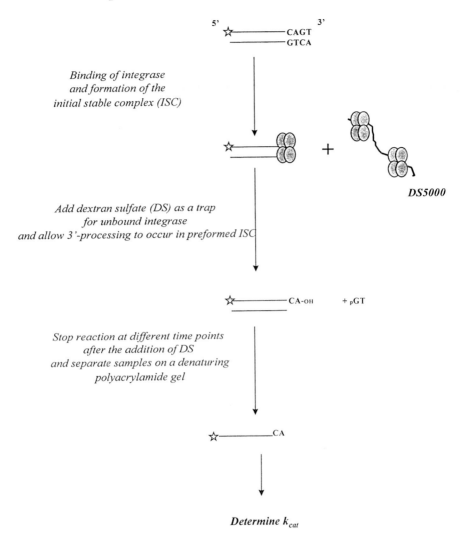

Fig. 2. Kinetic 3'-processing reaction. The principle is explained in (**A**). An example in the absence and presence of an inhibitor is shown in (**B**). IN is a positive control lane, with no dextran sulfate (DS) added. In C1, no integrase was added, and in C2, DS was added before the start of the reaction.

2. After 3 min, add 1 μL of excess target DNA T35/SK70 (final concentration 133 n*M*) with or without inhibitor, and incubate the samples at 37°C for 1 h. The excess target DNA blocks further binding of integrase to the viral DNA substrate.
3. Reactions are stopped and analyzed as described in **Subheading 3.1.2.**
4. Strand transfer products (upper DNA ladder) are quantified by phosphoimaging.

Note that inhibition of DNA-strand transfer is often evaluated in overall-integration assays (as explained in **Subheading 3.1.2.**) by quantifying the bands longer than the

## B

## Example of a kinetic 3'-processing assay

Reaction scheme :  $E + S \iff ES \xrightarrow{k_{cat}} P + E$

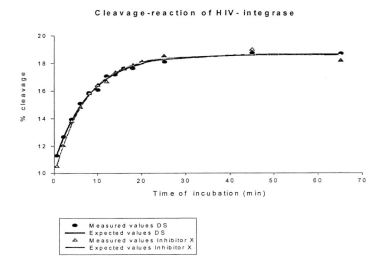

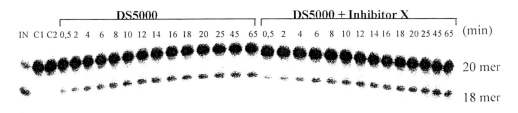

Fig. 2B.

initial substrate. However, since DNA-strand transfer depends on prior DNA binding and 3'-cleavage, the inhibition of strand transfer observed in this reaction cannot be determined independently. By confining strand-transfer to preformed integrase-DNA complexes, inhibition of this enzymatic step can be verified independently from prior DNA binding and 3'-cleavage.

### *3.1.5. Disintegration Reaction (**Fig. 4**)*

Recombinant HIV integrase can catalyze disintegration when using an oligonucleotide substrate that mimics the reaction intermediate formed by the initial cleavage-joining reaction *(2)*. Reaction conditions are as before.

## A    Principle

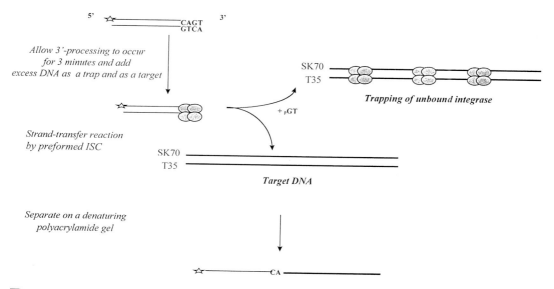

*Allow 3'-processing to occur for 3 minutes and add excess DNA as a trap and as a target*

*Strand-transfer reaction by preformed ISC*

*Trapping of unbound integrase*

*Target DNA*

*Separate on a denaturing polyacrylamide gel*

## B    Example

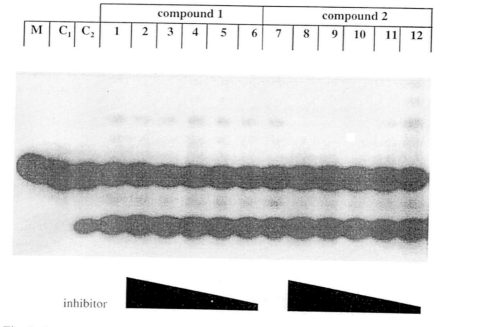

inhibitor

Fig. 3. Strand transfer reaction. The principle is explained in (**A**). An example in the absence (Lane 7) and presence of two inhibitors at varying concentrations (Lanes 1–6 and lanes 8–12) is shown in (**B**). In lane M, no enzyme was added; in lane C1, DS was added prior to the reaction; lane C2 shows the reaction product after the initial incubation.

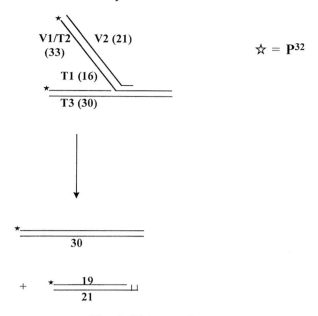

Fig. 4. Disintegration assay.

1. Label the 5'-end of T1 or V1/T2 with $^{32}$P.
2. Construct the Y-shaped substrate by annealing four oligonucleotides: T1, 16-mer; T3, 30-mer; V2, 21-mer; and the hybrid strand, V1/T2, 3-mer.
3. Production of a labeled fragment of 30 bp in the former case (labeling of T1) and of 19 bp in the latter case (labeling of V1/T2) can be detected after separation in a 15% native polyacrylamide gel using TBE buffer.

### 3.1.6. Integration Assays in Microtiter-Plate Format (*Fig. 5*)

High-throughput, nonradioactive microtiter-plate oligonucleotide-based assays have been designed for large-scale screening of integrase inhibitors. In one test, viral DNA containing a 5'-fluorescein isothiocyanate (FITC) group is coupled by IN to another oligonucleotide with a biotin group. Subsequently, the biotinylated oligonucleotides are bound to streptavidin-coated microtiter plates. FITC is detected using monoclonal anti-FITC alkaline phosphatase conjugate and chemiluminescence.

In our laboratory we use a modified version of the method developed by Hazuda et al. (*22*):

1. Immobilize double-stranded oligonucleotides INT3/INT4 onto NucleoLink polystyrene microtiter plates (NUNC) via a 5'-*O*-phosphoramidate function. In the condensation reaction INT3/INT4 (75 μL of 0.067 μM in 10 mM 1-methylimidazole, pH 7.0) reacts with 25 μL freshly prepared EDC (0.2 M 1-ethyl-3-(3-dimethylaminopropyl)-carbodiimide in 10 mM 1-methylimidazole, pH 7.0). Incubate at 50°C for 5 h. Remove the reaction medium and store plates at 4°C in blocking buffer (1% BSA and 0.2% sodium azide in PBS).
2. Plates can be stored for several months. Wash cups before use, once with 100 μL 5X SSC (sodium chloride/sodium citrate, *see* **ref. 24**), 0.25% SDS at 50°C and twice with 100 μL 75 mM NaCl, 20 mM Tris-HCl, pH 7.8.
3. Prepare the following reaction mixture (100 μL): 20 mM Tris-HCl, pH 7.8, 25 mM NaCl, 5 mM MnCl$_2$, 5 mM DTT, 10% DMSO, and 50 μg/mL BSA. Note that the reaction condi-

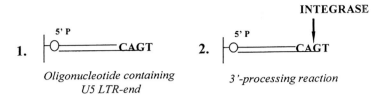

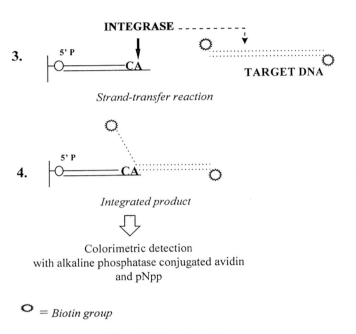

Fig. 5. Microtiter-plate assay.

tions are similar to those of integration reactions in solution, except for the presence of $Mn^{2+}$ instead of $Mg^{2+}$.

4. Add integrase at a final concentration of 300 n$M$.
5. After 30 min at 37°C, add 5 μL of 100 n$M$ biotinylated oligonucleotide INT5/INT6 as target DNA. Allow the strand transfer to proceed for 1 h at 37°C.
6. Wash the wells three times with 200 μL of PBS containing 0.05% Tween-20, and twice with PBS. Block plates with 1% BSA in PBS for 20 min.
7. Incubate with 100 μL of alkaline phosphatase-conjugated avidin (Dako, Glostrup, Denmark) for 30 min under gentle agitation. Wash once with 200 μL PBS, and twice with 200 μL alkaline phosphatase buffer. Add 200 μL of alkaline phosphatase substrate to each well (Sigma FAST para-nitrophenylphosphate [pNPP] substrate tablet). Incubate the plate in the dark for 30 min at 37°C, and determine the amount of *p*-nitrophenolate in solution spectrophotometrically at 405 nm.

### 3.1.7. Binding of HIV-1 Integrase to DNA (*Fig. 6*)

Several assays have been reported: a nitrocellulose filter binding assay *(26)*, UV crosslinking *(27)*, and a biosensor assay *(25)*. We describe the latter assay in more detail.

## A Principle

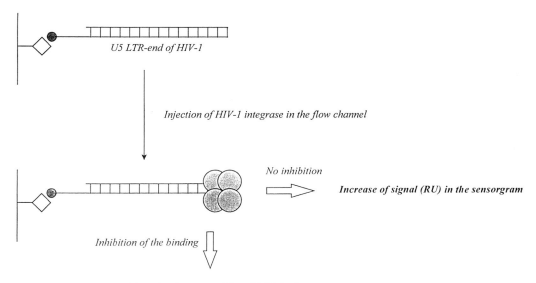

*U5 LTR-end of HIV-1*

*Injection of HIV-1 integrase in the flow channel*

*No inhibition*

*Increase of signal (RU) in the sensorgram*

*Inhibition of the binding*

*No or less increase of signal (RU) in the sensorgram*

## B Example of DNA binding inhibition in a sensorgram

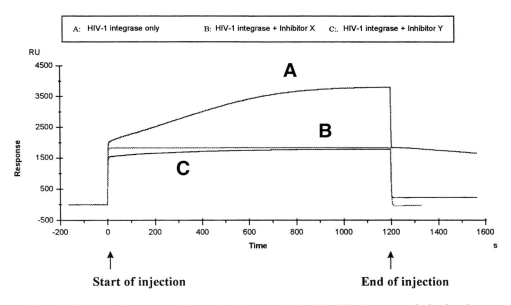

A: HIV-1 integrase only   B: HIV-1 integrase + Inhibitor X   C: HIV-1 integrase + Inhibitor Y

Start of injection

End of injection

Fig. 6. DNA-binding assay. The principle is explained in (**A**). An example in the absence and presence of two inhibitors is shown in (**B**). The response is expressed in resonance units (RU).

Binding experiments are carried out using biosensor technology on BIAcore2000 (Biacore) and sensor chips SA™ that contain preimmobilized streptavidin. The binding experiment is carried out at 37°C using binding buffer B (20 m*M* HEPES, pH 7.5, 10 m*M* MgCl$_2$, and 5 m*M* DTT).

1. Inject 100 µL of the 3'-biotinylated oligonucleotide 5'-ACTGCTAGAGATTTTCCACAC TGACTAAAAGGGTCAAAA (0.1 µg/mL) at a rate of 10 µL/min in channel 2.
2. Inject 100 µL of the complementary 35-mer 5'-GACCCTTTTAGTCAGTGTGGA AAATCTCTAGCAGT (0.5 µg/mL) in 150 m*M* NaCl at a rate of 10 µL/min in channel 2. Annealing will result in an integrase recognition sequence at the free end of the double-stranded oligonucleotide.
3. Inject 100 µL of HIV-1 integrase (33 µ*M*), diluted 10-fold in integrase dilution buffer, and another 10-fold in buffer B to a final concentration of 330 n*M*, at a flow rate of 5 µL/min in channel 1 (blank) and channel 2.
4. Varying concentrations of inhibitors can be added to integrase to quantify their effect. Dextran sulfate 5000 can be used as a control inhibitor of DNA binding. Coupling of a biotinylated oligonucleotide to a streptavidin-coated chip, followed by annealing of the complementary strand, works very well. This type of assay is highly suitable for analyzing DNA binding of proteins in general; for example, HIV-1 reverse transcriptase. Unfortunately, background binding of integrase to matrix in channel 1 and poor solubility of native integrase may interfere with quantitative data analysis. In a semiquantitative way, the assay can be used for analyzing inhibitors of the integrase–DNA interaction.

### 3.1.8. Analysis of Integrase Inhibitors

Candidate inhibitors are first tested in the integration assay, in radioactive (**Subheading 3.1.2.**) or microtiter plate (**Subheading 3.1.6.**) format. If inhibition is observed, analyze the mode of inhibition by testing the compound in a DNA-binding assay (**Subheading 3.1.7.**), the kinetic 3'-processing reaction (**Subheading 3.1.3.**), and the DNA strand transfer assay (**Subheading 3.1.4.**).

## 3.2. Preintegration-Complex Assay

The preintegration-complex (PIC) assay has been developed by Farnet and Haseltine *(3)*. Integrase present in the preintegration complexes (PICs), purified from the cytoplasm of acutely infected cells, catalyzes integration of viral retrotranscribed DNA into a heterologous target DNA. Integration is analyzed on Southern blots using HIV-specific probes. The assay is cumbersome in our hands, and requires optimal cell culture conditions and Molt/IIIB cells that produce sufficiently high titers of infectious virus particles.

1. Initiate infection by cocultivating human SupT1 cells ($2 \times 10^6$ cells/mL) with persistently infected Molt/IIIB cells at a ratio of 50:1. Pretreat Molt/IIIB cells for 24 h with 10 ng/mL phorbol myristate acetate.
2. Prepare cytoplasmatic extracts 4–6 h after infection. Wash cells twice with buffer K and lyse for 10 min on ice in the same buffer, but with 0.025% digitonin. Cell concentration during lysis is $33 \times 10^6$ cells/mL.
3. Centrifuge the lysate at 1000*g* for 3 min and discard the pellet containing intact nuclei.
4. Centrifuge at 8000*g* for 3 min, to remove additional debris, collect the supernatant (referred to as "cytoplasmatic extract") and freeze at –70°C. The extracts can be further purified by precipitation in low-ionic-strength solution *(28)*, although the crude extract contains suffiiciently high integrase activity to be used directly.
5. We use as a target for in vitro integration. Add 450 ng of target DNA (e.g., linearized PhiX174 RF DNA) to 100 µL of extract and incubate at 37°C for 1 h. Stop the reactions by adding 5 µL of 200 m*M* EDTA.

6. Deproteinate viral DNA by adding proteinase K to a final concentration of 1 mg/mL and SDS to a final concentration of 0.5% directly to the reaction mixture and incubating for 1 h at 55°C.
7. Extract samples once with an equal volume of phenol:chloroform:isoamylalcohol (25:24:1), and once with an equal volume of chloroform.
8. Precipitate DNA with ethanol, dissolve in water, and treat with RNase A (20 µg/mL) for 1 h at 37°C.
9. Viral DNA can be detected after Southern blotting of native agarose gels with an HIV-1-specific probe *(29)*.
10. Inhibitors are evaluated by preincubating the viral extract with the compound at room temperature for 10 min before adding target DNA *(28)*.

## *3.3. Mini-HIV-Based Assays*

These assays make use of long DNA substrates that are flanked by fragments of the LTR ends, and can be used to monitor concerted integration in genetic *(30,31)* or biochemical assays *(18)*. Integration of a mini-HIV DNA carrying a selectable marker gene can be scored by packaging the integration products into lambda phage heads in vitro and infecting a suitable *E. coli* strain. This assay is very sensitive, and can detect the low frequency of concerted integration events occurring in vitro with recombinant HIV-1 integrase *(30)*. The reaction products of the biochemical assay can be visualized by ethidium bromide staining after electrophoresis in agarose gels. The assay has been instrumental in obtaining electron-microscopic visualization of integrase-DNA complexes, and can also be used to analyze inhibitors of HIV-1 integration.

## 4. Notes

1. Since HIV-1 integrase is not highly soluble, use of a detergent (e.g., 7.5–10 m*M* CHAPS) is essential to prevent aggregation. No inhibition of integrase activity is observed at a final concentration of CHAPS up to 20 m*M*.
2. To prevent protein–DNA precipitation in reaction tubes after integrase reactions, include 0.2% SDS in formamide loading buffer and place samples in a boiling water bath for 5 min.
3. It is essential to use the *E. coli* strain BL21(DE3)(pLysS) Novagen (Madison, WI) BL21(DE3) strain for IN expression. Because the plasmid may be unstable in this host, backup stocks should therefore be maintained in HB101. The plasmid pLysS expresses low levels of T7 lysozyme, which inhibits the basal activity of T7 RNA polymerase before IPTG induction. In the absence of pLysS, the T7 promoter-based expression constructs are unstable because of the toxicity of integrase, which causes a loss of the integrase expression vector. An alternative way to circumvent plasmid instability is to introduce the more stringent kanamycin resistance gene into the expression construct.
4. The oligonucleotides used in the integrase assays must be purified by PAGE. In our experience, in-house purification yields better results than custom purification.

## References

1. Brown, P. O. (1997) Integration, in *Retroviruses* (Coffin, J. M., Hughes, S. H., and Varmus, H. E., eds.), Cold Spring Harbor Laboratory, Cold Spring Harbor, NY, pp. 161–203.
2. Chow, S. A., Vincent, K. A., Ellison, V., and Brown, P. O. (1992) Reversal of integration and DNA splicing mediated by integrase of human immunodeficiency virus. *Science* **255,** 723–726.

3. Farnet, C. M. and Haseltine, W. A. (1990) Integration of human immunodeficiency virus type 1 DNA in vitro. *Proc. Natl. Acad. Sci. USA* **87,** 4164–4168.
4. Engelman, A., Bushman, F. D., and Craigie, R. (1993) Identification of discrete functional domains of HIV-1 integrase and their organization within an active multimeric complex. *EMBO J.* **12,** 3269–3275.
5. van Gent, D. C., Vink, C., Oude Groeneger, A. A. M., and Plasterk, R. H. A. (1993) Complementation between HIV integrase proteins mutated in different domains. *EMBO J.* **12,** 3261–3267.
6. Pommier, Y., Pilon, A. A., Bajaj, K., Mazumder, A., and Neamati, N. (1997) HIV-1 integrase as a target for antiviral drugs. *Antiviral Chem. Chemother.* **8,** 463–483.
7. Katzman, M., Katz, R. A., Skalka, A. M., and Leis, J. (1989) The avian retroviral integration protein cleaves the terminal sequences of linear viral DNA at the in vivo sites of integration. *J. Virol.* **63,** 5319–5327.
8. Katz, R. A., Merkel, G., Kulkosky, J., Leis, J., and Skalka, A. M. (1990) The avian retroviral IN protein is both necessary and sufficient for integrative recombination in vitro. *Cell* **63,** 87–95.
9. Craigie, R., Fujiwara, T., and Bushman, F. D. (1990) The IN protein of Moloney murine leukemia virus processes the viral DNA ends and accomplishes their integration in vitro. *Cell* **62,** 829–837.
10. van Gent, D. C., Oude Groeneger, A. A. M., and Plasterk, R. H. A. (1992) Mutational analysis of the integrase protein of human immunodeficiency virus type 2. *Proc. Natl. Acad. Sci. USA* **89,** 9598–9602.
11. Vink, C., van der Linden, K. H., and Plasterk, R. H. A. (1994) Activities of the feline immunodeficiency virus integrase protein produced in Escherichia coli. *J. Virol.* **68,** 1468–1474.
12. Shibagaki, Y., Holmes, M. L., Appa, R. S., and Chow, S. A. (1997) Characterization of feline immunodeficiency virus integrase and analysis of functional domains. *Virology* **230,** 1–10.
13. Störmann, K. D., Schlecht, M. C., and Pfaff, E. (1995) Comparative studies of bacterially expressed integrase proteins of caprine arthritis-encephalitis virus, maedi-visna virus and human immunodeficiency virus type 1. *J. Gen. Virol.* **76,** 1651–1663.
14. Wright, M. (1971) Mutants of Escherichia coli lacking endonucleaseI, ribonuclease I, or ribonuclease II. *J. Bacteriol.* **107,** 87–94.
15. Cherepanov, P. P. and Wackernagel, W. (1995) Gene disruption in Escherichia coli: TcR and KmR cassettes with the option of Flp-catalyzed excision of the antibiotic-resistance determinant. *Gene* **158,** 9–14.
16. Goodarzi, G., Im, G.-J., Brackmann, K., and Grandgenett, D. (1995) Concerted integration of retrovirus-like DNA by human immunodeficiency virus type 1 integrase. *J. Virol.* **69,** 6090–6097.
17. Bushman, F. D. and Craigie, R. (1991) Activities of human immunodeficiency virus (HIV) integration protein in vitro: specific cleavage and integration of HIV DNA. *Proc. Natl. Acad. Sci. USA* **88,** 1339–1343.
18. Cherepanov, P., Surratt, D., Toelen, J., Pluymers, W., Griffith, J., De Clercq, E., and Debyser, Z. (1999) Activity of recombinant HIV-1 integrase on mini-HIV DNA. *Nucleic Acids Res.* **27,** 2202–2210.
19. Craigie, R., Mizuuchi, K., Bushman, F. D., and Engelman, A. (1991) A rapid in vitro assay for HIV DNA integration. *Nucleic Acid. Res.* **19,** 2729–2734.
20. Sherman, P. A. and Fyfe, J. A. (1990) Human immunodeficiency virus integration protein expressed in Escherichia coli possesses selective DNA cleaving activity. *Proc. Natl. Acad. Sci. USA* **87,** 5119–5123.

21. van Gent, D. C., Elgersma, Y., Bolk, M. W. J., Vink, C., and Plasterk, R. H. A. (1991) DNA binding properties of the integrase proteins of human immunodeficiency viruses types 1 and 2. *Nucleic Acids Res.* **19,** 3821–3827.
22. Hazuda, D. J., Hastings, J. C., Wolfe, A. L., and Emini, E. A. (1994) A novel assay for the DNA strand-transfer reaction of HIV-1 integrase. *Nucleic Acids Res.* **22,** 1121,1122.
23. Vink, C., Banks, M., Bethell, R., and Plasterk, R. H. A. (1994) A high-throughput, non-radioactive microtiter plate assay for activity of the human immunodeficiency virus integrase protein. *Nucleic Acids Res.* **22,** 2176,2177.
24. Sambrook, J., Fritsch, E. F., and Maniatis, T. (1989) *Molecular Cloning,* 2nd ed. Cold Spring Harbor Laboratory, Cold Spring Harbor, NY.
25. Cherepanov, P., Esté, J. A., Rando, R. F., Ojwang, J. O., Reekmans, G., Steinfeld, R., David, G., De Clercq, E., and Debyser, Z. (1997) Mode of interaction of G-quartets with the integrase of human immunodeficiency virus type 1 (HIV-1) *Mol. Pharmcol.* **52,** 771–780.
26. Lipford, J. R., Worland, S. T., and Farnet, C. (1994) Nucleotide binding by the HIV-1 integrase protein in vitro. *J. AIDS* **7,** 1215–1223.
27. Yoshinaga, T., Kimura-Ohtani, Y., and Fujiwara, T. (1994) Detection and characterization of a functional complex of human immunodeficiency virus type 1 integrase and its DNA substrate by UV cross-linking. *J. Virol.* **68,** 5690–5697.
28. Farnet, C. M. and Bushman, F. D. (1996) Differential inhibition of HIV-1 preintegration complexes and purified integrase protein by small molecules. *Proc. Natl. Acad. Sci. USA* **93,** 9742–9747.
29. Ausubel, F. M., et al. (ed.) *Current Protocols in Molecular Biology.* Wiley, New York.
30. Bushman, F. D., Fujiwara, T., and Craigie, R. (1990) Retroviral DNA integration directed by HIV integration protein in vitro. *Science* **249,** 1555–1558.
31. Fitzgerald, M. L., Vora, A. C., Zeh, W. G., and Grandgenett, D. P. (1992) Concerted integration of viral DNA termini by purified avian myeloblastosis virus integrase. *J. Virol.* **66,** 6257–6263.

# 11

## The Direction of Ribonucleases H by Antisense Oligodeoxynucleotides

**Richard V. Giles and David M. Tidd**

## 1. Introduction

Ribonuclease H (RNase H), first observed in extracts of calf thymus, is an apparently ubiquitous enzyme activity *(1)*, which catalyzes the hydrolysis of RNA in RNA-DNA heteroduplexes *(2)*. The two main classes (Type I and II) of eukaryotic RNase H have different biochemical requirements for maximal activity. A nuclear localization has been proposed *(3,4)*, although different subnuclear fractions for the separate types have been suggested (*see* **ref. 5**). The proposed physiological roles, DNA duplication (class I), and RNA replication (class II), are both nuclear processes (*see* **refs. 5** and **6**).

RNase H will cleave RNA when the RNA is hybridized to complementary beta-anomeric phosphodiester (PO) and phosphorothiodiester (PS) oligodeoxynucleotides (ON) *(7,8)*, but not alpha-anomeric *(9)* or methylphosphonodiester (PC) *(8,10)* analogs. In fact, RNase H appears to play a pivotal role in mediating the antisense effects obtained with PO and PS ON. Wheat-germ extracts provide reliable "hybrid arrest of translation" because of high levels of RNase H *(11)*, whereas effects in rabbit reticulocyte lysate systems are unreliable as a result of low but variable levels of RNase H *(12)*. RNase H-competent ON microinjected into *Xenopus* oocytes, eggs, and embryos *(13,14)* or delivered into human cells in culture *(15,16)* exert their antisense effects through RNase H-mediated destruction of target mRNA. Degradation of the target mRNA reduces expression of the encoded protein, as it decays with its normal half-life *(17)*.

With the use of ON, which do not direct RNase H, antisense effects may also be obtained. For example, morpholino structures have been shown to specifically inhibit translation *(18–20)*, splicing *(20)*, and induce mis-splicing following intracellular delivery. The β-globin mis-splicing that occurs in thalassemic cells can be corrected with 2'-*O*-modified oligoribonucleotides *(21,22)*. However, such compounds have a limited range of target sites within the selected RNA: the splice donor/acceptor/branch sequences to inhibit splicing or the 5'-untranslated region to inhibit translation. Fully formed ribosomes can unwind very stable secondary structures *(23,24)*, rendering sequences within the coding region ineffective target sites for ON that cannot direct RNase H.

From: *Methods in Molecular Biology, vol. 160: Nuclease Methods and Protocols*
Edited by: C. H. Schein © Humana Press Inc., Totowa, NJ

In contrast, RNase H-mediated cleavage of any accessible site within the target mRNA sequence may be obtained with PO and PS ON. However, PO ON are rapidly degraded by the nucleases present in biological fluids *(25–28)* and the released nucleosides and nucleotides may induce sequence-specific nonantisense effects *(29)*. On the other hand, nuclease-resistant PS ON may induce sequence-specific and -nonspecific nonantisense effects resulting from interaction with proteins (*see* **refs. 30–33**). Nevertheless, to date only one antisense compound has obtained FDA approval *(34)* for use as a drug and is a PS structure: Vitravene, (Isis Pharmaceuticals) a treatment for CMV retinitis.

Our research has focused on chimeric structures. The terminal PC sections confer resistance to the exonucleases found in culture medium, cell extracts, and within living cells *(25,35)*, whereas the central PO region directs RNase H. Appropriately configured chimeric PC/PO ON direct RNase H cleavage of target mRNA with enhanced efficiency and specificity in vitro *(10,36,37)* and in living cells in culture *(15,16,35)*, relative to PO or PS congeners.

Highly specific and efficient antisense effectors, such as the chimeric PC/PO structure, may eventually allow for enhanced chemotherapies for a range of diseases. Antisense ON are efficiently excluded from cells *(38–40)*, limiting therapeutic utility because intracellular delivery must be mediated by permeabilization or lipofection *(41–44)*. Antisense ON may instead provide a target-finding mechanism to precisely identify protein(s) critical for maintaining a given disease state. Standard drug discovery routes may then be employed to identify small molecule inhibitors of these proteins that can accumulate in cells in pharmacologically relevant concentrations.

In this chapter, we describe techniques to synthesize and purify chimeric PC/PO and PC/PS ON. Biochemical assays to investigate the specificity and efficiency of ON at directing the activity of commercially available *Escherichia coli* RNase H and human enzymes present in cell extracts and crude lysates are described. Identification of efficient target sites for ON is critical for antisense strategies, and current computer-based RNA secondary-structure prediction algorithms do not efficiently predict regions of RNA accessible to hybridization with antisense ON. We describe how the in vitro assays may be exploited to provide an empirical method supported by computer calculation, which identifies such regions.

Simple addition of ON to the culture medium of leukemia cells does not effect delivery of the ON to intracellular compartments. A procedure (streptolysin O reversible permeabilization) for delivering ON into living leukemia cells is provided. Control ON for cell work should be selected with care (*see* **Note 1** and **refs. 30–33,45–47**). Reduced expression of target mRNA is often observed following delivery of RNase H-directing ON into living cells. Short mRNA species—of the correct size to be fragments of the parent mRNA produced through RNase H cleavage at the site of ON hybridization—may be detected on Northern blots *(15,16)*. Such data, however, falls short of providing definitive proof of RNase H-mediated antisense effects in intact cells. Therefore, The RL-PCR procedure is described, which selectively amplifies and positively identifies the 3' product of RNase H-cleavage reactions.

## 2. Materials

### 2.1. Oligodeoxynucleotide Synthesis

1. DNA synthesizer: Model 381A Synthesizer (Applied Biosystems, Warrington, Cheshire, UK).
2. Methylphosphonamidite synthons: 100 m*M* A, C, and T 5'-*O*-dimethoxytrityl-2'-deoxynucleoside-3'-*O*-methylphosphonamidites (Glen Research Corporation, UK supplier Cambio Ltd, Cambridge) in anhydrous acetonitrile, 100 m*M* G synthon in 1:1 (v/v) anhydrous acetonitrile and anhydrous dichloromethane (Aldrich Chemical Co., Gillingham, Dorset, UK).
3. Amino linker: 100 m*M* β-cyanoethyl-protected 5'-Amino-Modifier C6-TFA (Note 2, Glen Research) in anhydrous acetonitrile.
4. Sulfurizing reagent: Tetraethylthiuram disulfide in acetonitrile (Applied Biosystems).
5. Carboxyfluorescein reagent: 12 μmol of the *N*-hydroxysuccinimide ester of 5(6)-carboxyfluorescein (Molecular Probes Inc., UK supplier, Cambridge BioScience, Cambridge, UK) in 50 μL dimethylformamide.
6. ON purification media: C18 Sep-Pak solid-phase extraction cartridges (Waters Chromatography Division of Millipore, Watford, Hertfordshire, UK). NAP 10 gel-filtration column (Pharmacia Biotech, St. Albans, Hertfordshire, UK).
7. Reverse-phase HPLC media: Brownlee Aquapore RP-300 7 micron (Applied Biosystems) and Asahipak C8P-50 (UK supplier Prolabo/Rhone-Poulenc, Manchester) columns.
8. Anion exchange HPLC media: Partisil-10 SAX column (Whatman International Ltd., Maidstone, UK).
9. HPLC precolumn: Column dry-packed with silica precolumn gel (37–53 μ*M*, Whatman, UK).
10. Repelcote(VS), BDH/Merck (Lutterworth, Leicestershire, UK).
11. Deprotection mixture: 45:45:10 (v/v) acetonitrile/ethanol/concentrated $NH_4OH$.
12. TEAA: 100 m*M* triethylamine (refluxed for 4 h over KOH pellets, then redistilled to remove UV-absorbing impurites) in water, pH to 8.0 with acetic acid.
13. Py/TEAA: 100 m*M* triethylammonium acetate pH 8.0 containing 0.1% pyridine (v/v).
14. Carbonate buffer: 600 m*M* $NaHCO_3$, pH to 8.5 with NaOH.
15. Anion-exchange HPLC buffers: Weak: 20 m*M* $KH_2PO_4$ (99+% A.C.S reagent grade, Aldrich), 50% formamide (v/v, spectrophotometric grade formamide, Aldrich), pH adjusted to 7.5. Strong: 1.5 *M* KCl (99+% A.C.S reagent grade, Aldrich) in weak buffer.

### 2.2. In Vitro RNase H Cleavage Reactions and Analysis

1. PhastSystem and "Homogenous 20" PhastGels (Pharmacia).
2. *E. coli* RNase H, Amersham International (Little Chalfont, Buckinghamshire) or Cambio.
3. RNase Block I, Stratagene (Cambridge, Cambridgeshire).
4. Nylon membranes for Northern blotting: Nytran (0.2-μm pore, Schleicher & Schuell, London, UK).
5. Bacteriophage RNA polymerases (Epicentre, UK Distributor Cambio).
6. LPA: 5 mg/mL linear polyacrylamide (Aldrich).
7. Alkaline phosphatase conjugated antidigoxigenin or antifluorescein, FAb fragments (Boehringer, Lewes, East Sussex, UK).
8. *E. coli* RNase H dilution buffer: 25 m*M* Tris-HCl, pH 7.5, 30 m*M* NaCl, 0.5 m*M* EDTA, 5 m*M* 2-mercaptoethanol, 10% glycerol (v/v).
9. *E. coli* RNase H 2X assay buffer: 80 m*M* Tris-HCl, 8 m*M* $MgCl_2$, 2 m*M* dithiothreitol (DTT), 0.006% bovine serum albumen (BSA), pH 7.6 at 37°C.
10. Phenol/chloroform: Phenol/chloroform/isoamyl alcohol 25:24:1 (v/v/v), saturated with 100 m*M* Tris-HCl, 1 m*M* EDTA, pH 8.0.
11. 20X MOPS: 400 m*M* MOPS, 100 m*M* Na acetate, 20 m*M* EDTA, pH 7.0.

12. Formaldehyde gel-loading buffer
    a. #1: 62.5% formamide (v/v), 1.25X MOPS buffer, 8.25% formaldehyde, 62.5 µg/mL ethidium bromide, 0.625 mg/mL orange G dye, and 8.25% sucrose (w/v).
    b. #2: 2X MOPS buffer, 13.2% formaldehyde, 100 µg/mL ethidium bromide, 1 mg/mL orange G, and 13.2% sucrose (w/v).
13. 5X in vitro transcription buffer: 200 m$M$ Tris-HCl, pH 7.5, 50 m$M$ NaCl, 10 m$M$ spermidine, 30 m$M$ MgCl$_2$.
14. NTP mixes:
    a. Nonlabeling NTP mix: 10 m$M$ ATP, 10 m$M$ CTP, 10 m$M$ GTP, 10 m$M$ UTP.
    b. 10% labeling NTP mix: 10 m$M$ ATP, 10 m$M$ CTP, 10 m$M$ GTP, 9 m$M$ UTP, and 1 m$M$ digoxigenin-11-UTP (Boehringer).
    c. 35% labeling NTP mix: 10 m$M$ ATP, 10 m$M$ CTP, 10 m$M$ GTP, 6.5 m$M$ UTP, 3.5 m$M$ digoxigenin-11-UTP (Boehringer).
15. Blot wash buffer: 100 m$M$ maleic acid, 150 m$M$ NaCl, pH 7.5, 0.3% (v/v) Tween-20.
16. Blot blocking buffer: 1% (w/v) blocking reagent (Boehringer) in wash buffer, freshly diluted from a diethylpyrocarbonate-treated 10% solution of blocking reagent in wash buffer.
17. Alkaline phosphatase buffer: 100 m$M$ diethanolamine, pH 10.0, 5 m$M$ MgCl$_2$.
18. Chromogen: 90 µL 37.5 mg/mL nitrobluetetrazolium chloride (in 70% dimethyformamide) and 70 µL 25 mg/mL 5-bromo-4-chloro-3-indolyl phosphate p-toluidine salt (in 100% dimethylformamide) per 10 mL alkaline phosphatase buffer.

## 2.3. Cell-Based Assays and Their Analysis

1. RPMI-1640 medium (Gibco-BRL, Paisley, Renfrewshire).
2. Fetal calf serum (FCS, SeraLab, Crawley Down, West Sussex, UK), heat-inactivated at 56°C for 30 min.
3. Cytoron Absolute flow cytometer (Ortho, High Wycombe, Buckinghamshire, UK).
4. Streptolysin O (SLO, Sigma, Poole, Dorset).
5. Gluteraldehyde-fixed chick red cells (Sigma).
6. Propidium iodide (PI, Sigma, 1 mg/mL) in water.
7. Phosphate-buffered saline (PBS): 8 g/L NaCl, 0.2 g/L KCl, 1.44 g/L Na$_2$HPO$_4$, 0.24 g/L KH$_2$PO$_4$, pH to 7.4 with HCl.
8. 2X intracellular buffer: 22 m$M$ KH$_2$PO$_4$, pH 7.4, 216 m$M$ KCl, 44 m$M$ NaCl, 2 m$M$ DTT, 6 m$M$ MgCl$_2$, 2 m$M$ ATP.
9. HEPES-buffered saline (HBS): 10 m$M$ HEPES, 137 m$M$ NaCl, 3 m$M$ MgCl$_2$, pH 7.4.
10. Cell lysis buffer for RNA: 4 $M$ guanidine thiocyanate, 5 m$M$ Na citrate pH 7.0, 0.5% Na sarkosyl, 200 m$M$ Na acetate, pH 4.2, and 100 m$M$ 2-mercaptoethanol.
11. Phenol (pH 4.3): Phenol saturated with 100 m$M$ Na citrate, pH 4.3.
12. 20X SSC: 300 m$M$ Na citrate, 3 $M$ NaCl, pH to 7.0.
13. Hybridization buffer: 50% formamide (v/v, deionized), 5X SSC, 5% blocking reagent (w/v, Boehringer), 0.1% Na $N$-laurylsarcosine (w/v), 0.02% SDS (w/v), 100 µg/mL yeast tRNA.
14. Stringency buffer: 0.1X SSC, 0.1% SDS (w/v), pH 7.0.

## 2.4. RL-PCR

1. T4 RNA ligase and supplied 10X reaction buffer (Boehringer).
2. Superscript reverse transcriptase (Gibco).
3. Light mineral oil (Aldrich).
4. "Treff" 1.5-mL microfuge tubes and disposable micropestles (distributed in the United Kingdom by Anachem-ScotLab, Luton, Bedfordshire, UK).

5. Pfu polymerase (native, not recombinant) and supplied 10X reaction buffer 1 (Stratagene).
6. fmol cycle sequencing kit (Promega, Southampton, Hampshire, UK).
7. Sequencing grade *Taq* polymerase, 5 U/μL (Promega).
8. Trans·Blot Semi-Dry cell (Bio-Rad, Hemel Hempstead, Hertfordshire, UK), *see* **Note 3**.
9. 0.2X TEN: 30 m$M$ NaCl, 2 m$M$ Tris-HCl, 0.2 m$M$ EDTA, pH 8.0.
10. Orange G gel-loading buffer: 5 mg/mL Orange G (Sigma), 67% (w/v) sucrose.
11. Modified lysis buffer: 25 μg/mL LPA in a 1:1 (v/v) mixture of cell-lysis buffer for RNA and phenol, pH 4.3.
12. dNTP mix: 10 m$M$ dATP, 10 m$M$ dCTP, 10 m$M$ dGTP, 10 m$M$ TTP.
13. RNase A: 1 ng/μL RNase A (Sigma) in 10 m$M$ Tris-HCl, 1 m$M$ EDTA, pH 8.0. Solution boiled (5 min) to destroy any contaminating DNases.
14. Linker-specific primer: 5'... GGG CAT AGG CTG ACC CTC GCT GAA A ...3'.
15. d/ddNTP mixes: All contain 20 μ$M$ dATP, dCTP, dTTP, and 7-deaza dGTP. The A mix also contains 350 μ$M$ ddATP, C mix 200 μ$M$ ddCTP, G mix 30 μ$M$ ddGTP, and the T mix 600 μ$M$ ddTTP.
16. 5X sequencing buffer: 250 m$M$ Tris-HCl, 10 m$M$ MgCl$_2$, pH 9.0.
17. Sequencing gel-loading buffer: 95% formamide (v/v), 10 m$M$ NaOH, 0.05% bromophenol blue, 0.05% xylene cyanol.
18. 0.5X TBE: 45 m$M$ Tris-borate, 1 m$M$ EDTA.

## 3. Methods

### 3.1. Synthesis, Purification, and Analysis of Fluorescein-Labeled Chimeric Oligodeoxynucleotides

It is desirable to have a readily detectable, fluorescent reporter group attached to ON for monitoring intracytoplasmic delivery and metabolism against a background of UV-absorbing biomolecules. Fluorescein adequately fulfills this function and, at the same time, does not appear to unduly affect the biochemical and biological properties of antisense ON. A robust fluorescein tag is provided by coupling carboxyfluorescein postsynthesis through an amide bond to a 5'-terminal alkylamino linker, incorporated onto the ON as the last cycle on the synthesizer.

#### 3.1.1. Synthesis of 5' Amino-Modified Chimeric Methylphosphonodiester/ Phosphodiester Oligodeoxynucleotides

Methylphosphonodiester ON analogs are synthesized in analogous fashion to normal PO ON on automated DNA synthesizers, except that the 5'-*O*-dimethoxytrityl-2'-deoxynucleoside-3'-*O*-(β-cyanoethyl)-phosphoramidites are replaced by 5'-*O*-dimethoxytrityl-2'-deoxynucleoside-3'-*O*-methylphosphonamidite synthons. The same cycle and reagents used for PO ON synthesis are employed. Methylphosphonamidites are less reactive than phosphoramidites, and coupling times need to be extended to 5 min. Optimal final yield of pure product from synthesis was achieved using the slower cycle, Version 1.23 software, 1 μmol cycle, and the "Improved 10-μmol synthesis cycle Model 381" (ABI User Bulletin No. 15, August 12, 1988). It is convenient to synthesize several different PC/PO chimeric ONs in a batch.

1. Install the methylphosphonamidite solutions on the 5-port synthesizer and synthesize the 3'-PC end section of the first ON.
2. Replace the column and synthesize the 3'-end section of the second ON; repeat until all 3' end sections have been synthesized.

3. Replace the methylphosphonamidites with phosphoramidite solutions and synthesize the central PO sections of the ON consecutively.
4. Replace the phosphoramidites with methylphosphonamidite solutions and synthesize the 5' PC section of first ON.
5. Couple the amino-linker to the ON. Confirm efficient coupling of amino-linker to the ON. Program an interrupt into the DNA synthesizer, immediately before the capping step of this cycle. The capping step is omitted by jumping to the next command. Execute two more cycles to couple an arbitrary deoxynucleoside phosphoramidite. The trityl fraction of the first supplementary cycle gives an indication of the efficiency of coupling of amino-linker, and ideally should be colorless. If coupling of the amino-modifier to the ON chain has been less than ideal, the detritylation step of the final cycle should be omitted.
6. Repeat **steps 4** and **5** until all ON are completed.

### 3.1.2. Synthesis of 5' Amino-Modified Chimeric Methylphosphonodiester/ Phosphorothiodiester Oligodeoxynucleotides

Chimeric PC/PS ON are synthesized in analogous fashion using the synthesizer manufacturer's recommended protocol for PS ON analog synthesis. Successful results were achieved using the sulfurising reagent and the Applied Biosystem's recommended PS cycle (ABI User Bulletin Number 58, February 1991).

### 3.1.3. Deprotection of 5'-Amino Modified Chimeric Oligodeoxynucleotides

Best results with PC containing chimeric ON are obtained using a modification of the one-pot deprotection procedure recommended by Hogrefe et al. (*48,49*).

1. Place the controlled pore glass support from a 1 μmol synthesis in a screw-cap vial.
2. Add 0.5 mL of deprotection mixture, seal, and incubate for 30 min at room temperature.
3. Add 0.5 mL ethylenediamine to the vial and incubate at room temperature for 72 h. The extended incubation time is required for complete removal of the trifluoroacetyl amino-protecting group.
4. Collect the supernatant.
5. Wash the support twice with 0.5 mL acetonitrile in water (1:1, v/v) and bulk the supernatant with that collected in **step 4**. Centrifuge to remove insoluble material. Discard the solid support.
6. Ethanol precipitate the ON. Add 0.1 vol of 3 *M* sodium acetate, 10 vol ethanol, and cool to –80°C for 1 h, or more. Pellet the precipitate and wash twice with 10 mL ethanol.
7. Resuspend the pellet in 10 mL Py/TEAA.

### 3.1.4. Removal of Trityl-ON Impurities by Reversed-Phase, Solid-Phase Extraction on C18 Sep-Pak Cartridges

1. Siliconize all glassware (Pasteur pipets, beakers, and other materials) and C18 Sep-Pak cartridges that contact the ON solutions during purification to prevent loss of ON through nonspecific adsorption. Treat with Repelcote(VS) for 5 min, followed by successive washes with dichloromethane, acetonitrile, and water. C18 Sep-Pak cartridges are given a final wash with Py/TEAA or TEAA as appropriate.
2. Apply the crude, deprotected ON solution in Py/TEAA to a C18 Sep Pak cartridge. Allow the ON solution to drip through under gravity, using a polypropylene disposable syringe barrel as reservoir (ON may adsorb to other types of plastic).
3. Take a 10-μL sample of the effluent for anion-exchange HPLC analysis to ensure that all full-length ON has bound to the cartridge.

4. In the event that not all ON has bound to the cartridge, wash the cartridge with 5 mL Py/TEAA, 10 mL 8% acetonitrile in Py/TEAA, 5 mL Py/TEAA. Dilute the initial effluent with 1 vol of Py/TEAA and reapply to the same cartridge.
5. When all ON is bound, wash the cartridge with 10 mL Py/TEAA.
6. Perform an exploratory step gradient of acetonitrile in Py/TEAA starting at 8% acetonitrile (v/v, 10 mL), increasing at 0.5% acetonitrile intervals.
    a. Monitor elution of 5'-amino ON by spectrophotometric examination (270–290 nm, away from the absorption maximum of pyridine) of cartridge effluents against a Py/TEAA blank.
    b. Take 10-μL samples of the effluents for anion-exchange HPLC, to confirm elution of the product.
    c. Full-length ON will begin to elute at around 10% acetonitrile, and the trityl-on oligomer from amino-modifier coupling failure will not appear in the effluent until 15% or higher acetonitrile (**Note 4**).
    d. Discard trityl-containing fractions, which develop an orange color when 50 μL is added to 1 mL 70% perchloric acid in water/absolute ethanol (1:1 v/v).
    End-capped ON synthesis failure sequences are removed from the full-length 5'-amino ON in **steps 7–9**. The ON is desalted in **steps 10–12**.
7. Bulk the 5'-amino ON-containing fractions, dilute with 1 vol Py/TEAA, and reapply to the same C-18 Sep Pak cartridge.
8. Wash the cartridge thoroughly with a concentration of acetonitrile in Py/TEAA just less than that required to remove the full-length product.
9. Elute the 5'-amino ON with an appropriate concentration of acetonitrile in Py/TEAA.
10. Dilute the ON-containing solution with 1 vol of TEAA and apply it to a new C18 Sep-Pak.
11. Wash the Sep-Pak thoroughly with water.
12. Elute the product from the Sep-Pak with 2 mL 50% acetonitrile/water and dry down in a Speed-Vac Concentrator.

### 3.1.5. Postsynthetic Derivitization of 5'-Amino Oligodeoxynucleotides with Fluorescein

Postsynthetic derivitization is achieved in **steps 1–6**, removal of the carboxyfluorescein impurity is achieved in **steps 7–10**, the counter-ion is changed to sodium in **step 11**, and the final product readied for storage in **steps 12** and **13**.

1. Dissolve 1 μmol of 5'-amino ON (approx 200 $A_{260}$ U of a 20-mer, **Note 5**) in 150 μL carbonate buffer.
2. Add 50 μL of carboxyfluorescein reagent.
3. Immediately check the pH of the solution with a semimicro electrode and readjust to 8.5 by addition of one or two grains of solid sodium carbonate, if necessary.
4. Reaction typically completes in <30 min at room temperature.
5. Remove a sample (1 μL) of the reaction mixture, precipitate by addition of 1 mL 90% ethanol, and analyze by anion-exchange HPLC. The reaction proceeds with the disappearance of the 5'-amino ON UV absorption peak and the appearance of a UV absorption/fluorescence peak at longer retention time, consistent with an increase in overall charge on the molecule because of the negative charges on fluorescein and removal of the positive charge at the amino function through formation of the amide.
6. If unreacted 5'-amino ON remains at 30 min, add more carboxyfluorescein ester solution to achieve further reaction (*see* **Note 6**).
7. Remove the bulk of the carboxyfluorescein impurity by adding 0.80 mL water, followed by ethanol precipitation (**Subheading 3.1.3., step 6**).

8. Redissolve the precipitate in 1 mL water and apply the solution to a NAP 10-column.
9. Elute the carboxyfluorescein-labeled ON from the column with 1.5 mL water.
10. Dilute the eluate with 10 mL TEAA and further purify the fluorescein-tagged product on a C18 Sep-Pak cartridge, using the reverse of the approach described in **Subheading 3.1.4.** (i.e., fluorescently labeled ON are retained on C18 Sep-Pak cartridges at higher concentrations of acetonitrile in water than unlabeled failure sequences).
11. For biological applications, the ON should be converted to the sodium salt, because the triethylamine salt is toxic. This may be achieved most simply by three consecutive ethanol precipitations (**Subheading 3.1.3., step 6**). Wash the final precipitate twice with ethanol.
12. Remove all residual small-molecule contaminants by passage through a NAP 10 column (*see* **steps 8** and **9** above).
13. Evaporate the final product to dryness in a Speed-Vac concentrator and store desiccated at –20°C.

### 3.1.6. Analysis of Chimeric Oligodeoxynucleotides by HPLC

HPLC column effluent is monitored with UV and fluorescence (excitation 494 nm, emission detection centered on 530 nm, Wratten 15 520 nm cut-off emission filter) detectors connected in series. Extend the lifetime of the silica-based columns by installing a precolumn upstream of the sample injector to saturate the buffer with silicate before contact with the analytical column.

1. Reverse-phase HPLC: A generally useful, steep gradient from 5 to 70% acetonitrile (HPLC grade) in TEAA in 20 min at a flow rate of 1 mL/min serves to separate 5'-dimethoxytrityl-on from 5'-dimethoxytrityl-off sequences in all applications. Shallower gradients over shorter concentration ranges may be used for more stringent analysis of final products.
2. HPLC column regeneration. The polymer-based C8P-50 column, but not silica-based columns, may be regenerated by flushing with a 0.2 *M* solution of sodium hydroxide followed by 0.2 *M* acetic acid. Therefore, this column is suitable for "dirty" applications, such as the analysis of ON in cell and tissue extracts. Periodically monitor column performance with a cocktail of ON standards. Regenerate when a deterioration in the separation is observed.
3. Anion-exchange HPLC. Excellent separations of chimeric ON are achieved by anion-exchange HPLC at a column temperature of 65°C, using a 60-min gradient from 0 to 100% strong buffer and a flow rate of 1 mL/min. Set the UV detector to 280 nm, off the absorption of formamide. In the case of chimeric ON carrying only four negative charges, the composition of the weak buffer is changed to 1 m*M* potassium phosphate in 50% formamide, pH 7.5, to achieve retention on the column.

### 3.1.7. Purification of Chimeric Oligodeoxynucleotides by HPLC

When highly pure chimeric ON are required, the product may be subjected to a final purification step on the analytical anion-exchange column. This is particularly appropriate for PS containing ON, where the preparation will contain substantial amounts of material seen as an apparent "*n* – 1" peak on anion-exchange HPLC analysis, but which coelutes with the true product on reverse-phase HPLC analysis and during C18 Sep-Pak purification. This impurity is a mixture of full-length ON in which one of the internucleoside linkages is PO rather than PS as a result of oxidation during deprotection or failure in the synthesis sulfurization step *(50)*.

1. Fit the injector with a large-volume sample loop.
2. Determine the nominal percentage of strong buffer at which the product elutes during an analytical separation.

3. Set the pumps to run isocratically at this constant eluent composition.
4. Inject an analytical sample of approx 0.05 $A_{260}$ U. Check that the ON elutes in the void volume of the column.
5. Reequilibrate the column at progressively reduced percentages of strong buffer until injection of an analytical sample of the ON produces a product peak retention time of about 20 min.
6. Set the HPLC UV detector to 290 nm and its range to 2 *A*U full scale.
7. Inject the ON in batches of 5–10 $A_{260}$ U and collect the required product as it elutes.
8. The separation may be fine-tuned by reducing the percentage of strong buffer still further during the preparative runs.
9. Rerun any inadequately purified fractions under the improved separation conditions after dilution in 5 vol of TEAA and concentration/desalting on C18 Sep-Pak cartridges, as described in **Subheading 3.1.4., steps 10–12**.
10. Product fractions are pooled and desalted on a C18 Sep-Pak cartridge (**Subheading 3.1.4., steps 10–12**) prior to NAP 10 gel filtration (**Subheading 3.1.5., steps 8** and **9**).

## 3.2. Direction of E. coli *RNase H by Antisense Oligodeoxynucleotides*

### 3.2.1. Oligomeric RNA Targets

A convenient assay for the intrinsic capacity of ON of given structure to direct RNase H cleavage of target RNA sequences is obtained using chemically synthesized RNA and *E. coli* RNase H.

1. To 10 μL 2X *E. coli* RNase H buffer, add 4 μL water, 2 μL of 100 μ*M* target oligomeric RNA and 2 μL 100 μ*M* antisense or control ON.
2. Place mixture at 37°C.
3. Dilute *E. coli* RNase H to 0.015 U/μL with dilution buffer.
4. Add 2 μL of the 0.015 U/μL RNase H solution to the RNA/ON solution. Incubate at 37°C.
5. Remove 2 μL samples at intervals (0–120 min) to 1 vol of ice-cold 8 m*M* EDTA (pH 8.0), 0.02% Bromophenol blue.
6. Fractionate 1 μL of the sample by electrophoresis through "Homogenous 20" PhastGels.
   a. Set electrophoresis temperature to 15°C.
   b. Program the PhastSystem: preseparation 400 V, 10 mA, 2.5 W, 100 $V_{hr}$; sample application 400 V, 1 mA, 2.5 W, 5 $V_{hr}$; separation 400 V, 10 mA, 2.5 W, 50–80 $V_{hr}$ (*51*). The values for voltage, current, and power represent upper limits, whereas "$V_{hr}$" represents the actual duration.
7. Visualize nucleic acids in the PhastGels by automated silver-staining. With 15-mer PO ON, this protocol has a sensitivity limit of 0.1 pmol and stains maximally with 15 pmol (*51*).
   a. Rinse in water for 0.1 min at 20°C.
   b. Fix gel with 20% trichloroacetic acid for 5 min at 20°C.
   c. Sensitize with 8.3% glutaraldehyde for 5 min at 50°C.
   d. Wash twice in water for 2 min at 50°C.
   e. Stain with 0.5% silver nitrate for 10 min at 40°C.
   f. Wash twice in water for 0.5 min at 30°C.
   g. Develop with 2.5% sodium carbonate, 0.04% formaldehyde for 0.5 min at 30°C.
   h. As in **step g**, but for 4.5 min at 30°C.
   i. Reduce background with 2.5% sodium thiosulfate (pentahydrate), 3.7% Tris-HCl for 1 min at 30°C.
   j. Stop further reaction with 5% acetic acid for 2 min at 50°C.
   k. Preserve gel with 10% acetic acid and 5% glycerol for 3 min at 50°C.
8. Quantify band intensities by densitometry.

### 3.2.1.1. THE "CATALYTIC" ACTIVITY OF OLIGODEOXYNUCLEOTIDES

The assay of **Subheading 3.2.1.** may be modified to test the capacity of ON to "catalytically" direct RNase H-mediated destruction of target RNA. Conditions are arranged to obtain a 100-fold molar excess of RNA over ON. A greatly increased concentration of RNase H is used.

1. To 10 µL 2X *E. coli* RNase H buffer, add 6 µL 1 m$M$ target oligomeric RNA and 2 µL 30 µ$M$ antisense or control ON.
2. Place mixture at 37°C.
3. Dilute *E. coli* RNase H to 1 U/µL with dilution buffer, if necessary.
4. Add 2 µL of the 1 U/µL RNase H solution to the RNA/ON solution. Incubate at 37°C.
5. Remove 1 µL samples at intervals (0–120 min) to 29 µL ice-cold 1 m$M$ EDTA (pH 8.0), 0.02% Bromophenol blue.
6. Fractionate 1 µL of the sample by electrophoresis through "Homogenous 20" PhastGels (**Subheading 3.2.1., step 6**).
7. Visualize nucleic acids in the PhastGels by automated silver-staining (**Subheading 3.2.1., step 7**).

### 3.2.2. Antisense Oligodeoxynucleotide Activity Against In Vitro Transcribed RNA Targets

The specificity of ON-directed RNase H-dependent scission of target RNA may be investigated in a similar manner to the assays described in **Subheadings 3.2.1.** and **3.2.1.1.**, but where in vitro transcribed RNA replaces oligomeric RNA. If near full-length in vitro transcript is used, these assays may also provide information regarding the influence of RNA secondary structure on the efficiency of the RNase H-cleavage reaction.

1. To 10 µL 2X RNase H digestion buffer on ice, add 2 µL digoxigenin labeled in vitro transcribed RNA (~250 ng/µL, **Subheading 3.2.2.1.**), 2 µL 10 µ$M$ ON, RNase Block I to a final concentration of 2 U/µL. Adjust volume to 18 µL.
2. Place reaction mixture at 37°C to equilibrate.
3. Dilute *E. coli* RNase H to 0.25 U/µL with dilution buffer.
4. Add 2 µL 0.25 U/µL RNase H to the reaction mixture and continue incubation at 37°C.
5. Remove 3 µL samples from the reaction at intervals (0–60 min) to 47 µL ice-cold 1 m$M$ EDTA, pH 8.0, containing 100 ng/µL *E. coli* ribosomal RNA.
6. Extract the RNA samples with phenol/chloroform to remove contaminating proteins.
   a. Add 1 vol of phenol/chloroform and vortex mix to obtain a milky suspension.
   b. Centrifuge (5 min at 12,000$g$) to separate the phases. Collect the upper aqueous phase.
7. Add 2 µL of the RNA sample to 8 µL of formaldehyde gel loading buffer #1. Store the remainder of RNA sample at –20°C.
8. Incubate the RNA/loading buffer sample at 65°C for 10 min then place on ice.
9. Fractionate substrate and product RNAs by electrophoresis through a formaldehyde–1% agarose gel (*see* **Note 7**). Capillary blot nucleic acids contained in the gel onto nylon membrane (*see* **Notes 8** and **9**).
10. Visualize digoxigenin-labeled RNA on the membrane immunologically (**Subheading 3.2.2.2.**).
11. Quantify band intensities by densitometry.

### 3.2.2.1. IN VITRO TRANSCRIPTION OF DIGOXIGENIN-LABELED TARGET RNA

Nonradioactively labeled target RNA is produced by replacing some of the UTP in standard in vitro transcription reactions with digoxigenin-11-UTP.

1. To 10 μL 5X in vitro transcription buffer, add 5 μL 100 m*M* DTT, 5 μL 10% labeling NTP mix. Adjust volume to 40 μL with water. Place on ice.
2. Add 100 U RNase Block I, 200 U of the appropriate bacteriophage RNA polymerase, approx 2 pmol of linear-template DNA and adjust volume to 50 μL.
3. Incubate at 37–40°C for 2 h.
4. Divide into 10-μL aliquots and store at –80°C. Do not remove template DNA by DNase I digestion. Further manipulation of the RNA product is not required for the subsequent assay.

### 3.2.2.2. IMMUNOLOGICAL DETECTION OF MEMBRANE-BOUND NONRADIOACTIVE NUCLEIC ACIDS

Digoxigenin or fluorescein-labeled nucleic acids may be detected on nylon membranes using antidigoxigenin, or antifluorescein, alkaline phosphatase-conjugated FAb fragments (Boehringer) essentially as described by the manufacturer (Boehringer. *DIG Nucleic Acid Detection Kit.*). **Steps 1–5** and **7** should be carried out at room temperature with constant gentle agitation.

1. Equilibrate the membrane for 5 min in blot wash buffer.
2. Block for 30 min to 1 h in blot-blocking buffer.
3. Dilute the antibody 1:5000 (v/v) in blocking buffer and apply to the membrane (20 mL/ 100 cm² of membrane) for 30 min to 1 h (*see* **Note 10**).
4. Wash the membrane once for 5 min and twice more for 15 min each in wash buffer. It is imperative to change the vessel between the first and second postantibody washes.
5. Equilibrate the membrane in alkaline phosphatase buffer for five min.
6. Drain the membrane well and add 10 mL of freshly prepared chromogen per 100 cm² of membrane. Place the chromogen-covered membrane in the dark at room temperature with minimal disturbance until the required bands reach the desired intensity.
7. Discard the chromogen. Wash the blot thoroughly in many changes of water. Dry and store in the dark.

### 3.2.2.3. COUPLED IN VITRO TRANSCRIPTION—RNASE H CLEAVAGE REACTIONS

*E. coli* RNase H is active under in vitro transcription reaction conditions. This assay may replace that described in **Subheading 3.2.2.**, in many circumstances. However, PS-containing ON (and 2'-*O*-allyl protected ribozymes) may not be used in this system, because minimal transcription occurs in the presence of such compounds.

1. To 10 μL 5X in vitro transcription buffer add 5 μL 100 m*M* DTT, 5 μL nonlabeling NTP mix, 100 U RNase Block I, 200 U bacteriophage RNA polymerase, approx 2 pmol of linear-template DNA and 1.25 U *E. coli* RNase H. Adjust volume to 45 μL with water and place on ice.
2. Place 1 μL 10 μ*M* ON into a fresh tube. Add 9 μL of the above mix. Five separate ON may be examined using the mix of **step 1**.
3. Incubate at 37°C for 60 min.
4. Stop reactions by adding 30 μL 1 m*M* EDTA, pH 8.0, and 1 μL LPA (coprecipitant) and placing on ice.
5. Phenol/chloroform extract (**Subheading 3.2.2., step 6**).
6. Ethanol-precipitate the nucleic acids by adding 1/10 vol of 3 *M* sodium acetate and 3 vol ethanol followed by incubation at –80°C for at least 1 h. Collect the precipitate by centrifugation (*see* **Notes 11** and **12**) 20 min at 12,000*g*.
7. Resuspend the pellet in 5 μL formamide by heating to 65°C and vortexing. Add 5 μL formaldehyde gel-loading buffer #2. Incubate at 65°C for 10 min; then place on ice.

8. Fractionate nucleic acids by formaldehyde–1% agarose gel electrophoresis (*see* **Note 7**).

9. Visualize abundant nucleic acid species (>approx 20 ng/band) in the gel by UV (312 nm) excitation of intercalated ethidium bromide *(52)*. Document the result by photography/ CCD camera video capture.

### 3.2.2.4. RESULTS

In these assays PO *(36)* and PS *(37)* ON induce substantial RNase H-dependent scission of nontarget RNA sequences. Progressively reduced undesired cleavage is observed using chimeric analogs with increasing PC substitution *(36,37)*. Comparable protocols to that described in **Subheading 3.2.2.**, which replace the in vitro transcribed RNA with 500 ng/µL total cellular RNA or 25 ng/µL A$^+$ selected RNA *(37)*, produce essentially identical results.

### 3.2.2.5. IDENTIFICATION OF OPEN-LOOP REGIONS OF RNA

The observation that RNase H cleavage occurs at nontarget RNA sequences may be exploited to identify open-loop regions of RNA. The sites at which nontargeted cleavage occur may represent regions of RNA that are particularly accessible to ON hybridization *(53,54)*.

1. Perform RNase H cleavage reactions and analysis as described in **Subheading 3.2.2.**
2. Identify fragments resulting from nontargeted RNase H action and estimate their lengths, by reference to RNA molecular weight standards.
3. Calculate all regions of contiguous partial complementarity between the ON and RNA used (*see* **Note 13**). From this data, calculate the two expected RNA fragment sizes that would result from cleavage at each position.
4. Fit observed fragments into the pool of hypothetical fragments to identify putative open-loop regions. You will notice that only a subset of predicted fragments are actually observed.

This system was investigated for the c-*myc* oncogene. Nontarget sites that were efficient substrates for cleavage were identified and complementary ON-synthesized. Parent and derived ON were assayed for antisense activity in living cells, following streptolysin O (SLO)-mediated delivery. A chimeric ON derived by this procedure was a much more potent effector than a chimeric ON with the parent sequence *(17)*.

## 3.3. Direction of Human RNases H by Antisense Oligodeoxynucleotides

### 3.3.1. Antisense Oligodeoxynucleotide Activity in Total Cell Protein Extracts with In Vitro Transcribed RNA Targets

The intracellular antisense efficacy of ON structures may be predicted by in vitro assays using whole-cell extracts containing human RNase H. It is important to use physiological salt conditions. Chimeric ON with a central region containing only two PO linkages efficiently direct the action of human-cell extract RNase H under low salt conditions (e.g., *E. coli* RNase H assay buffer). At physiological salt concentration, ON with central regions containing four or fewer PO linkages are essentially inactive, a finding that is consistent with observations following ON delivery into cells.

### 3.3.1.1. MAINTENANCE OF HUMAN LEUKEMIA CELL LINES

Maintain human leukemia cell lines (e.g., HL60, MOLT4, K562, and so on) in exponential growth between $2 \times 10^5$ and $1 \times 10^6$ cells/mL in 90% RPMI-1640 medium, 10% FCS by incubation at 37°C in an atmosphere of 5% $CO_2$, 95% air.

### 3.3.1.1.1. Total Cell Count by Flow Cytometry

The flow cytometric procedures described here used a Cytoron Absolute flow cytometer which possesses a 15-mW argon ion laser and records the variables of forward- and side-scatter. Green and red fluorescence data are collected from photomultiplier tubes equiped with 515–548 nm band pass and 620-nm long-pass filters, respectively.

1. Pass sample through the laser beam at the known rate of 1 μL/s. for a known duration (10 s).
2. Select forward vs side-scatter dot-plot output, and optimize forward- and side-scatter detector gains, using linear amplification, to obtain a tight population distribution somewhat away from both axes.
3. Apply a gate so that subsequent measurements only include data from single cells.
4. Obtain a simple cell count by taking an undiluted sample of cell culture through this protocol. Cells per milliliter in the original culture = **gated** cell count · (1000 μL/mL)/10 μL.

### 3.3.1.2. PREPARATION OF WHOLE-CELL EXTRACT *(55)*

Execute all procedures on ice and with ice-cold buffers.

1. Collect $2.5 \times 10^8$ exponentially growing cells and wash twice in PBS.
2. Resuspend in 4 mL 10 m$M$ Tris-HCl (pH 7.9), 1 m$M$ EDTA, 5 m$M$ DTT, and incubate on ice for 20 min.
3. Dounce homogenize the swollen cells (eight strokes, "B" pestle"). Transfer the lysate to a clean beaker.
4. Add 4 mL 50 m$M$ Tris-HCl (pH 7.9), 10 m$M$ MgCl$_2$, 2 m$M$ DTT, 25% sucrose (w/v), and 50% glycerol (v/v) to the suspension and mix by gentle stirring.
5. Continue the stirring while adding 1 mL saturated ammonium sulfate (pH 7.0), dropwise. Continue stirring for an additional 30 min.
6. Centrifuge the suspension for 3 h at 175,000$g$.
7. Collect the supernatant (~10 mL) and precipitate the remaining proteins and nucleic acids by adding 3.5 g of solid ammonium sulfate. When the ammonium sulfate has dissolved add 3.5 μL 1 $M$ NaOH and stir for 30 min.
8. Collect the precipitate by centrifugation (15,000$g$, 20 min) and resuspend in 500 μL 40 m$M$ Tris-HCl (pH 7.9), 0.1 $M$ KCl, 10 m$M$ MgCl$_2$, 0.2 m$M$ EDTA, 15% glycerol (v/v).
9. Dialyze the suspension against 100 mL of the same buffer overnight.
10. Divide the suspension (~800 μL) into 100-μL aliquots, snap freeze, and store in liquid nitrogen. The extracts retain RNase H activity through at least three rounds of thawing and snap freezing.

### 3.3.1.3. ASSAY FOR OLIGODEOXYNUCLEOTIDE ACTIVITY USING CELL-EXTRACT RNASE H

1. Place 15 μL 2X intracellular buffer in a tube on ice.
2. Add 3 μL 10 μ$M$ ON, 2 μL digoxigenin-labeled in vitro-transcribed RNA (~250 ng/μL) (**Subheading 3.2.2.1.**), yeast tRNA, and RNase Block I to final concentrations of 1 μg/μL, and 1 U/μL respectively. Adjust volume to 27 μL with water and equilibrate at 37°C.
3. Add 3 μL preheated (37°C) cell extract (**Subheading 3.3.1.2.**) and incubate at 37°C.
4. Remove 5-μL samples at intervals between 0 and 60 min.
5. Terminate reactions by extracting the RNA using the method of Chomczynski and Sacchi *(56)*.
   a. Add 500 μL of lysis buffer directly to the assay. Vortex to ensure complete mixing.
   b. Add 500 μL phenol (pH 4.3) and vortex to obtain a single clear phase. Add 130 μL chloroform:isoamyl alcohol (24:1 v/v), and vortex to obtain a milky suspension. Incubate on ice for 10 to 15 min.

c. Centrifuge (room temperature, 12,000*g* or higher, 10 min) to separate the phases.
d. Remove 400 μL of the upper, aqueous phase to a fresh tube (*see* **Note 14**) and precipitate the RNA by addition of 1 vol isopropanol and incubation at –20°C for 2 h or overnight.
e. Pellet the RNA by centrifugation (*see* **Notes 11** and **12**, room temperature, 12,000*g*, or higher, 30 min).

6. Resuspend the purified RNA (*see* **Note 15**) in 5 μL formamide, add 5 μL formaldehyde gel-loading buffer #2, and proceed with the analysis detailed in **Subheading 3.2.2., steps 8–11**.

## *3.3.2. Antisense Oligodeoxynucleotide Activity in Nonidet P-40 Lysed Cells*

Gentle removal of the cytoplasmic membrane using the nonionic detergent NP40 results in cell lysates that retain RNase H activity *(15)*. Such assays provide very good information regarding the efficiency and specificity that antisense ON would display on delivery to topologically intracellular locations. Successful results have been obtained with HL60, KYO1, LAMA84, and MOLT4 cell lines. However, not every cell-line produces results in this assay (*see* **Note 16**). The target RNA used in NP40 lysis RNase H assays is the endogenous transcript. Furthermore, the normal cellular compliment of proteins is retained. Therefore, the secondary and higher-order structure presented by the target RNA may accurately reflect the in vivo case. Thus, this assay may assist in identifying open-loop regions of RNA, using an approach similar to that described above (**Subheading 3.2.2.5.**).

1. Pellet $10^6$ exponentially proliferating cells by centrifugation (*see* **Note 11**) (800*g*, 4 min).
2. Wash by resuspension in an isotonic buffer, such as 1X intracellular buffer or HBS and recentrifugation.
3. To the cell pellet add 1 μL of a 50 μ*M* solution of ON and gently agitate to obtain a cell/ON "slurry."
4. Resuspend the cells in 50 μL of isotonic buffer (HBS, intracellular buffer) containing 10% (v/v) NP40 and 1 U/μL RNase Block I by gentle vortex mixing to form a nonviscous cloudy suspension.
5. Incubate at 37°C for 10 min.
6. Stop the reaction and take samples for RNA analysis as described for cell extract assays (**Subheading 3.3.1.3., steps 5** and **6**).
7. An additional Northern hybridization step (**Subheading 3.3.2.1.**), following capillary transfer and before immunological detection, will be required to detect specific RNA species.

### 3.3.2.1. NONRADIOACTIVE NORTHERN HYBRIDIZATION

1. Place the membrane in a preheated hybridization tube and add the tube manufacturer's recommended volume (usually 15–30 mL) of DEPC-treated hybridization buffer.
2. Incubate for 1 h. Ensure that no air bubbles become trapped between the tube and the membrane. Incubate for a further hour.
3. Add 1 μL in vitro transcribed nonradioactively labeled antisense RNA probe (**Subheading 3.3.2.2.**) per 10 mL of hybridization buffer. Incubate overnight. For pre- and hybridization temperature, *see* **Note 17**.
4. Wash the membrane three times in stringency buffer for 10 min at 65°C using at least 50 mL/100 cm$^2$ of membrane.

### 3.3.2.2. SYNTHESIS OF NONRADIOACTIVELY LABELED ANTISENSE RNA PROBE

1. Perform transcription reaction described in **Subheading 3.2.2.1., steps 1–3**, except use 35% labeling NTP mix instead of 10% labeling mix. Ensure that the antisense strand is synthesized.

2. Add 0.1 vol 10% SDS and 0.04 vol 2-mercaptoethanol to the transcription reaction, and incubate at 80°C for 5 min to stop transcription and preserve the RNA for subsequent use as a probe. The reaction products should then be stored at –20°C until needed. Multiple rounds of thawing and refreezing do not adversely affect the performance of the probe.

### 3.3.3. Antisense Oligodeoxynucleotide Effects in Living Cells

An overview of the procedure is as follows:

1. Check cell viability >95% (**Subheading 3.3.3.1.**).
2. Deliver ON into living cells by SLO-mediated reversible permeabilization (**Subheading 3.3.3.2.**).
3. Return cells to normal growth conditions (**Subheading 3.3.1.1.**).
4. Remove samples from the culture between one hour and 100 h (*see* **Note 18**) after ON delivery to assess the efficiency of the reversible permeabilization procedure by two-color flow cytometry (**Subheading 3.3.3.3.**) and to obtain a viable cell count (**Subheading 3.3.3.1.**).
5. Take samples, as in **step 4**, for RNA isolation (**Subheading 3.3.1.3.**, **step 5**).
6. Analyse purified RNA for evidence of ON-mediated inhibition of target and nontarget mRNA expression by formaldehyde gel electrophoresis, capillary blotting, Northern hybridization (**Subheading 3.3.2.1.**), and immunological visualization (**Subheading 3.2.2.2.**).
7. Perform suitable assay to investigate the biological response of the cells to antisense suppression of specific gene expression. The precise assay selected will vary with the target gene and cell line.

#### 3.3.3.1. VIABLE CELL COUNT BY FLOW CYTOMETRY

1. Take 1 mL of cell culture and place on ice. Incubate for 5 min.
2. Add 10 µL PI. Incubate the cells on ice for a further 5 min.
3. Pellet the cells by centrifugation (*see* **Note 11**, 800*g*, 4 min) and resuspend in 1 mL of ice-cold RPMI-1640.
4. Pass the cell sample through the flow cytometer using a modification of the total cell-count protocol (**Subheading 3.3.1.1.1.**).
   a. Use the forward/side-scatter-gated population to generate a red fluorescence histogram.
   b. Optimize the red fluorescence detector gain, using log amplification, to obtain good separation between the nonred (healthy, PI-excluding) and red (dead, PI-stained) sub-populations. Set regions which select these two populations. This returns the percentage and the number of the unstained and PI-stained cells.
5. Viable cells per milliliter in the original culture = PI unstained cell count · 100.

#### 3.3.3.2. STREPTOLYSIN ON REVERSIBLE PERMEABILIZATION OF CELLS

Barry et al. *(57)* first described antisense ON delivery by SLO reversible permeabilization of cells in culture. We adapted the method *(41,58)* to maximize the percentage of cells reversibly permeabilized and amount of ON delivered per cell and to minimize toxic side effects. SLO permeabilization will also deliver ON into cells of clinical origin *(59)*. SLO delivery has recently been used in a clinical trial *(60)*.

1. Resuspend SLO at 1000 U/mL in PBS containing 0.01% BSA.
2. Activate SLO by incubation at 37°C for 2 h in the presence of 5 m*M* DTT (freshly prepared).
3. Divide the solution into small aliquots (500–1000 U) and store at –20°C until needed. It is advisable to test the permeabilization activity of the SLO (*see* **Note 19**). Cycles of thawing and refreezing should be avoided as reduced SLO activity will be observed.

4. Wash $5 \times 10^6$ cells twice by centrifugation at 800$g$ for 5 min and resuspension in serum-free RPMI-1640 medium.
5. Pellet the cells and resuspend in 200 μL RPMI-1640.
6. Add the cell suspension to a mixture of SLO and sufficient fluorescently labeled ON to achieve the desired final concentration (suggest between 20 and 100 U SLO [*see* **Note 20**], and ~10 μ*M* ON). Gently agitate the suspension immediately.
7. Incubate at 37°C for 10 min. Agitate twice during this period.
8. Reseal the cells by adding 1 mL 90% RPMI-1640, 10% FCS, and incubating at 37°C for 20–30 min.
9. Transfer the cell suspension to flasks containing 9 mL of warmed and gassed normal growth medium.

### 3.3.3.3. TWO-COLOR FLOW CYTOMETRY

Two-color flow cytometry is used to calculate the percentages of a cell population which are living but did not take up the ON (not green or red); living and took up the ON (green but not red); and dead (red only or red and green) following SLO delivery of green fluorescent ON.

1. Perform **steps 1–3** of **Subheading 3.3.3.1.**
2. Pass the cells through the flow cytometer using a modification of the protocol in **Subheading 3.3.3.1., step 4**.
   a. Use the forward/side-scatter-gated population to generate green fluorescence vs red fluorescence dot-plot output. Red fluorescence detector gain and amplification identical to that optimized in **Subheading 3.3.3.1.** is used.
   b. Green fluorescence amplification is set to × 2 (linear) and detector gain is manipulated until the modal signal from a suspension of gluteraldehyde-fixed chick red cells, in PBS, falls into channel 50 (8-bit data collection). The green fluorescence amplification is then reset to log mode.
   c. Set-up electronic compensation to avoid green-channel signal spilling over into the red channel, and vice versa.
   d. Regions are set, using experimental samples of cells, which discriminate between PI-stained and unstained cells, as above, and green and nongreen cells. The green fluorescent/nonfluorescent boundary is set such that cells that have not received green fluorescent compound are observed to be 1–2% positive for green fluorescence.

### 3.3.3.4. RESULTS

A very wide range of cell lines are efficiently permeabilized by SLO. We have not identified any human suspension cells (established line or primary culture) that are refractory to SLO-reversible permeabilization. A variety of antisense ON analog structures targeted to a number of different mRNA species have been examined in this system including: PO, 3'-end protected PO, PS, C5-propyne PO, C5-propyne PS, chimeric PC/PO, PC/PS, PC/(PO – PS), 2' O-methyl PO/2'-deoxy PO, 2' O-allyl PO/2'-deoxy PO, 2' methoxyethoxy PO/2'-deoxy PO, 2' methoxyethoxy PO/2'-deoxy PS, 2' methoxytriethoxy PO/2'-deoxy PS, 2' methoxyethoxy PS/2'-deoxy PS, 2' methoxytriethoxy PS/2'-deoxy PS. Without exception, PC/PO chimeric ON (15–20-mers with 6–9 central PO) provided the most efficient and specific inhibition of target mRNA expression, when assayed 4 h postdelivery.

## 3.4. An Assay for RNase H Activity in Living Cells: RL-PCR

Reverse-ligation-mediated RT-PCR (RL-PCR), which amplifies and positively identifies RNA products, was reported as a method to identify in vivo protein-RNA interactions and ribozyme function *(61,62)*. We have adapted this method to search for products of RNase H cleavage reactions which occur within living cells *(15,16)*. The RNase H activities found in human cell extracts, NP40 lysates, and within intact cells in culture are demonstrable by this approach. It is highly likely that, in conjunction with a suitable biopsy procedure, this method could be used to search for ON-directed RNase H-cleavage fragments of target mRNA following delivery of ON in whole animals. To avoid false-positives in this case, it is imperative to ensure that RNA is extracted exclusively from living tissue (ON rapidly redistributes to intracellular locations when cells are lysed) *(63)* and that an RNA extraction procedure which efficiently denatures proteins is used (RNase H activity is retained when cells are lysed in nondenaturing conditions; *see* **Subheading 3.3.2.**).

### 3.4.1. RL-PCR Overview

1. Synthesize and purify a short RNA linker sequence (**Subheading 3.4.2.**).
2. Deliver ON into living cells by SLO reversible permeabilization (**Subheading 3.3.3.2., steps 4–9**) using conditions that minimize cell death.
3. Remove a sample ($0.5 \times 10^6$) of the cells 1 h (or later) following the initiation of permeabilization and purify RNA (**Subheading 3.3.1.3., step 5**).
4. Ligate the linker RNA (**step 1**) onto all available 5' phosphate groups in the RNA obtained in **step 3** (**Subheading 3.4.3.**) (*see* **Note 21**).
5. Reverse-transcribe target mRNA using a gene-specific primer in conditions that maximize specificity (**Subheading 3.4.4.**). Then degrade input RNA using RNase A (*see* **Note 22**).
6. Amplify reverse-transcribed RNA by PCR (*see* **Note 23**), using one primer specific for the linker RNA and one gene-specific primer (**Subheading 3.4.5.**).
7. Subamplify the product of the first PCR reaction using the same linker-specific primer and a nested, digoxigenin (or fluorescein) labeled, gene-specific primer, for sensitive detection of RNase H cleavage fragments (**Subheading 3.4.6.**).
8. Cycle-sequence the products of the first PCR using the nonradioactively labeled primer alone, to identify the precise site of target mRNA cleavage (**Subheading 3.4.7.**).
9. Nonradioactively labeled products from **steps 7** and **8** are fractionated by electrophoresis through denaturing polyacrylamide gels, transferred onto nylon membrane (detailed in **Subheading 3.4.6.** or **3.4.7.**), and immunologically detected (**Subheading 3.2.2.2.**)

### 3.4.2. SYNTHESIS AND PURIFICATION OF THE RNA LINKER

The high yield T7 RNA polymerase transcription conditions recommended by Milligan et al. *(64)* are used to synthesize a 25-mer linker RNA.

1. Incubate 10 nmol of both template and promoter ON (5' ... TTT CAG CGA GGG TCA GCC TAT GCC CTA TAG TGA GTC GTA TTA ...3' and 5'... TAA TAC GAC TCA CTA TAG ...3') in 250 µL 0.2X TEN at 95°C for 10 min. Cool to room temperature slowly to hybridize.
2. Check that complete hybridization has occurred by PhastGel "Homogenous-20" electrophoresis of a 1-µL sample (**Subheading 3.2.1., step 6**). Visualize nucleic acid bands by ethidium bromide-staining and UV illumination.
3. Dilute the hybrid to 5 µ*M* by addition of 1750 µL 0.2X TEN.

4. Place in a reaction tube in the following order: 153 μL water, 50 μL 0.1 *M* DTT, 7 μL 1 *M* MgCl$_2$, 100 μL 5X in vitro transcription buffer, 20 μL 100 m*M* ATP, 20 μL 100 m*M* CTP, 20 μL 100 m*M* GTP, 20 μL 100 m*M* UTP, 20 μL 5 μ*M* template hybrid, 20 μL RNase Block I (40 U/μL), and 50 μL T7 RNA polymerase (200 U/μL).
5. Incubate transcription reaction at 37°C for 6 h.
6. Stop reaction by adding 10 μL 0.5 *M* EDTA, pH 8.0, and placing on ice.
7. Add 50 μL Orange G gel loading buffer. Fractionate reaction template and products by electrophoresis through a 20% nondenaturing polyacrylamide gel (19:1 acrylamide:bis-acrylamide).
8. Identify the intensely UV-absorbing transcript band by UV-shadowing over a fluorescent thin-layer chromatography plate.
9. Excise the band, place in a "Treff" Eppendorf-type microfuge tube, and mash the gel slice using "Treff" micropestles.
10. Extract the linker RNA transcript from the gel into ~2 vol of cell lysis buffer for RNA by constant agitation overnight.
11. Pellet the mashed gel by centrifugation (*see* **Note 11**) (10,000–15,000*g*, 10 min) and collect the supernatant.
12. Reextract the mashed gel-slice pellet with the same volume of lysis buffer as in **step 10**, but for 2 h.
13. Pool the collected supernatant. Phenol extract, precipitate, and pellet the RNA as described in **Subheading 3.3.1.3., steps 5b–e**.
14. Resuspend the RNA pellet in 50 μL water. Reextract the RNA as if purifying total RNA (**Subheading 3.3.1.3., step 5**). Resuspend the resultant RNA pellet in 100 μL water and divide into 10 aliquots of 10 μL. Store at –20°C until needed.

### 3.4.3. RNA Ligation Reaction

1. Resuspend the total cellular RNA pellet in water at ~250 ng/μL.
2. Remove 4 μL of the total RNA to a fresh tube and add 1 μL gel-purified 25-mer RNA (~500 ng), 1 μL 10X T4 RNA ligase buffer, 1 μL RNase Block I (40 U), 3 U T4 RNA ligase (~0.5 μL), 0.5 MBU DNase I (~0.5 μL). Adjust volume to 10 μL with water.
3. Incubate overnight at 17°C followed by 30 min at 37°C (the latter to ensure complete DNase I digestion).
4. Stop the ligation/DNase digestion reaction by adding 5 μL 10 m*M* EDTA, pH 8.0, and 200 μL modified lysis buffer and vortex mix.
5. Separate the organic and aqueous phases by adding 25 μL chloroform and centrifuging at 12,000*g* for 10 min.
6. Recover the upper aqueous phase to a thermal cycler tube. Precipitate and pellet nucleic acids as described in **Subheading 3.3.1.3., steps 5d** and **e**.

### 3.4.4. Reverse Transcription

1. Resuspend the linker-ligated RNA precipitates in 4 μL water and place on ice until ready to proceed.
2. Make a mix with the following components: 0.9 μL water, 0.7 μL Pfu polymerase 10X reaction buffer 1, 0.7 μL 0.1 *M* DTT, 0.2 μL 10 μ*M* gene-specific RT primer, 0.5 μL RNase Block I. Place at 37°C.
3. Heat the linker-ligated RNA solution to 95°C for 5 min to denature RNA secondary structure; then immediately place the tube at 37°C. Incubate for 2–3 min to allow the temperature inside the tube to equilibrate.
4. Add the RT primer-containing mix to the RNA and incubate at 37°C for 45 min to hybridize.

5. Make a second mix containing 1.4 μL water, 0.3 μL Pfu polymerase 10X reaction buffer 1, 0.3 μL 0.1 *M* DTT, 0.5 μL dNTP mix, 0.5 μL Superscript reverse transcriptase (~100 U).
6. Add the second mix to the RNA/primer hybridization reaction and incubate at 37°C for 45 min more.
7. Stop the reaction by incubation at 95°C for 10 min, then place on ice.
8. Add 1 μL RNase A and incubate at 37°C for 20 min (*see* **Note 22**). Place reactions on ice.

### 3.4.5. First-Round PCR Amplification

1. Make a 10 μL reaction mix containing 4.8 μL water, 1 μL Pfu 10X reaction buffer 1, 2 μL 10 μ*M* linker-specific primer, 2 μL 10 μ*M* nested (position 5' to the RT primer) gene-specific primer, and 0.2 μL native Pfu polymerase (~0.5 U).
2. Add reaction mix to the RNase A-treated reverse transcription reaction: Overlay with 50 μL light mineral oil (if required by your PCR cycler) and pass through 20 rounds of 95°C for 1 min, 55–65°C (depending on the gene specific primer) for 1 min, and 72°C for 1 min.

### 3.4.6. Second-Round PCR Amplification

1. In a fresh thermal cycler tube place 9.5 μL water, 1.5 μL Pfu 10X reaction buffer 1, 1.5 μL 10 μ*M* linker-specific primer, 2 μL 10 μ*M* 5' nonradioactively labeled nested primer (position 5' to the first-round PCR primer), 0.3 μL dNTP mix, and 0.2 μL Pfu polymerase.
2. Add 5 μL of the first-round PCR to the mix, overlay with 50 μL light mineral oil, and perform 5–10 rounds of PCR using the conditions described in **Subheading 3.4.5., step 2**.
3. Remove 2 μL of the second-round PCR to 8 μL sequencing gel-loading buffer. Incubate at 95°C for 10 min to destroy secondary structure, and quench on ice.
4. Fractionate products by electrophoresis through a 7 *M* urea, 6–8% polyacrylamide (19:1 acrylamide:bis-acrylamide) gel.
5. Transfer the nucleic acids contained in the gel onto nylon membrane by semidry electroblotting using using a Trans·Blot Semi-Dry cell (Bio-Rad) (*see* **Notes 3** and **9**).
   a. Soak the gel for 15 min in 0.5X TBE still attached to one glass plate.
   b. Construct the transfer sandwich as described by the manufacturer, using 0.5X TBE as transfer buffer.
   c. Set the power pack to maximal current and voltage of 2 mA/cm$^2$ of membrane and 25 V, respectively.
   d. Transfer for 30 min.
6. Visualize the nonradioactively labeled products immunologically (**Subheading 3.2.2.2.**).

### 3.4.7. Nonradioactive Sequencing

  This procedure is an adaptation of the fmol cycle sequencing protocol.

1. Into the four base-specific chain-termination reaction tubes, add 2 μL dNTP/dideoxy-NTP mix.
2. In a separate tube place 5 μL 5X sequencing buffer, 3 pmol of 5' nonradioactively labeled nested gene specific primer (the same as used in **Subheading 3.4.6.**), 1 μL sequencing-grade *Taq* polymerase, and 1 μL of the first-round PCR product. Adjust volume to 17 μL with water.
3. Transfer 4 μL of the mix to each of the chain-termination reaction tubes.
4. Overlay with 25 μL light mineral oil and perform 30 cycles of PCR as described in **Subheading 3.4.5., step 2**.
5. Terminate reactions by adding 6 μL sequencing gel-loading buffer and heating to 95°C for 10 min.

6. Fractionate sequencing reaction products through a standard-sequencing polyacrylamide gel under standard conditions.

7. Contact blot nucleic acids from the polyacrylamide gel onto 0.2-μm pore Nytran membrane (*see* **Note 9**).

   a. Expose the gel by removing one of the plates.

   b. Pipet a small volume of 0.5X TBE onto the region of gel to be blotted.

   c. Wet a piece of Nytran membrane in 0.5X TBE, and place over the region to be blotted.

   d. Remove air bubbles and excess buffer by gently rolling a wallpapering edge roller over the membrane.

   e. Place a stack of 5–10 sheets of absorbent chromatography paper over the membrane and weigh down heavily.

   f. Allow the transfer to proceed for at least 2 h, but preferably overnight.

   g. Remove the membrane from the gel. Any fragments of polyacrylamide which adhere to the membrane should be removed before the membrane is dried, by briefly placing the membrane nucleic-acid side down on dry chromatography paper.

8. The sequence may be immunologically developed as described in **Subheading 3.2.2.2.**

## 4. Notes

1. Good controls for potential aptameric, sequence-specific, and base composition non-antisense effects may be provided by inverse antisense (reverse polarity antisense) and scrambled antisense sequences. In the case of chimeric ON, the base composition of the PC and PO (or PS) sections of the antisense molecule should be conserved in the control structures. Antisense ON that efficiently downregulated expression of other gene products provide convincing controls for an antisense mechanism of action. For example, bcr-abl, c-myc, and p53 chimeric PC/PO antisense effectors each inhibit expression of their target mRNA but not the two nontarget mRNAs *(15)*.

   In any event, detectable downregulation of the target mRNA (and protein) expression should precede any biological response of the cells. An absolute minimum of two appropriate control ON, of similar size and structure to the active compound, should be demonstrated to be inactive at inhibiting target expression and inducing the biological response *(47)*.

2. This amino-linker compound is compatible with the deprotection conditions for PC-containing chimeric ON. Other linkers may require extended deprotection with ammonia at room temperature leading to unacceptable loss of ON product through hydrolysis of PC internucleoside linkages.

3. Other types of semidry blotter, which possess carbon electrodes, give irreproducible results in this and similar applications.

4. A precise concentration of acetonitrile that elutes amino-modified product but permits retention of trityl-on impurities cannot be provided, because the degree of substitution of PO linkages with the more lipophilic, nonionic PC groups in the chimeric ON affect retention. Furthermore, the characteristics of C18 Sep-Pak cartridges vary slightly from batch to batch.

5. To convert $A_{260}$ U into approximate μmols of ON, calculate the mmolar absorptivity of an ON at 260 nm ($e_{260}$) by summing the values for each base *(65)*, 8.8 for T; 7.3 for C; 11.7 for G; 15.4 for A.

6. Failure to observe additional conversion of 5'-amino ON to product may mean that the reaction is inhibited by some form of secondary structure. In this case, complete reaction is secured by adding fresh carboxyfluorescein solution and heating the reaction mixture briefly to 90°C.

7. The resolution of the formaldehyde-agarose gel electrophoresis may be enhanced by addition of formaldehyde to 6.6% final concentration in the running buffer.

8. For subsequent immunological detection or Northern hybridization with nonradioactive probe, it is not necessary to wash the gel prior to capillary blotting, and neutral nylon membranes provide the lowest background.

9. Nucleic acids transferred to nylon membranes should be irreversibly crosslinked to the dried membrane by UV exposure for an optimized time, before any subsequent procedure.

10. Dilute antibody solution may be kept (at 4°C) for at least 24 h and used repeatedly within this time.

11. Always place Eppendorf-type tubes in the microcentrifuge in the same orientation, for example, hinges out. Supernatant may then be withdrawn from the opposite side of the tube without disturbing any pelleted material, even when the pellet is not readily visible.

12. A brief (10 s) second spin following removal of supernatant will collect residual fluid adhering to the tube walls for removal.

13. A computer-fitting algorithm that returns all regions of contiguous partial complementarity (of a user-designated minimum length; we suggest 6) between input RNA and ON sequences is available from our website (MSDOS executable, http://www.liv.ac.uk/~giles/).

14. If the temperature is permitted to drop below ~20°C, following centrifugal-phase separation and before recovering the aqueous RNA solution, gross DNA contamination of the RNA preparation may result. Such DNA contamination will interfere with subsequent processes.

15. Contamination of the RNA preparation by protein and guanidine thiocyanate and quantitation of RNA recovery may be assessed by spectrophotometric analysis of RNA, resuspended in aqueous solution in 10 m$M$ Tris-HCl, 1 m$M$ EDTA, pH 8.0, buffer, at 230, 260, and 280 nm. Typically, RNA extracted from MOLT4 cells by this method provides approx 10–15 µg/$10^6$ cells (using 1 $A_{260}$ ~ 40 µg/mL RNA in a 1-mL cuvet with a 1-cm path) (66) devoid of overt protein ($A_{260}/A_{280}$ > 1.8) or guanidine thiocyanate ($A_{260}/A_{230}$ >2.0) contamination.

16. Human histiocytic lymphoma U937 cells appear to release large quantities of nucleases and proteases during NP40 lysis, which result in extensively degraded RNA that cannot be interpreted by northern blotting. On the other hand, Veal et al. (67) recently reported antisense effects obtained in NP40 lysates of U937 cells. It is unclear whether the explanation for these disparate results is the analytical method selected (Veal et al., used the RNA degradation-tolerant Ribonuclease Protection Assay) or that our U937 cells differed.

17. Hybridization temperature is the most critical parameter for efficient and specific detection of RNA species. Deficiencies in specificity during hybridization cannot readily be rectified by increasing the stringency of the wash conditions. Higher than "textbook" temperatures may be required to obtain adequate stringency during hybridization. For example, a 1700 nt human c-*myc* antisense RNA probe is used in the described conditions at 80°C. Significantly lower hybridization temperatures result in mishybridization to the ribosomal (and preribosomal) RNAs.

18. Extended incubation times (>24 h) may require that the cells are fed to maintain them within acceptable cell density.

19. Many batches of commercial SLO appear to show essentially no activity in the described assay, even following activation. As a rule of thumb, most human leukemia cell lines used by us are optimally permeabilized using between 4 and 20 U of "good" SLO per $10^6$ cells.

20. The amount of SLO used in these experiments should be optimized for each cell line and batch of SLO. At optimal conditions we typically observe >90% of the population reversibly permeabilized to ON, with the remainder split equally between nonpermeabilized and dead.

21. This method detects 3' fragments of RNase H cleavage reactions. Intact target mRNA is not detected because of the 5' cap structure. The original method, described by Bertrand et al., called for a kinase step to precede the ligation step. We have empirically determined that

kinase treatment of the cellular RNA is not required, consistent with the report that RNase H produces 3' RNA fragments with 5' phosphate groups *(68)*.

22. Other results indicate that high levels (~1 µg) of undegraded RNA are capable of inhibiting PCR. We strongly recommend that the RNase A degradation step is retained.

23. PCR may produce "false-positive" results. A few suggestions for minimizing such problems follow:
    a. Prepare stocks of reagents for a project before commencing any PCR.
    b. Classify work as preamplification (RNA purification, reverse transcription, preparation of PCR reactions, which need to remain uncontaminated) and postamplification (such as gel electrophoresis, which does not).
    c. Use separate pipeters for pre- and postamplification work.
    d. Perform "preamplification" work in a cell-culture class II grade lamina flow cabinet and use barrier pipet tips.
    e. Prepare a stock of "Eppendorf" type tubes for preamplification work by systematically closing all the lids before opening the bag.

## Acknowledgments

The research described in this chapter is supported by the Leukemia Research Fund of Great Britain. RVG is supported by a grant from The Liposome Company Inc., Princeton, NJ.

## References

1. Crouch, R. J. and Dirksen, M. L. (1982) Ribonuclease H, in *Nucleases* (Linn, S. M. and Roberts, R. J., eds.), Cold Spring Harbor Laboratory, Cold Spring Harbor, NY, pp. 211–241.
2. Hausen, P. and Stein, H. (1970) Ribonuclease H. An enzyme degrading the RNA moiety of DNA-RNA hybrids. *Eur. J. Biochem.* **14,** 278–283.
3. Busen, W. (1980) Purification, subunit structure, and serological analysis of calf thymus ribonuclease H I. *J. Biol. Chem.* **255,** 9434–9443.
4. Sawai, Y., Kitahara, N., Thung, W. L., Yanokura, M., and Tsukada, K. (1981) Nuclear location of ribonuclease H and increased level of magnesium-dependent ribonuclease H in rat liver on thioacetamide treatment. *J. Biochem.* **90,** 11–16.
5. Busen, W. and Frank, P. (1998) Bovine ribonucleases H, in *Ribonucleases H* (Crouch, R. J. and Toulme, J. J., eds.), Les Editions Inserm, Paris, pp. 113–146.
6. Toulme, J. J., Frank, P., and Crouch, R. J. (1998) Human ribonucleases H, in *Ribonucleases H* (Crouch, R. J. and Toulme, J. J., eds.), Les Editions Inserm, Paris, pp. 147–162.
7. Stein, C. A., Subasinge, C., Shinozuka, K., and Cohen, J. S. (1988) Physicochemical properties of phosphorothioate oligodeoxynucleotides. *Nucleic Acids Res.* **16,** 3209–3221.
8. Furdon, P. J., Dominski, Z., and Kole, R. (1989) RNase H cleavage of RNA hybridized to oligonucleotides containing methylphosphonate, phosphorothioate and phosphodiester bonds. *Nucleic Acids Res.* **17,** 9193–9204.
9. Gagnor, C., Bertrand, J.-R., Thenet, S., Lemaitre, M., Morvan, R., Rayner, B., Malvy, C., Lebleu, B., Imbach, J. L., and Paoletti, C. (1987) α-DNA VI: comparative study of α- and β-anomeric oligodeoxyribonucleotides in hybridisation to mRNA and in cell free translation inhibition. *Nucleic Acids Res.* **15,** 10,419–10,436.
10. Giles, R. V. and Tidd, D. M. (1992) Enhanced RNase H activity with methylphosphonodiester/phosphodiester chimeric antisense oligodeoxynucleotides. *Anti-Cancer Drug Des.* **7,** 37–48.
11. Minshull, J. and Hunt, T. (1986) The use of single-stranded DNA and RNase H to promote quantitative "hybrid arrest of translation" of mRNA/DNA hybrids in reticulocyte lysate cell-free translations. *Nucleic Acids Res.* **14,** 6433–6451.

12. Walder, R. Y. and Walder, J. A. (1988) The role of RNase H in hybrid-arrested translation by antisense oligonucleotides. *Proc. Natl. Acad. Sci. USA* **85**, 5011–5015.

13. Shuttleworth, J. and Colman, A. (1988) Antisense oligonucleotide-directed cleavage of mRNA in *Xenopus* oocytes and eggs. *EMBO J.* **7**, 427–434.

14. Shuttleworth, J., Matthews, G., Dale, L., Baker, C., and Colman, A. (1988) Antisense oligodeoxyribonucleotide-directed cleavage of maternal mRNA in *Xenopus* oocytes and embryos. *Gene* **72**, 267–275.

15. Giles, R. V., Spiller, D. G., and Tidd, D. M. (1995) Detection of ribonuclease H-generated mRNA fragments in human leukemia cells following reversible membrane permeabilization in the presence of antisense oligodeoxynucleotides. *Antisense Res. Dev.* **5**, 23–31.

16. Giles, R. V., Ruddell, C. J., Spiller, D. G., Green, J. A., and Tidd, D. M. (1995) Single base discrimination for ribonuclease H-dependent antisense effects within intact human leukaemia cells, *Nucleic Acids Res.* **23**, 954–961.

17. Spiller, D. G., Giles, R. V., Broughton, C. M., Grzybowski, J., Ruddell, C. J., Clark, R. E., and Tidd, D. M. (1998) The influence of target protein half life on the effectiveness of antisense oligonucleotide analogue–mediated biological responses. *Antisense Nucl. Acid Drug Devel.* **8**, 281–293.

18. Partridge, M., Vincent, A., Matthews, P., Puma, J., Stein, D., and Summerton, J. (1996) A simple method for delivering morpholino antisense oligos into the cytoplasm of cells. *Antisense Nucl. Acid Drug Devel.,* **6**, 169–175.

19. Summerton, J., Stein, D., Huang, S. B., Matthews, P., Weller, D. D., and Partridge, M. (1997) Morpholino and phosphorothioate antisense oligomers compared in cell-free and in-cell systems. *Antisense Nucleic Acid Drug Devel.* **7**, 63–70.

20. Giles, R. V., Spiller, D. G., Clark, R. E., and Tidd, D. M. (1999) Antisense morpholino oligonucleotide analogue induces mis-splicing of c-*myc* mRNA. *Antisense Nucleic Acid Drug Devel.* **9**, 213–220.

21. Sierakowska, H., Sambade, M. J., Agrawal, S., and Kole, R. (1996) Repair of thalassemic human beta -globin mRNA in mammalian cells by antisense oligonucleotides. *Proc. Natl. Acad. Sci. USA* **93**, 12,840–12,844.

22. Sierakowska, H., Montague, M., Agrawal, S., and Kole, R. (1997) Restoration of beta-globin gene expression in mammalian cells by antisense oligonucleotides that modify the aberrant splicing patterns of thalassemic pre-mRNAs. *Nucleosides Nucleotides* **16**, 1173–1182.

23. Liebhaber, S. A., Cash, F. E., and Shakin, S. H. (1984) Translationally associated helix-destabilizing activity in rabbit reticulocyte lysate. *J. Biol. Chem.* **259**, 15,597–15,602.

24. Shakin, S. H. and Liebhaber, S. A. (1986) Destabilization of messenger RNA/complementary DNA duplexes by the elongating 80S ribosome. *J. Biol. Chem.* **261**, 16,018–16,025.

25. Tidd, D. M. and Warenius, H. M. (1989) Partial protection of oncogene, antisense oligodeoxynucleotides against serum nuclease degradation using terminal methylphosphonate groups. *Brit. J. Cancer* **60**, 343–350.

26. Woolf, T. M., Jennings, C. G. B., Rebagliati, M., and Melton, D. A. (1990) The stability, toxicity and effectiveness of unmodified and phosphorothioate antisense oligodeoxynucleotides in Xenopus oocytes and embryos. *Nucleic Acids Res.* **18**, 1763–1769.

27. Hoke, G. D., Draper, K., Freier, S. M., Gonzalez, C., Driver, V. B., Zounes M. C., and Ecker, D. J. (1991) Effects of phosphorothioate capping on antisense oligonucleotide stability, hybridization and antiviral efficacy versus herpes simplex virus infection. *Nucleic Acids Res.* **19**, 5743–5748.

28. Sburlati, A. R., Manrow, R. E., and Berger, S. L. (1991) Prothymosin alpha antisense oligomers inhibit myeloma cell division. *Proc. Natl. Acad. Sci. USA* **88**, 253–257.

29. Vaerman, J. L., Lammineur, C., Moureau, L. P., Lewalle, P., Deldime, P., Blumenfeld, M., and Martiat, P. (1995) BCR-ABL antisense oligodeoxyribonucleotides suppress the growth of leukemic and normal hematopoietic cells by a sequence-specific but non-antisense mechanism. *Blood* **86,** 3891–3896.

30. Stein, C. A. and Cheng, Y.-C. (1993) Antisense oligonucleotides as therapeutic agents— is the bullet really magical? *Science* **261,** 1004–1012.

31. Stein, C. A. (1995) Does antisense exist? *Nature Med.* **11,** 1119–1121.

32. Stein, C. A. (1996) Phosphorothioate antisense oligodeoxynucleotides: questions of specificity. *Trends Biotechnol.* **14,** 147–149.

33. Tidd, D. M. and Giles, R. V. (2000) Mechanisms of action of antisense oligonucleotides, in *Pharmaceutical Aspects of Oligonucleotides* (Couvreur, P. and Malvey, C., eds.), Taylor & Francis, London, UK, pp. 3–31.

34. Fisher, L. M. (1998) F. D. A. Approves a Drug That Blocks a Gene, *New York Times*, August 28.

35. Giles, R. V., Spiller, D. G., Green, J. A., Clark, R. E., and Tidd, D. M. (1995) Optimization of antisense oligodeoxynucleotide structure for targeting *bcr-abl* mRNA. *Blood* **86,** 744–754.

36. Giles, R. V. and Tidd, D. M. (1992) Increased specificity for antisense oligodeoxynucleotide targeting of RNA cleavage by RNase H using chimeric methylphosphonodiester/PO structures. *Nucleic Acids Res.* **20,** 763–770.

37. Giles, R. V., Spiller, D. G., and Tidd, D. M. (1993) Chimeric oligodeoxynucleotide analogues: enhanced cell uptake of structures which direct ribonuclease H with high specificity. *Anti-Cancer Drug Des.* **8,** 33–51.

38. Loke, S. L., Stein, C. A., Zhang, X. H., Mori, K., Nakanishi, M., Subasinghe, C., Cohen, J. S., and Neckers, L. M. (1989) Characterization of oligonucleotide transport into living cells. *Proc. Natl. Acad. Sci. USA* **86,** 3474–3478.

39. Yakubov, L. A., Deeva, E. A., Zarytova, V. F., Ivanova, E. M., Ryte, A. S., Yurchenko, L. V., and Vlassov, V. V. (1989) Mechanism of oligonucleotide uptake by cells: involvement of specific recptors? *Proc. Natl. Acad. Sci. USA* **86,** 6454–6458.

40. Spiller, D. G. and Tidd, D. M. (1992) The uptake kinetics of chimeric oligodeoxynucleotide analogues in human leukaemia MOLT-4 cells. *Anti-Cancer Drug Des.* **7,** 115–129.

41. Spiller, D. G. and Tidd, D. M. (1995) Nuclear delivery of antisense oligodeoxynucleotides through reversible permeabilization of human leukemia cells with streptolysin O. *Antisense Res. Dev.* **5,** 13–21.

42. Bergan, R., Connell, Y., Fahmy, B., and Neckers, L. (1993) Electroporation enhances c-*myc* antisense oligodeoxynucleotide efficacy. *Nucleic Acids Res.* **21,** 3567–3573.

43. Spiller, D. G., Giles, R. V., Grzybowski, J., Tidd, D. M., and Clark, R. E. (1998) Improving the intracytoplasmic delivery and molecular efficacy of antisense oligonucleotides in chronic myeloid leukaemia cells: a comparison of streptolysin O permeabilisation, electroporation and lipophillic conjugation. *Blood* **91,** 4738–4746.

44. Bennett, C. F., Chiang, M.-Y., Chan, H., Shoemaker, J. E. E., and Mirabelli, C. K. (1992) Cationic lipids enhance cellular uptake and activity of phosphorothioate antisense oligonucleotides. *Mol. Pharmacol.* **41,** 1023–1033.

45. Neckers, L. M. and Iyer, K. (1998) Nonantisense effects of antisense oligonucleotides, in *Applied Antisense Oligonucleotide Technology* (Stein, C. A. and Krieg, A. M., eds.), Wiley-Liss, New York, pp. 147–159.

46. Rando, R. F. and Hogan, M. E. (1998) Biological activity of guanosine quartet forming oligonucleotides, in *Applied Antisense Oligonucleotide Technology* (Stein, C. A. and Krieg, A. M., eds.), Wiley-Liss, New York, pp. 335–352.

47. Stein, C. A. and Krieg, A. M. (1994) Problems in interpretation of data derived from *in vitro* and *in vivo* use of antisense oligodeoxynucleotides. *Antisense Res. Dev.* **4,** 67–69.

48. Hogrefe, R. I., Vaghefi, M. M., Reynolds, M. A., Young, K. M., and Arnold, L. J., Jr. (1993) Deprotection of methylphosphonate oligonucleotides using a novel one-pot method. *Nucleic Acids Res.* **21,** 2031–2038.

49. Hogrefe, R. I., Reynolds, M. A., Vaghefi, M. M., Young, K. M., Riley, T. A., Klein, R. E., and Arnold, L. J., Jr. (1993) An improved method for the synthesis and deprotection of methylphosphonate oligonucleotides, in *Methods in Molecular Biology, Vol. 20: Protocols for Oligonucleotides and Analogs: Synthesis and Properties* (Agrawal, S., ed.), Humana Press, Totowa, NJ, pp. 143–164.

50. Zon, G. (1993) Oligonucleoside phosphorothioates, in *Methods in Molecular Biology, Vol. 20: Protocols for Oligonucleotides and Analogs: Synthesis and Properties* (Agrawal, S., ed.), Humana Press, Totowa, NJ, pp. 165–189.

51. Giles, R. V. (1993) Chimeric oligodeoxynucleotides: an *in vitro* investigation. Ph.D. Thesis, Liverpool University.

52. Kroczek, R. A. (1989) Immediate visualisation of blotted RNA in northern analysis. *Nucleic Acids Res.* **17,** 94–97.

53. Giles, R. V., Spiller, D. G., Grzybowski, J., Clark, R. E., Nicklin, P., and Tidd, D. M. (1998) Selecting optimal oligonucleotide composition for maximal effect following streptolysin O-mediated delivery into human leukaemia cells. *Nucleic Acids Res.* **26,** 1567–1575.

54. Giles, R. V., Spiller D. G., Clark, R. E., and Tidd, D. M. (1999) Identification of a good c-*myc* antisense oligodeoxynucleotide target site and the inactivity at this site of novel NCH triplet-targeting ribozymes. *Nucleosides Nucleotides* **18,** 1935–1944.

55. Manley J. L. (1984) Transcription of eukaryotic genes in a whole cell extract, in *Transcription and Translation: A Practical Approach* (Hames, B. D. and Higgins, S. J., eds.), IRL Press, Oxford, UK, pp. 71–88.

56. Chomczynski, P. and Sacchi, N. (1987) Single-step method of RNA isolation by acid guanidinium thiocyanate-phenol-chloroform extraction. *Anal. Biochem.* **162,** 156–159.

57. Barry, E. L. R., Gesek, F. A., and Friedman, P. A. (1993) Introduction of antisense oligonucleotides into cells by permeabilization with streptolysin O. *BioTechniques* **15,** 1016–1018.

58. Giles, R. V., Grzybowski, J., Spiller, D. G., and Tidd, D. M. (1997) Enhanced antisense effects resulting from an improved streptolysin-O protocol for oligodeoxynucleotide delivery into human leukaemia cells. *Nucleosides Nucleotides* **16,** 1155–1163.

59. Broughton, C. M., Spiller, D. G., Pender, N., Komorovskaya, M., Grzybowski, J., Giles, R. V., Tidd, D. M., and Clark, R. E. (1997) Preclinical studies of streptolysin-O in enhancing antisense oligonucleotide uptake in harvests from chronic myeloid leukaemia patients. *Leukemia* **11,** 1435–1441.

60. Clark, R. E., Grzybowski, J., Broughton, C. M., Pender, N., Spiller, D. G., Brammer, C. G. Giles, R. V., and Tidd, D. M. (1999) Transplantation of antisense oligonucleotide purged autografts in chronic myeloid leukaemia. *Bone Marrow Transplantation* **23,** 1305–1308.

61. Bertrand, E., Fromont-Racine, M., Pictet, R., and Grange, T. (1993) Visualisation of the interaction of a regulatory protein with RNA *in vivo*. *Proc. Natl. Acad. Sci. USA* **90,** 3496–3500.

62. Bertrand, E., Pictet, R., and Grange, T. (1994) Can hammerhead ribozymes be efficient tools to inactivate gene function? *Nucleic Acids Res.* **22,** 293–300.

63. Tidd, D. M. (1998) Ribonuclease H-mediated antisense effects of oligonucleotides and controls for antisense experiments, in *Applied Antisense Oligonucleotide Technology* (Stein, C. A. and Krieg, A. M., eds.), Wiley-Liss, New York, pp. 161–171.

64. Milligan, J. F., Groebe, D. R., Witherall, G. W., and Uhlenbeck, O. C. (1987) Oligoribonucleotide synthesis using T7 RNA polymerase and synthetic DNA templates. *Nucleic Acids Res.* **15,** 8783–8798.

65. Sproat, B. S. and Gait, M. J. (1984) Solid-phase synthesis of oligodeoxyribonucleotides by the phosphotriester method, in *Oligonucleotide Synthesis: A Practical Approach* (Gait, M. J., ed., IRL Press, Oxford, UK, pp. 83–114.

66. Sambrook, J., Fritsch, E. F., and Maniatis, T. (1989) *Molecular Cloning: A Laboratory Manual*, Cold Spring Harbor Laboratory, Cold Spring Harbor, NY.

67. Veal, G. J., Agrawal, S., and Bryn, R. A. (1998) Sequence-specific RNase H cleavage of *gag* mRNA from HIV-1 infected cells by an antisense oligonucleotide *in vitro. Nucleic Acids Res.* **26,** 5670–5675.

68. Eder, P. S., Walder, R. Y., and Walder, J. A. (1993) Substrate specificity of human RNase HI and its role in excision repair of ribose residues misincorporated in DNA. *Biochimie* **75,** 123–126.

# 12

# The 2-5A/RNase L Pathway and Inhibition by RNase L Inhibitor (RLI)

**Catherine Bisbal, Tamim Salehzada, Michelle Silhol, Camille Martinand, Florence Le Roy, and Bernard Lebleu**

## 1. Introduction

### 1.1. The 2-5A/RNase L Pathway and Its Physiological Relevance

Interferons (IFN) belong to a multigenic family of cytokines that activate the transcription of several genes through a signaling pathway involving cell-surface receptors (IFNAR), Janus kinases (JAK), and signal-transducing proteins (STAT), as comprehensively reviewed by Stark et al. *(1)*.

Although IFNs have been first characterized as broad-spectrum antiviral agents, they also exhibit growth-inhibitory and immunomodulatory activities which may be important for the control of various pathogens and for anticancer action (*see* **refs.** *2* and *3* for reviews).

The molecular basis of IFN-modulated antiviral activity has been the most widely studied and remains the best understood. Three major pathways based on dsRNA-dependent protein kinase (PKR), 2-5A/RNase L, and the Mx protein have been found.

The main features of the 2-5A/RNase L pathway (**Fig. 1**) were first delineated about 20 years ago through independent observations by the groups of P. Lengyel, who first described an IFN-modulated endoribonuclease (RNase L) activity *(4)*, and I. Kerr, who discovered and characterized the unusal 2',5'-oligoadenylates (2-5A) activators of this enzyme *(5)*. In brief, IFN-α,β induce the transcription of several 2-5A synthetase genes whose protein products are in turn activated by synthetic (poly (rI) · poly (rC), and so forth) or natural (such as picornavirus replicative intermediates) double-stranded RNA (dsRNA), and polymerize 2',5'-oligoadenylates (2-5A) from ATP. The 2-5A oligomers bind specifically to RNase L, which is activated to cleave single-stranded RNAs at UpX (mainly UpU and UpA) sequences (*see* **refs.** *6–8* for reviews).

RNase L was cloned and expressed in various cell types or extracts by R. Silverman and colleagues. These results have established the importance of this metabolic pathway for the antiviral and growth regulatory activities of IFN-α,β *(9–11)*.

A wider biological significance for the 2-5A/RNase L pathway has long been suspected. Biochemical and genetic tools are now available to determine the possible role

From: *Methods in Molecular Biology, vol. 160: Nuclease Methods and Protocols*
Edited by: C. H. Schein © Humana Press Inc., Totowa, NJ

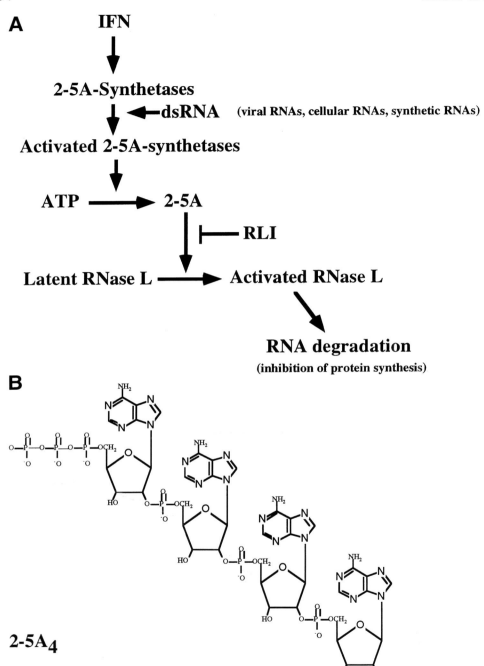

Fig. 1. (**A**) Outline of the 2-5A/RNase L pathway and (**B**) structure of the 5'-triphosphorylated 2-5A tetramer.

of RNase L in the control of cellular mRNAs metabolism, in the control of cell differentiation or in virus-induced apoptosis (reviewed in **ref. 8**).

## 1.2. RNase L Inhibitor (RLI), a New Component of the 2-5A/RNase L Pathway

The overall activity of the 2-5A/RNase L pathway was initially thought to be controlled solely by the availability of 2-5A, which in turn depended on the activity of 2-5A synthetases and of poorly characterized 2-5A degrading enzymes, such as phosphatases and phosphodiesterases *(12,13)*. Our group serendipitously cloned a new cDNA species whose protein product inhibited RNase L activity in cell-free extracts and in intact cells *(14)*.

Although a detailed mechanism for RNase L inhibitor (RLI) action has not been determined, we know that it associates reversibly with RNase L, thereby preventing 2-5A binding and endoribonuclease activation. RLI does not induce 2-5A degradation or sequestration. RLI may be instrumental in buffering RNase L- mediated RNA degradation and finely tuning the overall activity of the 2-5A/RNase L pathway *(14)*.

In this chapter, we will describe assays for RLI in whole cells or in cell extracts that rely on its capacity to interfere with the binding of 2-5A to RNase L, or on inhibition of RNase L biological activity.

## 1.3. Expression of RLI In Vivo and During Virus Infection

Recent studies in our group have documented an upregulation of RLI in cells infected with RNA viruses, such as picornaviruses *(15)* or HIV-1 *(16)*.

The stable transfection of cells with RLI sense or antisense-expressing plasmids has corroborated the hypothesis that viruses induce RLI to limit the antiviral activity of RNase L. Other physiological parameters might affect RLI expression and the overall activity of the 2-5A/RNase L pathway as suggested by recent studies in our group concerning the terminal in vitro differentiation of rodent myoblasts. We have observed an induction of RLI during the terminal differentiation of the murine myoblasts $C_2Cl_2$ in myotubes. This induction is concomitant to the decrease in RNase L activity (Bisbal et al., *Mol. Cell. Biol.*, in press).

The overexpression of RLI sense or antisense constructions in various cell lines (or possibly in vivo in transgenic animals) will provide additional tools to delineate the role of the 2-5A/RNase L pathway in cell physiology or in the control of virus multiplication. It is worth noting that the overexpression of a RLI antisense construction *(15,16)* or of RNase L *(17)* has led to essentially similar conclusions concerning the role of the 2-5A/RNase L pathway in cells infected with picornaviruses or with HIV-1.

## 2. Materials

## 2.1. Buffers and Solutions

1. Cell lysis hypotonic buffer: 0.5% (v/v) Nonidet P-40 (NP-40), 20 m$M$ HEPES, pH 7.5, 10 m$M$ KOAc, 15 m$M$ Mg(OAc)$_2$, 1 m$M$ phenylmethylsulfonyl fluoride (PMSF), 10 μg/mL aprotinin, 150 μg/mL leupeptin.
2. 19 m$M$ KCl, 31 m$M$ Mg(OAc)$_2$, 25 m$M$ HEPES/KOH, pH 7.4, 6 m$M$ ATP, 1.25 m$M$ dithiothreitol (DTT), 5 m$M$ fructose 1–6 diphosphate, and 25 μg/mL poly(rI)·poly(rC).
3. 125 m$M$ triethylammonium bicarbonate buffer, pH 8.5.
4. 450 m$M$ triethylammonium bicarbonate buffer, pH 8.5.
5. 50 m$M$ ammonium phosphate buffer, pH 7.2.
6. Methanol/water (1/1, v/v).

7. Ligase buffer: 100 m$M$ HEPES, pH 7.6, 15 m$M$ MgCl$_2$, 6.6 m$M$ DTT, 20% (v/v) dimethyl sulfoxide (DMSO). The ligase buffer is kept in aliquots at –20°C. DMSO and DTT are added immediately before use.
8. 100 m$M$ Tris-HCl, pH 8.5.
9. 20 m$M$ ethylenediamine tetraacetic acid disodium (EDTA).
10. 93.5 m$M$ sodium metaperiodate (NaIO$_4$) in 100 m$M$ Na(Oac), pH 4.75, prepared just before use.
11. 100 m$M$ HCl.
12. 100 m$M$ NaOH.
13. 100 m$M$ isopropylthio-β-D-galactoside (IPTG).
14. 6 $M$ guanidinium buffer, pH 7.8.
15. Denaturing binding buffer: 8 $M$ urea, 20 m$M$ NaH$_2$PO$_4$/Na$_2$HPO$_4$, pH 7.8, 500 m$M$ NaCl.
16. Denaturing wash buffer: 8 $M$ urea, 20 m$M$ sodium phosphate, pH 6.0, 500 m$M$ NaCl.
17. Denaturing elution buffer: 8 $M$ urea, 20 m$M$ NaH$_2$PO$_4$/Na$_2$HPO$_4$, pH 4.0, 500 m$M$ NaCl.
18. Dialysis buffer: 10 m$M$ HEPES, pH 7.5, 80 m$M$ KCl, 5 m$M$ Mg(OAc)$_2$, 1 m$M$ EDTA, 5%(v/v) glycerol.
19. Renaturation buffer: 0.5% (v/v) NP-40, 20 m$M$ HEPES, pH 7.5, 10 m$M$ KOAc, 15 m$M$ Mg(OAc)$_2$, 1 m$M$ PMSF, 10 μg/mL aprotinin, 150 μg/mL leupeptin.
20. TBE buffer: 50 m$M$ Tris-HCl, pH 8.0, 50 m$M$ boric acid, 1 m$M$ EDTA.
21. Phenol, dichloromethane.
22. 3 $M$ NaOAc.
23. TE buffer: 10 m$M$ Tris-HCl, pH 7.4, 0.1 m$M$ EDTA.
24. 5X transcription buffer: 200 m$M$ Tris-HCl, pH 7.5, 250 m$M$ NaCl, 40 m$M$ MgCl$_2$, 10 m$M$ spermidine.
25. RNTPs: 10 m$M$ rUTP, 10 m$M$ rCTP, 10 m$M$ rATP, 1 m$M$ rGTP, 750 m$M$ DTT.
26. 1.1 m$M$ amino-acid mixture.
27. L-($^{35}$S)methionine (1200 Ci/mmol) at 10 mCi/mL.
28. Poly(ethylene glycol) 6000 (25% [w/v]).
29. Carrier: bovine fetal serum (BFS) diluted two times in water.
30. 60 m$M$ sodium cyanoborohydride (NaBH$_3$CN) in 100 m$M$ NaH$_2$PO$_4$/Na$_2$HPO$_4$, pH 8.0, prepared just before use.
31. 100 m$M$ Tris-HCl, pH 8.9, 5% (w/v) sodium dodecyl sulfate (SDS), 20% (v/v) glycerol, 5% (v/v) mercaptoethanol.
32. 25 m$M$ Tris-HCl, pH 8.3, 192 m$M$ glycine, 20% (v/v) ethanol.
33. 10 m$M$ HEPES, pH 7.5, 80 m$M$ KCl, 5 m$M$ Mg(OAc)$_2$, 1 m$M$ EDTA, 5% (v/v) glycerol, 20 m$M$ mercaptoethanol.
34. Phosphate-buffered saline buffer (PBS): 140 m$M$ NaCl, 2 m$M$ KCl, 8 m$M$ Na$_2$HPO$_4$, 1.5 m$M$ KH$_2$PO$_4$, pH 7.4.
35. 20% (v/v) Tween 20.
36. Skimmed milk.
37. 100 m$M$ Acetate buffer (Na(OAc)), pH 4.5.
38. Storage buffer: 20 m$M$ Tris-HCl, pH 7.6, 5 m$M$ Mg(OAc)$_2$, 85 m$M$ KCl, and 5% (v/v) glycerol.
39. Cell lysis buffer: 5 m$M$ Tris-HCl, pH 7.6, 1.25% (v/v) glycerol, 20 m$M$ KCl.
40. 25 m$M$ Mg(OAc)$_2$.
41. Trypan blue 0.5% (w/v) in PBS.
42. Electrophoresis-loading buffer: 50% (v/v) glycerol, TBE buffer, and 0.25% (w/v) bromophenol blue.
43. Ethidium bromide: 10 mg/mL.
44. Geneticin (G418) 100 mg/mL.

45. Solution A: 40 μL DNA carrier (salmon sperm 500 μg/mL), 40 μL DNA plasmid (14 μg), 920 μL $H_2O$, 1 mL $CaCl_2$ (250 m$M$).
46. Solution B: 50 m$M$ HEPES/KOH, pH 7.1, 280 m$M$ NaCl, 1.5 m$M$ $Na_2HPO_4$. This solution is filtrated under sterile condition, aliquoted, and stored at –20°C. Thawed aliquots cannot be refrozen to avoid pH change.

## 2.2. Media

### 2.2.1. Cells

1. RPMI-1640 Glutamax medium (Gibco-BRL), 10% (v/v) fetal calf serum (FCS).
2. DMEM (Gibco-BRL), 10% (v/v) FCS.

### 2.2.2. Bacteria

1. SOB (for 1 L): 20 g tryptone, 5 g yeast extract, 0.5 g NaCl. Mix until dissolved. Add 10 mL 250 m$M$ KCl. Adjust the pH to 7.0 with 5 $N$ NaOH. Adjust the volume to 1 L and sterilize by autoclaving. After autoclaving, add 5 mL sterile 2 $M$ $MgCl_2$.

## 2.3. Enzymes and Proteins

1. Human leukocyte interferon (hu-IFN α), specific activity $2.10^6$ IU/mg. Aliquots of Hu-IFN are stored at –80°C. One possible source of IFN is PBL Biomedical laboratories (www.interferonsource.com/catalog.asp#top).
2. Trypsin solution 2.5% (w/v) (Gibco-BRL) diluted 1/10 in PBS.
3. T4 RNA ligase: Pharmacia (reference: 27-0883-01); other preparations may not provide a good yield of ligation.
4. *Escherichia coli* alkaline phosphatase Type III (BAP) (Sigma) 2.7 mg/mL, 64 U/mg.
5. RNA polymerase T7 (50 U/μL) (Promega).
6. RNA polymerase SP6 (20 U/μL) (Promega).
7. rRNAsin (RNase inhibitor) (40 U/μL) (Promega).
8. Proteinase K in water (1 mg/mL) (Boehringer).

## 2.4. Cell Lines, Plasmids, Nucleotides, and Kits

1. HeLa cells (human epithelial cells, ATCC CCL2) are plated at $10^4$ cells/mL in RPMI, 10% (v/v) FCS, grown at 37°C in 5% $CO_2$.
2. Daudi cells (human B lymphoblast cell line, ATCC CC213) are plated at $5 \times 10^4$ cells/mL in RPMI, 10% (v/v) FCS, grown at 37°C in 5% $CO_2$.
3. *E. coli* strain JM 109: recA1 supE44 endA1 hsdR17 gyrA96 relA1 thi Δ(lac-proAB) F' [traD36 proAB$^+$ lacl$^q$ lacZΔM15] (Invitrogen).
4. pRSET vector driven by a T7 promoter: the Xpress™ Protein Expression System (Invitrogen).
5. pcDNA3 (Invitrogen).
6. Rabbit reticulocyte lysate (Promega).
7. mCAP-mRNA capping kit from Stratagene.
8. L-($^{35}$S) methionine (1200 Ci/mmol, 10 mCi/mL), Amersham 5J1015 or AG1094.
9. ($^{32}$P) pCp (3000 Ci/mmol, 10 mCi/mL).
10. Chemoluminescence: Renaissance kit (NEN).

## 2.5. Equipment

1. DEAE-Trisacryl M (Pharmacia).
2. μBondapak C18 HPLC column 150 × 4.6 mm (Hypersil,ThermoQuest Corporation).

3. Zorbax GF-250 (DuPont Company).
4. Agarose-adipic acid hydrazide (27-5496-02) (Pharmacia).
5. Nickel agarose column (Invitrogen).
6. Protran Nitrocellulose Transfer Membrane (0.2 mm) (Schleicher & Schuell).
7. Electrotransfer (Western blotting) apparatus: Trans-blot™ Cell (Bio-Rad).

## 3. Methods
### 3.1. Enzymatic Synthesis of 2-5A Oligomers

2-5A$_n$ oligomers are synthesized enzymatically in extracts of IFN-treated human HeLa cells as described by Minks et al. *(18)*.

1. Prepare 10 (9-cm diameter) cell-culture plates containing 10 mL HeLa cell medium (**Subheading 2.2., item 1**). Dilute HeLa cells to $10^4$/mL.
2. After 20–24 h add 200 U/mL hu-IFN α and incubate 48 h more.
3. Take off the medium, rinse the cells with 10 mL PBS (**Subheading 2.1., item 34**), add 1 mL trypsin solution/dish (**Subheading 2.3., item 2**), and incubate 2–3 min at 37°C. Add 1 mL FCS and 9 mL PBS, collect the cells by pipeting in and out, and centrifuge 5 min at 800$g$.
4. Resuspend the cells in 2 vol of hypotonic buffer (**Subheading 2.1., item 1**), incubate 10 min on ice, disrupt with a Dounce homogenizer, and centrifuge 10 min at 10,000$g$.
5. Collect the supernatant, dilute with 4 vol of 2-5A synthesis buffer (**Subheading 2.1., item 2**) and incubate 2 h at 37°C.
6. Denature the proteins by incubation at 100°C for 5 min and centrifuge at 10,000$g$ for 10 min.
7. Dilute the supernatant with 3 vol water and adjust to pH 8.5 with 0.1 $M$ KOH before loading on a DEAE-Trisacryl M (Pharmacia) column equilibrated with buffer (**Subheading 2.1., item 3**). Wash the column with 1500 mL of this buffer.
8. Elute the 2-5A$_n$ oligomers with a linear gradient (1500 mL/1500 mL) of 125–450 m$M$ triethylammonium bicarbonate buffer pH 8.5 (**Subheading 2.1., item 4**).
9. Control the lengths of the 2-5A oligomers (usually 2–8 bases) by HPLC chromatography on a µBondapak C$_{18}$ column in ammonium phosphate buffer (**Subheading 2.1., item 5**). Equilibrate the column with buffer (**Subheading 2.1., item 5**) and inject the 2-5A$_n$, and elute over 25 min with 25 mL of a 0–50% linear gradient of methanol/water (1/1, v/v) *(19,20)*.
10. Combine the appropriate fractions of the DEAE-Trisacryl M column, concentrate *in vacuo* under reduced pressure and coevaporate with water several times to remove the triethylammonium bicarbonate buffer. Milligram amounts of each of the oligomer can be obtained in a pure form. 2–5A$_n$ oligomers can be stored for several years at –20°C in aqueous solution.

### 3.2. Synthesis of 2-5A$_n$-3'-($^{32}$P)pCp Radiolabeled Probe

The 2-5A$_n$-3'-($^{32}$P)pCp probe is synthesized by ligation of ($^{32}$P)pCp at the 3' end of 2-5A$_3$ or 2-5A$_4$ with T4 RNA ligase as originally described by Knight et al. *(21)*.

1. Mix 1 µL of 2-5A$_4$ (33 µ$M$ final concentration) with 9 µL of ATP (5 µ$M$ final concentration), 25 µL of ligase buffer (**Subheading 2.1., item 7**), 5 µL (60 U) of T4 RNA ligase and 10 µL (100 µCi) ($^{32}$P)pCp (3000 Ci/mmol). Incubate 48 h at 4°C.
2. Heat the reaction mixture at 90°C for 5 min and remove the precipitated T4 RNA ligase by centrifuging at 10,000$g$ for 10 min.
3. Separate the 2-5A$_4$-3'-($^{32}$P)pCp from excess 2-5A$_4$ and ($^{32}$P)pCp by HPLC chromatography on a µBondapak C$_{18}$ column in ammonium phosphate buffer (**Subheading 2.1., item 5**) as described in **Subheading 3.1, step 9**. Collect 500-µL fractions.

4. Count 5 μL of each fraction to identify the eluted peak corresponding to 2-5A$_4$-3'-($^{32}$P)pCp. In general 2-5A$_4$-3'-($^{32}$P)pCp elutes 12 min after the beginning of the gradient in four fractions.

5. Test each fraction for its capacity to bind to RNase L with the radiobinding assay described in **Subheading 3.7.** Daudi cell extracts (S10) are used as a source of natural RNase L, since these human lymphoblastoid cells are easy to grow and contain a high level of endogenous RNase L. The 2-5A$_4$-3'-($^{32}$P)pCp probe is used as such, or can be oxidized to be covalently bound to RNase L for denaturing gel analysis (SDS-PAGE) *(22)* as described in **Subheadings 3.3.** and **3.8.**

## 3.3. 2-5A$_4$-3'-($^{32}$P)pCp Oxidation

2-5A$_4$-3'-($^{32}$P)pCp are dephosphorylated under mild conditions to eliminate the 3' terminal phosphate and oxidize the 3' ribose residue *(23)*.

1. Mix 100 μL of 2-5A$_n$-3'-($^{32}$P)pCp, 50 μL of 100 m*M* Tris-HCl, pH 8.5 (**Subheading 2.1., item 8**) and 0.8 U BAP. Incubate 1 h in ice.

2. Inactivate BAP by adding 20 μL 20 m*M* EDTA (**Subheading 2.1., item 9**) and heating 5 min at 90°C.

3. Adjust the pH to 4.5 with 100 m*M* HCl.

4. Add 50 μL of freshly prepared NaIO$_4$ (**Subheading 2.1., item 10**) (10 m*M* final concentration) and incubate 1 h on ice in the dark.

5. Adjust the pH carefully to 8.0 with 100 m*M* NaOH. The oxidized 2-5A$_4$-3'-($^{32}$P)pCp can be stored at –20°C for several days.

## 3.4. Production of Recombinant RLI Protein in Bacteria

To produce large quantities of RLI protein for immunizing rabbits, the RLI coding sequence was cloned into a pRSET vector behind the T7 promoter of phage T7. The DNA insert is positioned downstream and in frame, with a sequence encoding a N-terminal fusion peptide. This sequence includes an ATG translation initiation codon and a tract of six histine residues which bind metals. The fusion protein can be purified in one step by chromatography on a nickel agarose column. The plasmid is propagated with the vector in JM109, which does not contain T7 polymerase. The transfected bacteria are then infected with an M13 phage containing a T7 polymerase gene driven by the *E. coli* lactose promoter. In the presence of IPTG, T7 polymerase is produced to transcribe the recombinant protein.

1. Infect 125 mL of the JM109 culture (DO = 0.3) (transfected with pRSET-RLI) with the M13/T7 phage (multiplicity of infection: 5 pfu/cell) in the presence of IPTG (final concentration 1 m*M*) for 6 h.

2. Centrifuge at 3000*g* for 5 min to collect the bacteria

3. Resuspend the pellet in 12 mL 6 *M* guanidinium buffer, pH 7.8 (**Subheading 2.1., item 14**), rock slowly the closed tubes for 5 min at room temperature, and sonicate six times with 5-s pulses. Store the lysate at –80°C.

4. Equilibrate 2 mL Nickel agarose in denaturing binding buffer (**Subheading 2.1., item 15**).

5. Load 5 mL of the bacterial extract on the column.

6. Wash the column four times with 10 ml of denaturing binding buffer (**Subheading 2.1., item 15**), and four times with 10 mL denaturing wash buffer (**Subheading 2.1., item 16**).

7. Elute the recombinant RLI protein with denaturing elution buffer (**Subheading 2.1., item 17**) and collect 1-μL fractions. The elution is monitored by following the absorbance at 280 nm.

8. Pool the RLI fractions and dialyse overnight at 4°C against $2 \times 1$ L of dialysis buffer (**Subheading 2.1., item 18**). The RLI protein precipitates during dialysis and is recovered by centrifugation at 6000*g* for 30 min at 4°C.
9. Resuspend the pellet in renaturation buffer (**Subheading 2.1., item 19**). Determine protein concentration by spectrophotometry *(24)* and its purity by SDS-PAGE *(22)*.

## 3.5. In Vitro Tranlation of Recombinant RLI Protein (see Note 1)

RLI can be produced in baculovirus-infected insect Sf9 cells *(15)* or in vitro in a wheat-germ cell extract (two biological systems in which RNase L and RLI are absent), but RLI produced by either method is biologically inactive *(14)*. This could reflect a defect in protein posttranslational modifications or the lack of a cofactor.

The most suitable system to produce biologically active RLI protein is the rabbit reticulocyte lysate. The RLI coding sequence is cloned in a pcDNA3 vector (Invitrogen) which allows in vitro transcription from T7 or SP6 promoters and in vivo expression from a CMV promoter after transfection in cultured cells.

1. Linearize 10 μg of DNA (RLI-pcDNA3) with a restriction enzyme that has a single cleavage site immediately downstream of the insert, in a final volume of 100 μL.
2. Apply 5 μL of the digest to an 0.8% (w/v) agarose gel in TBE buffer (**Subheading 2.1., item 20**) to check for complete plasmid DNA cutting.
3. Add 1 μg proteinase K (**Subheading 2.3., item 8**) and incubate 30 min at 37°C.
4. Add 100 μL water, 200 μL phenol, and 200 μL dichloromethane, mix with a Vortex, and centrifuge 10 min at 10,000*g*. Remove the top phase, repeat the phenol/dichloromethane extraction once, and follow by an extraction with dichloromethane alone.
5. Precipitate the template DNA with 2 vol 100% (v/v) ethanol, and 20 μL 3 *M* NaOAc.
6. Centrifuge 10 min at 10,000*g*. Eliminate the supernatant and rinse the pellet with 70% (v/v) ethanol. Centrifuge again, remove the supernatant, and dry air the pellet.
7. Resuspend the DNA in 20 μL RNase-free TE buffer (**Subheading 2.1., item 23**).
8. Mix 5 μL 5X transcription buffer (**Subheading 2.1., item 24**), 1 μg linearized DNA template, 1 μL rNTPs (**Subheading 2.1., item 25**), 2.5 μL cap analog (5'7meGppp5' G), 1 μL 750 m*M* DTT, 10 U T7 or SP6 RNA polymerase (depending on the cDNA orientation in pcDNA3) and RNase-free water to a final volume of 25 μL.
9. Incubate 1 h at 37°C.
10. Digest the DNA template by adding 10 U RNase-free DNase I to the transcription mixture and incubate for an additional 10 min at 37°C.
11. Purify the RNA by phenol/dichloromethane extraction (as in **step 4**) and precipitate with ethanol (as in **steps 5** and **6**).
12. Resuspend the RNA pellet in 25 μL RNase-free water.
13. Control the quality of the RNA by electrophoresis of a 2.5-μL aliquot in 1.2% (w/v) agarose gel in TBE buffer (**Subheading 2.1., item 20**). The amount of capped RNA synthesized is determined by diluting 2.5 μL of the RNA solution in 1 μL $H_2O$ (1/400 dilution) and measuring the optical density at 260 nm.
14. Mix 35 μL rabbit reticulocyte lysate, 11 μL RNase-free water, 1 μL RNasin ribonuclease inhibitor (40 U/mL), 1 μL 1 m*M* amino acid mixture, 2 μL (1 μg) mRNA (*see* **Note 2**), in a final volume of 50 μL. Incubate for 60 min at 30°C.
15. Run, in parallel as control, a smaller scale 25 μL translation reaction containing 2 μL L-($^{35}$S)methionine (1200 Ci/mmol) at 10 mCi/mL, 0.5 μL 1 m*M* amino-acid mixture minus methionine.
16. Analyze a 10-μL aliquot of this radioactive translation by SDS-PAGE *(22)* and autoradiography.

### 3.6. RLI Specific Antibodies

RLI is a highly conserved protein *(25)* and polyclonal antibodies are difficult to raise.

1. Mix $2 \times 50$ µg of the recombinant RLI protein produced in bacteria as described above (**Subheading 3.4.**) with Freund's adjuvant.
2. Inject one dose into a New Zealand rabbit. Inoculate two rabbits. Repeat every 2 wk until a rabbit has a positive antibody titer. The first time, we screened rabbit sera by Western blot against human RLI protein produced in bacteria, then contolled by Western blot with different cell extracts. (Six injections were needed and only one rabbit was positive. The titer of this polyclonal RLI antibody preparation remained low—it must be used at dilution 1/500°.)

### 3.7. Radiobinding Assay: Binding of Radiolabed 2-5A by RNase L in Solution

The specificity and the affinity of RNase L for 2-5A is very high, and the nuclease activity of RNase L is strictly dependent on its activation by 2-5A *(9,26)*. Measuring the binding of 2-5A to RNase L is therefore a good indicator of the presence of RNase L that can be activated in cell extracts. This quick and easy assay has been described by Knight et al. *(20)*, but we routinely make use of a slightly modified protocol *(23)*. This assay can be performed with S10 cell extracts of various origins to quantify RNase L. The presence and amount of RNase L inhibitors, such as RLI, can be determined in mixing experiments *(14)*. The appropriate cell extracts must be mixed before adding the $2\text{-}5A_4\text{-}3'\text{-}(^{32}P)pCp$ probe. As an example, adding increasing amounts of recombinant RLI to a reticulocyte lysate (which provides a source of RNase L) gives rise to a dose-dependent decrease of $2\text{-}5A_4\text{-}3'\text{-}(^{32}P)pCp$ binding in this radiobinding assay (*see* **Fig. 2**). This radiobinding assay has been used to demonstrate that human and mouse RLI are active on human, mouse, or rabbit RNase L (*14,25*, and unpublished results) and to establish the affinity of 2-5A analogs for RNase L *(12,27)*. In the latter case, the $2\text{-}5A_4\text{-}3'\text{-}(^{32}P)pCp$ probe and the 2-5A analog must be mixed and added together to the cell extract.

1. Mix 20,000 cpm of $2\text{-}5A_4\text{-}3'\text{-}(^{32}P)pCp$ with 200 µg of protein from S10 cell extract (**Subheading 3.1., step 4**).
2. Incubate for 60 min in ice.
3. Precipitate proteins at –20°C for 5 min by 300 µL poly(ethylene glycol) 6000 (25% [w/v]) after addition of 150 µL of bovine whole serum as a carrier.
4. Centrifuge 15 min at $10,000g$ at 4°C.
5. Measure the radioactivity of the pellet containing the $2\text{-}5A_4\text{-}3'\text{-}(^{32}P)pCp$ bound to RNase L.

### 3.8. Covalent Assay: Covalent Binding of Radiolabeled 2-5A by RNase L and Analysis by SDS-PAGE

This assay is based on the affinity of 2-5A for RNase L as well. Here, the radiolabeled $2\text{-}5A_4\text{-}3'\text{-}(^{32}P)pCp$ probe is oxidized at the 3' end, as described in **Subheading 3.3.** This modified probe can be covalently linked to amine groups on RNase L, and allows analysis by denaturing gel electrophoresis as initially described by Wreschner et al. *(28)*, and further modified in our group *(23)*.

1. Incubate 5 µL of the oxidized $2\text{-}5A_4\text{-}3'\text{-}(^{32}P)pCp$ probe with 200 µg of protein (S10 cell extract; **Subheading 3.1., step 4**) for 20 min in ice and for a further 20 min at room temperature with 5 µL sodium cyanoborohydride (**Subheading 2.1., item 30**) (20 m*M* final concentration) prepared just before use.

**Radiobinding assay**

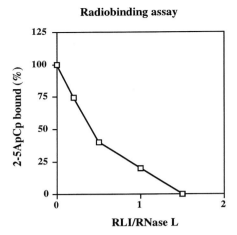

Fig. 2. Inhibition of 2-5A-binding to RNase L by RLI in the radiobinding assay. Increasing amounts of in vitro-translated recombinant RLI protein have been added to a reticulocyte lysate to increase the ratio between RLI and endogenous RNase L. The extracts are then tested in the 2-5A$_4$-3'-($^{32}$P)pCp binding assay. The results are expressed as percent of 2-5A$_4$-3'-($^{32}$P)pCp bound. 100% is the binding in the absence of added RLI.

**Covalent assay**

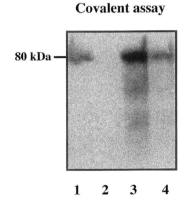

Fig. 3. Inhibition of 2-5A binding to RNase L by RLI in the covalent binding assay. Covalent 2-5A$_4$-3'-($^{32}$P)pCp binding was studied in a reticulocyte extract supplemented (Lanes 3 and 4) or not (Lanes 1 and 2) by recombinant RNase L with (Lanes 2 and 4) or without (Lanes 1 and 3) added recombinant RLI. After covalent binding of 2-5A$_4$-3'-($^{32}$P)pCp and RNase L, proteins are analyzed by SDS-PAGE. After drying, the gel is autoradiographed. An autoradiogram of the gel is presented.

2. Heat the incubation at 100°C for 3 min with 1 vol of denaturing gel electrophoresis buffer (**Subheading 2.1., item 31**) and analyze by 10% (v/v) SDS-PAGE *(22)*.
3. Dry the gel. Labeled proteins are visualized after autoradiography. As documented in **Fig. 3**, recombinant RLI inhibits 2-5A$_4$-3'-($^{32}$P)pCp binding to endogenous or overexpressed RNase L in this covalent assay.

### 3.9. Western Blot Assays with a Radiolabeled 2-5A Probe or with Specific Antibodies

The assays described in **Subheadings 3.7.** and **3.8.** are based on the binding of $2\text{-}5A_4\text{-}3'\text{-}(^{32}P)pCp$ to RNase L. Intracellular 2-5A, which competes with the $2\text{-}5A_4\text{-}3'\text{-}(^{32}P)pCp$ probe for binding to RNase L, or inhibitors, such as RLI, could inhibit the binding of $2\text{-}5A_4\text{-}3'\text{-}(^{32}P)pCp$ to RNase L, and therefore interfere with RNase L detection by these assays *(15,16,29)*. Changes in intracellular 2-5A concentration have indeed been observed during viral infection, cell growth, and cell differentiation, even in the absence of exogenous interferon (IFN) treatment *(7,30)*. Moreover, inhibitors of RNase L have previously been documented during virus infection *(7,31)*. More recently, we have reported an induction of RLI during encephalomyocarditis virus (EMCV) *(15)*, and human immunonodeficiency virus type 1 (HIV) infection *(16)*.

We have therefore developed a Western blot-based assay to detect RNase L and 2-5A binding proteins *(32)* in cells or tissue extracts independently of intracellular 2-5A levels or RNase L inhibitors. As documented in **Fig. 4**, an inhibitor of RNase L such as RLI can be easily dissociated from RNase L by this assay.

1. Mix 200 µg of S10 cell extract with an equal volume of denaturing gel electrophoresis buffer (**Subheading 2.1., item 31**) and analyze by 10% (w/v) SDS-PAGE *(22)*. The pH should be maintained at 8.9 all along the gel.
2. Electro-transfer the gel at 500 mA for 90 min to a nitrocellulose sheet (0.2 mm) in the electrotransfer buffer (**Subheading 2.1., item 32**) (*see* **Note 3**) *(33)*.
3. Incubate the nitrocellulose sheet with 200 mL of buffer (**Subheading 2.1., item 33**), 5% (v/v) skimmed milk at 4°C for 1 h.
4. Wash the nitrocellulose filter in buffer (**Subheading 2.1., item 33**) without milk.
5. Incubate the nitrocellulose filter overnight at 4°C with $10^6$ cpm $2\text{-}5A_4\text{-}3'\text{-}(^{32}P)pCp$ in 10 mL of the same buffer.
6. Wash the filter five times with 100 mL of buffer (**Subheading 2.1., item 33**), dry and expose to film to locate affinity-labeled RNase L.

For revelation with an antibody, after electrotransfer as described in **Subheading 3.9., steps 1** and **2**:

1. Incubate the nitrocellulose sheet for 1 h in PBS (**Subheading 2.1., item 34**) supplemented with 5% (w/v) skim milk.
2. Incubate overnight at 4°C with the antibody of interest (RNase L or RLI-specific antibodies, for instance).
3. Wash the filter with PBS supplemented with 0.05% (v/v) Tween 20.
4. Incubate the filter 1 h with PBS supplemented with 5% (w/v) skim milk
5. Incubate 1 h at room temperature with a second antibody conjugated to horseradish peroxidase in PBS supplemented with 5% (w/v) skim milk.
6. Chemoluminescence is generated by incubation of the nitrocelluse sheet with an equal volume of the oxidizing reagent and of the enhanced luminol reagent (1/1; v/v) from NEN renaissance kit (*see* **Note 4**).

As an example (**Fig. 4B**), RLI-specific antibodies have allowed us to document an increased expression of RLI during encephalomyocarditis virus infection *(15)*.

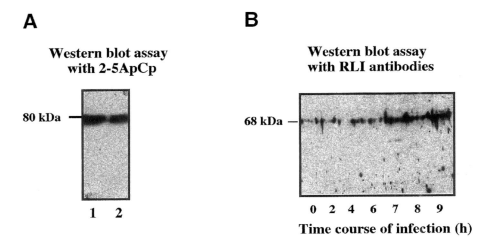

Fig. 4. Western blot analysis of RNase L and RLI. (**A**) A reticulocyte lysate has been comple-mented (Lane 2) or not (Lane 1) with recombinant RLI. Proteins are analyzed by SDS-PAGE, transferred to nitrocellulose sheet, and incubated with 2-5A$_4$-3'-($^{32}$P)pCp. An autoradiogram of the gel is shown. Contrary to the experience shown in **Fig. 3**, RNase L and RLI dissociate during SDS-PAGE, and no inhibition of 2-5A binding is observed. (**B**) HeLa cells were infected with EMCV (m.o.i.: 10) and harvested at the indicated times. For each time point, 100 µg proteins are fractionated by SDS/PAGE and Western blotting with a polyclonal antibody against RLI. The amount of RLI (which migrates as a 68 kDa polypeptide) increases after infection as reported in **ref. *15*.**

### 3.10. Assay of 2-5A-Dependent RNase L Cleavage Using 2-5A-Agarose (see Note 1)

RNase L can also be quantitated by its nuclease activity using poly(uridylic acid) (poly(U) or cellular rRNAs as substrates *(34–36)*.To avoid unspecific cleavage of poly(U), RNase L is purified on 2-5A-agarose according to the procedure described by Silverman *(37)* with slight modifications *(23)*. Poly(cytidylic acid) (poly(C)) is used as negative control, since it is not a substrate of RNase L.

1. Incubate 15 µL (300 µg) of poly(U) or poly(C) at 4°C for 48 h with 5 µL 1 m*M* ATP, 25 µL ligase buffer (**Subheading 2.1., item 7**), 5 µL (60 U) T4 RNA ligase, and 10 µL (100 µCi) ($^{32}$P)pCp.
2. Purify poly(U)-3'-($^{32}$P)pCp and poly(C)-3'-($^{32}$P)pCp by HPLC on a Zorbax GF-250 (DuPont Company) column equilibrated at 4°C in 0.1 *M* phosphate buffer, pH 7.0.
3. Collect 250 µL fractions every 0.5 min at 4°C and store labeled probes at –80°C.
4. Mix 2-5A$_6$ (0.5 mmol) with 0.5 mmol NaIO$_4$ in 0.1 *M* acetate buffer, pH 4.5 (**Subheading 2.1., item 37**) for 30 min on ice.
5. Add 10 µL adipic hydrazide-agarose (1:1 in suspension in 0.1 *M* acetate buffer, pH 4.5).
6. Incubate for 10 min at 25°C.
7. Wash the 2-5A$_6$-adipic hydrazide-agarose with 0.1 *M* acetate buffer, pH 4.5.
8. Eliminate the acetate buffer, and suspend the 2-5A$_6$-adipic hydrazide-agarose in the stor-age buffer (**Subheading 2.1., item 38**).
9. Store in aliquots at –20°C. About 99% of the 2-5A$_6$ becomes covalently linked to the agarose gel, as revealed by 260-nm *A* in the supernatant.

10. Incubate S10 cell extract (600 µg proteins) with 2-5A$_6$-adipic-hydrazid-agarose (25 µL) at 4°C for 1 h with stirring.
11. Wash the RNase L-agarose with RNase-free buffer (**Subheading 2.1., item 33**).
12. Incubate the RNase L-agarose at 35°C for 30 min with poly(U)-3'-($^{32}$P)pCp or with poly(C)-3'-($^{32}$P)pCp as a control (15,000 cpm in 40 µL buffer [**Subheading 2.1., item 33**]).
13. Separate uncleaved RNA material and breakdown products by precipitation with 2 vol ethanol, after addition of ammonium acetate to a final concentration of 2.5 *M*.
14. Centrifuge at 12,000*g* for 30 min and count radioactivity in the supernatant. Unspecific cleavage is determined in the same conditions by using a 2-5A$_6$-adipic-hydrazid-agarose without RNase L.

### 3.11. Assay of 2-5A-Dependent RNase L Cleavage of rRNAs (see Note 1)

1. Collect the cells.
2. Wash them once with PBS and centrifuge at 1000*g* for 5 min.
3. Add one cell volume of RNA cleavage cell lysis buffer (**Subheading 2.1., item 39**).
4. Resuspend the cells by pipeting and leave the tubes on ice for 10 min.
5. Pass the suspension at least 20 times through a 1-mL tuberculine syringe with a 0.5 × 16 mm needle. Cells breakdown must be controlled carefully under the microscope by adding Trypan blue (0.5% [w/v] in PBS) to an aliquot of cell extract, since breakdown varies significantly in different cell types.
6. Centrifuge the cell homogenate at 15,000*g* for 2 min.
7. Pool the supernatants and determine the protein concentration by spectrophotometry *(24)*.
8. Dispense the extracts in 100-µL aliquots in microcentrifuge tubes and keep at –80°C until use.
9. Incubate 200 µg of cell extract 30 min at 30°C with or without 1 m*M* 2-5An.
10. Extract RNAs with phenol-dichloromethane and precipitate them with 2 vol of ethanol and 300 m*M* NaOAc.
11. Centrifuge 10 min at 10,000*g*, discard the supernatant, rinse the pellet with 70% (v/v) ethanol and dry.
12. Resuspend RNA in 20 µL electrophoresis loading buffer (**Subheading 2.1., item 42**).
13. Analyze samples on a 1.2% (w/v) agarose gel (0.25 µg/mL ethidium bromide) in TBE buffer (**Subheading 2.1., item 20**) (0.25 µg/mL ethidium bromide).
14. Photograph the gel under an UV transilluminator and scan the photography with a densitometer. rRNAs and their characteristic degradation products (two bands between 28S and 18S, and 2–3 bands under 18S) are quantified by image analysis with the Intelligent Quantifier program (Bioimage Systems Corporation) or the NIH Image Program *(14,34)*.

### 3.12. Assays for RNase L and RNase L Inhibitor in Intact Cells

Studies of RLI/RNase L during virus infection *(15,16)* or cell differentiation (Bisbal et al., unpublished observation) have to be performed over periods of several days, which transient transfections do not allow. Stable cell lines were therefore established by transfection with calcium phosphate coprecipitation *(38)* of RLI and RNase L cDNA constructions in sense or antisense orientation in pcDNA3 vector (Invitrogen).

1. Test the cells for their resistance to the selection drug, e.g., geniticin (G418) for pcDNA3. Grow the cells in the presence of increasing concentrations of G418 during 10 d. G418 is toxic after 7 d at 800–1000 µg/mL for most cell lines.
2. Plate 5 × 10$^5$ cells in 10 mL of DMEM, 10% (v/v) FCS (*see* **Note 5**) in a cell-culture dish of 9-cm diameter.

3. The next day, homogenize solution A (**Subheading 2.1., item 45**) and add it dropwise to 2 mL of solution B (**Subheading 2.1., item 46**). A precipitate should form in 30 min.
4. Add 1 mL of A + B in a cell-culture dish of 9-cm diameter with 10 mL DMEM, 10% (v/v) FCS, and homogenize the medium.
5. After 24 h, wash the cell layer carefully with PBS, and add the usual cell-culture medium.
6. Tweny-four hours later, apply the selection with G418 at the concentration determined in **step 1**.
7. For nonadherent cell clones, count the cells with a hemocytometer and dilute to 1 cell/100 µL of medium in a 96-well dish. Reseed in fresh medium with G418 when a clone has grown to 20–30 cells, and continue with selection for 3 wk.
8. For adherent cells, plate the cells at normal density in the selection medium. When individual clones have 20–30 cells, remove the medium, wash the cells carefully with PBS, and add one drop of trypsin on each clone. Reseed each clone in a 96-well dish and carry on with the selection for 3 wk. All clones are amplified and frozen before screening for RNase L or RLI activity.

## 4. Notes

1. All dishes and reagents must be RNase-free. Wear gloves. Eliminate RNases from the gel electrophoresis apparatus by washing with 0.1 $N$ NaOH and abundant rinsing with RNase free water.
2. The template mRNA is heated at 67°C for 10 min and immediately cooled on ice to increase the efficiency of translation by destroying secondary structures.
3. The buffer and the transfer apparatus are kept ice-cold. Put the transfert apparatus in an ice bath. For efficient renaturation of RNase L activity, the temperature should not increase too much during transfer.
4. Another possibility is to use solutions prepared in the laboratory. Immediately before use mix: 100 m$M$ Tris-HCl, pH 8.5, 200 m$M$ coumaric acid, 1.25 m$M$ luminol, $H_2O_2$ 3/10$^4$ (v/v), and incubate on the membrane for 1 min. This is much less expensive, but the signal obtained is less stable.
5. For transfection with calcium phosphate coprecipitation, the cell-culture medium must be without phosphate. If your cells are grown in RPMI medium, for example, 48 h before the transfection, transfer your cells to DMEM medium, then follow the protocol. Twenty-four hours after the transfection, transfer the cells to the usual medium.

## References

1. Stark, G. R., Kerr, I. M., Williams, B. R., Silverman, R. H., and Schreiber, R. D. (1998) How cells respond to interferons. *Annu. Rev. Biochem.* **67,** 227–264.
2. Farrar, M. A. and Schreiber, R. D. (1993) The molecular cell biology of interferon-gamma and its receptor. *Annu. Rev. Immunol.* **11,** 571–611.
3. Boehm, U., Klamp, T., Groot, M., and Howard, J. C. (1997) Cellular responses to interferon-gamma. *Annu. Rev. Immunol.* **15,** 749–795.
4. Brown, G. E., Lebleu, B., Kawakita, M., Shaila, S., Sen, G. C., and Lengyel, P. (1976) Increased endonuclease activity in an extract from mouse Ehrlich ascites tumor cells which had been treated with a partially purified interferon preparation: dependence of double-stranded RNA. *Biochem. Biophys. Res. Commun.* **69,** 114–122.
5. Kerr, I. M. and Brown, R. E. (1978) pppA2"p5"A2"p5"A: an inhibitor of protein synthesis synthesized with an enzyme fraction from interferon-treated cells. *Proc. Natl. Acad. Sci. USA* **75,** 256–260.
6. Bisbal, C. (1997) RNase L: effector nuclease of an activatable RNA degradation system in mammals. *Prog. Mol. Subcell. Biol.* **18,** 19–34.

7. Silverman, R. H. (1997) 2-5A dependent RNase L: a regulated endoribonuclease in the interferon system, in *Ribonucleases: Structure and Functions* (D'Alessio, G. and Riordan, J. F., eds.), Academic Press, New York, pp. 515–551.

8. Player, M. R. and Torrence, P. F. (1998) The 2-5A system: modulation of viral and cellular processes through acceleration of RNA degradation. *Pharmacol. Ther.* **78,** 55–113.

9. Zhou, A., Hassel, B. A., and Silverman, R. H. (1993) Expression cloning of 2-5A-dependent RNase: a uniquely regulated mediator of interferon action. *Cell* **72,** 753–765.

10. Hassel, B. A., Zhou, A., Sotomayor, C., Maran, A., and Silverman, R. H. (1993) A dominant negative mutant of 2-5A-dependent RNase suppresses antiproliferative and antiviral effects of interferon. *EMBO J.* **12,** 3297–3304.

11. Zhou, A., et al. (1997) Interferon action and apoptosis are defective in mice devoid of 2',5'- oligoadenylate-dependent RNase L. *EMBO J.* **16,** 6355–6363.

12. Bayard, B., Bisbal, C., Silhol, M., Cnockaert, J., Huez, G., and Lebleu, B. (1984) Increased stability and antiviral activity of 2'-O-phosphoglyceryl derivatives of (2'-5')oligo-(adenylate). *Eur. J. Biochem.* **142,** 291–298.

13. Johnston, M. I. and Hearl, W. G. (1987) Purification and characterization of a 2'-phosphodiesterase from bovine spleen. *J. Biol. Chem.* **262,** 8377–8382.

14. Bisbal, C., Martinand, C., Silhol, M., Lebleu, B., and Salehzada, T. (1995) Cloning and characterization of a RNAse L inhibitor. A new component of the interferon-regulated 2-5A pathway. *J. Biol. Chem.* **270,** 13,308–13,317.

15. Martinand, C., Salehzada, T., Silhol, M., Lebleu, B., and Bisbal, C. (1998) RNase L inhibitor (RLI) antisense constructions block partially the down regulation of the 2-5A/RNase L pathway in encephalomyocarditis-virus-(EMCV)-infected cells. *Eur. J. Biochem.* **254,** 238–247.

16. Martinand, C., Montavon, C., Salehzada, T., Silhol, M., Lebleu, B., and Bisbal, C. (1999) RNase L Inhibitor is induced during human immunodeficiency virus type 1 infection and down regulates the 2-5A/RNase L pathway in human T cells. *J. Virol.* **73,** 290–296.

17. Maitra, R. K. and Silverman, R. H. (1998) Regulation of human immunodeficiency virus replication by 2,5'-oligoadenylate-dependent RNase L. *J. Virol.* **72,** 1146–1152.

18. Minks, M. A., Benvin, S., Maroney, P. A., and Baglioni, C. (1979) Synthesis of 2"5'-oligo(A) in extracts of interferon-treated HeLa cells. *J. Biol. Chem.* **254,** 5058–5064.

19. Brown, R. E., Cayley, P. J., and Kerr, I. M. (1981) Analysis of (2'-5')-oligo(A) and related oligonucleotides by high-performance liquid chromatography. *Methods Enzymol.* **79,** 208–216.

20. Knight, M., Cayley, P. J., Silverman, R. H., Wreschner, D. H., Gilbert, C. S., Brown, R. E., and Kerr, I. M. (1980) Radioimmune, radiobinding and HPLC analysis of 2-5A and related oligonucleotides from intact cells. *Nature* **288,** 189–192.

21. Knight, M., Wreschner, D. H., Silverman, R. H., and Kerr, I. M. (1981) Radioimmune and radiobinding assays for A2"p5"A2"p5"A, pppA2"p5"A, and related oligonucleotides. *Methods Enzymol.* **79,** 216–227.

22. Laemmli, U. K. (1970) Cleavage of structural proteins during the assembly of the head of bacteriophage T4. *Nature* **227,** 680–685.

23. Bisbal, C., Salehzada, T., Lebleu, B., and Bayard, B. (1989) Characterization of two murine (2'-5')(A)n-dependent endonucleases of different molecular mass. *Eur. J. Biochem.* **179,** 595–602.

24. Whitaker, J. R. and Granum, P. E. (1980) An absolute method for protein determination based on difference in absorbance at 235 and 280 nm. *Anal. Biochem.* **109(1),** 156–159.

25. Benoit De Coignac, A., Bisbal, C., Lebleu, B., and Salehzada, T. (1998) cDNA cloning and expression analysis of the murine ribonuclease L inhibitor. *Gene* **209,** 149–156.

26. Dong, B., Xu, L., Zhou, A., Hassel, B. A., Lee, X., Torrence, P. F., and Silverman, R. H. (1994) Intrinsic molecular activities of the Interferon-induced 2-5A-dependant RNase. *J. Biol. Chem.* **269,** 14,153–14,158.

27. Bisbal, C., Bayard, B., Silhol, M., Leserman, L., and Lebleu, B. (1985) 2"5' Oligo-adenylate analogues: synthesis, biological activities and intracellular delivery. *Prog. Clin. Biol. Res.* **202,** 89–95.

28. Wreschner, D. H., Silverman, R. H., James, T. C., Gilbert, C. S., and Kerr, I. M. (1982) Affinity labelling and characterization of the ppp(A2"p)nA-dependent endoribonuclease from different mammalian sources. *Eur. J. Biochem.* **124,** 261–268.

29. Martinand, C., Salehzada, T., Silhol, M., Lebleu, B., and Bisbal, C. (1998) The RNase L inhibitor (RLI) is induced by double-stranded RNAs. *J. Interferon Cytokine Res.* **18,** 1031–1038.

30. Sen, G. C. and Lengyel, P. (1992) The interferon system. A bird's eye view of its bio-chemistry. *J. Biol. Chem.* **267,** 5017–5020.

31. Silverman, R. H. and Cirino, N. M. (1997) RNA decay by the interferon-regulated 2-5A system as a host defense against virus, in *mRNA Metabolism and Post-Transcriptional Gene Regulation* (Hartford, J. B. and Morris, D. R., eds.), Wiley-Liss, New York, pp. 295–309.

32. Salehzada, T., Silhol, M., Lebleu, B., and Bisbal, C. (1982) Regeneration of enzyme activity after western blot: activation of RNase L by 2-5A on filter—importance for its detection. *J. Biol. Chem.* **257,** 12,739–12,745.

33. Towbin, H., Staehelin, T., and Gordon, J. (1979) Electrophoretic transfer of proteins from polyacrylamide gels to nitrocellulose sheets: procedure and some applications. *Proc. Natl. Acad. Sci. USA* **76,** 4350–4354.

34. Wreschner, D. H., James, T. C., Silverman, R. H., and Kerr, I. M. (1981) Ribosomal RNA cleavage, nuclease activation and 2-5A(ppp(A2"p)nA) in interferon-treated cells. *Nucleic Acids Res.* **9,** 1571–1581.

35. Wreschner, D. H., McCauley, J. W., Skehel, J. J., and Kerr, I. M. (1981) Interferon action—sequence specificity of the ppp(A2"p)nA-dependent ribonuclease. *Nature* **289,** 414–417.

36. Wreschner, D. H., James, T. C., Silverman, R. H., and Kerr, I. M. (1994) Ribosomal RNA cleavage, nuclease activation and 2-5A(ppp(A2"p)nA) in interferon-treated cells. *J. Interferon Res.* **14,** 101–104.

37. Silverman, R. H. (1985) Functional analysis of 2-5A-dependent RNase and 2-5a using 2',5'- oligoadenylate-cellulose. *Anal. Biochem.* **144,** 450–460.

38. Sambrook, J., Fritsch, E. F., and Maniatis, T. (1982) *Molecular Cloning: A Laboratory Manual.* Cold Spring Harbor Laboratory, Cold Spring Harbor, NY.

# III

# RELATING NUCLEASE STRUCTURE AND FUNCTION

# 13

## Crystallization and Crystal Structure Determination of Ribonuclease A-Ribonuclease Inhibitor Protein Complex

### Bostjan Kobe

## 1. Introduction

### 1.1. Ribonuclease Inhibitor Protein

Ribonuclease inhibitor (RI) is a cytoplasmic protein of ~50 kDa that tightly binds and inhibits ribonucleases (RNases) from the pancreatic superfamily *(1)*. Diverse RNases with very limited sequence similarities, including RNase A, angiogenin, RNase-2 (also known as eosinophil-derived neurotoxin (EDN) or placental RNase), and RNase-4 are inhibited with $K_i$ values between $10^{-14}$ and $10^{-16}$ $M$. These affinities are among the highest reported for noncovalent binding of proteins. The binding occurs with 1:1 stoichiometry. A subset of the pancreatic ribonuclease superfamily, including the amphibian ribonucleases such as frog-liver ribonuclease, sialic acid-binding lectin, and P-30 protein, are not inhibited by RI. The potent inhibitory activity of RI is believed to be utilized in RNA processing, angiogenesis, and protection of the cell from toxic ribonucleases.

The primary structure of RI consists of 15 repeated motifs termed the leucine-rich repeats (LRRs) *(2)*. LRRs have been found in a large number of proteins with diverse functions and cellular locations that all appear to be involved in protein–protein interactions *(3–5)*. These proteins participate in diverse biological processes, such as signal transduction, cell adhesion, DNA repair, RNA processing, and plant disease resistance.

RI contains a large number of cysteine residues (30 in porcine RI [pRI]) that are all found in a reduced state in the active native protein. Oxidative conditions lead to the formation of disulfide bridges and inactivation of the inhibitor.

### 1.2. X-Ray Crystallography

Although the three-dimensional structures of free pRI *(6)* and RNase A *(7,8)* had been determined earlier, and several biochemical and mutagenesis studies had addressed the interaction between them *(1)*, it was clear that understanding the atomic basis of the extremely tight interaction between RNases and RI, and the specificity of this association, required a three-dimensional structure of the complex. We chose the X-ray crystallography because it is the only suitable method for determining three-

From: *Methods in Molecular Biology, vol. 160: Nuclease Methods and Protocols*
Edited by: C. H. Schein © Humana Press Inc., Totowa, NJ

dimensional structures of proteins of the size of RNase-RI complex in the desired detail. The determination of a crystal structure comprises the following steps:

1. Protein purification;
2. Crystallization;
3. Measurement of diffraction data;
4. Phase determination;
5. Electron-density map interpretation and model building;
6. Crystallographic refinement.

## 1.3. Crystal Structure Determination of Ribonuclease A–Porcine Ribonuclease Inhibitor Protein Complex

We determined the crystal structure of the complex between pRI and bovine RNase A ($K_i = 5.9 \times 10^{-14}$ M; *9*), and refined it at 2.5 Å resolution (*10,11*). The reasons for choosing this particular complex were as follows. Porcine RI could be purified in large quantities from pig liver by a manageable procedure, and we had previously crystallized it on its own and determined its structure (*6,12*). Bovine RNase A could be purchased commercially, and its crystal structure had also been known for some time (*7,8,13*). Knowledge gained from the crystallization of free pRI could be used in developing strategies for screening for the crystals of the complex. Finally, the known three-dimensional structures of the two individual protein components could be used for determining the phases using the method of molecular replacement.

## 1.4. Structure of Ribonuclease A–Ribonuclease Inhibitor Protein Complex and the Mechanism of Ribonuclease Inhibition

The structure of the RNase A-pRI complex allowed us to assess the general properties of the ligand-binding functions of LRR-containing proteins (*10*), and to define the mechanism of the inhibition of ribonuclease activity by the inhibitor protein (*11*).

Two views of the structure of the RNase A-pRI complex are shown in **Fig. 1**. RI is a horseshoe-shaped α/β protein with a long parallel β-sheet lining its inner circumference, and α-helices lining its outer circumference. RNase A is kidney-shaped, built predominantly of antiparallel β-structure and stabilized by four disulfide bridges. The RNase active site is located in a cleft between the two lobes of RNase A. The enzyme binds to the inhibitor with its active site cleft covering the C-terminal corner of the horseshoe; one lobe inserts into the concave cavity, and the other lobe lies over the face of the horseshoe.

A very large accessible surface area (2550 Å$^2$), larger than is usual in protein–protein complexes (*14*), is buried in the interface between RI and RNase A in the complex. There are a large number of contacts between the two proteins (117 contacts at distances <4 Å, 18 hydrogen bonds at distances <3.5 Å), involving 28 residues of RI and 24 residues of RNase A. The degree of shape complementarity is slightly lower than usually observed in protein–protein complexes (*15*). In the RNase A-RI interface, the lesser complementarity is apparently compensated by the extremely large interface, which provides an opportunity for many interactions between the two molecules, and is thus probably one of the major factors for the observed tightness of the complex.

RI inhibits RNase by preventing the access of substrates to the active site by steric hindrance, and does not mimic the RNase-RNA interaction to a major extent. Several

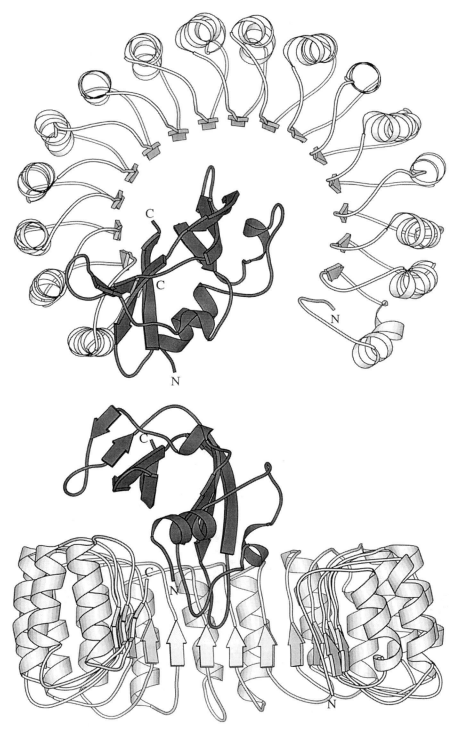

Fig. 1. Two orthogonal views of the structure of the RNase A-pRI complex, presented as a schematic ribbon diagram, and drawn with the program MOLSCRIPT *(32)*. RNase A, dark; pRI, light.

active-site residues do contact the inhibitor, but a solvent ion (most likely a sulfate) remains bound in the p1 phosphate-binding pocket of RNase A. The only major structural similarity between RNase-pRI and RNase-nucleotide binding appears to be the aromatic ring of Tyr-433 in pRI, which resembles the aromatic ring of a base in the B2 binding pocket of RNase A *(16,17)*.

The crystal structure of the RNase A-pRI complex shows that RNase A exploits the unusual structural features of pRI for tight binding to this protein. RI exhibits the flexibility of its structure by subtly but significantly conformationally adapting to the RNase A molecule upon binding. The structure suggests that the high affinity of binding to diverse RNases is most likely achieved by conformationally adjusting in a similar way to a particular RNase in the binding site, and by compensating a possibly lower degree of complementarity with an unusually large contact area.

The structure of RNase A-pRI complex enriches our basic understanding of protein–protein recognition, but also offers more practical applications. It can serve as a model for other LRR-containing, protein–ligand complexes, many of great biological importance *(3–5)*. It can also guide the design of inhibitors of angiogenesis and tumor growth *(18)*.

The methods used in the structural determination of RI-RNase A complex are described in detail here. Similar methods can be used, without major modifications, for structural determination of similar or more divergent proteins and complexes. For example, analogous methods have more recently been used to determine the structure of the complex of human angiogenin (Ang) and human placental RI (hRI) *(19)*. This structure underscores the major conclusions derived from the RNase A-pRI structure, but also emphasizes that many interactions in the two complexes are distinctive, indicating that the broad specificity of the inhibitor for diverse RNases relies on its capacity to recognize unique features of individual enzymes.

## 2. Materials

All reagents were of analytical grade. RNase A from bovine pancreas was purchased from Sigma (type XII-A) and used without further purification.

## 3. Methods
### 3.1. Purification of Ribonuclease Inhibitor from Pig Liver

RI was purified from porcine liver by ammonium sulfate fractionation followed by affinity chromatography on RNase A-Sepharose *(20)* (*see* **Note 1**). All steps were performed at 4°C. All buffers contained 1 m$M$ EDTA and 10 m$M$ freshly added dithiothreitol (DTT). Maintenance of reducing conditions was crucial to retain an active inhibitor.

1. Cut 4 kg liver (~two livers) fresh from slaughter into small pieces, and homogenize with a Polytron homogenizer 5 times for 1 min in 8 L 20 m$M$ Tris-HCl, pH 7.5, and 0.25 $M$ sucrose. (Some colleagues may prefer to avoid the cold room during this step of purification, so please leave a note on the door to alert them.)
2. Centrifuge homogenate for 1 h at 16,000$g$.
3. Add ammonium sulfate to the supernatant (slowly while stirring) to 35% saturation, leave for 1 h, and centrifuge as in **step 2**.
4. Add ammonium sulfate to the supernatant (slowly while stirring) to 60% saturation, leave for 1 h, and centrifuge as in **step 2**.

**A B C D**

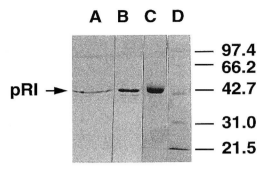

pRI ➡

— 97.4
— 66.2
— 42.7
— 31.0
— 21.5

Fig. 2. Purification of porcine RI. Proteins were separated by SDS-PAGE and stained by Coomassie blue. Lane A, fractions containing RI after affinity chromatography; Lane B, fractions containing RI after hydrophobic interaction chromatography; Lane C, pRI sample used for crystallization; Lane D, molecular-weight markers.

5. Dissolve pellets in 1.6 L 40 m$M$ potassium phosphate buffer (pH 6.5; buffer A) containing protease inhibitors (1 m$M$ PMSF, 1 µ$M$ pepstatin), and dialyze solution overnight against 16 L buffer A.
6. Add 20 mL RNase A-Sepharose to the dialyzed solution and stir overnight with an overhead stirrer. RNase A-Sepharose is prepared from bovine pancreatic RNase A and CNBr-activated Sepharose (Pharmacia) according to Pharmacia's instructions.
7. Collect affinity matrix on a glass sintered filter and wash the filter with 1/2 L buffer A, then 1/2 L buffer A supplemented with 1 $M$ NaCl, and collect in an empty column.
8. Elute RI with 100 mL borate buffer (pH 6.0), 4 $M$ NaCl, 15% glycerol, and protease inhibitors.
9. Dialyze eluate against 50 m$M$ potassium phosphate (pH 7.0), 1 $M$ ammonium sulfate, and load onto an FPLC Phenyl Superose column (hydrophobic interaction chromatography) equilibrated in the same buffer. Elute RI with a gradient from 1–0 $M$ ammonium sulfate.
10. Dialyze fractions containing RI against 40 m$M$ potassium phosphate buffer (pH 6.5) and load onto an FPLC Mono Q column (ion exchange chromatography) equilibrated in the same buffer. Elute with a gradient from 0–1 $M$ NaCl.
11. Dialyze fractions containing RI against 20 m$M$ potassium phosphate buffer (pH 7.0) containing 20 m$M$ DTT and no EDTA.
12. Concentrate in a Centricon-10 microconcentrator (Amicon) to 27 mg/mL (determined spectrophotometrically using a calculated molar extinction coefficient $\varepsilon_{280}$ = 33,110 $M^{-1}$/cm).
13. Store sample at 4°C for up to 1 mo, or freeze on dry ice and store at –70°C. The purity of pRI during the purification is shown in **Fig. 2**.

### *3.2. Crystallization*

Initial crystal growth conditions for RNase A-pRI complex were screened using an incomplete factorial search *(21)*. The variables were chosen based on the experience with crystallization of free pRI *(12)*. For general guidelines on crystallization screening strategies and methods used, *see* **ref. 22** and references therein.

The RNase A-pRI complex used in the initial screens was prepared by incubating pRI with excess RNase A, and purifying the complex on an FPLC Mono Q column *(21)*. It was subsequently found that the purification of the complex was unnecessary, as crystals could be obtained by mixing RNase A and pRI directly in the crystallization

Fig. 3. Crystallization drop containing crystals of RNase A-pRI complex.

drop. The prior purification of the complex may be a safer approach for the screening of crystal growth conditions of macromolecular complexes. As in the case of the RNase A-pRI complex crystallization, it can be established later whether it is possible for this step to be omitted.

The RNase A-pRI crystals used for structure determination were grown by the hanging drop-vapor diffusion technique using ammonium sulfate as the precipitating agent. The largest crystals grew to the size of $0.08 \times 0.08 \times 1$ mm (*see* **Note 2**).

1. Grease rims of individual compartments of a 24-well Linbro tissue culture plate with vacuum grease dispensed from a syringe.
2. Fill up the reservoir with 1 mL of reservoir solution (1.3 $M$ ammonium sulfate, 100 m$M$ sodium acetate (pH 5.0), and 20 m$M$ DTT).
3. Mix 4 µL reservoir solution with 4 µL pRI solution (27 mg/mL) in 20 m$M$ potassium phosphate (pH 7.0) and 20 m$M$ DTT and 1 µL RNase A solution (28 mg/mL [w/v] in water). Keep the droplet hemispherical and try not to introduce bubbles.
4. Invert the coverslip and gently press it onto the greased rim. Make sure it is thoroughly sealed.
5. Put plasticine in the corners of the plate to raise the plate cover above the cover slips, thus avoiding contact between the cover and cover slips. Leave the covered plate undisturbed at 21°C.
6. Crystals will appear as clusters of long needles in a few days, and can be inspected under the microscope (**Fig. 3**). Individual crystals for diffraction experiments can easily be isolated from the cluster.

### 3.3. Diffraction Data Collection

Protein crystals contain solvent, and therefore dehydrate if exposed to air. For diffraction experiments, they must be transferred into sealed capillaries.

1. Cleanly break a 0.7-mm diameter thin-walled glass or quartz capillary (to break the quartz capillary, a diamond knife is needed) at the thin end and attach with rubber tubing to a mouthpiece, syringe, or pipetman.

2. Under the microscope, draw the crystal into the capillary.
3. Position the plug of liquid containing the crystal ~1 cm from the cut end, and seal with wax.
4. Allow the crystal to sink to the meniscus closest to the sealed end, and remove the excess liquid from the capillary with a fine pipet (e.g., a Pasteur pipet drawn thin in a flame, or a capillary with a smaller diameter), or a filter-paper strip or wick, leaving only a small amount surrounding the crystal.
5. After adding some more reservoir solution into the capillary at some distance from the crystal, cut the other end of the capillary, and seal with wax.
6. Attach the capillary to the goniometer head with plasticine.

We collected diffraction data at room temperature as $1°$ oscillation images on an *R*-axis II image plate detector (Molecular Structure Corporation) using $CuK_\alpha$ radiation from a Rigaku RU-300 rotating anode generator. The generator was operated in fine focus mode at 50 kV and 100 mA. Data collected from two crystals to a resolution of 2.5 Å were combined to yield 111,046 measurements, which were subsequently reduced to 26,431 unique reflections with internal agreement $R_{merge} = 0.100$ (*see* **Note 3**). Crystals have the symmetry of the tetragonal space group I4 with $a = 133.3$ Å and $c = 86.7$ Å (*see* **Notes 4** and **5**).

### 3.4. Crystal Density Measurements

The measurement of crystal density is a simple method that yields invaluable information about the content of the crystals (e.g., the number of protein molecules in the crystal asymmetric unit), but is nevertheless rarely used. The measurement of the density of the RNase A-pRI complex crystals was instrumental in confirming that these crystals contained both RNase A and pRI. Free pRI crystallized under similar conditions (1.4 *M* ammonium sulfate, 20 m*M* potassium phosphate buffer (pH 7.0) and 20 m*M* DTT in the reservoir solution *[12]*), with the same crystal symmetry and similar unit-cell dimensions (a = 134.9 Å, c = 83.6 Å) as RNase A-pRI complex. Comparison of the crystal densities of free pRI and RNase A-pRI crystals confirmed that the latter did indeed contain a 1:1 complex of RNase A and pRI *(21)*.

1. Prepare a 0–50% Ficoll 400 gradient in 1.3 *M* ammonium sulfate and 100 m*M* sodium acetate (pH 5.0) in a transparent centrifuge tube by introducing layers of Ficoll (5% steps).
2. Calibrate the gradient by introducing droplets of water-saturated toluene/carbon tetrachloride mixtures of known density and centrifuge at 500 g (e.g., in a swinging-bucket rotor HB-4 [Sorvall]).
3. Introduce a crystal just under the meniscus, centrifuge as above, and at regular time intervals measure the crystal position using a ruler. Estimate the crystal density after it assumes a stable position in the gradient (*see* **Fig. 4**).
4. Calculate the mass of protein per asymmetric unit $M_p$ using the relation

$$M_p = [N(V/n)(\rho_c - \rho_w)]/(1 - v_p \rho_w),$$

where *N* is Avogadro's number, *V* is the unit-cell vol, *n* is the number of asymmetric U per unit cell, $\rho_c$ is the crystal density, $\rho_w$ is the solvent density (1.09 g/cm$^3$ for 1.3 *M* ammonium sulfate/100 m*M* sodium acetate), and $v_p$ is the partial specific vol of protein (0.74 cm$^3$/g).

### 3.5. Crystal Structure Determination and Refinement

Diffraction data are used to calculate an electron density map of the crystal, which describes the distribution of electrons in the crystal. This map can then be interpreted

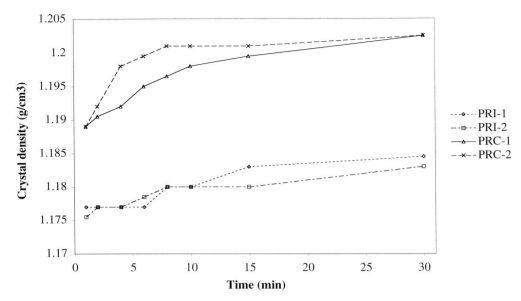

Fig. 4. Crystal density measurements as a function of time. PRI-1 and PRI-2 represent crystals of free pRI. PRC-1 and PRC-2 represent the crystals of RNase A-pRI complex.

as an atomic model of the molecules constituting the crystal, and the model refined for best fit with the experimental data. A diffraction experiment unfortunately provides only one part of the data required for the calculation of an electron density map. The missing part, the "phases," must be obtained by a different method.

In the case of the RNase A-pRI complex, the method of molecular replacement was used to obtain the starting phases. This was possible because the atomic models of free pRI (entry 1BNH in the Protein Data Bank) and ribonuclease A (entry 7RSA in the Protein Data Bank) were available. The starting phases were obtained by positioning these models of the individual components in the crystals of the complex.

Program X-PLOR *(23)* was used for molecular replacement calculations. The model of pRI was arbitrarily rotated before cross-rotation functions were calculated, using data between 15 and 4 Å resolution and a radius of 20 Å. The top 150 maxima of the rotation function were filtered by Patterson correlation (PC) refinement *(24)* consisting of 10 steps of conjugate gradient minimization of the molecular orientation. One outstanding peak was found with a correlation coefficient of 0.124 (signal-to-noise ratio 1.4), corresponding to a rotation with a height of 3.6 $\sigma$ in the cross-rotation function. A translation function search of the RI model in this orientation yielded the top solution of 17.3 $\sigma$. RI in this position produced an $R_{cryst}$ of 49.1% between 10 and 3.5 Å resolution and F >2$\sigma$(F) (*see* **Note 6**).

Cross-rotation and translation searches in various resolution ranges using RNase A as the search model did not yield a solution. Therefore, electron density maps calculated with coefficients 2|F$_{obs}$|–|F$_{calc}$| and |F$_{obs}$|–|F$_{calc}$| phased by only the properly positioned pRI molecule were inspected using the graphics program O *(25)*. Electron density resembling the RNase A molecule was clearly visible, and the model of RNase A was positioned into this electron density on the graphics display. The resulting model containing both pRI and RNase A had an $R_{cryst}$ of 49.0%, using data between 10 and 3.5 Å resolution and F >2σ(F).

Program X-PLOR *(23)* was used for crystallographic refinement. The inhibitor and the enzyme were first refined as rigid bodies resulting in an $R_{cryst}$ of 42.3% and $R_{free}$ of 43.0% between 10 Å and 3.5 Å resolution and F >2σ(F) (10% of data randomly omitted for the calculation of $R_{free}$). Further refinement of the inhibitor treating initially two halves, then four quarters, and finally eight-eighths of the sequence as rigid bodies resulted in an $R_{cryst}$ of 33.3% and $R_{free}$ of 34.2% for data between 10 Å and 3.5 Å resolution and F >2σ(F). Several cycles of positional refinement, refinement of individual isotropic B-factors, manual rebuilding, and addition of solvent molecules, resulted in $R_{cryst}$ of 19.4% and $R_{free}$ of 28.6% using data between 40 Å and 2.5 Å resolution, with a bulk solvent correction and F >σ(F). Although the inclusion of the low-resolution data and the weak reflections increased the R factors, there was a substantial improvement of the electron density maps. B-factors were restrained (target values for B-factor deviations were 1.5 Å$^2$ for bonded main-chain atoms, 2.0 Å$^2$ for bonded side-chain atoms and angle-related main-chain atoms, and 2.5 Å$^2$ for angle-related side-chain atoms). Candidate water molecules were selected as peaks in the electron density map calculated with coefficients |$F_{obs}$|–|$F_{calc}$| with a program written by S. R. Sprang, and kept in the model if they were within 3.5 Å of a polar atom and had a B-factor below 80 Å$^2$. Refinement by simulated annealing *(26)* caused an increase in $R_{free}$, and was therefore not used. In the final stage of refinement, we included the data thus far omitted from refinement for calculation of $R_{free}$, which resulted in a further improvement of the electron density maps and a final model with $R_{cryst}$ = 19.4%. To evaluate $R_{free}$ after the completion of refinement, 10% of data were again randomly omitted, and the model was subjected to a round of simulated annealing with an initial temperature of 3000 K, followed by 120 cycles of positional refinement. This resulted in an $R_{cryst}$ of 19.0% and $R_{free}$ of 26.1% for data between 40 and 2.5 Å resolution and F >σ(F).

State-of-the-art methods and software were used at the time of RNase A-pRI structure determination. Although the methodology of crystal-structural determination is an ever-developing field, similar methods are at present generally used, and program packages X-PLOR and O remain among the most popular. Common alternatives include the CCP4 suite *(27)* and CNS *(28)*; both contain programs covering most steps of structure determination.

## 4. Notes

1. Porcine RI can be expressed recombinantly in *S. cerevisiae (9)*. Although the effects of the protein from different sources on the crystallization of the RNase A-pRI complex are currently unknown, such preparations of pRI will probably perform equally well for crystallization.

2. The RNase A-pRI complex crystallized in presence of ammonium sulfate (as precipitant) and phosphate buffer, and is found to have a solvent ion, most likely a sulfate, bound in the active-site of RNase A. Correspondingly, the RNase active site residue Lys41 does not directly contact pRI. In the crystallization of Ang-hRI complex *(19)*, the concentrations of phosphate and sulfate were kept low, and the corresponding residue in Ang (Lys40) forms a hydrogen bond with hRI Asp435. Sulfate and other anions with affinity for the p1 site are not present under standard conditions used for kinetic characterization of RI complexes, and Asp435 mutations strongly affect the affinities for both RNase A and Ang *(29)*. The presence of salt also influences the affinity between RI and RNases *(30,31)*.

3. $R_{merge} = \sum hkl(\sum i(|I\ hkl,i-<I\ hkl>|))/\sum hkl,i\ <I\ hkl>$, where I *hkl,i* is the intensity of an individual measurement of the reflection with Miller indices h, k and l, and $<I_{hkl}>$ is the mean intensity of that reflection.

4. Data collection of crystals flash-cooled in a nitrogen stream at ~100 K generally improves the data quality by reducing radiation decay. Most protein crystals can be successfully flash-cooled by introducing them into a stabilizing solution supplemented with cryo-protectant (to prevent the solvent forming ice crystals, instead forming a glass-like solid phase) before subjecting them to the nitrogen stream. Common cryoprotectants are 13–30% glycerol, 11–30% ethylene glycol, 25–35% PEG 400, or 20–30% MPD (*see* **ref.** *22* and references therein). However, for the RNase A-pRI complex, no attempts have yet been made to establish conditions and perform cryogenic data collection.

5. Similar quality diffraction data could be obtained by using other commercially available image plate detectors, X-ray generators, and processing software. Synchrotron radiation would be expected to improve the quality and resolution of the diffraction data. The crystals of RNase A-pRI have not yet been exposed to synchrotron radiation.

6. $R_{cryst} = \sum_{hkl}(||Fobs_{hkl}|-|Fcalc_{hkl}||)/|Fobs_{hkl}|$, where $|Fobs_{hkl}|$ and $|Fcalc_{hkl}|$ are the observed and calculated structure factor amplitudes; $R_{free}$ is equivalent to $R_{cryst}$ but calculated with reflections omitted from the refinement process.

## References

1. Lee, F. S. and Vallee, B. L. (1993) Structure and action of mammalian ribonuclease (angiogenin) inhibitor. *Prog. Nucl. Acid Res. Mol. Biol.* **44**, 1–30.
2. Hofsteenge, J., Kieffer, B., Matthies, R., Hemmings, B. A., and Stone, S. R. (1988) Amino acid sequence of the ribonuclease inhibitor from porcine liver reveals the presence of leucine-rich repeats. *Biochemistry* **27**, 8537–8544.
3. Kobe, B. and Deisenhofer, J. (1994) The leucine-rich repeat: a versatile binding motif. *Trends Biochem. Sci.* **19**, 415–421.
4. Kobe, B. and Deisenhofer, J. (1995) Proteins with leucine-rich repeats. *Curr. Opin. Struct. Biol.* **5**, 409–416.
5. Buchanan, S. G. S. C. and Gay, N. J. (1996) Structural and functional diversity in the leucine-rich repeat family of proteins. *Prog. Biophys. Mol. Biol.* **65**, 1–44.
6. Kobe, B. and Deisenhofer, J. (1993) Crystal structure of porcine ribonuclease inhibitor, a protein with leucine-rich repeats. *Nature* **366**, 751–756.
7. Avey, H. P., Boles, M. O., Carlisle, C. H., Evans, S. A., Morris, S. J., Palmer, R. A., Woolhouse, B. A., and Shall, S. (1967) Structure of ribonuclease. *Nature* **213**, 557–562.
8. Kartha, G., Bello, J., and Harker, D. (1967) Tertiary structure of ribonuclease. *Nature* **213**, 862–865.
9. Vicentini, A. M., Kieffer, B., Matthies, R., Meyhack, B., Hemmings, B. A., Stone, S. R., and Hofsteenge, J. (1990) Protein chemical and kinetic characterization of recombinant porcine ribonuclease inhibitor expressed in *Saccharomyces cerevisiae*. *Biochemistry* **29**, 8827–8834.
10. Kobe, B. and Deisenhofer, J. (1995) A structural basis of the interactions between leucine-rich repeats and protein ligands. *Nature* **374**, 183–186.
11. Kobe, B. and Deisenhofer, J. (1996) Mechanism of ribonuclease inhibition by ribonuclease inhibitor protein based on the crystal structure of its complex with ribonuclease A. *J. Mol. Biol.* **264**, 1028–1043.
12. Kobe, B. and Deisenhofer, J. (1993) Crystallization and preliminary X-ray analysis of porcine ribonuclease inhibitor, a protein with leucine-rich repeats. *J. Mol. Biol.* **231**, 137–140.

13. Wlodawer, A., Svensson, L. A., Sjolin, L., and Gilliland, G. L. (1988) Structure of phosphate-free ribonuclease A refined at 1. 26 Å. *Biochemistry* **27,** 2705–2717.
14. Jones, S. and Thornton, J. M. (1996) Principles of protein-protein interactions. *Proc. Natl. Acad. Sci. USA* **93,** 13–20.
15. Lawrence, M. C. and Colman, P. M. (1993) Shape complementarity at protein/protein interfaces. *J. Mol. Biol.* **234,** 946–950.
16. Birdsall, D. L. and McPherson, A. (1992) Crystal structure disposition of thymidilic acid tetramer in complex with ribonuclease A. *J. Biol. Chem.* **267,** 22,230–22,236.
17. Fontecilla-Camps, J. C., De Llorens, R., Le Du, M. H., and Cuchillo, C. M. (1994) Crystal structure of ribonuclease A•d(ApTpApApG) complex. Direct evidence for extended substrate recognition. *J. Biol. Chem.* **269,** 21,526–21,531.
18. Polakowski, I. J., Lewis, M. K., Muthukkaruppan, V., Erdman, B., Kubai, L., and Auerbach, R. (1993) A ribonuclease inhibitor expresses anti-angiogenic properties and leads to reduced tumor growth in mice. *Am. J. Pathol.* **143,** 507–517.
19. Papageorgiou, A. C., Shapiro, R., and Acharya, K. R. (1997) Molecular recognition of human angiogenin by placental ribonuclease inhibitor—an X-ray crystallographic study at 2.0 Å resolution. *EMBO J.* **16,** 5162–5177.
20. Burton, L. E. and Fucci, N. P. (1982) Ribonuclease inhibitors from the livers of five mammalian species. *Int. J. Peptide Protein Res.* **19,** 372–379.
21. Kobe, B., Ma, Z., and Deisenhofer, J. (1994) Complex between bovine ribonuclease A and porcine ribonuclease inhibitor crystallizes in a similar unit cell as free ribonuclease inhibitor. *J. Mol. Biol.* **241,** 288–291.
22. Kobe, B., Gleichmann, T., Teh, T., Heierhorst, J., and Kemp, B. E. (1999) Crystallization of protein kinases and phosphatases, in *Protein Phosphorylation: A Practical Approach,* vol. 2 (Hardie, D. G., ed.), Oxford University Press, Oxford.
23. Brünger, A. T. (1996) *X-PLOR Version 3. 1.* Yale University, New Haven, CT.
24. Brünger, A. T. (1990) Extension of molecular replacement: a new search strategy base on Patterson correlation coefficient. *Acta Crystallogr.* **A46,** 46–57.
25. Jones, T. A., Zou, J.-Y., Cowan, S. W., and Kjeldgaard, M. (1991) Improved methods for building protein models in electron density maps and the location of errors in these models. *Acta Crystallogr.* **A47,** 110–119.
26. Brünger, A. T., Kuriyan, J., and Karplus, M. (1987) Crystallographic R factor refinement by molecular dynamics. *Science* **235,** 458–460.
27. CCP4 (1994) The CCP4 suite: programs for protein crystallography. *Acta Crystallogr.* **D50,** 760–763.
28. Brünger, A. T., Adams, P. D., Clore, G. M., DeLano, W. L., Gros, P., Grosse-Kunstleve, R. W., et al. (1998) Crystallography & NMR system: a new software suite for macromolecular structure determination. *Acta Crystallogr.* **D54,** 905–921.
29. Chen, C.-Z. and Shapiro, R. (1997) Site-specific mutagenesis reveals differences in the structural bases for tight binding of ribonuclease inhibitor to angiogenin and ribonuclease A. *Proc. Natl. Acad. Sci. USA* **94,** 1761–1766.
30. Lee, F. S., Shapiro, R., and Vallee, B. L. (1989) Tight-binding inhibition of angiogenin and ribonuclease A by placental ribonuclease inhibitor. *Biochemistry* **28,** 225–230.
31. Blackburn, P., Wilson, G., and Moore, S. (1977) Ribonuclease inhibitor from human placenta. *J. Biol. Chem.* **252,** 5904–5910.
32. Kraulis, P. (1991) MOLSCRIPT: a program to produce both detailed and schematic plots of protein structures. *J. Appl. Cryst.* **24,** 946–950.

# 14

## Methods for Studying the Interaction of Barnase with Its Inhibitor Barstar

### Gideon Schreiber

## 1. Introduction

Barnase is a 110-residue extracellular ribonuclease produced by *Bacillus amyloliquefaciens*. The same organism produces an 89-residue polypeptide, barstar, which binds tightly to barnase and inhibits its potentially lethal RNase activity inside the cell. The three-dimentional structures of barnase and barstar are known both from X-ray crystallography and NMR spectroscopy *(1–4)*. The crystal structure of the complex of barnase with the barstar (C40A/C82A) double mutant has been solved by Guillet et al. (1993) at 2.6 Å resolution and by Buckle et al. (1994) at 2.0 Å resolution *(5,6)*. Both barnase and barstar have proven to be excellent models for protein stability and folding studies *(7–11)*. The two proteins form a very tight complex, with $K_d = 10^{-14}$ $M$, half-life of about 2 d and an association rate constant of $4 \times 10^8$ $M^{-1}$/s. The kinetic and thermodynamics of the barnase-barstar interaction were investigated in detail *(12–20)*. High affinity is maintained at pH 7.0–9.0, while lowering the pH to under 4.5 eliminates binding completely. The plentiful structural and thermodynamic data available for this interaction, the ease of genetic manipulation and expression, and the excellent in vivo selection methods have made this interaction a target for many genetic selection techniques. Because free barnase is toxic, organisms expressing it can survive only if barstar is efficiently coexpressed. This positive selection system has been used as a convenient tool for transcriptional studies in plants. For example, the tissue specificity of the TA56 promoter was mapped by expressing a TA56/barnase fusion gene in tobacco plants and determining which tissues showed a requirement for barstar production *(21)*. Using a similar approach, a transgenic maize plant was developed where a plant with the *barnase* gene is pollen-sterile, whereas a plant with the *barnase* and *barstar* gene is pollen-fertile. Both genotypes are needed for the crossing system, thereby providing for the regular production of hybrid seeds *(22)*. The barnase–barstar system has been used to select affinity-compensating mutations at the interface of barnase–barstar *(23)*. A plasmid vector has been developed to allow positive selection and directional cloning based on the conditionally lethal barnase gene *(24)*.

From: *Methods in Molecular Biology, vol. 160: Nuclease Methods and Protocols*
Edited by: C. H. Schein © Humana Press Inc., Totowa, NJ

## 1.1. Structure-Function Analysis of the Barnase–Barstar Interaction

The binding site of a protein–protein complex can be characterized in a number of ways. The structure can be determined from X-ray crystallography. Here, one can determine the mode of binding and sort the interactions that occur at the interface, calculate buried surface area, and define interface water. However, this does not teach us much about the thermodynamics of the interaction, which can be determined by measuring binding affinities (from which the free energy of binding can be derived) and the enthalpy and entropy of binding. A more detailed energetic picture is obtained by introducing site-specific mutations, and varying solution and temperature conditions. These methods all measure binding only at equilibrium, while protein–protein interactions are a kinetic process involving the formation and dissociation of the complex. Association rate constants are dictated by random Brownian motion and long-range electrostatic forces, while dissociation rate constants are determined from the strength of short-range interactions formed between the proteins. Theoretical calculations may be used to explain the relation between structure, thermodynamics, and kinetics of an interaction. The interaction between barnase and barstar was investigated using all the different tools detailed above.

## 1.2. Barnase–Barstar Interaction and the Nature of the Interface

The mode of inhibition of barnase by barstar is very simple. The active site of barnase is sterically blocked by barstar $\alpha$-helix$_2$ and the loop connecting this to $\alpha$-helix$_3$ *(5,6)* (**Fig. 1**). Comparing the mode of barstar inhibition to that of substrate oligonucleotide binding into the barnase active site as seen in the barnase-3'GMP and barnase-d(CGAC) structures *(25,26)* reveal a number of structural parallels between the two. The most striking is that similar electrostatic interactions are formed between the phosphate oxygens of a RNA molecule and the phosphodiester cleavage site (the $p_1$ phosphate site) of the barnase active site as with the carboxyl side chain of Asp 39 of barstar. Thus, Asp 39 seems to mimic the mode of binding of the substrate. The structures of barnase and barstar in the complex are almost identical to that found in the free structures, except for possible movement of the 29–33 loop of barstar, which was observed in the NMR but not the crystal structure of unbound barstar *(4,6,27)*.

## 1.3. Thermodynamics of Binding

Upon formation of the barnase–barstar complex, 1590 Å$^2$ of solvent-accessible surface area is buried. Alanine-scanning mutagenesis of all residues at the interface has revealed that a small number of interface residues contribute most to the free energy of binding. On barstar, these residues are Tyr 29, Asp 35, and Asp 39 (3.4, 4.5, and 7.7 kcal/mol) *(15,17)*, which are all located on Helix$_2$ and the adjacent loop (**Fig. 1**). On barnase, a set of residues in the protein interface has been identified as being important for both binding and catalysis. Mutation of any one of these residues (Lys 27, Arg 59, Arg 83, Arg 87, and His 102) reduces the free energy of binding barstar by 5–6 kcal/mol *(13,16)*.

## 1.4. Double Mutant Cycle Analysis of the Barnase–Barstar Interface

The currently popular reductionist approach to study the energetic contribution of amino acids toward stability or binding involves the construction of single mutations.

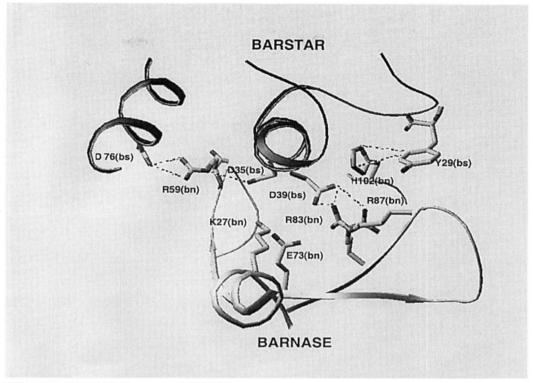

| Barnase | Barstar | $\Delta\Delta G_{int}$ (kcal mol$^{-1}$) | distance (Å) |
|---------|---------|------------------------------------------|--------------|
| K27A | D39A | 4.8 (0.21) | 4.5 (NZ↔OD1) |
| R59A | D35A | 3.4 (0.2) | 4.4 (NE↔OD1) |
| | | | 2.9 (N↔OD1) |
| R59A | E76A | 1.7 (0.24) | 3.0 (NH1↔OE1) |
| R83Q | D39A | 6.7 (0.25) | 2.5 (NH2↔OD1) |
| R87A | D39A | 6.1 (0.26) | 2.9 (NH2↔OD1) |
| H102A | Y29A | 3.3 (0.30) | 3.4 (◯↔CE1) |
| H102A | D39A | 4.9 (0.35) | 2.8 (NE2↔OD2) |

Fig. 1. The binding site between barnase and barstar, including the residues that contribute most to the binding energy. At the center of the binding site, Asp 39 (bs) binds to Arg 83 (bn), Arg 87 (bn), and His 102 (bn) with coupling energies of over 6 kcal mol$^{-1}$, making them among the strongest interactions yet reported. The adjacent table shows the interaction energies between some of the pairs of residues in the interface and the distances between them. Note that three of the interactions with coupling energies of more than 3.0 kcal/mol (Lys 27 [bn] with Asp 39 [bs], Arg 59 [bn] with Asp 35, and His 102 [bn] with Tyr 29 [bs]) were not expected to play an important part in binding according to structural analysis *(5,6)*.

However, the interpretation of such experiments is severely limited by the fact that the properties of a residue are a function of the entire system, and not necessarily the sum of its parts. As a result, the contribution of a particular residue to the energy of binding cannot be explained by physical analysis of the binding site *(17,19,28,29)*. A powerful experimental and theoretical approach in the study of protein–protein complexes is the use of double and higher-order mutant cycles, where interactions between amino acids are treated within their native context. Such cycles reveal whether the contributions from a pair of residues are additive or the effects of mutations are coupled. To measure the contribution of potential pairwise interactions to the stability of the complex *(17)*, we determined the pairwise interaction energies between most residues in the interface, regardless of the distance between them. Double mutant cycles were constructed between a subset of 5 barnase and 7 barstar residues, which were identified from structural and mutagenesis studies to be important in stabilizing the complex. The advantage of such a global approach is that it allows us to identify long-range interactions (which do not have to be direct), as well as standard short-range interactions, and to evaluate their contribution to the free energy of binding. In general, the coupling energy between pairs of residues was found to decrease as the distance between them increases *(17)*. To an approximation, residues separated by <7 Å can interact. At greater separations, the effects of mutation were additive; in other words the two residues affect binding independently of each other. The highest coupling energies were found between pairs of charged residues (1.6–7 kcal mol$^{-1}$).

## *1.5. Designing Peptides that Mimic Barstar*

As most barnase-binding residues on barstar are located on a single stretch of amino acids (residues 29–42) comprising a turn and helix, a peptide fragment of barstar should be sufficient to inhibit barnase. Further, residues 27 and 44 lie so close to each other in the three-dimentional structure that a disulfide bridge could be introduced between them to stabilize the correct peptide structure. A linear correlation between the helical amphipathy of peptides and their affinity for and ability to inhibit barnase has been observed. Despite the ample structural and thermodynamic information used to design them, the affinity of the best barstar peptide towards barnase is in the μ$M$ range, ~$10^8$ times lower than that of native barstar *(30)*. We are still determining what other factors could aid in the design of specific peptides as inhibitors.

## *1.6. Analyzing the Kinetics of Association*

Intermolecular hydrogen bonding plays an important role in barnase–barstar recognition. Almost one-half of the hydrogen bonds between barnase and barstar involve two charged residues, and an additional one-third involve one charged partner. Thus, the dominant forces in binding seem to be electrostatic in nature. On the whole, the interacting surfaces have a high degree of charge complementary. The negative electrostatic potential induced by Asp 35, Asp 39, and Glu 76 on the exposed face of barstar α-helix$_2$ and 4, interacts favorably with the electropositive base of the barnase active site, formed by the clustering of residues Lys 27, Arg 59, Arg 83, and Arg 87. This leads to a very high rate of association between the two proteins, on the order of $10^8$ $M^{-1}$/s under physiological conditions *(16)*. A similar charge complementarity was also observed for the interaction of porcine ribonuclease inhibitor (RI) with RNase A *(31)*. A number of theoretical studies, involving Brownian dynamic simulations of binding *(32)*, and electrostatic-free energy relations between the free and complexed proteins *(33–35)* were conducted to explain the extremely rapid association between barnase and barstar.

Structural studies, combined with thermodynamic and kinetic analysis of protein interactions and mutation analysis, have greatly advanced our understanding of the nature of a protein interface and the kinetic process of complexation. Still, a clear picture of intramolecular forces is not yet available, even for a well-studied case such as the interaction between barnase and barstar. In addition to the methods discussed in this chapter, the thermodynamics of this interaction were examined in detail ($\Delta H$ and $\Delta S$) *(12,14,36)*, and a number of theoretical studies have related the kinetics and thermodynamics of the barnase–barstar interaction to basic principles *(32,33,35,37)*. Many of the methods described here are not unique for investigating the barnase–barstar interaction but can be used for the investigation of other protein–protein interactions as well. Methods used to purify barnase and barstar, measure their activity, and determine the kinetics and thermodynamics of this tight, fast-associating interaction are summarized in the following sections.

## 2. Materials

### 2.1. Bacterial Strains and Plasmids

1. BL21DE3 strain of (*Escherichia coli*) (F-, *omp*T, $r_B$, $m_B$) contains the T7 polymerase under the control of the *lac UV5* promoter. This strain harbors the pLysS plasmid that confers chloramphenicol resistance and constitutively expresses the T7 lysozyme. T7 lysozyme is a specific inhibitor of T7 RNA polymerase, reducing the amount of active T7 RNA polymerase in the absence of IPTG induction.
2. The gene coding for wild-type barnase was cloned into the pUC19 plasmid *(38)*. A 1.4 kB *Eco*RI-*Hin*dIII restriction fragment of this plasmid, containing the structural gene for barnase fused to the promoter and signal sequence of the *E. coli* alkaline phosphatase gene, was subcloned into the pTZ18U plasmid (Pharmacia). This fragment also contains the gene for barstar under the control of its own promoter.
3. The gene coding for wild-type barstar was constructed by cloning the *Eco*RI-*Hin*dIII fragment of pMT410, which contains the barnase and barstar genes into the polylinker region of pTZ18U (Pharmacia). The gene-encoding barnase was removed by SacI digestion, resulting in the pML2bs plasmid, where barstar expression is under the control of a T7 promoter.
4. M13/T7 phage (available from Invitrogen) contain a 2.7 kB *Hin*dIII-*Eco*RI fragment that encodes T7 polymerase under the control of a lac promoter. T7 polymerase is introduced into the bacteria by infection of *E. coli* F' cells with M13/T7 phage at a titer of 5 PFU per cell (equal to $5 \times 10^9$ PFU/mL for a cell culture of density of $A_{600} = 1$).

### 2.2. Media

5. 2xTY: 16 g tryptone, 10 g yeast extract, and 5 g NaCl/L.
6. Low phosphate medium: 10X medium: 84 g MOPS, 8 g Tricine, 29.2 g NaCl, 16 g KOH, 5.1 g $NH_4Cl$, 20 mL solution O, 375 mg $K_2SO_4$, and 15 g casein hydrolysate, add water to 1 L. To make 1 L of 1X medium, take 100 mL 10X medium, 900 mL water, and 0.4 mL 1 $M$ $KH_2PO_4$ (0.4 m$M$ final). Following sterilization, 20 mL glucose (20% w/v) and 0.1 mL of 1% thiamine are added to each 1 L flask.
7. Solution O: 26.8 g $MgCl_2$, 0.08 mL HCl (conc), 0.058 g $FeCl_2 \cdot 6H_2O$, 1.84 mg $CaCl_2 \cdot 2H_2O$, 0.64 mg $H_3BO_3$, 0.4 mg $MnCl_2 \cdot 4H_2O$, 0.18 mg $CoCl_2 \cdot 6H_2O$, 0.04 mg $CuCl_2 \cdot 2H_2O$, 0.34 mg $ZnCl_2 \cdot$ 6.05 mg $Na_2MoO_4 \cdot 2H_2O/L$.
8. CAP and ampicillin solutions: CAP (chloramphenicol) stock solution is prepared in 100% ethanol at a concentration of 25 mg/mL, and ampicillin is prepared in water at a concentration of 100 mg/mL. Add 1 mL of either per liter of culture.

9. RNA-agar plates: Dissolve 15 g agar in 1 L 100 m*M* Tris-HCl, pH 8.4, and autoclave. Add 2 g/L total yeast RNA (BDH) and pour into standard 9-cm Petri dishes.

## 2.3. Buffers, Chemicals, and Other Stock Solutions

10. TE: 10 m*M* Tris-HCl, pH 8.0, 1 m*M* EDTA.
11. $G_pU_p$ - Guanylyl(3'→5')uridine 3'-monophosphate is from Sigma.
12. Dialysis tubing (Spectrum Medical Industries, Inc., Los Angeles, CA) with a cut-off of 3500 Dalton.
13. Acetate buffer: 50 m*M* Na-acetate, pH 4.5.
14. IPTG: 100 m*M* stock solution is prepared by dissolving 24 mg in 1 mL of water.
15. Tris buffer: 50 m*M* Tris-HCl buffer, pH 8.0.
16. TCA solution contains 10% (v/v) of trichloric acid in water.

## 3. Methods
## 3.1. Expression and Isolation of Barnase

1. Transform the BL21DE3 strain of *E. coli* with plasmid DNA *(39)*.
2. Inoculate 30 mL 2xTY + 100 mg/L ampicillin and 25 mg/L CAP with individual colonies of the transformation.
3. After overnight growth, use 5 mL of this culture to inoculate 1 L low-phosphate medium *(40)* containing the antibiotics from **step 2**. Phosphate starvation leads to activation of the alkaline phosphatase promoter and, consequently, to synthesis and secretion of barnase. Cells are grown at 37°C, with vigorous shaking for 20 h until the cell density at 600 nm levels off at about 1 OD. Alternatively, expression of some barnase mutants was higher when cells were grown for 48 h at 25°C.

Purification:

1. Chill the cell culture to about 4°C, slowly add 55 mL ice-cold glacial acetic acid per liter of culture and stir for 15 min. Acidification releases additional barnase from the periplasm into the medium.
2. Centrifuge for 10 min at 6000*g* to pellet the cells and decant the supernatant.
3. Add cation exchange resin (5 mL/L culture SP-trisacryl equilibrated with acetate buffer) while gently stirring. After 1 h stirring, the resin is allowed to settle and the supernatant is decanted.
4. Pour the resin and residual supernatant into a sintered glass filter and wash with acetate buffer until no *A* at 280 nm is detected. Barnase is eluted from the resin in a single step with acetate buffer containing 0.5 *M* NaCl, using three times the resin volume.
5. Dialyze the eluate containing barnase overnight against acetate buffer.
6. The dialyzed eluate is applied to a Pharmacia FPLC system using any standard cation exchange column. We used a mono-S 10/10 column and a flow rate of 1 mL/min. Barnase binds to the column at 50 m*M* Na-acetate (pH 4.5) and elutes at about 250 m*M* NaCl, using a linear gradient from 0–400 m*M* over 10 column volumes.
7. Average yields are 2–20 mg/L culture. Pure barnase is stable at 4°C for several days, or can be stored at −70°C after flash freezing in liquid nitrogen. Protein concentrations are determined from the UV *A* at 280 nm. An extinction coefficient of $\varepsilon_{280} = 27,400$ $M^{-1}$/cm was determined for barnase by the method of Gill and von Hippel *(41)*.

## 3.2. Expression and Isolation of Barstar

Barstar is expressed in *E. coli* TG1 cells containing the pML2bs plasmid.

1. Grow a 30-mL starter culture overnight in 2xTY + 100 mg/L ampicillin.
2. Inoculate 1 L 2xTY containing the above antibiotics with 5 mL of the overnight culture.
3. Add 0.2 m$M$ IPTG when the cell culture reaches $A_{600} = 0.3$.
4. After 1 h of vigorous shaking, add M13/T7 phage to a final MOI of 5 PFU/cell. Growth is continued for an additional 4 h before the cells are pelleted (*see* **Note 1**).
5. Alternatively, barstar can be produced in BL21(pLysE) cells containing the pML2bs plasmid (freshly transformed). An overnight starter culture starting from a single colony is grown, and 5 ml/L is used to inoculate 1 L 2xTY containing 100 mg/L ampicillin and 25 mg/L CAM. The bacterial culture is grown overnight without adding IPTG, and pelleted (15 min at 6000 rpm). The residual amount of T7 polymerase expressed from BL21(pLysE) is sufficient for barstar expression.
6. Resuspend the cell pellet in 25 mL/L TE + 1 m$M$ PMSF + 50 mg/mL lysozyme and freeze at –70°C. After thawing, sonicate cells for 3 min on ice in nine cycles of 20 s sonication, 40 s rest.
7. Centrifuge the solution in a Sorval SS34 rotor at 30,000$g$ for 30 min. Barstar is found in the soluble fraction.
8. Slowly add solid ammonium sulfate to 40% while stirring. The solution is centrifuged at 30,000$g$ for 30 min. Add additional solid ammonium sulfate to the eluate, bringing it to 80% saturation. At this concentration barnase precipitates. The solution is centrifuged as before. Dissolve the pellet in a minimal volume (5 mL/L culture is suitable) of Tris-HCl buffer + 100 m$M$ NaCl and dialyze against the same buffer.
9. Subject the protein solution to gel filtration on a Superdex™ 75 (26/60) column from Pharmacia at a flow rate of 4 mL/min. Fractions containing active barstar are detected using the RNA plate-inhibition assay (**Subheading 3.3.3.**).
10. Load the Superdex fractions containing barstar directly into any standard anion exchange column for final purification. We use a HiLoad 26/10 SP Sepharose HP column at a flow rate of 4 mL/min and elute with a salt gradient of 100–500 m$M$ NaCl in Tris buffer over 10 column vol. Barstar elutes at 300 m$M$ NaCl. One liter of culture yields 10–60 mg of pure protein (*see* **Notes 2** and **3**). An extinction coefficient of $\varepsilon_{280} = 22,100 \ M^{-1}$/cm was determined for barstar.

## *3.3. Activity Assays for Barnase and Barstar*

### *3.3.1. Rapid Assays of Barnase Activity*

RNA plate assay:

1. Drop 3 µL barnase protein solutions (0.01–0.1 µg/mL in 100 m$M$ Tris-HCl, pH 8.5) onto an RNA-agar plate and incubate for 10 min at 37°C.
2. Precipitate the RNA by pouring ~10 mL 10% TCA onto the plate. RNase activity results in the formation of clear plaques, whereas precipitated RNA gives the plate a whitish color. A concentration of 0.01 µg/µL of wild-type barnase is sufficient to see plaque formation. The plaque radius is proportional to the RNase activity.

Spectroscopic assay:

1. Dissolve 2 mg/mL total yeast RNA (BDH) in 100 m$M$ Tris-HCl, pH 8.5 at 25°C.
2. Add barnase to the reaction at time zero (0.01 µ$M$ wild-type barnase allows for at least 1–2 min of steady-state kinetics).
3. RNA hydrolysis is followed with a spectrophotometer as the decrease in A at 298 nm *(42)*. Enzyme activity is estimated from the initial rates of hydrolysis relative to wild-type barnase.

### 3.3.2. Assay of Barnase Activity Using $G_pU_p$ Dinucleotides

This assay allows the determination of values of $K_m$ and $V_{max}$ for the trans-esterification step from initial velocities by following the increase in $A$ at 275 nm *(42)*.

1. Dissolve 200 μ$M$ $G_pU_p$ in 100 m$M$ acetate/acetic acid buffer, pH 5.8, 100 m$M$ BSA at 25°C.
2. Add barnase to the reaction at time zero (0.025 μ$M$ wild-type barnase allows for at least 1–2 min of steady state kinetics).
3. $G_pU_p$ hydrolysis is followed with a spectrophotometer as the increase in $A$ at 280 nm *(42)* with time. $K_m$ and $k_{cat}$ for wild-type barnase are 19.9 μ$M$ and 53 s$^{-1}$, respectively *(43)*.

pH 5.8 has been shown to be optimum for the transesterification of dinucleotides, and 8.5 is optimal for RNA degradation.

### 3.3.3. Rapid Assay of Barnase Inhibition by Barstar

This assay is based on the RNA plate assay (**Subheading 3.3.1.**).

1. Prepare 3 μL barnase solutions at concentrations of 0.01, 0.1, and 1 μg/mL in 100 m$M$ Tris-HCl, pH 8.5 at room temperature.
2. Add 3 μL barstar at an approximate concentration of 0.1 μg/mL incubate 1 min at room temperature.
3. Put the mixture on a RNA-agar plate and continue as in **Subheading 3.3.1.** Barstar activity reduces the size of clear plaques made by barnase alone on the plate.
4. Barstar activity can also be monitored by its ability to inhibit RNA or $G_pU_p$ degradation by barnase monitored as change in absorbance (see **Subheadings 3.3.1.** and **3.3.2.**).

## 3.4. Direct Measurements of Barnase–Barstar Binding

Binding is measured by following a specific signal, which allows the complexed and free form of the proteins to be distinguished. Probes for measuring binding can be divided into two main groups. The first involves measurements in real time; for example, monitoring the change of a spectroscopic probe as fluorescence, circular dichroism, or surface-plasmon-resonance. Such methods have the ability to measure both kinetics and thermodynamic parameters. A second type of "readout" requires longer measurement times, and thus is applicable either for slow kinetics or for equilibrium measurements. This group include enzyme-activity inhibition assays, radioactive tracer techniques (either in equilibrium or pulse-chase experiments), isothermal titration calorimetry, and other methods. In **Subheading 3.4.1.**, we will discuss the various methods used for measuring binding affinities, thermodynamics, and kinetics of the barnase–barstar interaction.

### 3.4.1. Quenching of Tryptophan Fluorescence

Barnase, which contains three tryptophans, has a fluorescence spectrum with an emission maximum at 337 nm upon excitation at 280 nm. Barstar has a maximum at 332 nm, and the complex has an emission maximum of fluorescence at 335 nm *(16)*. The fluorescence of the barnase–barstar complex is lower by about 20% than the sum of fluorescence of barnase and barstar measured separately. Interestingly, for some mutant proteins the difference in fluorescence intensity between the free and complexed proteins is larger at excitation at 230 nm *(15)*. This phenomenon was used extensively in studying the kinetics of complex formation.

### 3.4.2. Measuring Association Rate Constants

Barnase–barstar association was measured by monitoring dequenching of tryptophan fluorescence upon interaction, using an Applied Photophysics Bio Sequential DX-17MV stopped-flow apparatus. Second-order conditions with equal concentrations of both proteins (0.25 μ$M$) were used for most experiments. Using these conditions, virtually all of the protein is bound within the first 0.5 s of the reaction. The association is effectively irreversible at these concentrations, i.e., [E]+[I] → [E·I], where [E] is the barnase concentration, [I] the barstar concentration and [E·I] the complex concentration. The analytical solution for second-order conditions, where $[E]_o = [I]_o$, is:

$$1/([E]_o - [I·E]) - 1/[E]_o = k_1 t \ (44) \tag{1}$$

Only the first 80% of the reaction amplitude should be used for fitting the data to a second-order equation, because small errors in measuring the initial concentrations of barnase or barstar transform the second-order reaction to a pseudo-first-order process toward the later part of the association curve. For slower-associating and faster-dissociating mutants, pseudo-first-order conditions may be applied, with one of the proteins at a fivefold excess. Here, the rates of the interaction $k_s$ are determined from an exponential fit of the data. The association-rate constant $k_{on}$ are determined from a plot of $k_s$ as a function of protein concentration by linear regression *(17,18)* (*see* **Note 4**).

### 3.4.3. Determination of Dissociation Rate Constants

Wild-type and mutant complexes with a dissociation-rate constants slower then $5 \times 10^{-3}$/s are measured manually by forming a complex with one of the proteins radioactively labeled with [$^3$H], and then measuring the rate at which the labeled protein is chased off using a large excess of cold protein (*see* **Note 4**).

1. Form a barnase–barstar complex using 20,000 cpm of [$^3$H]barstar and unlabeled barnase, with barnase being in excess concentration of about 1 μ$M$, so that all of the [$^3$H] barstar will be in complex. A total volume of 1 mL is convenient for tacking 20 time-point measurements.
2. At time zero, "chase" off [$^3$H] barstar from the complex using unlabeled barstar at 10× the concentration of barnase (10 μ$M$).
3. Complexed and noncomplexed [$^3$H] barstar are separated using an anion exchange resin (DE52 from Whatman). We found that the most convenient method for separation is to fill approx 20 μL of DE52 into an Eppendorf tip, and to use this as a mini ion-exchange column. The separation procedure is based on the observation that barstar binds to DE52 at salt concentrations of up to 300 m$M$ NaCl, the complex elutes at about 70 m$M$ NaCl, and barnase does not bind at all.
4. Measuring residual binding of [$^3$H] barstar to barnase: at various time points after "chasing": Inject about 1000 cpm from the solution into the Eppendorf tip containing DE52 (using a standard pipettor). Wash the resin with 2 vol of 150 m$M$ NaCl. The [$^3$H] barnase-barstar complex elutes, while free [$^3$H] barstar remains bound to the column. Remove the free [$^3$H] barstar using 2 column volumes of buffer, including 500 m$M$ NaCl and measure [$^3$H] barstar using a scintilation counter *(15,16)*.
5. The same procedure with small modifications can be done using [$^3$H] barnase instead of [$^3$H] barstar. Here we use 25 m$M$ Tris-HCl, pH 8.0, to load the solution containing the complex into the eppendorf tip containing DE52. At these buffer conditions, free [$^3$H] barnase will not bind the column, but the complex does. The complex can be eluted using 150 m$M$ NaCl.

6. Dissociation-rate constants greater then $5 \times 10^{-3}$/s may be measured with a stopped-flow fluorimeter, using wild-type barstar to chase off mutant barstar from the barnase–barstar complex. This method takes advantage of the different fluorescence spectrum of the wild-type and mutant complexes *(15,16)*. Both "chasing" procedures gave similar results for complexes having "off rates" of approx $5 \times 10^{-3}$/s. Dissociation-rate constants are calculated by fitting the data to a single exponential equation.

### 3.4.4. Determination of Binding Energies

Barnase and barstar form a very tight complex, with an equilibrium constant ($K_d$) of approx $10^{-14}$ $M$ *(16)*. Equilibrium constants of this magnitude are difficult to measure directly, but can be obtained from the ratio $K_d = k_{on}/k_{off}$ (*see* **Note 4**). When the affinity is low enough ($>10^{-9}$ $M$) $K_d$ can be determined directly using Isothermal Titration Calorimetry. A good agreement was found between the values determined from kinetic measurements and those determined from equilibrium titration *(12)*. The binding energy is calculated from: $\Delta G = -RT\ln K_d$, and the change in binding energy after introducing a mutation is $\Delta\Delta G_{wt\text{-}mut} = \Delta G_{wt} - \Delta G_{mut}$.

## 3.5. Site-Directed Mutagenesis

We suggest that the most efficient way to perform site-directed mutagenesis on small genes, as in the case of barnase or barstar, is by performing complimentary PCR.

1. Two primers of 20-nucleotides are synthesized. The mutagenic primer has the mutated codon placed at its 5'-end, followed by 6 codons downstream of the point of mutation in the gene sequence. The other primer is composed of the sequence complementary to the ~7 codons immediately upstream from the site of mutation.
2. Standard PCR conditions are used, with PWO (Boehringer Mannheim), or PFU (Strategene) used as high-fidelity polymerase enzymes. The product of the PCR reaction consists of the whole linearized plasmid, with blunt ends on both sides.
3. The PCR product is run on a standard agarose gel, purified, and resuspended in 20 µL water.
4. The purified PCR fragment is phosphorylated using $T_4$ polynucleotide kinase (1 U of enzyme in standard buffer including 1 m$M$ ATP, for 30 min at 37°C) and heated for 10 min at 70°C for enzyme inactivation.
5. 5 µL of phosphorilated PCR fragment is ligated in a total of 15 µL, using 1 U ligase in appropriate buffer, for 1 h at room temperature, after which the plasmid is transformed into JM109 or TG1 cells.
6. Mutations are confirmed by direct sequencing of the gene, which should be done with care, as random mutations can be incorporated into the gene during the PCR reaction. The efficiency of site-specific mutagenesis is about 90%, with the occurrence of random mutations in 20% of the genes analyzed. This method is especially suitable for small genes, where the rate of random mutations will be low. The total length of the plasmid, including the gene, should be kept to minimum, as the whole plasmid is amplified.

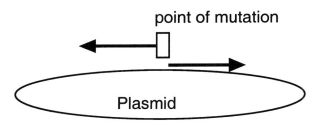

### 3.6. Double-Mutant Cycle Analysis

The now conventional procedure for experimentally analyzing interactions is by site-directed mutagenesis, usually through single mutations. If a mutated residue is directly or indirectly involved in binding, the mutation can lead to a change in the binding energy of the protein–protein complex. A change in free energy following the mutation of a residue X to A, $\Delta\Delta G_{X\to A}$, is not, in general, the intrinsic binding energy of the side-chain of X that is mutated, but is simply a quantitative measure of the relative binding energy of X rather than A being present, the *specificity* of the interaction. A more advanced procedure for mutagenesis is to mutate, both singly and doubly, pairs of residues (X and Y): the double mutant-cycle method (*see* **Subheading 2.3.**) *(8,17,45)*. This gives a coupling energy, $\Delta\Delta G_{int}$, defined by:

$$\Delta\Delta G_{int} = \Delta\Delta G_{X\to A, Y\to B} - \Delta\Delta G_{X\to A} - \Delta\Delta G_{Y\to B} \qquad (2)$$

where $\Delta\Delta G_{Y\to B}$ is the change in binding energy on mutation of Y to B, $\Delta\Delta G_{X\to A}$ is the change in binding energy on mutation of X to A, and $\Delta\Delta G_{X\to A, Y\to B}$ the change on simultaneously mutating X to A and Y to B. $\Delta\Delta G_{int}$ is a measure of the cooperativity of interaction of the two components that are mutated. If the effects of the mutations are independent (noncooperative), the change in free energy for the double mutant is the sum of those for the two single mutations, but if the mutated residues are coupled, then the change in free energy for the double mutant differs from the sum of the two single mutants. In some circumstances, $\Delta\Delta G_{int}$ can reduce to the interaction energy between two residues *(8,45)*.

Much of the mutational analysis is currently performed by mutating residues to Ala, regardless of the original residue. This is well justified if we want to delete a certain side chain. However, it is possible to make more subtle mutations, such as mutating Glu to Gln, keeping the stereospecificity of the side-chain while omitting the negative charge. Replacement mutations should be analyzed with caution, as the new residue may introduce new interactions, alterations in interface water or changing desolvation energies *(19)*.

## 4. Notes

1. The advantage of using M13/T7 as source of T7 polymerase is that any F' *E. coli* strain can be used, and high yields of expression are achieved more regularly.
2. Pure barstar protein will precipitate if dialyzed against water, or if the pH of the solution is near the isoelectric point of this protein, which is about pH 4.0.
3. Wild-type barstar has two cysteine residues that do not form a disulfide bridge *(46)*. DTT is not necessary to avoid formation of intermolecular disulfide bonds. However, 5 m*M* DTT must be added to unfolded barstar protein before refolding to avoid oxidation. Replacing the two Cys residues (40 and 82) with Ala causes only a small decrease in protein stability, with no significant structural change. Thus, some of the investigations were done with this double-mutant used as pseudo-wild-type.
4. The use of surface plasmon resonance (SPR) for the determination of binding constants has become very popular in recent years, with most work done on BIAcore instruments (Pharmacia). Because of technical limitations, it is possible to use a BIAcore to determine association-rate constants of up to $5 \times 10^6$ $M^{-1}$/s, dissociation-rate constants faster then $10^{-5}$/s, and binding affinities lower then $10^{11}$/*M*. As the rate of association of barnase to barstar is over $10^8$/$M^{-1}$/s, the rate of dissociation is $4 \times 10^{-6}$/s, and the affinity is in the

range of 10–14 *M*, methods based on SPR technology are not well-suited to investigate this interaction. To experimentally determine whether this will be the case, we bound barnase to a CM5 sensor chip through amine coupling (barstar does not bind to the negatively charged dextran surface of the CM5 sensor chip because its surface is strongly negatively charged). As expected, association was instantaneous, and no dissociation phase could be observed because binding is too tight.

## References

1. Bycroft, M., Ludvigs, S., Fersht, A. R., and Poulsen, F. M. (1991) Determination of the three-dimensional solution structure of barnase using nuclear magnetic resonance spectroscopy. *Biochemistry* **30,** 8697–8701.
2. Lubienski, M. J., Bycroft, M., Freund, S. M., and Fersht, A. R. (1994) Three-dimensional solution structure and 13C assignments of barstar using nuclear magnetic resonance spectroscopy. *Biochemistry* **33,** 8866–8877.
3. Mauguen, Y., Hartley, R. W., Dodson, E. J., Dodson, G. G., Bricogne, G., Chothia, C., and Jack, A. (1982) Molecular structure of a new family of ribonucleases. *Nature* **29,** 162–164.
4. Ratnaparkhi, G. S., Ramachandran, S., Udgaonkar, J. B., and Varadarajan, R. (1998) Discrepancies between the NMR and X-ray structures of uncomplexed barstar: analysis suggests that packing densities of protein structures determined by NMR are unreliable. *Biochemistry* **37,** 6958–6966.
5. Buckle, A. M., Schreiber, G., and Fersht, A. R. (1994) Protein-protein recognition: crystal structural analysis of a barnase–barstar complex at 2.0-A resolution. *Biochemistry* **33,** 8878–8889.
6. Guillet, V., Lapthorn, A., Hartley, R. W., and Maugen, Y. (1993) Recognition between a bacterial ribonuclease, barnase and its natural inhibitor, barstar. *Structure* **1,** 165–177.
7. Fersht, A. R. (1993) Protein folding and stability: the pathway of folding of barnase. *FEBS Lett* **325,** 5–16.
8. Fersht, A. R., Matouschek, A., and Serrano, L. (1992) The folding of an enzyme i: theory of protein engineering analysis of stability and pathway of protein folding. *J. Mol. Biol.* **224,** 771–782.
9. Nath, U., Agashe, V. R., and Udgaonkar, J. B. (1996) Initial loss of secondary structure in the unfolding of barstar [letter]. *Nature Struct. Biol.* **3,** 920–923.
10. Nolting, B., Golbik, R., Neira, J. L., Soler, G. A., Schreiber, G., and Fersht, A. R. (1997) The folding pathway of a protein at high resolution from microseconds to seconds. *Proc. Natl. Acad. Sci. USA* **94,** 826–830.
11. Zaidi, F. N., Nath, U., and Udgaonkar, J. B. (1997) Multiple intermediates and transition states during protein unfolding. *Nature Struct. Biol.* **4,** 1016–1024.
12. Frisch, C., Schreiber, G., Johnson, C. M., and Fersht, A. R. (1997) Thermodynamics of the interaction of barnase and barstar: changes in free energy versus changes in enthalpy on mutation. *J. Mol. Biol.* **267,** 696–706.
13. Hartley, R. W. (1993) Directed mutagenesis and barnase-barstar recognition. *Biochemistry* **32,** 5978–5984.
14. Martinez, J. C., Filimonov, V. V., Mateo, P. L., Schreiber, G., and Fersht, A. R. (1995) A calorimetric study of the thermal stability of barstar and its interaction with barnase. *Biochemistry* **34,** 5224–5233.
15. Schreiber, G., Buckle, A. M., and Fersht, A. R. (1994) Stability and function: two constraints in the evolution of barstar and other proteins. *Structure* **2,** 945–951.
16. Schreiber, G. and Fersht, A. R. (1993) Interaction of barnase with its polypeptide inhibitor barstar studied by protein engineering. *Biochemistry* **32,** 5145–5150.

17. Schreiber, G. and Fersht, A. R. (1995) Energetics of protein-protein interactions: analysis of the barnase–barstar interface by single mutations and double mutant cycles. *J. Mol. Biol.* **248,** 478–486.
18. Schreiber, G. and Fersht, A. R. (1996) Rapid, electrostatically assisted association of proteins. *Nature Struct. Biol.* **3,** 427–431.
19. Schreiber, G., Frisch, C., and Fersht, A. R. (1997) The role of Glu73 of barnase in catalysis and the binding of barstar. *J. Mol. Biol.* **270,** 111–122.
20. Yakovlev, G. I., Moiseyev, G. P., Protasevich, I. I., Ranjbar, B., Bocharov, A. L., Kirpichnikov, M. P., Gilli, R. M., et al. (1995) Dissociation constants and thermal stability of complexes of Bacillus intermedius RNase and the protein inhibitor of Bacillus amyloliquefaciens RNase. *FEBS Lett.* **366,** 156–158.
21. Beals, T. P. and Goldberg, R. B. (1997) A novel cell ablation strategy blocks tobacco anther dehiscence. *Plant Cell.* **9,** 1527–1545.
22. Mariani, C., Gossele, V., De Beuckeleer, M., De, B. M., Goldberg, R. B., De Greef, W., and Leemans, J. (1992) A chimaeric ribonuclease-inhibitor gene restores fertility to male sterile plants. *Nature* **357,** 384–387.
23. Jucovic, M. and Hartley, R. W. (1996) Protein-protein interaction: a genetic selection for compensating mutations at the barnase-barstar interface. *Proc. Natl. Acad. Sci. USA* **93,** 2343–2347.
24. Deyev, S. M., Yazynin, S. A., Kuznetsov, D. A., Jukovich, M., and Hartley, R. W. (1998) Ribonuclease-charged vector for facile direct cloning with positive selection. *Mol. Gen. Genet.* **259,** 379–382.
25. Buckle, A. M. and Fersht, A. R. (1994) Subsite binding in an RNase: structure of a barnase-tetranucleotide complex at 1.76 Å resolution. *Biochemistry* **33,** 1644–1653.
26. Guillet, V., Lapthorn, A., and Mauguen, Y. (1993) Three-dimensional structure of a barnase-3'GMP complex at 2.2A resolution. *FEBS Lett.* **330,** 137–140.
27. Wong, K. B., Fersht, A. R., and Freund, S. M. (1997) NMR 15N relaxation and structural studies reveal slow conformational exchange in barstar C40/82A. *J. Mol. Biol.* **268,** 494–511.
28. Bogan, A. A. and Thorn, K. S. (1998) Anatomy of hot spots in protein interfaces. *J. Mol. Biol.* **280,** 1–9.
29. Clackson, T. and Wells, J. A. (1995) A hot spot of binding energy in a hormone-receptor interface. *Science* **267,** 383–386.
30. Soler-Gonzalez, A. S. and Fersht, A. R. (1997) Helix stability in barstar peptides. *Eur. J. Biochem.* **249,** 724–732.
31. Kobe, B. and Deisenhofer, J. (1995) A structural basis of the interactions between leucine-rich repeats and protein ligands. *Nature* **374,** 183–186.
32. Gabdoulline, R. R. and Wade, R. C. (1997) Simulation of the diffusional association of barnase and barstar. *Biophys. J.* **72,** 1917–1929.
33. Janin, J. (1997) The kinetics of protein-protein recognition. *Proteins* **28,** 153–161.
34. Selzer, T. and Schreiber, G. (1999) Predicting the rate enhancement of protein complex formation from the electrostatic energy of interaction. *J. Mol. Biol.* **287,** 409–419.
35. Vijayakumar, M., Wong, K. Y., Schreiber, G., Fersht, A. R., Szabo, A., and Zhou, H. X. (1998) Electrostatic enhancement of diffusion-controlled protein-protein association: comparison of theory and experiment on barnase and barstar. *J. Mol. Biol.* **278,** 1015–1024.
36. Makarov, A. A., Protasevich, I. I., Lobachov, V. M., Kirpichnikov, M. P., Yakovlev, G. I., Gilli, R. M., et al. (1994) Thermostability of the barnase-barstar complex. *FEBS Lett.* **354,** 251–254.
37. Covell, D. G. and Wallqvist, A. (1997) Analysis of protein-protein interactions and the effects of amino acid mutations on their energetics. The importance of water molecules in the binding epitope. *J. Mol. Biol.* **269,** 281–297.

38. Paddon, C. J. and Hartley, R. W. (1987) Expression of bacillus amyloliquefaciens extracellular ribonuclease (barnase) in Escherichia coli following an inactivating mutation. *Gene* **53,** 11–19.

39. Studier, F. W. and Moffatt, B. A. (1986) Use of bacteriophage T7 RNA polymerase to direct selective high-level expression of cloned genes. *J. Mol. Biol.* **189,** 113–130.

40. Serpesu, E. H., Shortle, D., and Mildvan, A. S. (1986) Kinetic and magnetic resonance studies of effects of genetic substitutions of a $Ca^{++}$ liganding amino acid in *Staphysococcal Nuclease. Biochemistry* **25,** 67–87.

41. Gill, S. C. and von Hippel, P. H. (1989) Calculation of protein extinction coefficients from amino acid sequence data. *Anal. Biochem.* **182,** 319–326.

42. Mossakowska, D. E., Nyberg, K., and Fersht, A. R. (1989) Kinetic characterization of the recombinant ribonuclease from Bacillus amyloliquefaciens (barnase) and investigation of key residues in catalysis by site-directed mutagenesis. *Biochemistry* **28,** 3843–3850.

43. Day, A. G., Parsonage, D., Ebel, S., Brown, T., and Fersht, A. R. (1992) Barnase has subsites that give rise to large rate enhancements. *Biochemistry* **31,** 6390–6395.

44. Fersht, A. R. (1985) *Enzyme Structure and Mechanism,* W. H. Freeman, New York.

45. Horovitz, A., Serrano, L., Avron, B., Bycroft, M., and Fersht, A. R. (1990) Strength and cooperativity of contributions of surface salt bridges to protein stability. *J. Mol. Biol.* **216,** 1031–1044.

46. Frisch, C., Schreiber, G., and Fersht, A. R. (1995) Characterization of in vitro oxidized barstar. *FEBS Lett.* **370,** 273–277.

# 15

## Assaying In Vitro Refolding of RNases by Mass Spectrometry

## Gennaro Marino, Piero Pucci, and Margherita Ruoppolo

## 1. Introduction
### 1.1. The Folding Problem

Although evidence has been mounting since the 1930s that proteins could be reversibly denatured, it was not until the 1960s that the elegant experiments of Anfinsen on bovine pancreatic Ribonuclease A (RNase A) established that all the information needed for a protein to acquire three-dimentional structure is contained in its amino-acid sequence *(1)*. However, the code that yields a fully folded protein from a string of amino acids has not been deciphered yet.

Folding of disulfide-containing proteins is thermodynamically associated with formation of the unique set of native disulfide bonds *(2)*. As a result, many theoretical and experimental studies have addressed the characterization of disulfide-bonded intermediates in order to elucidate protein-folding pathways by determining the appearance of intermediates and their interconversion rates *(3,4)*. Methods used to determine the various intermediates present at different refolding times, such as those of Creighton *(5)* on Bovine Pancreatic Trypsin Inhibitor (BPTI), trap the intermediates with thiol-blocking reagents. The trapped intermediates are then separated, and the cysteine residues paired in disulfide bonds are identified by using different analytical methods.

### 1.2. Folding Studied by Mass Spectrometric Methodologies

Mass spectrometry is used in protein-folding studies to characterize the population of species formed on the folding pathway, and the disulfide bridges present at different times of the process *(6,7)*. **Figure 1** outlines the general strategy. Aliquots withdrawn at different times during the refolding process are trapped by alkylation of the free thiol and analyzed by electrospray mass spectrometry (ESMS). The alkylation reaction prevents reoxidation and increases the molecular mass of the intermediates by a fixed amount for each reacted free SH group. Intermediates containing different numbers of disulfide bonds are separated by mass, and their relative abundance in the sample is determined. The nature and the quantitative distribution of the disulfide-bonded species present at a given time can be used to establish the refolding pathway of the protein and, in some cases, to develop a kinetic analysis of the process. The method can also be used to determine the effect of folding catalysts on each step of the refolding

From: *Methods in Molecular Biology, vol. 160: Nuclease Methods and Protocols*
Edited by: C. H. Schein © Humana Press Inc., Totowa, NJ

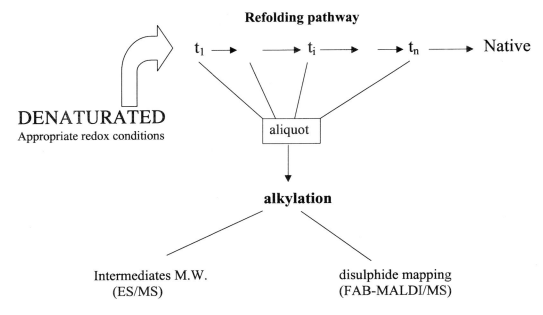

Fig. 1. Outline of the general strategy to study protein folding by extensive use of mass spectrometry.

pathway. Any alteration in the relative distribution of the disulfide-bonded species present at a given time because of the catalytic action can be identified and quantitated.

Once the content of disulfide bonds of the refolding intermediates have been characterized by ESMS, the next step is the structural assignment of the various disulfide bonds formed at different times in the entire process by high-field, fast-atom bombardment mass spectrometry (FABMS) or matrix-assisted laser desorption ionization mass spectrometry (MALDI-MS). The experimental approach is based upon determination of the masses of disulfide-linked peptides by direct FAB-MS or MALDI-MS analysis of unfractionated or partially purified proteolytic digests of refolding aliquots in mapping experiments. Both the S-S bond assignment strategy itself, and the concept of its use in the study of the protein-folding problem, have been described by the authors in previous articles *(8,9)*. The refolding intermediates should be cleaved at points between the potentially bridged cysteine residues under conditions known to minimize disulfide reduction and reshuffling, using aspecific enzymes or a combination of enzymes, in an attempt to isolate any cysteine residue within an individual peptide. The FAB-MS or MALDI-MS mapping approach is then used to search for any S-S bridged peptides, which are characterized by their unique masses. The interpretation is then confirmed by performing reduction or Edman reactions followed by rerunning the MS spectrum *(8)*. The choice of using FAB-MS or MALDI-MS depends on the availability of sample, MALDI is much more sensitive (FAB-MS requires 1 nmol of peptide mixture and MALDI-MS requires 10 pmol). However, FAB-MS is still used for historical reasons. In fact, it was the first mass spectrometry methodology to be developed to solve problems linked to peptides and protein characterization.

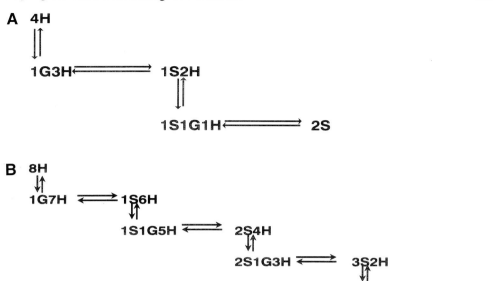

Fig. 2. Proposed mechanism of refolding of RNase T1 (**A**) and RNase A (**B**).

## 1.3. Folding of RNases Studied by Mass Spectrometric Methodologies

Here we describe the application of mass spectrometry to the analysis of the refolding pathways of RNase T1 and RNase A *(6,7,9–12)*. Refolding was started from either fully reduced protein or a mixed disulfide formed by treating the protein with glutathione, and the relative abundance of the refolding intermediates determined for each starting state. Our MS time-course analysis indicated that in quasi-physiological conditions, the refolding of single-domain disulfide-containing proteins occurs through reiteration of two sequential steps: formation of a mixed disulfide with glutathione, and internal attack of a free SH group to form an intramolecular disulfide bond (*see* **Fig. 2**). We detected only a limited number of intermediates, and found that isomerization between species with the same number of disulfides is only extensive at late stages of the process, where slow conformational transitions become significant. Theoretically, the number of all the possible intermediates during the refolding process should increase exponentially with the increase in the number of cysteine residues. However, the number of the populations of intermediates increases linearly with the increase in the number of disulfide bonds. The amount and accumulation rate of individual intermediates varies with each protein.

## 1.4. Folding of RNases in the Presence of Folding Catalysts Studied by Mass Spectrometric Methodologies

When the refolding was carried out in the presence of Protein Disulfide Isomerase (PDI), the refolding pathway of reduced RNases and their GS-derivatives was unchanged, but the relative distribution of the various populations of intermediates was altered. All the experiments suggest that PDI catalyzes the formation of mixed disulfides with glutathione, the reduction of mixed disulfides, and the formation of intramolecular disulfide bonds. These results are not surprising, considering the broad range of activities shown by PDI *(13–15)*.

The structural characterization of intermediates present in the Glutaredoxin-assisted refolding showed, for the first time, that Glutaredoxin catalyzes both formation and reduction of mixed disulfides involving glutathione. Synergism between Glutaredoxin and PDI *(13)* has been explained through the demonstration that Glutaredoxin catalyzes formation or reduction of mixed disulfides, which are then rapidly converted into intramolecular disulfides by PDI. These catalytic steps repeat sequentially throughout the refolding pathway, thus resulting in an immediate formation of fully oxidized species even at the early stages of the process.

In conclusion, we propose a sequential mechanism to account for the limited population of isomers in the folding pathway. Conformational and/or kinetic constraints severely restrict the number of SS bonds and the possible folding pathways of disulfide-containing proteins. The extremely high rate and efficiency of the folding process in vivo may be a result of the folding factors further restricting the number of isomers.

## 2. Materials

### 2.1. Reagents

1. *Escherichia coli* Grx was the gift of Drs. O. Björnberg and J. Bushweller (Karolinska Institute, Stockholm, Sweden).
2. RNase T1 was a gift from N. Pace (Texas A & M University, Health Science Center).
3. PDI was purified from bovine liver as previously described *(13)*.
4. DTNB, DTT, EDTA, GSH, GSSG, guanidinium chloride, hen egg-white lysozyme, Iodoacetic acid, Pepsin, Trypsin, bovine pancreatic RNase A, and horse-heart myoglobin were obtained from Sigma Chemical Co.
5. Iodoacetamide (IAM) and Tris were purchased from Fluka.
6. Prepacked Sephadex G-25M PD10 columns were acquired from Pharmacia.
7. Endoproteinase Asp-N was acquired from Boehringer Mannheim GmbH.
8. All other reagents were HPLC grade from Carlo Erba.

### 2.2. Solutions

1. Reduction and denaturation buffer: 0.1 $M$ Tris-HCl, 1 m$M$ EDTA containing 6 $M$ guanidinium chloride, pH 8.5, stable at room temperature for up to 1 mo.
2. Refolding buffer: 0.1 $M$ Tris-HCl, 1 m$M$ EDTA, pH 7.5; stable at room temperature for up to 1 mo.
3. Reduced glutathione stock solution: 50 m$M$ reduced glutathione in refolding buffer; prepare fresh daily.
4. Oxidized glutathione stock solution: 50 m$M$ oxidized glutathione in refolding buffer; prepare fresh daily.
5. Iodoacetamide solution: 2.2 $M$ Iodoacetamide solution in refolding buffer; prepare fresh daily. IAM is freshly dissolved in refolding buffer at 65°C and cooled to room temperature before use. During preparation of the reagents, the solutions should be protected from light to minimize iodine production, which is a very potent oxidizing agent for thiols.
6. DTNB solution: 20 m$M$ DTNB in 0.3 $M$ Tris-HCl, pH 7.5, 1 m$M$ EDTA containing 6 $M$ guanidinium chloride; prepare fresh daily.
7. 1 $M$ hydrochloric acid.
8. 0.01 $M$ hydrochloric acid.

9. 5% acetic acid.
10. 5% formic acid.
11. HPLC solvent A: 0.1% trifluoroacetic acid in degassed water, stable at room temperature for up to 1 wk
12. HPLC solvent B: 0.07% Trifluoroacetic acid in 95% acetonitrile/5% degassed water, stable at room temperature for up to 1 mo.
13. Electrospray solution: Degassed water/acetonitrile (50/50) containing 1% acetic acid, stable at room temperature for up to 1 wk.
14. Trypsin digestion buffer: 0.4% ammonium bicarbonate, pH 8.5, prepare fresh daily.
15. Endoproteinase Asp-N digestion buffer: 100 m$M$ ammonium bicarbonate, pH 8.0, containing 10% acetonitrile; prepare fresh daily
16. MALDI solution: Ethanol/acetonitrile/0.1% trifluoroacetic acid 1:1:1 (v:v:v).

## 2.3. Protein Extinction Coefficient

The concentrations of solutions of native RNase T1 and its modified forms were determined using a molecular absorption coefficient $\varepsilon_{278} = 18{,}370\ M^{-1}/cm$ *(16)*. The concentration of solutions of native RNase A and its modified forms ($\varepsilon_{278} = 9591\ M^{-1}/cm$ *[17]*), of Protein Disulfide Isomerase ($\varepsilon_{280} = 47{,}300\ M^{-1}/cm$ *[13]*) and of Glutaredoxin ($\varepsilon_{280} = 12{,}500\ M^{-1}/cm$ *[18]*) were determined similarly.

## 2.4. Computer Program

1. Mass Lynx software provided by Micromass (Manchester, UK).
2. Program MatLab (Teoresi, Torino, Italy).

## 2.5. Equipment

1. HPLC analysis: System Gold Beckman equipped with an UV detector with double-wavelength, controlled by an IBM software.
2. ES-MS analysis: BIO-Q triple-quadrupole mass spectrometer from Micromass equipped with an electrospray ion source.
3. FAB-MS analysis: Mass spectra were recorded on three types of instruments at different times with equivalent results: a VG ZAB HF operating with a M-Scan xenon ion gun at 10 kV (10 mA), a VG ZAB 2SE operating with a VG cesium ion gun at 25 kV (2 µA) and a Finnegan MAT 90 operating with an Ion Tech xenon atom gun at 10 kV (2 µA). **Subheading 3.10.** describes sample preparation.
4. MALDI-MS analysis: Peptide mixtures were analyzed by MALDI-TOF mass spectrometry using a Voyager DE mass spectrometer (PerSeptive Biosystems, Boston, MA). **Subheading 3.11.** describes sample preparation.

# 3. Methods
## 3.1. Preparation of Fully Reduced Samples

1. Reduce RNase A or RNase T1 in Reduction and denaturation buffer by incubation with reduced DTT (DTT mol/S-S mol = 50/1) for 2 h at 37°C, under nitrogen atmosphere.
2. Add 0.2 vol 1 $M$ HCl.
3. Desalt the reaction mixture on a prepacked PD10 column (Pharmacia) equilibrated and eluted with 0.01 $M$ HCl.
4. Recover the protein fraction, test for the SH content (described in **Subheading 3.2.**), lyophilize, and store at –20°C.

### 3.2. Test for the SH Content

1. Add reduced or GS-RNases (about 0.1 mg) to 20 m$M$ DTNB solution.
2. Measure the formation of the 2-nitro-5-thiobenzoate dianion at 412 nm ($\varepsilon$ = 13,600 $M^{-1}$/cm).
3. Calculate the number of SH groups using the formula: $n$ = mol wt of the protein $\times$ A$_{412}$ $\times$ vol (mL)/13,600 $\times$ mg of the protein.

### 3.3. Preparation of Fully Glutathionylated Samples

1. Treat reduced proteins (2 mg/mL), prepared as described in **Subheading 3.1.**, with oxidized glutathione stock solution containing 6 $M$ guanidinium chloride at room temperature for 3–10 h, under nitrogen atmosphere. Use a 1000-fold molar excess of GSSG to protein.
2. Acidify the reaction mixture with 0.2 vol of 1 $M$ HCl
3. Desalt on a Sephadex G-25 column eluted with 0.01 $M$ HCl.
4. Recover the protein fraction, test for SH content (described in **Subheading 3.2.**) and lyophilize.
5. Store the preparation (*see* **Note 1**) at –20°C.

### 3.4. Refolding Reactions

1. Dissolve lyophilized rd or GS-RNases to approx 3 mg/mL in 0.01 $M$ HCl and then dilute into the refolding buffer to a final concentration of 1 mg/mL.
2. Add the desired amounts of GSH and GSSG stock solutions to initiate refolding; typically, final concentrations of the glutathione species were 2 m$M$ GSH/1.5 m$M$ GSSG for RNase T1 and 1.5 m$M$ GSH/0.3 m$M$ GSSG for RNase A (*see* **Note 2**). Adjust the pH of the solution to 7.5 with Tris-base and incubate at 25°C under nitrogen atmosphere.
3. Monitor the refolding processes by removing 50–100 mL samples of the refolding mixture at appropriate intervals.
4. Alkylate the protein samples as described in **Subheading 3.6.**
5. Purify from the excess of blocking reagent by rapid HPLC desalting as described in **Subheading 3.7.**
6. Recover the protein fraction and lyophilize.

### 3.5. Use of Folding Catalysts

1. Dissolve lyophilized protein disulfide isomerase and/or glutaredoxin, in refolding buffer.
2. Preincubate PDI and/or Grx in the presence of 2 m$M$ GSH/1.5 m$M$ GSSG or 1.5 m$M$ GSH/0.3 m$M$ GSSG (*see* **Note 2**) for 10 min at 25°C. Add this mixture to reduced proteins or GS-derivatives and continue the refolding at 25°C under nitrogen atmosphere, as described in **Subheading 3.4.** Typically the Glutaredoxin concentration was fixed to 0.5 μ$M$ and the PDI concentrations were either 1 or 10 μ$M$.

### 3.6. Alkylation of the Refolding Aliquots

1. Add the refolding aliquots (50–100 μL) to an equal volume of a 2.2 $M$ idoacetamide solution. Perform the alkylation for 30 s, in the dark at room temperature under nitrogen atmosphere as described *(6)* according to Gray *(19)*.
2. Add 100 μL of 5% trifluoroacetic acid, vortex the aliquots, and store on ice prior to loading on the HPLC for desalting described in **Subheading 3.7.**

## 3.7. Desalting by HPLC

1. Load the alkylated refolding samples on a Vydac TP 214 reversed-phase C4 column (0.46 × 25 cm). Elute with a linear gradient of solvent B from 15–95% at flow rate of 1 mL/min, monitoring protein at 220 nm (*see* **Note 3**).
2. Recover the protein fractions and lyophilize.

## 3.8. Electrospray Analysis of Refolding Intermediates

1. Perform mass-scale calibration by means of multiply charged ions from a separate injection of hen egg-white lysozyme (average mol mass 14,305.99 Da) or of horse-heart myoglobin (average molecular mass 16,951.5 Da) (*see* **Note 4**).
2. Dissolve the protein samples in a mixture of $H_2O/CH_3CN$ (50/50) containing 1% acetic acid.
3. Inject the protein samples (10 µL) in concentrations ranging from 10 to 20 pmol/µL into the ion source via loop injection at a flow rate of 10 µL/min.
4. Record the spectra by scanning the quadrupole at 10 s/scan. Data are acquired and analyzed by the MassLynx software (*see* **Note 5**).
5. Accurately quantify each population of intermediates by measuring the total ion current produced by each species, provided that the different components are endowed with comparable ionization capabilities. Each set of refolding data should be obtained as the mean of two independent folding experiments. The differences between refolding experiments performed completely independent of each other should be about 2% (*see* **Note 6**).
6. Perform kinetic analysis (*see* **Note 7**).

## 3.9. Proteolytic Hydrolysis

1. Hydrolyze the protein samples withdrawn at different incubation times of refolding with pepsin and trypsin or with trypsin, and endoproteinase Asp-N.
2. Hydrolyze with pepsin in 5% formic acid at 37°C for 6 h, using an enzyme to substrate ratio of 1:50 (w/w).
3. Digest with trypsin in 0.4% ammonium bicarbonate, pH 8.5, at 37°C for 8 h, using an enzyme-to-substrate ratio of 1:50.
4. Digest with endoproteinase Asp-N in 100 m*M* ammonium bicarbonate, pH 8.0, using 10% acetonitrile as activator, at 37°C for 18 h using an enzyme-to-substrate ratio of 1:100 (w/w).

## 3.10. Fast Atom Bombardment (FAB)-MS Analysis

1. Dissolve samples in 5% acetic acid and load onto a probe tip coated with either glycerol or nitrobenzyl alcohol. In the mixed matrix experiment samples are added to a probe tip coated with a mixture of nitrobenzyl alcohol and glycerol.
2. Add a trace of thioglycerol just before inserting the probe into the ion source to enhance sensitivity.

## 3.11. Matrix-Assisted Laser Desorption Ionization (MALDI)-MS Analysis

1. Dissolve samples in 0.1% trifluoroacetic acid at 10 pmol/µL.
2. Apply 1 µL sample to a sample slide and allow to air-dry.
3. Apply 1 µL bovine insulin to the sample slide and allow to air-dry.
4. Apply 1 µL α-cyano-4-hydroxycinammic acid (10 mg/mL) in ethanol/acetonitrile/0.1% trifluoroacetic acid 1:1:1 (v:v:v) and allow to air-dry.
5. Collect spectra. Mass spectra are generated from the sum of 50 laser shots.
6. Calibrate the mass range using bovine insulin (average mol mass 5734.6 Da) and a matrix peak (379.1 Da) as internal standards.

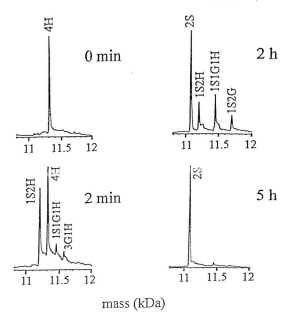

Fig. 3. ES spectra of the mixtures of species sampled at time 0, 2 min, 2 h, and 5 h from the refolding mixture of RNase T1.

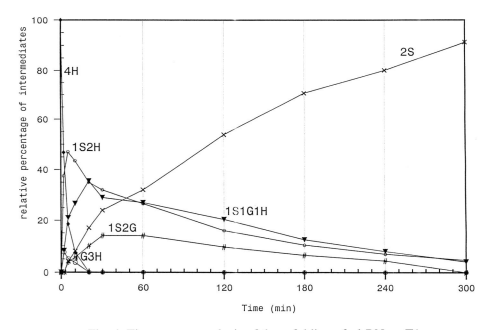

Fig. 4. Time-course analysis of the refolding of rd-RNase T1.

## 4. Notes

1. The GS-derivatives were structurally characterized by ES-MS analysis. The ES-MS spectrum of GS-RNase T1 showed a major component with a molecular mass of 12,307.15 ±

1.28 Da in agreement with the expected value (12,309.91 Da). The ES-MS spectrum of GS-RNase A showed the presence of an equimolar mixture of two molecular species, the GS-RNase A (with eight glutathione moieties) and a derivative with one intramolecular disulfide bond and six glutathione moieties. The former was then purified by HPLC using a Vydac TP 214 reversed-phase C4 column (0.46 × 25 cm) using a linear gradient of solvent B from 24 to 29% at flow rate of 1 mL/min. The first eluted peak was indeed the pure GS-RNase A product, as confirmed by ESMS analysis (measured molecular weight = 16,132.6 ± 1.35 Da; expected molecular weight = 16,132.7 Da).

2. The choice of the concentrations of reduced and oxidized glutathione depend on the protein under investigation, and is based on conditions giving the highest yield of native protein at the end of the refolding process.

3. When the refolding was carried out in the presence of PDI or Glutaredoxin, the catalysts were eluted in the void volume and did not interfere with the subsequent analysis of the refolding intermediates.

4. Hen egg-white lysozyme is used to calibrate the mass range for the acquisition of the spectra of RNase T1 samples and horse-heart myoglobin for RNase A samples.

5. **Figure 3** shows, as an example, the deconvoluted ES spectra of the mixtures of species sampled at time 0, 2 min, 2 h, and 5 h from the refolding mixture starting from reduced RNase T1. The different populations of disulfide intermediates present were identified on the basis of their molecular mass. Each population of trapped intermediates is characterized by a different number of intramolecular disulfide bonds (indicated as nS), mixed disulfides with the exogenous glutathione (nG) and carboxyamidomethyl groups (nCAM). The number of CAM groups corresponds to the number of free thiols present in the refolding intermediates and are therefore indicated as nH.

6. The time course of refolding of reduced RNase T1, plotted in **Fig. 4**, shows that intermediates 1S2H and 1S1G1H are predominant from the beginning of the reaction up to about 1 h, when the relative concentration of the species 2S increases. A steady state is established between intermediates 1S2H and 1S1G1H; from 30 min onward, the concentration of these species is approximately equal. The component 1G3H is present only in the first 20 min of the reaction, and it rapidly disappears. The species 1S2G is present at a very low concentration at the early stages of the reaction, reaching a constant concentration (<20%) and disappearing at the late stages of the refolding. Finally, after 5 h of refolding, only component 2S was detected in the mixture.

7. In the analysis of RNase T1 refolding, kinetic analyses have been performed using the Runge-Kutta method *(20)* and the Symplex algorithm *(21)*. Both methods have been implemented in the program MatLab (Teoresi, Torino, Italy). Kinetic analysis of the refolding experiments of rd or GS-RNase T1 shows that in both cases the formation of component 2S represents the rate-limiting step of the whole process.

## References

1. Anfinsen, C. B. (1961) Studies on the reduction and re-formation of protein disulfide bonds. *J. Biol. Chem.* **236,** 1361–1363.

2. Creighton, T. E. (1983) Pathways and energetics of protein disulphide formation, in *Functions of Glutathione: Biochemical, Physiological, Toxicological and Chemical Aspects.* (Larsson, A., Orrenius, S., Holmgren, A., and Mannervik, R., ed.), Raven Press, New York, pp. 205–211.

3. Creighton, T. E. (1992) Folding pathways determined using disulphide bonds, in *Protein Folding* (Creighton, T. E., ed.), W. H. Freeman and Company, pp. 301–351.

4. Gilbert, H. F. (1994) The formation of native disulphide bonds, in *Mechanism of Protein Folding* (Pain, R. H., ed.), Oxford University Press, pp. 104–136.

5. Creighton, T. E. (1977) Conformational restrictions on the pathway of folding and unfolding of the pancreatic trypsin inhibitor. *J. Mol. Biol.* **113,** 275–280.

6. Torella, C., Ruoppolo, M., Marino, G., and Pucci, P. (1994) Analysis of RNase A refolding intermediates by electrospray/mass spectrometry. *FEBS Lett.* **352,** 301–306.

7. Ruoppolo, M., Torella, C., Kanda, F., Panico, M., Pucci, P., Marino, G., and Morris, H. R. (1996) Identification of disulfide bonds in the refolding of bovine pancreatic RNase A. *Folding Design* **1,** 381–390.

8. Morris, H. R. and Pucci, P. (1985) A new method for rapid assignment of S-S bridges in proteins. *Biochem. Biophys. Res. Commun.* **126,** 122–128.

9. Morris, H. R., Pucci, P., Panico, M., and Marino, G. (1990) Protein folding/refolding analysis by mass spectrometry. *Biochem. J.* **268,** 803–806.

10. Ruoppolo, M., Freedman, R. B., Pucci, P., and Marino, G. (1996) The glutathione dependent pathways of refolding of RNase T1 by oxidation and disulfide isomerization. Catalysis by protein disulfide isomerase. *Biochemistry* **35,** 13,636–13,646.

11. Ruoppolo, M., Lundstrom-Ljung, J., Talamo, F., Pucci, P., and Marino, G. (1997) Effect of glutaredoxin and protein disulfide isomerase on the glutathione-dependent folding of Ribonuclease A. *Biochemistry* **36,** 12,259–12,267.

12. Ruoppolo, M., Moutiez, M., Mazzeo, M. F., Pucci, P., Mènez, A., Marino, G., and Quèmèneur, E. (1998) The length of a single turn controls the overall folding rate of "three-fingered" snake toxins. *Biochemistry* **37,** 16,060–16,068.

13. Lundström, J. and Holmgren, A. (1995) Glutaredoxin accelerates glutathione-dependent folding of reduced Ribonuclease A together with protein disulfide isomerase. *J. Biol. Chem.* **270,** 7822–7828.

14. Freedman, R. B. (1984) Native disulfide bond formation in protein biosynthesis: evidence for the role of protein disulfide isomerase. *Biochem. Soc. Trans.* **12,** 929–932.

15. Freedman, R. B., Hirst, T. R., and Tuite, M. F. (1994) Protein disulphide isomerase: building bridges in protein folding *Trends Biochem. Sci.* **19,** 331–336.

16. Pace, N. C. and Grimsley, G. R. (1988) Ribonuclease T1 is stabilized by cation and anion binding *Biochemistry* **27,** 3242–3246.

17. Schaffer, S. W., Ahmed, A. K., and Wetlaufer, D. B. (1975) Salt effects in the glutathione-facilitated reactivation of reduced bovine pancreatic ribonuclease. *J. Biol. Chem.* **250,** 8483–8486.

18. Björnberg, O. and Holmgren, A. (1993) A putative glutathione binding site in T4 glutaredoxin investigated by site-directed mutagenesis. *J. Biol. Chem.* **266,** 16,105–16,110.

19. Gray, W. A. (1993) Disulfide structures of highly bridged peptides: a new strategy for analysis. *Prot. Sci.* **2,** 1732–1748.

20. Forsythe, G. E. and Malcolm, M. A. (1977) *Computer Methods for Mathematical Computation,* Prentice Hall.

21. Nelder, J. A. and Mead, R. (1965) A Symplex method for function minimization. *Comp. J.* **7,** 308–313.

# 16

## Dissecting Nucleases into Their Structural and Functional Domains

*Mapping the RNA-Binding Surface of RNase III by NMR*

**Andres Ramos and Annalisa Pastore**

## 1. Introduction
### *1.1. Sequence and Modular Structure of RNase III*

The enzyme RNase III catalyzes the hydrolysis of the phosphodiester bond between two nucleotides. The protein was discovered in 1968, when Robertson and coworkers isolated a double-stranded RNA (dsRNA) cleaving activity in *Escherichia coli* cells *(1)*. A few years later, specific targets for this enzyme were identified *(2)*. Proteins analogous to RNase III have been isolated in many organisms, both eubacteria *(3,4)* and eukaryotes *(5,6)* suggesting the importance of the regulatory function of RNase III. The amount of RNase III present in the cell is downregulated by specific cleavage of its own messenger RNA. Recombinant RNase III is a dimer under physiological conditions *(7–11)*.

RNase III is a small protein 226 amino acids long with a molecular weight of 25.6 kDa. Sequence analysis and mutational studies indicate the presence in the protein of two distinct regions. The first 150 amino acids are responsible for catalytic function and dimerization, whereas the last 70 amino acids are necessary and sufficient for RNA binding *(12–15)*. The C-terminal domain contains a sequence motif, named double-stranded RNA Binding Domain (dsRBD). This motif has been identified in many proteins with diverse architecture *(14,15)*. Lack of sequence conservation in the regions flanking the dsRBD in the family suggested that the two functional domains of RNase III could fold independently.

### *1.2. RNase III Activity*

In vitro, RNase III cleaves double-stranded RNA helices of over 20 nucleotides in length with structural specificity, but no apparent sequence specificity *(1)*. However, the secondary structure of specific cellular targets cleaved by RNase III range from a regular double helix to internal loops or bulges (reviewed **ref.** *15*). Some specific targets of RNase III, such as the cleavage sites for the processing of preribosomal mRNA

From: *Methods in Molecular Biology, vol. 160: Nuclease Methods and Protocols*
Edited by: C. H. Schein © Humana Press Inc., Totowa, NJ

and a processing site within the RNA transcript of the T7 phage (reviewed in **refs. *15*** and ***16***), have a putative double-helical conformation. These templates are cleaved on both strands with a frameshift of two or three nucleotides between the two hydrolysed bonds. Other target sites fold in non-A-form secondary structures as internal loops or bulges. In internal loops the cleavage site has been found at specific positions and involves only one strand. The recognition mechanism of these motifs is still unclear: Although several T7 sites are near bulges, insertion of a bulge hinders RNase III cleavage of the MS2 RNA *(17)*. The role of secondary structure elements in recognition is still unclear. However, an element common to all RNase III targets is a long A-form double-helical region in the proximity of the binding site. Such a structure seems to be the leading motif in RNase III-RNA interaction.

RNase III cellular targets include its own mRNA *(18)*, the mRNA of polynucleotide phosphorylase *(19)* and the preribosomal RNA of *E. coli (2)*. RNase III target sites are often located 5' of the mRNA coding sequence; the cleavage of the double-helical structure leads to a much shorter mRNA half-life. This mechanism is responsible for RNase III self-regulation. Processing of pre-ribosomal RNA is necessary to obtain ribosomal RNAs of the correct size. In cases of very intense protein synthesis, most of RNase III protein is bound to ribosomal RNA; this results in a larger portion of uncut rnc mRNA and in overproduction of RNase III. RNase III is also involved in the processing of the RNA transcript of the T3, T7, and lambda phages in infected *E. coli* cells and in many other posttranscriptional and posttranslational regulation mechanisms *(15,18,20)*.

Systematic attempts were made to elucidate whether any sequence specificity existed in the dsRNA targets. Krinke and Wulff analyzed 20 sequences of double-stranded regions of RNase III targets *(21)*. They found sequence preference at a few positions of the RNA double-helical strand. In a more recent study based on a similar analysis, Zhang and Nicholson proposed that sequence antideterminants located in the RNA double-helix regulate binding and catalysis *(22)*. A series of RNA mutants was produced to investigate the effect of forbidden nucleotides in crucial positions. The mutations impaired binding and thus cleavage. The sequence antideterminants could be grouped into two "boxes" (proximal and distal box).

## *1.3. Function of the RNA-Binding Motif in the Context of the Full-Length Protein*

The RNase III dsRBD plays a major role in the functions of the enzyme, suggesting that this region is functionally independent from the catalytic domain. Regulation by binding of the enzyme without RNA cleavage has been observed in a specific case *(23)*. Genetic uncoupling of the catalytic and RNA-binding function has been demonstrated in vitro and in vivo *(24,25)*. Mutations of the dsRBD impair binding and therefore enzymatic activity. Mutations in the catalytic domain impair RNA cleavage, but do not detectably change the RNA-binding properties. From this evidence, it is expected that the catalytic domain is necessary to position the groups involved in the catalytic reaction and to stabilize the reaction intermediate. However, these contacts should not provide a significant contribution to the binding affinity or recognition of a single sequence or secondary structure. The nonsequence specific-binding of RNase III to A-form RNA helices is achieved instead by the dsRBD, which is therefore thought to provide the necessary binding affinity and be crucial for the modulation of the binding

*(15,24,25)*. Incorrect RNA recognition would interfere with the catalytic reaction and result in impairment of the cleavage.

How this binding is achieved is not known precisely. Since RNA-DNA hybrids are not bound by RNase III, the geometry of an RNA helix and possibly the presence of 2'-hydroxyl groups should be relevant for the interaction. Selection of the RNA targets could occur also by structural deformations linked to specific sequences. Recognition mechanisms based on nonsequence-specific interactions occur in other proteins which rely on dsRBD motifs for RNA recognition. Such a recognition mechanism is, in any case, expected to be quite different from the ones described for other RNA-protein interactions *(26–28)*, where the protein interacts with functional groups of exposed RNA bases. Structural information is therefore essential to investigate this recognition mechanism. For this reason we solved the solution structure of the dsRBD of RNase III using the strategy described **ref.** *29* and reviewed in the following sections. A similar strategy is in principle applicable to other nuclease enzymes.

## 1.4. Using NMR to Study Isolated Domains in Multidomain Proteins

Nuclear magnetic resonance (NMR) is a technique widely used, together with X-ray crystallography, to determine the structure of proteins and protein complexes. NMR is ideally suited for small molecular weight molecules. In the last 20 years, since the first protein structure determined by this technique *(30)*, solution NMR has been applied successfully to solve the structure of proteins up to 35 kDa. Recent developments *(31)* have at least partially overcome the limitations arising from resonance broadening, and the increasing complexity of the spectra typical of large systems, pushing the size of molecules that can be studied by NMR to >40,000. On the other hand, many proteins are assembled from smaller, independently folding units or domains (on average, 40–300 residues) *(32)*. This suggests an alternative strategy: structure determination of multidomain proteins can be approached by dissecting them into individual domains more easily studied by NMR. The overall shape and structure of the protein can then be reconstructed by combining information on domain interfaces with low-resolution information. The study of protein-nucleic-acid recognition, one of the fundamental problems in molecular biology, has strongly benefited from this approach. Essential protein-nucleic-acid interactions often involve multidomain proteins in which one or more domains specialize in providing the interaction. Structure determination of these domains is the prerequisite for mapping the interaction surface and understanding the binding mechanism. The simple modular architecture of the dsRBD of RNase III provides a good paradigm of the strategy described above and of its application in predicting the RNA interaction surface from structural information.

## 1.5. Prediction of a Putative RNA-Binding Surface on the dsRNA from the dsRBD Structure

The structure of the dsRBD of RNase III was solved in 1995 by the combined use of homonuclear NMR and a novel approach based on iterative assignment of ambiguous NOE crosspeaks. The structure consists of an alpha–beta fold with two alpha-helices packing on one face of a three antiparallel strands beta-sheet (**Fig. 1**). Details about the structure are described in **ref.** *29*. The structure of the dsRBD, together with biochemical and genetic information on other dsRBD-containing proteins, was then used for

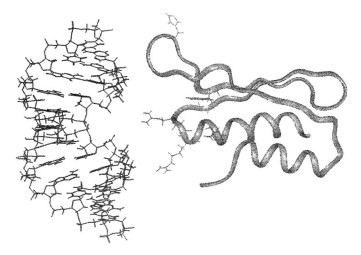

Fig. 1. Representation of the RNase III dsRBD/RNA complex. The NMR structure of RNase III dsRBD is represented as a ribbon that follows the backbone. The sidechains of the residues His 32, Phe 36, Gly 53, Lys 56 and 57, and Glu 60, conserved in the dsRBD family, are displayed. The protein faces an A-form RNA helix 12 base-pair long.

defining the RNA binding surface of the protein. The hypotheses drawn by the structural analysis of the dsRBD of RNase III were also supported by the structure of a dsRBD from the *staufen* protein of *Drosophila* published at the same time *(33)*.

Analysis of amino acids that are exposed in the structure and conserved in the dsRBD family identifies a positively-charged cluster on the dsRBD surface which includes loops 2 and 4 and the amino-terminal part of helix 2. Residues 32, 34, and 35 and residues 56, 57, 58, and 60, are positively charged, and are located respectively on loops 2 and 4. Of these, 32, 56, and 57 are also very conserved in the whole dsRBD family. Biochemical tests carried out on protein A, VvE3 protein, and PKR *(34–36)* have demonstrated that mutations in residues aligning to positions 56, 57, and 60 lead to loss of binding. A similar result was obtained for histidine 32 in the case of protein A. RNA–protein interactions are often mediated by long, positively charged side chains. These residues can interact with the RNA phosphate groups both through salt bridges and/or H-bonds (direct or water-mediated). The high conservation and the functional importance of these residues cannot be explained by their structural role in the protein fold, since these residues are not buried in the hydrophobic core of the protein, but exposed in solution. This strongly suggests that they play a more direct role in protein–RNA interaction.

A model of the complex made from NMR of the dsRBD structure of RNase III and a model dsRNA suggests further details of how the two components could interact (**Fig. 1**). The distance separating the two loops in the free protein allows a simultaneous interaction with the RNA major and minor grooves. Experimentally, RNA 2'-hydroxyl groups appear to be necessary for interaction: blocking of 2'-hydroxyl and phosphate groups in the nucleotides of the so-called proximal and distal box results in impaired cleavage *(25)*. The positively charged groups on the side chains of the RNase III dsRBD loops 2 and 4 can in principle make contacts with the RNA phosphate moieties across the major groove. Contacts could also be achieved by other hydrogen-bond donors in the same loops (serines 54, 55, and glutamate 61). The relative orientation of the positively charged cluster is determined by a small hydrophobic patch composed by phenylalanine 36, which is sandwiched between the two protein loops, and glycine 53. Since these two positions are extremely conserved in the dsRBD family, it is likely that the main features of the RNA-dsRBD interactions will be common to the entire dsRBD family. The precise positioning of the protein loop 2 and 4 could fit the required geometry of the RNA groove(s). This hypothesis, first inferred from the structure of isolated dsRBD, was more recently confirmed by the structure of the complexes between *Xenopus laevis* RNA-Binding Protein (xrbp) and dsRNA *(37)* and between the third dsRBD of staufen protein of *Drosophila* and two different RNA templates *(38)*. In the latter dsRBD, a beta + bulge is present in strand 1 defining the start of loop 2 that should determine the precise conformation of loop 2. The interaction specificity of a dsRBD could then be easily modulated by few additional secondary structure elements.

The predictions made on the basis of the structural information and modeling provide the first step in the description of the mode of action of RNase III. The exact prediction of the contacts attributable to the single RNase III monomers and the precise geometry of the interaction becomes difficult, since the protein binds as an homodimer. Their validity needs to be checked by the direct determination of the RNase III dsRBD/RNA complex structure and by judiciously chosen site-directed mutagenesis experiments both on the protein and on the RNA.

## 2. Materials

### 2.1. Sequence Analysis

1. BLAST *(39)* or equivalent programs (website address: http://www.embl-heidelberg.de/Services/index.html).
2. CLUSTALX *(40)* (available through the web from http://www-igbmc.u-strasbg.fr/BioInfo/ClustalX/Top.html).

### 2.2. Sample Preparation

1. PCR primers (5'-TAT ATT CAT GAG TCA TCA CCA TCA CCA TCA GTC CAT GGG TCA AAA AGA TCC GAA AAC GCG CTT GCA AGA A 3' and 5'-ATA TAA GCT TAT TCC AGC TCC AGT TTT TTC AAC GCC TGT TCG-3'.
2. pET3d *E. coli* vector *(41)* and *E. coli* strain BL21 (Pharmacia).
3. Buffers and solutions: 100 mL of bis-Tris propane buffer 5 m$M$ (pH 6.0 and 9.0); NaCl solution.
4. Enzymes and inhibitors: DNase I (Boheringer Mannheim), protease inhibitor phenylmethyl sulfonylfluoride (PMSF) (Sigma).
5. Q-Sepharose and monoQ columns (Pharmacia).

### 2.3. Software for NMR Structure Determination

1. AURELIA (Bruker) or equivalent programs for spectral analysis.
2. Xplor *(42)* (for the licence, write to the author) or equivalent sofware.

### 2.4. Software for Structure Display and Modeling

1. InsightII (MSI) or equivalent graphic programs.

## 3. Methods

An overview of the whole procedure used for the structure determination of the dsRBD of RNase III is summarized in **Fig. 2**.

### *3.1. Identification of Domain Boundaries*

1. Make a sequence Database search of the given sequence motif (here the dsRBD) in the database by BLAST or one of its derivatives (e.g., BLASTPSI, BLAST2).
2. Extract the sequences found with the highest score. When more than one copy of the motif is found in the same sequence, make sure that you extract all repeats. For example, this can be done by following the strategy described in **ref. *43***.
3. Align the sequences with a multiple alignment program, such as CLUSTALX. Then identify the boundaries with a method that preserves conserved patterns.

### *3.2. Sample Preparation of the dsRBD in* E. coli

A successful protocol for preparation of the dsRBD of RNase III is the following, as described in **ref. *29***.

1. Amplify the *E. coli* DNA sequence coding for residues 153–226 of RNase III by PCR using *E. coli*. Design the PCR primers to fuse fMet with the *E. coli* sequence. As extra nucleotides are necessary to produce the desired restriction site, add a Glycine N-terminal to the construct.
2. Clone the amplified fragment into a modified expression vector pET3d *(41)*. Cleave the plasmid and dephosphorilate it. Heat-inactivate the enzyme if possible, and purify on a 1% agarose gel. Phosphorilate the amplified DNA, heat-inactivate the phosphorylase, and ligate to the plasmid. Transform into *E. coli* BL21. All these procedures are standard, as described in **ref. *43***. Express the protein inducing at OD = 0.6 with IPTG to a final concentration of 0.4 m$M$. Four liters of *E. coli* growth are sufficient to obtain an NMR sample.
3. Suspend the pellet in 100 mL bis-Tris propane buffer 5 m$M$ (pH 6.0), 0.11 μ$M$ PMSF. Lysate the cells by French press at 4°C.
4. Add DNase (20 mg/L) and MgCl$_2$ (3 m$M$) to reduce the viscosity of the suspension. Centrifuge the suspension for 1 h at a RCF of 117,000$g$ and 4°C. Collect and filter the supernatant.
5. Purify the filtrate by applying it to a Q-Sepharose FastFlow anion-exchange column equilibrated with bis-Tris propane 5 m$M$ (pH 6.0). Elute the protein with a linear NaCl gradient from 0 to 0.5 $M$ NaCl. The protein elutes at 0.26 $M$ NaCl. Pool the fractions of the protein peak. Desalt and dilute with bis-Tris propane 5 m$M$ (pH 9.0) until the pH of the protein solution is close to 9.0.
6. Load the protein on a monoQ column equilibrated with the bufferdescribed in **step 5**. The construct will be recovered in flowthrough.
7. Dialyze the protein against 100 m$M$ NaCl solution (pH 6.0). Concentrate it to a final concentration of 1–2 m$M$. Check the mass by electronspray mass spectroscopy. The mass of the construct should be of 8350.2 Da (or 8482.2 Da if the initial methionine is not cleaved).

### *3.3. Structure Determination*

1. Record preliminary 1D and 2D NOESY and TOCSY NMR spectra to ensure that the protein is folded and monomeric. A good dispersion of the protein resonances in the 1D spectrum are diagnostic for folded proteins. Efficient transfer both in NOESY and TOCSY 2D experiments suggest that proteins of this size are monomeric and well-behaved.
2. Optimize the conditions (temperature, salt content, buffer, pH) for spectra recording in order to improve the spectral resolution and the transfer. The temperature of the experi-

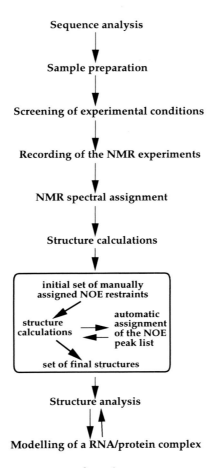

Fig. 2. Flowchart of the relevant steps to get from the sequence to a protein structure determination.

ments must be chosen as a compromise between a temperature that allows the protein to still be folded (i.e., well below the protein melting point) and the resonance narrowing usually observed by increasing the temperature. Since the stability and the protein behavior is strongly dependent on the salt, buffer, and pH, these three parameters might be systematically screened to obtain the best experimental conditions.

3. Record 2D and 3D spectra to achieve full resonance assignment as well as distance and angular information. Different bacteries of standard experiments may be applied in a protein-structure determination by NMR according to the isotopical state of the sample. A comprehensive guide to the most widely used 2D–4D experiments can be found in **ref. 44**. The strategy of NMR sequence assignment is designed according to the degree of spectral overlap. It is now common practice to label the samples with at least one of the 15N, 13C, and 2H isotopes, provided that the yield of the expression is adequate.

4. Perform structure calculations using the NMR observables as experimental restraints. This is a self-correcting iterative procedure; new distance-based information is added at each iteration after careful reinspection of the compatibility of the resulting set of structure with the experimental restraints. For a semiautomatic assignment of the spectra, use one of the existing packages *(45–48)*. If using the ARIA protocol (website http://www.nmr.embl-heidelberg.de/nmr/nilges/), you must optimize a number of parameters.

Of these, we discuss only the most critical ones that can be useful in a general case. Spectral assignment requires a tolerance of the chemical shift values (PPMD parameter). This is chosen according to the intrinsic resolution of individual nuclei and to the spectral resolution. For a homonuclear experiment, typical values are 0.010–0.015/ppm and 0.020–0.25/ppm in the direct and indirect dimensions, respectively. A larger tolerance is expected for 3D experiments. Different PPMD values might be chosen for each of the NOESY-like experiments chosen for use in the calculations. At each iteration, the assignment is performed using the $N$ lowest energy structures from the previous iteration (Assignstruct parameter). A restraint is not used in the next iteration if it was violated by more than a target value in at least one-half of the $N$ structures. In practice, for runs that will produce 20–50 structures at each iteration, it seems an appropriate choice to consider one-third to one-fifth of the total number of structures *(7–10)*. The violation tolerance (violtoler parameter), the tolerance parameter that excludes systematically violated restraints in the converged structures, is usually decreased slowly at each iteration. An appropriate choice is to choose a large number in the first iteration (e.g., 100.0) and decrease the values to reach 0.00 A in the last iteration (e.g., if eight iterations are done, use 1000, 5.0, 1.0, 0.5, 0.1, 0.1, 0.01, and 0). A dynamic cutoff parameter for the assignment of ambiguous NOEs is also needed. At the beginning of the iterative procedure, all peaks contribute to the assigned distance. Therefore, the Ambigcutoff parameter is usually set to 1.0. In the following iterations, increasingly smaller Ambigcutoffs are used (between 0.9999 and 0.0 at the last iteration).

This automatic procedure should always be checked manually to ensure that the automatic assignment is correctly performed (peaks might be misassigned if, for instance, overlap alters the center of the peak). The number of peaks excluded after each iteration and the reasons of their exclusion must also be verified. In practice, the whole procedure will only be semiautomatic, because human intervention is needed to verify robustness of the result. For small modifications of the protocol, the fold of the protein should remain stable.

### 3.4. Modeling the RNA-dsRBD Complex

1. To visualize how the dsRBD could interact with RNA, it is necessary to obtain a 3D model of the complex. To achieve this, build the structure of the RNA target by any of the available graphics programs with builder facilities (e.g., InsightII). Display both components through a suitable graphics program.
2. Dock the two molecules, taking into account low-resolution structural information (e.g., crosslinking data, chemical and enzymatic probing, data derived from low-resolution spectroscopic techniques) as well as genetic and functional information when available. An excellent book to guide beginners through structural analysis and molecular graphics is published by A. M. Lesk *(49)*.

## 4. Notes

1. Sequence analysis: Identification of domain boundaries is a difficult but necessary part of a structural determination when working with protein fragments rather than with the whole protein. A number of semirational approaches have been developed to perform this task, ranging from limited proteolitic hydrolysis to identification of domains based on sequence analysis. In multidomain proteins, where the same-sequence motif can be present in several copies either within the same or in different proteins, it has become common practice to assume that conserved patterns, as identified in a multiple alignment, could correspond to an independently folding unit. This assumption must be tested experimentally.

The assumption that identifiable sequence motifs constitute a structurally independent unit is right as often as it is wrong. Other regions from the same or different molecules can be necessary for the stability of the motif. To ensure that regions proximal to the motif are not "what is missing", it is advisable to produce constructs that extend on both sides of the conserved motif. The RNase III dsRBD was a very favorable case: Since the C-terminus of the dsRBD corresponds to the C-terminus of the protein, uncertainty is limited to the N-terminus of the domain. Extensive multiple alignments had already been reported in the literature together with wide direct evidence of a functional independence of the two RNase III domains *(12–15)*.

2. The cloning, overexpression, and purification of the protein were carried out using standards techniques. The construct has a calculated isoelectric point of 7.35. The purification procedure is based on two anion-exchange columns run at very different pH values. The protein binds to the first column, which is run at a pH of 6.0, whereas the second column, run at a pH of 3.0, removes protein with a strong positive charge.

3. Description of NMR experiments: The size of the RNase III dsRBD (74 amino acids) could in principle allow structure determination without having to go through the expensive step of isotopic labeling, although this procedure is necessary for high precision structures. Sequential assignment was performed by standard homonuclear 2D techniques performed at different temperatures. The NMR experiments were performed on either a Bruker AMX-500 or an AMX-600 spectrometer equipped with z-shielded gradient coils, using 1.0–1.5-m$M$ samples in 90% $H_2O$/10% $D_2O$ or in $D_2O$ at pH 6.0. In all measurements, the TPPI method was employed to obtain quadrature detection in the indirectly detected dimensions mode (TPPI). Water suppression was achieved with WATERGATE *(50)* 2D TOCSY, NOESY, and DQF-COSY spectra were recorded at temperatures between 280 and 310 K, with 512 increments in t1 and 2048 data points in t2. 2D TOCSY spectra were acquired using the TOWNY composite pulse cycle to minimize cross-relaxation effects. Mixing times used were in the 30–70 ms range for the TOWNY and 50–160 ms for the NOESY experiments. The spectra were processed on a Bruker-X32 workstation using standard Bruker UXNMR and AURELIA softwares. Prior to Fourier transformation, the time-domain data were multiplied by suitable weighting functions and zero-filled in both dimensions. Baseline correction was applied in each dimension using a third-order polynomial. Nevertheless, despite the small size of the RNase III dsRBD, severe overlap, especially at the level of contacts involving resonances of aromatic residues, would impair unique assignment of some of the key contacts. Without the use of the possibility of taking into account ambiguous restraints in a automatic and iterative way *(45)*, it would have been very difficult to resolve the ambiguities of the homonuclear spectra.

4. Structure calculation: RNase III was one of the first applications of a novel approach, later developed into the ARIA protocol, that provides a set of routines for the automatic assignment of a NOESY spectrum on the basis of NOE and chemical shift lists *(45,51,52)*. Structural calculations of RNase III dsRBD were based on a 2D NOESY recorded at 300 K and 60 ms mixing time. NOE peak intensities were converted into distance restraints from volume integrated by the AURELIA program (400 interresidual restraints). Twenty percent correction was added to the distances as upper bounds. Additional hydrogen bonds were imposed on elements of secondary structures. Fifty structures were calculated and resubmitted to an iterative procedure. At each iteration, the ambiguous NOEs were reassigned by keeping only those possibilities that were fulfilled in the 10 best structures in terms of lowest energy.

5. Modeling of the RNA/dsRBD complex: Modeling of the complex was attempted as described in **ref. *29*** and further extended for the present review to take into account also the most recent literature on RNase III. Modeling was supported by experimental evi-

dence. Low-resolution structural information is available for an RNA target of the enzyme. NMR data, as well as an enzymatic and chemical mapping, are available for the R1.1 site of the T7 phage mRNA *(53)*. The dsRBD-binding region of this RNA forms a double helix *(22,25)* up to the RNA internal loop. The R1.1 site was therefore modeled with reasonable accuracy from high-resolution structures of A-form RNA double helical available in the Protein Data Bank. The RNA fragment was built using the Biopolymer module of InsightII. The two molecules were displayed using the same program, and the groups thought to be involved in the interaction were highlighted (e.g conserved exposed residues, positively charged, negatively charged, H-bond acceptor, and others). Manually docking of the two molecules was achieved, taking into account genetic and functional information. No further refinement was attempted. The complex was then displayed by one of the several graphics packages available.

# References

1. Robertson, H. D., Webster, R. E., and Zinder, N. D. (1968) Purification and properties of ribonuclease III from Escherichia coli. *J. Biol. Chem.* **243,** 82–91.
2. Dunn, J. J. and Studier, F. M. (1973) T7 early RNAs and Escherichia coli ribosomal RNAs are cut from large precursors RNAs in vivo by ribonuclease III. *Proc. Natl. Acad. Sci. USA* **70,** 3296–3300.
3. Conrad, C., Rauhut, R., and Klug, G. (1998) Different cleavage specificities of RNAses III from Rhodobacter capsulatus and Escherichia coli. *Nucleic Acid Res.* **26,** 4446–4453.
4. Oguro, A., Kakeshita, H., Nakamura, K., Yamane, K., Wang, W., and Bechofer, D. H. (1998) Bacillus subtilis RNase III cleaves both 5'- and 3'-sites of the small cytoplasmatic RNA precursor. *J. Biol. Chem.* **273,** 19,542–19,547.
5. Allmang, C. and Tollervey, D. (1998) The role of the 3' external transcribed spacer in yeast pre-rRNA processing. *J. Mol. Biol.* **278,** 67–78.
6. Elela, S. A., Igel, H., and Ares, M., Jr. (1996) RNase III cleaves eukariotic preribosomal RNA at a U3 snoRNP-dependent site. *Cell* **85,** 115–124.
7. Watson, N. and Apirion, D. (1985) Molecular cloning of the gene for the RNA-processing enzyme RNase III of Escherichia coli. *Proc. Natl. Acad. Sci. USA* **82,** 849–853.
8. Chen, S. M., Takiff, H. E., Barber, A. M., Dubois, G. C., Bardwell, J. C., and Court, D. L. (1990) Expression and characterization of RNase III and Era proteins. Products of the rnc operon of Escherichia coli. *J. Biol. Chem.* **265,** 2888–2895.
9. March., P. E., Ahnn, J., and Inouye, M. (1985) The DNA sequence of the gene (rnc) encoding ribonuclease III of Escherichia coli. *Nucleic Acid Res.* **13,** 4677–4685.
10. Chen, S. and Court, D. L. (1992) Overexpression of rnc gene and purification. *Chin. J. Biotec.* **8,** 82–91.
11. March, P. E. and Gonzales, M. A. (1990) Characterisation of the biochemical properties of recombinant ribonuclease III. *Nucleic Acid Res.* **18,** 3293–3298.
12. Nicholson, A. W. (1995) Structure, reactivity and biology of double stranded RNA in *Progr. Nucleic Acid Res. Mol. Biol.* **52,** 1–65.
13. Iino, Y., Sugimoto, A., and Yamamoto, M. (1991) S. pombe pac1+, whose overexpression inhibits sexual development, encodes a ribonuclease III-like RNase. *EMBO J.* **10,** 221–226.
14. St Johnston, D., Brown, N. H., Gall, J. G., and Jantsch, M. (1992) A conserved double-stranded RNA-binding domain. *Proc. Natl. Acad. Sci. USA* **89,** 10,979–10,983.
15. Court, D. C. (1993) RNA processing and degradation by RNase III, in *Control of mRNA Stability* (Belasco, J. G. and Brawerman, G., eds.), Academic Press, New York, pp. 71–116.
16. Dunn, J. J. (1982) Ribonuclease III, in *The Enzymes*, vol. 15, part B, pp. 485–490.
17. Klovins, J., van Duin, J., and Olsthoorn, R. L. C. (1997) Rescue of the RNA phage genome from RNase III cleaveage. *Nucleic Acid Res.* **25,** 4201–4208.

18. Bardwell, J. C. A., Regneir, P., Chen, S., Nakamura, Y., Grunberg-Manago, M., and Court, D. L. (1989) Autoregulation of RNase III operon by mRNA processing. *EMBO J.* **8,** 3401–3407.

19. Meur, M. R. and Portier, C. (1994) Polynucleotide phosphorylase of Escherichia coli induces the degradation of its RNase III processed messenger by preventing its translation. *Nucleic Acid Res.* **22,** 397–403.

20. Mayer, J. E. and Schweiger, M. (1983) RNase III is positively regulated by T7 protein kinase. *J. Biol. Chem.* **258,** 5340–5343.

21. Krinke, L. and Wulff, D. L. (1990) The cleavage specificity of RNase III. *Nucleic Acid Res.* **18,** 4809–4815.

22. Zhang, K. and Nicholson, A. (1997) Regulation of ribonucleaseIII processing by double helix sequence antideterminants. *Proc. Natl. Acad. Sci. USA* **94,** 13,437–13,441.

23. Oppenheim, A. B., Kornitzer, D., Altuvia, S., and Court, D. L. (1993) Posttranscriptional control of the lysogenic pathway in bacteriophage lambda. *Prog. Nucleic Acid Res. Mol. Biol.* **46,** 37–49.

24. Dasgupta, S., Fernandez, L., Kameyama, L., Inada, T., Nakamura, Y., Pappas, A., and Court, D. L. (1998) Genetic uncoupling of the dsRNA binding and RNA cleavage activities of the Escherichia coli endoribonuclease RNase III—the effect of dsRNA binding on gene expression. *Mol. Microb.* **28,** 629–640.

25. Li, H. and Nicholson, W. A. (1996) Defining the enzyme binding domain of a ribonuclease III processing signal. Ethylation interference and hydroxyl radical footprinting using catalytically inactive RNase III mutants. *EMBO J.* **15,** 1421–1433.

26. Nagai, K. (1996) RNA-protein complexes. *Curr. Opin. Struct. Biol.* **6,** 53–61.

27. Varani, G. (1997) RNA-protein intermolecular recognition. *Acc. Chem. Res.* **26,** 1345–1351.

28. Cusack, S. (1999) RNA protein complexes. *Curr. Opin. Struct. Biol.* **9,** 66–73.

29. Kharrat, A., Macias, M. J., Gibson, T. J., Nilges, M., and Pastore, A. (1995) Structure of the dsRNA binding domain of E. coli RNase III. *EMBO J.* **14,** 3572–3584.

30. Braun, W., Bosch, C., Brown, L. R., Go, N., and Wuthrich, K. (1981) Combined use of proton-proton Overhauser enhancements and a distance geometry algorithm for determination of polypeptide conformations. Application to micelle-bound glucagon. *Biochim. Biophys. Acta* **667,** 377–396.

31. Pervushin, K., Riek, R., Wider, G., and Wuthrich, K. (1997) Attenuated T2 relaxation by mutual cancellation of dipole-dipole coupling and chemical shift anisotropy indicates an avenue to NMR structures of very large biological macromolecules in solution. *Proc. Natl. Acad. Sci. USA* **94,** 12,366–12,371.

32. Hegyi, H. and Bork, P. (1997) On the classification and evolution of protein modules. *J. Protein Chem.* **16,** 545–551.

33. Bycroft, M., Grunert, S., Murzin, A. G., Proctor, M., and St. Johnston, D. (1995) NMR solution structure of a dsRNA binding domain from Drosophila staufen protein reveals homology to the N-terminal domain of ribosomal protein S5. *EMBO J.* **14,** 3563–3571.

34. Krovat, B. C. and Jantsch, M. F. (1996) Comparative mutational analysis of the double stranded RNA binding domain of Xenopus laevis RNA-binding protein A. *J. Biol. Chem.* **271,** 28,112–28,119.

35. McMillan, N. A. J., Carpick, B. W., Hollis, B., Toone, W. M., Zamanian-Daryoush, M., and Williams, B. R. G. (1995) Mutational analysis of the double-stranded RNA (dsRNA) binding domain of the dsRNA-activated protein kinase, PKR. *J. Biol. Chem.* **270,** 2601–2606.

36. Ho, C. K. and Shuman, S. (1996) Mutational analysis of the vaccinia virus E3 protein defines amino acid residues involved in E3 binding to double-stranded RNA. *J. Virol.* **70,** 2611–2614.

37. Ryter, J. M. and Schultz, S. C. (1998) Crystal structure of methionyl-tRNA transformylase complexed with the initiator formyl-methionyl-tRNA. *EMBO J.* **17,** 6816–6826.

38. Ramos, A., Gunert, S., Adams, J., Micklem, D., Proctor, M. R., Freund, S., Bycroft, M., St. Johnston, D., and Varani, G. (2000) RNA recognition by a Staufen double-stranded RNA-binding domain. *EMBO J.* **19,** 997–1009.

39. Altschul, S. F., Madden, T. L., Schaffer, A. A., Zhang, J., Zhang, Z., Miller, W., and Lipman, D. J. (1997) Gapped BLAST and PSI-BLAST: a new generation of protein database search programs. *Nucleic Acids Res.* **25,** 3389–3402.

40. Thompson, J. D., Gibson, T. J., Plewniak, F., Jeanmougin, F., and Higgins, D. G. (1997) The CLUSTALX windows interface: flexible strategies for multiple sequence alignment aided by quality analysis tools. *Nucleic Acids Res.* **25,** 4876–4882.

41. van der Oost, J., Lappalainen, P., Mussacchio, A., Warne, A., Lemieux, L., Rumbley, J., Gennis, R. B., Aasa, R., Pascer, T., Malmstrom, B. G., and Saraste, M. (1992) Restoration of a lost metal-binding site. Construction of two different copper sites into a sub-unit of the E. coli cytochrome O quinol oxidase complex. *EMBO J.* **11,** 3209–3217.

42. Brünger, A. T., ed. (1992) *X-PLOR. A System for X-Ray Crystallography and NMR.* Yale University Press, New Haven, CT.

43. Sambrook, K. J., Fritsch, E. F., and Maniatis, T. (1989) *Molecular Cloning.* Cold Spring Harbor Laboratory Press, Cold Spring Harbor, NY.

44. Canavagh, J., Fairbrother, W. J., Palmer, A. G., III and Skelton, N. J. (1996) *Protein NMR Spectroscopy. Principle and Practice.* Academic Press.

45. Nilges, M. (1995) Calculation of protein structures with ambiguous distance restraints. Automated assignment of ambiguous NOE crosspeaks and disulphide connectivities. *J. Mol. Biol.* **245,** 645–660.

46. Xu, Y., Schein, C. H., and Braun, W. (1999) Combined automated assignment of NMR spectra and calculation of three-dimensional protein structures, in *Biological Magnetic Resonance, vol. 17, Structure Computation and Dynamics in Protein NMR* (Krishna and Berliner, eds.), Kluwer Ac. /Plenum Publishers, New York.

47. Xu, Y., Wu, J., Gorenstein, D., and Braun, W. (1999) Automated 2D NOESY assignment and structure calculation of crambin(S22/I25) with the self-correcting distance geometry based NOAH/DIAMOND programs. *J. Magn. Res.* **136,** 76–85.

48. Guntert, P., Mumenthaler, C., and Wuthrich, K. (1997) Torsion angle dynamics for NMR structure calculation with the new program DYANA. *J. Mol. Biol.* **273,** 283–298.

49. Lesk, A. M. (1991) *Protein Architecture: A Practical Approach.* IRL Press, Oxford.

50. Piotto, M., Saudek, V., and Sklenar, V. (1992) Gradient-tailored excitation for single-quantum NMR spectroscopy of aqueous solutions. *J. Biomol. NMR* **2,** 661–665.

51. Nilges, M., Macias, M., O'Donoghue, S. I., and Oschkinat, H. (1997) Automated NOESY interpretation with ambiguous distance restraints in the refined NMR solution structure of the pleckstrin homology domain from beta-spectrin. *J. Mol. Biol.* **269,** 408–422.

52. Nilges, M. and O'Donoghue, S. I. (1998) Ambiguous NOEs and automated NOE assignment. *Progr. NMR Spectr.* **32,** 107–139.

53. Schweisguth, D. C., Chelladurai, B. S., Nicholson, A. W., and Moore, P. B. (1994) Structural characterization of a ribonuclease III processing signal. *Nucleic Acid Res.* **22,** 604–612.

# 17

## Using Electrostatics to Define the Active Site of *Serratia* Endonuclease

### Kurt L. Krause and Mitchell D. Miller

## 1. Introduction

The location of an enzyme's active site is usually known prior to the determination of its three-dimensional structure. Analysis of sequence homology may indicate residues that are near an enzyme's active site. Catalytic residues are often identified when site-directed mutagenesis experiments or chemical modification of residues lead to an alteration in enzyme activity. The enzyme's three-dimensional (3D) structure is then used to supplement and interpret preceding experimental data. Occasionally, a new protein structure is solved before the active site has been located through conventional methods. Using the extracellular endonuclease from *Serratia marcescens* as an example, we show here how the active site of an enzyme can sometimes be determined from electrostatic analysis of its 3D structure *(1)*.

### 1.1. Background on Serratia Nuclease

*Serratia marcescens* is a Gram-negative pathogen that secretes, among other proteins, a very active endonuclease with broad specificity *(2)*. The presence of this endonuclease is one characteristic that distinguishes *S. marcescens* from other Enterobacteriaceae *(3)*. The *Serratia* nuclease cleaves double- and single-stranded DNA and RNA without sequence specificity. The enzyme cleaves nucleic acids very rapidly, with a catalytic rate almost 15 times faster than DNase I *(4)*. It shows long-term stability at room temperature and is active in the presence of many reducing and chaotropic agents *(5,6)*. For example, rate enhancement is observed in 4 *M* urea. These characteristics make it a useful enzyme in biotechnological and pharmaceutical applications, and it is marketed under the trade name Benzonase (Merck). In order to help understand this enzyme's mechanism of action, we undertook a crystal-structure determination. At the time the structure was initially solved, comparative sequence data, mutagenesis data, and chemical modification data were not available. Our search of GENBANK revealed only one other related protein, NUC1, an uncharacterized yeast endonuclease.

From: *Methods in Molecular Biology, vol. 160: Nuclease Methods and Protocols*
Edited by: C. H. Schein © Humana Press Inc., Totowa, NJ

## 1.2. Approach to Locating the Serratia Nuclease Active Site

After we completed the building and refinement of the *Serratia* nuclease structure, visual inspection failed to reveal the active site location. Because the fold of the *Serratia* nuclease is unique, no information could be obtained from comparison to other nuclease structures. Therefore, we attempted to analyze the structure in terms of surface charges. We reasoned that in order to bind a strongly negatively charged nucleic-acid substrate, the enzyme would need several positively charged residues on its surface. Although there was no cluster of charged surface residues to indicate a DNA binding site, electrostatic calculations described in **Subheading 1.3.** clearly indicated the location of the *Serratia* nuclease active site.

## 1.3. Background on Electrostatic Field Calculations

Scientists have long understood the importance of electrostatics in explaining how enzymes and substrates are drawn to each other *(7,8)*. Electrostatic forces dominate long-range intermolecular interactions because the potential function, which determines the strength of the electrostatic interaction between two molecules, has a $1/r$ dependence, where $r$ is the distance between the two molecules under study. In practice, this potential has effects that reach several Å out into solution from the protein's surface. In contrast, other important noncovalent interactions have smaller long range effects, such as the attractive part of a van der Waals potential, which has a $1/r^n$ dependence where $n$ is often 6. This potential function diminishes rapidly with distance.

Advances in algorithm speed and computing power have improved the ability to perform detailed and accurate electrostatic calculations on proteins. The math behind these methods ranges from fairly simple to complicated, but it is easy to survey the relevant equations *(9)*. For example, Coulomb's law mathematically describes the force between point charges in a vacuum (**Eq. 1**):

$$F = kq_1q_2/r^2 \qquad (1)$$

In this equation $F$ is the force on two atoms with charge $q$, $r$ is the distance between these two atoms, and k is the proportionality constant. These pairwise interactions are additive within a protein. The integral form of Coulomb's law is Gauss's equation, which is still relatively simple (**Eq. 2**):

$$\Phi = 4\pi\int\rho dv \qquad (2)$$

Here $\Phi$ is the total flux or total electrostatic potential, $\rho$ is the charge density, and $v$ is the vol. If this integral equation is transformed into its differential form, it can be used to derive the Poisson equation (**Eq. 3**). This equation relates the second partials of the electric potential to the charge density.

$$\frac{\partial^2\phi}{\partial x^2} + \frac{\partial^2\phi}{\partial y^2} + \frac{\partial^2\phi}{\partial z^2} = -4\pi\rho \text{ or more commonly } \nabla^2\Phi = -4\pi\rho \qquad (3)$$

Performing these calculations on proteins and small molecules in solution is more complicated because the effects of solvent and ionic strength must be included. Altering the Poisson equation to include these additional effects results in the Poisson-Boltzmann (PB) equation (**Eq. 4**) *(10,23)*.

$$\nabla \cdot [\varepsilon(r) \cdot \nabla \cdot \phi(r)] - \kappa(r)^2 \sinh[\phi(r)] = -4\pi\rho(r) \tag{4}$$

In this equation, $\phi$ is the electrostatic potential, which is usually given in units of $kT/q$ where k is Boltzmann's constant, $T$ is temperature in Kelvin, and $q$ is the charge of a proton. $r$ is the real space vector; $\varepsilon(r)$ is the spatially varying dielectric constant; $\rho(r)$ is the charge distribution function; and $\kappa(r)$ is a modified Debye-Hückel parameter included to account for the effects of salt. When the dielectric is 1 and the ionic strength is zero, i.e., a protein in a vacuum, this equation reduces to the PB equation. It is worth noting that publications on protein electrostatic calculations often include slightly different, but equivalent, forms of the PB equation.

The PB equation is a nonlinear partial differential equation which is costly to solve computationally. Several different methods have been developed to provide solutions to the PB equation, and each has particular advantages and disadvantages. Often a linearized PB equation is used because it is easier and faster to solve, especially using the finite difference method. This equation's derivation is based on the first-order Taylor series approximation to the hyperbolic sine function. Today, modern computers can determine solutions to either the PB equation or the linearized PB equation using a variety of well-known algorithms and approaches.

## 1.4. Performing Electrostatic Calculations

Solving the PB equation yields the electrostatic potential at any point in space. This potential can be mapped onto the molecular surface of the protein and into solvent regions many angstroms away from the protein. Several software packages for displaying the results of these calculations are available. We have experience with the Graphical Representation and Analysis of Protein Structures (GRASP) *(11)* and the University of Houston Brownian Dynamics (UHBD) packages *(12,13)*. There are a number of other packages available which can perform similar calculations (*see* **Table 1** and **Note 1**). These generally give comparable answers, but since GRASP is a little easier for beginners, we will use it for the calculations described in the next section. These types of calculations have shown that enzymes often attract substrate or other proteins several angstroms away from their surface. Several reviews of enzymes and electrostatic calculations have been published *(10,14–16)*.

Electrostatic calculations for protein structures are analyzed by studying electrostatic surface maps and electrostatic contour maps. An electrostatic surface is created when electrostatic potential is mapped onto the molecular surface of a protein molecule. It can be analyzed for areas of complementary shape and charge between interacting molecules such as a substrate and its enzyme, two interacting proteins, or a protein and a nucleic acid. An electrostatic contour map is an isopotential surface that reveals how the potential penetrates into solution. Contour maps can be used to propose long-range electrostatic effects. Both types of maps are displayed in units of kT/e, and the results are often color-coded, with red representing negative potential and blue representing positive potential. This color convention opposes the usual physics and electronics convention where the anode is always red, but does make sense in terms of CPK coloring where electronegative oxygens are red and electropositive nitrogens are blue. Both types of calculations have become ubiquitous in published descriptions of newly solved crystal structures.

**Table 1**
**Computer Software for Protein Electrostatic Calculations**

| Software title | Program source | World Wide Web site |
|---|---|---|
| GRASP | Columbia University | http://honiglab.cpmc.columbia.edu/grasp/ |
| MOLMOL | Eidgenössische Technische Hochschule, Switzerland | http://www.mol.biol.ethz.ch/wuthrich/software/molmol/ |
| UHBD | University of Houston | http://glycine.ncsa.uiuc.edu/~dlivesay/TEST/bd_test.html or http://chemcca10.ucsd.edu/uhbd.html |
| DelPhi | Molecular Simulations Inc. | http://www.msi.com/life/products/insight/modules/DelPhi_page.html |
| QUANTA | Molecular Simulations Inc. | http://www.msi.com/life/products/quanta/ |
| Insight II | Molecular Simulations Inc. | http://www.msi.com/life/products/insight/ |
| O | Uppsala University, Sweden | http://imsb.au.dk/~mok/o/ |

In **Subheading 2.** we will calculate the electrostatic potential around the *Serratia* nuclease, using the finite difference method to solve the PB equation (FDPB). We will calculate a contour map and also display the potential mapped onto the protein surface. From analysis of the potential maps, we will discover that the enzyme presents a remarkable electrostatic surface that clearly identifies its active site. The electrostatic potential has a geometry that attracts DNA toward its proper binding configuration and directs it away from nonproductive binding orientations.

### 1.5. Electrostatic Analysis of the Serratia Nuclease-Active Site

Inspection of electrostatic maps calculated for the *Serratia* nuclease reveals a broad area of positive electrostatic potential (blue) that surrounds a narrow cleft containing negative electrostatic potential (red) (**Fig. 1B**). This area of positive density is of the appropriate dimension to bind to double-stranded DNA. Note that the strip of negative electrostatic potential is caused by Glu 127, an active-site residue that aids in binding the catalytic magnesium water cluster *(17)*.

To confirm our identification of the active site, we docked a canonical B-DNA dodecamer into this broad area of positive potential. We used FRODO, a crystallographic model building program, to fit a wireframe DNA model to a dot surface of the *Serratia* nuclease generated at two times the van der Waals radii of its fixed protein atoms *(1,18)*. The results of this docking are shown in **Fig. 1C**. This docking exercise is used to illustrate that the surface cleft on *Serratia* nuclease is the appropriate size to accommodate double-stranded DNA, but is not described in **Subheading 2.** because docking methods that are more accurate and advanced are currently available.

### 1.6. Confirmation and General Applicability of the Method

Our location of the active site of *Serratia* nuclease is in agreement with all other experimental analyses which have appeared including those based on multiple-sequence alignments, kinetics and mutagenesis studies *(4,19)*. In fact, the residues we cited in our electrostatic analysis were shown to be mechanistically important. These residues include His 89, Glu 127, and Asn 119, and each dramatically affects *Serratia* endonuclease activity when mutated to alanine.

To test the general applicability of this method of identifying nuclease active sites, we applied it to on several other nuclease structures, including staphylococcal nuclease (PDB identifier 3nuc), DNase I (3dni), I-*Ppo*I (1a73), and *Eco*RI (1eri). The results from staphylococcal nuclease and DNase I were difficult to interpret, as both enzymes are highly charged using the current program defaults. Staphylococcal nuclease has a net charge of +8, whereas DNase I has a net charge of –8. The excess charge dominated the display, engulfing most of the protein. These last two cases might benefit from a different charge model (*see* **Note 6**). The method worked well for the intron-encoded homing endonuclease I-*Ppo*I and the restriction endonuclease *Eco*RI. The resulting positive field contours projecting out from the DNA-binding clefts guided us to their active sites (*see* **Fig. 2**).

## 2. Materials

GRASP (Graphical Representation and Analysis of Structural Properties) is a program developed in Professor Barry Honig's laboratory at Columbia University for the calculation and display of electrostatic surfaces *(11)*. The program is only available for Silicon Graphics computers. The current version requires IRIX operating system version 5.x or 6.x; however, there is also a version available for systems running IRIX version 4.x. A demo version of the program can be downloaded from http://honiglab. cpmc.columbia.edu/grasp/. For $500 (at the time of this writing) a release key can be obtained for academic users. You need a working knowledge of basic UNIX commands, although once you are inside GRASP this is not really necessary.

Complete information about other programs which can perform similar electrostatic analyses can be found at the following Web sites (*see* **Note 1** and **Table 1**).

## 3. Methods

GRASP is easy to run. Most of the input controls are menu-driven and intuitive, and informative results can be obtained quickly. The default values for most parameters are reasonable and can be used at least initially (*see* **Note 5**). Commands are input through the menus, command line, control keys, script files, or via the panels interface. With a little patience, it is possible to exploit the program's flexibility and create more complicated figures.

### 3.1. Obtain PDB

To repeat the calculations on the *Serratia* nuclease described here, download the PDB file "1smn" from the protein data bank (*http://www.rcsb.org/pdb/cgi/explore. cgi?pdbId=1smn*) *(20,21)*. It is possible to control exactly which atoms are included in the surface and electrostatics calculations from within GRASP, but it is simpler for beginners to edit the PDB file to delete all atoms they do not want in the model. In our example, this involves creating a PDB file which contains only one monomer and no solvent molecules. This can be done interactively using a text editor or via the following UNIX command. (GRASP commands and UNIX commands are shown in bold.)

```
grep ATOM 1smn.pdb | grep ' A ' > 1smn-a.pdb
```

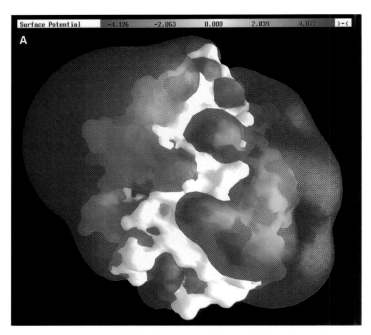

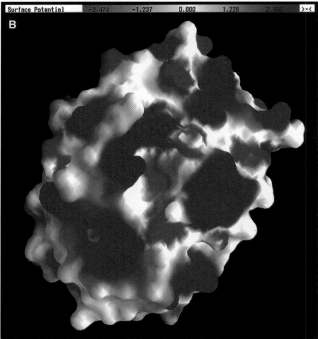

Fig. 1. *Serratia* nuclease. (**A**) Electrostatic contour map at +/– 1.2 kT/e for the monomer. Note the prominent dipole with the positive (blue) field projecting into solution on one side and a negative (red) contour on the opposite side. (**B**) Looking at the molecular surface in the region of the positive electrostatic contour, we see that there is a large cleft with a patch of negative electrostatic potential surround by a ridge of positive potential. The active site is at the base of this cleft (*see text*). This view is rotated approx 90° relative to (A).   *(continued on next page)*

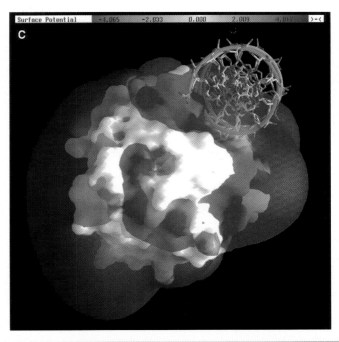

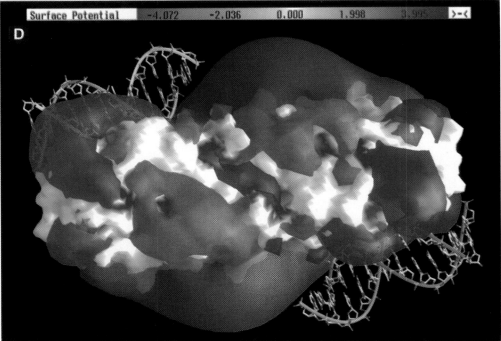

Fig. 1. *(continued).* Our model with B-DNA docked in the active site is shown in **(C)**. This model illustrates that the cleft is the correct size to accommodate B-DNA *(1)*. **(D)** Electrostatic map calculated for the physiological dimer. The DNA strands were not included in the map calculation, but were added to help orient the reader to the location of the two independent active sites in the dimer.

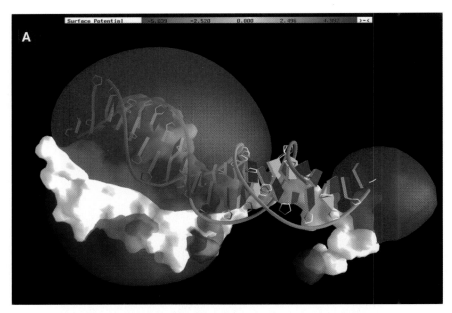

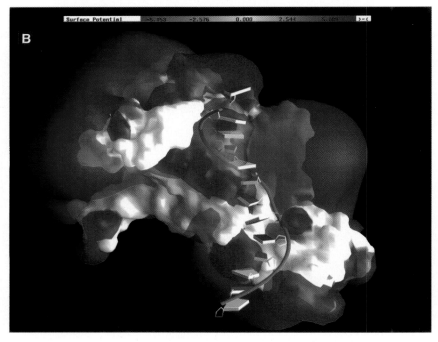

Fig. 2. Electrostatic potential map calculated as described in the text for (**A**) the I-*Ppo*I monomer and (**B**) the *Eco*RI monomer. The DNA shown in both figures identifies the location of the active site and was not included in the map calculation.

## 3.2. Run Grasp

1. Start GRASP. If the executable version of GRASP is in a directory called "/programs" then to run GRASP you would type:

`/programs/grasp`

2. Read in the PDB file. Use the right mouse button to bring up the menus. Select **Read > PDB file > show list** from the list, choose the 1smn-a.pdb file we just created containing monomer A. You will now have a display of the atoms from this monomer.

3. Assign charges to the atoms. Using the right mouse button to bring up the menus. Select: **Read > Radius/Charge file (+ Assign) > full.crg**. It is assumed that the C-terminus and all Asp, Glu, Arg, and Lys residues are fully ionized, while the N-terminus and all His residues are set to neutral. In many cases these assumptions are not valid. Other charge schemes will be discussed below (*see* **Note 6**).

4. Build a molecular surface of the molecule. From the menus select: **Build > molecular surface > all atoms.**

5. We can turn off the atom display to clean things up a bit. From the menus select: **Display > hide > atoms.**

6. Use the FDPB (finite difference Poisson Boltzmann) solver in GRASP to calculate an electrostatic potential map. From the menus select: **Calculate > new potential map.**

7. Map the potential map onto the molecular surface. From the menus select: **Calculate > Pot. Via Map at Surfaces/Atoms > all atoms > all surfaces**. Now the electrostatic charges you have calculated in **step 6** are displayed on the surface of your protein scaled by color.

8. Adjust the scale. You can now adjust the scale by clicking on the scale bar values. The left mouse button lowers the scale limits, and the middle mouse button raises the scale limits. In our example, we will make the maximum red and blue values correspond to +/– 4 kT/e.

9. Build a three-dimentional contour from the potential map. From the menus select: **Build > contour > 3-D**. Enter the values for contouring and colors. +/– 1.2 kT/e is a good place to start (*see* **Note 4**). Now enter contour values: **1.2, –1.2** and contour colors: **4,2** (this sets the positive contour blue and the negative contour red using GRASP's default color scheme with 4 = blue and 2 = red).

10. Make the surface semitransparent. From the menus select: **Display > alter > contours > transparent.** The semitransparent surface allows ready visualization of both the electro-static surface potentials and the contour map.

### 3.3. Localization of the Serratia Nuclease-Active Site

1. Inspect the contour map. Note the presence of two prominent potential clouds that project from the molecular surface. They indicate a prominent electrostatic dipole around the *Serratia* monomer (**Fig. 1A**). The positive electrostatic potential (blue) potential would attract DNA, while the negative electrostatic potential (red) would repel DNA.

2. Study the electrostatic map projected onto the molecular surface. Locate the area of posi-tive electrostatic potential underneath the contour cloud. This broad slightly curved area dominates the molecular surface.

3. Locate the narrow cleft containing negative electrostatic potential (red) (**Fig. 1B**). This cleft is located in the middle of the DNA binding region and is caused by Glu 127, an active-site residue which aids in binding the catalytic magnesium water cluster (*17*).

4. Observe the shape of the large positive potential region. The shallow, flat, slightly curved surface suggests that it might be complementary to a B-DNA molecule. Docking experi-ments described above confirmed this assumption for us (**Fig. 1C**).

## 4. Notes

1. Other programs: We chose GRASP for these calculations for its ease of use and at the time of our initial analysis it was the only program available that combined an FDPB solver

with a graphical interface designed to examine surface properties. There are many other programs which can be used in the analysis of electrostatic properties (*see* **Table 1**). For example, MOLMOL *(22)* implements the FDPB algorithm developed by the authors of GRASP *(23)* and can display electrostatic contours and map electrostatic values to the surface of the molecule. UHBD *(12,13)* and DelPhi *(10)* can both be used to calculate maps for display in GRASP or other modeling programs, e.g., QUANTA and InsightII (Molecular Simulations Inc.), and O *(24)*. We have found that UHBD produces maps with similar qualitative results to GRASP.

2. Our configuration: For this calculation we ran GRASP version 1.3.6 on a Silicon Graphics $O_2$ operating under IRIX 6.3 with 64 MB RAM. Calculations from other versions may produce slightly different results. Our original published calculations were performed on a Silicon Graphics 4D/320VGX using GRASP version 1.0.3.

3. World Wide Web: There is now a Web-based GRASP server, GRASS (Graphical Representation and Analysis of Structures Server), that simplifies many operations through a World Wide Web interface *(25)*. The results can be displayed with MDL's Chime plug-in *(26)*, a VRML plug-in, e.g., SGI's cosmo player *(23,27,28)*, and GRASP. The Chime and VRML plug-ins are available for a wide array of platforms, including UNIX, MacOS and Windows, and are used in conjunction with a web browser, e.g., Netscape's communicator version 4.x. Display using GRASP requires a Silicon Graphics machine running IRIX 4.x or later. The server web site (*http://trantor.bioc.columbia.edu/GRASS/surfserv_enter.cgi*) contains detailed descriptions and hypertext links which will guide a beginner through any software and browser configurations necessary.

4. Contouring: It is informative to vary the contour levels and the surface color scale and examine all sides of the structure. Depending on the hardware system you are using, it is possible to use hardware or side-by-side stereo views to help improve visualization of the structure. GRASP has many features that can aid the production of final figures for publication. These include changing the buffer depth to increase the number of colors accessible on 8-bit graphics subsystems and calculating antialiased polygons and lines. While the rotation of such modified displays is not recommended, it allows for the creation of high-quality images for publication purposes.

5. Electrostatic parameters: It is possible to alter several different parameters used in the calculation of electrostatic potentials within GRASP. These include the inner and outer dielectric constant, and the salt concentration. Probe radii used to calculate molecular and accessible surfaces can be altered. When starting out, keeping the default values is the easiest methods, and will yield useful information.

   The default values for these parameters are:

   | | |
   |---|---|
   | interior dielectric | 2.0 |
   | exterior dielectric | 80.0 |
   | water probe radius | 1.4 |
   | ionic probe radius | 2.0 |
   | salt concentration | 0.0 |

6. Charge sets: Since many of the side-chain atoms in proteins are charged and many others contain permanent dipoles, the charging model used in the calculations can significantly affect the results. In our example we used the GRASP distributed full.crg file. This is a simple charge model which assigns full charges to side chains, e.g., Glu has a charge of –0.5 assigned to each carboxylate oxygen. There are several other more detailed charges sets available based on the CHARMM *(29)*, OPLS *(30)*, and AMBER *(31)* force fields. In addition, it is possible to develop charging models for a specific case. For example, it is possible to include information about experimentally determined side-chain ionization

constants. We have found that changing the charging model does affect the details of the potential map, but the global picture remains similar.

The GRASS web server offers two charging models: the formal charge set, which is similar to the full.crg set we used in our example, and the PARSE charge set, which was developed to provide more accurate solvation energies *(32)*. Note that the partial charge PARSE set requires *all* atoms to be present in the PDB file, including hydrogens. A modified PDB database has been created on the GRASS server with these additional atoms already built.

7. The *Serratia* endonuclease is a physiological dimer: After our initial structural communication, we learned that the *Serratia* nuclease was a homodimer under physiological conditions *(33)*. We repeated the electrostatic calculations with the dimer structure to look at the field contour and cleft orientation. We discovered that the clefts were located on opposite sides of the dimer and the dimer interface does not obscure the entrance to the active site. A positive electrostatic field still projects out into solvent from the active-site cleft (**Fig. 1D**).

## Acknowledgments

We would like to thank Anna K. Dvorak for helpful discussion of this manuscript. This work was supported by The Robert A. Welch Foundation, the National Institutes of Health, the State of Texas, and the W. M. Keck Foundation.

## References

1. Miller, M., Tanner, J., Alpaugh, M., Benedik, M., and Krause, K. (1994) 2.1 Å structure of *Serratia* endonuclease suggests a mechanism for binding to double-stranded DNA. *Natural Struct. Biol.* **1,** 461–468.

2. Benedik, M. J. and Strych, U. (1998) *Serratia marcescens* and its extracellular nuclease. *FEMS Microbiol. Lett.* **165,** 1–13.

3. Eisenstein, B. (1995) Enterobacteriaceae, in *Principles and Practice of Infectious Diseases* 4th ed. (Mandell, G. L., Bennett, J. E., and Dolin, R., eds.), Churchill Livingstone, New York, pp. 1964–1980.

4. Friedhoff, P., Kolmes, B., Gimadutdinow, O., Wende, W., Krause, K., and Pingoud, A. (1996) Analysis of the mechanism of the *Serratia* nuclease using site-directed mutagenesis. *Nucleic Acids Res.* **24,** 2632–2639.

5. Filimonova, M. N., Balaban, N. P., Sharipova, F. P., and Leshchinskaia, I. B. (1980) Isolation and physico-chemical properties of homogenous nuclease from *Serratia marcescens. Biokhimiia* **45,** 2096–2104.

6. Benzon (1993) Product Specifications for Benzonase, the first industrial endonuclease. Benzon Pharma A/S, Helseholmen 1, P.O. Box 1185, DK-2650, Hvidovre, Denmark.

7. Feynman, R. P., Leighton, R. B., and Sands, M. (1963) *The Feynman Lectures on Physics.* Addison-Wesley, Reading, MA.

8. Lipscomb, W. N. (1982) Acceleration of reactions by enzymes. *Accts. Chem. Res.* **15,** 232–238.

9. Purcell, E. M. (1965) *Electricity and Magnetism.* McGraw Hill, New York.

10. Honig, B. and Nicholls, A. (1995) Classical electrostatics in biology and chemistry. *Science* **268,** 1144–1149.

11. Nicholls, A., Sharp, K. A., and Honig, B. (1991) Protein folding and association: insights from the interfacial and thermodynamic properties of hydrocarbons. *Proteins* **11,** 281–296.

12. Madura, J. D., Briggs, J. M., Wade, R. C., Davis, M. E., Luty, B. A., Ilin, A., Antosiewicz, J., Gilson, M. K., Bagheri, B., Scott, L. R., and McCammon, J. A. (1995) Electrostatics

and diffusion of molecules in solution: simulations with the University of Houston Brownian Dynamics program. *Comput. Phys. Commun.* **91,** 57–95.

13. Davis, M. E., Madura, J. D., Luty, B. A., and McCammon, J. A. (1991) Electrostatics and diffusion of molecules in solution: simulations with the University of Houston Brownian Dynamics program. *Comput. Phys. Commun.* **62,** 187–197.

14. Davis, M. E. and McCammon, J. A. (1990) Electrostatics in biomolecular structure and dynamics. *Chem. Rev.* **90,** 509–521.

15. Gilson, M. K. (1995) Theory of electrostatic interactions in macromolecules. *Curr. Opin. Struct. Biol.* **5,** 216–223.

16. Warshel, A. and Papazyan, A. (1998) Electrostatic effects in macromolecules: fundamental concepts and practical modeling. *Curr. Opin. Struct. Biol.* **8,** 211–217.

17. Miller, M. D., Cai, J., and Krause, K. L. (1999) The active site of *Serratia* endonuclease contains a conserved magnesium-water cluster. *J. Mol. Biol.* **288,** 975–987.

18. Jones, T. A. (1978) A graphics model building and refinement system for macromolecules. *J. Appl. Crystallogr.* **11,** 268–272.

19. Friedhoff, P., Gimadutdinow, O., and Pingoud, A. (1994) Identification of catalytically relevant amino acids of the extracellular *Serratia marcescens* endonuclease by alignment-guided mutagenesis. *Nucleic Acids Res.* **22,** 3280–3287.

20. Abola, E. E., Bernstein, F. C., Bryant, S. H., Koetzle, T. F., and Weng, J. (1987) Protein Data Bank, in *Crystallographic Databases–Information Content, Software Systems, Scientific Applications* (Allen, F. H., Bergerhoff, G., and Sievers, R., eds.), Data Commission of the International Union of Crystallography, Bonn/Cambridge/Chester, pp. 107–132.

21. Bernstein, F. C., Koetzle, T. F., Williams, G. J. B., Meyer, E. F., Brice, M. D., Rodgers, J. R., Kennard, O., Shimanouchi, T., and Tasumi, M. (1977) The protein data bank: a computer-based archival file for macromolecular structures. *J. Mol. Biol.* **112,** 535–542.

22. Koradi, R., Billeter, M., and Wüthrich, K. (1996) MOLMOL: a program for display and analysis of macromolecular structures. *J. Mol. Graphics* **14,** 51–55.

23. Nicholls, A. and Honig, B. (1991) A rapid finite difference algorithm, utilizing successive over-relaxation to solve the Poisson-Bolzmann equation. *J. Comp. Chem.* **12,** 435–445.

24. Jones, T. A., Zou, J.-Y., and Cowan, S. W. (1991) Improved methods for building protein models in electron density maps and the location of errors in these models. *Acta Crystallographica* **A47,** 110–119.

25. Nayal, M., Hitz, B., and Honig, B. (1999) GRASS Browser: Graphical Representation and Analysis of Structure Server. Columbia University, New York. http://trantor.bioc.columbia.edu/GRASS/surfserv_enter.cgi.

26. MDL (1999) MDL Worldwide Services: Chime Free Support. MDL Information Systems, San Leandro, CA. http://www.mdli.com/support/chime/chimefree.htm.

27. Hartman, J. and Wernecke, J. (1996) *The VRML 2.0 Handbook: Building Moving Worlds on the Web.* Addison-Wesley, Reading, MA.

28. SGI (1999) SGI–Cosmo Products References. Silicon Graphics, Inc., Mountain View, CA. http://www.sgi.com/software/cosmo/redirect.html.

29. Brooks, B. R., Bruccoleri, R. E., Olafson, B. D., States, D. J., Swaminathan, S., and Karplus, M. (1983) CHARMM: a program for macromolecular energy, minimization, and dynamics calculations. *J. Comput. Chem.* **4,** 187–217.

30. Jorgensen, W. L. and Tirado-Rives, J. (1988) The OPLS potential function for proteins. Energy minimizations for crystals of cyclic peptides and crambin. *J. Am. Chem. Soc.* **110,** 1657–1666.

31. Weiner, S. J., Kollman, P. A., Nguyen, D. T., and Case, D. A. (1986) An all atom force field for simulations of proteins and nucleic acids. *J. Comput. Chem.* **7,** 230.

32. Sitkoff, D., Sharp, K. A., and Honig, B. (1994) Accurate calculation of hydration free energies using macroscopic solvent models. *J. Phys. Chem.* **98,** 1978–1988.

33. Miller, M. and Krause, K. (1996) Identification of the *Serratia* endonuclease dimer: structural basis and implications for catalysis. *Protein Sci.* **5,** 24–33.

# 18

# Homology Modeling and Simulations of Nuclease Structures

## Kizhake V. Soman, Catherine H. Schein, Hongyao Zhu, and Werner Braun

## 1. Introduction

### 1.1. Nuclease Sequences Can Be Grouped into Families

As discussed in other chapters in this book (see especially Chapters 6, 7, 9, 23, and 24), many nucleases were first identified as such by their sequence identity to known nucleases. Several of these were isolated because they exhibited a seemingly unrelated activity (protein kinase, angiogenesis, eosinophil activation, tumor-cell growth inhibition). Multiple-sequence alignments of many proteins in a family can reveal a consistently occurring pattern which is not easily detected by pairwise alignments (1,2). Even proteins with extremely low overall sequence identity (such as human angiogenin) to the pancreatic RNase (ribonuclease) sequences can be recognized by a pattern of highly conserved amino acids in a multiple-sequence alignment (3). The sequence of six RNases (RNase A and two mutants thereof, bovine seminal RNase, angiogenin, and onconase) from those whose structures are available in the Protein Data Bank (PDB) (4) are aligned in **Fig. 1**. The "RNase mask" (asterisks in **Fig. 1**) for this sequence family consists of <20 absolutely conserved residues, most of which are involved in disulfide linkages or catalysis. The largest block of identity in the series is CKXXNTF. Because insertions and deletions alter the distance between other conserved residues throughout the alignment, the mask, although useful for classifying members of the family, cannot be used to search sequences for new RNase homologs in a straightforward way.

### 1.2. Proteins with a High Degree of Sequence Identity Share a Common Three-Dimentional-Fold

However, once a novel protein has been identified as a member of this family, an alignment such as the one in **Fig. 1** can be used to predict the three-dimensional (3D) structure from template structures. A good 3D protein structure or model provides invaluable help in studying a protein's function and specific binding to other proteins, activators, or substrates. Several chapters in this section illustrate how protein structures determined by X-ray (Chapters 13, 17, 19, and 20) or NMR (Chapter 16) can be used to identify functional areas of a protein and to design site-directed mutagenesis experiments.

From: *Methods in Molecular Biology, vol. 160: Nuclease Methods and Protocols*
Edited by: C. H. Schein © Humana Press Inc., Totowa, NJ

```
1ONC_P30      -----DWLTFQKKHITNT------RDVDCDNIMS---TNLFHCKDKNTFIYSRPEPVKAIC
1AGI_Angi     AQDDYRYIHFLTQHYDAK-PKGRNDEYCFNMMKNRRLTRP-CKDRNTFIHGNKNDIKAIC
3RN3_RN       --KETAAAKFERQHMDSSTSAASSSNYCNQMKSRNLTKDRCKPVNTFVHESLADVQAVC
1A5P_RNCC     --KETAAAKFERQHMDSSTSAASSSNYCNQMKSRNLTKDRAKPVNTFVHESLADVQAVC
1BSR_BS-RN    --KESAAAKFERQHMDSGNSPSSSSNYCNLMMCCRKMTQGKCKPVNTFVHESLADVKAVC
1A5Q_RNP93A   --KETAAAKFERQHMDSSTSAASSSNYCNQMKSRNLTKDRCKPVNTFVHESLADVQAVC
                                F    H        C   M        K  NTF        A C
                                *    *        *   *  *      * .  ***.    .  .*.*

1ONC_P30      KG-IIASKN------VLTTSEFYLSDCNVT---SRP-CKYKLKKSTNKFCVTCENQ--APV
1AGI_Angi     EDRNGQPYRG--DLRISKSEFQITICKHKGGSSRPPCRYGATEDSRVIVVGCENG--LPV
3RN3_RNmut    SQKNVACKNGQTNCYQSYSTMSITDCRETGSSKYPNCAYKTTQANKHIIVACEGNPYVPV
1A5P_RNmut    SQKNVACKNGQTNCYQSYSTMSITDCRETGSSKYPNAAYKTTQANKHIIVACEGNPYVPV
1BSR_BS-RN    SQKKVTCKNGQTNCYQSKSTMRITDCRETGSSKYPNCAYKTTQVEKHIIVACGGKPSVPV
1A5Q_RNmut    SQKNVACKNGQTNCYQSYSTMSITDCRETGSSKYANCAYKTTQANKHIIVACEGNPYVPV
                         S        C        Y     C    Y     C    PV
                         *          * ..     *       *  *  *   . **

1ONC_P30      HFVGVGSC---
1AGI_Angi     HFDESFITPRH
3RN3_RNmut    HFDASV-----
1A5P_RNmut    HFDASV-----
1BSR_BS-RN    HFDASV-----
1A5Q_RNmut    HFDASV-----
              HF
              **
```

Fig. 1. Multiple sequence alignment of Onconase (P30 protein, PDB structure file 1ONC), angiogenin (1AGI), bovine RNase A (3RN3), bovine seminal RNase (1BSR, only one monomer is used), and two site mutants (sites underlined) of RNase A that have profound effects on the secondary structure of the molecule (1A5P:Cys40 and 95 to Ala; 1A5Q:Pro93 to Ala). The residues in bold indicate the pancreatic RNase "mask" of conserved residues.

264

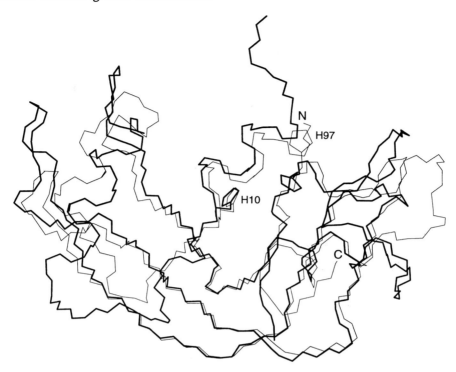

Fig. 2. Similarity of the onconase structure (PDB file 1ONC) to that of RNase A (3RN3; thick line). Most of the secondary structural elements overlap. Despite the low sequence identity (29 residues of onconase are identical to RNase A if seven gaps are introduced) between these two distantly related proteins, the two catalytic histidines in each molecule are similarly located (shown for both). The N- and C-terminus and the catalytic histidines of ONC (10 and 97) are labeled.

High-resolution methods for determining structures require large amounts of pure protein and months or even years of data accumulation and analysis to solve a structure. In the meantime, other biophysical techniques may yield some information on protein structure *(5)* which can be used in combination with homology modeling *(6)* and protein-fold recognition methods *(7,8)*. As **Fig. 2** shows, even proteins as distantly related by sequence identity as onconase and RNase A have a common fold, and essentially identical catalytic centers. This information can be used to build a 3D model as a structural framework to help rationalize experimental findings, design further experiments, and even assist structure solution by X-ray crystallography or multidimensional NMR spectroscopy. The methods presented in this chapter make use of statistical surveys of the influence of protein sequence on protein secondary structure and previously determined structures to predict the structure of a new nuclease protein. For example, a reasonable model of the 3D-structure of a stylar RNase could be built based on sequence identity to RNase Rh *(9)*. Comparing multiple aligned sequences (*see* **Subheading 3.2.**) can yield a wealth of information about areas of proteins essential for retaining structure or function, as will be discussed in our description of MASIA (*see* **Subheading 3.6**).

Although *ab intio* modeling of proteins, using various statistical prediction methods, has made some progress *(10–12)*, models based on a high-resolution 3D structure of a reference protein closely related in sequence are the most reliable. Coordinates that define most of the high-resolution protein structures solved are now available on the World Wide Web in the PDB (URL: http://www.rcsb.org/pdb/). Analysis of these data has shown that proteins with a high degree of identity at the primary structure level have the same overall fold *(13)*. The root mean square deviation (RMSD; *see* **Note 1**) was ≤1.5 Å from the template when the residue identity of the pair was ≥30%. At lower levels of identity, the RMSD between the structures became progressively higher *(14)*.

## 1.3. Homology Modeling Is a Reliable Way to Predict the 3D Structure of a Protein

For "homology modeling," the structure of the reference protein is used as a "template" on which to fold the unknown protein. Suitable template structures are identified by searching the Protein Data Bank with FASTA *(15)* or another sequence search program. The FASTA search results are expressed in a table that tells how closely each of the sequences matches the probe sequence. The model and template sequences are then aligned in the most probable fashion using an alignment program, such as Clustal W (*see* **Subheading 3.2.**). The next phase is to determine inter-atomic distances and dihedral angles in the template and force the corresponding residues of the model to assume similar distances and angles.

As described in **Subheading 3.3.**, our in-house program EXDIS is then used to extract "distance constraints" and "dihedral angle constraints" from the template file in the PDB that reflect the distance between and the orientation of the residues in the template file. EXDIS calculates the distances of a specified number of other atoms, chosen at random, from each atom in the structure. Each distance, plus a tolerance value, is used as an upper and a lower distance constraint for that atom pair in the model. At each position, if the amino acids in the model protein and the template are identical, EXDIS includes all the dihedral angles between them in the angle constraint list. When the residues are different, only the backbone dihedral angles are included. Once the dihedral angle file for generating a starting structure and distance and dihedral constraint files have been generated, they can either be used as input for the distance geometry program DIAMOD *(10,16,17)* or the "energy minimization" program FANTOM *(19,20,37)* to build the model. During much of energy minimization, the model will be kept "restrained" to the template by imposing the interatomic distances and dihedral angles extracted from the template.

In energy minimization programs, a series of formulae are used to represent various contributions to the conformational energy of a structure. Different force fields are alternate methods of accounting for these energies. For example, in the CHARMm force field *(18)* used by XPLOR, one can choose a "united" atom model, where hydrogens are implicit in the parameters for the heavier atoms to which they are attached, or an "all-atom" representation (more precise but more expensive computationally), where every atom including hydrogen is explicitly represented. In our CPU-intensive molecular dynamics (MD) simulations using XPLOR, we resorted to an intermediate solution where the polar hydrogen atoms (in addition to all non-hydrogens) are represented explicitly (*see* **Subheading 3.8.**). FANTOM uses the ECEPP/2 force field

*(19)*, which allows such an increase in calculation efficiency that all hydrogen atoms can be explicitly represented. Following ECEPP philosophy, bond lengths and bond angles are kept fixed, while dihedral angles are allowed to rotate to generate the lowest energy structure. This reduces the number of variables compared to a representation in Cartesian coordinates *(20)*. The total energy calculated by FANTOM is the sum of the *conformational energy* (electrostatic + hydrogen bond + van der Waals + torsional energies) and the *constraint energy* (weighted penalties for violations of dihedral angles + upper + lower distance constraints).

First, FANTOM constructs a starting structure of the protein, taking standard geometry from a library and dihedral angles from the angle file generated with EXDIS. A "constraint energy" penalizes deviations from the allowed ranges for the distances or dihedral angles. Only violations of distances or dihedral angles that exceed the tolerance values contribute to the constraint energy. The contribution to the constraint energy depends on the severity of the violation. The user can choose the functional form and the scale factor for distance and dihedral angle constraints, thus selecting the relative weights of their contributions. For example, a fourth power energy function can be selected for distance constraints, which adds kT/2 to the constraint energy for a violation of 0.2 Å. The dihedral angle constraint function can be instructed to add an energy of, e.g., 10.0*kT/2, for every 5° violation. The sum of these violation energies is the total constraint energy for that conformation. At this stage, called "regularization," conformational energy is not calculated. The total constraint energy is minimized, leading to a crude model. This starting model would have the overall shape (fold) of the template, but would be stereochemically unacceptable. In the next stage, the full energy function is turned on and minimized.

One can choose between two minimization algorithms in FANTOM: *quasi-Newton* and *Newton-Raphson*. As the minimization progresses, constraints to the template are made looser, until at the end, a short "free," or unrestrained minimization, is run. The resulting model will be largely consistent with the constraints and free of short contacts. EXDIS and FANTOM modeling of a structure for the free PBCV-1 glycosylase is demonstrated in **Subheadings 3.3.** and **3.4.**

Alternatively, after EXDIS, a distance geometry program, such as our group's DIAMOD *(10,11,16,17)* can be used to build a series of models. In DIAMOD, the calculation starts with the protein in a random conformation. A specified number (~50) of conformers are generated by minimizing a "target function." The target function for a given conformation is the sum of penalties for violations of distance and dihedral angle constraints, and van der Waals clashes approximated by a "hard-sphere" function. The advantage of the distance geometry approach is that a large number of conformers can be generated and analyzed, and selected ones can be energy minimized with FANTOM. The DIAMOD approach, illustrated in **Subheading 3.7.**, is especially suited for modeling a complex of two or more molecules, when intermolecular distances are known.

## 1.4. Glycosylase and Endonuclease Modeling and Simulations

Our model of the UV-light-damage-specific DNA glycosylase from Chlorella virus strain PBCV-1 (pyrimidine dimer glycosylase; PDG) in complex with its substrate, DNA, containing UV-induced thymidine dimers, illustrates the use of several programs

*(21)*. A better understanding of the enzymes involved in repair of damaged DNA is useful for developing treatments for diseases in which cellular repair mechanisms are absent or impaired *(22)*. Precise definition of their substrate-binding mechanism can aid in designing inhibitors of these enzymes to use when their activity limits the effectiveness of chemotherapy *(23)*.

While this example was first used to illustrate how easily functional homologues of newly sequenced proteins can be identified (*see* **Subheading 3.1.**), **Subheadings 3.7.** and **3.8.** outline how EXDIS and DIAMOD can be used to model protein/nucleic acid complexes, and how molecular dynamics simulations can be used to follow protein/substrate interactions. The amino-acid sequence of PDG is ~41% identical (~65% similar) to that of T4 endonuclease V (Endo V), one of the best-studied DNA repair enzymes. High-resolution crystal structures have been determined for the free Endo V and its complex with damaged DNA *(24–27)*. As has been determined for several other DNA modifying enzymes *(28,29)*, Endo V "flips out" an adenine base from the ordered helical structure on the opposite strand from the damaged bases. This space in the DNA is occupied by protein residues that cleave the N-C-1' glycosidic bond joining the base to the sugar phosphate backbone. The two recombinant proteins both recognize *cis-syn* cyclobutane pyrimidine dimers (PD) and effect glycolytic cleavage via formation of an imino (Schiff base) intermediate with damaged DNA that can be trapped by adding $NaBH_4$. They differ in that the PDG enzyme is also active on DNA substrates containing a trans-syn II PD, which is not recognized by Endo V, and that PDG binds its substrate better at a higher salt concentration *(30)*.

Despite the high degree of identity between the two proteins, building a homology model of the complex was not straightforward. Quantum chemical calculations were used to determine the optimum geometry and charges of the damaged base pair. This data, together with normal EXDIS files from the Endo V structure file, were used in DIAMOD to generate a PDG/DNA model whose backbone RMSD to the Endo V/DNA structure was 0.5, 0.5, and 0.8Å for the DNA, the protein, and the whole complex, respectively. To determine structural details that could account for differences in activity of the two enzymes, molecular dynamics simulations were used to follow protein-DNA interactions in an aqueous environment. The simulations of the Endo V/DNA complex indicated new roles for Arg22 and Arg26 in the active site in recognizing irregular pairing and maintaining the strand separation needed for incision of the damaged bases. The model for the PDG/DNA complex and simulations thereof indicate a similar mechanism for DNA binding by this enzyme despite significant differences in residues maintaining the flipped-out adenine and strand separation in the area of damage. The increased affinity for substrate as well as PDG's activity on *trans-syn II* cyclobutane dimers could be explained by differences in the surface charge and active site.

## 2. Materials

SGI/R10000 workstations and a Cray J90 were used for these simulations. *See* **Tables 1** and **2** for the software used.

**Table 1**
**In-House Software Used**

| Software | Used for | Reference |
|---|---|---|
| MASIA | Prediction of secondary structures and exposed/buried residues from sequence; analysis of alignments | *(11,17)* |
| EXDIS | Extract distances and dihedral angles from 3-D structure | *(19,20)* |
| FANTOM | Energy minimization; Monte Carlo simulation | |
| DIAMOD | Self-correcting distance geometry (SECODG) calculations | *(10,16,17)* |

**Table 2**
**External Software Used**

| Software | Web site URL |
|---|---|
| MOLMOL2.6 *(31)* | www.mol.biol.ethz.ch/wuthrich/software/molmol |
| XPLOR3.1 *(32)* | atb.csb.yale.edu/xplor-info/ |
| MOPAC93 6.0 *(33)* | www.psc.edu/general/software/packages/mopac/mopac.html |
| ClustalW *(34)* | www2.ebi.ac.uk/clustalw/ |
| GAUSSIAN94 *(35)* | www.gaussian.com |

## 3. Methods

### 3.1. Sequence Search of the Protein Data Bank Using the Program FASTA to find the Best Template for the Homology Modeling of PBCV-1 UV Glycosylase

1. Go to the PDB web site http://www.rcsb.org/pdb > *Search Fields* > check *FASTA Search* ON.
2. Cut and paste the PBCV-1 sequence into the input box in *FASTA* format (*see* **Fig. 3**).
3. Press *Search* button; this will run a *FASTA* search using the default parameters.
4. Select the template structure: The output of the search command is a list of similar sequences. As of Aug. 19, 1999, seven proteins were listed (this may increase as more structures are solved and deposited in the PDB). Look for the one with the lowest E value (*see* **Note 2**). For the example chosen, Endonuclease V, with an E value of $8.1 \times 10^{-16}$ and ~40% sequence identity over a 132 residue region, is clearly the best template.
5. Download the PDB file (code = 2END) with coordinates by clicking successively on: EXPLORE > Download File > PDB format.
6. Save the downloaded PDB file as *2end.pdb*.

### 3.2. Align the Sequences of PBCV-1 and 2END Using ClustalW

1. Prepare an input file for alignment, with PBCV-1 as the first sequence, as shown in **Fig. 4**.
2. Run ClustalW to align the sequences. ClustalW can be accessed on the web, e.g., at www2.ebi.ac.uk/clustalw/. Cut and paste the sequence pair shown above into the input box, and run ClustalW without changing any of the options. The result is the alignment shown in **Fig. 5** where the asterisks indicate residue identity and the dots show residue similarity. Save this result into a file (*pbcv1_2end.aln*).
3. This file can be edited to optimize the alignment for extraction of secondary structure. Although there is only one additional residue between the sequences in this example, alignments having a lower sequence identity will probably require more gapping.

```
> PBCV-1 Glycosylase sequence in FASTA format:
TRVNLVPVQELADQHLMAEFRELKMIPKALARSLRTQSSEKILKKIPSKFTLNTGHVLF
FYDKGKYLQQRYDEIVVELVDRGYKINVDAKLDPDNVMTGEWYNDYTPTEDAFNIIRAR
IAEKIAMKPSFYRFTKAKTSNN
```

Fig. 3. The amino-acid sequence of PBCV-1 glycosylase (PDG) in "FASTA" format.

```
>PBCV-1;
TRVNLVPVQELADQHLMAEFRELKMIPKALARSLRTQSSEKILKKIPSKF
TLNTGHVLFFYDKGKYLQQRYDEIVVELVDRGYKINVDAKLDPDNVM
TGEWYNDYTPTEDAFNIIRARIAEKIAMKPSFYRFTKAKTSNN
>2END;
MTRINLTLVSELADQHLMAEYRELPRVFGAVRKHVANGKRVRDFKISPTF
ILGAGHVTFFYDKLEFLRKRQIELIAECLKRGFNIKDTTVQDISDIPQEF
RGDYIPHEASIAISQARLDEKIAQRPTWYKYYGKAIYA
```

Fig. 4. Sequences of PBCV-1 (model) and 2END (template) proteins in ClustalW input format.

## 3.3. Run EXDIS

1. Download FANTOM/EXDIS from our web site: http://www.scsb.utmb.edu/fantom and install the program suite on your workstation following the detailed instructions given there. The installation procedure produces two executable programs, *exdis* and *fantom*.
2. Go to the directory where the input files (template structure and sequence alignment) are. To begin the session, type "exdis." EXDIS will prompt the reader to supply the names of the alignment and template's pdb file, the two user-specified parameters (interval for dihedral angle constraints and the number of distance constraints per atom) as input and the output file names. **Figure 6** shows a typical EXDIS session. At the end of the session, EXDIS writes the dihedral angle constraints into the file *pbcv1.aco* (**Fig. 7**), the upper/lower distance constraints into the *pbcv1.upl* (**Fig. 8**) and the dihedral angles necessary to generate an initial conformation into *pbcv1.seq* (**Fig. 9**). These files are required for the FANTOM session that follows.

## 3.4. Run FANTOM to Build and Refine a Model for PBCV-1

FANTOM needs these three types of file as input. The parameters needed to control the regularization and minimization tasks can be taken from the default values, which may need to be modified for a given job. In general, these parameters are specified in command (*.com*) files, which FANTOM reads as input. Some sample input files are supplied with the FANTOM package, and usually the best strategy is to edit one to suit your purpose.

### 3.4.1. Structure Regularization

To start the regularization session, type "fantom." The program prompts the user for the names of the command file and names of two output files. A sample command file suitable for PBCV-1 is shown in **Fig. 10** (*regu.com*). A limited explanation is provided in the figure; full explanations of all the parameters are available in the FANTOM

```
CLUSTAL W(1.60) multiple sequence alignment

PBCV_1      TRVNLVPVQELADQHLMAEFRELKMIPKALARSLRTQSSEKILKKIPSKFTLNTGHVLF
2END        TRINLTLVSELADQHLMAEYRELPRVFGAVRKHVAN-GKRVRDFKISPTFILGAGHVTF
            ** ** * ******** ***    . * . . .      **   * * .. *** *

PBCV-1      FYDKGKYLQQRYDEIVVELVDRGYKINVDAKLDPDNVMTGEWYNDYTPTEDAFNIIRARI
2END        FYDKLEFLRKRQIELIAECLKRGFNIKDTTVQDISDIPQ-EFRGDYIPHEASIAISQARL
            ****  .  .. *  . *  * .  **.  *       *   .* *  * * . ** .

PBCV-1      AEKIAMKPSFYRFTKAKTSNN
2END        DEKIAQRPTWYKYYGKAIYA-
            **** .  . *  * . .
```

Fig. 5. The sequences of PBCV-1 and 2END aligned by ClustalW. This file, named *pbcv1_2end.aln*, will be used as input for EXDIS.

```
exdis
exdis> homology
Sequence alignment file: pbcv1_2end.aln
PDB file of the template molecule: 2end.pdb
Library file: (supply name of the FANTOM library file here)

Define fragments of model sequence that will be     (See Note 3)
used to generate dihedral and distance constraints.
Format: residue_number <space> residue_number
Type "0 0" to take entire sequence into account,
or to end fragment definition.
Enter pairs of numbers:
0 0
Type core name of the output file.
Default extensions will be provided by program: pbcv1
Interval for dihedral angle constraints (deg): 15  (See Note 4)
Dihedral angle constraints file pbcv1.aco
Number of distance constraints per atom: 10          (see Note 5)
exdis> end
```

Fig. 6. A typical EXDIS session. User response is in bold letters and comments are in italics.

```
2 THR    CHI1      -77.634     -47.634
2 THR    PSI       117.043     147.043
3 ARG    OMEGA    -185.000    -175.000
3 ARG    PHI      -150.854    -120.854
3 ARG    CHI1     -193.635    -163.635
3 ARG    CHI2      163.226     193.226
3 ARG    CHI3       51.850      81.850
3 ARG    CHI4     -156.809    -126.809
3 ARG    CHI5      -13.138      16.862
3 ARG    PSI       113.626     143.626
```

Fig. 7. Part of the dihedral angle constraint file (*.aco*) for PBCV-1 homology modeling from EXDIS.

```
1 THR
          N        11 LEU   O        18.32
          N        28 LYS   C        13.10
          N        34 LEU   CA       20.21
          N        44 LYS   CA       17.64
          N        62 ASP   N        16.02
          N        69 GLN   N        18.40
          N        83 TYR   C        22.39
          N        98 THR   CA       24.14
          N       121 GLU   N        16.75
          N       131 ARG   O        19.78
          N       135 ALA   C        25.35
```

Fig. 8. Part of a distance constraint file (*.upl*) for PBCV-1 homology modeling from EXDIS.

```
#PBCV-1_
# Angle file pbcv1.seq generated by exdis
#141                     Energy:           0.0
   1 MET    PHI   -180.000 CHI1  -180.000 CHI2  -180.000 CHI3  -180.000
           CHI4   -180.000 PSI   -180.000
   2 THR    OMEGA -180.000 PHI   -180.000 CHI1   -62.634 CHI21 -180.000
           CHI22  -180.000 PSI    132.043
   3 ARG    OMEGA -180.000 PHI   -135.854 CHI1  -178.635 CHI2   178.226
           CHI3     66.850 CHI4  -141.809 CHI5     1.862 CHI61 -180.000
           CHI62  -180.000 PSI    128.626
   4 VAL    OMEGA -180.000 PHI   -105.445 CHI1  -180.000 CHI21 -180.000
           CHI22  -180.000 PSI    129.426
   5 ASN    OMEGA -180.000 PHI   -103.619 CHI1   -55.832 CHI2   -54.694
           CHI32  -180.000 PSI    164.818
```

Fig. 9. Part of an angle (*.seq* or *.ang*) file for PBCV-1 homology modeling from EXDIS. This and the previous two figures are formatted as FANTOM input files.

```
# Fantom command file (regu.com) for regularization of PBCV-1 ----------
#
library             (specify here the location of the FANTOM library file)
group               NH3+ CPDB-
cutoff              distance 8.0 90
cutoff              surface 100.0 list 0
smooth              2.0
dielectric          proportional
construct           pbcv1.seq
dhconstraint        pbcv1.aco
diconstraint        upper pbcv1.upl lower pbcv1.upl
pdb                 pbcv1_0.pdb              (initial structure)
physical            none             (no conformational energy)
energy
summary             1
demin               0.0 0.0
gradient            0.01
lambda              500.0 50.0
sigma               0.2
rho                 0.2
tau                 0.01
#----Phase I of regularization------
switch              quasi info
loweight            4 0.2
upweight            4 0.2
dhweight            10.0 5.0
minimize            200
summary             1
#----Phase II of regularization-----
switch              newton 20
minimize            500
summary             1
sequence            pbcv1_1.ang             (regularized structure)
pdb                 pbcv1_1.pdb             (regularized structure)
#----End of regularization    -----
end
```

Fig. 10. A typical command file (*regu.com*) for structure regularization with FANTOM with explanations for some of the parameters.

manual. A typical session is reproduced in **Fig. 11**. Details of the run are written into the *ovw*, *pcl*, and *summary* files, and the structure at the end of regularization into *pbcv1_1.ang* (the molecule in terms of dihedral angles) and equivalently, into

```
%fantom
                 ***************
                 * F A N T O M *
                 ***************
        Refinement and Monte Carlo Simulations of Proteins
              Program Version 4.2 - Double Precision

        name of command file  (CR for interactive use): regu.com
        name of overview file (CR for standard output): regu.ovw
        name of protocol file (CR if not needed): regu.pcl
```

Fig. 11. A typical FANTOM session, to launch a job for regularization of a structure, as seen on the computer screen.

*pbcv1_1.pdb* (Cartesian coordinates). The *ang* file is used to continue the FANTOM calculation, and the *pdb* file can be opened in graphics programs, such as *MOLMOL*, to view the structure. The regularization or minimization of total constraint energy is done by quasi-Newton minimization followed by Newton-Raphson minimization. The computer time needed will depend, primarily, on the number of residues in the protein, and on the parameters chosen. The run specified by the command file of **Fig. 10** will need about 1 h of CPU time on an SGI Indigo2 workstation.

### 3.4.2. Constrained Minimization

Run a FANTOM session similar to that in **Subheading 3.4.1.**; but specify in the command file (e.g., **Fig. 12**) the parameters for constrained minimization. This *.com* file runs constrained minimization in three stages, progressively reducing the weights on the constraints. A maximum of 50 steps of quasi-Newton and 200 steps of Newton-Raphson minimization are run at each stage. Minimization at any stage would also be stopped if the gradient of total energy (with respect to dihedral angles) falls below 0.01, even if the number of maximum steps has not exceeded. The CPU time for this example is ~2 h. The minimized structure will be written into *pbcv1_4.ang*. See the file *min1.com* (**Fig. 12**) for details.

### 3.4.3. Unconstrained Minimization

Remove the constraints to the template and allow the structure to minimize freely in the final FANTOM run. A minimization protocol, core name *min2,* is shown in **Fig. 13**. This run will need about 1 h of CPU time. The final minimized structure is written into *pbcv1_5.ang* and *pbcv1_5.pdb*. The three FANTOM runs can also be combined into one large job using the UNIX script shown in **Fig. 14**.

The two candidate models for pbcv-1 are in the files *pbcv1_4.pdb* and *pbcv1_5.pdb*. The energy contributions and the values of gradients are written at specified intervals during FANTOM runs into *summary* files, usually of the form *min1.su1*. Check these summary files to insure that the values for the various energy contributions are decreasing, and that the gradient is low and leveling off to convergence. The pdb files are convenient for visualizing the model with graphics programs, making plots, and for analyzing the robustness of the model. Our FANTOM calculations were run on an SGI O2 workstation. **Figure 15** compares the structure of the PBCV-1 model to the Endo V template.

### 3.5. Analysis of a 3D Model

#### 3.5.1. Structural Validation

The model should be structurally validated and its robustness verified (this is a different issue than determining whether the model is "correct" from a functional standpoint). Use the program PROCHECK *(36)* to ensure that the backbone dihedral angles ϕ and ψ and the peptide dihedral ω (that controls the planarity of the peptide unit) at each residue are within the allowed ranges, and that there are no interatomic clashes. The summary files from PROCHECK can be plotted. Then use our group's program GETAREA *(37)*, available on the web at http://www.scsb.utmb.edu/getarea, to ensure the exposed hydrophobic surface area of the protein is low, as would be expected for a correctly folded protein.

#### 3.5.2. Test for Consistency of the Model with Experimental Data

Open the model in a graphics program and check for consistency with all known experimental data to verify functional viability. For example, are known glycosylation sites, epitopes, and intermolecular interaction sites surface-exposed? Are residues implicated in catalysis near each other in space?

### 3.6. Using MASIA to Predict the Secondary Structure of the Stylar RNases

MASIA is a program that interprets a set of aligned sequences to predict secondary structures and exposed and buried residues. The dihedral angle constraints and distance constraints extracted from MASIA results can be used in preparing a 3D model of a protein for which no template exists by running DIAMOD in a self-correcting mode. In this example, we use a list of amino-acid sequences available for the plant stylar RNases (*see* Chapters 6 and 7 for more information on these proteins). The C-terminal sequences of 65 different stylar alleles and related proteins were obtained from Dr. Adam Richman *(38)*.

1. Prepare a ClustalW alignment (or Pileup alignment if using the GCG package) as input for MASIA. The complete sequence alignment, essentially identical to that in **ref. *38***, is not shown here.
2. Open the MASIA program at the web site and use the "Browse" key to open the aligned sequence file in the program. MASIA will prepare a colored alignment of the sequences; the consensus sequence and predicted secondary structure for the alignment can be obtained by using the appropriate buttons in the bottom half of the screen in the graphic user interface. **Figure 16** shows the printout with the secondary structure prediction for the stylar example.
3. Compare the predictions for secondary structure with any biophysical data available (e.g., circular dichroism or FT-IR *[39]*) before going on to build a model. The bottom line of **Fig. 16** shows the secondary structure elements of the corresponding sequences from the X-ray crystal determination for RNase Rh *(40)*, which shares sequence identity with the stylar family especially around the active site and the Cys residues involved in crosslinks. MASIA predicted all the major helical areas and some β-strands that are not seen in the RH crystal structure. Note that especially in cases such as this one where there are many homologues with widely differing sequences, MASIA detects areas of stable structural elements well. However, the secondary structure predictions of the program should, where possible, be combined with other biophysical data before building a model.

```
#----Fantom command file min1.com for restrained energy minimization
library          (specify the location of the FANTOM library file)
group            NH3+ CPDB-
cutoff           distance 10.0 50
smooth           2.0
dielectric       proportional
construct        pbcv1_1.ang                    (regularized strucure)
dhconstraint     pbcv1.aco
diconstraint     upper pbcv1.upl lower pbcv1.upl
demin            0.0 0.0
gradient         0.01
lambda           500.0 50.0
sigma            0.2
rho              0.2
tau              0.01
physical         ecepp
energy           cutoff
summary          1
energy           full
summary          2
# ---------step 1 -----------------------------
loweight         4 0.5
upweight         4 0.5
dhweight         1.0 5.0
switch           quasi info
minimize         50
energy           cutoff
summary          1
energy           full
summary          2
# ---------step 2 -----------------------------
switch           newton 20
minimize         200
energy           cutoff
summary          1
energy           full
summary          2
sequence         pbcv1_2.ang                    (partially minimized)
pdb              pbcv1_2.pdb
# ---------step 3 -----------------------------
loweight         4 1.0
upweight         4 1.0
dhweight         1.0 5.0
switch           quasi info
minimize         50
energy           cutoff
summary          1
energy           full
summary          2
```

Fig. 12. A sample command file for restrained energy minimization by FANTOM (*min1.com*).

### 3.7. Building a Homology Model of Chlorella Glycosylase Using DIAMOD

Our distance geometry program DIAMOD *(10,11)* was used to construct the homology model of PDG in complex with UV-damaged DNA. The template was the crystal structure of Endo V Glu23Gln mutant complexed with a duplex DNA substrate con-

```
# ----------step 4 ------------------------------
switch               newton 20
minimize             200
energy               cutoff
summary              1
energy               full
summary              2
sequence             pbcv1_3.ang          (partially minimized)
pdb                  pbcv1_3.pdb
# ----------step 5 ------------------------------
loweight             4 1.0
upweight             4 1.0
dhweight             1.0 10.0
switch               quasi info
minimize             50
energy               cutoff
summary              1
energy               full
summary              2
# ----------step 6 ------------------------------
switch               newton 20
minimize             500
energy               cutoff
summary              1
energy               full
summary              2
sequence             pbcv1_4.ang          (minimized structure)
pdb                  pbcv1_4.pdb          (minimized structure)
end
```

Fig. 12. Part II.

taining a *cis-syn* cylcobutane pyrimidine dimer determined at 2.75Å resolution (entry *1vas* in the PDB *[4]*).

1. Align the amino-acid sequences with the program ClustalW *(34)* or other alignment program such as the Pileup program from the GCG packet. PDG from *Chlorella* virus *(41)* aligns with EndoV (Glu23Gln) with only two small gaps.

2. Decide on a numbering scheme that will allow easy comparison of the model to the template structure and determine how to handle gaps in the sequences (here, PDG residues were numbered according to the EndoV structure, with two additional residues in PDG numbered Lys45' and Val86'). DNA should be numbered from 5' to 3', so here the strand containing the pyrimidine dimer was numbered 201–213 and the opposite strand was numbered 214–226. For complexes of proteins with nucleic acids, use the one-letter code followed by the residue number (e.g., A221 is the "flipped out" adenine) for the nucleic acids and the three letter code for the proteins.

3. Decide what constraints can be derived from experimental evidence. In this example, residues known to be in the active site of Endo V (particularly the N-terminal residue Thr2 and the area around Glu23) and areas critical for the recognition of the cyclobutane pyrimidine dimer are highly conserved in PDG.

4. Then use EXDIS (**Subheading 3.3.**) to extract distance and dihedral angle constraints from the X-ray crystal coordinates. In this example, the DNA distance constraints were extracted between all the heavy atoms of a base and the heavy atoms in bases ≥4 bases away within each strand and between P, C4, C1', and C4' of each base pair between strands.

**#----Fantom command file <u>min2.com</u> for unrestrained minimization**

```
library              (specify the location of the FANTOM library file)
group                NH3+ CPDB-
cutoff               distance 10.0 50
cutoff               surface 50.0 list 0
smooth               2.0
dielectric           proportional
construct            pbcv1_4.ang                    (minimized structure)
demin                0.0 0.0
gradient             0.01
lambda               500.0 50.0
sigma                0.2
rho                  0.2
tau                  0.01
physical             ecepp
energy               cutoff
summary              1
energy               full
summary              2
# -------step 7 ----------------------------------------------------
switch               quasi info
minimize             50
energy               cutoff
summary              1
energy               full
summary              2
# -------step 8 ----------------------------------------------------
switch               newton 20
minimize             500
energy               cutoff
summary              1
energy               full
summary              2
# ------------------------------------------------------------------
sequence             pbcv1_5.ang                    (fully minimized structure)
pdb                  pbcv1_5.pdb                    (fully minimized structure)
end
```

Fig. 13. A sample command file for unrestrained energy minimization by FANTOM (*min2.com*).

For the pyrimidine dimer PD, the distances between the C5 and C6 of one thymine and O2, O4, C5, and C6 of the other thymine were extracted. The X-ray coordinates of PD in the crystal structure of Endo V were added to the library of DIAMOD to reproduce that local conformation in the model.

5. Extract distance constraints for the side chains for the regions that are ~50% identical in the two proteins. For this example, there were three regions of high identity. Thr2-Arg36 (51%), Ile46-Ile86 (48%), and Asp103-Ala126 (50%). The resulting distance constraints include the $C_\alpha$-$C_\alpha$ distances for residues ≥2 apart and the $C_\beta$-$C_\beta$ distances between residues ≥3 apart in each of the three regions, and the $C_\alpha$-$C_\alpha$ distances between residues (in different regions), which are identical in the two proteins. Distances ≤16 Å between the P atoms of the DNA and $C_\alpha$ atoms of the protein within the three regions of the PDG sequence were used to constrain the relative position of protein and DNA substrate in the complex. The resulting ~6000 distance constraints were used to set the limits of permis-

```
#!/bin/csh
fantom <<
EOF
regu.com
regu.ovw
regu.pcl
EOF
fantom <<
EOF
min1.com
min1.ovw
min1.pcl
EOF
fantom <<
EOF
min2.com
min2.ovw
min2.pcl
EOF
```

Fig. 14. A *unix* script for running the regularization and the constrained and unconstrained FANTOM minimizations of PBCV-1 together as one job in batch mode.

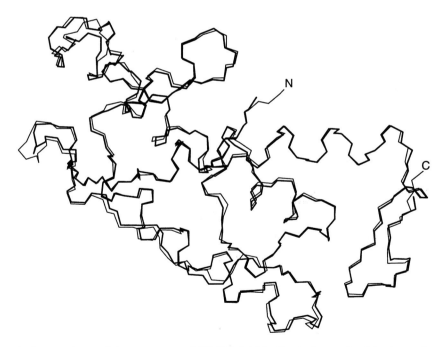

Fig. 15. Comparison of the structure of PBCV-1 (thin line), determined by homology modeling (**Subheading 3.4.**), with the crystal structure of Endo-V (PDB file 2END). To prepare the drawing shown, both files were opened in MOLMOL and were superimposed by specifying the two fragments used in modeling (RMSD = 0.41 Å). The line width of the Endo V structure was specified as 2, whereas that for the model was left at the default value of 1. The N- and C-termini of the model are marked.

```
sequence WPQGSG--TSLTNCPQ--GSPFDITK--ISHLQSQLTLWPNVLRANN-----QQFWSHEW
beta     B           BBB B       BB B      B   B    BBBB B       B BB  BB
                                                      BBBBBBB
alph     H H      HH         HHH    HHH  H H H HHH  HH       H HHH H
                             HHHHHHHHHHHHHHHHHHHHHHHHHHH      HHHHHHHHH
turn       TTT T    TTTT   TTT     TT   TT TT      T T        T  TT T
                    TTTT
insd     i         ii  i         i      ii  i  i  i i          ii
         i         ii  i         i      ii  i  i  i i          ii
outs       ooo o  o     oo  ooo oo o  oo   oo oo   o o ooo    oo   o o
           ooo o  o     oo  ooo oo o  oo   oo oo   o o ooo    oo   o o
---------------------------------------------------------------------
insd     i         ii  i         i      ii  i  i  i i          ii
outs       ooo o  o     oo  ooo oo o  oo   oo oo   o o ooo    oo   o o
SSP                 TTTT    HHHHHHHHHHHHHHHHHHHHHHHHHHH      HHHHHHHHH
sequence WPQGSG--TSLTNCPQ--GSPFDITK--ISHLQSQLTLWPNVLRANN-----QQFWSHEW
                               α₁       α₂                      α₃
```

```
sequence TKHGTCSE-STFNQAAYFKLVDMRNN-YDIIGALRPHAAGPNGR-TKSRQAIKGFLKAKF
beta        BBBB  BB B BBB   B B   B B  B      BB      BBB         BB
            BBBBBB    BBBBB
alph     HH      H H   H H H HHH HH  H HH  H    H        H HHH H H
         HHH           HHHHHHHHHHHHHHHHHHHHH          HHHHHHHHHH
turn      T T TT   T TT T T        TT  T       T T  TT T  T  T    T T
insd        ii  i        i i i i   i ii  i      ii        i   i   i i
            ii  i        i i i i   i ii  i      ii        i   i   i i
outs     oo         o  ooo      o oo  o  oo oo    o o    o  o o  oo o
         oo         o  ooo      o oo  o  oo oo    o o    o  o o  oo o
---------------------------------------------------------------------
insd        ii  i        i i i i   i ii  i      ii        i   i   i i
outs     oo         o  ooo      o oo  o  oo oo    o o    o  o o  oo o
SSP      HHHBBBBBB    HHHHHHHHHHHHHHHHHHHHHHHHH        HHHHHHHHHH
sequence TKHGTCSE-STFNQAAYFKLVDMRNN-YDIIGALRPHAAGPNGR-TKSRQAIKGFLKAKF
                     α₄              α₅                      α₆
```

```
sequence G-KFPGLRC
beta             B
alph     H    HH
turn          TTT TT
insd              i
                  i
outs      o  ooo o
          o  ooo o
-------------------
insd              i
outs      o  ooo o
SSP
sequence G-KFPGLRC
```

Fig. 16. MASIA prediction of secondary structural elements from the aligned stylar RNase structures of Richman et al. *(38)*. The positions of alpha-helical regions of the RNase Rh structure, predicted by alignment with that protein, are indicated. SSP: prediction of secondary structure, where *H* is α-helical, *B* is β-sheet, and *T* is turn.

sible distances between the residues involved. In addition, dihedral angle constraints were extracted for each strand of the DNA substrate and those three regions of the protein.

6. Prepare the list of constraints to start DIAMOD distance geometry calculations. In this example, a modified target function for a violated distance constraint was constructed to prevent large violations of distance constraints from dominating the nonlinear fit procedure *(11)*. DIAMOD calculates structures from random initial conformations. For the PDG example, minimizations were performed with 37 levels for the variable target function *(16,42)* where the final minimization level covers all distance constraints.

## 3.8. Molecular Dynamics Simulations of the Model and the Template Crystal Structure

Molecular dynamics simulations of both the crystal structure of the Endo V complex and our model structure were then used to determine residues potentially responsible for the glycosylase's distinct substrate specificity. These "molecular movies" permit a more detailed analysis of catalysis than a crystal structure can provide, and offer provocative clues to how these enzymes establish their catalytic platform within damaged DNA.

1. Perform energy minimization on the model resulting from the DIAMOD calculations with the CHARMm polar hydrogen force-field parameters *(18)* in vacuum for 500 steps, keeping backbone atoms fixed in position, followed by another 500 steps without restraints. The structure can be used for subsequent MD simulation in a water environment. For this example, the PDG/DNA complex was solvated first in a cubic box of water. Water molecules with close contacts to the solute atoms (within 2.8 Å) were excluded from the solution. All water molecules beyond 10 Å from the heavy solute atoms were deleted. The initial configuration of the system consisted of 2015 atoms from the protein/DNA complex and 2485 water molecules.

2. Calculate geometries and charges by quantum chemistry for any irregular bases or amino acids which are not available in the structure file. For the *cis-syn* UV-induced PD, modified atom types, partial charges, and force-field parameters were prepared for the simulations. The geometry of an *N*-methanolated model molecule with PD conformation was optimized with AM1 using MOPAC93 *(33)* followed by Hartree-Fock calculations with the 6-31G* basis set *(35)*. Water molecules were treated as TIP3P residues *(43)*.

3. Use the X-PLOR program *(32)* with the CHARMM polar hydrogen force-field parameters *(18)* for molecular dynamics simulations of the solvated complex.

4. Minimize the assemblies prior to dynamics runs to relax the local strain in the initial configurations. For this example, 500 steps of conjugated gradient minimization were performed with all heavy atoms fixed, and then 500 minimization steps followed without constraints. Nonbonding interactions were cut off at 10 Å. To prevent water evaporation during the MD simulation runs, a harmonic potential with a 2.0 kcal/mol force constant was used to restrain the water molecules in the 4 Å outer shell. All simulations used a 1 *fs* time step and assumed a dielectric constant of 1.0. All covalent bonds to hydrogen were constrained by the SHAKE algorithm *(44)* with a tolerance of $1 \times 10^{-6}$ nm. The MD simulations on the entire system started with assigned random velocities, and gradually warmed up to 300 K using 20 K temperature steps of 30 ps. The temperature was held constant by means of the Berendsen coupling algorithm *(45)* with a coupling constant of 0.2 ps$^{-1}$. An additional 50 ps equilibration followed, to stabilize the system. During heating and equilibration runs for PDG/DNA system, hydrogen-bonding distance constraints for base pairs in DNA were applied with square-well potential at $2.0 \pm 0.3$ Å, and then were removed during the MD simulations. No such constraints were used during simulations of the Endo V com-

plex. A 300 ps simulation was performed at 300 K as production run. Configurations were recorded every 0.5 ps (500 steps). For MD simulations of PDG-DNA alone in an aqueous environment, the starting configuration was that in the EndoV crystal complex, with A221 flipped out. A 15 Å water shell with 5 Å outer layer of constrained water molecules was used for 500 ps of simulations at 300 K. Configurations were recorded every 1 ps.

5. Obtain the dynamic averaged structure by averaging all frames of 150~300 ps trajectory, followed by 200 energy minimization steps with all heavy atoms fixed, 200 steps with backbone atoms fixed, and 1000 steps with no constraints. In this example, for the dynamic averaged structures, all water molecules were removed, and a dielectric constant of 80 was used for the energy minimizations.

6. Calculate RMSD values with MOLMOL *(31)* or a similar graphics program, to create molecular graphics and calculate contact distances between parts of the complex.

## 4. Notes

1. Comparing structures: The most common measure for comparing structures (e.g., a model with the experimentally determined structure, or models among themselves) is the root mean square deviation (RMSD) of atomic coordinates. An RMSD can be calculated using any selected number of equivalent atoms from the two structures being compared. The RMSD of backbone atoms will tell us how close the folds of the proteins are, and that of all heavy atoms indicates overall structural differences. The RMSD of a model from its native structure can in favorable cases be roughly estimated from the degree of identity between the two proteins *(12–14)*. A generally accepted view of the assessment of a model by RMSD values is that if the RMSD is <2 Å, the model can be used for functional studies, <4 Å, the structure can show what groups are on which side of the protein and can help in planning experiments, whereas a 6 Å model provides only the rough path of the polypeptide chain *(12)*.

   Two caveats should be noted when using RMSD values as a measure of how two structures differ from each other. A low RMSD between two structures, even of the order of 1–2 Å, does not guarantee the correct formation of α helices and β sheets. This is because their characteristic hydrogen bonds depend on the precise orientation of the hydrogen-bonding groups, which may not significantly affect RMSDs. An easy way to look for the presence of secondary structures is to use a graphics program. A second word of caution relates to high RMSDs, i.e., of the order of 5 Å or more. Despite the high RMSD, the two structures being compared may still have the same fold. Especially when there are large gaps between the sequences, a linear overlap of the two structures may be poor. In this case, "contact maps," which can be made automatically, for example, in MOLMOL *(31)* provide a more reliable indication of structural relatedness. A contact map is a plot of the interatomic distances for all pairs of atoms (e.g., using only α-carbons). The closer a contact, the darker its square in the plot. Thus comparing the contact maps of structures can reveal structural similarity, even when the RMSD between them is high.

2. The meaning of the "E" value in the table generated by FASTA: The FASTA sequence comparison program reports, in addition to the "raw" similarity score for each alignment, an E or "expected" value. E is an assessment of the statistical significance of the alignment and depends on its score (which varies depending on the gapping of the alignment and the number of identical and similar residues), the length of the query sequence and the size of the database searched. It represents the number of distinct alignments with equivalent or better score that can be expected to occur purely by chance in a database of that size *(46)*. Hence, an E value of 5 is not significant, but a value of 0.01 would be. To locate very short peptide sequences (5–6 amino acids long) in a database using a FASTA search, raise

the E-value allowed in the search from the default, as there is a high probability in a large database that these sequences will occur several times.

3. Analyzing and editing the ClustalW alignment results: The gaps in the alignment file produced by ClustalW correspond to residue insertions or deletions between the model sequence and the template. Gaps thus lead to "fragments" of aligned regions. We recommend leaving one or two residues out, i.e., making such fragments shorter, at each end, to account for the inevitable conformational differences between the model and template in these regions. The EXDIS session (**Fig. 6**) shows how to define fragments. One must occasionally manually edit the ClustalW alignment file before running EXDIS, e.g., to align known active-site residues. Other justifiable changes in a given alignment would be to improve the matching of Cys residues known to form disulfide bridges. **Note 4** describes how some changes must be made in the EXDIS file when Pro residues are not well-aligned.

4. Readying the EXDIS output files for FANTOM run: After running EXDIS, depending on the protein being modeled, it may be necessary to edit the output *aco* (dihedral angle constraint file) or *seq* (angle file), before using them as input for FANTOM. To keep the peptide units planar, allow ω, the peptide torsion angle, to deviate within a smaller range (such as ± 5°) than the other dihedral angles (where deviation can be, for example, ± 15°).

   Proline residues are a special problem, as they can exist in a *cis* or *trans* configuration. If the template contains an X-*cis*-Pro peptide bond, EXDIS will recognize this and set the corresponding ω to 0° in the *seq* file and to –5° to +5° in the *aco* file. However, if there is a Pro in the test sequence but not in the corresponding position in the template, the user can choose to define ω to 0° or to 180°, depending on what other evidence is available about this residue. For example, there may be a structure for a protein more distantly related to the test sequence which may contain a *cis*-Proline at this position. The user can also choose to remove angle constraints completely for the Pro and allow the angles in the structure to be determined by the program. Finally, the number of disulfide bridges must be specified in each *com* file used for running FANTOM. The command for this, *disulfide*, should be issued before the *construct* command.

5. Number of distance constraints per atom assigned in EXDIS: More distance constraints lead to a better constrained model, but will also increase the computation time needed. We typically use 10 distance constraints per atom.

## Acknowledgments

We acknowledge the help of Cynthia J. Orlea in the preparation of this manuscript. This work was supported by the National Science Foundation (DBI-9714937), the Department of Energy (DE-FG03-96ER62267), and the John Sealy Memorial Endowment Fund.

## References

1. Higgins, D. G., Bleasy, A. J., and Fuchs, R. (1992) CLUSTAL_W: improved software for multiple sequence alignment. *Comp. Appl. Biosc.* **8,** 189–191.
2. Baxevanis, A. D. (1998) in *Bioinformatics: A Practical Guide to the Analysis of Genes and Proteins* (Baxevanis, A. D. and Ouellette, B. F. F., eds.), Wiley-Interscience, New York, pp. 172–188.
3. Schein, C. H. (1997) From housekeeper to microsurgeon: the diagnostic and therapeutic potential of ribonucleases. *Nature Biotechnology* **15,** 529–536.
4. Bernstein, F., Koetzle, T., Williams, G., Meyer, E., Jr., Brice, M., Rodgers, J., Kennard, O., Shimanouchi, T., and Tasumi, M. (1977) The Protein Data Bank: a computer-based archival file for macromolecular structures. *J. Mol. Biol.* **112,** 535.

5. Schein, C. H. (1999) Protein aggregation degradation, in *Encyclopedia of Bioprocess Technology: Fermentation, Biocatalysis, and Bioseparation* (Flickinger, M. C. and Drew, S. W., eds.), Wiley, New York, pp. 2134–2155.

6. Sanchez, R. and Sali, A. (1997) Evaluation of comparative structure modeling by MODELLER-3. *Proteins Suppl.* **1,** 50–58.

7. Johnson, M. S., May, A. C. W., Rodionov, M. A., and Overington, J. P. (1996) in *Computer Methods for Macromolecular Sequence Analysis* (Doolittle, R. F., ed.), Academic Press, San Diego, CA, pp. 575–598.

8. Bowie, J. U., Zhang, K., Wilmanns, M., and Eisenberg, D. (1996) in *Computer Methods for Macromolecular Sequence Analysis* (Doolittle, R. F., ed.), Academic Press, San Diego, CA, pp. 598–616.

9. Parry, S., Newbigin, E., Craik, D., Nakamura, K. T., Bacic, A., and Oxley, D. (1998) Structural analysis and molecular model of a self-incompatibility RNase from wild tomato. *Plant Physiol.* **116,** 463–469.

10. Mumenthaler, C. and Braun, W. (1995) Predicting the helix packing of globular proteins by self-correcting distance geometry. *Protein Sci.* **4,** 863–871.

11. Hänggi, G. and Braun, W. (1994) Pattern recognition and self-correcting distance geometry calculations applied to myohemerythrin. *FEBS Lett.* **344,** 147–153.

12. Koehl, P. and Levitt, M. (1999) A brighter future for protein structure prediction, *Nat. Struct. Biol.* **6,** 108–110.

13. Abagyan, R. A. and Batalov, S. (1997) Do aligned sequences share the same fold? *J. Mol. Biol.* **273,** 355–368.

14. Chothia, C. and Lesk, A. M. (1986) The relation between the divergence of sequence and structure in proteins. *EMBO J.* **5,** 823–826.

15. Pearson, W. R. and Lipman, D. J. (1988) Improved tools for biological sequence analysis. *Proc. Natl. Acad. Sci. USA* **85,** 2444–2448.

16. Güntert, P., Braun, W., and Wüthrich, K. (1991) Efficient computation of three-dimensional protein structures in solution from NMR data using the program DIANA and the supporting programs CALIBA, HABAS, and GLOMSA. *J. Mol. Biol.* **217,** 517–530.

17. Zhu, H. and Braun, W. (1999) Sequence specificity, statistical potentials and 3-D structure prediction with self-correcting distance geometry calculations of beta-sheet formation. *Protein Sci.* **8,** 1–17.

18. Brooks, B., Bruccoleri, R., Olafson, B., States, D., Swaminathan, S., and Karplus, M. (1983) CHARMM: a program for macromolecular energy, minimization, and molecular dynamics calculations. *J. Comp. Chem.* **4,** 187–217.

19. Abe, H., Braun, W., Noguti, T., and Go, N. (1984) Rapid calculation of first and second derivatives of conformational energy with respect to dihedral angles for proteins. General recurrent equations. *Comp. Chem.* **8,** 239–247.

20. Schaumann, T., Braun, W., and Wuthrich, K. (1990) The program FANTOM for energy refinement of polypeptides and proteins using a Newton-Raphson minimizer in torsion angle space. *Biopolymers* **29,** 679–694.

21. Zhu, H., Schein, C. H., and Braun, W. (1999) Homology modeling and molecular dynamics simulations of PBCV-1 glycosylase with UV-damaged DNA. *J. Mol. Modeling* **5,** 302–316.

22. Lee, B. J., Sakashita, H., Ohkubo, T., Ikehara, M., Doi, T., Morikawa, K., Kyogoku, Y., Osafune, T., Iwai, S., and Ohtsuka, E. (1994) Nuclear magnetic resonance study of the interaction of T4 endonuclease V with DNA. *Biochemistry* **33,** 57–64.

23. Koc, O. N., Phillips, W. P. J., Lee, K., Liu, L., Zaidi, N. H., Allay, J. A., and Gerson, S. L. (1996) Role of DNA repair in resistance to drugs that alkylate O6 of guanine. *Cancer Treat. Res.* **87,** 123–146.

24. Lloyd, R. S. and Cheng, X. (1997) Mechanistic link between DNA methyltransferases and DNA repair enzymes by base flipping, *Biopolymers* **44,** 139–151.

25. Lloyd, R. S. (1998) Base excision repair of cyclobutane pyrimidine dimers. *Mutation Res.* **408,** 159–170.

26. Vassylyev, D. G., Kashiwagi, T., Mikami, Y., Ariyoshi, M., Iwai, S., Ohtsuka, E., and Morikawa, K. (1995) Atomic model of a pyrimidine dimer excision repair enzyme complexed with a DNA substrate: structural basis for damaged DNA recognition. *Cell* **83,** 773–782.

27. Vassylyev, D. G. and Morikawa, K. (1997) DNA-repair enzymes. *Curr. Opin. Struct. Biol.* **7,** 103–109.

28. Cheng, X. (1995) DNA modification by methyltransferases. *Curr. Opin. Struct. Biol.* **5,** 4–10.

29. Cheng, X. and Blumenthal, R. M. (1996) Finding a basis for flipping bases. *Structure* **4,** 639–645.

30. McCullough, A. K., Romberg, M. T., Nyaga, S., Wei, Y., Wood, T. G., Taylor, J. S., Van Etten, J. L., Dodson, M. L., and Lloyd, R. S. (1998) Characterization of a novel *cis-syn* and *trans-syn-II* pyrimidine dimer glycosylase/AP lyase from a eukaryotic algal virus, *Paramecium bursaria chlorella* virus-1. *J. Biol. Chem.* **273,** 13,136–13,142.

31. Koradi, R., Billeter, M., and Wüthrich, K. (1996) MOLMOL: a program for display and analysis of macromolecular structures. *J. Mol. Graphics* **14,** 51–55.

32. Brünger, A. T. (1992) *X-PLOR, Version 3.1. A System for X-ray Crystallography and NMR*, Yale University Press, New Haven, CT.

33. Steward, J. J. P. (1993) MOPAC Manual. Fujitsu Limited, Tokyo, Japan.

34. Thompson, J. D., Higgins, D. G., and Gibson, T. J. (1994) CLUSTAL W: improving the sensitivity of progressive multiple sequence alignment through sequence weighting, positions-specific gap penalties and weight matrix choice. *Nucleic Acids Res.* 4673–4680.

35. Frisch, M. J., Trucks, G. W., Schlegel, H. B., Gill, P. M. W., Johnson, B. G., Robb, M. A., et al. (1995) Gaussian Inc., Pittsburgh, PA.

36. Laskowski, R. A., MacArthur, M. W., Moss, D. S., and Thornton, J. M. (1993) PROCHECK: a program to check the stereochemical quality of protein structures. *J. Appl. Crystallogr.* **26,** 283–291.

37. Frackiewicz, R. and Braun, W. (1998) Exact and efficient analytical calculation of the accessible surface areas and their gradients for macromolecules. *J. Comput. Chem.* **19,** 319–333.

38. Richman, A. D., Uyenoyama, M. K., and Kohn, J. R. (1996) Allelic diversity and gene genealogy at the self-incompatibility locus in the Solanaceae. *Science* **273,** 1212–1216.

39. Schein, C. H. (1990) Solubility as a function of protein structure and solvent components. *Bio/Technology* **8,** 308–317.

40. Kurihara, H., Nonaka, T., Mitsui, Y., Ohgi, K., Irie, M., and Nakamura, K. T. (1996) The crystal structure of the ribonuclease from Rhizopus Niveus at 2.0A resolution. *J. Mol. Biol.* **255,** 310–320.

41. Lu, Z., Li, Y., Zhang, Y., Kutish, G. F., Rock, D. L., et al. (1995) Analysis of 45 kb of DNA located at the left end of the chlorella virus PBCV-1 genome. *Virology* **206,** 339–352.

42. Braun, W. and Go, N. (1985) Calculation of protein conformations by proton-proton distance constraints. *J. Mol. Biol.* **186,** 611–626.

43. Jorgensen, W., Chandrasekar, J., Madura, J., Impey, R., and Klein, M. (1983) Comparison of simple potential functions for simulating liquid water. *J. Chem. Phys.* **79,** 926–935.

44. Ryckaert, J.-P., Ciccotti, G., and Berendsen, H. J. C. (1977) Numerical-integration of cartesian equations of motion of a system with constraints—molecular-dynamics of N-Alkanes. *J. Comput. Phys.* **23,** 327–341.
45. Berendsen, H. J. C., Postma, J. P. M., van Gunsteren, N. F., DiNola, A., and Haak, J. R. (1984) Molecular dynamics with coupling to an external bath. *J. Chem. Phys.* **81,** 3684–3690.
46. Altschul, S. F. (1998) Fundamentals of database searching, in *Trends Guide to Bioinformatics* (Patterson, M. and Handel, M., eds.), Elsevier Science, New York, pp. 6–8.

# 19

## Methods for Determining Activity and Specificity of DNA Binding and DNA Cleavage by Class II Restriction Endonucleases

### Albert Jeltsch and Alfred M. Pingoud

### 1. Introduction to the Structure and Function of Class II Restriction Enzymes

Restriction endonucleases coupled with DNA methyltransferases form the restriction-modification (RM) systems that occur ubiquitously among bacteria. They protect bacterial cells against bacteriophage infection by cleaving incoming foreign DNA highly specifically if it contains the recognition sequence. Cellular DNA is protected from cleavage by a specific methylation within the recognition sequence, which is introduced by the methyltransferase (for review, *see* **refs.** *1,2*). Restriction endonucleases recognize palindromic recognition sites, 4–8 base pairs in length. These enzymes are indispensable tools for genetic engineering. The biology and biochemistry of type II restriction endonucleases has been reviewed recently *(3,4)* and will be summarized only briefly here. Type IIS restriction enzymes differ from type II enzymes in that they recognize an asymmetric recognition sequence (for review, *see* **ref.** *5*). Monomeric in solution, these enzymes consist of a DNA recognition domain and a catalytic domain *(6)*.

Restriction endonucleases initiate DNA cleavage by binding nonspecifically to DNA. Association of the enzymes to DNA is diffusion controlled, with rate constants in the order of $10^7$–$10^8$ $M^{-1}s^{-1}$ *(7,8)*. In a series of dissociation and association steps and/or sliding along the DNA, the enzyme scans the DNA and searches for the recognition site in a one-dimensional diffusion process. This process permits much faster target site location than would be possible via a three-dimensional search *(9–15)*. During recognition, conformational changes in the enzyme-DNA complex lead to the activation of the catalytic centers and cleavage of the DNA. Subsequently, the products are released, allowing for a new reaction cycle to take place. Often product release is the rate limiting step for multiple turnover cleavage reactions.

### 1.1. Structures of Restriction Endonucleases

The crystal structures of *Eco*RI *(16)*, *Eco*RV *(17)*, *Pvu*II *(18,19)*, *Bam*HI *(20,21)*, *Cfr*10I *(22)*, and *Bgl*I *(23)* have been determined. These enzymes all contain a central

From: *Methods in Molecular Biology, vol. 160: Nuclease Methods and Protocols*
Edited by: C. H. Schein © Humana Press Inc., Totowa, NJ

mixed five-stranded β-sheet flanked by two α-helices. Whereas the structural similarities between *Eco*RI and *Eco*RV are difficult to recognize *(24)*, they are quite pronounced between *Eco*RI and *Bam*HI *(20)*, *Eco*RV and *Pvu*II *(18,19)*, and *Eco*RV and *Bgl*I *(23)*. Moreover, the catalytic domain of the type IIS restriction enzyme *Fok*I also has this common fold *(25)*, and forms a dimer upon interaction with DNA *(26)*. Intriguingly, the catalytic centers of all restriction enzymes analyzed thus far are very similar with respect to the amino-acid residues involved, as well as secondary-structure elements on which these residues are located (for review, *see* **refs. *4,27***). However, the dimerization interfaces of the subunits of *Eco*RI and *Bam*HI, *Eco*RV and *Pvu*II, and *Eco*RV and *Bgl*I vary greatly, probably reflecting the structural constraints needed to position two active sites so that 5' overhangs (*Eco*RI and *Bam*HI), blunt ends (*Eco*RV and *Pvu*II), or 3' overhangs (*Bgl*I) are generated upon cleavage. These differences in their cleavage mode may also explain why *Eco*RI *(16,28)* and *Bam*HI *(21)* approach the DNA from the major groove but *Eco*RV *(17)*, *Pvu*II *(19)*, and *Bgl*I *(23)* contact the DNA via the minor groove.

### 1.2. DNA Recognition

One of the most fascinating aspects of restriction enzymes is their extraordinary specificity. Under optimal conditions, sites differing in only one base pair from the recognition sequence are cleaved by several orders of magnitude more slowly than the canonical site *(29–31)*. This specificity results from preferential interactions between the enzymes and substrates, including a multitude of contacts of the enzymes to the bases as well as to the phosphodiester backbone of the DNA (for review, *see* **ref. *4***) which differ qualitatively and quantitatively from the contacts to DNA seen in nonspecific complexes.

The crystal structures of the specific *Eco*RI, *Eco*RV, *Bam*HI, *Pvu*II, and *Bgl*I DNA complexes demonstrate that structural elements responsible for specific binding are interwoven with those involved in catalysis. The transition from nonspecific to specific binding leads to a highly cooperative restructuring of the protein–DNA interface. In the course of a mutual induced fit of enzyme and substrate, all groups participating in catalysis are brought into the required proximity and precise orientation to activate the catalytic centers. Thus, specific binding is coupled to catalysis in a transient process that can only be observed through kinetic experiments *(15,32–35)*.

### 1.3. Mechanism of DNA Cleavage

Cleavage of DNA by restriction endonucleases yields 3'-OH and 5'-phosphate ends *(36)*. Hydrolysis of the phosphodiester bonds occurs with inversion of configuration at the phosphorous atom *(37,38)*, suggesting an attack of a water molecule *in line* with the 3'-OH leaving group. In general, hydrolysis of phosphodiester bonds requires three functional entities *(39)*: a general base that activates the attacking nucleophile; a Lewis acid that stabilizes the extra negative charge in the pentacovalent transition state; and an acid that protonates or stabilizes the leaving group. The individual importance of each function depends on the actual mechanism of the reaction (associative vs dissociative or intermediate). The active sites of *Eco*RI and *Eco*RV, the first two restriction endonucleases for which the structure of an enzyme-DNA complex was determined, are very similar, and both contain a PD...(D/E)XK motif *(40,41)*. This motif, with modifications, can be found in other restriction endonucleases and related enzymes. The mechanism of DNA cleavage has not been elucidated unequivocally for any restriction endonuclease, but several different models have been proposed *(21,23,25, 39,42–44)*. In all of these models, the acidic amino-acid residues within the motif chelate a metal ion at the active center of the enzyme. Subsequent mechanistic details are less clear, and it has been speculated that additional metal ions, other amino-acid residues, and/or phosphate groups of the DNA may be actively involved in catalysis.

To study the molecular enzymology of restriction endonucleases, generally all steps of the cleavage reaction must be investigated. Methods to study the thermodynamics and kinetics of DNA binding (**Subheading 3.1.**) and the kinetics of DNA cleavage (**Subheading 3.2.**) are thus required. If small differences in the cleavage rates of different substrates must be detected, e.g., to measure linear diffusion of restriction enzymes on DNA, competitive cleavage assays are required (**Subheading 3.3.**).

## 2. Materials
### 2.1. Assays to Measure the Thermodynamics and Kinetics of Nonspecific and Specific Binding of Proteins to DNA
#### 2.1.1. Nitrocellulose Filter Assay

2.1.1.1. EQUIPMENT

1. Phosphorimager or scintillation counter.
2. Filtration device (we recommend a microtiter-plate format device, e.g., Bio-Dot, Biorad).
3. Microtiter plate.
4. Nitrocellulose membrane (e.g., Protran BA 85 Cellulosenitrat(E), Schleicher & Schuell, Dassel, Germany).
5. Protein at sufficient concentration and amount.

2.1.1.2. SOLUTIONS

1. Labeled DNA at concentrations between 0.1 p$M$ and 10 n$M$, depending on the protein concentrations to be used (either oligonucleotides labeled with polynucleotide kinase or PCR products prepared using $\alpha$-[$^{32}$P]ATP can be used. Product can be purified using kits such as QIAquick PCR purification™ or QIAquick Nucleotide Removal™ [Qiagen]).
2. 1X assay buffer (without BSA).
3. 10X assay buffer (containing approx 10–200 ng/µL BSA). The composition of the assay buffer depends on the protein to be investigated and must be optimized (start with: Tris-HCl or HEPES at 20–50 m$M$; pH 7.0–8.0; NaCl at 0–100 m$M$, $\beta$-ME or dithiothreitol (DTT) at 0–100 µ$M$). Some restriction enzymes require $Ca^{2+}$ for specific DNA binding.

#### 2.1.2. Electrophoretic Mobility Shift Assay

2.1.2.1. EQUIPMENT

1. Power supply.
2. PAGE chamber.
3. Polyacrylamide gel (typically 4–10% in 0.5X TTE).
4. Labeled DNA (either oligonucleotides labeled with polynucleotide kinase or PCR products prepared using $\alpha$-[$^{32}$P]ATP can be used) at concentrations between 0.1 p$M$ and 10 n$M$, depending on the protein concentrations to be used. The shift substrate must be free of contaminations. PCR products can be purified by electrophoretic separation on a polyacrylamide gel run in TPE (80 m$M$ Tris, 20 m$M$ EDTA, adjusted to pH 8.0 with $H_3PO_4$).

Excise product containing bands and elute the DNA with water. If necessary, the product can be concentrated, e.g., by ultrafiltration using centricon-30 tubes (Amicon), and desalted by gel filtration (e.g., NAP-columns, Pharmacia) or using a PCR purification kit (e.g., Qiaquick, Qiagen).

5. Protein at sufficient concentration and amount.

### 2.1.2.2. Solutions

1. 0.5X TTE buffer: 100 m$M$ Tris-HCl, 29 m$M$ taurine, 1.25 m$M$ EDTA, pH not adjusted.
2. 10X incubation buffer, depending on the protein to be investigated: Tris-HCl or HEPES at 20–50 m$M$, pH 7.0–8.0; NaCl at 0–100 m$M$; β-ME or DTT at 0–100 μ$M$. One may add BSA (about 100 ng/μL) and spermine (approx 2 m$M$). Some restriction enzymes need $Ca^{2+}$ for specific DNA binding.
3. 5X gel-loading buffer: 50% glycerol, 0.25% (w/v) xylene cyanol, 0.15% (w/v) chromotrope FB in 1X incubation buffer.

## 2.2. Assays to Measure the Kinetics of DNA Cleavage by Class II Restriction Enzymes

### 2.2.1. Cleavage of Macromolecular DNA

#### 2.2.1.1. Equipment

1. Power supply.
2. Agarose gel electrophoresis chamber.
3. Agarose gel (0.5–2%, depending on the size of the DNA fragments to be separated).
4. Substrate DNA (either λ-DNA or plasmid DNA at >0.1 μg/μL).

#### 2.2.1.2. Solutions

1. Running buffer, e.g., TPE: 80 m$M$ Tris, 20 m$M$ EDTA, adjusted to pH 8.0 with $H_3PO_4$.
2. 10X reaction buffer (as recommended by the supplier of the restriction enzyme).
3. 5X gel-loading buffer: 0.25 $M$ EDTA, 25% (v/v) Ficoll, 0.2% (w/v) xylene cyanol, 0.2% (w/v) bromophenol blue, 1% (w/v) SDS. For details concerning preparation, handling, and staining of agarose gels, refer to handbooks of molecular biology techniques.

### 2.2.2. Cleavage of Oligonucleotide Substrates Studied by Homochromatography

#### 2.2.2.1. Equipment

1. Phosphorimager.
2. Hair dryer.
3. Thin-layer chromatography buffer tank.
4. DEAE cellulose thin-layer plates (20 × 20 cm works well) (e.g., Polygram CEL 300 DEAE/HR-2/15 Macherey-Nagel, Düren, Germany).
5. Oligonucleotide substrate.

#### 2.2.2.2. Solutions

1. Running buffer (RNA lysate).
2. 10X reaction buffer (as recommended by the supplier of the restriction enzyme).

### 2.2.3. Analysis of Oligonucleotide Cleavage Reactions Studied by Denaturing Gel Electrophoresis

1. Phosphorimager.
2. PAGE chamber.

3. Power supply.
4. 20 × 20 cm denaturing polyacrylamide gel.
5. Oligonucleotide substrate.

### 2.2.3.1. SOLUTIONS

1. 0.5X TTE.
2. 10X reaction buffer.
3. Gel-loading buffer: 0.5X TTE, 250 m$M$ EDTA, 89.5% formamide, 0.5% (w/v) chemotrope FB.

## 2.3. Competitive Cleavage Assays

### 2.3.1. The Biotin-ELISA Competitive Cleavage Assay

#### 2.3.1.1. EQUIPMENT

1. PCR cycler.
2. Microtiter plate (suitable for adsorption of proteins).
3. Microtiter-plate ELISA reader.
4. Microtiter-plate washer is optional.
5. Avidin (10 µg/mL in water).
6. 1 mg TMB (3,3',5,5'tetramethylbenzidin) tablets.
7. α-fluorescein-POD and α-digoxygenin-POD antibody conjugates

#### 2.3.1.2. SOLUTIONS

1. PBST: 140 m$M$ NaCl, 2.7 m$M$ KCl, 4.3 m$M$ Na$_2$HPO$_4$, 1.4 m$M$ K$_2$HPO$_4$, 0.05% (v/v) Tween 50, pH 7.2.
2. PBSTM: PBST containing 1% (w/v) skim milk powder, PBSTM must be freshly prepared each day.
3. 200 m$M$ NaHCO$_3$ buffer (pH 9.6).
4. 50 m$M$ Na-acetate/citric-acid buffer (pH 4.9).
5. Reaction buffer.
6. DMSO.
7. 30% (w/v) H$_2$O$_2$.
8. TMB solution: dissolve 1 mg TMB in 1 mL DMSO and add 9 mL 50 m$M$ Na-acetate/citric acid, pH 4.9. Add 1–5 µL 30% (w/v) H$_2$O$_2$. This solution must be freshly prepared for each experiment.
9. 2 $M$ H$_2$SO$_4$.

### 2.3.2. Competitive Cleavage Assay Using Capillary Electrophoresis

#### 2.3.2.1. EQUIPMENT

1. PCR cycler.
2. Capillary electrophoresis system.
3. DNA template.
4. PCR primers (one of the two primers for each reaction should be labeled with a characteristic fluorophor).

#### 2.3.2.2. SOLUTIONS

1. Genetic analysis buffer™ (Perkin-Elmer) supplemented with 1 m$M$ EDTA.
2. Template suppression reagent™ (Perkin-Elmer).
3. 10X reaction buffer.

## 3. Methods

### *3.1. Assays to Measure the Thermodynamics and Kinetics of Nonspecific and Specific Binding of Proteins to DNA*

DNA-binding assays can be divided into two groups: homophasic assays based on the observation of a property of the enzyme or DNA that changes during complex formation, and heterophasic assays, which require separation of free enzyme, DNA, and the protein–DNA complex. In principle, most assays can be used for thermodynamic or kinetic analyses, although some tests are more suitable for one application or the other. Here, two common heterophasic assays will be described in detail (nitrocellulose filter-binding assay and gel retardation assay) and an overview and introduction into homophasic assays based on spectrophotometry will be given. Details concerning the analysis of the binding specificity of DNA-binding proteins are provided in **Note 1**. For advantages and disadvantages of the individual methods, refer to **Table 1**.

### *3.1.1. Nitrocellulose Filter Assay*

#### 3.1.1.1. PRINCIPLE

In nitrocellulose filter assays, protein-bound and free DNA molecules are quantified after they are physically separated *(45–49)* (**Fig. 1**). The basis of this assay is as follows: when a solution containing a mixture of labeled DNA and protein is passed through a nitrocellulose membrane filter, protein and protein-bound DNA are retained, but free DNA is not. The nitrocellulose filter-binding assay is one of the oldest methods to analyze protein–DNA interactions, and it still offers a convenient solution for many experimental problems.

#### 3.1.1.2. MEASUREMENT OF DNA-BINDING CONSTANTS

Typically, in nitrocellulose filter-binding experiments, the concentration of the labeled DNA is held constant. In titration experiments, the protein concentration should be varied over a range of 0.2–2X the presumptive $K_{Diss}$ at a constant DNA concentration of <0.02× $K_{Diss}$ (if you do not have an estimate for $K_{Diss}$, start with $10^{-7}$ *M*); BSA (10–50 ng/µL) should be included in the incubation buffer to prevent adsorption of the protein to surfaces.

1. Mix DNA, buffer, BSA, and protein solutions in a total volume of 50 µL in wells of a microtiter plate, and incubate for 30–60 min.
2. Incubate the nitrocellulose membrane for at least 30 min in assay buffer, tightly fix it in the dot blot device, and wash the membrane in the device with 200 µL assay buffer/well.
3. Transfer the samples with a multichannel pipet into the wells of the dot blot device and immediately suck the liquid through the membrane.
4. Wash the membrane in the device at least three times using 200-µL assay buffer/well.
5. Air-dry the membrane and analyze the bound radioactivity using a phosphorimager. Alternatively, the membrane can be cut into pieces corresponding to the spots where the samples were applied, and the radioactivity can be measured by scintillation counting.
6. To calculate the DNA binding constant ($K_{Ass}$) of the enzyme, analyze the experimental data (cpm) using **Eq. 1**, which is derived from the law of mass action applied to a bimo-

**Table 1**
**Advantages and Disadvantages of the Assays to Measure DNA Binding**

|  | Advantages | Disadvantages |
|---|---|---|
| Nitrocellulose-filter binding | The assay is easy to use and very fast. The experimental setup is simple. Many samples can be processed in parallel. | The assay is heterophasic and discontinuous. |
|  | The experimental conditions can be easily varied over wide ranges of pH and salt concentrations. | The assay cannot discriminate between DNA species with one, two, or more protein molecules bound. |
|  | Fast separation of free and bound DNA allows detection of weak protein–DNA complexes with high dissociation-rate constants. |  |
|  | The assay is well-suited for kinetic studies, as free and bound DNA are rapidly separated. |  |
| Electrophoretic mobility-shift assay | Discrimination between DNA species with one, two, or more proteins bound is possible. | Heterophasic assay. |
|  | Both free and bound DNA are observed, reducing the risk of misinterpretation because of artifacts. | Gel conditions can have a pronounced impact on the apparent specificity and $K_{Diss}$ values of the protein *(56)*. |
|  | The assay provides information on the conformation of the protein–DNA complex, e.g., DNA bending can be studied *(52–55)*. | The composition of the gel-loading buffer can alter apparent binding constants by up to one order of magnitude *(57,58)*. |
|  |  | Experimental conditions are difficult to control and can only be varied within certain limits. |
|  |  | Weak complexes often cannot be detected. |
|  |  | The assay is not well-suited for kinetic analyses. |
| Spectrophotometric assays | Homophasic assays determine true thermodynamic binding constants. | Relatively large amounts of pure protein are required. |
|  | Continuous data acquisition is possible, making these assays well-suited for kinetic purposes. | Requires advanced technical equipment. |
|  | Fast data acquisition for transient kinetic analyses is possible. | The relation of signal changes to physical processes must be established. |

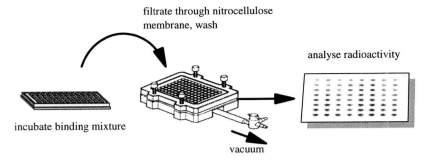

Fig. 1. Setup of a nitrocellulose filter binding experiment to measure DNA-binding constants.

lecular binding equilibrium, by a least squares fitting procedure, e.g., using the Solver function of Excel (Microsoft):

$$\text{cpm}_{\text{theo}} = A + B \times \tfrac{1}{2}\left[(c_{E,\text{tot}} + c_{D,\text{tot}} + 1/K_{\text{Ass}}) - \sqrt{(c_{E,\text{tot}} + c_{D,\text{tot}} + 1/K_{\text{Ass}})^2 - 4 \times c_{E,\text{tot}} \times c_{D,\text{tot}}}\,\right] \quad (1)$$

where: $A$ = background correction; $B$ = filter binding efficiency; $c_{E,\text{tot}}$ = total concentration of enzyme; $c_{D,\text{tot}}$ = total concentration of DNA.

### 3.1.1.3. ASSOCIATION KINETICS

To measure the kinetics of DNA association, the DNA and the protein are used at fixed concentrations.

1. Incubate the nitrocellulose membrane for at least 30 min in assay buffer, tightly fix it in the dot blot device, and wash the membrane in the device with 200-µL assay buffer/well.
2. Prepare a stock solution of buffer and DNA and distribute it into the wells of a microtiter plate.
3. Add protein to these solutions at defined times (e.g., after 0, 15, 30, 40, 50, 55, 58, and 59 min) yielding a total volume of 50 µL.
4. After 60 min, transfer all samples to the nitrocellulose membrane and immediately suck the liquid through the membrane.
5. Wash the membrane in the device at least three times using 200-µL assay buffer/well.
6. Air-dry the nitrocellulose membrane and analyze the bound radioactivity.

The concentrations of enzyme and DNA must be carefully optimized, because at concentrations too low no binding will be observed because of the law of mass action, while at too high, the reaction may proceed too rapidly. If, for example, the protein and DNA are used at equimolar concentrations ($c_0$), the half-life ($t_{1/2}$) of the bimolecular association reaction ($A + B \rightarrow AB$) depends on the association-rate constant (k) as given in **Eq. 2**.

$$t_{1/2} = 1/(k \times c_0) \quad (2)$$

This means that with an association-rate constant of $k = 1 \times 10^9\ M^{-1}\text{s}^{-1}$, which is in the range often observed for diffusion-controlled protein–DNA association reactions, the reaction has a half-life of only 1 s at 1 n$M$ protein and DNA, which increases to 16.7 min at 1 p$M$. Clearly, the latter concentrations are to be preferred for a kinetic analysis. If, however, the binding constant ($K_{\text{Ass}}$) is $10^7\ M$, as is often observed for nonspecific binding of restriction enzymes to oligonucleotides, only $10^{-17}\ M$ complex will be present at equilibrium (at DNA and protein concentration of 1 p$M$), which is certainly below the detection limit. For these reasons, only complex formation that is slow and/ or leads to stable complexes can be investigated with this technique.

### 3.1.1.4. DISSOCIATION KINETICS

To measure the kinetics of dissociation of a protein–DNA complex, the DNA and the protein are used at fixed concentrations. To allow for significant complex formation, use protein concentrations in the range of the value of $K_{Diss}$ (= $1/K_{Ass}$).

1. Prepare a stock solution (20–50 µL) containing DNA and protein in buffer.
2. Incubate the nitrocellulose membrane for at least 30 min in assay buffer, tightly fix it in the dot blot device, and wash the membrane in the device with 200-µL assay buffer/well.
3. At defined time intervals (e.g., after 0, 15, 30, 40, 50, 55, 58, and 59 min) add 2–5 µL of the binding mixture to 50–100 µL of buffer containing a large excess of unlabeled oligonucleotide (at a concentration at least 100-fold higher than the concentration of the labeled substrate and at least five- to 10-fold higher than the total protein concentration).
4. After 60 min, transfer all samples to the nitrocellulose membrane, and immediately suck the liquid through the membrane.
5. Wash the membrane in the device at least three times using 200 µL assay buffer/well.
6. Air-dry the nitrocellulose membrane and analyze the bound radioactivity.

## 3.1.2. Electrophoretic Mobility Shift Assay

### 3.1.2.1. PRINCIPLE

Electrophoretic mobility shift assays are often used for the study of DNA binding by proteins. In this assay, bound and free DNA are separated on native polyacrylamide gels. DNA bound to protein molecules migrates slower in the gel than free DNA *(50,51)* (**Fig. 2**). This assay is particularly well-suited to investigate the specificity of the protein-DNA interaction, because under conditions of specific binding, a single shifted band is observed if the DNA has one recognition site for the protein. In contrast, under conditions of nonspecific DNA binding, a ladder of shifted bands appears corresponding to DNA molecules with one, two, three, or more protein molecules bound. Complexes with much shorter half-lives in solution than the duration of the gel run can be detected, because the gel matrix prevents the separation of molecules after dissociation. This leads to rapid reassociation and much higher apparent binding constants in the gel ("cage effect").

### 3.1.2.2. EXPERIMENTAL PROCEDURE

Typically, the concentration of the labeled DNA is held constant. In titration experiments one should try to vary the protein concentration over a range from 0.2 to 2 times the presumptive $K_{Diss}$ at a constant DNA concentration of <0.02X $K_{Diss}$. DNA, buffer and protein are mixed and incubated for 30–60 min. A polyacrylamide gel is prepared (any size might be used; however, the length of the gel limits the resolution). Gel-loading buffer is added to the samples, the samples are transferred into the slots of the gel, and electrophoresis is started immediately (in general, low voltage is preferrable because it leads to less heating of the gel, and aggregation of protein during entry of the gel is less often observed). After electrophoresis, the gel is dried and the radioactivity analyzed, e.g., using a phosphorimager.

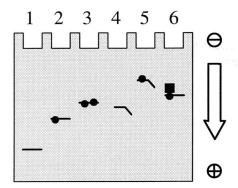

Fig. 2. Principle of electrophoretic mobility shift experiments. Free DNA (lane 1) runs faster in polyacrylamide gels than DNA bound to a protein (lane 2) or curved DNA (lane 4). If two proteins bind to one DNA molecule, or if a second protein binds to the first one, migration is even slower (compare lanes 2 and 3 or lanes 2 and 6).

### 3.1.3. Spectrophotometric Assays

A variety of spectrophotometric assays have been used to study protein–DNA binding interactions. In principle, DNA binding can be analyzed employing any signal change, such as fluorescence intensity, fluorescence depolarization or fluorescence resonance energy transfer (FRET). In the first assay, a change of the fluorescence properties of tryptophan, or less frequently tyrosine, residues of the protein upon DNA binding is measured. In the second assay, a fluorescent probe is attached to the oligonucleotide-binding substrate. Upon DNA-protein complex formation, the effective molecular mass of the fluorescent entity changes, leading to altered kinetics of fluorescence depolarization. In the last assay, a fluorescent probe is attached to the oligonucleotide that can form a FRET pair with a tryptophan residue. Upon formation of the protein–DNA complex, the fluorescent probe may come close to tryptophan residues, which leads to a FRET signal *(59)*. Whereas changes of the fluorescence intensity of the protein upon DNA binding and fluorescence energy transfer from the protein to a suitably positioned acceptor group in the oligonucleotide substrate are observed in only some systems, observation of changes of the fluorescence depolarization upon protein binding is a very general method. Homophasic binding constants of proteins to DNA molecules can also be measured by fluorescence correlation spectroscopy. With this technique, changes of the diffusion constant of the DNA upon protein binding are detected. As a comprehensive description of these techniques is beyond the scope of this chapter, a biophysical textbook should be consulted for details *(74)*.

The relationship between signal changes observed and physical processes, such as complex formation or dissociation, conformational changes or DNA cleavage, must be established for each system by careful control experiments. In general, spectrophotometric assays have the advantage of being homophasic. Therefore, true thermodynamic binding constants can be determined. This aspect is particularly important if the heterophasic assays described here produce inconsistent results. For example, depending on the conditions, gel shift and nitrocellulose filter binding experiments have shown that the restriction enzyme *Eco*RV binds completely nonspecifically to DNA *(54,60–63)* or prefers binding to substrates containing a recognition site about 10-fold *(56)*. This discrepancy could be resolved using different homophasic techniques that have confirmed that *Eco*RV binds nonspecifically to DNA in the absence of $Mg^{2+}$ *(59)*.

## 3.2. Assays to Measure the Kinetics of DNA Cleavage by Class II Restriction Enzymes

Restriction endonucleases are among the most specific enzymes known (*see* **Note 2**). Sites differing in only one bp from the canonical site (so-called "star sites") are cleaved at rates several orders of magnitude lower, and sites differing in more than 1 bp are not cleaved at all (*see* **Note 2**). The kinetics of DNA cleavage and the specificity of DNA cleavage by restriction enzymes have been carefully studied for many restriction endonucleases. Different substrates can be used in cleavage assays. Assays using macromolecular substrates provide information about the cleavage site of the enzyme as well as qualitative data on its specificity, while assays using oligonucleotide substrates are needed for mechanistic studies or systematic approaches. For advantages and disadvantages of the individual methods, refer to **Table 2**.

### 3.2.1. Cleavage of Macromolecular DNA

In this assay, macromolecular DNA is incubated with the restriction enzyme for a defined period of time, and the distribution of cleaved and uncleaved DNA fragments is analyzed by agarose (or polyacrylamide) gel electrophoresis. The assay is one of the oldest still in use, and is the most common assay to analyze DNA cleavage by restriction endonucleases because it is inexpensive, does not require any high-tech materials or equipment, and is easy to perform. Moreover, it provides detailed, although qualitative, information on the specificity of DNA cleavage, by far the most important property of these enzymes. The most popular substrates are phage λ-DNA and plasmid-DNA.

Because of its length (48,502 bp) λ-DNA usually contains several cleavage sites and many star-sites for most enzymes with 4-, 5-, and 6-bp recognition sites. In cleavage reactions with λ-DNA, specific nucleolytic activity results in a defined cleavage pattern. Under certain conditions, for example, at very high concentrations of enzyme, slow cleavage at some noncanonical sites occurs and additional bands appear on the agarose gel. A smear on the agarose gel indicates that nonspecific nucleolytic activity is present in the enzyme preparation. Such activity can be caused by a contamination or the enzyme itself. Because of its availability, λ-DNA has become the reference substrate for many restriction enzymes. Usually, the time required to completely digest the DNA is measured, with 1 U for most enzymes being defined as the amount of enzyme required to digest 1 μg λ-DNA in 1 h. However, such endpoint measurements are not particularly useful for mechanistic studies, where initial cleavage rates are to be determined.

For many purposes, plasmids are preferable to λ-DNA as substrates. Cleavage of substrates containing only one (or at most two) cleavage sites can be analyzed with respect to initial cleavage rates. Moreover, if only one site is present on a superhelical plasmid, the transition of the superhelical to the open-circle form represents cleavage of one strand of the DNA, and the transition of the open-circle form to linear DNA represents cleavage of the second strand. Thus, this assay allows to analyze if cleavage of both strands of the DNA occurs in a concerted manner *(33,64)*. If no accumulation of open-circle intermediates is observed, the cleavage events on both strands of the

**Table 2**
**Advantages and Disadvantages of the Assays to Measure DNA Cleavage**

| | Advantages | Disadvantages |
|---|---|---|
| Cleavage of macro-molecular DNA | Very simple, only low-tech and low-cost materials required. | Difficult to analyze quantitatively if more than two cleavage sites are present. |
| | Very efficient method to determine enzyme specificity. | Substrate concentrations can be varied only over a limited range. |
| | Detection of cleavage of both strands of the DNA is easy if the substrate is a superhelical plasmid. | Only cleavage of both strands of the DNA can be detected with linear substrates. Multistep kinetic mechanism obscures the relation of apparent-rate constants to the elementary reaction steps. |
| Cleavage of oligo-nucleotide substrates (homo-chromatography) | Substrates can be designed at will. Modified bases can be incorporated into the substrates. A wide range of substrate concentrations can be employed ($10^{-10}$–$10^{-4}$ $M$). Fast, convenient, easy to handle. Well-suited for quantitative analysis. | Concerted cleavage of both strands of the DNA is difficult to detect. |
| Cleavage of oligo-nucleotide substrates (denaturing polyacryl-amide-gel electro-phoresis | Substrates can be designed at will. Modified bases can be incorporated into the substrates. A wide range of substrate concentrations can be employed ($10^{-10}$–$10^{-4}$ $M$). Single nucleotide resolution. | Concerted cleavage of both strands of DNA is difficult to detect. |
| Biotin–avidin competitive cleavage assay | Competitive cleavage of different substrates is possible. Does not require gel electrophoresis; thus substrates of very different length can be analyzed. | Only disapperance of uncleaved substrate can be detected. |
| Competitive cleavage assay using capillary electrophoresis | Competitive cleavage of different substrates is possible. With current capillary systems up to four substrates can be analyzed in one run. Uncleaved and cleaved substrates are observed. | Requires capillary electrophoresis system or automatic sequencer. |

target site are coupled, and the second strand is cleaved before the enzyme dissociates from the DNA after cleavage of the first strand.

1. Mix DNA (20–100 ng/µL) and enzyme in cleavage buffer. Incubate at reaction temperature.
2. After appropriate incubation times, remove 10-µL aliquots from the reaction mixture,

immediately add 5 μL gel-loading buffer, mix, and store on ice. Because λ-DNA has cohesive ends, samples should be heated for 5 min to 65°C before electrophoresis, and then kept on ice.

3. Transfer the samples into the slots of an agarose gel and run the gel. Be careful to avoid excessive heating of the gel.
4. Stain the gel by soaking in ethidium bromide solution and analyze by transillumination with UV light (312 nm).

### 3.2.2. Cleavage of Oligonucleotide Substrates Studied by Homochromatography

The analysis of cleavage reactions using oligonucleotides is the most important tool in the study of the mechanism of DNA recognition and cleavage by restriction enzymes. Because the nucleotide sequence can be chosen at will, modified nucleotides can be introduced, and the substrate concentrations can be varied over a wide range. Moreover, the costs of custom-synthesized oligonucleotides have decreased dramatically over the past few years. One of the most convenient (although not widely used) assays to analyze cleavage of oligonucleotide substrates is thin-layer chromatography (TLC) on DEAE cellulose plates under denaturing conditions: homochromatography *(65)*. This assay allows study of the cleavage of each strand of the substrate separately. The only two disadvantages are that it is difficult to measure cleavage of both strands of the DNA, and resolution may not be adequate. Denaturing polyacrylamide gel electrophoresis described below provides a higher resolution than homochromatography.

#### 3.2.2.1. PREPARING AN RNA LYSATE AS RUNNING BUFFER

1. Dissolve 20 g yeast RNA (Boehringer Mannheim) in 80 mL water and 20 mL 5 *M* KOH.
2. Incubate at 37°C for 24 h. The incubation time with KOH depends on the quality of the RNA preparation, and needs to be optimized (2–48 h). An RNA hydrolysate generated by 24 h incubation of the RNA with KOH should separate oligonucleotides ranging from 5 to 25 nucleotides.
3. Neutralize with HCl to pH 7.0, and add 420 g urea and water to a final volume of 1 L. Store the RNA hydrolysate at 4°C.

#### 3.2.2.2. DNA CLEAVAGE REACTION

1. Mix DNA (0.05–20 μ*M*) and enzyme in 1X reaction buffer. It is most convenient to work at ambient temperature.
2. After appropriate incubation times, remove 0.5–2-μL aliquots from the reaction mixture and immediately spot them onto a DEAE cellulose plate to stop DNA cleavage.
3. Preincubate a thin-layer chromatography tank with RNA hydrolysate running buffer for at least 30 min at 60–70°C.
4. Run the thin-layer chromatography plate in a separate tank with water until the water front has reached a height of at least 2 cm above the positions of the spots. The plate does not need to be dried after this step.
5. Transfer the thin layer chromatography plate into the thin-layer chromatography tank with RNA hydrolysate running buffer, and develop the plate with running buffer at 60–70°C until the solvent front has almost reached the top end of the plate.
6. Dry the thin-layer chromatography plate with a hair-dryer and analyze radioactivity, using a phosphorimager. Note that synthetic oligonucleotides are often not cleaved to 100% by restriction endonucleases.

### 3.2.3. Analysis of Oligonucleotide Cleavage Reactions Studied by Denaturing Gel Electrophoresis

1. Prepare a $20 \times 20$ cm polyacrylamide gel containing 7 $M$ urea. Appropriate polyacrylamide concentrations are 15% for separation of 30–100-mers, 17% for 10–60-mers, and 20% for 5–30-mers if acrylamide and *bis*-acrylamide are used in a 19:1 ratio.
2. DNA cleavage is performed as described in **Subheading 2.2.2.2.**, except that sample volumes between 0.5 and 10 µL are possible.
3. Remove aliquots from the reaction mixture and immediately add 5 µL gel-loading buffer to stop the cleavage reaction.
4. Prerun the gel at 50 V/cm until it reaches a temperature of approx 50°C.
5. Heat the samples for 5 min to 95°C in loading buffer and place on ice.
6. Load the samples on the gel and run the gel at 50 V/cm. It is important to maintain a temperature of about 50°C.
7. Dry the gel and analyze the radioactivity, using a phosphorimager.
8. Alternatively, any convenient nonradioactive detection method may be used (e.g., DNA sequencing technology, digoxygenin labeling).

### 3.2.4. Analysis of DNA Cleavage by Restriction Endonucleases Using Fast Kinetics Systems

Single-turnover DNA cleavage reactions using stopped flow or quenched flow devices are needed for most mechanistic studies. If spectroscopic properties change during the reaction, transient conformational changes of the enzyme-DNA complex can be detected by stopped-flow experiments, where a fast mixing device is coupled to a spectrophotometer or a spectrofluorimeter (*see* **Subheading 3.1.3.**). To investigate a single turnover of DNA cleavage, quenched-flow experiments must be carried out, where enzyme, DNA and $Mg^{2+}$ are rapidly mixed and the reaction is quenched after a very short incubation period by addition of excess EDTA (for review, *see* **refs. *66,67***). A detailed description of these techniques is beyond the scope of this chapter (*see* **refs. *7,8,68–70*** for examples of application of these methods to restriction endonucleases).

### 3.3. Competitive Cleavage Assays

Competitive cleavage experiments can be useful to measure small differences in the cleavage rates of different substrates. The advantage of this experimental design is that such factors as errors made during pipetting and variations of the quality of the enzyme preparation cancel, because they affect the cleavage rates of both substrates to the same extent. In addition, competitive cleavage assays are also particularly suited to study linear diffusion of restriction enzymes on DNA. This process leads to a faster association of restriction enzymes to recognition sites embedded in longer DNA substrates. However, under many conditions DNA cleavage by restriction enzymes is limited by product release, which would obscure any rate differences in target-site location. Moreover, higher amounts of nonspecific DNA in cleavage reactions with large substrates reduce the cleavage rate by trapping the enzyme, and, thus, titrating the free enzyme out. Both of these problems can be circumvented by performing the cleavage reactions of different substrates in competition, i.e., in one reaction tube.

## 3.3.1. The Biotin-ELISA Competitive Cleavage Assay

In this assay, two DNA substrates produced by PCR are used. In each case, one primer of the PCR carries a biotin label on its 5' end. One of the second primers is labeled with fluorescein, and the other with digoxygenin. These modifications are commercially available. After incubation of both substrates with the enzyme in one reaction mixture, cleaved and uncleaved substrates are separated on an avidin-coated microtiter plate, and both substrates are quantitated separately using two ELISA reactions directed against fluorescein and digoxygenin (**Fig. 3**). If linear diffusion is to be investigated, it is advisable to amplify both substrates using the same template. Then, only one biotinylated primer is required, which also introduces the cleavage site into the PCR products. The second primers should be chosen to amplify DNA pieces of different lengths, which of course must not contain a second recognition site *(14)*. The resulting DNA substrates differ only in length, but not in the sequence of the DNA immediately flanking the cleavage site *(71)*.

### 3.3.1.1. PERFORMING THE DNA CLEAVAGE REACTION

1. Mix both substrates in cleavage buffer.
2. Start the reaction by adding the enzyme and mixing. It is convenient to work at ambient temperature.
3. After appropriate incubation times, transfer aliquots containing at least 100 fmol of each substrate DNA into tubes containing enough EDTA to stop the reaction.

### 3.3.1.2. ELISA

Note that for each time point, two separate ELISA reactions must be carried out—one for each substrate. In addition, all ELISA experiments should be carried out at least in duplicate.

1. Coat the microtiter plates with avidin by adding 100 μL of 10 μg/μL avidin in 200 m$M$ NaHCO$_3$ (pH 9.6) into each well of the plate, and incubating overnight at ambient temperature. Coated plates can be stored at 4°C for several days.
2. Wash the plate three times with PBST.
3. To block nonspecific adsorption, incubate the plate for 30 min with 200 μL/well PBSTM.
4. Wash again three times with PBST.
5. Pipet 50 μL PBST into each well and add the sample. The optimal amount of each substrate must be tested; start with 1–10 fmol. It is reasonable to use different amounts of each substrate in the ELISA to get similar signals with both antibodies.
6. Incubate the microtiter plate for 45 min.
7. Wash six times with 100 μL/well PBST.
8. Add the antibodies in 100 μL PBSTM. The optimal amounts of each antibody must be tested; refer to the technical data provided by the supplier.
9. Incubate the plate for 30 min.
10. Wash it six times with PBST.
11. If antibody-POD conjugates are used, pipet 50 μL TMB solution into each well.
12. Allow the staining reaction to proceed for 2–15 min, and stop it by addition of 50 μL 2 $M$ H$_2$SO$_4$.
13. Measure the optical density at 450 n$M$ with an ELISA reader. Determine the background of the reaction (use completely cleaved substrates), and normalize the reaction progress curves using the background signal *(71)*.

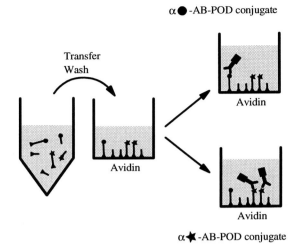

Fig. 3. Principle of the biotin-ELISA competitive cleavage assay. Two biotin-labeled (○) DNA molecules with different haptens (● or ★) are cleaved in competition. After defined times, aliquots of the reaction mixture are transferred to an avidin-coated microtiter plate and analyzed by an ELISA for intact DNA molecules.

### 3.3.2. Competitive Cleavage Assay Using Capillary Electrophoresis

The DNA substrates for this assay can be generated by PCR reactions. Up to four DNA fragments can be cleaved in competition. Each fragment must be labeled with one characteristic fluorophore which is introduced by one of the PCR primers. After cleavage in competition, the reaction mixtures are subjected to capillary electrophoresis, which allows for the simultaneous detection of the cleavage of all four substrates.

1. Mix both substrates in cleavage buffer. Start the reaction by adding the enzyme and mixing. It is most suitable to work at ambient temperature.
2. After appropriate incubation times, transfer aliquots containing at least 10 fmol of each substrate DNA into tubes containing EDTA to stop the reaction.
3. Heat aliquots of the reaction mixtures containing about 5 fmol of each oligonucleotide for 5 min at 95°C in template suppression reagent (Perkin-Elmer) and cool on ice.
4. Analyze on ABI PRISM 310 genetic analyzer (Perkin-Elmer) capillary electrophoresis system using a 47 cm capillary (inner diameter: 50 μm) containing the POP-4 polymer in 8 $M$ urea. Electroinject the samples for 5 s at 15,000 V and run for 10–30 min at 60°C and 15,000 V using genetic analysis buffer (Perkin-Elmer) supplemented with 1 m$M$ EDTA as electrode buffer.
5. Use the Genescan software (Perkin-Elmer) to quantify the amounts of the substrate and product *(72)*.

An automatic DNA sequencer may be employed instead of the capillary electrophoresis system (*see* **ref. 73**).

## 4. Notes

1. Determination of the DNA binding specificity of DNA-binding proteins: Specificity in DNA binding is defined as the preference of a protein for binding to its recognition site compared to other "nonspecific" sites. When comparing binding to substrates A and B,

which should be of identical sequence except that one substrate should contain a recognition site and the other an unrelated sequence, specificity can be described by the difference in the free energy of binding to both substrates ($\Delta\Delta G°_{A/B,Ass}$) as given in **Eq. 3**:

$$\Delta\Delta G°_{A/B,Ass} = \Delta G°_{Ass}(A) - \Delta G°_{Ass}(B) = -RT \ln[K_{Ass}(A)/K_{Ass}(B)] \qquad (3)$$

To analyze whether proteins specifically bind to DNA, different experimental approaches can be used:

a. Binding constants for several substrates that only differ in one or few bases within or adjacent to the recognition site can be measured. This approach yields the most accurate data, yet, some precautions must be followed:

 i. If oligonucleotides are used, the length of the substrate is very important. It is good practice to use substrates that are at least 3–5 bp longer on both sides than the region covered by the enzyme. If footprint analysis (*see* Chapter 30) has not been conducted, 20-mers should be used to study type II restriction endonucleases (this is not true for type IIS enzymes).

 ii. If longer substrates, e.g., PCR products, are used, the large number of nonspecific-binding sites may obscure specific binding.

b. Competition experiments can be used to compare binding constants for different substrates: protein and labeled DNA$_A$, both at constant concentrations, are incubated with increasing amounts of unlabeled DNA$_B$ to displace the labeled DNA from the enzyme. This experimental setup can result in considerable savings of protein.

c. Using longer substrates, specific binding can be easily demonstrated by analyzing the concentration of DNA substrates carrying zero, one, two, three, and more, protein molecules bound at different protein concentrations by gel-shift analyses. If the protein binds nonspecifically to DNA, these complexes are populated according to a Poisson-distribution (**Eq. 4**).

$$P(x) = (e^{-\lambda}\lambda^x)/x! \qquad (4)$$

with: $x$ = number of proteins bound to the DNA; $P(x)$ = probability of occurrence of DNA having $x$ proteins bound; $\lambda$ = average number of proteins bound to the DNA.

 For specific binding, only the first (one-to-one) complex is populated over a large range of protein concentrations. In the case of little specificity, many bands are observed; however, the first complex is more highly populated than expected on the basis of **Eq. 4**. This approach was first used to show that the restriction enzyme *Eco*RV binds nonspecifically to DNA in the absence of $Mg^{2+}$ *(60)*. In contrast, this enzyme binds to DNA with high specificity when $Mg^{2+}$ or $Ca^{2+}$ were added *(54,61)*. However, an *Eco*RV mutant displays specificity even in the absence of $Mg^{2+}$ as indicated by slight deviations from the expected distribution of bands demonstrating approximately a 10-fold preference of this mutant for binding to its recognition site under these experimental conditions *(62)*.

2. Specificity of DNA cleavage by restriction endonucleases: The preference ($\sigma$) of a restriction enzyme for one substrate (substrate A) over another (substrate B) is given by the relation of its specificity constants ($k_{cat}/K_m$) for each substrate (**Eq. 5a**). This specificity can be related to the difference in the free energy of activation for both reactions, $\Delta\Delta G^{\#}_{AB}$ (**Eq. 5b**):

$$\sigma = [k_{cat}(A)/K_m(A)]/[k_{cat}(B)/K_m(B)] \qquad (5a)$$

$$\Delta\Delta G^{\#}_{AB} = -RT \ln \sigma \qquad (5b)$$

 The specificity of a restriction enzyme is defined as the preference of the enzyme to cleave at its recognition sequence faster than at any other site. In theory, this means that specificity can only be conclusively determined by comparing cleavage rates of a restriction enzyme at any possible site to that observed at the canonical site. In practice, the most convenient way to demonstrate DNA cleavage specificity is to incubate macromolecular DNA with high concentrations of the restriction enzyme and analyze the cleavage products on an agarose or polyacrylamide gel. In the case of relaxed activity, additional bands will appear after all canonical sites have been cleaved. High relaxed activity or cleavage at completely noncognate sites are indicated by a smear. Usually, no additional bands will appear under optimal reaction conditions, even at moderate to high enzyme concentration. However, such experiments cannot be rigorously analyzed in quantitative terms. Only oligonucleotide cleavage experiments yield quantitative data. Thus far, there are three data sets available, in which cleavage rates for all sites differing in one base pair from the recognition site of a restriction enzyme have been analyzed *(29–31)*. These demonstrate that the rate of cleavage in one DNA strand at noncognate sites usually is reduced by at least two to three orders of magnitude, and that many sites are cleaved even more slowly. Cleavage in both strands of the DNA occurs with $10^5$–$10^6$-fold reduced rates. It has never been observed that a restriction enzyme cleaves DNA at sites differing from the recognition site by more than 1 bp.

3. Star conditions: Under certain reaction conditions ("star" conditions), restriction endonucleases cleave DNA with a markedly reduced specificity (for review, *see* **ref. *4***), cutting sites that differ from the recognition site in one base pair at rates similar to those for the canonical site. It is interesting that different enzymes lose specificity under the same star conditions, such as:
   a. The presence of $Mn^{2+}$ (usually at 1–5 m$M$) instead of $Mg^{2+}$ in the reaction buffer.
   b. The presence of osmolytes such as organic solvents (e.g., DMSO or glycerol at concentrations above 10%) or sucrose.
   c. Alkaline pH (e.g., pH 8.5) and/or low salt concentrations (e.g., 1 m$M$ $MgCl_2$, 20 m$M$ Tris-HCl, no NaCl).

   Star cleavage of a macromolecular substrate results in the appearance of many additional bands upon gel electrophoresis. To distinguish star cleavage from a nucleolytic contamination of the enzyme preparation, use the following considerations:
   a. Star cleavage should be observed only under certain buffer conditions.
   b. Star activity copurifies with the canonical activity throughout the purification.
   c. Star cleavage only occurs at sites that differ in 1 bp from the recognition site.
   d. Contaminating endo/exo-nucleases usually will produce a smear on agarose or polyacrylamide gels and and no distinct cleavage pattern.

# References

1. Noyer-Weidner, M. and Trautner, T. A. (1993) Methylation of DNA in prokaryotes, in *DNA Methylation: Molecular Biology and Biological Significance* (Jost, J. P. and Saluz, H. P., eds.), Birkhauser Verlag, Basel, pp. 40–108.
2. Cheng, X. (1995) Structure and function of DNA methyltransferases. *Annu. Rev. Biophys. Biomol. Struct.* **24**, 293–318.
3. Roberts, R. J. and Halford, S. E. (1993) Type II restriction enzymes, in *Nucleases* (Linn, S. M., Lloyd, R. S., and Roberts, R. J., eds.), Cold Spring Harbor Laboratory, Cold Spring Harbor, NY, pp. 35–88.
4. Pingoud, A. and Jeltsch, A. (1997) Recognition and cleavage of DNA by type-II restriction endonucleases. *Eur. J. Biochem.* **246**, 1–22.
5. Szybalski, W., Kim, S. C., Hasan, N., and Podhajska, A. J. (1991) Class-IIS restriction enzymes—a review. *Gene* **100**, 13–26.

6. Li, L., Wu, L. P., and Chandrasegaran, S. (1992) Functional domains in *Fok*I restriction endonuclease. *Proc. Natl. Acad. Sci. USA* **89,** 4275–4279.

7. Alves, J., Urbanke, C., Fliess, A., Maass, G., and Pingoud, A. (1989) Fluorescence stopped-flow kinetics of the cleavage of synthetic oligodeoxynucleotides by the *Eco*RI restriction endonuclease. *Biochemistry* **28,** 7879–7888.

8. Baldwin, G. S., Vipond, I. B., and Halford, S. E. (1995) Rapid reaction analysis of the catalytic cycle of the *Eco*RV restriction endonuclease. *Biochemistry* **34,** 705–714.

9. Jack, W. E., Terry, B. J., and Modrich, P. (1982) Involvement of outside DNA sequences in the major kinetic path by which *Eco*RI endonuclease locates and leaves its recognition sequence. *Proc. Natl. Acad. Sci. USA* **79,** 4010–4014.

10. Terry, B. J., Jack, W. E., and Modrich, P. (1985) Facilitated diffusion during catalysis by *Eco*RI endonuclease: nonspecific interactions in *Eco*RI catalysis. *J. Biol. Chem.* **260,** 13,130–13,137.

11. Ehbrecht, H.-J., Pingoud, A., Urbanke, C., Maass, G., and Gualerzi, C. (1985) Linear diffusion of restriction endonucleases on DNA. *J. Biol. Chem.* **260,** 6160–6166.

12. Jeltsch, A., Alves, J., Wolfes, H., Maass, G., and Pingoud, A. (1994) Pausing of the restriction endonuclease *Eco*RI during linear diffusion on DNA. *Biochemistry* **33,** 10,215–10,219.

13. Jeltsch, A., Wenz, C., Stahl, F., and Pingoud, A. (1996) Linear diffusion of the restriction endonuclease *Eco*RV on DNA is essential for the in vivo function of the enzyme. *EMBO J.* **15,** 5104–5111.

14. Jeltsch, A. and Pingoud, A. (1998) Kinetic characterization of linear diffusion of the restriction endonuclease *Eco*RV on DNA. *Biochemistry* **37,** 2160–2169.

15. Schulze, C., Jeltsch, A., Franke, I., Urbanke, C., and Pingoud, A. (1998) Crosslinking the *Eco*RV restriction endonuclease across the DNA-binding site reveals transient intermediates and conformational changes of the enzyme during DNA binding and catalytic turnover. *EMBO J.* **17,** 6757–6766.

16. Kim, Y., Grable, J. C., Love, R., Greene, P. J., and Rosenberg, J. M. (1990) Refinement of *Eco*RI endonuclease crystal structure: a revised protein chain tracing. *Science* **249,** 1307–1309.

17. Winkler, F. K., Banner, D. W., Oefner, C., Tsernoglou, D., Brown, R. S., Heathman, S. P., Bryan, R. K., Martin, P. D., Petratos, K., and Wilson, K. S. (1993) The crystal structure of *Eco*RV endonuclease and of its complexes with cognate and non-cognate DNA fragments. *EMBO J.* **12,** 1781–1795.

18. Athanasiadis, A., Vlassi, M., Kotsifaki, D., Tucker, P. A., Wilson, K. S., and Kokkinidis, M. (1994) Crystal structure of *Pvu*II endonuclease reveals extensive structural homologies to *Eco*RV. *Nat. Struct. Biol.* **1,** 469–475.

19. Cheng, X., Balendiran, K., Schildkraut, I., and Anderson, J. E. (1994) Structure of *Pvu*II endonuclease with cognate DNA. *EMBO J.* **13,** 3927–3935.

20. Newman, M., Strzelecka, T., Dorner, L. F., Schildkraut, I., and Aggarwal, A. K. (1994) Structure of restriction endonuclease *Bam*HI and its relationship to *Eco*RI. *Nature* **368,** 660–664.

21. Newman, M., Strzelecka, T., Dorner, L. F., Schildkraut, I., and Aggarwal, A. K. (1995) Structure of *Bam*HI endonuclease bound to DNA: partial folding and unfolding on DNA binding. *Science* **269,** 656–663.

22. Bozic, D., Grazulis, S., Siksnys, V., and Huber, R. (1996) Crystal structure of *Citrobacter freundii* restriction endonuclease *Cfr*10I at 2.15 Å resolution. *J. Mol. Biol.* **255,** 176–186.

23. Newman, M., Lunnen, K., Wilson, G., Greci, J., Schildkraut, I., and Phillips, S. E. V. (1998) Crystal structure of restriction endonuclease *Bgl*I bound to its interrupted DNA recognition sequence. *EMBO J.* **17,** 5466–5476.

24. Venclovas, C., Timinskas, A., and Siksnys, V. (1994) Five-stranded beta-sheet sandwiched with two a-helices: a structural link between restriction endonucleases *Eco*RI and *Eco*RV. *Proteins* **20,** 279–282.

25. Wah, D. A., Hirsch, J. A., Dorner, L. F., Schildkraut, I., and Aggarwal, A. K. (1997) Structure of the multimodular endonuclease *Fok*I bound to DNA. *Nature* **388,** 97–100.

26. Wah, D. A., Bitinaite, J., Schildkraut, I., and Aggarwal, A. K. (1998) Structure of *Fok*I has implications for DNA cleavage. *Proc. Natl. Acad. Sci. USA* **95,** 10,564–10,569.

27. Aggarwal, A. K. (1995) Structure and function of restriction endonucleases. *Curr. Opin. Struct. Biol.* **5,** 11–19.

28. Rosenberg, J. M. (1991) Structure and function of restriction endonucleases. *Curr. Opin. Struct. Biol.* **1,** 104–113.

29. Lesser, D. R., Kurpiewski, M. R., and Jen-Jacobson, L. (1990) The energetic basis of specificity in the *Eco*RI endonuclease-DNA interaction. *Science* **250,** 776–786.

30. Thielking, V., Alves, J., Fliess, A., Maass, G., and Pingoud, A. (1990) Accuracy of the *Eco*RI restriction endonuclease: Binding and cleavage studies with oligodeoxynucleotide substrates containing degenerate recognition sequences. *Biochemistry* **29,** 4682–4691.

31. Alves, J., Selent, U., and Wolfes, H. (1995) Accuracy of the *Eco*RV restriction endonuclease: Binding and cleavage studies with oligodeoxynucleotide substrates containing degenerate recognition sequences. *Biochemistry* **34,** 11,191–11,197.

32. Jeltsch, A., Alves, J., Oelgeschlager, T., Wolfes, H., Maass, G. M., and Pingoud, A. (1993) Mutational analysis of the function of Gln115 in the *Eco*RI restriction endonuclease, a critical amino acid for recognition of the inner thymidine residue in the sequence -GAATTC- and for coupling specific DNA binding to catalysis. *J. Mol. Biol.* **229,** 221–234.

33. Stahl, F., Wende, W., Jeltsch, A., and Pingoud, A. (1996) Introduction of asymmetry in the naturally symmetric restriction endonuclease *Eco*RV to investigate intersubunit communication in the homodimeric protein. *Proc. Natl. Acad. Sci. USA* **93,** 6175–6180.

34. Stahl, F., Wende, W., Jeltsch, A., and Pingoud, A. (1998) The mechanism of DNA cleavage by the type II restriction enzyme *Eco*RV: Asp36 is not directly involved in DNA cleavage but serves to couple indirect readout to catalysis. *Biol. Chem.* **379,** 467–473.

35. Stahl, F., Wende, W., Wenz, C., Jeltsch, A., and Pingoud, A. (1998) Intra- vs intersubunit communication in the homodimeric restriction enzyme *Eco*RV: Thr37 and Lys38 involved in indirect readout are only important for the catalytic activity of their own subunit. *Biochemistry* **37,** 5682–5688.

36. Smith, H. O. and Wilcox, K. W. (1970) A restriction enzyme from *Haemophilus influenzae*. I. Purification and general properties. *J. Mol. Biol.* **51,** 379–391.

37. Connolly, B. A., Eckstein, F., and Pingoud, A. (1984) The Stereochemical course of the restriction endonuclease *Eco*RI-catalyzed reaction. *J. Biol. Chem.* **259,** 10,760–10,763.

38. Grasby, J. A. and Connolly, B. A. (1992) Stereochemical outcome of the hydrolysis reaction catalyzed by the *Eco*RV restriction endonuclease. *Biochemistry* **31,** 7855–7861.

39. Jeltsch, A., Alves, J., Maass, G., and Pingoud, A. (1992) On the catalytic mechanism of *Eco*RI and *Eco*RV. A detailed proposal based on biochemical results, structural data and molecular modelling. *FEBS Lett.* **304,** 4–8.

40. Thielking, V., Selent, U., Köhler, E., Wolfes, H., Pieper, U., Geiger, R., Urbanke, C., Winkler, F. K., and Pingoud, A. (1991) Site-directed mutagenesis studies with *Eco*RV restriction endonuclease to identify regions involved in recognition and catalysis. *Biochemistry* **30,** 6416–6422.

41. Winkler, F. K. (1992) Structure and function of restriction endonucleases. *Curr. Opin. Struct. Biol.* **2,** 93–99.

42. Jeltsch, A., Alves, J., Wolfes, H., Maass, G., and Pingoud, A. (1993) Substrate-assisted catalysis in the cleavage of DNA by the *Eco*RI and *Eco*RV restriction enzymes. *Proc. Natl. Acad. Sci. USA* **90,** 8499–8503.

43. Vipond, I. B., Baldwin, G. S., and Halford, S. E. (1995) Divalent metal ions at the active sites of the *Eco*RV and *Eco*RI restriction endonucleases. *Biochemistry* **34,** 697–704.
44. Kostrewa, D. and Winkler, F. K. (1995) Mg$^{2+}$ binding to the active site of *Eco*RV endonuclease: a crystallographic study of complexes with substrate and product DNA at 2 Angstrom resolution. *Biochemistry* **34,** 683–696.
45. Riggs, A. D., Suzuki, H., and Bourgeois, S. (1970) Lac repressor-operator interaction. I. Equilibrium studies. *J. Mol. Biol.* **48,** 67–83.
46. Riggs, A. D., Bourgeois, S., and Cohn, M. (1970) The lac repressor-operator interaction. 3. Kinetic studies. *J. Mol. Biol.* **53,** 401–417.
47. Hinkle, D. C. and Chamberlin, M. J. (1972) Studies of the binding of *Escherichia coli* RNA polymerase to DNA. II. The kinetics of the binding reaction. *J. Mol. Biol.* **70,** 187–195.
48. Hinkle, D. C. and Chamberlin, M. J. (1972) Studies of the binding of *Escherichia coli* RNA polymerase to DNA. I. The role of sigma subunit in site selection. *J. Mol. Biol.* **70,** 157–185.
49. Wong, I. and Lohman, T. M. (1993) A double-filter method for nitrocellulose-filter binding: Application to protein-nucleic acid interactions. *Proc. Natl. Acad. Sci. USA* **90,** 5428–5432.
50. Fried, M. and Crothers, D. M. (1981) Equilibria and kinetics of lac repressor-operator interactions by polyacrylamide gel electrophoresis. *Nucleic Acids Res.* **9,** 6505–6525.
51. Garner, M. M. and Revzin, A. (1981) A gel electrophoresis method for quantifying the binding of proteins to specific DNA regions: application to components of the *Escherichia coli* lactose operon regulatory system. *Nucleic Acids Res.* **9,** 3047–3060.
52. Stöver, T., Köhler, E., Fagin, U., Wende, W., Wolfes, H., and Pingoud, A. (1993) Determination of the DNA bend angle induced by the restriction endonuclease *Eco*RV in the presence of Mg$^{2+}$. *J. Biol. Chem.* **268,** 8645–8650.
53. Withers, B. E. and Dunbar, J. C. (1993) The endonuclease isoschizomers, *Sma*I and *Xma*I bend DNA in opposite orientations. *Nucleic Acids Res.* **21,** 2571–2577.
54. Vipond, I. B. and Halford, S. E. (1995) Specific DNA recognition by *Eco*RV restriction endonuclease induced by calcium ions. *Biochemistry* **34,** 1113–1119.
55. Cal, S. and Connolly, B. A. (1996) The *Eco*RV modification methylase causes considerable bending of DNA upon binding to its recognition sequence GATATC. *J. Biol. Chem.* **271,** 1008–1015.
56. Engler, L. E., Welch, K. K., and Jen-Jacobson, L. (1997) Specific binding by *Eco*RV endonuclease to its DNA recognition site GATATC. *J. Mol. Biol.* **269,** 82–101.
57. Zhang, X., Duggan, L. J., and Gottlieb, P. A. (1993) Loading dyes used in gel electrophoresis alter the apparent thermodynamic equilibrium of the lac repressor-operator complex. *Anal. Biochem* **214,** 580–582.
58. Hassanain, H. H., Dai, W., and Gupta, S. L. (1993) Enhanced gel mobility shift assay for DNA-binding factors. *Anal. Biochem* **213,** 162–167.
59. Erskine, S. G. and Halford, S. E. (1998) Reactions of the *Eco*RV restriction endonuclease with fluorescent oligodeoxynucleotides: identical equilibrium constants for binding to specific and non-specific DNA. *J. Mol. Biol.* **275,** 759–772.
60. Taylor, J. D., Badcoe, I. G., Clarke, A. R., and Halford, S. E. (1991) *Eco*RV restriction endonuclease binds all DNA sequences with equal affinity. *Biochemistry* **30,** 8743–8753.
61. Thielking, V., Selent, U., Köhler, E., Landgraf, Z., Wolfes, H., Alves, J., and Pingoud, A. (1992) Mg$^{2+}$ confers DNA binding specificity to the *Eco*RV restriction endonuclease. *Biochemistry* **31,** 3727–3732.
62. Jeltsch, A., Maschke, H., Selent, U., Wenz, C., Köhler, E., Connolly, B. A., Thorogood, H., and Pingoud, A. (1995) DNA binding specificity of the *Eco*RV restriction endonuclease is increased by Mg$^{2+}$ binding to a metal ion binding site distinct from the catalytic center of the enzyme. *Biochemistry* **34,** 6239–6246.

63. Szczelkun, M. D. and Connolly, B. A. (1995) Sequence-specific binding of DNA by the *Eco*RV restriction and modification enzymes with nucleic acid and cofactor analogues. *Biochemistry* **34,** 10,724–10,733.

64. Halford, S. E. and Goodall, A. J. (1988) Modes of DNA cleavage by the *Eco*RV restriction endonuclease. *Biochemistry* **27,** 1771–1777.

65. Brownlee, G. G. and Sanger, F. (1969) Chromatography of $^{32}$P-labeled oligonucleotides on thin layers of DEAE- cellulose. *Eur. J. Biochem.* **11,** 395–3999.

66. Fierke, C. A. and Hammes, G. G. (1995) Transient kinetic approaches to enzyme mechanisms. *Methods Enzymol.* **249,** 3–37.

67. Johnson, K. A. (1995) Rapid quench kinetic analysis of polymerases, adenosinetriphosphatases, and enzyme intermediates. *Methods Enzymol.* **249,** 38–61.

68. Zebala, J., Choi, J., and Barany, F. (1992) Characterization of steady state, single-turnover, and binding kinetics of the *Taq*I restriction endonuclease. *J. Biol. Chem.* **267,** 8097–8105.

69. Groll, D. H., Jeltsch, A., Selent, U., and Pingoud, A. (1997) Does the restriction endonuclease *Eco*RV employ a two-metal-ion mechanism for DNA cleavage? *Biochemistry* **36,** 11,389–11,401.

70. Erskine, S. G., Baldwin, G. S., and Halford, S. E. (1997) Rapid-reaction analysis of plasmid DNA cleavage by the *Eco*RV restriction endonuclease. *Biochemistry* **36,** 7567–7576.

71. Jeltsch, A., Fritz, A., Alves, J., Wolfes, H., and Pingoud, A. (1993) A fast and accurate enzyme-linked immunosorbent assay for the determination of the DNA cleavage activity of restriction endonucleases. *Anal. Biochem.* **213,** 234–240.

72. Wenz, C., Hahn, M., and Pingoud, A. (1998) Engineering of variants of the restriction endonuclease *Eco*RV that depend in their cleavage activity on the flexibility of sequences flanking the recognition site. *Biochemistry* **37,** 2234–2242.

73. Glasner, W., Merkl, R., Schmidt, S., Cech, D., and Fritz, H. J. (1992) Fast quantitative assay of sequence-specific endonuclease activity based on DNA sequencer technology. *Biol. Chem. Hoppe Seyler* **373,** 1223–1225.

74. van Holde, K. E., Johnson, W. C., and Ho, P. S. (1998) *Principles of Physical Biochemistry*. Prentice Hall, Upper Saddle River, NJ.

# 20

# Engineered Properties and Assays for Human DNase I Mutants

## Clark Q. Pan, Dominick V. Sinicropi, and Robert A. Lazarus

## 1. Introduction

Human deoxyribonuclease (DNase I) is an important clinical agent currently used in the treatment of cystic fibrosis (CF) patients (*1*). It is inhaled into the airways, where it degrades DNA to lower molecular weight fragments, thus reducing the viscoelasticity of CF sputum and improving lung function (*2,3*). In engineering DNase I, our goal was to address several properties that could not only provide more biochemical and pharmacological understanding, but might also result in a more efficacious clinical agent. This goal included engineering DNase I to improve its catalytic activity and to increase its resistance to inhibition by both G-actin and salt.

Although the enzymatic hydrolysis of DNA is widely accepted as the mechanism of action for decreasing the viscoelasticity, an alternative mechanism is possible because DNase I can depolymerize F-actin, another viscous polymer found in CF sputum (*4*). Since G-actin is a potent inhibitor ($K_i$ ~1 n$M$) of DNase I hydrolytic activity (*5*), actin could also affect the biological activity of DNase I. Therefore, based on the pharmacological and clinical significance of human DNase I as well as the structural and functional properties of the enzyme, we recently engineered several different classes of DNase I mutants that affect either actin binding or DNA catalytic activity (*6–10*). The latter mutants are altered at residues that are either directly involved in catalysis or that bind DNA or metal ions. Crystal structures of both bovine and human DNase I, either free (*11,12*), complexed with DNA (*13,14*), or complexed with G-actin (*15*), have been essential for the site-directed mutagenesis approach we have taken.

In order to elucidate the basic mechanism of action and assess the significance of G-actin as an inhibitor of human DNase I in CF sputum (*6*), we generated and characterized two classes of mutants: actin-resistant variants that catalyze DNA hydrolysis comparable to wild-type, but are no longer inhibited by actin, and active site variants which bind actin with affinity comparable to wild-type, but no longer catalyze DNA hydrolysis (*see* **Subheading 3.1.1.**). This protein engineering strategy showed that the reduction of viscoelasticity in CF sputum by DNase I resulted from DNA hydrolysis and not from depolymerization of F-actin, that G-actin is a significant inhibitor of

From: *Methods in Molecular Biology, vol. 160: Nuclease Methods and Protocols*
Edited by: C. H. Schein © Humana Press Inc., Totowa, NJ

DNase I in CF sputum, and that actin-resistant variants are 10- to 50-fold more potent than wild-type in CF sputum *(6,10)*.

DNase I is an endonuclease that catalyzes the hydrolysis of double-stranded DNA predominantly by a single-stranded nicking mechanism under physiological conditions when both $Ca^{2+}$ and $Mg^{2+}$ ions are present *(16)*. To improve the enzymatic activity, i.e., more effective reduction of DNA length under physiological conditions, we assessed the importance of various residues at the DNA-binding interface and engineered hyperactive variants that utilize a more efficient functional mechanism involving processive nicking of DNA, resulting in more double-stranded breaks *(7,8)*. The hyperactive variants were created by introducing basic amino acids into DNase I at selected positions in the DNA-binding interface to generate attractive interactions with the negatively charged phosphates on the DNA backbone; increased activities up to 10,000-fold relative to wild-type were observed. Furthermore, unlike wild-type, the hyperactive variants were no longer inhibited by physiological saline.

In order to carry out the studies described above, we constructed site-directed human DNase I mutants, expressed them in human 293 cells (*see* **Subheading 3.1.2.**), and characterized the cell-culture media in a variety of assays. DNase I concentration was measured by a two-site enzyme-linked immunoassay *(17)*, using homogeneously pure recombinant human DNase (Pulmozyme® rhDNase, Genentech) as the standard (*see* **Note 1**) (*see* **Subheading 3.1.3.** and **Fig. 1**). Kinetic parameters of the DNase I variants were determined using a Kunitz-based hyperchromicity kinetic assay *(18)* (*see* **Subheading 3.2.1.**). A colorimetric DNA-methyl green assay *(19)* was used for high-throughput screening of cell-culture supernatants (*see* **Subheading 3.2.2.** and **Fig. 2**). Supercoiled and linear plasmid digestion assays *(7–10)* were used to measure the degree of hyperactivity and to evaluate the proportion of single- vs double-stranded cleavage (*see* **Subheading 3.2.3.** and **Fig. 3**). A *p*-nitrophenyl phenylphosphonate assay *(20)* was used to evaluate the hydrolysis of a low-molecular weight substrate (*see* **Subheading 3.2.4.**). Actin binding (*see* **Subheading 3.3.1.** and **Fig. 4**) and actin inhibition (*see* **Subheading 3.3.2.**) were determined using ELISA and kinetic methods respectively *(6)*.

## 2. Materials

### 2.1. DNase I Mutant Design, Mutagenesis, and Expression

#### 2.1.1. Mutant Design, Mutagenesis, and Expression

1. InsightII program (Molecular Simulation, Inc., San Diego, CA).
2. Silicon Graphics workstation.
3. pRK.DNase expression vector *(21)*.
4. *Escherichia coli* XL1 blue MRF' (Stratagene, La Jolla, CA).
5. Qiagen tip-500 columns (Qiagen Inc., Valencia, CA).
6. Human embryonic kidney: 293 cells (ATCC CRL 1573; ATCC, Rockville, MD).
7. Cell-culture media (per 1 L): 900 mL of 50:50 F12 Nutrient Mixture (Ham's):D-MEM (low glucose), 10 mL of 200 m*M* glutamine, 10 mL of 1 *M* HEPES, pH 7.2, and either 100 mL 10% fetal bovine serum serum-containing media) or 100 mL of 50 µg/mL insulin (serum-free media). Media components can be obtained from Life Technologies (Gaithersburg, MD).
8. Adenovirus DNA *(22)*.
9. Centriprep 10 concentrators (Amicon, Beverly, MA).

## 2.1.2. DNase I ELISA

1. Polystyrene microtiter plates (Nunc, Rochester, NY).
2. Polyclonal antibodies (anti-DNase I) used in the assay were generated in rabbits, affinity-purified using standard procedures, and labeled with biotin-X-NHS (Research Organics, Inc., Cleveland, OH).
3. 0.05 $M$ Sodium carbonate, pH 9.6.
4. Wash buffer: Phosphate-buffered saline (PBS) containing 0.05% polysorbate 20.
5. Pulmozyme® rhDNase or diluted cell-culture media from 293 cells transfected with wild-type or variant DNase I.
6. ELISA buffer: 25 m$M$ HEPES-NaOH, 4 m$M$ CaCl$_2$, 4 m$M$ MgCl$_2$, 0.1% BSA, 0.05% polysorbate 20, 0.03% Proclin 300 (a preservative available from Supelco, Inc., State College, PA), pH 7.5.
7. Streptavidin-β-galactosidase (Roche Molecular Biochemicals, Indianapolis, IN).
8. 4-Methylumbelliferyl-β-D-galactopyranoside (Molecular Probes, Eugene, OR).
9. Substrate buffer: 0.1 $M$ sodium phosphate, 1 m$M$ MgCl$_2$, pH 7.3.
10. 0.15 $M$ Glycine, pH 10.5.
11. Fluorescence microtiter plate reader (96-well).

## 2.2. DNase I Activity Assays

### 2.2.1. Kunitz Hyperchromicity Assay

1. Calf thymus DNA (Sigma, St. Louis, MO).
2. Diluted cell-culture media from 293 cells transfected with wild-type or variant DNase I.
3. Assay buffer: 25 m$M$ HEPES, pH 7.0, 1 m$M$ MgCl$_2$, 2.5 m$M$ CaCl$_2$, and 150 m$M$ NaCl.
4. UV-readable 96-well microtiter plates (96-well quartz plate or disposable SPECTRAplates; Molecular Devices, Sunnyvale, CA).
5. Molecular Devices Spectra Max 250 spectrophotometer (Molecular Devices).
6. Kaleidagraph v 3.0.8 c (Synergy Software, Reading, PA).

### 2.2.2. DNA-Methyl Green Assay

1. Polystyrene microtiter plates (Nunc, Rochester, NY).
2. DNA from salmon testes (Sigma) is prepared at a target concentration of 2.5 mg/mL in HEPES/EDTA buffer (25 m$M$ HEPES-NaOH, 1 m$M$ EDTA, pH 7.5) by gentle orbital mixing in an Erlenmeyer flask for 3–4 d at room temperature until a homogeneous solution is obtained. The concentration of DNA is measured by absorbance at 260 nm and adjusted to 2 mg/mL with HEPES/EDTA buffer. Store at 4°C.
3. 0.4% Methyl green in 20 m$M$ acetate-NaOH, pH 4.2. The solution is repeatedly extracted with chloroform until the organic layer is colorless. The upper aqueous layer is then separated and stirred in a fume hood for 3 h to evaporate excess chloroform. Store at 4°C.
4. Assay buffer: 25 m$M$ HEPES-NaOH, 4 m$M$ CaCl$_2$, 4 m$M$ MgCl$_2$, 0.1% BSA, 0.01% Thimerosal (optional), 0.05% polysorbate (Tween) 20, pH 7.5. A high-grade preparation of BSA free of nuclease contamination is recommended.
5. Pulmozyme® rhDNase or diluted cell-culture media from 293 cells transfected with wild-type or variant DNase I.
6. Microtiter plate reader (96-well).

### 2.2.3. Plasmid Digestion Assay

1. Supercoiled pBR322 plasmid DNA (New England Biolabs, Beverly, MA).
2. *Eco*RI (New England Biolabs).

3. Pulmozyme rhDNase or diluted cell-culture media from 293 cells transfected with wild-type or variant DNase I.
4. Incubation buffer: 25 m$M$ HEPES, pH 7.0, 100 μg/mL BSA, 1 m$M$ MgCl$_2$, 2.5 m$M$ CaCl$_2$.
5. Agarose quench buffer: 25 m$M$ EDTA, 6% glycerol, xylene cyanol, and Bromophenol blue.
6. Acrylamide quench buffer: 20 m$M$ EDTA, xylene cyanol, and Bromophenol blue in deionized formamide.
7. TBE: 90 m$M$ Tris-borate and 2 m$M$ EDTA.
8. Ethidium bromide.
9. Model 575 FluorImager (Molecular Dynamics).
10. Agarose NA (Pharmacia, Piscataway, NJ).
11. Acrylamide.
12. Multichanel-compatible combs (Life Technologies, Gaithersburg, MD).
13. Klenow fragment of DNA polymerase I (Amersham Pharmacia Biotech, Piscataway, NJ).
14. α-$^{32}$P-dATP (Amersham Pharmacia Biotech).
15. Fuji phosphoimager (BAS2000) and phosphoimaging plate (type BAS-III) (Fuji Medical Systems Inc., Stamford, CT).

### 2.2.4. p-Nitrophenyl Phenylphosphonate Assay

1. NPPP: *p*-nitrophenyl phenylphosphonate (Sigma).
2. Pulmozyme rhDNase or diluted cell-culture media from 293 cells transfected with wild-type or variant DNase I.
3. Reaction buffer: 10 m$M$ NPPP in 60 m$M$ HEPES, pH 7.0, 50 m$M$ MgCl$_2$, and 50 m$M$ CaCl$_2$.

## 2.3. DNase I/Actin Assays

### 2.3.1. Actin-Binding ELISA

1. MaxiSorp plates (Nunc).
2. Gc globulin (Calbiochem, La Jolla, CA).
3. Buffer A: 25 m$M$ HEPES, pH 7.5, 4 m$M$ CaCl$_2$, 4 m$M$ MgCl$_2$, 0.1% BSA, 0.5 m$M$ ATP, 0.01% thimerosal, and 0.05% Tween 20.
4. Wash buffer: PBS containing 0.05% Tween 20.
5. Rabbit skeletal muscle G-actin (1 mg/mL, prepared by the method of Pardee and Spudich *[23]* or obtained from Sigma).
6. Pulmozyme rhDNase or diluted cell-culture media from 293 cells transfected with wild-type or variant DNase I.
7. Antihuman DNase I rabbit polyclonal antibody-horseradish peroxidase conjugate.
8. Substrate reagent: Sigma Fast o-phenylenediamine and urea/H$_2$O$_2$ reagent (prepared according to Sigma).
9. 4.5 $N$ H$_2$SO$_4$.
10. Microtiter-plate reader (96-well).

### 2.3.2. Actin Inhibition Assay

1. Pulmozyme rhDNase or diluted cell-culture media from 293 cells transfected with wild-type or variant DNase I.
2. Buffer I: 25 m$M$ HEPES, pH 7.5, 4 m$M$ CaCl$_2$, 4 m$M$ MgCl$_2$, 0.1% BSA, 0.5 m$M$ ATP, 0.5 m$M$ β-mercaptoethanol, 150 m$M$ NaCl 0.01% thimerosal, and 0.05% Tween 20.
3. $^{33}$P-labeled M13 DNA (*see* Chapter 21).
4. Salmon testes DNA (Sigma).
5. 25 m$M$ EDTA.

6. Trichloroacetic acid (TCA).
7. Kaleidagraph v 3.0.8 c (Synergy Software).

## 3. Methods
### 3.1. DNase I Mutant Design, Mutagenesis, and Expression
#### 3.1.1. DNase I Mutant Design

1. Obtain the coordinates for different DNase I structures from the Protein Data Bank (PDB; the website address is http://www.rrcsb.org). These include 3DNI, free DNase I at 2 Å *(11)*, 1ATN, the G-actin DNase I complex at 2.8 Å *(15)*, and 2DNJ and 1DNK, DNase I complexes with octomeric DNA at 2.0 and 2.3 Å, respectively *(13,14)*.
2. For the actin-resistant variants *(6)*, model appropriate mutations at residues at the actin-binding site of DNase I (1ATN). These include His44, Asp53, Tyr65, Val67, Glu69, and Ala114.
3. For the active-site variants *(6,8)*, alter the key residues at the active site (His134, His 252, Asp212, or Glu78) to Ala; His or Glu residues can also be changed to Gln.
4. For the hyperactive variants *(7,9)*, to identify possible positions for basic residue substitutions, only the $C_\alpha$ and $C_\beta$ atoms of DNase I and DNA backbone were displayed. The sidechains of Lys or Arg were modeled into 11 DNA-proximal positions (Gln9, Thr10, Gly12, Glu13, Thr14, His44, Asn74, Ser75, Asn110, Thr205, and Thr207) with $C_\alpha$ to $C_\beta$ trajectories directed toward the DNA backbone. The positively-charged amines of the engineered Arg or Lys can be placed to within ~3 Å of the negatively-charged phosphates in the DNA backbone (2DNJ and 1DNK).

#### 3.1.2. Mutagenesis and Expression of DNase I

1. Make the desired mutant DNase I in the pRK.DNase expression vector *(21)* using site-directed mutagenesis according to the method of Kunkel et al. *(24)*. Multiple mutations were introduced into DNase I by including several single mutant primers in one hybridization reaction.
2. Verify the desired mutation by the dideoxy sequencing method *(25)*.
3. Purify mutant DNA from 500-mL cultures of transformed *E. coli* XL1 blue MRF' using Qiagen tip-500 columns.
4. Grow 293 cells in 25 mL serum-containing cell-culture media in 150-mm plastic Petri dishes.
5. When cells are in log phase (approx 50% confluent), transiently cotransfect with 22.5 µg DNase variant DNA and 17 µg adenovirus DNA *(22)*.
6. About 16 h after transfection, carefully suck off media, wash the cells with 15 mL PBS, and change to 25 mL serum-free media.
7. Harvest the cell-culture supernatant after 72–96 h, concentrate approx 10-fold using Centriprep 10 concentrators, and store at 4°C (*see* **Note 2**).

#### 3.1.3. DNase I ELISA (*Fig. 1*)

Unless indicated otherwise, all incubations were carried out at room temperature with rotary agitation of the microtiter plates.

1. Coat wells of microtiter plates overnight at 4°C with 100 µL of 62.5 ng/mL anti-DNase I in 0.05 *M* sodium carbonate, pH 9.6.
2. Wash plates with wash buffer and block nonspecific binding sites by incubation for 1 h with 200 µL of ELISA buffer.

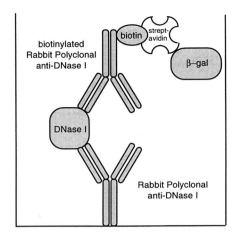

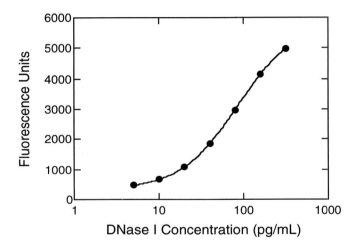

Fig. 1. DNase I ELISA. Diagram of the two-site ELISA used to quantify the concentration of DNase I and DNase I variants **(top)**. Typical calibration curve generated using human DNase I as the standard **(bottom)**.

3. Incubate 100 μL of DNase I standards and samples diluted in ELISA buffer for 2 h.
4. Wash plates with wash buffer and incubate 100 μL of biotinylated anti-DNase I at a concentration of 200 ng/mL for 2 h.
5. Wash plates with wash buffer and incubate 100 μL of streptavidin-β-galactosidase, diluted 10,000-fold in ELISA buffer, for 1 h.
6. Wash plates with wash buffer and incubate 50 μL of 4-methylumbelliferyl-β-D-galacto-pyranoside at a concentration of 1 m$M$ in substrate buffer in the dark for 24 h.
7. Stop the reaction by the addition of 200 μL of 0.15 $M$ glycine, pH 10.5, and read the fluorescence at an excitation wavelength of 360 nm and an emission wavelength of 460 nm.
8. Interpolate the sample concentration from DNase I standard curve (5–320 pg/mL) fitted to a 4-parameter logistic equation using the nonlinear regression method of Marquardt *(26)*.

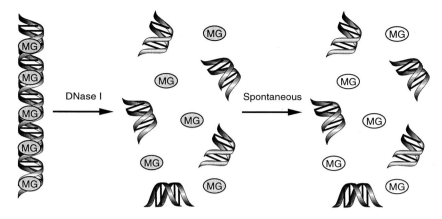

Fig. 2. DNA-methyl green assay. The diagram shows that a complex of DNA with methyl green releases free methyl green when DNA is digested by DNase I. Initially, the free methyl green retains its color but fades to a colorless compound in a subsequent spontaneous reaction at neutral pH.

## 3.2. DNase I Activity Assays

### 3.2.1. Kunitz Hyperchromicity Assay

1. Dilute culture media from 293 cells transfected with wild-type or variant DNase I by two- to 50-fold into a solution containing 10 μg/mL to 400 μg/mL calf thymus DNA in assay buffer at room temperature in a final volume of 150 μL in a 96-well microtiter plate.
2. Initiate the DNA cleavage reactions simulaneously for all samples using a multichannel pipet-aid to add diluted DNase I to DNA-containing Assay buffer.
3. Measure the increase in $A$ at 260 nm for 6 min.
4. Plot the initial rates of the increase in $A_{260}$ vs DNA concentration, and fit the data to the Michaelis-Menten equation using nonlinear regression methods to generate apparent $K_m$ and $V_{max}$ kinetic values (*see* **Note 3**).

### 3.2.2. DNA-Methyl Green Assay (**Fig. 2**)

1. Preparation of DNA-methyl green substrate: 192 mL of 2 mg/mL DNA is combined with 11.5 mL of 0.4% methyl green and 46 mL of assay buffer. The mixture is stored overnight at room temperature before use to permit equilibration and fading of unbound methyl green. Thereafter, the mixture is stored at 4°C for up to 30 d.
2. Prepare Pulmozyme rhDNase standards (usually 7) in assay buffer between 5.7 and 119 ng rhDNase/mL.
3. Prepare multiple dilutions of samples to be analyzed in Assay Buffer.
4. DNA-methyl green substrate (100 μL) is added to the wells of a microtiter plate.
5. Standards and samples (100 μL) are added to individual wells, and the contents are mixed thoroughly by trituration (8–12 cycles of aspiration/dispensing with a multichannel pipet) (*see* **Note 4**).
6. Seal the plates with adhesive film and incubate at 37°C for 6 h (*see* **Note 5**).
7. Measure the absorbance of each well in a microtiter plate reader at 620 nm. Optionally, a reference $A$ at 492 nm can be measured and subtracted from the $A$ at 620 nm.
8. Plot the $A$ of the standards vs the logarithm of the rhDNase concentration and fit the data to a 4-parameter logistic curve by nonlinear regression *(26)*.

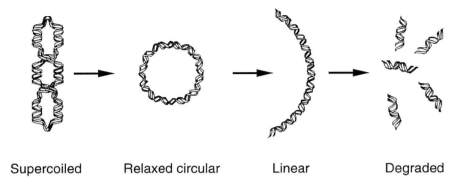

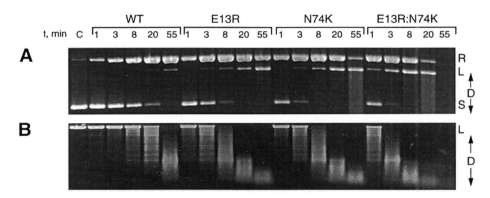

Fig. 3. Plasmid digestion assay in native agarose gel. **(Top)** DNase I catalyzes the conversion of supercoiled pBR322 plasmid DNA to relaxed circular plasmid by cleavage of a single strand, to linearized plasmid by cleavage of both strands, and to smaller DNA fragments by subsequent cleavage. **(Bottom) (A)** Supercoiled pBR322 substrate with a low level of relaxed circle background (first lane, "C" for control) is incubated with 1 ng/mL of wild-type, E13R, N74K, and E13R:N74K DNase I in a solution containing 150 m$M$ NaCl from 1 to 55 min. The linear product (L) runs in between the slower-moving relaxed circular product (R) and the faster-moving supercoiled substrate (S). Smaller degraded forms of DNA (D) run as a smear. **(B)** *Eco*RI-linearized pBR322 substrate is degraded by 10 ng/mL of DNase I in the presence of 150 m$M$ NaCl.

9. Interpolate the activity of samples from the standard curve in units of DNase concentration. A methyl green unit of activity (MG-U) is defined as the activity equivalent to 1 µg of the rhDNase standard.

### 3.2.3. Plasmid Digestion Assay in Native Agarose Gel *(Fig. 3)*

1. Incubate 30 µg/mL of supercoiled pBR322 plasmid DNA or 25 µg/mL of *Eco*RI linearized pBR322 (purified by phenol-chloroform extraction and purified by ethanol precipitation) with DNase I or ~1000-fold diluted cell-culture media from human DNase I-transfected 293 cells in the presence of incubation buffer in a final volume of 160 µL at room temperature (*see* **Note 6**).
2. Initiate the DNA cleavage reactions simultaneously for all samples using a multichannel pipet-aid to add diluted DNase I to DNA-containing incubation buffer (*see* **Note 7**).

3. At various time intervals, quench 30-μL aliquots of the reaction mix with 7 μL agarose quench buffer using a multichannel pipet-aid (*see* **Note 8**).
4. Load 30 μL directly onto a 0.8% agarose gel casted with multichannel-compatible combs using a multichannel pipet-aid (*see* **Note 9**).
5. Run the agarose gel overnight at ~1 V/cm in TBE and stain with ethidium bromide (*see* **Note 10**).
6. Quantify the individual bands (supercoiled, relaxed, or linearized) with a Molecular Dynamics Model 575 FluorImager.
7. Measure the overall activity as the initial rate of disappearance of supercoiled or linear DNA. The linear to relaxed ratio can be determined from the initial rates of appearance of linear and relaxed products (*see* **Note 11**).

### 3.2.4. p-*Nitrophenyl Phenylphosphonate Assay*

1. Add DNase I or cell-culture media from human DNase I-transfected 293 cells to reaction buffer using a multichannel pipet-aid to start all reactions at the same time.
2. Monitor the increase in the *A* at 405 nm as a result of the generation of *p*-nitrophenol over time.
3. Determine the specific activity for the enzyme by dividing the initial linear change in $A_{405}$/min by the concentration in the assay (*see* **Note 12**).

## 3.3. DNase I/Actin Assays

### 3.3.1. Actin Binding ELISA (*Fig. 4*)

1. Coat plates with 100 μL human Gc globulin at 10 μg/mL in 25 m*M* HEPES, pH 7.2, 4 m*M* $MgCl_2$, 4 m*M* $CaCl_2$, at 4°C for 16–24 h.
2. Discard the Gc globulin by banging the plates dry and block nonspecific binding sites by incubation for 1 h with 200 μL of buffer A (*see* **Note 13**).
3. Wash with wash buffer and incubate 100 μL of 50 μg/mL G-actin in buffer A for 1 h (*see* **Note 14**).
4. Wash with wash buffer and incubate 100 μL of DNase I or cell-culture harvest at various dilutions for 1 h.
5. Wash with wash buffer and incubate 100 μL of antihuman DNase I rabbit polyclonal antibody-horseradish peroxidase conjugate (19 ng/mL) for 1 h.
6. Wash with wash buffer and add 100 μL/well of Substrate reagent for 5–30 min.
7. Stop the reaction by adding 100 μL/well 4.5 *N* $H_2SO_4$ and read the *A* at 492 nm.
8. Plot the $A_{492}$ vs the DNase I concentration and fit the resultant sigmoidal curve to a 4-parameter equation by nonlinear regression analysis to determine the the $EC_{50}$ value (*26*); the $EC_{50}$ value is the DNase I concentration that produces a half-maximal signal.

### 3.3.2. Actin Inhibition Assay

1. Incubate varying concentrations of G-actin in duplicate for 15 min at room temperature with 0.54 n*M* DNase I or DNase I variant in buffer I.
2. Add $^{33}$P-labeled M13 DNA and salmon testes DNA to a final concentration of 4.1 μg/mL and incubate at room temperature for 2 h.
3. Quench with 25 m*M* EDTA and cold TCA (6.7% final concentration) and incubate samples on ice for 10 min.
4. Centrifuge at 9300*g* for 5 min at 4°C and determine the acid-soluble counts.
5. Plot the fractional activity (CPM inhibited/CPM uninhibited) versus the G-actin concentration and fit the curve using nonlinear regression analysis to the following equation to determine the $K_i$ value.

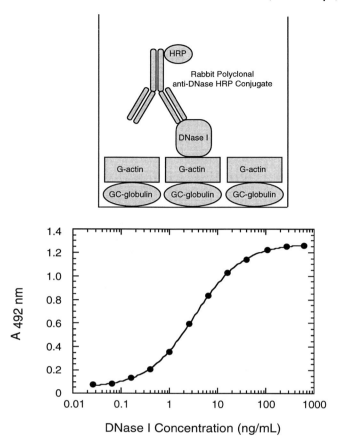

Fig. 4. Actin-binding ELISA. (**Top**) Diagram of the actin-binding ELISA where G-actin is bound to GC-globulin coated in the wells of microtiter plates. Samples containing DNase I or DNase I variants are subsequently incubated in the wells, and the bound variants are detected with a polyclonal antibody to DNase I. (**Bottom**) Data and nonlinear regression fit of DNase I binding to G-actin; the calculated $EC_{50}$ is 3 ng/mL.

$$\text{Fractional activity} = 1 - \{[E_o] + [I_o] + K_i - \sqrt{([E_o] + [I_o] + Ki)^2 - (4 \cdot [E_o] \cdot [I_o])}\}/2 \cdot [E_o] \quad (1)$$

where $[E_o]$ is the DNase I concentration, and $[I_o]$ is the total G-actin concentration.

## 4. Notes

1. Pulmozyme rhDNase is available for research use upon written request to Genentech, Inc., Research Contracts and Reagents, M/S #50, 1 DNA Way, South San Francisco, CA 94080, or via the internet at www.gene.com.
2. The concentrated cell-culture harvest generally contained 10–100 µg of DNase I variant; the concentration was generally >10 µg/mL.
3. No activity was detected in 1:2 diluted mock-transfected media. Plots of the initial rate vs DNA concentration were hyperbolic for most variants; several mutants could not be fitted to the Michaelis-Menten equation. Wild-type DNase I has a $K_m$ of 230 µg/mL DNA and $V_{max}$ of 170 $A_{260}$ U/min/mg DNase I in assay buffer. For experiments to measure inhibi-

tion by NaCl, the DNA concentration was fixed at 300 µg/mL and the amount of NaCl varied from 0 to 400 m$M$.

4. The final concentration of reagents in the assay mixture is 0.77 mg DNA-methyl green/mL, 24 m$M$ HEPES, 2.4 m$M$ MgCl$_2$, 2.4 m$M$ CaCl$_2$, 0.06% BSA, 0.006% Thimerosal, 0.03% Tween 20, 0.46 m$M$ acetate, 0.4 m$M$ EDTA, pH 7.5.

5. The incubation time and concentration of standards can be adjusted for samples outside the quantitative range *(19)*.

6. The amount of DNase I varies depending on the mutant and the concentration of NaCl or various metal ions. Under the conditions stated, approx 1 ng/mL DNase I is suitable. When only Mg$^{2+}$ or Mn$^{2+}$ ions are present, 0.4 m$M$ EGTA was added to the reaction mix to sequester trace Ca$^{2+}$ ions present in solution. For reactions in the absence of metal ions, DNase I was mixed with plasmid DNA and 1 m$M$ EDTA for ~40 min prior to quenching, followed by immediate loading on the 0.8% agarose gel. The effect of G-actin as an inhibitor of DNase I can also be determined using this assay *(27)*.

7. *Eco*RI-linearized pBR322 DNA at ng/mL concentrations can be used if it is 3'-end radiolabeled with α-$^{32}$P-dATP and Klenow. Shorter DNA fragments can also be generated by digesting the end-labeled pBR322 with other restriction enzymes and purifying the resulting fragments by polyacrylamide gel electrophoresis *(7,9)*.

8. For small DNA substrates, quench the reaction with an equal volume of acrylamide quench buffer.

9. For small-size DNA substrates, samples are loaded onto the appropriate % denaturing polyacrylamide gel. Single-stranded nicking of linearized plasmid can also be monitored on a 0.8% denaturing agarose gel in 50 m$M$ NaOH by adding NaOH to the quenched reaction to a final concentration of 50 m$M$ *(7,9)*.

10. For radiolabeled DNA substrates, dry gel and incubate with a phosphoimaging plate, which is then scanned with a Fuji phosphoimager.

11. Wild-type human DNase I cleaves linear plasmid pBR322 with a specific activity of 23 mg DNA/min/mg DNase I in incubation buffer containing 150 m$M$ NaCl. For digestion of supercoiled DNA by human DNase I in this buffer, the specific activity for the overall loss of supercoiled substrate DNA is 1200 mg DNA/min/mg DNase I and the linear/relaxed product ratio is 0.010.

12. The activity of DNase I-transfected media is corrected for background activity detected in mock-transfected media; the specific activity for human DNase I in this assay is 1.0 A$_{405}$ min/mg DNase I.

13. Buffer A was used in all dilution steps unless otherwise noted; incubations were at room temperature for 1 h with shaking.

14. G-actin (1 mg/mL; prepared by the method of Pardee and Spudich *(23)* or obtained from Sigma) was dialyzed overnight at 4°C against 5 m$M$ HEPES, pH 7.2, 0.2 m$M$ CaCl$_2$, 0.5 m$M$ ATP, 0.5 m$M$ β-mercaptoethanol. After centrifugation at 13,000g for 5 min, the amount of G-actin·ATP was quantitated by measuring the absorbance at 290 nm, using $\varepsilon_{290} = 28.3$ m$M^{-1}$/cm *(28)*.

# References

1. Ramsey, B. W. (1996) Management of pulmonary disease in patients with cystic fibrosis. *N. Engl. J. Med.* **335,** 179–188.
2. Ramsey, B. W., Astley, S. J., Aitken, M. L., Burke, W., Colin, A. A., Dorkin, H. L., Eisenberg, J. D., Gibson, R. L., Harwood, I. R., Schidlow, D. V., Wilmott, R. W., Wohl, M. E., Meyerson, L. J., Shak, S., Fuchs, H., and Smith, A. L. (1993) Efficacy and safety of short-term administration of aerosolized recombinant human deoxyribonuclease in patients with cystic fibrosis. *Am. Rev. Respir. Dis.* **148,** 145–151.

3. Fuchs, H. J., Borowitz, D. S., Christiansen, D. H., Morris, E. M., Nash, M. L., Ramsey, B. W., Rosenstein, B. J., Smith, A. L., and Wohl, M. E. (1994) Effect of aerosolized recombinant human DNase on exacerbations of respiratory symptoms and on pulmonary function in patients with cystic fibrosis. *N. Engl. J. Med.* **331,** 637–642.

4. Vasconcellos, C. A., Allen, P. G., Wohl, M. E., Drazen, J. M., Janmey, P. A., and Stossel, T. P. (1994) Reduction in viscosity of cystic fibrosis sputum in vitro by gelsolin. *Science* **263,** 969–971.

5. Lazarides, E. and Lindberg, U. (1974) Actin is the naturally occurring inhibitor of deoxyribonuclease I. *Proc. Natl. Acad. Sci. USA* **71,** 4742–4746.

6. Ulmer, J. S., Herzka, A., Toy, K. J., Baker, D. L., Dodge, A. H., Sinicropi, D., Shak, S., and Lazarus, R. A. (1996) Engineering actin-resistant human DNase I for treatment of cystic fibrosis. *Proc. Natl. Acad. Sci. USA* **93,** 8225–8229.

7. Pan, C. Q. and Lazarus, R. A. (1997) Engineering hyperactive variants of human deoxyribonuclease I by altering its functional mechanism. *Biochemistry* **36,** 6624–6632.

8. Pan, C. Q., Ulmer, J. S., Herzka, A., and Lazarus, R. A. (1998) Mutational analysis of human DNase I at the DNA binding interface: implications for DNA recognition, catalysis, and metal ion dependence. *Protein Sci.* **7,** 628–636.

9. Pan, C. Q. and Lazarus, R. A. (1998) Hyperactivity of human DNase I variants: dependence on the number of positively charged residues and concentration, length, and environment of DNA. *J. Biol. Chem.* **273,** 11,701–11,708.

10. Pan, C. Q., Dodge, T. H., Baker, D. L., Prince, W. S., Sinicropi, D. V., and Lazarus, R. A. (1998) Improved potency of hyperactive and actin-resistant human DNase I variants for treatment of cystic fibrosis and systemic lupus erythematosus. *J. Biol. Chem.* **273,** 18,374–18,381.

11. Oefner, C. and Suck, D. (1986) Crystallographic refinement and structure of DNase I at 2 Å resolution. *J. Mol. Biol.* **192,** 605–632.

12. Wolf, E., Frenz, J., and Suck, D. (1995) Structure of human pancreatic DNase I at 2.2 Å resolution. *Protein Eng.* **8 (Suppl.),** 79.

13. Lahm, A. and Suck, D. (1991) DNase I-induced DNA conformation: 2 Å structure of a DNase I-octamer complex. *J. Mol. Biol.* **221,** 645–667.

14. Weston, S. A., Lahm, A., and Suck, D. (1992) X-ray structure of the DNase I-d(GGTATACC)$_2$ complex at 2.3 Å resolution. *J. Mol. Biol.* **226,** 1237–1256.

15. Kabsch, W., Mannherz, H. G., Suck, D., Pai, E. F., and Holmes, K. C. (1990) Atomic structure of the actin:DNase I complex. *Nature* **347,** 37–44.

16. Campbell, V. W. and Jackson, D. A. (1980) The effect of divalent cations on the mode of action of DNase I. *J. Biol. Chem.* **255,** 3726–3735.

17. Prince, W. S., Baker, D. L., Dodge, A. H., Ahmed, A. E., Chestnut, R. W., and Sinicropi, D. V. (1998) Pharmacodynamics of recombinant human DNase I in serum. *Clin. Exp. Immunol.* **113,** 289–296.

18. Kunitz, M. (1950) Crystalline desoxyribonuclease I. Isolation and general properties, spectrophotometric method for the measurement of desoxyribonuclease activity. *J. Gen. Physiol.* **33,** 349–362.

19. Sinicropi, D., Baker, D. L., Prince, W. S., Shiffer, K., and Shak, S. (1994) Colorimetric determination of DNase I activity with a DNA-methyl green substrate. *Anal. Biochem.* **222,** 351–358.

20. Liao, T. H. and Hsieh, J. C. (1988) Hydrolysis of *p*-nitrophenyl phenylphosphonate catalysed by bovine pancreatic deoxyribonuclease. *Biochem. J.* **255,** 781–787.

21. Shak, S., Capon, D. J., Hellmiss, R., Marsters, S. A., and Baker, C. L. (1990) Recombinant human DNase I reduces the viscosity of cystic fibrosis sputum. *Proc. Natl. Acad. Sci. USA* **87,** 9188–9192.

22. Gorman, C., Gies, D. R., and McCray, G. (1990) Transient production of proteins using an adenovirus transformed cell line. *DNA and Prot. Eng. Tech.* **2,** 3–10.

23. Pardee, J. D. and Spudich, J. A. (1982) Purification of muscle actin. *Methods Enzymol.* **85,** 164–181.

24. Kunkel, T. A., Roberts, J. D., and Zakour, R. A. (1987) Rapid and efficient site-specific mutagenesis without phenotypic selection. *Methods Enzymol.* **154,** 367–382.

25. Sanger, F., Nicklen, S., and Coulson, A. R. (1977) DNA sequencing with chain-terminating inhibitors. *Proc. Natl. Acad. Sci. USA* **74,** 5463–5467.

26. Marquardt, D. W. (1963) An algorithm for least squares estimation of non linear parameters. *J. Soc. Indust. Appl. Math.* **11,** 431–441.

27. Baron, W. F., Pan, C. Q., Spencer, S. A., Ryan, A. M., Lazarus, R. A., and Baker, K. P. (1998) Cloning and characterization of an actin-resistant DNase I-like endonuclease secreted by macrophages. *Gene* **215,** 291–301.

28. Mannherz, H. G., Goody, R. S., Konrad, M., and Nowak, E. (1980) The interaction of bovine pancreatic deoxyribonuclease I and skeletal muscle actin. *Eur. J. Biochem.* **104,** 367–379.

# IV

# NUCLEASES IN THE CLINIC

# 21

# Assays for Human DNase I Activity in Biological Matrices

## Dominick V. Sinicropi and Robert A. Lazarus

## 1. Introduction

Pulmozyme® recombinant human deoxyribonuclease I (DNase I) is currently used as a therapeutic for cystic fibrosis (CF) *(1)* and may be effective in the treatment of systemic lupus erythematosus (SLE) *(2,3)*. As described in Chapter 20, degradation of high-molecular-weight DNA following inhalation of DNase I as an aerosol decreases the viscoelasticity of CF sputum and improves lung function. In SLE, antibodies to DNA and DNA/anti-DNA immune complexes have been implicated as the principal causative factor underlying clinical pathogenesis *(4–7)*. Degradation of extracellular DNA or DNA/anti-DNA immune complexes may be of clinical benefit in SLE by two mechanisms. First, hydrolysis of the DNA component of membrane deposited DNA/anti-DNA immune complexes may reduce inflammation in the affected tissues. Second, hydrolysis of DNA and/or DNA/anti-DNA immune complexes in the circulation may elicit a decrease in the production of antibodies to DNA over time by reducing the antigen load *(8)*.

To evaluate the mechanism of action of DNase I in CF and SLE, and to compare the potency of actin-resistant and hyperactive variants of DNase I (*see* Chapter 20), we developed methods to measure DNase I activity in both CF sputum and human serum. The development of methods to quantify DNase I activity in complex biological matrices, such as sputum and serum, is complicated by the presence of factors that alter the catalytic activity of the enzyme. In CF sputum, these include high concentrations of DNA (up to 20 mg/mL) and the presence of both proteases and G-actin, a potent inhibitor of DNase I *(9)*. In serum, other factors, such as the ionic milieu, the presence of DNA-binding proteins, DNase I inhibitors, and other biological polymers, can affect the catalytic activity of DNase I *(8)* and several DNase I-like endonucleases *(10)*. For example, endogenous DNase I (approx 3.2 ng/mL in serum determined by ELISA) does not catalyze detectable hydrolysis of DNA because of an unfavorable ionic milieu and the presence of DNase I inhibitors in serum *(8)*. Therefore, in developing assays for DNase I activity in biological matrices, it is important that the assay conditions are representative of the environment in vivo. To preserve the sample composition and environmental conditions found in vivo, the assay method should not require extraction, fractionation, or significant dilution of the biological matrix. The importance of

From: *Methods in Molecular Biology, vol. 160: Nuclease Methods and Protocols*
Edited by: C. H. Schein © Humana Press Inc., Totowa, NJ

maintaining the integrity of the biological matrix is illustrated by our observation that as little as a twofold dilution of serum is sufficient to eliminate the inhibition of endogenous DNase I in serum *(8)*. Therefore, we have chosen to develop assays that minimize the dilution and alteration of the composition of biological samples.

To quantify DNase I activity with minimal disruption of the biological matrix, we have described two types of assays, those that measure the degradation of either exogenous or endogenous DNA. Degradation of exogenous synthetic or natural DNA added in a small volume to the biological matrix is the approach used in the [33]P-DNA digestion assay *(8)* (*see* **Subheading 3.2.2.** and **Fig. 3**), the end-labeled oligonucleotide assay *(3)* (*see* **Subheading 3.2.1.** and **Fig. 2**), and the chromatin degradation assay *(8)* (*see* **Subheading 3.2.3.**). Degradation of endogenous DNA in the biological matrix is quantified by pulsed-field gel electrophoresis *(9)* (*see* **Subheading 3.1.2.**) and the resulting decrease in sputum viscoelasticity is measured in the compaction assay *(11,12)* (*see* **Subheading 3.1.1.** and **Fig. 1**). Other methods to evaluate DNase I activity in biological matrices include a radial enzyme diffusion assay, which is based on the degradation of exogenous DNA *(13)* and a PCR-based assay, which measures degradation of endogenous DNA *(9)*.

## 2. Materials

### 2.1. Assays for DNase I Activity in CF Sputum

#### 2.1.1. Compaction Assay

1. Pulmozyme® recombinant human DNase I was used as a standard for quantitation of DNase I activity (*see* **Note 1**).
2. Melting point capillary tubes (Kimax-51, $1.5–1.8 \times 90$ mm, Kimble Products, Vineland, NJ).
3. Positive displacement pipets (Gilson Microman pipets, Rainin, Emeryville, CA).
4. Microcap capillary tubes, 50 μL (Drummond Scientific, Broomall, PA).
5. Biofuge A equipped with a horizontal hematocrit-type rotor (Heraeus Instruments, Hanau, Germany).
6. Sputum samples obtained from CF patients were frozen immediately after collection in 50-mL polypropylene centrifuge tubes and stored at –70°C.

#### 2.1.2. Pulsed-Field Electrophoresis of Sputum DNA

1. Sputum samples were obtained from CF patients (*see* **Subheading 2.1.1., step 6**).
2. Positive displacement pipets (Gilson Microman pipets, Rainin, Emeryville, CA).
3. Fluorimager model 545 (Molecular Dynamics, Sunnyvale, CA).
4. Foto/Eclipse UV image digitizer with Collage image analysis software (Fotodyne, Inc., Hartland, WI).
5. DNA size markers: 0.05–1 mb, λ ladder; 0.2–2.2 mb, *S. cerevisiae* (Bio-Rad, Hercules, CA) and 125 bp–23.1 kb λDNA/*Hin*dIII fragments (Life Technologies, Gaithersburg, MD).
6. Pulsed-field certified agarose and low-melting temperature agarose (Bio-Rad).
7. Miscellaneous reagents: Proteinase K was obtained from Roche Molecular Biochemicals, Indianapolis, IN, and ethidium bromide, 10 mg/mL, was obtained from Sigma (St. Louis, MO).
8. 10X TBE stock solution: 890 m*M* Tris base, 890 mM boric acid, 20 m*M* EDTA, pH 8.0.
9. Pulmozyme DNase I (*see* **Note 1**) and DNase I variants (*see* Chapter 20) were produced at Genentech.
10. CHEF Mapper pulsed-field electrophoresis system (Bio-Rad).

## 2.2. Assays for DNase I Activity in Serum

### 2.2.1. End-Labeled Oligonucleotide Assay

1. Synthetic oligonucleotides: Two complementary 25-base oligonucleotides were synthesized with either biotin or dinitrophenol (DNP) on the 5' end. The oligonucleotide sequences were 5'-biotin-GTA-ATG-CAA-CCG-AGA-CTT-GAC-TGC-A-3' and 5'-DNP-TGC-AGT-CAA-GTC-TCG-GTT-GCA-TTA-C-3'.
2. Affinity-purified rabbit anti-DNP antibody was produced at Genentech. Anti-DNP antibodies are commercially available from several suppliers. Streptavidin-horseradish peroxidase conjugate was obtained from Roche Molecular Biochemicals (Indianapolis, IN).
3. ELISA buffer: 25 m$M$ HEPES, 1 mg/mL BSA, 0.05% polysorbate 20, 0.01% thimerosal, pH 7.5.
4. DNase buffer: In addition to the components of ELISA buffer, DNase buffer contains 2.5 m$M$ $CaCl_2$, 1 m$M$ $MgCl_2$, and 150 m$M$ NaCl.
5. Wash buffer: Phosphate-buffered saline (PBS) containing 0.05% polysorbate 20.
6. OPD substrate: 2 m$M$ o-phenylenediamine in PBS, pH 7.4, with 0.012% (w/v) $H_2O_2$.
7. Microtiter plates and plate reader.

### 2.2.2. $^{33}$P-DNA Digestion Assay

1. The following items were obtained from Life Technologies: M13mp18 single-stranded DNA, 200–300 μg/mL (cat. no. 18218-016); M13 Forward Primer, 70 μg/mL (cat. no. 18431-015); DNA polymerase I, large fragment, 3–9 U/mL (cat. no. 18012-021).
2. The following items were purchased from Amersham Pharmacia Biotech (Piscataway, NJ): 5'-[α-$^{33}$P]dATP (*see* **Note 2**), 10 mCi/mL (cat. no. AH9904); 100 m$M$ dNTP Set (cat. no. 27-2035-01).
3. 10X Klenow buffer: 0.5 $M$ Tris-HCl, 0.1 $M$ $MgCl_2$, 10 m$M$ DTT, 0.5 mg/mL BSA, pH 7.5.
4. Salmon testes DNA (Sigma, St. Louis, MO).
5. Quick Spin Sephadex G-50 columns (cat. no. 1273965, Roche Molecular Biochemicals).
6. Scintillation counter: Model LS3801 (Beckman-Coulter, Palo Alto, CA).

### 2.2.3. Chromatin Digestion Assay

1. Human chromatin was isolated from human leukocytes by a detergent lysis method *(8)*. The DNA concentration of the preparation was quantified by the diaminobenzoic acid method *(14)* using salmon testes DNA as the standard.
2. Agarose gel electrophoresis apparatus.
3. Proteinase K (Roche Molecular Biochemicals).
4. MetaPhor agarose (FMC, Rockland, ME).
5. FotoAnalyst UV image analysis system with Collage software (Fotodyne, Hartland, WI).

## 3. Methods

## 3.1. Assays for DNase I Activity in CF Sputum

### 3.1.1. Compaction Assay (*Fig. 1*)

1. Aliquot 50 μL of sputum to microcentrifuge tubes using a positive displacement pipet.
2. Prepare serial dilutions of DNase I or DNase I variant in formulation buffer.
3. Add 50 μL of each DNase I dilution to separate tubes containing 50 μL of sputum. Mix thoroughly by vortexing for 15 s.
4. Incubate the tubes for 20 min at room temperature.
5. Load the reaction mixtures into melting-point capillary tubes using 50-μL Microcap capillary tubes.

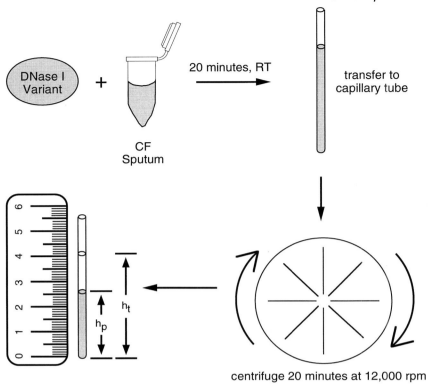

Fig. 1. Compaction assay. CF sputum incubated with DNase I or DNase I variants is centrifuged in microcapillary tubes. The ratio of the pellet to the total height is proportional to the viscoelasticity of the sputum.

6. Centrifuge the capillary tubes in horizontal heamatocrit-type rotor for 20 min at 12,000 rpm.
7. Measure the total height of the reaction mixture in the capillary ($h_t$) and the height of the pellet ($h_p$) with a metric ruler.
8. Calculate the percent compaction as $100 \times (h_p/h_t)$.

### 3.1.2. Pulsed-Field Gel Electrophoresis of Sputum DNA

1. Mix 100 µL of sputum with 11 µL of DNase I or DNase I variant in a microcentrifuge tube and incubate at 37°C for 30 min (*see* **Note 3**).
2. Quench nuclease activity by addition of 19.8 µL of 167 m*M* EDTA, 6.7% SDS.
3. Add 7 µL of 20 mg/mL proteinase K to each tube and incubate 14–18 h overnight at 50°C (*see* **Note 4**).
4. Prepare 1% agarose gel: Dissolve 1.5 g pulsed-field grade agarose in 0.5X TBE in microwave oven at 50% power. Pour gel and allow to polymerize for 1 h at 4°C.
5. Add sample (approx 67 µg DNA/lane) or DNA length markers to each well (*see* **Note 5**). Overlay the samples and DNA length markers with 1% low melting temperature agarose and allow to polymerize for 1 h at room temperature.
6. Transfer the gel to the pulsed-field electrophoresis unit containing 0.5X TBE prechilled to 14°C. Electrophoresis is carried out under the following conditions: run time = 20 h, 18 min; angle between pulses = 120°; initial switching time = 0.5 s, final switching time = 17.4 s, voltage = 6 V/cm; amperage = 130–150 mA.

7. Dilute ethidium bromide to 0.5 µg/mL in water. After electrophoresis, incubate the gel in diluted ethidium bromide for 1 h at room temperature with gentle rotary agitation. Decant the ethidium bromide solution, rinse the gel with deionized water, and incubate for 1–3 h in fresh deionized water at room temperature with gentle agitation.

8. Digitize the fluorescence image of the gel on Fluorimager using 488 nm excitation and 620 nm emission filters. Quantify the fluorescence intensity in the region of each lane corresponding to DNA longer than 23 kb. The percent DNA hydrolyzed in each lane is expressed as the fluorescence intensity in this high molecular weight region divided by that of a control lane containing no DNase I.

## 3.2. Assays for DNase I Activity in Serum

### 3.2.1. End-Labeled Oligonucleotide Assay (*Fig. 2*)

1. Reconstitute the synthetic nucleotides in 10 m$M$ Tris-HCl, 1 m$M$ EDTA, pH 8.0, at a concentration of 500 µg/mL.

2. Hybridize the labeled oligonucleotides at a concentration of 100 µg/mL of each strand in the same buffer containing 100 m$M$ NaCl and dilute the double-stranded oligonucleotide to 16 µg/mL (*see* **Note 6**).

3. For quantification of exogenous nuclease activity, add 1 µL of DNase I or DNase I variant to 18 µL of each undiluted serum sample. For quantification of endogenous nuclease activity, add 1 µL of DNase buffer to 18 µL of each undiluted serum sample.

4. Add 1 µL of 16 µg/mL double-stranded oligonucleotide to each tube and incubate at room temperature for 1 h.

5. Stop the reaction by adding 1 µL of 0.5 $M$ EDTA and dilute each sample 1:40 and 1:80 in ELISA buffer to quantify the concentration of hydrolyzed substrate by ELISA.

6. Coat the wells of microtiter plates overnight at 4°C with 100 µL of affinity-purified anti-DNP antibody at a concentration of 2 µg/mL in 0.05 $M$ sodium carbonate buffer, pH 9.6.

7. Wash with wash buffer and block for 1 h at room temperature with 200 µL of ELISA buffer. Incubate 100 µL of the standards (*see* **Note 7**) and diluted samples in the wells for 2 h at room temperature with agitation.

8. After washing the wells with wash buffer, incubate 100 µL of a 1:2000 dilution of the streptavidin-HRP conjugate for an additional 2 h at room temperature with agitation.

9. Wash the wells with wash buffer and incubate 100 µL of OPD substrate for 10–12 min at room temperature.

10. Stop the chromogenic reaction by the addition of 100 µL of 4.5 $N$ H$_2$SO$_4$ and measure the absorbance of the wells at 492 nm with a reference wavelength of 405 nm.

11. Generate standard curves by fitting the data to a four-parameter logistic function *(15)*. The concentration of unhydrolyzed substrate in the samples is determined by interpolation from the curves. Calculate the concentration of hydrolyzed substrate by subtraction from the total concentration of oligonucleotide added to the serum samples. Report the DNase I activity in units of ng DNA hydrolyzed/min/mL serum.

### 3.2.2. [33]P-DNA Digestion Assay (*Fig. 3*)

1. Prepare [33]P-M13 DNA reaction mixture: 28 µg/mL M13mp18 single-stranded DNA, 1.4 µg/mL M13 forward primer, approx 0.25 U/µL Klenow fragment, 0.077 m$M$ dNTPs, including 130 µCi/µL [33]P-dATP in 1X Klenow buffer. After incubation at 37°C for 30 min, add an additional 0.077 m$M$ unlabeled dNTPs, and incubate for an additional 15 min at 37°C. Remove unincorporated nucleotides by passing the mixture through a spin column.

2. Prepare substrate: 80 µg/mL salmon testes DNA, 0.26 µCi/µL [33]P-M13 DNA, 25 m$M$ HEPES, 150 m$M$ NaCl, 2.5 m$M$ CaCl$_2$, 1 m$M$ MgCl$_2$, 0.1% BSA, 0.01% thimerosal, pH 7.5.

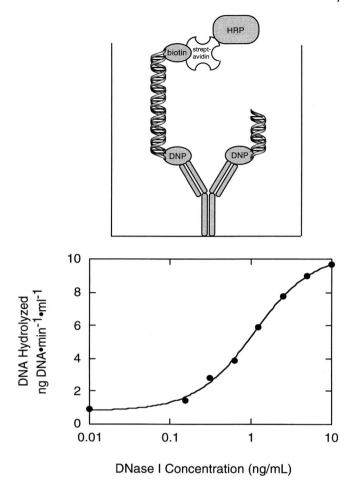

Fig. 2. End-labeled oligonucleotide assay. (**Top**) Diagram of the end-labeled oligonucle-otide assay where intact, double-stranded oligonucleotide is captured by the antibody to dini-trophenol (DNP) coated in the wells of microtiter plates and is detected by binding of streptavidin-HRP to the biotin-labeled strand. (**Bottom**) Typical dose response curve for the hydrolysis of the oligonucleotide by DNase I.

3. For quantification of exogenous nuclease activity, add 5 μL of DNase I or DNase I variant to 90 μL of each undiluted serum sample. For quantification of endogenous nuclease activity, add 5 μL of DNase buffer to 90 μL of each undiluted serum sample.
4. Add 5 μL of DNA substrate to the above mixture and incubate at 37°C for 2 h.
5. Stop the reaction by sequential addition of 100 μL of 50 m*M* EDTA and 100 μL of ice-cold 20% TCA. Incubate tubes on ice for at least 10 min.
6. Centrifuge at 12,000 rpm in a microcentrifuge and count 100 μL of the supernatant in 10 mL of scintillation cocktail.
7. Determine the total CPM in 5 μL of DNA substrate. Calculate the fraction of total DNA solubilized, which is equal to (3 times the CPM obtained for sample)/total CPM. The μg of DNA solubilized is equal to the fraction of total DNA solubilized times 0.4 μg DNA. Reac-tion velocity is expressed as the μg of DNA solubilized $\cdot$ min$^{-1}$ $\cdot$ mL$^{-1}$ serum (*see* **Note 8**).

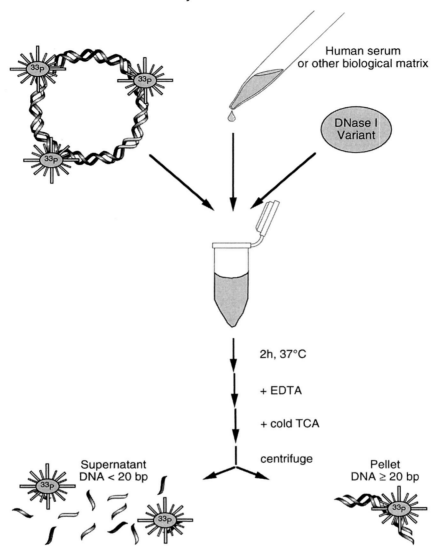

Fig. 3. [33]P-DNA digestion assay. DNase I or DNase I variants and [33]P-labeled M13 DNA are added to serum or other biological matrix in a small volume to minimize dilution of the sample. After incubation and precipitation of large DNA, soluble radioactivity associated with DNA shorter than 20 bp is quantified in a scintillation counter.

### 3.2.3. Chromatin Digestion Assay

1. Add 10 µL of 1.2 mg/mL chromatin DNA to 80 µL of human serum in a microcentrifuge tube and mix.
2. Add 10 µL of DNase I or DNase I variant to be tested (*see* **Note 3**), mix, and incubate at 37°C for 2 h.
3. Stop the enzymatic reaction by the addition of EDTA to a final concentration of 25 m*M*.
4. Prepare the samples for gel electrophoresis by addition of (final concentrations) 2% SDS and 1 mg/mL Proteinase K followed by incubation overnight at 50°C.

5. Analyze the samples and DNA length standards (75–1020 bp) by electrophoresis on 2.7% MetaPhor agarose/0.2% SDS gels in TBE buffer followed by poststaining with 1 μg/mL ethidium bromide.
6. Digitize the gel images with the FotoAnalyst system. Visual comparison of the gels is usually sufficient for comparison of the relative potency of the DNase I variants (*see* **Note 9**).

## 4. Notes

1. DNase I is available for research use upon written request to Genentech, Inc. at Research Contracts and Reagents, M/S 50, 1 DNA Way, South San Francisco, CA 94080) or via the internet at www.gene.com. DNase I is formulated at a concentration of 1 mg/mL in 150 m$M$ NaCl, 1 m$M$ CaCl$_2$, pH 7.4, and was diluted in the same buffer. Hyperactive and actin-resistant DNase I variants were produced as described by Pan et al. (*see* Chapter 20).
2. We chose to use $^{33}$P-dATP as an alternative to $^{32}$P-dATP because of its longer half-life and lower energy emission; however, either isotope can be used.
3. The DNase I or DNase I variant should be tested at several concentrations to produce dose-response data.
4. At this stage, samples may be stored at –20°C for analysis at a later date.
5. The λ ladder and yeast DNA-length markers are supplied as agarose plugs. A small piece of each plug type is inserted in separate sample wells. Three microliters of the λDNA *Hin*dIII fragments can be added to the well containing the yeast DNA markers if desired.
6. The hybridized strands are stable and can be stored at 4°C for several months; the strands are diluted to 16 μg/mL on the day the assay is performed.
7. Prepare a standard curve by dilution of the 16 μg/mL double-stranded oligonucleotide in ELISA buffer to the following concentrations: 0, 0.2, 0.039, 0.78, 1.56, 3.13, 6.25, and 12.5 ng/mL.
8. Reaction velocity is constant if the fraction of total DNA solubilized is <0.4. If the fraction of total DNA solubilized is >0.4, the sample must be reassayed with a shorter incubation period.
9. After electrophoresis, the majority of intact chromatin is not resolved and migrates slowly, corresponding to DNA longer than 1020 bp; minor polynucleosome bands correspond to DNA lengths as short as 180 bp. Increasing concentrations of DNase I produce a decrease in the fluorescence >1020 bp and an increase in the polynucleosomal bands. As an alternative to visual comparison, fluorescence intensity in the region corresponding to high-mol-wt chromatin can be quantified with the Collage software.

## References

1. Ramsey, B. W. (1996) Management of pulmonary disease in patients with cystic fibrosis. *New Engl. J. Med.* **335,** 179–188.
2. Lachmann, P. J. (1996) The *in vivo* destruction of antigen-a tool for probing and modulating an autoimmune response. *Clin. Exp. Immunol.* **106,** 187–189.
3. Macanovic, M., Sinicropi, D., Shak, S., Baughman, S., Thiru, S., and Lachmann, P. J. (1996) The treatment of systemic lupus erythematosus (SLE) in NZB/W F1 hybrid mice; studies with recombinant murine DNase and with dexamethasone. *Clin. Exp. Immunol.* **106,** 243–252.
4. Lefkowith, J. B. and Gilkeson, G. S. (1996) Nephritogenic autoantibodies in lupus: current concepts and continuing controversies. *Arthritis Rheum.* **39,** 894–903.
5. Koffler, D., Agnello, V., and Kunkel, H. G. (1973) Polynucleotide immune complexes in serum and glomeruli of patients with systemic lupus erythematosus. *Am. J. Pathol.* **74,** 109–124.

6. terBorg, E. J., Horst, G., Hummel, E. J., Limburg, P. C., and Kallenberg, C. G. M. (1990) Measurement of increases in anti-double-stranded DNA antibody levels as a predictor of disease exacerbation in systemic lupus erythematosus. *Arthritis Rheum.* **33,** 634–643.

7. Yamada, A., Miyakawa, Y., and Kosaka, D. (1982) Entrapment of anti-DNA antibodies in the kidney of patients with systemic lupus erythematosus. *Kidney Int.* **22,** 671–676.

8. Prince, W. S., Baker, D. L., Dodge, A. H., Ahmed, A. E., Chestnut, R. W., and Sinicropi, D. V. (1998) Pharmacodynamics of recombinant human DNase I in serum. *Clin. Exp. Immunol.* **113,** 289–296.

9. Pan, C. Q., Dodge, A. H., Baker, D. L., Prince, W. S., Sinicropi, D. V., and Lazarus, R. A. (1998) Improved potency of hyperactive and actin-resistant human DNase I variants for treatment of cystic fibrosis and systemic lupus erythematosus. *J. Biol. Chem.* **273,** 18,374–18,381.

10. Baron, W. F., Pan, C. Q., Spencer, S. A., Ryan, A. M., Lazarus, R. A., and Baker, K. P. (1998) Cloning and characterization of an actin-resistant DNase I-like endonuclease secreted by macrophages. *Gene* **215,** 291–301.

11. Daugherty, A. L., Patapoff, T. W., Clark, R. C., Sinicropi, D. V., and Mrsny, R. J. (1995) Compaction assay: A rapid and simple in vitro method to assess the responsiveness of a biopolymer matrix to enzymatic modification. *Biomaterials* **16,** 553–558.

12. Ulmer, J. S., Herzka, A., Toy, K. J., Baker, D. L., Dodge, A. H., Sinicropi, D., Shak, S., and Lazarus, R. A. (1996) Engineering actin-resistant human DNase I for treatment of cystic fibrosis. *Proc. Natl. Acad. Sci USA* **93,** 8225–8229.

13. Macanovic, M. and Lachmann, P. J. (1997) Measurement of deoxyribonuclease I (DNase) in the serum and urine of systemic lupus erythematosus (SLE)-prone NZB/NZW mice by a new radial enzyme diffusion assay. *Clin. Exp. Immunol.* **108,** 220–226.

14. Kissane, J. M. and Robins, E. (1958) The fluorometric measurement of deoxyribonucleic acid in animal tissues with special reference to the central nervous system. *J. Biol. Chem.* **233,** 184–188.

15. Marquardt, D. W. (1963) An algorithm for least squares estimation of non linear parameters. *J. Soc. Indust. Appl. Math.* **11,** 431–441.

# 22

# Evaluation of Retroviral Ribonuclease H Activity

## Jennifer T. Miller, Jason W. Rausch, and Stuart F. J. Le Grice

## 1. Introduction

Reverse transcription is the process whereby the single-stranded RNA genome of a retrovirus is converted into double-stranded DNA by the DNA polymerase and ribonuclease H (RNase H) activities of virus-coded reverse transcriptase (RT) *(2,16)*. The former activity of reverse transcriptase has been studied extensively, both as a potential target for antiviral therapy and a tool for molecular biology. However, increasing attention has been directed recently toward understanding the roles of RNase H activity, both nonspecific and highly specialized, in reverse transcription (reviewed in **ref. 10**). A complete summary of the process is depicted in **Fig. 1**. Quantitatively, the most important RNase H function is to degrade the RNA strand of ~10,000 bp RNA–DNA replication intermediate, eliminating the need for strand-displacement during synthesis of the second or (+) strand. This degradation is largely nonspecific, yet several cleavage events demand a high degree of precision, such as scission of the RNA strand at specific positions near the 3' termini of small, purine-rich stretches of RNA called polypurine tracts (PPTs). These RNAs serve as primers for (+) strand synthesis, but must also be removed by the enzyme once initiation has occurred. The (–) strand primer, a host-derived tRNA, is removed after its 3'-terminus is copied during (+) strand synthesis. If the (+) and (–) strand primers are not precisely removed, the resulting errors would alter the 5'- and 3'-termini of viral DNA (**Fig. 1E**). This would inhibit integration of viral DNA into the host-cell genome and perhaps adversely affect the viability of the virus at later stages of its life cycle. Finally, RNase H activity has been shown to increase greatly the efficiency of (–) and (+) strand transfer events, both of which are essential for completion of reverse transcription.

Understanding of these "simple" and "complex" RNase H-mediated events has been aided by the development of a wide variety of methodologies using substrates ranging from randomly labeled, M13-derived RNA–DNA duplexes to highly specific, uniquely end-labeled heteropolymeric substrates that mimic replication intermediates. All of the techniques and protocols required for these analyses will be described in detail in this chapter, together with a recent technology in which $Fe^{2+}$ is substituted for $Mg^{2+}$ in the RNase H active center, allowing RNase H-mediated cleavage of duplex DNA *(8)*. For reference, several reviews are available describing the retroviral enzyme and relation-

From: *Methods in Molecular Biology, vol. 160: Nuclease Methods and Protocols*
Edited by: C. H. Schein © Humana Press Inc., Totowa, NJ

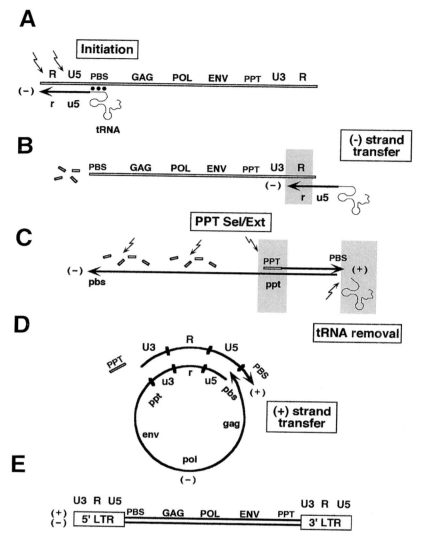

Fig. 1. RNase H-mediated events during retrovirus replication. (**A**) (−) strand DNA synthesis (filled line), initiated from a cellular tRNA hybridized to the primer binding site (PBS), continues to the 5' terminus of viral RNA (open line), during which the resulting RNA–DNA hybrid is hydrolyzed via a "synthesis-dependent" RNase H activity. (**B**) "Synthesis-independent" RNase H activity hydrolyzes the replicative intermediate within 8–10 nt of the template 5' terminus permitting transfer of nascent (−) strand DNA to the 3' end of the RNA genome via complementarity of repeat (R) sequences. (−) strand synthesis continues along the RNA genome. Concomitantly, RNase H activity continues to hydrolyze RNA of the replicative intermediate. The exception to this is the polypurine tract (PPT), which resists hydrolysis to serve as primer for (+) strand synthesis (**C**). (+) strand DNA is copied on the (−) strand DNA template up to and including 18 nucleotides of the tRNA primer, where transient pausing of RT at this base leads to RNase H-mediated excision of the tRNA primer. (**D**) Homology between PBS regions of (−) and (+) DNA permits strand transfer and relocation of nascent (+) strand DNA to the 3' end of the fully elongated (−) strand. Although an intramolecular event is depicted here, intermolecular strand switching is also possible. (**E**) Following second-strand transfer, DNA-dependent DNA polymerase activity completes synthesis of (−) and (+) strands to yield a double-stranded preintegrative intermediate.

ship of its catalytic centers *(1,3,7,9,15)*. With the exception of *in situ* analysis of RNase H activity, the methodologies described herein assume that a highly purified enzyme is available.

## 2. Materials

### 2.1. Supplies

1. Chromatography media: $Ni^{2+}$-NTA resin (Qiagen, Chatsworth CA), Sephadex G-50, S-Sepharose and DEAE-Sepharose (Amersham-Pharmacia Biotech).
2. Silica based-TLC plate with UV indicator, DE 81 ion-exchange filters (Whatman).
3. Glogos markers (Stratagene)
4. Biomax MS film (Kodak)
5. Imidazole, PMSF (Roche Molecular Biochemicals), IPTG (Research Products International), $ddH_2O$, nuclease/metal free (Sigma-W4502).

### 2.2. Reagents

1. Buffer A/78: 50 m$M$ $Na_2HPO_4/NaH_2PO_4$, pH 7.8.
2. Buffer A/60: 50 m$M$ $Na_2HPO_4/NaH_2PO_4$, pH 6.0, 0.3 $M$ NaCl, 10% (v/v) glycerol.
3. Buffer D: 50 m$M$ Tris-HCl, pH 7.0, 25 m$M$ NaCl, 1 m$M$ EDTA, 0.5 m$M$ DTT (added fresh), 10% (v/v) glycerol.
4. 4X lower Tris: 1.5 $M$ Tris-HCl, pH 8.8, 0.4% (w/v) SDS.
5. 4X upper Tris: 0.5 $M$ Tris-HCl, pH 6.8, 0.4% (w/v) SDS.
6. 10X Tris/Borate/EDTA (TBE): 0.89 $M$ Tris base, pH 8.3, 0.89 $M$ boric acid, 20 m$M$ $Na_2$-EDTA.
7. Acrylamide solution: 40% (w/v) acrylamide/*bis*-acrylamide (19:1).
8. 10X SDS running buffer (Tris/glycine/SDS): 250 m$M$ Tris, pH 8.3, 2.0 $M$ glycine, 1% (w/v) SDS.
9. SDS sample buffer: 1X upper Tris, 20% (v/v) glycerol, 3% (w/v) SDS, 3% (v/v) β-mercaptoethanol, 0.02% (w/v) bromophenol blue.
10. SDS/PAGE stain: 15% (v/v) methanol, 10% (v/v) acetic acid, 0.2% (w/v) Coomassie blue.
11. SDS/PAGE Destain: 5% (v/v) methanol, 4.2% (v/v) acetic acid.
12. Urea sample buffer (USB): 0.1% (w/v) bromophenol blue, 0.1% (w/v) xylene cyanol, 7 $M$ urea, 1X TBE.
13. 10X Tris/glycine buffer: 250 m$M$ Tris-HCl, 1.92 $M$ glycine, 10 m$M$ EDTA.
14. Nondenaturing gel-loading buffer: 30% (v/v) glycerol, 0.1% (w/v) xylene cyanol, 0.1% (w/v) bromophenol blue.
15. 10X T7 RNAP salts: 400 m$M$ Tris-HCl, pH 7.5, 130 m$M$ $MgCl_2$, 100 m$M$ NaCl.
16. 10X NTP: 4 m$M$ ATP, 4 m$M$ CTP, 4 m$M$ GTP, 4 m$M$ UTP.
17. 10X Transcription salts: 400 m$M$ Tris-HCl, pH 7.9, 100 m$M$ $MgCl_2$, 1.5 $M$ NaCl, 1 m$M$ EDTA, 10 m$M$ DTT.
18. 10X NTP solution for internal labeling: 2.5 m$M$ ATP, 2.5 m$M$ GTP, 2.5 m$M$ CTP, 0.5 m$M$ UTP.
19. 10X Bacterial alkaline phosphatase (BAP) buffer: 100 m$M$ Tris-HCl, pH 8.0.
20. 10X Polynucleotide kinase buffer: 500 m$M$ Tris-HCl, pH 7.6, 100 m$M$ $MgCl_2$, 100 m$M$ DTT.
21. 10X RNA ligase buffer: 500 m$M$ Tris-HCl, pH 7.5, 100 m$M$ $MgCl_2$, 100 m$M$ DTT, 10 m$M$ ATP, 100 mg/mL BSA.
22. 10X Annealing buffer: 100 m$M$ Tris, pH 7.5, 250 m$M$ NaCl.
23. Gel elution buffer: 0.5 $M$ sodium acetate, pH 5.2, 0.2% (w/v) SDS, 1 m$M$ EDTA.
24. 10X RNase H buffer: 0.5 $M$ Tris-HCl, pH 8.0, 50 m$M$ DTT, 0.8 $M$ NaCl, 80 m$M$ $MgCl_2$.

25. 10X SSC: 0.15 $M$ Na$_3$-citrate, pH 7.6, 1.5 $M$ NaCl.
26. 10X Renaturation buffer: 0.5 $M$ Tris-HCl, pH 8.0, 0.5 $M$ NaCl, 0.1 $M$ MgCl$_2$, 50 m$M$ DTT.
27. 10X Cleavage buffer: 0.1 $M$ Tris-HCl, pH 7.5, 0.25 $M$ NaCl.
28. Formamide loading buffer: 98% formamide, 10 m$M$ EDTA, 0.1% (w/v) xylene cyanol, 0.1% (w/v) Bromophenol blue.
29. 10X Reaction buffer: 0.1 $M$ Tris-HCl, pH 8.0, 0.8 $M$ NaCl, 50 m$M$ DTT, 60 m$M$ MgCl$_2$.
30. 10X DNA polymerase buffer: 500 m$M$ Tris-HCl, pH 8.0, 500 m$M$ KCl, 80 m$M$ MgCl$_2$, 20 m$M$ DTT.
31. 10X RNase H cleavage buffer: 500 m$M$ Tris-HCl, pH 7.5, 500 m$M$ KCl, 80 m$M$ MgCl$_2$, 20 m$M$ DTT.
32. dNTP solution: 10 m$M$ dATP, dGTP, dCTP, and TTP in 1X DNA polymerase buffer.
33. 5X Fe cleavage buffer: 0.5 $M$ NaCl, 0.8 $M$ HEPES, pH 7.2.
34. Stop solution: 0.1 $M$ thiourea, 10.0 m$M$ EDTA, 0.6 $M$ sodium acetate, pH 5.2.

## 3. Methods
### 3.1. Expression and Purification of Retroviral RT/RNase H

Most of these RNase H assays require enzymes purified to near homogeneity. In this chapter, we present the technique used in our laboratory. Most commonly, recombinant enzymes are overexpressed in *Escherichia coli* and purified using a combination of metal chelate, ion exchange, and/or hydrophobic interaction chromatography. We express the enzymes from a vector that encodes a short polyhistidine extension at the N-terminus of either the p66 or p51 subunit *(8)*. The enzyme is rapidly purified from the high-speed supernatant of a bacterial homogenate via sequential passage through Ni$^{2+}$-NTA, S- and DEAE-sepharose resins. Although we have not noted any adverse effects of the polyhistidine extension, some researchers may prefer to use enzymes which more closely resemble the natural form. In this case, Fletcher et al. *(4)* have published a rapid purification protocol utilizing strong anion, followed by strong cation-exchange chromatography, which produces an enzyme lacking the polyhistidine extension. If either of these approaches prove unsatisfactory or time-consuming, purified recombinant enzymes may also be obtained from the NIH AIDS Reagent Repository or the author of this article.

Our method is to purify HIV RT purification from 1- to 3-L cultures. Running conditions for column chromatography have been given in column volumes/h, to allow scale up to purification of 100–200 mg enzyme from 100 L of culture volume. We use NTA-sepharose. Although it is possible to use chelating sepharose (Pharmacia), the running and column regeneration conditions differ from those of NTA-sepharose.

### 3.1.1. Induction of Protein Production

1. Allow cells to grow in antibiotic-supplemented L-Broth until OD$_{600}$ = 0.7 (225 rpm at 37°C).
2. Induce with IPTG at 200 mg/mL and continue agitation for 3 h.
3. Harvest biomass by centrifugation at 12,000$g$ for 15 min, decant supernatant.
4. Wash once with RTA 7/8 to remove residual medium components and spin as above. The cells may be frozen at –20°C/–80°C until ready to use.

### 3.1.2. Cell Breakage

1. Resuspend cells in 2 vol buffer A/78 containing 1 m$M$ PMSF (fresh)/g wet weight and stir at 4°C until an even suspension is obtained.

2. Add lysozyme (in buffer A/78) to an end concentration of 0.5 mg/mL and stir until the suspension becomes viscous, approx 15 min.
3. Add NaCl to 300 m$M$ and sonicate the suspension (200–300 W) 3 × 1-min (pulsating) with 1-min cooling periods.
4. Centrifuge the sonicated cells at 100,000$g$ at 4°C for 30 min and carefully decant the high-speed supernatant.
5. Pass the supernatant twice through an 18-gage needle to reduce viscosity.

### 3.1.3. Ni²⁺-NTA-Sepharose Metal Chelate Affinity Chromatography (Note 1)

NTA resin is stripped of nickel and recharged before each run (*see below*) and then equilibrated in Buffer A/78 + 300 m$M$ NaCl.

1. After the high speed supernatant is loaded (at a speed of 1–2 column volumes/h), the column is washed with 5 vol buffer A/78 + 300 m$M$ NaCl.
2. Following the column wash, elute loosely bound proteins by application of ~10 vol buffer A/60.
3. Elute RT with a gradient (usually 10 column volumes) of 0–0.5 $M$ imidazole in buffer A/60 (**Note 2**).
4. Check RT-containing samples (eluting at approx 0.16 $M$ imidazole) immediately for purity by SDS-PAGE (*see below*) on Bio-Rad mini Protean II gels (10% gel). Do not boil samples that contain imidazole before applying them to the gel, as it tends to hydrolyze labile bonds. Instead, heat the sample at 37°C for 5 min before loading the gel. Pool fractions containing RT>70% pure and dialyze overnight against 100 vol of buffer D.
5. To regenerate the Ni²⁺-NTA column, first wash with 0.05 $M$ EDTA until it turns white, then with 100 m$M$ NaOH to remove traces of contamination. If material still remains, repeat wash with 0.5 $M$ NaOH. Neutralize by washing extensively with water until pH is near 7.0. Re-nickel with 2–3 column volumes of 2% NiSO₄, and remove unbound nickel by washing with 1 column volumes of 0.2 $M$ acetic acid. Equilibrate immediately with desired running buffer.

### 3.1.4. DEAE-Sepharose Chromatography

1. Pass RT-containing fractions from the NTA column over a DEAE-sepharose column equilibrated in buffer D.
2. Pool the column flowthrough containing RT, and then apply directly to S-sepharose.

### 3.1.5. S-Sepharose Chromatography

1. Load pooled fractions from DEAE-sepharose onto an S-sepharose column equilibrated in buffer D until E$_{280}$ is <0.1.
2. Develop column with a 25–500 m$M$ NaCl gradient in buffer D. HIV-1 RT normally elutes between 0.2 and 0.3 $M$ NaCl.
3. Concentrate pooled fractions by dialysis against buffer D containing 50% glycerol and store at –20°C.

## 3.2. SDS-Polyacrylamide Gel Electrophoresis (SDS-PAGE)

SDS-PAGE is the method of choice for fractionating protein samples on the basis of molecular weight. It may be used to evaluate sample purity and estimate the molecular weight and concentration of a protein (by comparison to appropriate standards). SDS may be eluted from the gels to permit protein refolding and recovery of activity in samples containing denatured enzyme. This technique may be used to evaluate RNase H assay *in situ*, and is described in detail in **Subheading 3.9.2.**

Although the SDS-PAGE protocol described here was developed using the Bio-Rad Protean II and Mini-Protean II Electrophoresis Systems equipped with 0.5-mm-thick combs and spacers, it should prove easily adaptable to other systems.

1. For a 40-mL resolving gel, mix 10 mL of 4X lower Tris with acrylamide and ddH$_2$O to obtain a gel percentage appropriate for separation of your protein. Add 200 µL 10% APS and 40 µL TEMED, and carefully fill the gel cassette with a syringe. Fill only 75% of the gel cassette, and overlay with H$_2$O-saturated butanol.
2. The gel should polymerize in 5–10 min, evidenced by a sharp line at the surface. Discard the H$_2$O-saturated butanol layer, and rinse thoroughly with distilled water.
3. Prepare a 3.3% stacking gel (for use with a 40-mL resolving gel we make up 15 mL) and layer onto the resolving gel to within 0.5 cm of the top of the cassette. Carefully insert the comb and allow 5–10 min for polymerization.
4. Mix samples with at least an equal volume of SDS sample buffer, heat to 95°C for 5 min, and cool to room temperature prior to loading.
5. Carefully remove the comb, rinse out sample wells, and load samples (2–30 µL, depending on well size) using a Hamilton syringe or micropipettor with gel-loading tip. If sample concentration is known, add no more than 1–2 µg of a single protein (no more than 10 µg of a mixture) per lane for best resolution.
6. Run in 1X SDS running buffer at constant current (15 mA/gel for mini Protean II cells, 25 mA/gel for standard Protean II cells) until the tracking dye enters the resolving gel. At this point, increase the current (25 mA/50 mA per gel) and run until the tracking dye reaches the bottom of the gel (or as dictated by the mol wt of the protein and desired resolution).
7. Incubate gel in SDS-PAGE stain for 15–30 min with agitation. Transfer to SDS-PAGE destain and agitate until the background is clear. Soak in water to remove acetic acid and methanol, then dry or photograph for permanent storage.

### 3.3. Denaturing PAGE

Denaturing PAGE is the method of choice for separating single-stranded nucleic acids according to size. High concentrations of urea are included in both the sample buffer and gel solution to ensure that fragments do not form inter- or intrastrand secondary structures that might affect their electrophoretic migration rates. In the context of this review, denaturing PAGE is used in RNA/DNA purification (**Subheading 3.8.**) and for fractionating RNase H cleavage products in various assays (*see below*).

Solution volumes and procedures are specific for the Bio-Rad Sequi-Gen Sequencing Cell, 21 × 40 cm plates, and 0.4-mm-thick spacers and combs, but may be scaled/modified to fit the specifications of other systems.

1. Add 500 µL APS and 25 µL TEMED to 50 mL of the appropriate percentage acrylamide/7 *M* urea/1X TBE solution.
2. Insert the comb at an angle (inverted if of the sharkstooth variety). Allow 20 min for polymerization.
3. Transfer the gel to the electrophoresis unit, filling the upper and lower chambers with 1X TBE. Sharkstooth combs should be removed and the well rinsed, while flat combs should remain in place. Prerun gels at a constant 1700–2400 V until the temperature reaches 55°C (larger and higher-percentage gels require higher voltage). As this typically requires 30 min to 1 h, prerunning provides an excellent opportunity for final preparation of samples or markers. When the prerun is complete, remove flat combs and flush the wells with 1X TBE using a syringe and narrow-gage needle.

4. Place sharkstooth combs "teeth down" in the well immediately before loading, and mix all samples with at least 1 vol of urea sample buffer (USB). Heat samples to 95°C for 2 min, then chill quickly, rinse sample wells, and load immediately.

5. Run gels at sufficient voltage to maintain a temperature of 55–60°C until the marker dyes have migrated a to the appropriate position. Use the prerun voltage to start, and adjust as necessary.

6. Following electrophoresis, disassemble the unit and transfer the gel onto Whatman #1 chromatography paper. Cover with Saran wrap. Dry the gel with a standard gel drier and process by autoradiography or phosphorimaging.

## 3.4. Nondenaturing PAGE

This protocol is specific for the Bio-Rad Model V16 vertical gel apparatus, $15 \times 17$ cm plates, 0.4-mm-thick combs and spacers, but may be scaled/modified to fit the specifications of other systems.

1. Add 400 μL APS and 20 μL TEMED to 40 mL of an appropriate percentage polyacrylamide/1X Tris/glycine solution.

2. Insert the comb and allow 20 min for polymerization.

3. Precool the gel apparatus and the running buffer (1X Tris/glycine) at 4°C. Transfer the gel to the electrophoresis unit, and prerun at 4°C (3 V/cm for at least 30 min). Do not remove the comb.

4. Prior to loading, gently mix samples with 0.2 vol nondenaturing gel-loading buffer at 4°C. Rinse wells very thoroughly, load samples, and run gel at 3 V/cm at 4°C until tracking dye has reached the appropriate position.

5. Following electrophoresis, process, dry, and analyze as in **Subheading 3.3.**

## 3.5. Preparation of Unlabeled RNA by In Vitro Transcription

In vitro transcription permits synthesis of large amounts of RNA 50 bases to several kb in length from a linear, double-stranded DNA template. This approach may be used to generate the RNA portions of specialized RNase H substrates requiring specific sequences. In addition, in vitro transcribed RNA's may be uniquely 5' or 3' end-labeled, as discussed in subsequent sections.

1. The template for in vitro transcription is usually a plasmid containing the sequence of interest inserted downstream from the bacteriophage T7 promoter, linearized by restriction digestion to produce an appropriate termination site. Alternatively, for smaller probes, the sequence to be transcribed can be prepared by the polymerase chain reaction (PCR), incorporating the T7 promoter sequence into the 5' primer.

2. To assemble the transcription reaction (300 μL), add the reagents in the following order at room temperature to prevent precipitation of the DNA template by spermidine: ddH$_2$O (**Note 3**), $10 \times$ T7 RNA polymerase (RNAP) salts to 1X, DTT to 5 m$M$, 10X NTP to 1X, RNasin to 1 U/μL, Triton X-100 to 0.5%, BSA to 0.5 mg/mL, spermidine to 1 m$M$, DNA template to 0.1 mg/mL, T7 to 0.67 U/μL, and inorganic pyrophosphatase to 1 μg/mL (**Note 4**).

3. Incubate 1 h at 37°C, then supplement with 5 μL T7 RNAP and 6 μL 0.1 $M$ GTP. Supplement again after 2 h.

4. After 3 h, add 10 μL RNase-free DNase I and incubate at 37°C for 15 min.

5. Extract RNA with an equal volume of phenol:chloroform:isoamyl alcohol (25:24:1) to remove the protein components of the reaction. Transfer the aqueous phase to an empty microtube, and add an equal volume of USB, heat to 95°C for 1 min and purify over a preparative denaturing polyacrylamide gel (**Subheading 3.8.2.**).

6. Quantify RNA after gel purification by UV absorption (E$_{260}$ 1.0 = 37 μg/mL).

## 3.6. Preparation of Internally Labeled RNA–DNA Hybrids

In the absence of a specific substrate, retroviral RNase H activity may be quantified via release of acid-precipitable counts from a randomly labeled RNA–DNA hybrid (**Subheading 3.9.1.**). A substrate for this assay may be conveniently generated using a single-stranded phage DNA and bacterial RNA polymerase. In the absence of specific "promoters" or recognition elements, RNA is nonspecifically transcribed to create a randomly labeled RNA–DNA hybrid.

1. Prepare 50 μL transcription mix containing 1X transcription Salts, 1X NTP solution for internal labeling, 100 mCi α-[$^{32}$P]-UTP (800 Ci/mmol), 2–3 μg M13 mp18 single-stranded DNA and 25 U *E. coli* RNA polymerase and incubate 1 h at 37°C. [$^3$H], [$^{35}$S], or otherwise-labeled NTPs may be substituted for [$^{32}$P]-labeled UTP if desired.
2. Extract with an equal volume of phenol:chloroform:isoamyl alcohol (25:24:1) to remove protein.
3. Apply sample to a sephadex G-50 spin column and centrifuge at 1000g for 1 min to remove unincorporated nucleotides from aqueous phase. Calculate specific activity of labeled RNA via UV absorption (A$_{260}$) and scintillation counting.

## 3.7. Preparation of Uniquely End-Labeled RNA–DNA Hybrids

RNA–DNA duplexes containing a uniquely end-labeled RNA strand are used frequently to evaluate RNase H cleavage specificity. In these reactions, only one fragment of a bisected RNA strand is detectable by autoradiography, which greatly facilitates identification of the cleavage site. When an RNA strand is cleaved multiple times, both the directionality and temporal sequence of cleavage events may be determined by comparing cleavage profiles generated using substrate containing 5'-end-labeled RNA to those generated from the 3' end-labeled substrate.

To prepare an RNA–DNA hybrid containing uniquely end-labeled RNA, the RNA strand is conventionally generated by in vitro transcription, end-labeled, and annealed to a synthetic or recombinant single-stranded DNA. Methods for radiolabeling RNA with [$^{32}$P] are described in **Subheadings 3.7.1.** and **3.7.2.** Prior to 5' end-labeling, RNA must be dephosphorylated to eliminate 5'-terminal phosphates from the 5' terminus. Both 5' and 3' end-labeled RNAs must be purified when labeling is complete.

### 3.7.1. 5'-End Labeling

1. To dephosphorylate the RNA, incubate with 150 U BAP (**Note 5**) in 1X BAP buffer for 1 h at 65°C.
2. Add 1 μL 0.5 *M* EDTA and incubate 10 min at 50°C to terminate the reaction.
3. Extract sample once with an equal volume of phenol:chloroform:isoamyl alcohol (25:24:1) and again with chloroform:isoamyl alcohol (24:1).
4. Add 0.1 vol 3.0 *M* sodium acetate (pH 5.2), 0.5 μL glycogen (10 mg/mL), and 3 vol cold 95% (v/v) EtOH to precipitate RNA. Mix thoroughly and incubate for 10 min on dry ice. Pellet by centrifugation for 10 min at 15,000 rpm. Wash with 70% (v/v) EtOH, dry *in vacuo*, and resuspend RNA in ddH$_2$O.
5. Quantify by spectrophotometry (E$_{260}$ 1.0 = 37 μg/mL).
6. Incubate dephosphorylated RNA (20 pmol) for 1 h at 37°C in 1X polynucleotide kinase (PNK) buffer containing 50 mCi γ-[$^{32}$P]-ATP (3000 Ci/mmol) and 10 U T4 PNK (V$_{tot}$ = 20 μL) to add the label, and then extract with an equal volume of phenol:chloroform: isoamyl alcohol as described above.

7. Apply sample to a sephadex G-50 spin column and centrifuge at 1000$g$ for 1 min.
8. Add an equal volume of USB to the eluate, denature by heating to 95°C for 1 min and fractionate by preparative, denaturing high-voltage electrophoresis (**Subheading 3.8.1.**).

### 3.7.2. 3'-End Labeling

1. Heat 20 pmol RNA in 1X RNA ligase buffer, 10% dimethylsulfoxide (DMSO), 10 µg/mL BSA, and 50 mCi [$^{32}$P]-pCp (3000 Ci/mmol) ($V_{tot}$ 23 µL), to 95°C for 3 min and quickly cool on ice in order to expose 3'-ends masked by RNA secondary structure. Spin tubes quickly in a microcentrifuge to remove any condensation on the lid.
2. Add 4 U T4 RNA ligase (2 µL), and allow ligation to proceed ~16 h at 4°C.
3. Increase the reaction volume to 50 µL and extract RNA with an equal volume of phenol:chloroform:isoamyl alcohol. Apply the aqueous phase to a sephadex G-50 spin column, centrifuge 1000$g$ for 1 min, and mix the eluate with an equal volume USB. Denature RNA at 95°C for 5 min, and then apply to a preparative denaturing polyacrylamide gel (**Subheading 3.8.1.**).

### 3.7.3. Preparation of RNA–DNA Hybrids

Subsequent to extraction from the gel (**Subheading 3.8.**) and quantitation, mix radiolabeled RNA with a twofold molar excess of unlabeled complementary DNA oligonucleotide in 1X annealing buffer and heat to 95°C followed by cooling slowly to room temperature. Store at –20°C until ready for use in the assays below.

## 3.8. Preparative Denaturing PAGE

### 3.8.1. Purification of Radiolabeled Products

1. Precipitate nucleic acid samples, if necessary, in 3 vol of ethanol and 0.1 vol 3 $M$ sodium acetate (pH 5.2). Dry briefly *in vacuo,* resuspend samples in an appropriate volume of USB, denature by heating 2 min at 95°C and quickly cool on ice.
2. Load the samples on a polyacrylamide/7 $M$ urea gel for purification (1-mm-thick with 1-cm wells). The polyacrylamide concentration can be varied according to the size of nucleic-acid fragments, although exceeding 10% tends to reduce recovery. Run gels in 1X TBE at a constant voltage of 1000 V. Perform subsequent steps behind a plexiglass shield.
3. Remove the top glass plate from the gel, apply Glogos markers (Stratagene) directly to the gel, and cover with Saran wrap. Expose the gel to Biomax MS film (Kodak) for 1–2 min, and develop the film.
4. Using the markers for positioning, place the film beneath the lower glass plate and remove the Saran wrap. Use the film as a guide to excise the desired product from the gel.
5. Place the gel slice in a 1.5-mL microcentrifuge tube and crush with a pipet tip. Add 400 µL gel elution buffer and incubate overnight at 37°C.
6. Apply the gel slurry to a sephadex G-50 spin column and spin at 1000$g$ for 1 min in a microcentrifuge to remove any particulate matter.
7. Precipitate nucleic acids by adding 0.1 vol 3 $M$ sodium acetate (pH 5.2) and 3 vol of 95% (v/v) ethanol, and incubate 1 h at –20°C. Spin at 4°C for 10 min at 15,000 rpm. Remove the supernatant and wash the pellet with 500 µL 70% (v/v) ethanol. Dry and resuspend the pellet in a small volume of ddH$_2$O.

### 3.8.2. Purification of Nonradiolabeled Products

1. Following electrophoresis, remove top glass plate from the gel and transfer gel to plastic wrap (**Note 6**).

2. Place the gel on a silica based-TLC plate with UV indicator protected by the plastic wrap, and visualize nucleic acid by exposure to a short-wave (254 n$M$) UV light. Excise the desired product and extract nucleic acids as described above. Resuspend nucleic acid duplexes in a small volume of 10 m$M$ Tris-HCl, pH 8.0, 1 m$M$ EDTA.

### 3.9. Evaluation of RNase H Activity–Nonspecific Substrates

### 3.9.1. RNase H-Mediated Hydrolysis of an Internally Labeled RNA–DNA Hybrid

An RNA–DNA hybrid containing randomly labeled RNA is the simplest RNase H substrate. Activity is determined through loss of radioactivity recovered on ion-exchange membranes. However, this method will not distinguish between hydrolysis caused by the retroviral RT/RNase H and its bacterial counterpart. Although *E. coli* RNase H may be present in trace amounts, it is considerably more active than the retroviral enzyme. It is therefore recommended that this assay be combined with an *in situ* evaluation in order to determine contamination (*see* **Subheading 3.9.2.**).

1. Incubate ~50,000–100,000 cpm of randomly labeled RNA–DNA hybrid with 50–100 ng RT in 1X RNase H buffer at 37°C to initiate hydrolysis.
2. Remove aliquots from the reaction (time points ranging from 5 s to 10 min) and pipet onto prelabeled Whatman DE 81 ion-exchange filters. Incubate 5–10 min at room temperature to facilitate drying.
3. Wash DE 81 filters 3 × 10 min with 1X SSC, 2 × 10 min with 95% ethanol, and then air-dry. The amount of RNA remaining in hybrid form is determined by scintillation counting.

### 3.9.2. In Situ *Analysis of RNase H Activity*

When evaluating retroviral RT/RNase H, it is imperative to determine to what extent the enzyme is free of contaminating nucleases, especially the RNase H of *E. coli*, the primary source from which recombinant enzymes are purified. Although the bacterial and retroviral enzymes can be distinguished by their divalent metal ion requirement, a simple and more informative approach is the *in situ* analysis. The enzyme preparation is separated in a SDS-polyacrylamide gel matrix containing a radiolabeled RNA–DNA *(12)*. The proteins migrate faster than the nucleic-acid duplexes, which can essentially be regarded as immobilized. Following electrophoresis, the gel is soaked in a SDS-free buffer containing the monovalent and divalent cations necessary for RNase H activity. RNase H renatures and hydrolyzes the radiolabeled substrate, whose fragments are leeched from the gel. An autoradiograph of the gel after incubation shows a "zone of clearing" on an otherwise dark background at a migration position corresponding to the molecular weight of the protein. The RNase H activity of recombinant HIV-1 RT, associated with the 66 kDa subunit of the p66/p51 heterodimer (the p51 subunit lacks an RNase H domain), is readily separated from its 17 kDa *E. coli* counterpart (**Fig. 2**).

Although this method has been used to study RNase H activity of multiple RT mutants on a single gel and directly in bacterial lysates *(6,14)*, its limitations should be pointed out. First, differences in RNase H activity can result from altered catalysis or altered renaturation kinetics. As single-point mutations can alter the level to which a protein is expressed, a second gel is needed to quantitate the amount of recombinant enzyme.

1. Cast SDS/polyacrylamide gels as outlined in **Subheading 3.2.**, but include 2 × 10$^5$ cpm of randomly labeled RNA (5 × 10$^6$ cpm/μg, *see* **Subheading 3.6.**)/DNA hybrid/mL of gel.

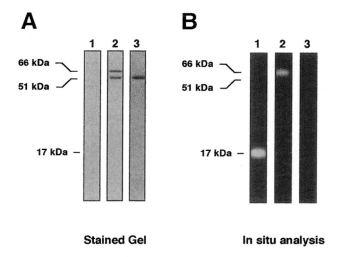

**Stained Gel**          **In situ analysis**

Fig. 2. *In situ* evaluation of bacterial and retroviral RNase H activities. **(A)** Coomassie Blue stained gel containing *E. coli* RNase H (Lane 1), the purified p66/p51 HIV-1 RT/RNase H heterodimer (Lane 2), and its purified p51 component (Lane 3). **(B)** The same proteins were fractionated through a denaturing polyacrylamide gel containing a radiolabeled RNA–DNA hybrid, renatured by removal of SDS and evaluated for RNase H activity via a "zone of clearing." Of the two HIV-1 RT polypeptides, RNase H activity is associated exclusively with p66, since this domain is proteolytically excised to generate p51. **(B)** Lane 1 illustrates that equivalent levels of RNase H activity can be demonstrated with ~1000-fold less *E. coli* RNase H, which can potentially contaminate preparations of recombinant RT. Thus, when embarking on studies with RT/RNase H, such an assay is useful to unequivocally determine the source of activity. Reprinted from Smith et al., 1993 *(14)*, with permission.

For simultaneous analysis of both retroviral (Mr ~ 66 kDa) and bacterial RNases H (Mr 17–30 kDa), a 10–15% gel is recommended.

2. Prepare crude bacterial lysates by 5 alternate incubations of 1 min each in a dry ice/ethanol bath followed by 37°C. Mix 1:1 with SDS sample buffer. Alternatively, incubate ~1 mg of purified protein at 37°C for 10 min in SDS sample buffer. Perform electrophoresis as described in **Subheading 3.2.**

3. Following electrophoresis, incubate gels in 1X renaturation buffer (~40-fold excess) with gentle rocking at room temperature for 3 d. Change renaturation buffer every 24 h to facilitate removal of SDS and RNase H renaturation.

4. Renatured RNase H will cleave the immobilized RNA–DNA hybrid into fragments small enough to elute from the gel matrix. Then stain the gel with SDS–PAGE stain, destain, dry, and subject to autoradiography as described in **Subheading 3.2.** RNase H activity will appear as bands of clearing against a black background. By comparing the location of these clear zones to the proteins on the stained gel itself, one may determine the size of the active RNase H species.

## *3.10. Evaluation of RNase H Activity-Specific RNA–DNA Hybrids*

### *3.10.1. Cleavage of Uniquely End-Labeled RNA–DNA Hybrids*

The strategy discussed in **Subheading 3.9.1.** is useful for quantitative evaluation of RNase H activity, yielding some measure of the number of cleavage events on an RNA–

DNA duplex per unit time. However, nothing is revealed about the location or specificity of cleavage. For this information, the lengths of RNA fragments generated by RNase H cleavage must be unambiguously determined. This can be achieved by subjecting a defined RNA-DNA duplex whose RNA moiety is uniquely end-labeled to cleavage by the RNase H to be evaluated, and evaluating reaction products via high resolution polyacrylamide gel electrophoresis and autoradiography. A model RNA–DNA hybrid and the specificity with which it is hydrolyzed are illustrated in **Fig. 3**.

When a uniquely end-labeled RNA is cleaved multiple times, only cleavage closest to the end-label is identified by autoradiography, regardless of the temporal sequence of hydrolysis. Therefore, with prolonged incubation, a reduction in product size provides an approximation of cleavage toward the radiolabel. By evaluating 5' and 3'-end-labeled RNAs, the directionality of processing relative to either terminus of the RNA template can be assessed. An example of this is indicated in **Fig. 3**, indicating that the retroviral enzyme can rebind a nicked template and initiate a new round of hydrolysis *(11)*.

1. To generate the RNA-DNA hybrid, mix the 5' or 3' end-labeled RNA (RNA prepared by either in vitro transcription [**Subheading 3.5.**] or by chemical synthesis, and end-labeled as per **Subheading 3.6.**) with a twofold excess of complementary DNA in 1X annealing buffer. Then heat the mixture to 95°C and cool slowly to room temperature. Typically, we anneal 20 pmol RNA template with 40 pmol DNA primer in a final volume of 10 μL.
2. Incubate RT (0.1–0.2 pmol) with radiolabeled RNA–DNA hybrid (0.1 pmol) in 1X cleavage buffer for 2 min for 37°C to form a binary complex. Initiate hydrolysis by addition of MgCl$_2$ to a final concentration of 10 m$M$ followed by thorough mixing and terminate by addition of an equal volume of formamide-loading buffer.
3. Apply reaction products directly to a high-resolution denaturing polyacrylamide gel, fractionate by high-voltage electrophoresis (**Subheading 3.3.**) and evaluate either by autoradiography or phosphorimaging. For analysis of product sizes, fractionate an alkaline hydrolysate of the single-stranded RNA template in parallel.
4. To prepare an alkaline hydrolysate of the RNA template, treat ~20,000–50,000 cpm of RNA (1–2 pmol) with 0.1 $M$ NaOH at room temperature in a final volume of 12.5 μL. Remove 2.5-μL aliquots between 20 and 100 s and mix with an equal volume of 0.1 $M$ HCl and 20 μL of formamide-loading buffer. Pool the combined reaction products to give a total volume of 125 μL, 10 ml of which (~5000 cpm) is applied to the gel.

### 3.10.2. Polypurine Tract Selection and Removal from (+) Strand DNA

In addition to generalized digestion of the retroviral RNA genome following (–) strand DNA synthesis, RT-associated RNase H is required for several highly specialized events during the course of reverse transcription. One such event is initiation of

---

Fig. 3. HIV-1 RT/RNase H-mediated cleavage of a model heteropolymeric RNA-DNA duplex. (**Top**) Schematic representation of the assay system. Substrate is a 90-nt RNA with a 36-nt DNA is hybridized at its 3' terminus. Positioning the primer 3' OH in the DNA polymerase catalytic center locates the C-terminal RNase H domain around position –17 of the duplex for initial endonucleolytic cleavage and release of 70–73-nt fragments, corresponding to hydrolysis between positions –16 and –19. Subsequently, a "directional" processing activity hydrolyzes toward the template 5' terminus as far as position –8, generating a 62-nt hydrolysis product (Scheme **A**). Creating this "gapped" duplex makes the recessed RNA 5' terminus avail-

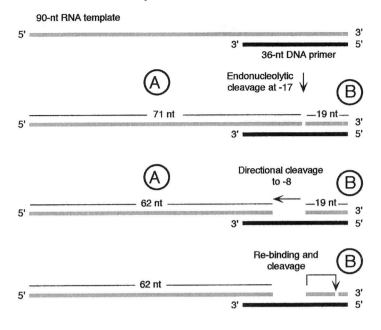

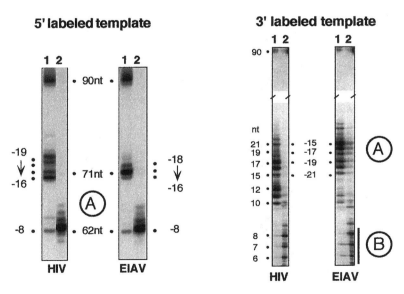

able for rebinding and an equivalent round of endonucleolytic and directional processing (Scheme **B**). Although initial, endonucleolytic cleavage will be detected with a 5' or 3'-end-labeled RNA, subsequent events are only revealed by labeling a specific terminus. **(Bottom)** Experimental evaluation of RNase H activity using substrates labeled at either the 5' (left) or 3' terminus of the RNA template (right). With the 3' labeled substrate, longer fragments represent the products of initial RT binding to the DNA 3' OH, whereas shorter fragments represent products resulting from later rebinding to the recessed 5' terminus of the RNA template. The activities of HIV-1 and EIAV RT are presented to indicate subtle differences in hydrolysis profiles generated by structurally similar enzymes.

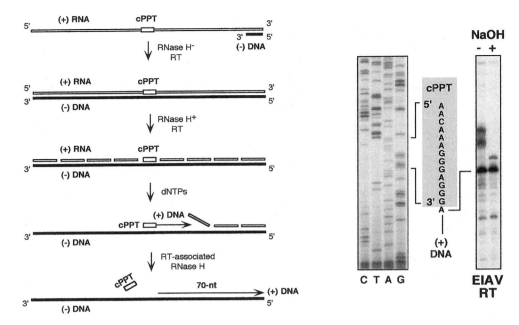

Fig. 4. RNase H-mediated selection of the polypurine-tract RNA primer and its excision from nascent (+) strand DNA. (**Top**) schematic representation of the assay system. RNA–DNA hybrid, prepared by reverse-transcribing PPT-containing RNA with an RNase H-deficient RT, is purified by nondenaturing polyacrylamide gel electrophoresis and incubated with wild-type RT in the presence of deoxynucleoside triphosphates, one of which is radiolabeled. Initially, The PPT primer, selected through its resistance to hydrolysis, is extended into (+) strand DNA. As polymerization proceeds, RNase H activity is required to cleave at the junction between the (+) strand RNA primer and nascent DNA. (**Bottom**) Selection of the EIAV central PPT (cPPT) by purified, recombinant p66/p51 EIAV RT. (+) strand products in the absence of NaOH treatment evaluate the capacity of the retroviral enzyme to remove the PPT from nascent DNA. Sequencing of the DNA strand of the duplex in parallel is used to accurately define the initiation site for (+) strand synthesis, as well as the size of the PPT RNA primer. The sequence of the EIAV cPPT has been indicated within the shaded area, adjacent to which is the first nucleotide of the (+) strand.

(+) strand DNA synthesis, wherein the enzyme must first cleave at a specific site within highly conserved, 15–20 nt purine-rich regions of the RNA genome referred to as polypurine tracts (PPTs). Subsequently initiation of DNA synthesis must occur from the RNA primer, and finally this primer must be removed from the nascent (+) strand DNA. To evaluate PPT selection and (+) strand initiation, highly specialized in vitro assays were required, as in the example developed by Rausch et al. (*11*) described and depicted in **Fig. 4**.

### 3.10.2.1. PREPARATION OF HYBRID DUPLEXES

1. Anneal a 20 nt DNA primer to a 126 nt RNA generated via in vitro transcription (**Subheading 3.5.**), and extend by an RNase H deficient RT(13) in 1X reaction buffer containing 200 μ*M* dNTPs for 1 h at 37°C. Approximately 20 pmol of RNA–DNA hybrid should be generated in each reaction.

2. Terminate DNA synthesis by extraction with an equal volume of 25:24:1 phenol/chloroform/isoamyl alcohol. Precipitate reaction products with 3 vol of ethanol, 0.1 vol of 3 $M$ NaOAc (pH 5.2), dry, and resuspend in nondenaturing gel-loading buffer. Fractionate through a 10% nondenaturing polyacrylamide gel in 1X TBE at 4°C (**Subheading 3.4.**).

3. Following electrophoresis, place gel on a silica-based TLC plate (protected by plastic wrap), and visualize the 126-bp hybrid by exposure to short-wave (254 n$M$) UV light. Perform excision and purification as described in **Subheading 3.8.** Resuspend the purified RNA–DNA hybrid in a small volume of ddH$_2$O (~20 µL).

### 3.10.2.2. EVALUATION OF PPT SELECTION AND PLUS-STRAND SYNTHESIS

1. Incubate the RNA–DNA hybrid (50 n$M$) with RT (85 n$M$), 50 m$M$, and 0.8 m$M$ α-[$^{32}$P]-dATP (3000 Ci/mmol) in 1X reaction buffer for 60 min at 37°C. Under these conditions, RNase H activity selects the PPT for subsequent extension into (+) strand DNA.

2. Terminate reactions by thermal denaturation (90°C, 2 min) and incubate on ice.

3. Mix 50% of each reaction mixture with 0.3 vol 1 $N$ NaOH and incubate at 65°C for 20 min to achieve complete RNA hydrolysis. Compare alkali-treated and -untreated products to determine the extent to which the PPT primer is removed by RT following plus-strand DNA synthesis.

4. Precipitate reaction products (treated with alkali or untreated) with 3 vol of ethanol and 0.1 vol of 3 $M$ NaOAc (pH 5.2), wash with 70% (v/v) ethanol, vacuum-dry and resuspend in 10 mL formamide-loading buffer. Fractionate the fragments by high-voltage gel electrophoresis by 10% denaturing PAGE, with a sequencing ladder in parallel for precise analysis of product sizes.

### 3.10.2.3. DISSECTION OF PPT SELECTION AND PLUS-STRAND SYNTHESIS

A modification of this protocol was developed in which RNase H-mediated cleavage of the PPT and plus-strand DNA synthesis occur in separate reactions, thereby permitting independent analysis of the two processes.

1. Incubate enzymes to be evaluated for PPT selection with RNA–DNA hybrid as above, but omit dNTPs to prevent primer extension, i.e., under these conditions, RNase H activity is evaluated.

2. Terminate RNase H activity by addition of an equal volume of 25:24:1 phenol/chloroform/isoamyl alcohol, and precipitate reaction products (0.1 vol 3 $M$ sodium acetate, pH 5.2, 3 vol 100% ethanol), wash with 70% ethanol, vacuum-dry, and resuspend in a small volume of 2X reaction buffer.

3. Establish reaction conditions capable of supporting exclusively DNA synthesis by addition of dNTPs (including α-[$^{32}$P]-dATP) and HIV-1 p66$^{E478}$Q/p51 RT (*13*) under the conditions described above (37°C, 30 min). Because this RNase H-deficient enzyme is incapable of further processing the PPT, direct evaluation of plus-strand primers generated by the "selecting" enzymes is possible.

4. Terminate DNA synthesis, partition the reactions for alkali treatment, and prepare the reaction products for high-voltage denaturing electrophoresis as described above.

## 3.10.3. RNase H-Mediated tRNA Primer Release

Upon initiation of (+) strand synthesis from the 3' polypurine tract, RT first utilizes the (–) strand DNA, and subsequently the first 18 nucleotides of the tRNA replication primer, as a template (**Fig. 1**). At this stage, RT is transiently halted as the methylated tRNA base Me-A$^{58}$ cannot be copied. Consequently, the C-terminal RNase H domain

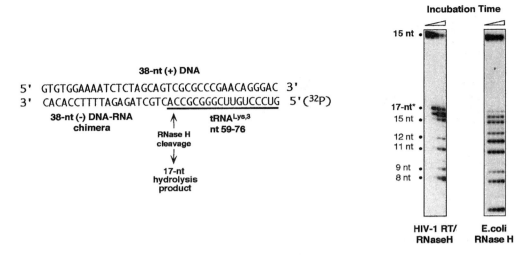

Fig. 5. RNase H-mediated excision of the retroviral tRNA from (+) strand DNA prior to the second-strand switch. (**Top**) Model substrate mimicking (+) strand DNA synthesized on a (–) strand DNA-RNA chimeric template; its RNA portion represents the 3' 18 nucleotides of tRNA$^{Lys,3}$, the HIV replication primer (*see* **Fig. 1**). In vivo, a methylated tRNA base (Me-A$^{58}$) base stalls the replication machinery, positioning the RNase H domain over the junction of the DNA-RNA chimera. Preparing a substrate whose RNA component is 5'-end-labeled allows its fate to be followed. Endonucleolytic cleavage by HIV-1 enzyme at the penultimate ribonucleotide bond, followed by "directional processing" activity, results in release of radiolabeled fragments varying in length from 17 to 8 nt. Although not shown here, in vitro experiments indicate that the oligoribonucleotide can substitute for the intact tRNA molecule without loss of RNase H specificity. (**Bottom**) Hydrolysis of the chimeric substrate by either HIV-1 RT/ RNase H (left) or *E. coli* RNase H (right). Significant differences in their hydrolysis profiles are clearly evident, providing a convenient means of evaluating "pure" preparations of retroviral RT for contamination with a bacterial nuclease.

is positioned immediately adjacent to the (–) strand DNA/tRNA junction and poised to remove the (-) strand replication primer. This mandatory event in retroviral replication permits "strand switching," which relocates nascent DNA to the 5' end of the genome for continued (+) strand synthesis. Although cleavage at the DNA–RNA junction might be predicted, the HIV-1 enzyme has been shown to hydrolyze the penultimate phosphodiester bond, leaving a single ribonucleotide on the (+) strand DNA. **Figure 5** depicts a model substrate to study these events, and its cleavage by retroviral and bacterial RNases H. Shown in the right panel are the strikingly different hydrolysis profiles exhibited by the two enzymes, and a comparison of these reveals the level to which *E. coli* RNase H contaminates a "pure" preparation of recombinant RT. The substrate for this experiment is a 38-nt DNA template to which an 18-nt radiolabeled RNA primer is hybridized at the 3' end and subsequently extended by an exonuclease-deficient Klenow fragment of DNA polymerase I. Template and primer are prepared by chemical synthesis.

1. In a preliminary step, label a 40 pmol oligoribonucleotide representing the 3' terminus of tRNA$^{Lys,3}$ at its 5' terminus with γ-[$^{32}$P]-ATP and polynucleotide kinase according to **Sub-**

**heading 3.7.1.** Include the RNase H inhibitor RNasin in the labeling reaction at a final concentration of 1.5 U/mL.

2. Unless the RNA is chemically synthesized, purify the labeled oligoribonucleotide by denaturing polyacrylamide gel electrophoresis (**Subheading 3.8.1.**) and resuspend in ddH$_2$O.

3. Hybridize the labeled RNA with 40 pmol of the 38-nt DNA template presented in **Fig. 5** by heating to 95°C and slow cooling in 1X DNA polymerase buffer, in a final volume of 20 µL. Subsequently, add dNTPs to a final concentration of 0.4 m*M*, together with 8 U of Exo (–) Klenow fragment. Allow DNA synthesis to proceed 5 min on ice, 5 min at room temperature, and 1 h at 37°C.

4. Separate annealed, extended RNA–DNA substrate from prematurely terminated products by nondenaturing polyacrylamide gel electrophoresis (**Subheading 3.4.**), purify and resuspend in a small volume (2 µL) of ddH$_2$O.

5. Incubate 5–10 pmol of annealed, extended RNA–DNA substrate with 0.1–0.2 pmol of RT in 1X RNase H cleavage buffer at 37°C. Withdraw samples at the appropriate times, mix with an equal volume of urea sample buffer, and evaluate by high-voltage denaturing polyacrylamide gel electrophoresis (usually a 20% polyacrylamide gel, as the cleavage products are small) and autoradiography.

## 3.11. RNase H-Mediated Cleavage of Duplex DNA

RNase H cleavage of the defined heteropolymeric substrate described in **Fig. 3** indicates cleavage of the RNA moiety of an RNA–DNA hybrid approx 17 bp behind the DNA polymerase catalytic center. Recently, Götte and colleagues *(5)* have demonstrated an elegant modification of this by substituting Fe$^{2+}$ for Mg$^{2+}$ in the RNase H catalytic center, thereby permitting hydroxyl radical-mediated cleavage of duplex DNA in the immediate vicinity of the bound metal, i.e., at position –17. An example of this is presented in **Fig. 6**. Although a nonenzymatic cleavage, this unique method has been exploited to demonstrate subtle differences in the manner in which RNA–DNA and DNA–DNA duplexes are accommodated in the nucleic-acid binding cleft, and will undoubtedly find further use in providing high-resolution data on the architecture of the RNase H domain. Since the specificity of this approach is defined by binding of the DNA polymerase catalytic center to the primer 3' OH, the method can be applied to any heteropolymeric template/primer. Although Mg$^{2+}$ is coordinated at both the DNA polymerase and RNase H catalytic centers, Fe$^{2+}$-mediated cleavage is restricted to the latter.

1. Prepare substrate by mixing 5'-end-labeled template with primer as in **Subheading 3.7.3.**

2. Prepare binary complexes at room temperature in a total volume of 7 µL. The reaction should contain 50 n*M* template/primer and 1 µ*M* RT in 1X Fe cleavage buffer.

3. Incubate 5 min at room temperature. Then pipet the following reagents onto the side of the reaction tube: 1.0 µL 50 m*M* DTT, 1 µL 0.5% H$_2$O$_2$ (prepared fresh), and 2 µL 2 m*M* Fe(NH$_4$)$_2$(SO$_4$)$_2$·6H$_2$O.

4. Gently close tubes and spin in a microcentrifuge to initiate Fe-mediated cleavage. Incubate a further 5 min at room temperature.

5. Terminate cleavage and precipitate the nucleic-acid substrate by adding 40 µL stop solution, 1 µL glycogen, and 200 µL 95% (v/v) ethanol.

6. After thorough mixing, place reaction tubes on dry ice for 10 min. Precipitate nucleic acids by centrifugation at 13,000 rpm in a refrigerated microcentrifuge for 10 min.

7. Decant supernatant and carefully wash precipitated nucleic acids with cold 70% (v/v) ethanol. Repeat centrifugation as above, decant the supernatant, and dry the nucleic acids *in vacuo.*

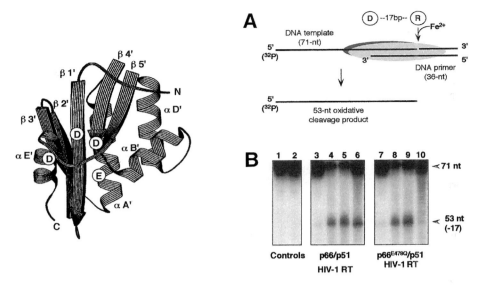

Fig. 6. Hydroxyl radical cleavage of duplex DNA through replacement of $Mg^{2+}$ in the RNase H catalytic center with $Fe^{2+}$. **(Top)** Structure of the C-terminal RNase H domain of HIV-1 RT illustrating the positions of residues implicated in binding divalent cation, i.e., $Asp^{443}$ ($\alpha$-strand 1'), $Asp^{478}$ ($\alpha$-helix A') $Glu^{498}$ ($\alpha$-strand 4'), and $Asp^{549}$ ($\alpha$-helix E'). Although the exact mechanism is unclear, $Fe^{2+}$ can replace $Mg^{2+}$ to promote hydroxyl radical-mediated cleavage of DNA at a position defined by the spatial relationship between the DNA polymerase and RNase H catalytic centers. **(Bottom) (A)** Schematic representation of the assay. Substrate is a 71-nt $[^{32}P]$-5' labeled DNA template to which a 36-nt primer is hybridized. Localization of RT over the primer 3' OH positions the RNase H domain (R) ~17 bp behind the DNA polymerase catalytic center (D). Using this template-primer (TP), $Fe^{2+}$-mediated cleavage originating within the RNase H domain releases a 53-nt radiolabeled fragment. **(B)** $Fe^{2+}$-mediated cleavage, catalyzed by wild-type HIV-1 RT (Lanes 3–6) and an RNase H-deficient mutant (Lanes 7–10). Lanes 1, 71-nt template; Lane 2, TP/reactants/no RT; Lanes 3, 7, TP/RT, no $Fe^{2+}$; Lanes 4, 8, TP/RT/$Fe^{2+}$; Lanes 5, 9, TP/RT/$Fe^{2+}$/100 *M* EDTA; Lanes 6, 10, TP/RT/$Fe^{2+}$/400 *M* EDTA. Note that high concentrations of EDTA can remove $Fe^{2+}$ from the RNase H domain of mutant RT, causing loss of cleavage.

8. Dissolve the nucleic acids in 5 µL USB, fractionate by denaturing PAGE, and analyze by autoradiography or phosphorimaging.

## 4. Notes

1. Never use high (>1 m*M*) concentrations of DTT, as this will reduce the nickel ions, and therefore the NTA-sepharose. If you need to use a reductant, you may add up to 20 m*M* β-mercaptoethanol. To reduce nonspecific binding, do not use too large a column for isolation. In general, we use a 3-mL column for RT purification for between 3 and 10 g cells (corresponding to 5–20 mg RT).

2. This is a linear gradient set up with equal volumes of each solution. The imidazole solution must be fresh, as even a slightly yellow solution will complicate $OD_{280}$ measurements.

3. When working with RNA, it is important to use $ddH_2O$ that is free of nucleases and contaminating metals. We prefer to purchase previously tested $ddH_2O$ from Sigma chemicals molecular biology reagents (*see* **Subheading 2.1.**). See protocols from Kormanec (Chapter 31) or Schein (Chapter 9) for preparing nuclease-free water.

4. Inorganic pyrophosphatase prevents product inhibition of RNA synthesis by eliminating pyrophosphate.
5. The BAP should include $Zn^{2+}$ in the enzyme preparation; thus, no $Zn^{2+}$ is needed in 10X BAP buffer.
6. For thin gels, cover an old film slightly larger than the gel with plastic wrap, lay on top of the gel plastic wrap side down, and flip the sandwich over. Gradually let the gel drop onto the covered film as you slid the gel off the edge of the lab counter (support the film with one hand). When the gel is completely transferred, lay the film flat on the counter. Cover the gel with plastic wrap and flip it onto a silica plate with UV indicator. Remove the film and visualize as in **Subheading 3.8.2.**

# References

1. Arts, E. J. and Le Grice, S. F. (1998) Interaction of retroviral reverse transcriptase with template-primer duplexes during replication. *Prog. Nucleic Acid Res. Mol. Biol.* **58,** 339–393.
2. Baltimore, D. (1970) RNA-dependent DNA polymerase in virions of RNA tumor viruses. *Nature* **226,** 1209–1211.
3. Ding, J., Hughes, S. H., and Arnold, E. (1997) Protein-nucleic acid interactions and DNA conformation in a complex of human immunodeficiency virus type 1 reverse transcriptase with a double-stranded DNA template-primer. *Biopolymers* **44,** 125–138.
4. Fletcher, R. S., Holleschak, G., Nagy, E., Arion, D., Borkow, G., Gu, Z., Wainberg, M. A., and Parniak, M. A. (1996) Single-step purification of recombinant wild-type and mutant HIV-1 reverse transcriptase. *Protein Expr. Purif.* **7,** 27–32.
5. Gotte, M., Maier, G., Gross, H. J., and Heumann, H. (1998) Localization of the active site of HIV-1 reverse transcriptase-associated RNase H domain on a DNA template using site-specific generated hydroxyl radicals. *J. Biol. Chem.* **273,** 10,139–10,146.
6. Hizi, A., Hughes, S. H., and Shaharabany, M. (1990) Mutational analysis of the ribonuclease H activity of human immunodeficiency virus 1 reverse transcriptase. *Virology* **175,** 575–580.
7. Hottiger, M. and Hubscher, U. (1996) Human immunodeficiency virus type 1 reverse transcriptase. *Biol. Chem. Hoppe Seyler* **377(2),** 97–120.
8. Le Grice, S. F. and Gruninger-Leitch, F. (1990) Rapid purification of homodimer and heterodimer HIV-1 reverse transcriptase by metal chelate affinity chromatography. *Eur. J. Biochem.* **187,** 307–314.
9. Le Grice, S. F., Naas, T., Wohlgensinger, B., and Schatz, O. (1991) Subunit-selective mutagenesis indicates minimal polymerase activity in heterodimer-associated p51 HIV-1 reverse transcriptase. *EMBO J.* **10,** 3905–3911.
10. Rausch, J. W. and Le Grice, S. F. (1997) Reverse transcriptase-associated ribonuclease H activity as a target for antiviral chemotherapy. *Antivir. Chem. Chemother.* **8,** 173–185.
11. Rausch, J. W. and Le Grice, S. F. (1997) Substituting a conserved residue of the ribonuclease H domain alters substrate hydrolysis by retroviral reverse transcriptase. *J. Biol. Chem.* **272,** 8602–8610.
12. Rucheton, M., Lelay, M. N., and Jeanteur, P. H. (1979) Evidence from direct visualization after denaturing gel electrophoresis that RNase H is associated with MSV-MuLV reverse transcriptase. *Virology* **97,** 221–223.
13. Schatz, O., Cromme, F. V., Gruninger-Leitch, F., and Le Grice, S. F. (1989) Point mutations in conserved amino acid residues within the C-terminal domain of HIV-1 reverse transcriptase specifically repress RNase H function. *FEBS Lett.* **257,** 311–314.
14. Smith, J. S. and Roth, M. J. (1993) Purification and characterization of an active human immunodeficiency virus type 1 RNase H domain. *J. Virol.* **67,** 4037–4049.

15. Tarrago-Litvak, L., Andreola, M. L., Nevinsky, G. A., Sarih-Cottin, L., and Litvak, S. (1994) The reverse transcriptase of HIV-1: from enzymology to therapeutic intervention. *FASEB J.* **8,** 497–503.
16. Temin, H., and Mizutani, S. (1970) RNA-directed DNA polymerase in virions of Rous sarcoma virus. *Nature* **226,** 1211–1213.

# 23

## Assays for Detection of RNase A Superfamily Ribonucleases

### Helene F. Rosenberg and Joseph B. Domachowske

## 1. Introduction

Ribonuclease A (RNase A), isolated from bovine pancreas, maintains a well-deserved place in the history of modern biochemistry, as many of the earliest studies on amino-acid sequencing, protein crystallography, and protein folding were performed on this stable and abundant protein *(1)*. The existence of what is currently known as the ribonuclease (RNase) A superfamily (proteins with specific elements of shared sequence despite >50% amino-acid sequence divergence) emerged in the mid 1980s. There are currently six known human members of this superfamily, including pancreatic ribonuclease (RNase 1), eosinophil-derived neurotoxin (RNase 2), eosinophil cationic protein (RNase 3), RNase 4, angiogenin (RNase 5), and RNase 6 *(2,3)*. Although elevated levels of serum ribonuclease activity were once thought to be diagnostic of pancreatic cancer *(4)*, subsequent work demonstrated that serum ribonuclease activity was most closely correlated with declining renal function regardless of the underlying disease process *(5,6)*. While not diagnostic of any specific disease or condition, ribonuclease activity has been used as a marker for the presence and activation state of human eosinophilic leukocytes both in vitro *(7,8)* and in vivo *(9,10)*. For example, Harrison and colleagues *(9)* demonstrated that eosinophils recruited into the lower airways in response to infection with respiratory syncytial virus were activated and degranulating, as demonstrated by the presence of immunoreactive and ribonucleo-lytically active eosinophil-derived neurotoxin (EDN/RNase 2) and eosinophil cationic protein (ECP/RNase 3).

More recently, the focus in the RNase A field has been on therapeutics. To date, the best-characterized therapeutic application of RNase A ribonucleases is Onconase, an atypical RNase A family member isolated from oocytes of the bullfrog *Rana pipiens* *(11)*. Onconase is toxic to malignant cells in vitro, and is currently in phase III trials for the treatment of mesothelioma (http://www.alfacell.com). Angiogenin (RNase 5) is a target of anticancer therapy, as the antiangiogenin monoclonal antibody 26-2F has been shown to prevent the growth and development of metastases of the colon carcinoma cell-line HT-29 and the formation of human breast cancer xenografts in athymic mice *(12–14)*.

From: *Methods in Molecular Biology, vol. 160: Nuclease Methods and Protocols*
Edited by: C. H. Schein © Humana Press Inc., Totowa, NJ

As analysis and detection of RNase A superfamily ribonucleases are not among the tests used as part of the diagnosis of human disease, the protocols to follow were designed for research use only. There are two basic types of assay for these ribonucleases. The first is based on enzymatic activity, and thus detects one or more of these proteins in a fairly nonspecific manner based on their ability to degrade standard substrates. The second type of assay detects ribonucleases on the basis of their specific amino-acid or nucleotide sequences. These include immunologic assays, such as radioimmunoassays, Western blotting, and immunohistochemistry, and detection of specific mRNAs via Northern blotting. We will profile specific examples of each technique, as each has been used to specifically detect RNase A superfamily ribonucleases and genes. Although not included among the protocols, we would like to refer the reader to the ribonuclease inhibitor-competition assay designed by Bond *(15)*. This is a clever and unusual variation on an immunoassay that was developed specifically for angiogenin, but has potential for use in the detection of other RNase A superfamily members.

## 2. Materials

### *2.1. Materials Required*

1. Substrates: yeast tRNA (Sigma Aldrich, St. Louis, MO); dinucleotides CpN and UpN (Sigma Aldrich); poly C and poly U (Amersham Pharmacia, Piscataway, NJ); human multitissue Northern blots (Clontech, Palo Alto, CA); random prime labeling kits (Boehringer Mannheim, Indianapolis, IN).
2. Equipment: gamma-detecting scintillation counter (LS6000IC, Beckmann, Fullerton, CA); UV-VIS spectrophotometer (Beckmann); 14% tris-glycine polyacrylamide gels, polyacrylamide gel electrophoresis and Western blotting chamber (Novel Experimental Technologies, San Diego, CA); Mini-Protean II Dual slab gel apparatus ($0.8 \times 8 \times 10$ cm, Bio-Rad, Hercules, CA); X-ray processor and film (Kodak, Rochester, NY), pH meter (Accumet model 15, Fisher Scientific), 256 nm UV lamp (UVP transilluminator, Upland, CA), F254 fluorescent indicator (Merck, Whitehouse Station, NJ); multi-temperature rotating water bath (Fisher Scientific), microcentrifuge (Sorvall, Newtown, CT), quartz cuvets (Fisher Scientific), electrophoresis power source (Amersham Pharmacia), TLC plates.
3. Biochemicals (*sources listed for items with limited availability only): sodium phosphate monobasic, sodium phosphate dibasic, sodium azide, Tris (Tris[hydroxymethyl]aminomethane) HCl, Tris base, lanthanum nitrate (Sigma Aldrich), 60% perchloric acid (Fisher Scientific), sodium chloride, MOPS (3-[*N*-morpholino]propane sulfonic acid), acrylamide, *bis*-acrylamide, Bromophenol blue, glycerol, protamine sulfate, magnesium sulfate heptahydrate, 20X SSC (Biofluids, Rockville, MD), Ficoll 400 (Sigma Aldrich), polyvinylpyrrolidone (Sigma Aldrich), bovine serum albumin (BSA), Tween-20 (Sigma Aldrich), formamide, sodium dodecyl sulfate, sodium dextran sulfate, NBT (nitroblue tetrazolium, Bio-Rad), BCIP (5-bromo-4-chloro-3-indolyl phosphate, Bio-Rad), nonfat dry milk, gelatin (Bio-Rad), *N, N* dimethylformamide (Sigma Aldrich), acetic acid, methanol, isopropanol, toluidine blue (Sigma Aldrich) Coomassie blue R-250 (Sigma Aldrich), salmon sperm DNA (3 prime-5 prime, Boulder, CO), nitrocellulose filters (Schleicher & Schuell, Keene, NH).
4. Radiochemicals: $\alpha^{32}$P-dCTP (1.0 mCi; Amersham Pharmacia).
5. Antibodies: monoclonals EG1 and EG2 (Amersham Pharmacia); alkaline phosphatase-conjugated goat antimouse IgG (Bio-Rad).

## 2.2. Reagents

1. Basic RNase assay buffer: 100 m*M* sodium phosphate, pH 7.0.
2. RNase quenching solution: 50 m*M* lanthanum nitrate and 3% perchloric acid.
3. Dinucleotide RNase assay buffer: 100 m*M* MOPS, pH 7.0, and 100 m*M* sodium chloride.
4. Zymogram stacking gel: 3% acrylamide and 0.08% *bis*-acrylamide.
5. Zymogram separating gel: 12, 15, or 18% acrylamide and 0.3% *bis*-acrylamide with 0.3 mg/mL poly C or poly U substrate.
6. Zymogram sample buffer: 50 m*M* Tris-HCl, pH 6.8, 2% SDS, 0.001% Bromophenol blue, and 5–15% glycerol.
7. Zymogram wash buffer: 20 m*M* Tris-HCl, pH 8.0 and 20% isopropranol.
8. Zymogram incubation buffer: 100 m*M* Tris-HCl, pH 7.5–8.0.
9. Zymogram staining solution: 0.2% toluidine blue in 10 m*M* Tris-HCl, pH 7.5–8.0.
10. Coomassie blue staining solution: 0.1% Coomassie blue in 50% methanol and 10% acetic acid.
11. Destain solution: 7% acetic acid.
12. Western-blotting antibody dilution solution: 50 m*M* Tris-HCl, pH 8.0, 150 m*M* sodium chloride, 0.05% Tween-20, and 1% gelatin.
13. Western-blotting wash buffer: 50 m*M* Tris-HCl, pH 8.0, 150 m*M* sodium chloride, and 0.05% Tween-20.
14. Western-blotting blocking buffer: 50 m*M* Tris-HCl, pH 8.0, 150 m*M* sodium chloride, 0.05% Tween-20, and 5% nonfat dry milk.
15. Western-blotting developing buffer: 50 m*M* Tris-HCl, pH 9.5, and 0.01 mg/mL magnesium sulfate heptahydrate.
16. Western-blotting developing reagents: 3 mg/mL NBT in 70% *N,N*-dimethylformamide; 1 mg/mL BCIP in *N,N*-dimethylformamide.
17. Denhardt's solution (100X): 10 g Ficoll 400, 10 g polyvinylpyrolidone, 10 g BSA (fraction V), and distilled water to 500 mL, filtered at stored at –20°C.
18. Northern-blotting prehybridization solution: 5X SSC, 50% formamide, 5X Denhardt's solution, 0.05 *M* sodium phosphate, pH 6.5, 0.01% SDS, and 0.1 mg/mL sheared denatured salmon-sperm DNA.
19. Northern-blotting hybridization solution: prehybridization solution containing 6% dextran sulfate.
20. Northern-blotting washing solution A: 2X SSC and 0.1% SDS.
21. Northern-blotting washing solution B: 1X SSC and 0.1% SDS.
22. Northern-blotting washing solution C: 0.2X SSC and 0.1% SDS.

## 3. Methods

### 3.1. Basic Ribonuclease Assay

This assay is based on the one originally described by Anfinsen and colleagues (*16*) and modified by Slifman and colleagues (*17*) to assay purified EDN and ECP. Once time points within the linear portion of the curve have been identified, single time points can be used for comparative analysis. Variations in this basic assay include the use of wheat-germ RNA and total yeast RNA as substrates. This assay has also been used to determine the relative activities of various ribonucleases against defined nucleotide substrates, such as poly (C) and poly (U).

1. Dilute the ribonuclease assay buffer to 40 m*M* with distilled water to a final volume of 800 µL in each microfuge tube; prepare triplicate samples for each time point.
2. Add dilutions of body fluid (serum, urine, CSF, and others) in volumes of 20 µL or less. Add 500 µL cold quenching solution (500 µL) to one tube to serve as a time = 0 control, or blank.

3. At time = 0, add 10 μL of a 4 mg/mL solution of yeast tRNA substrate to the blank and to each test sample. At various time points (3, 5, 10, and 20 min) add 500 μL cold quenching solution to each of the test samples. The test samples and blank are held on ice for 10 min.
4. Precipitate the acid-insoluble, undigested tRNA by centrifugation for 5 min at room temperature in a microcentrifuge.
5. Read the optical density (OD) at wavelength 260 nm of the supernatant containing the solubilized ribonucleotides. The OD units generated per unit time can be converted into nmol ribonucleotides using conversion factors mol wt tRNA ~28,100; 1 OD 260 nm = 40 μg RNA *(18)*.

## 3.2. Detection of Ribonuclease Activity Using Dinucleotide Substrates

This method was used by Sorrentino and Libonati *(19,20)* for comparative analysis of RNase A superfamily ribonucleases. Although this assay is generally used to compare the activities of known ribonucleases against standard substrates, it requires fewer steps than the standard assay described in **Subheading 3.1.**

1. Dilute the sample (final volume of <20 μL) in a quartz cuvet containing 1 mL dinucleotide assay buffer at room temperature (25°C). Do the assay in triplicate for each sample. Include a control without sample, also in triplicate.
2. Initiate the reaction by adding 10 μL dinucleotide substrate to 0.1 m*M* final concentration; mix sample by covering and inverting end-over-end.
3. Monitor the degradation of the dinucleotide substrate continuously as the reduction in OD at 280–290 nm/U time.
4. Subtract rates observed in controls without sample (i.e., spontaneous hydrolysis of substrate) from the observed rate.
5. Express data as the change in OD per time. Determine the total measurable change in OD and calculate change in OD per time/total measurable change in OD.

## 3.3. Zymogram Detection

Hodes and colleagues *(21–23)* described methods for both positive and negative staining zymograms using substrate-impregnated agarose overlays or acrylamide gels for detecting ribonuclease activity in electrophoretically separated proteins. The method described here is that of Bravo and colleagues *(24)*, in which the substrate is included in the acrylamide-gel matrix. The ribonucleases separated in the gel remain inactive in the presence of SDS, and are activated after its removal. We suggest reference to the original article for more extensive instruction in this technique. The authors note that they can detect as little as 1 pg of RNase A by this technique. The direct visualization method permits the operator to follow the extent of substrate degradation as the assay progresses.

1. Pour the acrylamide gel via standard procedures *(25* and Chapter 22), selecting an acrylamide concentration that permits optimal separation of the proteins of interest. Add poly (C) or poly (U) to a final concentration of 0.3 μg/mL in the gel to the water for the separating gel. Heat to 50°C to enhance dissolution, before combining with the acrylamide. Wear gloves to avoid contamination with ribonucleases from the skin surface.
2. Apply samples diluted in sample buffer and run gel at 30–40 mA current until the tracking dye reaches the separating gel; then reduce the current to 20 mA.
3. After electrophoresis, remove the SDS with two 15-min washes in zymogram wash buffer (enough to completely cover the gel) with continuous shaking at room temperature. Assess

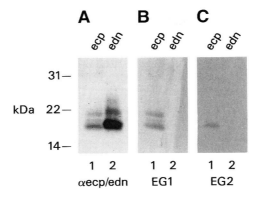

Fig. 1. Western blot. Degranulated ribonucleases EDN/RNase 2 and ECP/RNase 3 detected with (**A**) rabbit polyclonal anti-EDN/ECP, (**B**) MAb EG1, and (**C**) MAb EG2. Denaturing, nonreducing gel probed with 1:10 dilutions (10 μg/mL) monoclonal antibodies, followed by alkaline-phosphatase-linked secondary goat antimouse IgG. Reprinted with permission from Rosenberg, H. F. and Tiffany, H. L. (1994) *J. Leukoc. Biol.* **56,** 502–506.

ribonuclease activity after a 15–30-min incubation at room temperature in zymogram developing buffer. The areas of activity can be visualized by placing the gel on a silica TLC plate with an F254 fluorescent indicator, and irradiating from above with a 256-nm UV source.

4. Stain proteins with Coomassie blue stain and wash with destain to the desired degree of contrast. Areas of ribonuclease activity can also be assessed by staining the gel (even after staining with Coomassie blue) by staining with toluidine blue, which will stain the undigested substrate.

## 3.4. Western Blotting

Here the commercially-available MAbs EG1 and EG2 are used to detect EDN/RNase 2 and ECP/RNase 3 isolated from human eosinophils (**Fig. 1**) *(26)*. An important point to remember is that neither EG1 nor EG2 will detect reduced proteins.

1. Eosinophil extracts, degranulation products, and/or bronchoalveolar lavage fluids are subjected to nonreducing gel electrophoresis and transferred to nitrocellulose membranes using standard techniques *(25)*.
2. Block nonspecific binding to the nitrocellulose membrane by a 5–30-min incubation in blocking buffer (an amount sufficient to cover the membrane) at room temperature with constant agitation. Wash the nitrocellulose membrane with Western blotting wash buffer three times, at least 5 min per wash, with amounts sufficient to cover the membrane, at room temperature.
3. Immerse the membrane in a 1:10 dilution of MAb EG1 or EG2 in antibody dilution buffer using an amount sufficient to cover the membrane, and let incubate at room temperature for 1 h. Wash the membrane three times again, as in **step 2**.
4. Immerse the membrane in 20 mL of a 1:1000 dilution of alkaline-phosphatase conjugated goat antimouse IgG, and let incubate at room temperature for 1 h. Wash the membrane three times again, as in **step 2**.
5. After the final wash, immerse the membrane in 30 mL developing solution. Add 0.3 mL each of NBT and BCIP solutions. Immunoreactive proteins emerge as purple-tinged bands, generally within 30 min.
6. Stop the developing process by rinsing the membrane several times in distilled water.

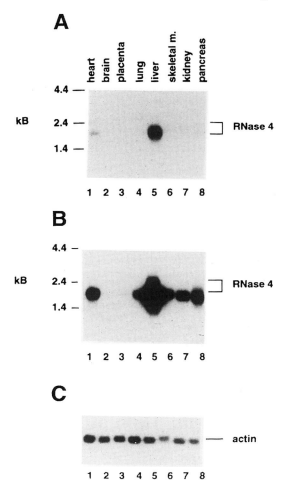

Fig. 2. Northern blot. Poly A+ RNA from human tissues probed with the full-length coding sequence of RNase 4 generated by PCR from a plasmid and $^{32}$P-radiolabeled by random priming using the kit described above. (**A**) and (**B**) are 4- and 24-h exposures, respectively; (**C**) is probed with a cDNA encoding human beta-actin. Reprinted with permission from Rosenberg, H. F. and Dyer, K. D. (1995) *Nucleic Acids Res.* **23**, 4290–4295.

### 3.5. Northern Blotting

One can distinguish among the six known human RNase A ribonucleases that have been cloned and sequenced by this technique. An example of the results of this type of analysis is shown in **Fig. 2**; semiquantitative comparisons can be achieved by scanning densitometry with normalization of each test signal to that obtained by probing with the actin-encoding cDNA, as shown. Futami and colleagues *(27)* have done an extensive analysis of the tissue-specific expression of each of these ribonucleases in humans; the following are the specific hybridization conditions used in their report.

1. Prehybridize prepared blots (**Note 1**) 3 h, 42°C in prehybridization solution and hybridize for 12 h at 42°C with individual $^{32}$P-radiolabeled probes, $10^6$ cpm/mL, prepared using the

random priming kit available from Boehringer Mannheim and $\alpha$-$^{32}$P-dCTP. In this specific case, full-length cDNA probes were used; the GenBank accession numbers for these cDNAs are included in **Note 2**.

2. Wash the blots twice at 42°C with Northern-blotting wash solution A with 5 min per wash, and then twice at 42°C with Northern-blotting wash solution B, 5 min per wash.

3. Wash further, once at 60°C with Northern-blotting wash solution C for 30 min, followed by a final wash at 65°C in wash solution B for 5 min. Analyze the washed blots by autoradiography *(25)*.

## 4. Notes

1. This protocol makes use of membranes containing RNA from human organs prepared by Clontech Laboratories. This procedure can also be used for probing membranes prepared in research laboratories.

2. The GenBank Accession numbers of the human RNase A superfamily ribonucleases include: pancreatic ribonuclease (RNase 1), X79235; eosinophil-derived neurotoxin (EDN/RNase 2), M24157; eosinophil cationic protein (ECP/RNase 3), X15161; ribonuclease 4, U36776; angiogenin, (RNase 5), M11567; ribonuclease 6, U64998.

## References

1. Raines, R. T. (1998) Ribonuclease A. *Chem. Rev.* **98,** 1045–1066.
2. D'Alessio, G. and Riordan, J. F. (1997) *Ribonucleases: Structures and Functions.* Academic Press, San Diego.
3. Beintema, J. J. (1998) Ribonucleases. A multi-author review. *Cell. Mol. Life Sci.* **54,** 761–832.
4. Reddi, K. K. and Holland, J. F. (1976) Elevated serum ribonuclease in patients with pancreatic cancer. *Proc. Natl. Acad. Sci. USA* **73,** 2308–2310.
5. Shenkin, A., Citrin, D. L., and Rowan, R. M. (1976) An assessment of the clinical usefulness of plasma ribonuclease assays. *Clin. Chim. Acta* **72,** 223–231.
6. Karpetsky, T. P., Humphrey, R. L., and Levy, C. C. (1977) Influence of renal insufficiency on levels of serum ribonuclease in patients with multiple myeloma. *J. Natl. Cancer Inst.* **58,** 875–880.
7. Domachowske, J. B. and Rosenberg, H. F. (1997) Eosinophils inhibit retroviral tranduction of human target cells by a ribonuclease-dependent mechanism. *J. Leukoc. Biol.* **62,** 363–368.
8. Domachowske, J. B., Dyer, K. D., Bonville, C. A., and Rosenberg, H. F. (1998) Recombinant human eosinophil-derived neurotoxin/RNase 2 functions as an effective antiviral agent against respiratory syncytial virus. *J. Infect. Dis.* **177,** 1458–1464.
9. Harrison, A. M., Bonville, C. A., Rosenberg, H. F., and Domachowske, J. B. (1999) Respiratory syncytial virus-induced chemokine expression in the lower airways: eosinophil recruitment and degranulation. *Am. J. Resp. Dis. Crit. Care Med.* **159,** 1918–1924.
10. Domachowske, J. B., Bonville, C. A., Dyer, K. D., Easton, A. J., and Rosenberg, H. F. (1999) Pulmonary eosinophilia is a prominent response to infection with pneumonia virus of mice. In review.
11. Youle, R. J. and D'Alessio, G. (1997) Antitumor ribonucleases, in *Ribonucleases: Structures and Functions.* (D'Alessio, G. and Riordan, J. F., eds.), Academic Press, San Diego, CA.
12. Olson, K. A., French, T. C., Vallee, B. L., and Fett, J. W. (1994) A monoclonal antibody to human angiogenin suppresses tumor growth in athymic mice. *Cancer Res.* **54,** 4576–4579.
13. Olson, K. A., Fett, J. W., French, T. C., Key, M. E., and Vallee, B. L. (1995) Angiogenin antagonists prevent tumor growth in vivo. *Proc. Natl. Acad. Sci. USA* **92,** 442–446.

14. Piccoli, R., Olson, K. A., Vallee, B. L., and Fett, J. W. (1998) Chimeric anti-angiogenin antibody cAb 26-2F inhibits the formation of human breast cancer zenografts in athymic mice. *Proc. Natl. Acad. Sci. USA* **95,** 4579–4583.

15. Bond, M. D. (1988) An in vitro binding assay for angiogenin using placental ribonuclease inhibitor. *Anal. Biochem.* **173,** 166–173.

16. Anfinsen, C. B., Redfield, R. R., Choate, W. L., Page, J., and Carroll, W. R. (1954) Studies on the gross structure, cross-linkages, and terminal sequences in ribonuclease. *J. Biol. Chem.* **207,** 201–208.

17. Slifman, N. R., Loegering, D. A., McKean, D. J., and Gleich, G. J. (1986) Ribonuclease activity associated with human eosinophil-derived neurotoxin and eosinophil cationic protein. *J. Immunol.* **137,** 2913–2917.

18. Rosenberg, H. F. and Dyer, K. D. (1995) Eosinophil cationic protein and eosinophil-derived neurotoxin: evolution of novel function in a primate ribonuclease gene family. *J. Biol. Chem.* **270,** 21,539–21,544.

19. Sorrentino, S. and Libonati, M. (1994) Human pancreatic-type and nonpancreatic-type ribonucleases: a direct side-by-side comparison of their catalytic properties. *Arch. Biochem. Biophys.* **312,** 340–348.

20. Sorrentino, S. and Libonati, M. (1997) Structure-function relationships in human ribonucleases: main distinctive features of the major RNase types. *FEBS Lett.* **404,** 1–5.

21. Thomas, J. M. and Hodes, M. E. (1981) Isozymes of ribonuclease in human serum and urine: methodology and a survey of a control population. *Clin. Chim. Acta* **111,** 185–197.

22. Thomas, J. M. and Hodes, M. E. (1981) Improved method for ribonuclease zymogram. *Anal. Biochem.* **113,** 343–351.

23. Karn, R. C., Crisp, M., Yount, E. A., and Hodes, M. E. (1979) A positive zymogram method for ribonuclease. *Anal. Biochem.* **96,** 464–468.

24. Bravo, J., Fernandez, E., Ribo, M., de Llorens, R., and Cuchillo, C. M. (1994) A versatile negative-staining ribonuclease zymogram. *Anal. Biochem.* **219,** 82–86.

25. Ausubel, F. M., Brent, R., Kingston, R. E., Moore, D. D., Seidman, J. G., Smith, J. A., and Struhl, K. (1992) *Short Protocols in Molecular Biology*, 2nd ed., Wiley, New York.

26. Rosenberg, H. F. and Tiffany, H. L. (1994) Characterization of the eosinophil granule proteins recognized by the activation-specific antibody EG2. *J. Leukoc. Biol.* **56,** 502–506.

27. Futami, J., Tsushima, Y., Murato, Y., Tada, H., Sasaki, J., Seno, M., and Yamada, H. (1997) Tissue-specific expresson of pancreatic-type RNases and RNase inhibitor in humans. *DNA Cell Biol.* **16,** 413–419.

# 24

# Assay for Antitumor and Lectin Activity in RNase Homologs

## Kazuo Nitta

## 1. Introduction

Lectins, proteins that specifically bind carbohydrates, are widely distributed in animal and plant species. Functions of animal lectins include uptake of serum asialoglycoproteins into liver, self-defense mechanisms, and modulation of cell–cell interactions during differentiation *(1)*.

   Lectins are found in the eggs of many frog species *(2–6)*. Two sialic acid-binding lectins (SBLs), isolated from the eggs of *Rana catesbeiana* (SBL-C) and *R. japonica* (SBL-J), agglutinate a large variety of tumor cells *(3,4)*, and are cytotoxic against some tumor cell lines *(7,8)*. These SBLs are carbohydrate-free single-chain basic proteins consisting of 111 amino-acid residues with four disulfide bonds *(9,10)*. Since amino acid sequences of both SBLs are highly homologous to that of a pancreatic RNase family and act as RNases *(11,12)*, they are classed as members of the RNase superfamily, and designated as leczymes *(13)*. This bifunctional nature distinguishes SBLs from other known lectins, which generally have no apparent catalytic activity.

### 1.1. Biological Action of RNase-Like Lectins

SBLs have antitumor effects in vivo as well as in vitro *(3,7,11)*. Both their tumor cell-agglutinating activity and RNase activity were inhibited by sialomucin, heparin, and nucleotides, but not by the RNase inhibitor, RNasin, from the human placenta *(7)*.

### 1.2. Sequence Comparison Between RNase-Like Lectins and the RNase Superfamily

**Figure 1** shows covalent structures of SBLs and related proteins, such as liver RNase from *R. catesbeiana* *(14)*, antitumor protein, onconase from *R. pipiens* oocytes *(15)*, angiogenic protein, angiogenin originally from human tumor cells *(16)*, and pancreatic RNase *(17)*. Structural homologies were observed between SBLs and RNases, in terms of position of disulfide linkages, and conservation of the amino-acid residues (Gln-11, His-12, Lys-41, Thr-45, His-119, and Phe-120) in the active site of RNase *(11,17,18)*.

   Substitution of the N-terminal pyroglutamic acid residue of onconase with other amino-acid residues drastically reduced its RNase activity *(19)*. Since frog proteins,

From: *Methods in Molecular Biology, vol. 160: Nuclease Methods and Protocols*
Edited by: C. H. Schein © Humana Press Inc., Totowa, NJ

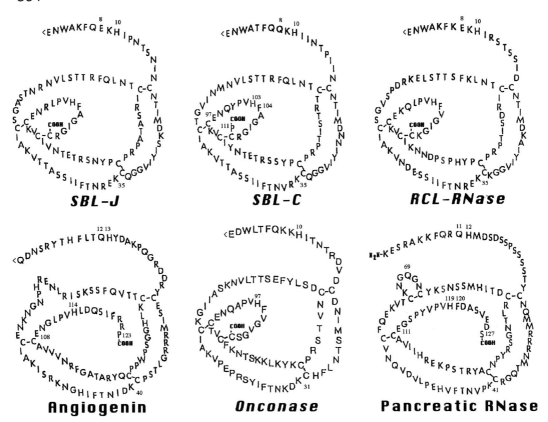

Fig. 1. Covalent structures of frog proteins (SBLs *[9,10]*, liver RNase *[14]* and onconase *[15]*), human angiogenin *(16)*, and human pancreatic RNase *(17)*.

SBLs, liver RNase and onconase have pyroglutamic acid residue in common at the N-termini, the residue may be included in the region of the RNase active site *(20)*. In contrast, frog liver RNase and onconase did not show lectin activity, but SBLs showed strong pyrimidine-base specific RNase activity *(11,12)*.

## 1.3. Relationship of RNase Activity and Antitumor Activity of RNase-Like Lectins

Protein toxins have been successfully used to generate active immunotoxins and chimeric toxins for targeting to tumor cells. Ribotoxins (ribonucleolytic toxins), such as restrictocin and α-sarcin, are ribosome-inactivating proteins produced by *Aspergillus*, and have antitumor activity *(21,22)*. They are homologous to RNase U$_2$ from *Ustilago sphaerogena (21,22)*. Restrictocin cannot enter into a tumor cell on its own, and must be introduced inside the cell by artificial means to manifest its effect *(23)*, e.g., as the toxin moiety of an immunotoxin or a chimeric toxin. The choice of the ligand for specific receptors on the tumor-cell surface is extremely important in the construction of immunotoxins.

SBL binds to the *O*-linked carbohydrate chain(s) of SBL receptor glycoprotein(s) at the tumor-cell surface. There are two hypotheses for SBL antitumor activity: RNA degradation, an apoptosis signal transduction theory, or both *(13)*. In the first theory, internalized SBL-C, acting as an RNase, promotes RNA degradation. RNA degradation affects cell proliferation, leading to cell death via inhibition of protein synthesis. SBL-C-treated P388 cells showed extensive RNA degradation over the course of 1 h. In contrast, SBL-C-treated RC150 cells showed no RNA degradation even after 24 h *(13)*.

In the apoptosis signal transduction theory, binding of SBL-C to tumor cells causes tumor-cell agglutination, leading to alteration of normal signaling pathway, activation of caspase, and induction of apoptotic cell death. To determine whether SBL-C affects growth stimulatory or inhibitory signals, we compared formation of a ladder of fragmented DNA, caspase activity, concentration of intracellular $Ca^{2+}$, and protein kinase A, G, and C activities. Treatment of P388 cells with SBL-C leads to DNA fragmentation, increased caspase-3 activity (unpublished observation), decreased concentration of intracellular $Ca^{2+}$, decreased protein kinase A activity, and increased protein kinase G activity *(13)*. Treatment of RC150 cells with SBL-C had no effect on any of these parameters. These results suggest that the cytotoxicity of SBL-C correlates with an increased rate of hydrolysis of RNA as well as degradation of DNA.

## 1.4. Possible Evolutionary Role of RNase-Like Lectins

Pyrimidine base-specific RNases, including the frog proteins described above, have been found only in vertebrates. The phylogenetic tree constructed by the probability inference method of Adachi and Hasegawa *(24)* showed that SBLs are most closely related to frog liver RNase, and that frog proteins appeared before mammalian proteins did *(25)*.

Although SBLs, onconase (frog proteins), angiogenin, eosinophil-cationic protein *(26)*, and eosinophil-derived neurotoxin (EDN) *(27)* (mammalian proteins) are classified as members of RNase A family, and all have some RNase activity, each protein has divergent functions. SBL exerts cytotoxic effect against many tumor-cell lines originated from various tissues through the interaction of SBL with sialoglycoconjugates generally expressed on tumor-cell surfaces *(3)*. In the process of molecular evolution, mammalian proteins with special biological actions should be endowed with binding capacity to specific binding protein (angiogenin–angiogenin binding protein [α-actin]) or to specific target tissue (EDN-myelinated neurons).

## 2. Materials
## 2.1. Enzymes and Reagents

1. Bovine pancreatic RNase A (type IIIA), RNasin, and Torula yeast RNA (type VI) were obtained from Sigma Chemical Co. (St. Louis, MO).
2. Neuraminidase from *Arthrobacter ureafaciens* was purchased from Nacalai Tesque (Kyoto, Japan).
3. Neuraminidase inhibitor (2,3-dehydro-2-deoxy-*N*-acetylneuraminic acid) was obtained from Boehringer Mannheim (Mannheim, Germany).
4. Sephadex G-75, heparin-Sepharose, S-Sepharose and Mono-S were obtained from Amersham Pharmacia Biotech (Uppsala, Sweden).
5. Hydroxyapatite was obtained from Seikagaku Kogyo (Tokyo, Japan).

## 2.2. Cell Lines

1. P388 and L1210 murine leukemia cells were obtained from the Japanese Cancer Research Resources Bank (Tokyo, Japan).
2. SBL-C-resistant mutant RC150 was established by prolonged exposure of P388 cells to SBL-C (*8*).
3. Rat (Donryu) ascites hepatoma AH109A cells were originally donated from the Institute of Development, Aging and Cancer, Tohoku University (Sendai, Japan).
4. Sarcoma 180 cells and Ehrlich cells were propagated as ascites in ddY mice. These cell lines were also donated from the Cancer Cell Repository, the Institute of Development, Aging and Cancer, Tohoku University.

## 2.3. Solutions and Media

### 2.3.1. Solutions

#### 2.3.1.1. EXTRACTION AND COLUMN CHROMATOGRAPHY

1. Extraction solution: 0.15 $M$ NaCl.
2. Sephadex G-75 equilibration and elution solution: 0.15 $M$ NaCl.
3. DEAE column equilibration buffer: 1 m$M$ Na$_2$HPO$_4$, 1 m$M$ KH$_2$PO$_4$, pH set to 6.8.
4. DEAE column elution buffer: 1 m$M$ Na$_2$HPO$_4$, 1 m$M$ KH$_2$PO$_4$, pH set to 6.8.
5. Hydroxyapatite column equilibration buffer: 1 m$M$ Na$_2$HPO$_4$, 1 m$M$ KH$_2$PO$_4$, pH set to 6.8.
6. Hydroxyapatite column elution buffer: 10 m$M$ Na$_2$HPO$_4$, 10 m$M$ KH$_2$PO$_4$, pH set to 6.8.
7. CM column equilibration buffer: 10 m$M$ Na$_2$HPO$_4$, 10 m$M$ KH$_2$PO$_4$, pH set to 6.8.
8. CM column elution buffer: 100 m$M$ Na$_2$HPO$_4$, 100 m$M$ KH$_2$PO$_4$, pH set to 6.8.
9. Heparin column equilibration buffer: 50 m$M$ CH$_3$COONa, pH set to 6.0 with acetic acid.
10. Heparin column elution buffer: 0.8 $M$ NaCl, 50 m$M$ CH$_3$COONa, pH set to 6.0 with acetic acid.
11. S-Sepharose column equilibration buffer: 50 m$M$ CH$_3$COONa, pH set to 5.0 with acetic acid.
12. S-Sepharose column elution buffer: 0.8 $M$ NaCl, 50 m$M$ CH$_3$COONa, pH set to 5.0 with acetic acid.

#### 2.3.1.2. ASSAY

1. RNase activity assay buffer: 50 m$M$ Tris-HCl, 0.5 m$M$ EDTA, pH set to 7.5 with HCl, or 50 m$M$ CH$_3$COONa, 0.5 m$M$ EDTA, pH set to 6.0 with acetic acid (for measurement of absorbance); 100 m$M$ Tris-HCl, 10 m$M$ EDTA, pH set to 7.6 with HCl (for electrophoresis).
2. Reaction termination solution: 5% perchlorate containing 0.25% uranyl acetate (for measurement of absorbance); 0.5% SDS containing 10 μg of proteinase K (for electrophoresis).
3. Gel loading solution: 80% formamide containing 0.05% bromophenol blue and 0.05% xylene cyanol.

### 2.3.2. Media

Culture medium of P388, L1210, and RC150 cells: RPMI-1640 supplemented with 10% fetal calf serum.

## 3. Methods

### 3.1. Preparation of Lectins

### 3.1.1. Purification of RNase-Like Lectins

SBL-C can be isolated from *R. catesbeiana* eggs according to the procedure of sequential chromatography on Sephadex G-75, DEAE-cellulose, hydroxyapatite, and CM-cellulose columns (*3*), and further purified by Mono-S cation-exchange chromatography *(11)* or by heparin-Sepharose affinity chromatography *(7)*. The subsequent protocol revision aids in obtaining a lectin preparation free of contamination by foreign RNase(s).

### 3.1.1.1. EXTRACTION AND SEPHADEX G-75 SIZE EXCLUSION CHROMATOGRAPH

1. *R. catesbeiana* eggs are homogenized in 20 vol of ice-cold acetone, and the acetone-insoluble residue is filtered and dried.
2. The acetone powder is placed in the homogenizing vessel and homogenized in 0.15 *M* NaCl (1 g acetone powder/20 mL).
3. The homogenate is centrifuged at 9000*g* for 30 min at 4°C.
4. The supernatant is carefully decanted through cotton loosely packed in the stem of a funnel.
5. The filtrate is then dialyzed exhaustively against distilled water and lyophilized (saline-extracted fraction, SEF).
6. SEF (1.0 g in weight/20 mL) is separated by Sephadex G-75 (column size, 4.0 × 35 cm) equilibrated with 0.1 *M* NaCl.
7. The column is run at room temperature and flow rate of 80 mL/h.

### 3.1.1.2. DEAE-CELLULOSE ANION-EXCHANGE CHROMATOGRAPHY

1. Peak 3 fraction (1.0 g) obtained by Sephadex G-75 gel filtration is loaded onto a column of DEAE-cellulose (column size, 2.2 × 35 cm) equilibrated with 1 m*M* phosphate buffer (pH 6.8).
2. The column is eluted with a step gradient from 1 to 400 m*M* phosphate buffer (pH 6.8). Flow rate is 30 mL/h.

### 3.1.1.3. HYDROXYAPATITE CHROMATOGRAPHY

1. DEAE-fraction 1 (eluates of 1 m*M* phosphate buffer, 1.0 g) is loaded onto a column of hydroxyapatite (column size, 2.2 × 30 cm) equilibrated with 1 m*M* phosphate buffer (pH 6.8).
2. The column is eluted with the same step gradient as described in **Subheading 3.1.1.2.** Flow rate is 30 mL/h.

### 3.1.1.4. CM-CELLULOSE CATION-EXCHANGE CHROMATOGRAPHY

1. Hydroxyapatite-fraction 2 (eluates of 10 m*M* phosphate buffer, 1.0 g) is loaded onto a column of CM-cellulose (column size, 2.2 × 35 cm) equilibrated with 10 m*M* phosphate buffer (pH 6.8).
2. After washing with 120 mL of equilibrating buffer, the column is developed with a linear gradient of 10 to 100 m*M* phosphate buffer (total volume, 600 mL).
3. SBL-C can be eluted between 70–80 m*M* phosphate buffer.

### 3.1.1.5. HEPARIN-SEPHAROSE AFFINITY CHROMATOGRAPHY

1. SBL-J can be obtained from *R. japonica* eggs by an improved 4-Step method using Sephadex G-75, DEAE-cellulose, heparin-Sepharose, and S-Sepharose columns in series *(7)*.
2. **Subheadings 3.1.1.1.** and **3.1.1.2.** are performed by the same procedure as described in preparation of SBL-C.
3. DEAE-fraction 1 from *R. japonica* eggs can be further fractionated by heparin-Sepharose and S-Sepharose columns.
4. DEAE-fraction 1 (eluates of 1 m*M* phosphate buffer, 100 mg protein/5 mL) is loaded onto a column of heparin-sepharose (column size, 1.0 × 20 cm) equilibrated with 50 m*M* acetate buffer (pH 6.0).
5. After washing with 300 mL of equilibrating buffer, the column is developed with a linear 0–0.8 *M* NaCl gradient in the same buffer (total volume, 600 mL).

### 3.1.1.6. S-Sepharose Cation-Exchange Chromatography

1. Fraction 2 (eluates in the range of 0.22–0.26 $M$ NaCl, 10 mg protein/2 mL) from the heparin column is applied to an S-Sepharose column ($1.0 \times 30$ cm) equilibrated with 50 m$M$ acetate buffer (pH 5.0), and eluted with a linear 0–0.8 $M$ NaCl gradient in the same buffer (total volume, 600 mL). SBL-J can be eluted between 0.33 and 0.35 $M$ NaCl.
2. Lyophilized preparations of SBLs can be stored for several years at 4°C or –20°C without detectable loss of activity. SBLs are heat stable (75°C for 5 min) at pH 2.0–8.0. In addition, no activity of SBLs is lost at 85°C for 5 min at pH 6.0 *(12)*.

## 3.1.2. Preparation of Modified RNase-Like Lectins

1. Prepare amino group-modified SBL-Cs according to acetylation, succinylation, maleylation, and carbamylation by the method of Nitta et al. *(3)*.
2. Prepare chemically modified SBL-Cs according to reductive methylation and 2,3-butanedione modification as described by Nitta et al. *(3,11)*.

## 3.2. Assays for Activities

### 3.2.1. Lectin Agglutination and Inhibition Assay

Tumor-cell agglutinating activities are determined by counting particle number under a microscope and calculating the percentage of total particle number (%TPN). When the %TPN falls below 90 (the initial particle number, $2 \times 10^5$ cells/40 µL; 100 %TPN) because of the formation of cell aggregates, the sample is evaluated as having positive agglutinating activity *(3)*.

#### 3.2.1.1. Agglutination Assay

Tumor-cell agglutination is determined using Ehrlich cells, AH109A cells, or P388 cells.

1. Dilute a solution of SBL in 0.15 $M$ NaCl from a concentration of 500 µg/100 µL to 0.1 µg/100 µL.
2. Add 100 µL of tumor-cell suspension ($10^7$ cells/mL) in 0.15 $M$ NaCl to each small test tube.
3. Incubate the tubes under gentle rotation for 10 min at room temperature.
4. Allow to stand for 30 min without rotation.

#### 3.2.1.2. Inhibition Assay

Inhibition of lectin-induced AH109A agglutination is determined based on effects of serial dilutions of each potential inhibitor. RNasin (1000 U/80 µL) is also serially diluted with PBS containing 1 m$M$ DTT.

1. Incubate each tube with three agglutination units (1.5 µg/20 µL) of SBL-C (total volume; 100 µL) for 30 min at room temperature.
2. Add 100 µL of tumor cell suspension to the mixture.
3. Incubate the tubes under gentle rotation for 10 min at room temperature.
4. Allow to stand for 30 min without rotation.

### 3.2.2. RNase Activity and Enzyme Inhibition Assay

#### 3.2.2.1. RNase Activity

##### 3.2.2.1.1. Release of Acid-Soluble Nucleotides

The standard assay is performed according to the method of Reddi *(28)* with some modifications.

1. The reaction mixture (200 μL) is contained 100 μg of homopolynucleotide or 200 μg of yeast RNA, 50 m$M$ Tris-HCl (pH 7.5) or 50 m$M$ acetate buffer (pH 6.0), 0.5 m$M$ EDTA, and SBL.
2. Incubate for 30 min at 37°C.
3. Terminate the reaction by addition of 200 μL of ice-cold 5% perchlorate containing 0.25% uranyl acetate.
4. Cool the mixture in an ice-bath for 30 min.
5. Centrifuge at 2000$g$ for 10 min.
6. Dilute a 300-μL aliquot of the supernatant with 1.2 mL of distilled water.
7. Measure absorbance at 260 nm (yeast RNA, poly[A], poly[I], poly[U]) or at 280 nm (poly[C]). Number of moles of mononucleotide equivalents formed by the action of SBL on the homopolynucleotide is calculated from the molar extinction coefficient (*12,29*).

### 3.2.2.1.2. Evaluation of Degradation Product Using Gel Electrophoresis

The effect of SBL on RNA is examined by polyacrylamide gel electrophoresis under denaturing conditions as described by Maniatis et al. (*30*).

1. Incubate rRNA with SBL at 37 or 0°C in 100 m$M$ Tris-HCl, pH 7.6, containing 10 m$M$ EDTA in a total volume of 5 μL, for 60 min in the presence (40 U) or absence of RNasin.
2. Terminate the reaction by addition of 100 μL of 0.5% SDS solution containing 10 μg of proteinase K, for 15 min at 37°C.
3. Add 10 μL of 3 $M$ sodium acetate and 250 μL of ethanol.
4. After 15 min at –80°C, centrifuge and dry the precipitate in a Speed-Vac Concentrator.
5. Dissolve the pellet in 5 μL of 0.05% Bromophenol blue and 0.05% xylene cyanol in 80% formamide.
6. Heat at 70°C for 5 min, load onto a 4% polyacrylamide gel containing 7 $M$ urea, and electrophorese at a constant voltage of 150 V. RNA is visualized with ethidium bromide by soaking the gel for 60 min in an ethidium bromide solution (0.00005%, 25 μL of 10 μg/μL ethidium bromide in 500 mL distilled water) (*11*).

### 3.2.2.2. Enzyme Inhibition Assay

Enzyme inhibition assay is performed as follows.

1. Add inhibitors to the enzyme solution.
2. Incubate for 30 min at room temperature.
3. Measure residual activity as described in **Subheading 3.2.2.1.1.** (*7*).

## 3.2.3. In Vitro and In Vivo Antitumor Activity Assay

### 3.2.3.1. In Vitro Antitumor Activity Assay

P388, L1210, and RC150 cells are cultured in RPMI-1640 supplemented with 10% fetal calf serum. Neuraminidase-treated cells are prepared by adding *A. ureafaciens* neuraminidase (0.1 U) to 1 mL of the cell suspension (~5 × 10⁵ cells/mL), and treating for 30 min at 37°C. The basic protocol for in vitro antitumor activity assay is summarized in **Table 1**.

1. Dilute the cells with medium, and seed 2 × 10⁵ cells/well in 96-well plates.
2. Add SBL-C or SBL-J, and incubate in a CO$_2$ incubator for 48 h.
3. Count the number of cells in each well according to trypan blue dye-exclusion assay method.
4. Determine cell proliferation as percentage increase in cell numbers: ([number of cells at 48 h – number of cells at time zero]/number of cells at time zero) × 100.

**Table 1**
**Basic Protocol for In Vitro Antitumor Activity Assay**

Seed $2 \times 10^5$ cells/well in 96-well plates
↓
Add SBL, and incubate in $CO_2$ incubator

| for 48 h | for 44 h |
|---|---|
| ↓ | ↓ |
| Count the number of cells in each well | Incubate with alamar Blue™ for 4 h in $CO_2$ incubator |
| ↓ | ↓ |
| Calculate % increase in cell nos.: [(no. of cells at 48 h – no. of cells at time zero)/ no. of cells at time zero] × 100 | Determine optical density at 570 nm/500 nm |
| | ↓ |
| | Calculate % cytotoxicty |
| *Trypan blue dye-exclusion assay* | *Alamar Blue assay* |

5. Normalize growth of cultures without SBL to 100% (0% inhibition), and express proliferation in experimental cultures as a percentage of this control growth. The number of cells at 48 h in all cultures without SBL is at least four times that at 0 h *(7)*.
6. In a separate experiment, SBL-treated cells are incubated with alamar Blue™ for 4 h.
   a. Determine optical density using a microplate reader MPR-A4i (Tosoh, Tokyo, Japan) at 570 nm/600 nm.
   b. Calculate the percent cytotoxicity.

### 3.2.3.2. In Vivo Antitumor Activity Assay

Mice are given intraperitoneal injections of Sarcoma 180 cells or Ehrlich cells ($1 \times 10^6$ cells/mouse) on d 0.

1. Divide the Sarcoma 180-bearing ddY mice into two groups, one given a single ip injection of SBL-C on d 1 and the other given repeated injections of SBL-C every 10 d from d 1 to d 10.
2. Give the Ehrlich-bearing ddY mice injections of modified SBL-C on d 1, 2, and 3.
3. Incubate Sarcoma 180 cells ($1 \times 10^6$ cells/200 µL) in the presence of SBL-C (1000, 100, or 10 µg) at 37°C for 30 min, and transplant ip into ddY mice.
4. Record survival time in days *(8)*.

## 4. Notes

1. Caution for preparation of acetone powder: The acetone-insoluble residue is rapidly filtered under vacuum through Toyo filter paper (No. 2) on Buchner funnel. The residue is air-dried. This filtration and drying are done in a draft chamber at room temperature. The powder is carefully scraped into containers and stored at –20°C for later use. If we inhale the minute acetone powder, we have a cough.
2. Separation of SBL-C from anti-A lectin: Peak 3 fraction obtained by Sephadex G-75 contains two lectins, SBL-C and anti-A lectin agglutinated specifically human type A erythrocytes. They can be separated by chromatography on a DEAE cellulose with a step

gradient 1, 10, 50, 100, 200, and 400 m*M* phosphate buffer (pH 6.8), i.e., SBL-C is eluted in DEAE-fraction 1, and anti-A lectin in the third fraction (eluates of 50 m*M* phosphate buffer).

3. Agreement of elution position: The peak of chromatographic profile of protein content is concordant with that of RNase activity in tumor-cell agglutination-positive fractions which are eluted from heparin-Sepharose or S-Sepharose column.

4. Preparation of cells for agglutination assay: Ascites fluids are collected at the fifth or sixth day from Donryu rats transplanted AH109A cells and ddY mice transplanted Ehrlich cells, respectively. Tumor cells are harvested and washed with 0.15 *M* NaCl by repeated centrifugation. We can obtain the vigorous cells from the fifth day's ascites fluid. P388 cells at the logarithmic growth phase ($5 \times 10^5$/mL, 42% confluence) suit the agglutination assay, because SBL-C agglutinates the logarithmic phase cells rather than confluent cells.

## Acknowledgments

This work was supported in part by grant (09240228) from the Ministry of Education, Science and Culture of Japan.

## References

1. Barondes, S. H. (1981) Lectins: their multiple endogenous cellular functions. *Annu. Rev. Biochem.* **50,** 207–231.
2. Kawauchi, H., Sakakibara, F., and Watanabe, K. (1975) Agglutinins of frog eggs: a new class of proteins causing preferential agglutination of tumor cells. *Experientia* **31,** 364,365.
3. Nitta, K., Takayanagi, G., Kawauchi, H., and Hakomori, S. (1987) Isolation and characterization of *Rana catesbeiana* lectin and demonstration of the lectin-binding glycoprotein of rodent and human tumor cell membranes. *Cancer Res.* **47,** 4877–4883.
4. Sakakibara, F., Kawauchi, H., Takayanagi, G., and Ise, H. (1979) Egg lectin of *Rana japonica* and its receptor glycoprotein of Ehrlich tumor cells. *Cancer Res.* **39,** 1347–1352.
5. Ozeki, Y., Matsui, T., Nitta, K., Kawauchi, H., Takayanagi, Y., and Titani, K. (1991) Purification and characterization of β-galactoside binding lectin from frog (*Rana catesbeiana*) eggs. *Biochem. Biophys. Res. Commun.* **178,** 407–413.
6. Ahmed, H., Pohl, J., Fink, N. E., Strobel, F., and Vasta, G. R. (1996) The primary structure and carbohydrate specificity of a β-galactosyl-binding lectin from toad (*Bufo arenarum* Hensel) ovary reveal closer similarities to the mammalian galectin-1 than to the galectin from the clawed frog *Xenopus laevis*. *J. Biol. Chem.* **271,** 33,083–33,094.
7. Nitta, K., Ozaki, K., Ishikawa, M., Furusawa, S., Hosono, M., Kawauchi, H., Sasaki, K., Takayanagi, Y., Tsuiki, S., and Hakomori, S. (1994) Inhibition of cell proliferation by *Rana catesbeiana* and *R. japonica* lectins belonging to the ribonuclease superfamily. *Cancer Res.* **54,** 920–927.
8. Nitta, K., Ozaki, K., Tsukamoto, Y., Furusawa, S., Ohkubo, Y., Takimoto, H., Murata, R., Hosono, M., Hikichi, N., Sasaki, K., Kawauchi, H., Takayanagi, Y., Tsuiki, S., and Hakomori, S. (1994) Characterization of a *Rana catesbeiana* lectin-resistant mutant of leukemia P388 cells. *Cancer Res.* **54,** 928–934.
9. Titani, K., Takio, K., Kuwada, M., Nitta, K., Sakakibara, F., Kawauchi, H., Takayanagi, G., and Hakomori, S. (1987) Amino acid sequence of sialic acid binding lectin from frog (*Rana catesbeiana*) eggs. *Biochemistry* **26,** 2189–2194.
10. Kamiya, Y., Oyama, F., Oyama, R., Sakakibara, F., Nitta, K., Kawauchi, H., Takayanagi, Y., and Titani, K. (1990) Amino acid sequence of a lectin from Japanese frog (*Rana japonica*) eggs. *J. Biochem.* **108,** 139–143.

11. Nitta, K., Oyama, F., Oyama, R., Sekiguchi, K., Kawauchi, H., Takayanagi, Y., Hakomori, S., and Titani, K. (1993) Ribonuclease activity of sialic acid-binding lectin from *Rana catesbeiana* eggs. *Glycobiology* **3**, 37–45.

12. Okabe, Y., Katayama, N., Iwama, M, Watanabe, H., Ohgi, K., Irie, M., Nitta, K., Kawauchi, H., Takayanagi, Y., Oyama, F., Titani, K., Abe, Y., Okazaki, T., Inokuchi, N., and Koyama, T. (1991) Comparative base specificity, stability, and lectin activity of two lectins from eggs of *Rana catesbeiana* and *R. japonica* and liver ribonuclease from *R. catesbeiana*. *J. Biochem.* **109**, 786–790.

13. Nitta, K., Ozaki, K., Tsukamoto, Y., Hosono, M., Ogawa-Konno, Y., Kawauchi, H., Takayanagi, Y., Tsuiki, S., and Hakomori, S. (1996) Catalytic lectin (leczyme) from bullfrog (*Rana catesbeiana*) eggs: mechanism of tumoricidal activity. *Int. J. Oncol.* **9**, 19–23.

14. Nitta, R., Katayama, N., Okabe, Y., Iwama, M., Watanabe, H., Abe, Y., Okazaki, T., Ohgi, K., and Irie, M. (1989) Primary structure of a ribonuclease from bullfrog (*Rana catesbeiana*) liver. *J. Biochem.* **106**, 729–735.

15. Ardelt, W., Mikulski, S. M., and Shogen, K. (1991) Amino acid sequence of an antitumor protein from *Rana pipiens* oocytes and early embryos. Homology to pancreatic ribonucleases. *J. Biol. Chem.* **266**, 245–251.

16. Strydom, D. J., Fett, J. W., Lobb, R. R., Alderman, E. M., Bethune, J. L., Riordan, J. F., and Vallee, B. L. (1985) Amino acid sequence of human tumor derived angiogenin. *Biochemistry* **24**, 5486–5494.

17. Beintema, J. J., Wietzes, P., Weickmann, J. L., and Glitz, D. G. (1984) The amino acid sequence of human pancreatic ribonuclease. *Anal. Biochem.* **136**, 48–64.

18. Blackburn, P. and Moore, S. (1982) Pancreatic ribonuclease, in *The Enzymes*, 3rd ed., Vol. 15 (Boyer, P. D., ed.), Academic, New York, pp. 317–433.

19. Boix, E., Wu Y., Vasandani, V. M., Saxena, S. K., Ardelt, W., Ladner, J., and Youle, R. J. (1996) Role of the N terminus in RNase A homologues: differences in catalytic activity, ribonuclease inhibitor interaction and cytotoxicity. *J. Mol. Biol.* **257**, 992–1007.

20. Youle, R. J. and D'Alessio, G. (1997) Antitumor RNases, in *Ribonucleases: Structures and Functions* (D'Alessio, G. and Riordan, J. F., eds.), Academic Press, New York, pp. 491–514.

21. Lopez-Otin, C., Barber, D., Fernandez-Luna, J. L., Soriano, F., and Mendez, E. (1984) The primary structure of the cytotoxin restrictocin. *Eur. J. Biochem.* **143**, 621–634.

22. Wool, I. G. (1997) Structure and mechanism of action of cytotoxic ribonuclease α-sarcin, in *Ribonucleases: Structures and Functions* (D'Alessio, G. and Riordan, J. F., eds.), Academic Press, New York, pp. 131–162.

23. Rathore, D. and Batra, J. K. (1996) Generation of active immunotoxins containing recombinant restrictocin. *Biochem. Biophys. Res. Commun.* **222**, 58–63.

24. Adachi, J. and Hasegawa, M. (1992) Program for molecular phylogenetics I. PROTML: maximum likelihood inference of protein phylongeny. Institute of Statistical Mathematics, Tokyo. *Comput. Sci. Monographs* **27**, 1–77.

25. Irie, M. (1997) Structures and functions of ribonuclease. *Yakugaku Zasshi (in Japanese)* **117**, 561–582.

26. Rosenberg, H. F., Ackerman, S. J., and Tenen, D. G. (1989) Human eosinophil cationic protein. Molecular cloning of a cytotoxin and helminthotoxin with ribonuclease activity. *J. Exp. Med.* **170**, 163–176.

27. Barker, R. L., Loegering, D. A., Ten, R. M., Hamann, K. J., Pease, L. R., and Gleich, G. J. (1989) Eosinophil cationic protein cDNA. Comparison with other toxic cationic proteins and ribonucleases. *J. Immunol.* **143**, 952–955.

28. Reddi, K. K. (1975) Nature and possible origin of human serum ribonuclease. *Biochem. Biophys. Res. Commun.* **67,** 110–118.

29. Kumagai, H., Igarashi, K., Takayama, T., Watanabe, K., Sugimoto, K., and Hirose, S. (1980) A microsomal endoribonuclease from rat liver. *Biochim. Biophys. Acta* **608,** 324–331.

30. Maniatis, T., Fritsch, E. F., and Sambrook, J. (1982) in *Molecular Cloning. A Laboratory Manual*, Cold Spring Harbor Laboratory, Cold Spring Harbor, NY.

# 25

# Isolation and Enzymatic Activity of Angiogenin

## James F. Riordan and Robert Shapiro

## 1. Introduction

Angiogenin is one of several unusual members of the pancreatic ribonuclease super-family (for reviews *see* **refs.** *1,2*). It was first isolated as a 14 kDa soluble protein from culture medium conditioned by human colon carcinoma (HT-29) cells, and identified as an angiogenic substance based on its capacity to induce blood-vessel formation on the chorioallantoic membrane of the chicken embryo *(3)*. Although angiogenin is secreted by most tumor cells, and has been shown to be essential for tumor growth *(4)*, it is not a tumor-specific protein. It is present at a concentration of 250–360 ng/mL in normal human plasma. Higher or lower concentrations have been seen in a variety of conditions including endometrial cancer, pregnancy, and renal dialysis, but thus far its plasma level has not been shown to have diagnostic relevance. Recently, a protein that inhibits the degranulation of polymorphonuclear leukocytes was isolated from plasma ultrafiltrates of patients with uremia and shown to be identical to angiogenin *(5)*.

Protein and cDNA sequencing of the HT-29 protein demonstrated its relationship to the pancreatic ribonucleases, and although its enzymatic activity was not immediately apparent, later mutagenesis studies showed that it not only catalyzed the hydrolysis of RNA, but that this property was essential for its angiogenic activity. Replacement of any of the active-site residues eliminates both activities. It cleaves preferentially on the 3' side of pyrimidines to generate a cyclic phosphate intermediate that is subsequently hydrolyzed. The overall rate is 4–6 orders of magnitude less than that of human pancreatic ribonuclease. The 123-residue protein contains a catalytic active site, and also has an endothelial cell-binding site (which is believed to recognize an as-yet uncharacterized angiogenin receptor), an actin-binding site, and a nuclear localization signal.

In addition to its roles as an enzyme, angiogenin is pleiotropic toward its target cells: it binds to endothelial and smooth-muscle-cells (SMC) to induce second messengers, activates cell-associated proteases, stimulates cell migration and invasion of the extra-cellular matrix, supports cell adhesion, promotes tube formation by cultured endothe-lial cells, and induces endothelial cell proliferation in sparse cultures. It is also internalized by endothelial cells and subsequently translocated to the nucleus, where it accumulates in the nucleolus.

From: *Methods in Molecular Biology, vol. 160: Nuclease Methods and Protocols*
Edited by: C. H. Schein © Humana Press Inc., Totowa, NJ

Angiogenin, like other members of the pancreatic ribonuclease family, is inhibited by a 50 kDa protein originally isolated from human placenta (*see* Chapter 13). Indeed, the binding of angiogenin to this ribonuclease inhibitor is extremely tight, and the $K_i$, $7 \times 10^{-16}$ $M$, is one of the lowest values ever reported for the interaction of two proteins. The overall binding of angiogenin to the inhibitor resembles that of RNase A, but the majority of the interactions are distinctive, indicating that the inhibitor has the capacity to recognize features unique to each protein.

Subsequent to its isolation from tumor-conditioned medium, angiogenin was found to be a circulating plasma protein. Sequence analysis showed that the material purified from normal human plasma was identical to that produced by tumor cells. Indeed, the yield from blood plasma, 60–150 µg/L, was more than 100-fold greater than could be obtained from HT-29 cell-conditioned medium. To obtain even greater amounts of protein, recombinant angiogenin was expressed first in transformed BHK cells, and later in *Escherichia coli* using the pAng2 vector under the control of the *E. coli trp* promoter *(6)*. In the latter case, the system generates cytoplasmic inclusion bodies that require solubilization by reduction and reoxidation. The solubilized product has an additional Met residue at the N-terminus, but it can be converted to the native pyroGlu form (<Glu-1) by treatment with *Aeromonas* aminopeptidase. A further improvement *(7)* introduced the pAng3 vector in which a DNA sequence encoding the *E. coli pho*A signal sequence replaces the initiator Met codon of pAng2 at the N-terminus of angiogenin and, thus allows the <Glu-1 protein (actually Gln-1) to be expressed directly into the periplasm. This method of preparation is described in detail in this chapter. The preparation of bovine angiogenin is also described because of the ready availability of the starting material.

Numerous potential assay methodologies can be used to monitor the isolation and purification process. Among them, the hydrolysis of tRNA, described here, is the most convenient and direct approach. It is not foolproof, however, since the very low level of ribonucleolytic activity of angiogenin necessitates rather extreme precautions to avoid contamination by pancreatic-type RNases. These common reagents are notorious for their stability, ubiquity, and persistence once they have been brought into a laboratory environment. For example, the slightest trace of RNase A can overwhelm the catalytic activity of angiogenin. It is possible, although tedious, to distinguish the ribonucleolytic activity of angiogenin from that of RNase A by using rRNA (18S and 28S) as substrate and agarose gel electrophoresis to analyze products *(8)*. The major degradation products generated by angiogenin, even after >4 h digestion, are 100–500 nucleotides long, whereas those resulting from just a 3-min exposure to RNase A under the same conditions are quite small and migrate ahead of the dye front. The basis for this apparent difference in specificity is unknown, but is not seen when 5S rRNA or tRNA is the substrate.

An alternative approach to monitor isolation and measure activity, which obviates the problem of contaminating RNases, is to assay for angiogenesis. This was the basis on which angiogenin was purified in the first place. Unfortunately, a simple, reproducible, quantitative angiogenesis method still awaits development, and undoubtedly this has been a major limitation to progress in the field. Perhaps the most widely employed method to assess angiogenic activity has been endothelial-cell proliferation. But this is a nonspecific assay. Angiogenin, like several other angiogenic factors, has a relatively weak proliferative activity, although it is quite potent when tested with in vivo assays.

Another increasingly popular method, often referred to as in vitro angiogenesis, examines the formation of capillary-like tubes in matrigel, collagen, or laminin upon addition of an angiogenic factor to endothelial cells in culture.

Certainly the most dramatic evidence for induction of capillary growth is seen with the rabbit corneal assay. By properly implanting an angiogenic substance into the cornea, which is normally avascular, blood vessels can be induced to emerge from the limbus, penetrate into the cornea, and migrate toward the source of stimulation. Clearly this method has serious disadvantages, and not just for the rabbit. It is difficult to quantify, and requires an investigator skilled in microsurgery with an ability to insert samples reproducibly at precisely the right distance from the corneoscleral junction. Obviously, it is not amenable to screening many samples and, despite the best technical expertise, it is highly variable.

One of the more classic angiogenesis methods involves placing the material to be tested onto the chorioallantoic membrane of the chicken embryo and, after 2–3 d, looking for a characteristic "spoke-wheel" pattern of vessels directed toward the sample. Many variations of this assay have been developed in an effort to establish it as the standard of reference, but it remains tedious, difficult to quantify, susceptible to artifacts, poorly reproducible, and dependent on subjective evaluation. Even in our own laboratory, where thousands of eggs have been used over the years, virtually no aspect of the technique has survived unmodified. For this reason, and because their vagaries have been elaborated upon in a review *(9)*, we have not attempted to include a detailed protocol for this or any other angiogenic assay in this chapter.

## 2. Materials

All solutions should be prepared in disposable, sterile plastic tubes, or in 0.1 *N* NaOH-treated plastic or glassware. All reagents should be "certified" grade or higher.

### *2.1. Isolation of PyroGlu Angiogenin*

1. Centrifuge bottles (250 mL) and caps.
2. Plastic pipets (10 mL).
3. Plastic spatula.
4. Polypropylene test tubes (10 × 75 mm).
5. Amicon concentrator (400 mL) equipped with a YM3 membrane.
6. Sorvall centrifuge (or equivalent) and GSA rotor.
7. Branson Model 350 Sonifier (or equivalent).
8. Micropipets.
9. Eppendorf tubes (1.5 mL).
10. Vortex mixer.
11. Polypropylene conical tubes (15 mL).
12. Stirring bars.
13. Incubator (37°C).
14. Syringes (10 and 60 cc).
15. Incubator/shaker (37°C).
16. Sterile filters (0.45 or 0.22 μm).
17. HPLC system.
18. Synchropak C18 HPLC column (Synchrom, RP-P 250 × 4.6-mm, or equivalent).
19. Mono S column (Pharmacia, 50 × 5 mm).

20. Nylon66 filters (Rainin).
21. Sep-Pak C18 cartridge (Waters, follow manufacturer's instructions for use).
22. Sep-Pak filtered water.
23. MilliQ water.
24. 0.02% azide in filtered water.
25. Culture flasks (2.8 L).
26. Ehrlenmeyer flask (1 L).
27. Snap-top tubes (15 mL).
28. Microfuge.
29. 10X M9 salts *(10)*: 60 g $Na_2HPO_4$ (anhydrous), 30 g $KH_2PO_4$ (anhydrous), 5 g NaCl, 10 g $NH_4Cl$. Dissolve in ~850 mL water and titrate to pH 7.4 with 10 $N$ NaOH. Adjust volume to 1 L. Sterilize by autoclaving at 121°C, 15 lb/in.$^2$, 20 min.
30. 1 $M$ $MgSO_4$ (8 mL required for each 4 L of culture). Prepare 100 mL and autoclave.
31. 100 m$M$ $CaCl_2$ (4 mL required for each 4 L of culture). Prepare 50 mL and autoclave.
32. 20% glucose (126 mL required for each 4 L of culture). Prepare 500 mL and sterile filter.
33. 10% casamino acids (Difco), prepared fresh each time (240 mL required for each 4 L of culture). Prepare 250 mL and sterile filter.
34. LB medium *(10)*: 10 g bacto-tryptone, 10 g bacto-yeast extract, 5 g NaCl. Dissolve in water and adjust volume to 1 L (12 mL required for each 4 L of culture). Autoclave. To make LB-amp, add ampicillin to 50 μg/mL just prior to use.
35. Ampicillin (sodium salt) 25 mg/mL in water (8 mL required for each 4 L culture, 24 μL are used in smaller cultures). Prepare 50 mL, sterile filter, and store at –20°C.
36. Indole-3-acrylic acid (Aldrich), 20 mg/mL in 100% ethyl alcohol. Prepare ~5 mL just before use. Two batches will be needed.
37. 0.5 $M$ $Na_2EDTA$, pH 8.0 (8 mL required for each 4 L of culture). Prepare 100 mL and autoclave.
38. 5 $M$ NaCl (not sterile, 16 mL required for each 4 L of culture). Make 250 mL.
39. TSOPB (not sterile; **store at 4°C**): 20 m$M$ Tris-HCl, 10% sucrose, 10 m$M$ 1,10-phenanthroline (Sigma) (*see* **Note 1**), 0.25 m$M$ benzamidine (Sigma), 0.5 m$M$ EDTA (from 0.5 $M$ stock). Adjust pH to 9.0. (375 mL required for each 4 L of culture). Make 1 L.
40. Lysozyme: dissolve hen egg-white lysozyme (Sigma, L-7001), 2 mg/mL, in TSOBP. Prepare ~6 mL fresh and keep on ice.
41. PMSF, 0.45 $M$ in 10% dry DMF (dissolve in this first)/ 90% isopropanol. Use a polypropylene tube. Keep at room temperature. Prepare ~7 mL for 4 L of culture.
42. SP-sepharose buffers:
    a. 10 m$M$ Tris-HCl, 0.2 $M$ NaCl, pH 8.0 (not sterile). Make 2 L.
    b. 10 m$M$ Tris-HCl, 0.6 $M$ NaCl, pH 8.0 (not sterile). Make 500 mL.
43. Mono S buffers:
    a. 10 m$M$ Tris-HCl, pH 8.0 (Nylon 66 filter, degas). Make 1 L.
    b. 10 m$M$ Tris-HCl, 1 $M$ NaCl, pH 8.0 (Nylon 66 filter, degas). Make 0.5–1 L.
44. C18 solvents:
    a. 0.1% TFA in water (Nylon 66 filter, degas). Make 1 L.
    b. 0.08% TFA in 2-propanol:acetonitrile:water, 3:2:2 (Nylon66 filter, degas) Separately measure 300-, 200-, and 200-mL solvents, mix, and add 560 μL TFA.
45. 0.4 $M$ potassium phosphate, pH 7.2 (Sep-Pak filter/sterile filter 5–10 mL required for each 4 L of culture).
46. Glycerol (~1 mL required for each 4 L of culture). Autoclave.
47. Bacterial culture "seeds": Grow an overnight culture (starting with cryopreserved W3110 cells containing the pAng3 vector) in 2–3 mL LB medium containing 50 μg/mL ampicillin. Add 1 mL of this to 1 mL of 60% LB-amp/40% glycerol. Freeze 300-μL aliquots in sterile Eppendorf tubes (dry ice/propanol) and store at –70°C.

48. SP-sepharose column: (Preparation of this column can be done during d 1 or 2). The column should be ~25 mL with a diameter of 1.7–2.4 cm. The flow rates for all of the following steps can be up to 3 mL/min, except for the NaOH wash, which should be ~0.5 mL/min. If the resin has been used previously, it should be washed (in the column) with ~50 mL of 1 *N* NaOH (contact time 1–2 h) followed by several column volumes of buffer B. Then equilibrate with at least 5 column volumes of buffer A. All of this is done at room temperature. If the resin is new, it should be equilibrated with Mono S buffer A and then 25 mL of 2 mg/mL lysozyme in the same buffer should be run through it. Then run another 25 mL of Mono S buffer A, followed by 100 mL of Mono S buffer B. Finally, equilibrate with at least 5 column volumes of (SP-sepharose) buffer A. This preconditioning with lysozyme greatly reduces irreversible adsorption of angiogenin to the resin.
49. Angiogenin standard: 40 µg/mL in Sep-Pak water. This can be purchased from R and D Systems, Inc. in the Met- (-1) form.

## 2.2. Isolation of Bovine Angiogenin from Milk

1. ~15 L 1% fat milk (*see* **Note 2**).
2. 5X CPD: 131.5 g sodium citrate dihydrate, 16.35 g citric acid monohydrate, 11.1 g monobasic sodium phosphate monohydrate, 116.2 g anhydrous dextrose per liter. Filter through Whatman filter paper before use.
3. Whatman No. 1 filter paper.
4. CM 52 cellulose ion-exchange resin preswollen (Whatman, Ltd.).
5. CM 52 loading buffer: 50 m*M* dibasic sodium phosphate heptahydrate, titrated with HCl to pH 6.6 (*see* **Note 3**). Prepare 10–12 L. Store at 4°C.
6. CM 52 elution buffer: loading buffer plus 1 *M* NaCl. Prepare 2 L. Store at 4°C.
7. CM 52 washing buffer: loading buffer plus 3 *M* NaCl. Prepare 2 L.
8. Glass test tubes, 13 × 100 mm.
9. 2-L sintered glass funnel.
10. 4-L flask.
11. Amicon 400 mL concentrator, with YM3 membrane.
12. Eppendorf tubes.
13. Microfuge.
14. Chromatography column, 5 × 50 cm.
15. Mono S column (50 × 5 mm).
16. Mono S buffers: *see* **Subheading 2.1., item 43**.
17. C18 HPLC column: 4.6 × ~250 mm Synchropak (or equivalent).
18. C18 solvents: *see* **Subheading 2.1., item 44**.
19. Centrex filters, 0.45 µm, cellulose acetate (Schleicher and Schuell).

## 2.3. tRNA Assay for Ribonucleolytic Activity of Angiogenin (see *Note 4*)

1. Eppendorf tubes (1.5 mL).
2. Microfuge.
3. 37°C water bath.
4. Micropipets.
5. Spectrophotometer.
6. Sep-Pak C18 cartridge (Waters).
7. Syringes (10 and 60 cc).
8. Falcon tubes, 50 mL.
9. Assay buffer: 0.1 *M* HEPES-NaOH, 0.1 *M* NaCl, pH 7.0. Sep-Pak filter and then sterile filter. Store at room temperature (short-term). Make 50 mL.

10. Sep-Pak filtered water. Store at room temperature.
11. BSA (RNase-free, Worthington), 1 mg/mL in Sep-Pak water. Aliquot and store at –20°C. Keep on ice while setting up assays.
12. tRNA (Type X, Sigma), 10 mg/mL in Sep-Pak water. Store at –20°C for up to 1 mo. Keep on ice while seting up assays.
13. Perchloric acid, 3.4%. Store at 4°C.

## 3. Methods

### 3.1. Isolation of PyroGlu-Angiogenin

Briefly, *E. coli* W3110 cells harboring the pAng3 expression plasmid are cultured in LB broth for about 6 h and then added to $4 \times 1$ L of nutrient-supplemented M9 medium. Cells are grown overnight, harvested, lysed, and centrifuged. Angiogenin is isolated from the supernatant by successive chromatographic fractionations on S-sepharose, Mono S, and C18 columns. The following is a detailed, day-by-day, step-by-step description of the isolation of angiogenin by this procedure.

Certain general precautions should be noted:

1. Extreme care must be taken to avoid contamination by even trace amounts of RNases (e.g., RNase A, a common laboratory reagent) if the angiogenin preparation is to be used for enzymatic assays, since the RNase activity of angiogenin itself is extremely weak. The various base treatments described are effective in removing such contaminants.
2. Angiogenin binds to polystyrene, so this type of plastic must be avoided.

### 3.1.2. Day 1

1. Autoclave four 2.8-L culture flasks, each containing 900 mL water.
2. Six hours prior to starting the large cultures, thaw a bacterial culture "seed" of the W3110 cells containing the pAng3 vector, and add 30 µL to 3 mL of LB-amp in each of four 15 mL snap-top tubes. Incubate at 37°C in a shaker.
3. To each 2.8-L flask add (after cooling the flask to <40°C) 100 mL 10X M9 salts, 2 mL 1 *M* MgSO₄, 1 mL 100 m*M* CaCl₂, 21 mL 20% glucose, and 40 mL 10% casamino acids.
4. Shortly before starting the large cultures, prepare the indole-3-acrylic acid solution.
5. At the end of the day, add to each 2.8-L flask 2 mL of 25 mg/mL ampicillin, 1 mL of the indoleacrylic acid solution, and 3 mL of the "seed" culture.
6. Place flasks in 37°C incubator/shaker.

### 3.1.2. Day 2

1. Approximately 16 h after starting the large cultures, add 10 mL of 20% glucose, 20 mL 10% casamino acids, and 1 mL 20 mg/mL indoleacrylic-acid solution.
2. Shake the cultures at 37°C for another 2 h.
3. Shortly before the end of the incubation, prepare lysozyme and PMSF solutions (*see* **Sub-heading 2.1., items 40** and **41**).
4. Pour the contents of the culture flasks into six 250-mL centrifuge bottles, place in Sorvall GSA rotor, and spin for 15 min at 5800*g*. Three spins are required. Do not remove pellets. Keep the culture flasks in the cold room while waiting to centrifuge their contents. After the last centrifugation, invert the bottles, and allow them to drain onto paper towels for 5 min. Place bottles on ice.
5. To each tube add 62 mL cold TSOPB plus 330 µL PMSF and mix.
6. Detach the pellets with a plastic spatula and resuspend cells by repeated pipetting with a 10-mL plastic pipet.

7. Combine into two bottles.
8. Add to each bottle (mixing after each addition) 4 mL 0.5 $M$ EDTA, 8 mL 5 $M$ NaCl, 2.5 mL 2 mg/mL lysozyme, and 1 mL 0.45 $M$ PMSF.
9. Incubate 1 h in an ice-water bath.
10. Add 1 mL PMSF to each bottle. Sonicate 3 × 40 s (power setting 5, 70% cycle) in cold room. Store in ice water between rounds. Temperature should not rise above ~25°C.
11. Centrifuge in Sorvall GSA rotor for 30 min at 23,400$g$.
12. Pour supernatants into another pair of centrifuge bottles (chilled). Freeze at –70°C. Transfer to –20°C freezer the night before thawing.

### 3.1.3. Day 3

1. Thaw supernatants in 37°C water bath with frequent mixing to prevent temperature from going above ~20°C.
2. Centrifuge in Sorvall GSA rotor 30 min at 23,400$g$.
3. During the centrifugation, start running the SP-sepharose column with buffer A, flow rate ~2–3 mL/min.
4. Pour supernatants into a 1-L flask and keep on ice.
5. Load onto column at ~2–3 mL/min.
6. After loading is complete, run buffer A through column until $A_{280}$ <0.1.
7. Elute angiogenin with buffer B. Collect fractions of ~4 mL each in plastic test tubes (20 will probably be sufficient).

### 3.1.4. Day 4

#### 3.1.4.1. SAMPLE PREPARATION

1. Soak a 76 mm-diameter YM3 Amicon membrane in MilliQ water for ~1 h with several changes (this can be done the previous day).
2. Add all SP-sepharose fractions with $A_{280}$ >0.07 to the Amicon 400-mL concentrator fitted with the YM 3 membrane.
3. Concentrate to <10 mL under $N_2$ in the cold room.
4. Add 50 mL of cold Mono S buffer A.
5. Reconcentrate to ~5–7 mL.
6. Remove sample from concentrator with a syringe and transfer to a 15-mL tube. Wash out concentrator with 2 × 1–2 mL of buffer A and add this to the rest of the sample.
7. Mix and transfer aliquots to 1.5-mL Eppendorf tubes. Centrifuge for 15 min at 4°C.

#### 3.1.4.2. MONO S CHROMATOGRAPHY

1. Clean the column and HPLC system by injecting ~2 mL of 2 $N$ NaOH and running for 30 min on 50% Mono S buffer B, flow rate 0.2 mL/min.
2. Increase the flow to 0.8 mL/min, and begin the wash program: 0–100% buffer B in 15 min, followed by 15 min at 100%.
3. Equilibrate the system with 15% B for 25 min at 0.8 mL/min.
4. Inject ~2 µg of an angiogenin standard in water (50–100 µL) and run a 50-min gradient from 15–55% B at 0.8 mL/min. (Flow rate for all Mono S and C18 chromatography described below is 0.8 mL/min. All gradients are linear.) Monitor absorbance at 214 nm, 0.05 $AU$ full-scale.
5. Angiogenin should elute at 30–36 min in a peak ~3–4 min wide; no other significant peaks should be present except from the flowthrough. Proceed with a pilot run of the preparation after allowing the column to re-equilibrate at 15% B for at least 25 min.
6. The pilot preparation should be ~0.5% of the total (for a 4-L preparation). Centrifuge this material in an Eppendorf tube for 15 min at 4°C before loading. Run the pilot sample

exactly as the standard, but monitor absorbance with 1.0 AU as full scale. Angiogenin will elute as a doublet (representing Gln-1 and <Glu-1 protein).

7. Estimate the yield from the pilot run by comparing the combined area of the two peaks to that of the standard. Determine how many preparative runs will be required (the maximum amount that can be chromatographed at one time is ~4 mg), and divide the total accordingly. Preparative runs are made using the same gradient as for the standard and pilot, but monitor at 254 or 280 nm (full-scale absorbance = 1) and collect 1-min fractions (*see* **Note 5**). At the end of each run increase % B to 100 for 10 min or until a peak elutes. Then go to initial conditions and equilibrate for 25 min before starting the next run.
8. To each fraction containing angiogenin, add 0.8 mL of 0.4 *M* potassium phosphate, pH 7.2, and vortex briefly. Incubate the tubes at 37°C for 18–20 h (*see* **Note 6**).
9. Store Mono S column in 0.02% azide in water or 20% ethanol.

### 3.1.5. Day 5

#### 3.1.5.1. SAMPLE PREPARATION

1. Add Mono S fractions containing potassium phosphate to concentrator and concentrate to ~5–7 mL.
2. Add 50 mL of Mono S buffer B and reconcentrate.
3. Add another 50 mL of Mono S buffer B and reconcentrate to ~5–7 mL.
4. Remove from concentrator as above, washing with Mono S buffer B.
5. Centrifuge for 15 min at 4°C in Eppendorf tubes.

#### 3.1.5.2. C18 HPLC

1. Wash the HPLC system and the Synchropak C18 column with >20 mL water.
2. Run the following wash program at 0.8 mL/min: 0–100 % B in 15 min, followed by 15 min at 30% B.
3. Equilibrate for 25 min at 30% B.
4. Run an angiogenin standard as for the Mono S column, but with a 25-min gradient from 30 to 50% B followed by 5 min at 50% B. The standard should elute between 20 and 27 min with an area as above.
5. Run an aliquot of the preparation as a pilot run (again, ~0.5% of a 4-L prep), with the same gradient as for the standard. Set full-scale absorbance based on expected yield. Up to 2 µg can be run on 0.05 AU full scale.
6. If the material looks clean, run up to 4 mg at a time on the same gradient and collect 1-min fractions. Monitor at 254 or 280 nm as above. If the pilot run revealed extra peaks eluting close to angiogenin, and these are more than a few percent of the total, use a shallower gradient for the main prep (e.g., 30–50% B in 50 min) (*see* **Note 5**).
7. Pool fractions containing angiogenin and
   a. Dialyze vs MilliQ water (>200 vol; 2×) or
   b. Add 1 vol Sep-Pak water to dilute the organic solvents, lyophilize, and reconstitute with Sep-Pak water.
   Quantitate by absorbance at 280 nm: 1 mg/mL reads 0.89.
8. Aliquot, store at 4°C (up to several months) or at –70°C (stable for >5 yr).

### 3.2. Isolation of Bovine Angiogenin from Milk

Angiogenin has been found in the plasma of nonhuman species, including pig, rabbit, mouse, and bovine *(11)*. Noteworthy is the presence of the protein in bovine milk where it occurs at a concentration that is 10 times higher than in bovine plasma *(12)*. Recently, a heparin-binding growth factor purified from bovine colostral fat-globule membranes was identified as angiogenin *(13)*. Since the protein can be readily isolated from milk obtained from a supermarket, this becomes a convenient source to obtain reasonable amounts, particularly for structural and functional studies. All of the following steps prior to Mono S chromatography are performed at 4°C (*see* **Note 7**).

1. Before starting the purification, prepare 600 mL of CM 52 ion-exchange resin. Suspend the resin in 2 L loading buffer, allow to settle, and pour off fines. Resuspend in ~600 mL of loading buffer and transfer to a 2-L sintered glass funnel attached to a 4-L filter flask. Wash under house- or water-vacuum with 2 L of 3 *M* NaCl in loading buffer and then equilibrate with loading buffer.
2. To ~15 L milk containing no more than 1% fat (*see* **Note 2**) add 28 mL/L of 5X CPD (which is the standard citrate-phosphate-dextrose anticoagulant used to make blood plasma). Check the pH of the milk-CPD mixture to make sure that it is no higher than that of the loading buffer.
3. Apply the milk mixture semibatchwise to the resin at a flow rate of ~3 L/h. This flow rate is achievable with a water-vacuum and can be maintained throughout the procedure if 1–1.5 L of loading buffer is passed through the resin (with resuspension of the resin near the end of the wash) after each 4–5 L of milk.
4. After all the milk has been loaded, wash the resin extensively with at least 5–6 L loading buffer, with occasional resuspension, and then transfer it to a 5 × 50 cm chromatography column. Allow the column to pack under flow for about 1 h and then let it sit overnight.
5. The next morning, replace the top buffer (which is cloudy) with fresh buffer and start the flow again (~5 mL/min). When the $A_{280}$ of the effluent is <0.05 (usually 1–2 column volumes), replace the buffer with 1 *M* NaCl in loading buffer. After 200 mL passing through the column, begin collecting 8-mL fractions in glass test tubes (*see* **Note 8**). Pool fractions having an $A_{280}$ of at least 0.08.
6. Concentrate the pooled fractions by ultrafiltration in an Amicon 400-mL device. When the volume has been reduced to <10 mL, add 5 vol of water, reconcentrate to <10 mL, add another 5 vol water, and concentrate to ~10 mL. Rinse the concentrator twice with 1–2 mL water, and add the washings to the concentrate. Centrifuge in Eppendorf tubes for 15 min at 4°C immediately before loading onto a Mono S column.
7. Proceed to purify the bovine angiogenin by Mono S and C18 chromatography as described above, except that the gradient for Mono S chromatography is 10–45% B in 60 min. Bovine angiogenin elutes at ~35 min. Elution time during C18 HPLC is ~24 min.
8. Lyophilize or dialyze as in **Subheading 3.1.5.2., step 7**.
9. Store as in **Subheading 3.1.5.2., step 8**.

### 3.3. tRNA Assay for Ribonucleolytic Activity of Angiogenin

When the sequence of angiogenin was first determined and its homology to pancreatic ribonuclease became apparent, it was surprising to find that it lacked any apparent ribonuclease activity, especially since it contained all the residues considered important for catalysis. Eventually, as a result of persistent and intensive investigations, its enzymatic capability was established and, although extremely low compared to RNase A, it was shown to be essential for angiogenic activity *(8)*. Numerous assay procedures have since been developed, the most convenient of which is based on the hydrolysis of tRNA and is suitable for routine monitoring of ribonuclease activity *(14)*. To ensure that reagents are RNase-free, they should be prepared in Sep-Pak filtered water. The HEPES buffer can also be filtered through a Sep-Pak cartridge.

1. To each of a series of 1.5-mL Eppendorf tubes, add in this order: $110 - x$ μL Sep-Pak filtered water, 100 μL assay buffer (do not mix), 30 μL BSA solution (mix by repeated pipetting) and $x$ μL of sample to be assayed (mix several times) (*see* **Note 9**).
2. At time 0, add 60 μL of tRNA solution to tube 1, mix extensively by pipetting, and place in 37°C water bath.
3. At convenient (20 or 30 s) intervals add 60 μL of tRNA solution to the remaining tubes, mixing extensively after each addition, and place tubes in a 37°C water bath.
4. Terminate the reactions after an appropriate incubation time (*see* **Note 10**) by adding 700 μL of ice-cold 3.4% perchloric acid at the same intervals used for the tRNA additions.
5. Vortex each tube throughly and place in ice or ice water for 10 min.
6. Centrifuge each tube at 4°C for 10 min and put tubes in ice (*see* **Note 11**).
7. Determine the absorbance of each supernatant at 260 nm (zero the instrument with water) (*see* **Note 12**).

## 4. Notes

1. Use the monohydrochloride, monohydrate form of 1,10-phenanthroline, which is more soluble than the other forms.
2. Milk with a higher fat content will clog the CM 52 resin.
3. If you use 50 m*M* of the monobasic sodium phosphate and titrate with NaOH, too much nonangiogenin protein will bind to the resin.
4. Wear powder-free gloves while preparing all solutions, and store all solutions in sterile plasticware.
5. If you have more than ~1 mg of total protein per sample, it will probably elute somewhat earlier than it did during the pilot run.
6. This incubation is included in the protocol to allow all N-terminal Gln residues to cyclize to pyroGlu, which is the native form of angiogenin found in conditioned medium or blood plasma.
7. Bovine angiogenin prepared by this method is suitable for biological and biochemical studies, but not for measurement of intrinsic ribonucleolytic activity since trace contaminants (<0.01%) of RNase A are usually present.
8. It is recommended that a UV monitor **not** be used for this since during the latter part of the protein peak a great deal of insoluble material (probably calcium phosphate) elutes, and this will clog the system. Reading absorbancies is a problem because of the precipitate. Only read every fifth fraction or so, and Centrex filter to clarify those that have precipitate.
9. For maximum efficiency, use one pipette tip for all assay water aliquots, one for all buffer aliquots, and individual tips for each BSA sample, sample, and tRNA addition.
10. Typical angiogenin standards are 5, 10, 20, 30, 40, and 50 μL of 40 μg/mL in Sep-Pak water and the incubation time is 2 h. It is necessary to use a range of concentrations because the standard curve is not linear. The blank obtained with no sample (do in duplicate) should read between 0.15 and 0.3. The top of the standard curve will be 0.9–1.1. The incubation time and concentration of standards can be adjusted for convenience. For example, a 4-h incubation time with 5- to 50-μL aliquots of 20 μg/mL angiogenin is often better to use if the test sample is dilute or little is available. The blank will not increase too much with this extended incubation time.
11. If samples cannot be read within 30 min, transfer each supernatant to a fresh tube. Prolonged contact with perchloric acid will result in degradation of the RNA in the pellet, and increased absorbance. The potential action of perchloric acid on RNA also makes it critical that samples be kept at 0–4°C once the acid is added.
12. To obtain activities of test samples, first plot a standard curve of absorbancies of angiogenin standards (minus the blank) vs angiogenin concentrations. Calculate from the angiogenin concentration required to produce the same absorbance measured in your sample.

## References

1. Riordan, J. F. (1997) Structure and function of angiogenin, in *Ribonucleases, Structures and Functions* (D'Alessio, G. and Riordan, J. F., eds.), Academic, New York, pp. 445–489.
2. Vallee, B. L. and Riordan, J. F. (1997) Organogenesis and angiogenin. *Cell. Mol. Life Sci.* **53,** 803–815.
3. Fett, J. W., Strydom, D. J., Lobb, R. R., Alderman, E. M., Bethune, J. L., Riordan, J. F., and Vallee, B. L. (1985) Isolation and characterization of angiogenin, an angiogenic protein from human carcinoma cells. *Biochemistry* **24,** 5480–5486.
4. Olson, K. A., Fett, J. W., French, T. C., Key, M. E., and Vallee, B. L. (1995) Angiogenin antagonists prevent tumor growth *in vivo. Proc. Natl. Acad. Sci. USA* **92,** 442–446.
5. Tschesche, H., Kopp, C., Horl, W. H., and Hempelmann, U. (1994) Inhibition of degranulation of polymorphonuclear leukocytes by angiogenin and its tryptic fragment. *J. Biol. Chem.* **269,** 30,274–30,280.
6. Shapiro, R., Harper, J. W., Fox, E. A., Jansen, H.-W., Hein, F., and Uhlmann, E. (1988) Expression of Met (-1) angiogenin in *Escherischia coli*: conversion to the authentic <Glu-1 protein. *Anal. Biochem.* **175,** 450–461.
7. Shapiro, R. and Vallee, B. L. (1992) Identification of functional arginines in human angiogen by site-directed mutagenesis. *Biochemistry* **31,** 12,474–12,485.
8. Shapiro, R., Riordan, J. F., and Vallee, B. L. (1986) Characteristic ribonucleolytic activity of human angiogenin. *Biochemistry* **25,** 3527–3532.
9. Auerbach, R., Auerbach, W., and Polakowski, I. (1991) Assays for angiogenesis: a review. *Pharmac. Ther.* **51,** 1–11.
10. Sambrook, J., Fritsch, E. F., and Maniatis, T. (1989) *Molecular Cloning. A Laboratory Manual*, 2nd ed., Cold Spring Harbor Laboratory, Cold Spring Harbor, N.Y.
11. Bond, M. and Strydom, D. J. (1989) Amino acid sequence of bovine angiogenin. *Biochemistry* **28,** 6110–6113.
12. Maes, P., Damart, D., Rommens, C., Montreuil, J., Spik, G., and Tartar, A. (1988) The complete amino acid sequence of bovine milk angiogenin. *FEBS Lett.* **241,** 41–45.
13. Hironaka, T., Ohishi, H., Araki, T., and Masaki, T. (1997) Identification of a heparin-binding growth factor isolated from bovine colostral fat globule membrane. *Milchwissenschaft-Milk Sci. Int.* **52,** 508–510.
14. Shapiro, R., Weremowicz, S., Riordan, J. F., and Vallee, B. L. (1987) Ribonucleolytic activity of angiogenin: Essential histidine, lysine and arginine residues. *Proc. Natl. Acad. Sci. USA* **84,** 8783–8787.

# 26

# Preparation and Preclinical Characterization of RNase-Based Immunofusion Proteins*

## Dianne L. Newton and Susanna M. Rybak

## 1. Introduction

Toxins from plants and bacteria have been coupled to antibodies to produce selective cytotoxic agents (immunotoxins) *(1–3)*. However, toxic side effects and immunogenicity have presented major obstacles to the successful clinical application of these proteins *(4–9)*. Production of humanized antibodies has alleviated the immunogenicity of the targeting component *(10)*. Substituting plant and bacterial toxins with members of the pancreatic RNase A superfamily (for review *see* **ref.** *11*) presents a solution to problems associated with the effector portion of immunotoxins *(12)*.

With regard to immunotoxin construction, noncytotoxic RNases, such as pancreatic RNase, eosinophil-derived RNase (EDN), or angiogenin, become as or more cytotoxic to cells than the classical plant and bacterial toxins when directly introduced into the cytosol *(13)*. Chemically linking or fusing RNases to targeting agents that internalize them into the cytosol results in RNA degradation and inhibition of protein synthesis *(14,15)*. RNase-based immunotoxins exhibit potent antitumor activity in vitro and in vivo *(16–24)*.

**Subheading 3.1.** describes the synthesis of an RNase gene using only a single strand of DNA. This method reduces the cost of synthesizing a gene by 50% and is as effective as the double-stranded method. **Subheadings 3.2.** and **3.3.** describe the preparation of an RNase immunofusion protein. RNase-sFv fusion proteins as well as cytokine (RNase-IL2), growth factor (RNase-EGF) or protein ligand (RNase-CD4) fusion proteins have been successfully prepared using the methods described (D.L.N. and S.M.R., unpublished data). **Subheadings 3.4.** and **3.5.** describes characterization of the activities of both moieties of the fusion protein. This section describes methods to determine the effectiveness of the targeting agent to bind to its receptor and the enzymatic activity of the RNase. Also included are methods to determine the interaction of the RNase

*This project has been funded in whole or in part with Federal funds from the National Cancer Institute, National Institutes of Health, under Contract No. No1-CO-56000. The content of this publication does not necessarily reflect the views or policies of the Department of Health and Human Services, nor does mention of trade names, commercial products, or organizations imply endorsement by the US Government.

From: *Methods in Molecular Biology, vol. 160: Nuclease Methods and Protocols*
Edited by: C. H. Schein © Humana Press Inc., Totowa, NJ

inhibitor (RI) with the RNase. Numerous studies have shown that cytotoxic activity depends on expressed ribonuclease activity (*14,25,26*) and sensitivity to RI may affect intracellular RNase cytotoxicity (*27–29*). **Subheadings 3.6.** and **3.7.** describe methods for the preclinical evaluation of RNase immunotoxins both in vitro and in vivo. Our results demonstrate that targeted RNases are as efficacious as classical immunotoxins in the treatment of disseminated human lymphoma in mice with severe combined immunodeficiency disease (SCID). Yet targeted RNases do not cause nonspecific toxic side effects in mice at doses 90% greater than that needed for efficacy (D.L.N. and S.M.R., unpublished data).

## 2. Materials

### 2.1. PCR-Based Synthesis of RNase Genes Using Only One of the Strands

1. Oligonucleotides (95–105 nucleotides in length) encoding the gene of interest, purified on oligonucleotide purification cartridges (OPC cartridges, Cruachem, Inc., Dulles, VA). Purify oligonucleotides and primers to remove any short failure sequences. Unpurified oligonucleotides can introduce nucleotide deletion(s).
2. Junction oligonucleotides (18 nucleotides in length) complementary to the junction of the two-template-strand oligonucleotides purified on OPC cartridges (**Table 1**). *See* **Fig. 1A** and **Subheading 3.1., step 2** for discussion of the design.
3. Primers containing flanking restriction enzyme sites purified on OPC cartridges (**Table 1**). *See* **Fig. 1A** and **Subheading 3.1., step 3** for discussion of the design.
4. Thermal cycling (PCR) machine and reagents for performing PCR, GeneAmp PCR Reagent Kit (Perkin-Elmer Corp., Foster City, CA).
5. Agarose and NuSieve 3:1 agarose (FMC BioProducts, Rockland, ME) and appropriate gel electrophoresis apparatus.
6. Gene Clean II (Bio 101 Inc., La Jolla, CA).

### 2.2. Construction of the RNase-sFv Fusion Gene

1. Gene encoding the single chain antibody (sFv) or binding ligand of choice.
2. Oligonucleotides (primers) purified on OPC cartridges (*see* **Subheading 2.1., item 1**).
3. Vector, pET-11d, or appropriate pET vector (Novagen, Madison, WI) (*see* **Note 4**).
4. Appropriate restriction enzymes.
5. Rapid DNA Ligation Kit (Boerhinger, Mannheim, Indianapolis, IN).
6. PCR machine and reagents (*see* **Subheading 2.1., item 4**).
7. Agarose and electrophoresis apparatus (*see* **Subheading 2.1., item 5**).
8. Gene Clean II (*see* **Subheading 2.1., item 6**).
9. Sequenase II Kit (U.S. Biochemical Corp., Cleveland, OH) and sequencing apparatus.

### 2.3. Purification of a RNase-sFv Recombinant Fusion Protein

1. Competent bacteria, BL21(DE3) (Novagen).
2. Luria Broth (LB)/amp plates: 10 g tryptone, 5 g yeast extract, 5 g NaCl, and 16 g agar in 1 L $H_2O$; autoclave 20 min, let cool to 55°C, and add 1 mL 100 mg/mL ampicillin. Pour 20–25 mL/100 mm petri dish.
3. Isopropyl-β-D-thiogalactopyranoside (IPTG), 120 mg/mL (0.5 *M*) in $H_2O$ (Gibco-BRL, Grand Island, NY).
4. Superbroth: 12 g tryptone, 24 g yeast extract, 6.3 g glycerol, 12.5 g $K_2HPO_4$, 3.8 g $KH_2PO_4$ in 1 L $H_2O$; autoclave 20 min. After cooling, add 20 mL of 25% glucose (250 g/L $H_2O$, sterile-filtered) and 10 mL 80 m*M* $MgSO_4$ (9.75 g $MgSO_4 \cdot 7H_2O$/500 mL $H_2O$, sterile-filtered).

**Table 1**
**Sequence of Junction and Primer Oligonucleotides**
**Encoding the _Rana catesbiana_ Gene (32)**

| Primer | Sequence |
|---|---|
| a (Primer) | 5'-ATATA**TCTAGA**-_AATAATTTTGTTTAACTTTAAGAAGGAGATATACAT_-<br>ATGcagaactgggctactttccag-3' |
| d (Primer) | 5'-CGCGC**GGATCC**-_CTACTA_ -cgggcaacgaccgataccagc-3' |
| b (Junction) | 5'-gttcatgttgataacacc-3' |
| c (Junction) | 5'-ggtgttatcaacatgaac-3' |

Lower case letters, gene sequence; uppercase letters, clamp for later restriction enzyme digest; bold uppercase letters, restriction enzyme sites (_Xba_I [Primer a] and _Bam_HI[Primer d]); italicized letters, modified cloning vector sequence; underlined uppercase letters, ATG start site; underlined italicized letters, two stop signals.

5. Sucrose buffer: 30 m$M$ Tris-HCl, pH 7.5, containing 20% sucrose (200 g/L $H_2O$) and 1 m$M$ EDTA.
6. Janke & Kunkel polytron tissuemizer (Janke & Kunkel, GMBH, KG Staufen, West Germany) or similar model with a 100-mm long × 10-mm OD shaft.
7. Tris-EDTA buffer: 50 m$M$ Tris-HCl, pH 7.5, containing 20 m$M$ EDTA.
8. Lysozyme (Sigma, St. Louis, MO), 5 mg/mL in $H_2O$, prepared just before use.
9. Triton X-100 (Sigma), 25% solution in $H_2O$; use low heat to solubilize.
10. 5 $M$ NaCl, 292.25 g/L; use low heat to solubilize.
11. Solubilization buffer: 0.1 $M$ Tris-HCl, pH 8.0, containing 6 $M$ guanidine-HCl (573.2 g/L) and 2 m$M$ EDTA.
12. Dithioerythritol (DTE) (Sigma).
13. Renaturation buffer: 0.1 $M$ Tris-HCl, pH 8.0, containing 0.5 $M$ L-arginine-HCl (105 g/L) (Sigma), 8 m$M$ glutathione, oxidized (GSSG) (Boehringer Mannheim, Indianapolis, IN) (4.9 g/L) and 2 m$M$ EDTA, prechilled to 10°C.
14. Dialysis buffer; 10X concentration containing 0.2 $M$ Tris-HCl, pH 7.5, and 1 $M$ urea (60 g/L) prepared ≤ 2–3 h before use. Because of the formation of cyanate ions, which can then carbamylate amino groups, do not prepare the 10X dialysis solution more than a few hours in advance. Dilute to 1X just before needed.
15. Appropriate chromatography columns for purification of the fusion protein (Sephadex ion exchangers, heparin Sepharose, affinity columns, Mono S or Q HR 5/5) and the appropriate equipment and buffers to use with the columns.

## 2.4. Characterization of the Targeting Component of the RNase Immunofusion Protein

1. Ice-cold phosphate-buffered saline (PBS).
2. Ice-cold 1% BSA in PBS (a squirt bottle containing this solution is very convenient).
3. Unlabeled and [$^{125}$I] labeled antibody (_see_ **Note 9**).
4. Cells containing the appropriate antibody binding site.
5. Suction apparatus.
6. Gamma counter.

## 2.5. Characterization of the Enzymatic Component of the RNase Immunofusion Protein

The following are used for determining enzymatic activity using tRNA:

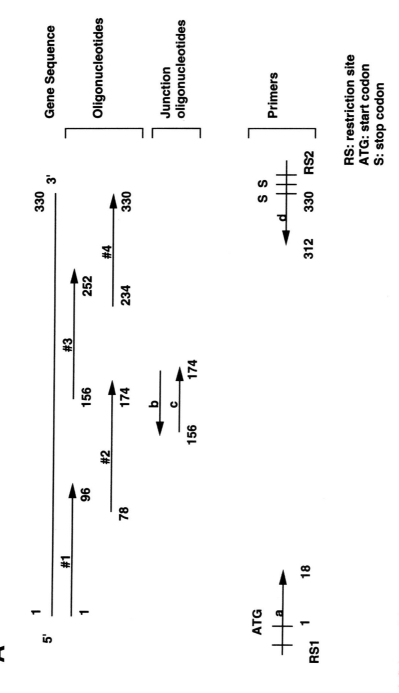

Fig. 1. Scheme of the strategy used for the synthesis of an RNase gene containing 330 bp. (A) Components required for the synthesis of the gene. Oligonucleotides labeled #1–4 are 95–105 bp in length and encode one strand of the *Rana catesbeiana* RNase gene (*32*). *b* and *c* are the junction oligonucleotides, 18 bp in length, and are complementary to the junction of oligonucleotides #2 and #3. *b* is in the antisense direction; *c* is in the sense direction. *a* and *d* are the primer oligonucleotides and contain the flanking restriction enzyme sites and the first 18 bp of oligonucleotide #1 (*a*) and the last 18 bp of the oligonucleotide #4 (*d*). *a* is in the sense and *d* is in the antisense direction. **Table 1** depicts the actual bp sequences. *See* **Subheading 3.1.** for details on the construction of the RNase gene.

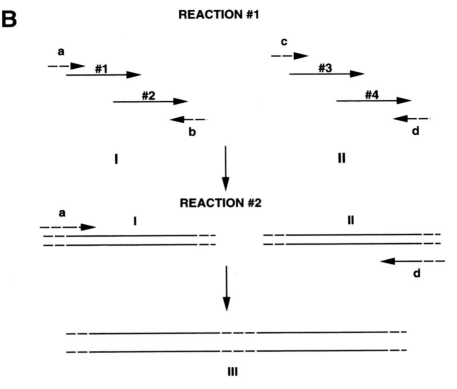

Fig. 1. **(B)** PCR strategy used for the synthesis of the gene. I and II depict the synthesis of each half of the gene. III depicts the total cDNA resulting from the joining of the two halves, I and II, together.

1. RNase-free tubes and pipet tips (*see* **Note 10**).
2. RNase-free $H_2O$ (*see* **Note 10**).
3. tRNA (Sigma) in RNase-free $H_2O$.
4. Assay buffer: for pH 7.5 or 8.0, 0.5 *M* Tris-HCl, pH 7.5 or 8.0, containing 5 m*M* EDTA and 0.5 mg/mL HSA; for pH 6.0, 30 m*M* HEPES, pH 6.0, containing 0.5 mg/mL HSA.
5. Dilution buffer, 0.5 mg/mL human serum albumin (HSA) (Sigma) in RNase-free $H_2O$. HSA is a very good diluent for RNases and RNase fusion proteins, as very low blank values can be obtained in its presence and RNases maintain full activity at very low concentrations *(30)*.
6. $H_2O$ bath or incubator at 37°C.
7. Ice-cold perchloric acid, 3.4%: 2.4 mL 70% perchloric acid diluted to 50 mL with $H_2O$.
8. Centrifuge: Eppendorf or equivalent model.
9. UV absorbance spectrophotometer.
10. RI, ribonuclease inhibitor: RNasin Ribonuclease Inhibitor (Promega).
11. Dithiothreitol (DTT) (Sigma).

The following are used to determine cell free-protein synthesis inhibition:

12. Rabbit reticulocyte lysate (Promega). Do not freeze and thaw the rabbit reticulocyte lysate more than twice.
13. Brom mosaic viral mRNA, 500 µg/mL (Promega).

14. Amino acid mixture minus methionine, 1 m*M* (Promega).
15. [$^{35}$S]methionine (>1000 Ci/mmol, 10 mCi/mL) (NEN Life Science Products, Boston, MA) (*see* **Note 13**).
16. Nuclease-free water (*see* **Note 10**).
17. Water bath at 30°C.
18. Whatman GF/A filter paper (Thomas, Swedesboro, NJ).
19. 10% (w/v) trichloroacetic acid (TCA) (stored at 4°C).
20. Absolute ethanol.
21. Scintillation fluid.
22. Scintillation counter.

## 2.6. In Vitro Analysis of RNase Immunofusion Protein Potency

1. 96-well microtiter plates (Nunc, Inc., Gaithersburg, MD or equivalent).
2. Millipore Millex-HV filters (Millipore Products Division, Bedford, MA) (*see* **Note 14**).
3. Serum- and leucine-free RPMI (Gibco-BRL).
4. [$^{14}$C] leucine, 0.1 mCi in 10 µL PBS (NEN Life Science Products) or WST-1 (Boehringer Mannheim, Indianapolis, IN).
5. Glass fiber filters (Brandel Inc., Gaithersburg, MD).
6. PHD cell harvester (Cambridge Technology Inc., Watertown, MA).
7. Scintillation fluid.
8. Scintillation counter for measuring [$^{14}$C] leucine incorporation or microtiter-plate reader (MR4000; Dynatech Laboratories, Chantilly, VA or equivalent) for WST-1 assay.

## 2.7. In Vivo Analysis of RNase Immunofusion Proteins

The following are used for in vivo toxicity studies:

1. Outbred male and female mice, 3/group, 6–8 wk of age, approx 20 g weight. Animal care procedures should be in accordance with standards described in the National Institutes of Health Guide for Care and Use of Laboratory Animals.
2. RNase immunofusion, free RNase, appropriate control buffer (*see* **Note 8**).
3. Millipore Millex-HV filters (Millipore Products Division, Bedford, MA) (*see* **Note 14**).
4. 1-mL syringes and 25-gage needles.
5. Scale to weigh animals.

The following are used for in vivo tumor treatment models:

6. Nude female mice, 8–10 wk old, approx 20 g weight (mice should be within a 5 g range), 8–10 animals per group. Use only one sex, since the development of the tumor as well as the response to the drug may vary between sexes because of hormonal differences *(31)*.
7. Anti-asialo GM1 (rabbit) (ASGM-1) (Wako Pure Chemical Industries Ltd. (Richmond, VA).
8. Tumor cells.
9. Trypan blue stain, 0.4% (Gibco-BRL).
10. Dulbecco's phosphate-buffered saline (DBSS) or equivalent (Gibco-BRL).
11. Sterile RNase immunofusion, RNase, antibody, appropriate control buffer (*see* **Notes 8, 14,** and **16**).
12. 1-mL syringes and 25-gage needles.
13. 1X Trypsin-EDTA (Gibco-BRL).
14. Falcon sterile cell strainer, 100-µm nylon (Becton Dickinson Labware (Franklin Lakes, NJ).
15. Mark I [$^{137}$]Cesium Irradiator (J. L. Shepherd and Associates, Glendale, CA).

16. Fixatives; 10% formalin solution (Fisher Scientific, Pittsburgh, PA), Bouins fixative solution (Ricca Chemical Co., Arlington, TX).

## 3. Methods

### 3.1. PCR-Based Synthesis of RNase Genes Using Only One of the Strands

The genes for most RNase proteins range from 312 to 402 basepairs (bp) in length, and can be synthesized with 4 oligonucleotides of 90–105 bp. For example, in the construction of the frog RNase, *Rana catesbiana (32)* which is 330 bp long, four oligonucleotides each approx 96 bp in length were prepared (*see* **Fig. 1A**).

1. Design oligonucleotides corresponding to the sense strand of the gene with an 18–20 bp overlap between each pair of segments (*see* **Fig. 1A**). Analyze each for the restriction enzyme sites needed for insertion of the gene into the vector.
2. Design the junction oligonucleotides to contain 18 bp of overlap sequence from each of the two template strands to be joined together. Make these in both the sense and antisense directions (*see* **Fig. 1A** and **Table 1**).
3. Design the 5' and 3' primers to have an 18–21 bp overlap with the gene, to contain the ATG start codon and the termination signal at the 5' and 3' ends, respectively, as well as the appropriate restriction enzyme sites at the 5' and 3' ends to facilitate cloning (*see* **Table 1**). The use of two different restriction enzyme sites for cloning facilitates the directional insertion of the gene into the vector. One can add nucleotides encoding a tag, such as 3–6 histidyl residues for affinity purification, into the primer if they are not already in the vector.
4. Set up PCR reaction mixture #1 to contain the following per 100 µL reaction volume (*see* **Fig. 1B**):

    Final Concentration

    1X reaction buffer (10 m$M$ Tris-HCl, pH 8.3, 50 m$M$ KCl, 1.5 m$M$ MgCl$_2$)
    100 µ$M$ of each of the nucleotides
    0.5 µ$M$ junction oligonucleotide *b* or *c*
    0.5 µ$M$ primer *a* or *d*
    0.4–0.5 µg of each of the oligonucleotides 1 and 2 or 3 and 4

    Two reactions are prepared, each containing the oligonucleotides, junction oligonucleotides, and primers appropriate for one-half of the gene (*see* **Fig. 1B**). The following PCR conditions have been found to be optimal for many RNase genes: 94°C, 5 min hot start before beginning the program; add 1.0 µL AmpliTaq DNA Polymerase; then 94°C, 1 min (denaturation); 55°C, 2 min (annealing); 72°C, 2 min (extension) for 20 cycles (*see* **Note 1**).

5. Analyze the PCR reaction by NuSieve 3:1 Agarose to determine the size of the PCR product (*see* **Note 2**).
6. Purify the DNA using Gene Clean II following the manufacturer's instructions to remove the junction oligonucleotides. Performing the second PCR without prior purification may result in a decrease in full-length clones *(33)*. The final product should be in a volume of 15 µL.
7. Perform PCR reaction #2 to attach the first half of the gene to the second half using the same PCR conditions as described for reaction #1 (*see* **Fig. 1B**) (final volume 100 µL):

    Final Concentration
    1X reaction buffer (10 m$M$ Tris-HCl, pH 8.3, 50 m$M$ KCl, 1.5 m$M$ MgCl$_2$)
    100 µ$M$ of each of the nucleotides

1.0 µL AmpliTaq DNA Polymerase (added after 5 min at 94°C as described in reaction #1)
0.5 µ*M* primer a and d
3.2 µL (approx 20% of the 15 µL from PCR reaction #1) of each half of the DNA
purified from PCR reaction #1 (Fig. 1B, I and II)

8. Repeat **steps 5** and **6** above.
9. The DNA (**Fig. 1B**, III) is now ready to be restricted and inserted into the vector of choice for sequencing and expression or spliced onto another gene as described below in **Subheading 3.2.**

## 3.2. Construction of the RNase-sFv Fusion Gene (see Note 3)

Before constructing the RNase-sFv fusion construct, the orientation of the RNase to the sFv should be identified, i.e., should the sFv be attached to the 5' or 3' end of the RNase. For some RNases, the N or C terminus may be involved in the active center. For example, the last 3 amino-acid residues of the C-terminus of the human RNase, angiogenin, have been shown to be an active-center subsite *(34)* and attachment of an sFv to this end of the gene results in loss of enzymatic activity *(35)*. In another case, the N-terminus of the amphibian RNase, onconase, contains a pyroglutamic acid residue which forms part of the enzyme active site *(36)*. Expression of this protein with a Met as the first amino acid results in a dramatic loss of RNase and cytotoxic activity *(37,38)*.

1. Design primers to incorporate the following:
   a. A spacer, such as amino-acid residues 48–60 of fragment B of staphylococcal protein A to separate the RNase and sFv genes *(39)*. RNase fusion proteins without a spacer have lower enzymatic, receptor binding and cytotoxic activities *(35)*.
   b. 3–6 histidyl residues or other peptide tag for affinity purification at either the 5' or 3' end of the entire construct (*see* **Note 4**).
   c. A termination signal at the 3' end of the total construct.
   d. Appropriate restriction enzyme sites to facilitate cloning into the vector of choice.
2. Modify the two genes by PCR using the primers described in **step 1** and splice the two modified genes together using the PCR technique of splicing by overlap extension *(40)* following the detailed protocol described in **ref. *41***.
3. Purify the DNA using the Gene Clean II procedure according to the manufacturer's instructions, extracting the resin with 15 µL sterile $H_2O$.
4. Restrict the spliced DNA and ligate the DNA and the vector together using the Rapid DNA Ligation Kit, according to the manufacturer's instructions.
5. Identify those clones expressing protein of the appropriate molecular weight as described in **ref. *41*** and sequence the DNA to ensure that no PCR errors have occurred.

## 3.3. Purification of a RNase-sFv Recombinant Fusion Protein

The denaturation and renaturation of fusion proteins prepared from inclusion bodies have been carefully optimized *(42)*. The functional domains of the fusion proteins fold independently of each other, and the peptide connecting the two together influences the folding rate *(43)*. The optimal length of refolding is empirical for each construct, and must be determined for that protein. Aggregation caused by incorrect folding will occur during the renaturation step resulting in the formation of a precipitate.

1. Transform competent bacteria (BL21[DE3]) with the vector containing the fused gene and grow on a LB/amp plate overnight at 37°C.
2. Scrape the colonies from the plate into 1 L superbroth, grow overnight at 27°C, and induce the next morning with 0.8 m*M* IPTG for 1.5–3 h (*see* **Note 5**).

3. Centrifuge the bacteria at 8000*g* 10–20 min at 4°C and store the pellet at –20°C if not ready to begin the extraction of the protein.

4. Resuspend the pellet with ice-cold sucrose buffer (200 mL/1 L bacterial culture), incubate on ice 10 min, and pellet by centrifugation at 8000*g* at 4°C for 20 min.

5. Carefully pour off the supernatant, resuspend the pellet with ice-cold $H_2O$ (200 mL/1 L bacterial culture), place into eight polypropylene tubes (25 mL/tube), and incubate on ice for 20 min.

6. Centrifuge at 17,000*g* for 20 min at 4°C. Carefully pour off the supernatant (periplasm). The pellet (spheroplast) may be stored at –70°C until ready to process further.

7. Resuspend the pellet in each tube with 9 mL Tris-EDTA buffer using a Janke & Kunkel polytron tissuemizer (or similar model) and combine two tubes together so that each liter of the original bacterial culture is now contained in four tubes. Let the tubes stand at room temperature for 30 min.

8. Add 0.9 mL of 5 mg/mL lysozyme (final concentration, 240 µg/mL)/18 mL and incubate at room temperature for 60 min, shaking occasionally.

9. Add 2.5 mL 5 *M* NaCl and 2.5 mL 25% Triton X-100 per tube, shaking after each addition. Incubate at room temperature for 30 min, shaking intermittently.

10. Homogenize the samples just before centrifugation and centrifuge 17,000*g* for 40 min at 4°C.

11. Carefully pour off supernatant, resuspend the pellet in 15 mL Tris-EDTA buffer with the tissuemizer, and centrifuge again as in **step 10** for 30 min. Repeat the homogenization and centrifugation three times more, decreasing the centrifugation time to 20 min for the last two centrifugations. The washed inclusion bodies may be stored at –70°C until ready for solubilization.

12. Resuspend the purified inclusion bodies in 10–20 mL solubilization buffer so that the solution is free-flowing and not too viscous, and incubate for 2 h at room temperature. Be careful not to dilute too much, as final protein concentration should be 8 mg/mL (*see* **step 14**).

13. Centrifuge 20 min at 4°C at 17,000*g*.

14. Determine the protein concentration of a 1:10 dilution of the supernatant and adjust the solution to a final protein concentration of 8 mg/mL with solubilization buffer. Add dry DTE to a final concentration of 0.3 *M*. Incubate for 2 h or more at room temperature.

15. Renature the protein by diluting 100 times with renaturation buffer at 10°C (*see* **Note 6**) and incubating for 2–3 d at 10°C.

16. Before application to the first chromatography column, dialyze the refolded protein extensively at 4°C (10 vol of renaturation solution each time, total of at least two changes) (*see* **Note 7**).

17. Apply the eluate of the first column to a final affinity column (such as $Ni^{2+}$-NTA agarose (Qiagen, Chatsworth, CA) for proteins containing the His tag) or to FPLC or HPLC to obtain the purified RNase-sFv (*see* **Note 8**). For an extensive discussion of the purification of RNase-sFv fusion proteins, the reader is referred to **ref. *41***.

## *3.4. Characterization of the Targeting Component of the RNase Immunofusion Protein*

The binding assay compares the amount of the RNase fusion protein required to displace 50% of the [$^{125}I$] labeled form of the intact IgG antibody ($EC_{50}$) from cells expressing the appropriate receptor with that of the unlabeled form of the same antibody. Before determining the binding activity of the fusion protein, the $EC_{50}$ of the nonfused antibody should be known. It can be determined as follows:

1. Wash cells twice with ice-cold PBS to remove any dead cells or nonadherent cells that may interfere with the binding assay. Keep the cells cold throughout the procedure to

prevent or slow down the rate of internalization of the cell-surface receptors. Resuspend nonadherent cells in 1% BSA/PBS; use 100 μL/1.5-mL Eppendorf tube. For adherent cells plated into a 24-well plate the day before, add 500 μL of 1% BSA/PBS to each well.

2. Prepare a standard curve of unlabeled antibody over a wide range of concentrations. Apply 10 μL of the diluted antibody solutions or buffer (control to determine the maximum amount of binding of labeled antibody) to the appropriate wells or tubes containing the cells.
3. Apply 10 μL of the [$^{125}$I] labeled antibody to the cells (5000–10,000 cpm/reaction).
4. Incubate the cells on ice for 2 h.
5. Wash the cells three times (an aspirator can be used to remove the liquid, but care should be taken not to disturb the cells) with ice-cold 1% BSA/PBS, and count the pellet in the tip of the tube (either count the entire tube or with a scapel carefully remove the tip of the tube and count the tip), or lyse the cells by the addition of 0.1 $N$ NaOH followed by an incubation for 30 min at 37°C, and count the entire lysate.
6. Determine the percentage of maximal binding as follows:

$$\text{(cpm of reaction containing cold antibody)}/ \text{(cpm of reaction containing labeled antibody alone)} \times 100 \qquad (1)$$

Plot the percentage of maximal binding on the $y$-axis vs the concentration of cold antibody on the $x$-axis. Calculate the EC$_{50}$ by determining that concentration of cold antibody required to inhibit the binding of labeled antibody by 50%. Repeat the binding experiment described above with the RNase fusion protein. A preincubation step of 15–30 min may be incorporated at **step 2** to help a weaker binding antibody fusion protein, such as a monovalent sFv or Fab fusion protein, bind to the receptor before the bivalent intact labeled antibody is added to the cells. Bivalent antibodies or antibody fragments usually bind with higher affinities than monovalent Fab or sFv fragments *(44)*.

## 3.5. Characterization of the Enzymatic Component of the RNase Immunofusion Protein

Many substrates can be used to measure the enzymatic activity of RNases: Some are specific substrates for a particular RNase, whereas others, such as yeast RNA, are more general substrates. The assay described here uses tRNA, which is a convenient substrate for many of the pancreatic type RNases for the following reasons:

1. The blank value which increases with increasing incubation time is low. This is important if extremely long incubation times are required because of low enzymatic activity.
2. The assay is not difficult to perform, and many samples can be processed simultaneously.
3. No special equipment is needed. Some substrates require an HPLC *(45)* or a computer-linked spectrophotometer equipped with a temperature controlled cuvet compartment to monitor the very small changes in absorbance *(45)*.

Before any of the kinetic values for the RNase are determined, the optimal pH for ribonuclease activity should be determined as follows:

1. Set up numbered reaction tubes on ice in replicate and add 100 μL of each of the following solutions: tRNA (stock concentration, 1.0 mg/mL), HSA (stock concentration, 0.5 mg/mL), buffer (stock concentration, 0.5 $M$ Tris-HCl, pH 7.5, 5 m$M$ EDTA, 0.5 mg/mL HSA, or the same buffer at pH 8.0 or 30 m$M$ HEPES, pH 6.0, 0.5 mg/mL HSA).
2. Add 10 μL of either dilution buffer or RNase (*see* **Note 11**).
3. Initiate the reaction by placing the tubes at 37°C. Incubation time varies with the RNase; EDN, 15 min; angiogenin, 18 h. To shorten the incubation time, the concentration of the enzyme can be increased (*see* **Note 11**).

4. To terminate the reaction, place tubes on ice and add 700 μL of 3.4% ice-cold perchloric acid. Allow the tubes to sit on ice for 10 min before centrifuging for 10 min in an Eppendorf centrifuge at full speed.
5. Read the absorbance of the supernatant at $OD_{260nm}$.
6. Average the replicate values and subtract the blank (those tubes with dilution buffer and no enzyme) from those assays containing the enzymes. Choose the buffer at the pH yielding the greatest RNase activity for further enzymatic studies.

To determine the $K_m$ (the substrate concentration at which the reaction velocity is half maximal) of the RNase, run preliminary assays as described above at the appropriate pH with a wide range of substrate concentrations. Five different substrate concentrations spanning 0.5–5 times the approximate $K_m$ value will usually give a reproducible straight line on the double reciprocal plot (*see* **step 7**) of the following sequence. Substitute the following for **step 1** of the previous section; 100 μL of each solution:

1. tRNA (stock concentration, vary from 0.25–10 mg/mL), HSA (stock concentration, 0.5 mg/mL), buffer (buffer at optimal pH (*see* **step 1** above for description).
   Then follow **steps 2–6** described above, using blanks for each substrate concentration because perchloric acid-soluble nucleotides change with different tRNA concentrations.
7. Prepare a double-reciprocal (Lineweaver-Burk) plot by plotting $1/s$, where $s$ is the substrate concentration, on the *x*-axis vs $1/v_o$, the adjusted $OD_{260nm}$ value obtained for each $s$ (*see* **step 6** above), on the *y*-axis. Determine the $V_{max}$ (maximal velocity, the intercept on the $1/v_o$ axis is $1/V_{max}$) and $K_m$ (approximate dissociation constant of the enzyme-substrate complex; the $K_m$ is equal to the substrate concentration in mol/L that results in one-half the maximum velocity of $V_{max}$ and is obtained from the intercept on the $1/s$ axis, which is $-1/K_m$) from the plots and calculate the $K_{cat}$ (catalytic constant or turnover number) and $K_{cat}/K_m$ (indicator of catalytic efficiency) as follows; using the following approximation, convert $K_m$ from mg/mL to a molar value using the $M_r$ for tRNA of 28,100 *(46)*. Convert $V_{max}$ from $OD_{260nm}$ to a molar value by dividing $OD_{260nm}$ value obtained by $7 \times 10^5$ ($OD_{260nm}$ of 1.0 equals 40 μg of RNA [*47*], thus from the equation e = A/C, e = $7 \times 10^5$). $K_{cat}$ can be obtained from the equation, $K_{cat}(E_o) = V_{max}/t$ where $Eo$ is the molar concentration of enzyme used and $t$ is the incubation time in seconds, the final units are $s^{-1}$. The efficiency of the enzyme is obtained from $K_{cat}/K_m$, the units of $K_{cat}/K_m$ are $M^{-1}S^{-1}$.

To determine whether ribonuclease inhibitor (RI) binds to the enzyme and inhibits RNase activity, perform the assay as described above, preparing two sets of identical tubes. Add RI to one set (RI or RI diluent can be included in the buffer). RI binds noncovalently to RNases in a 1:1 ratio with an association constant $>10^{14}$ *(48,49)*, thus the concentration of RI should be adjusted to be two- to 10-fold higher than the concentration of enzyme (*see* **Note 12**). Include 1.0 m*M* DTT in the assay (*see* **Note 12**). Choose enzyme and substrate concentrations that will yield an $A_{260}$ nm between 0.5 and 1.0 in the absence of RI to provide an accurate assessment of partial or complete inhibition. This assay is not meant to determine the $K_i$, the inhibition constant for the ribonuclease inhibitor. The reader is referred to the literature for more in depth coverage of enzyme kinetics of inhibitors *(50,51)*.

The following method is used to determine cell free-protein synthesis inhibition. The incubation of rabbit reticulocyte lysate, mRNA, and other components necessary for protein synthesis results in the incorporation of added [$^{35}$S] methionine into newly synthesized protein. Inclusion of RNases in the incubation mix results in a potent inhibition of the translational capacity of this system.

1. Thaw the rabbit reticulocyte lysate on ice and the other components at room temperature, storing them on ice as soon as thawed (*see* **Note 10**).
2. Prepare the following reaction components on ice in a sterile 1.5-mL screw-cap tube in a final volume of 25 µL.
   a. Control reaction #1 (100% protein synthesis): 7.5 µL rabbit reticulocyte lysate; 7.5 µL reaction mixture containing 0.14 µg Brom mosaic virus mRNA; 0.3 nmol amino acids minus methionine, 1–2 µCi of [$^{35}$S]methionine; 10 µL diluent buffer
   b. Control reaction #2 (background): 7.5 µL rabbit reticulocyte lysate; 7.5 µL reaction mixture containing no mRNA, 0.3 nmol amino acids minus methionine, 1–2 µCi of [$^{35}$S]methionine; 10 µL diluent buffer.
   c. Reaction mixtures: 7.5 µL rabbit reticulocyte lysate; 7.5 µL reaction mixture containing 0.14 µg Brom mosaic virus mRNA, 0.3 nmol amino acids minus methionine, 1–2 µCi of [$^{35}$S]methionine; 10 µL appropriate dilution of RNase/RNase-sFv.
3. Briefly centrifuge to return the reaction components to the bottom of the tube and gently shake the tubes before incubating them for 60 min at 30°C.
4. Terminate the reaction by placing the tubes on ice. The reaction tubes can be frozen at this point and completed at a later date.
5. Prepare a grid (0.5 × 0.5 in. squares) on Whatman GF/A filter paper. With a number 2 pencil, label each square with a number corresponding to a reaction tube.
6. Spot 5 µL of each reaction in the middle of the appropriate square.
7. Place the spotted filter into a tray containing 10% ice-cold TCA, cover, and shake gently for 5 min. Pour off the TCA into a designated radioactive container. Repeat the TCA wash two times more.
8. Rinse the filter in absolute ethanol and air dry for 30 min.
9. Cut the filter grid into the individual squares and place into appropriately numbered scintillation vials, add scintillation fluid, and count in a liquid scintillation counter.
10. Determine the percentage of control protein synthesis as follows:

$$\text{(cpm of reaction mixture } - \text{ cpm of control reaction \#2)/}$$
$$\text{(cpm of control reaction \#1 } - \text{ cpm of control reaction \#2)} \times 100 \qquad (2)$$

The interaction of RI with the RNase can also be determined from this assay. Prepare two sets of identical reaction mixtures, one containing RI and the other containing RI diluent buffer. Perform the assay as described above.

## 3.6. In Vitro Analysis of RNase Immunofusion Protein Potency

In contrast to the assay described in **Subheading 3.5.**, which measures protein synthesis inhibition in a cell-free system, this assay determines the activity of the RNase/RNase immunofusion on intact cells. For the RNases to be cytotoxic to cells, they must be internalized into the cell. Some RNases, such as the amphibian RNase onconase *(52)*, or the newly described human RNase, (-4)rhEDN *(53)*, contain motif(s) that allow them to target and kill cells, whereas other RNases, such as the human RNases, EDN, angiogenin, and pancreatic RNase, require the use of a targeting moiety such as antibodies to aquire cytotoxic properties. Many different cell lines should be tested to identify those cells most sensitive to the RNase immunofusion of interest, especially since it has been shown that different RNases use different pathways and/or mechanisms to kill cells *(26)*. Nontarget cell lines, i.e., cells that do not express a relevant antigen or receptor, should be included as a control to demonstrate the specificity of the targeting ligand.

1. Plate 2500 or 10,000 cells (in 0.1 mL) per well of a 96-well microtiter plate for adherent or nonadherent cells, respectively, the day before treatment. The plating media should be the media in which the cells are normally grown.
2. On the day of treatment, prepare dilution curves of the control and test samples (*see* **Note 14**). Dilutions are made in sterile dilution buffer (*see* **Subheading 2.5., item 5**). As some cells are very sensitive to the glycerol and imidazole in the storage buffer, include buffer controls in the assay. Apply 10 µL to the appropriate wells. Test each sample at least in triplicate.
3. Incubate the plates at 37°C in a humidified $CO_2$ incubator for 1–3 d. Shorter incubation times require higher concentrations of immunofusion proteins. We usually incubate the cells for 3 d before termination.
4. To determine protein synthesis, replace the serum-containing medium with 100 µL of serum- and leucine-free RPMI. Add 0.1 mCi of [$^{14}$C] leucine (in 10 µL) and continue incubation for another 2–4 h at 37°C. Cell viability can also be determined by using the colorimetric assay, WST-1, and following the manufacturer's instructions. If cell viability is determined, proceed to **step 6**.
5. Harvest the cells onto glass fiber filters using a PHD cell harvester, wash with $H_2O$, dry with ethanol, and count in a scintillation counter.
6. Express the results as percentage of buffer-treated wells, calculated as follows:

$$[\text{cpm (or absorbance for WST-1) of sample-treated cells}]/$$
$$[\text{cpm (or absorbance for WST-1) of buffer-treated cells}] \times 100 \qquad (3)$$

Plot protein synthesis (% of control) on the *y*-axis vs sample concentration on the *x*-axis. Determine the $IC_{50}$, the concentration of test sample which inhibits protein synthesis or cell viability by 50%.

### 3.7. In Vivo Analysis of RNase Immunofusion Proteins

The following describes in vivo toxicity studies.

The maximum amount of drug that can be administered without encountering lethality is determined before the evaluation of the antitumor effects *(54)*. To evaluate the toxicity of the RNase immunofusion protein, the dosage is increased until the occurrence of lethal side effects.

1. Prepare sterile solutions containing the RNase immunofusion protein, the RNase alone, and the buffer control (*see* **Notes 8** and **14**).
2. Inject the mice intraperitoneally with 0.2 mL of the solutions noted in **step 1** (inject the RNase immunofusion protein at the highest concentration achievable).
3. Monitor the weight of the animals daily, and look for other signs of toxicity, such as coat ruffling, changes in gait, and so forth. Death is the final endpoint.
4. If no toxicity is observed, increase the number of injections per day (*see* **Note 15**). Continue increasing the number of daily injections and/or the number of days the sample is injected until toxicity is observed.
5. Report the data as $LD_{50}$ (mg test substance/kg body weight of animal) that caused death in 50% of the animals.

The following describes in vivo tumor treatment models.

The selective cytotoxicity observed in **Subheading 3.6.** controls the choice of the in vivo model. Only those cell lines sensitive should be further evaluated in animals. Two models are presented in this section: iv injection of tumor cells and iv injection of the RNase immunofusion protein with either survival or colony counts in a target tissue as the endpoint; and subcutaneous implant of the tumor cells into the flank of the animal with direct intratumoral injection of the RNase immunofusion protein with evaluation of tumor size as the end point.

### 3.7.1. Model 1: Survival

1. Forty-eight hours before the injection of cells, inject (iv) 0.2 mL ASGM-1 (1:20 dilution in sterile $H_2O$).
2. The next day, irradiate the animals with 200 rads for 0.9 s. The further immunosuppression caused by irradiation will promote survival of many cell types in the nude mouse *(55)*.
3. On the third day prepare cells as follows (for the best results, the cells should be in log phase of growth):
   a. Remove the media from ten flasks and rinse the cells twice with DBSS.
   b. Apply 7 mL 1X trypsin-EDTA solution (for T-150 cm² flasks), coat the cells evenly with the trypsin solution, and remove 5 mL.
   c. Allow the cells to sit in the hood for 30 s before tapping the flask to dislodge the cells.
   d. Add 10 mL of serum-containing media to the first flask and resuspend the cells. Use this same solution to resuspend the cells from a total of 10 flasks. Repeat this procedure for each set of 10 flasks.
   e. Add media to bring the volume to 50 mL for each set of 10 flasks.
   f. Centrifuge the cells for 5 min at 2300 rpm.
   g. Resuspend all the cells in a final volume of 40 mL DBSS.
   h. Repeat **steps f** and **g**.
   i. Centrifuge as described in **step f** and resuspend the cells in a final volume of 10 mL.
   j. Pour the cells through a cell strainer and count viable cells using the trypan blue exclusion assay per the manufacturer's instructions.
   k. Adjust the cell number with DBSS for 0.2-mL injection volume.
4. Prepare sterile solutions of the RNase immunofusion protein, antibody alone, RNase alone, the combination of antibody and RNase, or vehicle in an injection volume of 0.2 mL. Begin treatment 3–10 d after the injection of cells (*see* **Notes 14** and **16**).
5. Treat the animals twice weekly for 3 wk (*see* **Note 17**).
6. Weigh the animals twice weekly during the treatment period to assess toxicity and once weekly thereafter.
7. Monitor the animals for death, or terminate the experiment at a designated time, fix the tissue of interest, such as lung or liver, in 10% formalin or Bouin's fixative, and count the number of tumor nodules present in the tissue.

### 3.7.2. Model 2: Flank Model

Follow **steps 1–6** of **Subheading 3.7.1.** with the following exception: cells are injected subcutaneously into the flank of the mouse. Growth of the tumor is monitored using *in situ* caliper measurements to determine tumor mass, using the following equation to calculate tumor volume; $TV = 1/2LW^2$ where $TV$ = tumor volume, $L$ = length, and $W$ = perpendicular diameter. Treatment begins when the tumor reaches 3–5 mm². Animals are treated twice weekly for 3 wk (*see* **Note 18**). The size of the tumor is calculated after termination of the experiment.

## 4. Notes

### 4.1. PCR-Based Synthesis of RNase Genes Using Only One of the Strands

1. Keep the cycle number to a minimum to avoid amplification of errors. The final constructed gene should always be sequenced to verify that no insertions, deletions, or muta-

tions have occurred either during the synthesis of the oligonucleotides or during the multiple PCR reactions. Note that the primers and/or junction oligonucleotides can be used as sequencing primers.

2. Because DNA <500 bp gives a faint band on 1% agarose, use NuSieve 3:1 agarose which is capable of finely resolving DNA fragments ranging from 30 to 1000 bp.

## 4.2. Construction of the RNase-sFv Fusion Gene

3. This section is described in great detail in **ref. *41*** and is briefly described here.

4. A variety of pET vectors containing different peptide tags for affinity purification are available. These tags can then be removed after purification with proteases if desired. The pET vector system (*56*) was designed for the expression of toxic genes. Several pET vectors for cloning and bacterial hosts that suppress basal expression levels of toxic genes are available (Novagen).

## 4.3. Purification of a RNase-sFv Recombinant Fusion Protein

5. The $OD_{600nm}$ should be ≥1.0 in the superbroth medium, since it is an enriched medium, and the bacteria can grow to a higher density than in a less enriched medium, such as LB.

6. The final concentration of protein in the renaturation buffer affects the final yield of protein; final concentration of 80 µg/mL seems to work well for many constructs. The optimal DTE to GSSG ratio has been carefully optimized (*42*) to give a redox system of reduced and oxidized glutathione. To maintain this redox activity, the protein solution is diluted exactly 100 times.

7. Since RNases are basic proteins, many, such as pancreatic RNase, angiogenin, or onconase, will adhere to CM-sephadex, but the majority of contaminating bacterial proteins will not. Heparin sepharose is an appropriate first column for EDN. Prepare 4-mL columns for 2 L renatured protein (160 mg of protein).

8. The final buffer is very important for maintaining the solubility of the fusion protein; 20 m*M* Tris-HCl, pH 7.5, containing 10% glycerol and 200 m*M* imidazole or 100 m*M* NaCl is a very good storage buffer for RNase fusion proteins. Samples should be stored at 4°C and freezing should be avoided, since this increases aggregation.

## 4.4. Characterization of the Targeting Component of the RNase Immunofusion Protein

9. There are several easy methods for labeling antibodies with iodine, such as chloramine T, iodogen, Bolton-Hunter, and others. The chloramine T method generally yields a protein of higher specific activity (*57*). However, some antibodies may be damaged by the introduction of a bulky iodine atom onto a tyrosine group and thus another method, such as the Bolton-Hunter method, which labels lysines, may be attempted.

## 4.5. Characterization of the Enzymatic Component of the RNase Immunofusion Protein

10. Extra precaution must be taken against contaminating RNases found on hands and in the environment. Gloves should always be worn when handling the tubes, tips, buffers, and enzyme solutions. Disposable presterilized tubes and tips may be used without autoclaving. DEPC-treated $H_2O$ is not required; deionized $H_2O$ that is autoclaved for 30 min is sufficient.

11. The concentration of the enzyme should be in the linear range of the assay. Substrate should not be limiting (50–100 times the enzyme concentration). Since RNases are generally diluted severalfold with dilution buffer before assaying, interference caused by imi-

dazole or NaCl present in the storage buffer is eliminated. If the sample cannot be diluted, include the buffer components in the standards to control for buffer effects.

12. Because RI is costly, keep the enzyme concentration low, yet still high enough to give an $A_{260\ nm}$ of 0.5–1.0. The assay time may be increased to achieve this. RI is very labile *(58–60)*, requires a minimum of 1 m$M$ DTT to maintain activity, and should be stored at –70°C. When removed from the freezer, it should immediately be stored on ice.

13. Translation grade [$^{35}$S] methionine is recommended to reduce background labeling *(61)*. [$^{35}$S]methionine should be aliquotted and stored at –70°C. Repeated thawing and freezing leads to oxidation, and this can interfere with the assay.

## 4.6. In Vitro Analysis of RNase Immunofusion Protein Potency

14. To sterilize RNase fusion proteins, the Millipore Millex-HV (Millipore products Division, Bedford, MA) filters have less protein loss than other filters tested, including the Millex-GV.

## 4.7. In Vivo Analysis of RNase Immunofusion Proteins

15. RNase immunofusion proteins are nontoxic to mice. We have yet to achieve a maximum tolerated dose (the highest concentration of an anti-CD22 antibody-RNase conjugate administered has been 1.2 mg/d given as 300-μg injections four times per day for 5 d).

16. The antibody should be included as one of the controls because of its possible antitumor effects *(62–64)*. The combination of antibody and free RNase is included to demonstrate that the activity observed with the immunofusion protein is caused by the covalent coupling between antibody and RNase, and not to an additive or synergistic effect between the two proteins.

17. Various dosing schedules should be tried. For the first evaluation, the RNase immunofusion can be administered 24 h after the administration of the tumor cells. If under these conditions there is activity, begin treatment at later times after tumor-cell injection, i.e., 3, 5, or 7 d later. In some experiments the fusion protein is injected daily for 5 d or twice weekly as described here.

18. Water-soluble agents can be injected iv eliminating the need for absorption of the agent into the blood stream. Although iv injection is most similar to the clinical condition *(54)*, it requires more technical skill, is more time-consuming, and causes tail damage after multiple injections. In contrast, intraperitoneal and subcutaneous injections can be accomplished quickly and allow for the evaluation of water-insoluble agents.

## References

1. Brinkmann, U. and Pastan, I. (1994) Immunotoxins against cancer. *Biochim. Biophys. Acta* **1198,** 27–45.
2. Thrush, G. R., Lark, L. R., Clinchy, B. C., and Vitetta, E. S. (1996) Immunotoxins: an update. *Ann. Rev. Immunol.* **14,** 49–71.
3. Pastan, I. (1997) Targeted therapy of cancer with recombinant immunotoxins. *Biochim. Biophys. Acta* **1333,** C1–C6.
4. Rybak, S. M. and Youle, R. J. (1991) Clinical use of immunotoxins: monoclonal antibodies conjugated to protein toxins. *Immunol. Allergy Clin. North Am.* **11,** 359–380.
5. Soler-Rodriguez, A. M., Ghetie, M.-A., Oppenheimer-Marks, N., Uhr, J. W., and Vitetta, E. S. (1993) Ricin A-chain and ricin A-chain immunotoxins rapidly damage human endothelial cells: implications for vascular leak syndrome. *Exp. Cell Res.* **206,** 227–234.

6. Vitetta, E. S., Thorpe, P. E., and Uhr, J. W. (1993) Immunotoxins: magic bullets or misguided missiles. *TiPS* **14,** 148–154.
7. Sawler, D. L., Bartholomew, R. M., Smith, L. M., and Dillman, R. (1985) Human immune response to multiple injections of murine monoclonal IgG. *J. Immunol.* **135,** 1530–1535.
8. Schroff, R. W., Foon, K. A., Beatty, S. M., Oldham, R., and Morgan, A. (1985) Human anti-murine immunoglobulin response in patients receiving monoclonal antibody therapy. *Cancer Res.* **45,** 879–885.
9. Harkonen, S., Stoudemire, J., Mischak, R., Spitler, L., Lopez, H., and Scannon, P. (1987) Toxicity and immunogenicity of monoclonal antimelanoma antibody-ricin A chain immunotoxins in rats. *Cancer Res.* **47,** 1377–1385.
10. Khazaeli, M. B., Conry, R. M., and LoBuglio, A. F. (1994) Human immune response to monoclonal antibodies. *J. Immunother.* **15,** 42–52.
11. Schein, C. H. (1997) From housekeeper to microsurgeon: the diagnostic and therapeutic potential of ribonucleases. *Nature Biotechnol.* **15,** 529–536.
12. Rybak, S. M. and Newton, D. L. (1999) Immunoenzymes, in *Antibody Fusion Proteins* (Chamow, S. M. and Ashkenazi, A., eds.), Wiley, New York, pp. 53–110.
13. Saxena, S. K., Rybak, S. M., Winkler, G., Meade, H. M., McGray, P., Youle, R. J., and Ackerman, E. J. (1991) Comparison of RNases and toxins upon injection into Xenopus oocytes. *J. Biol. Chem.* **266,** 21,208–21,214.
14. Wu, Y. N., Mikulski, S. M., Ardelt, W., Rybak, S. M., and Youle, R. J. (1993) Cytotoxic ribonuclease: a study of the mechanism of Onconase cytotoxicity. *J. Biol. Chem.* **268,** 10,686–10,693.
15. Lin, J. J., Newton, D. L., Mikulski, S. M., Kung, H. F., Youle, R. J., and Rybak, S. M. (1994) Characterization of the mechanism of cellular and cell free protein synthesis inhibition by an anti-tumor ribonuclease. *Biochem. Biophys. Res. Commun.* **204,** 156–162.
16. Rybak, S. M., Saxena, S. K., Ackerman, E. J., and Youle, R. J. (1991) Cytotoxic potential of ribonuclease and ribonuclease hybrid proteins. *J. Biol. Chem.* **266,** 21,202–21,207.
17. Rybak, S. M., Hoogenboom, H. R., Meade, H. M., Raus, J. C., Schwartz, D., and Youle, R. J. (1992) Humanization of immuntoxins. *Proc. Natl. Acad. Sci. USA* **89,** 3165–3169.
18. Newton, D. L., Ilercil, O., Laske, D. W., Oldfield, E., Rybak, S. M., and Youle, R. J. (1992) Cytotoxic ribonuclease chimeras: targeted tumoricidal activity *in vitro* and *in vivo*. *J. Biol. Chem.* **267,** 19,572–19,578.
19. Newton, D. L. and Rybak, S. M. (1996) Single-chain immunofusions engineered with human RNases (Abstract), in *Exploring and Exploiting Antibody and Ig Superfamily Combining Sites*. Taos, NM, pp. 20.
20. Newton, D. L., Pearson, J. W., Xue, Y., Smith, M. R., Fogler, W. E., Mikulski, S. M., Alvord, W. G., Kung, H. F., Longo, D. L., and Rybak, S. M. (1996) Anti-tumor ribonuclease combined with or conjugated to monoclonal antibody MRK16, overcomes multidrug resistance to vincristine in vitro and in vivo. *Int. J. Oncol.* **8,** 1095–1104.
21. Zewe, M., Rybak, S. M., Dubel, S., Coy, J. F., Welschof, M., Newton, D. L., and Little, M. (1997) Cloning and cytotoxicity of a human pancreatic RNase immunofusion. *Immunotechnology* **3,** 127–136.
22. Deonarain, M. P. and Epenetos, A. A. (1998) Design, characterization and antitumor cytotoxicity of a panel of recombinant, mammalian ribonuclease-based immunotoxins. *Br. J. Cancer* **77,** 537–546.
23. Psarras, K., Ueda, M., Yamamura, T., Ozawa, S., Kitajima, M., Aiso, S., Komatsu, S., and Seno, M. (1998) Human pancreatic RNase1-human epidermal growth factor fusion: an entirely human immunotoxin analog with cytotoxic properties against squamous cell carcinomas. *Prot. Eng.* **11,** 1285–1292.

24. Yoon, J. M., Han, S. H., Kown, O. B., Kim, S. H., Park, M. H., and Kim, B. K. (1999) Cloning and cytotoxicity of fusion proteins of EGF and angiogenin. *Life Sci.* **64,** 1435–1445.

25. Rybak, S. M., Pearson, J. W., Fogler, W. F., Volker, K., Spence, S. E., Newton, D. L., Mikulski, S. M., Ardelt, W., Riggs, C. W., Kung, H. F., and Longo, D. L. (1996) Enhancement of vincristine cytotoxicity in drug-resistant cells by simultaneous treatment with Onconase, an antitumor ribonuclease. *J. Natl. Cancer Inst.* **88,** 747–753.

26. Smith, M. R., Newton, D. L., Mikulski, S. M., and Rybak, S. M. (1999) Cell cycle-related differences in susceptibility of NIH/3T3 cells to ribonucleases. *Exp. Cell Res.* **247,** 220–232.

27. Leland, P. A., Schultz, W., Kim, B.-M., and Raines, R. T. (1998) Ribonuclease A variants with potent cytotoxic activity. *Proc. Natl. Acad. Sci. USA* **95,** 10,407–10,412.

28. Suzuki, M., Saxena, S. K., Boix, E., Prill, R. J., Vasandani, V. M., Ladner, J. E., Sung, C., and Youle, R. J. (1999) Engineering receptor-mediated cytotoxicity into human ribonucleases by steric blockade of inhibitor interaction. *Nat. Biotechnol.* **17,** 265–270.

29. Rybak, S. M. and Newton, D. L. (1999) Uncloaking RNases. *Nat. Biotechnol.* **17,** 408.

30. Bond, M. D. (1988) An in vitro binding assay for angiogenin using placental ribonuclease inhibitor. *Anal. Biochem.* **173,** 166–173.

31. Gart, J., Krewski, D., Lee, P., Tarone, R., and Wahrendorf, J. (1986) The design and analysis of long-term animal experiments, in *Statistical Methods in Cancer Research,* Vol. 3, International Agency for Research on Cancer, New York, pp. 10–26.

32. Huang, H.-C., Wang, S.-C., Leu, Y.-J., Lu, S.-C., and Liao, Y.-D. (1998) The Rana catesbeiana rcr gene encoding a cytotoxic ribonuclease. *J. Biol. Chem.* **273,** 6395–6401.

33. Jayaraman, K. and Puccini, C. J. (1992) A PCR-mediated gene synthesis strategy involving the assembly of oligonucleotides representing only one of the strands. *BioTechniques* **12,** 392–398.

34. Russo, N., Nobile, V., DiDonato, A., Riordan, J. F., and Valee, B. L. (1996) The C- terminal region of human angiogenin has a dual role in enzymatic activity. *Proc. Natl. Acad. Sci. USA* **93,** 3243–3247.

35. Newton, D. L., Xue, Y., Olson, K. A., Fett, J. W., and Rybak, S. M. (1996) Angiogenin single chain immunofusions: influence of peptide linkers and spacers between fusion protein domains. *Biochemistry* **35,** 545–553.

36. Mosimann, S. C., Ardelt, W., and James, M. N. G. (1994) Refined 1.7 A X-ray crystallographic structure of P-30 protein, an amphibian ribonuclease with anti-tumor activity. *J. Mol. Biol.* **236,** 1141–1153.

37. Boix, E., Wu, Y., Vasandani, V. M., Saxena, S. K., Ardelt, W., Ladner, J., and Youle, R. J. (1996) Role of the N terminus in RNase A homologues: differences in catalytic activity, ribonuclease inhibitor interaction and cytotoxicity. *J. Mol. Biol.* **257,** 992–1007.

38. Newton, D. L., Xue, Y., Boque, L., Wlodawer, A., Kung, H. F., and Rybak, S. M. (1997) Expression and characterization of a cytotoxic human-frog chimeric ribonuclease: potential for cancer therapy. *Protein Eng.* **10,** 463–470.

39. Tai, M. S., Mudgett-Hunter, M., Levinson, D., Wu, G.-M., Haber, E., Oppermann, H., and Huston, J. S. (1990) A bifunctional fusion protein containing Fc-binding fragment B of staphylococcal protein A amino terminal to antidigoxin single-chain Fv. *Biochemistry* **29,** 8024–8030.

40. Horton, R. M., Cai, Z. L., Ho, S. N., and Pease, L. R. (1990) Gene splicing by overlap exension: tailor made genes using the polymerase chain reaction. *BioTechniques* **8,** 528–535.

41. Newton, D. L. and Rybak, S. M. (2000) Preparation of recombinant RNase single chain antibody fusion proteins, in *Methods in Molecular Medicine, vol. 25: Drug Targeting* (Francis, G. E. and Delgado, C., eds.), Humana Press, Totowa, NJ, pp. 77–96.

42. Buchner, J., Pastan, I., and Brinkmann, U. (1992) A method for increasing the yield of properly folded recombinant fusion proteins: single-chain immunotoxins from renaturation of bacterial inclusion bodies. *Anal. Biochem.* **205,** 263–270.

43. Brinkmann, U., Buchner, J., and Pastan, I. (1992) Independent domain folding of *Pseudomonas* exotoxin and single-chain immunotoxins: influence of interdomain connections. *Proc. Natl. Acad. Sci. USA* **89,** 3075–3079.

44. Crothers, D. M. and Metzger, H. (1972) The influence of polyvalency on the binding properties of antibodies. *Immunochem.* **9,** 341–357.

45. Shapiro, R., Fett, J. W., Strydom, D. J., and Vallee, B. L. (1986) Isolation and characterization of a human colon carcinoma-secreted enzyme with pancreatic ribonuclease-like activity. *Biochemistry* **25,** 7255–7264.

46. Rosenberg, H. F. and Dyer, K. D. (1995) Eosinophil cationic protein and eosinophil-derived neurotoxin. Evolution of novel function in a primate ribonuclease gene family. *J. Biol. Chem.* **270,** 21,539–21,544.

47. Sambrook, J., Fritsch, E. F., and Maniatis, T. (1989) *Molecular Cloning: A Laboratory Manual* (Ford, N., Nolan, C., and Ferguson, M., eds.), Cold Spring Harbor Laboratory, Cold Spring Harbor, NY.

48. Blackburn, P., Wilson, G., and Moore, S. (1977) Ribonuclease inhibitor from human placenta. Purification and properties. *J. Biol. Chem.* **252,** 5904–5910.

49. Shapiro, R. and Vallee, B. L. (1991) Interaction of human placental ribonuclease with placental ribonuclease inhibitor. *Biochemistry* **30,** 2246–2255.

50. Fromm, H. J. (1995) Reversible enzyme inhibitors as mechanistic probes. *Meth. Enzymol.* **249,** 123–143.

51. Hofsteenge, J. (1997) Ribonuclease inhibitor, in *Ribonucleases: Structures and Functions* (D'Allessio, G. and Riordan, J. F., eds.), Academic Press, San Diego, CA, pp. 621–658.

52. Ardelt, W., Mikulski, S. M., and Shogen, K. (1991) Amino acid sequence of an anti-tumor protein from *Rana pipiens* oocytes and early embryos. *J. Biol. Chem.* **266,** 245–251.

53. Newton, D. L. and Rybak, S. M. (1998) Unique recombinant human ribonuclease and inhibition of kaposi's sarcoma cell growth. *J. Natl. Cancer Inst.* **90,** 1787–1791.

54. Corbett, T., Valeriote, F., LoRusso, P., Polin, L., Panchapor, C., Pugh, S., White, K., Knight, J., Demchik, L., Jones, J., Jones, L., and Lisow, L. (1997) In vivo methods for screening and preclinical testing. Use of rodent solid tumors for drug discovery, in *Anticancer Drug Development Guide* (Teicher, B. A., ed.), Humana Press, Totowa, NJ, pp. 75–99.

55. Giovanella, B. C., Stehlin, J. S., Shepard, R. C., and Williams, L. J. (1979) Hyperthermic treatment of human tumors heterotransplanted in nude mice. *Cancer Res.* **39,** 2236–2241.

56. Studier, F. W., Rosenberg, A. H., Dunn, J. J., and Dubendorff, J. W. (1990) Use of T7 RNA polymerase to direct expression of cloned genes. *Methods Enzymol.* **185,** 60–89.

57. Harlow, E. and Lane, D. (1988) *Antibodies. A Laboratory Manual.* Cold Spring Harbor Laboratory, Cold Spring Harbor, NY, pp. 319–358.

58. Roth, J. S. (1958) Partial purification and characterization of a ribonuclease inhibitor in rat liver supernatant fraction. *J. Biol. Chem.* **231,** 1085–1095.

59. Shortman, K. (1961) Studies on cellular inhibitors of ribonuclease. The assay of the ribonuclease-inhibitor system, and the purification of the inhibitor from rat liver. *Biochim. Biophys. Acta* **1961,** 37–49.

60. Gagnon, C. and Lamirande, G. D. (1973) A rapid and simple method for the purification of rat liver RNase inhibitor. *Biochem. Biophys. Res. Comm.* **51,** 580–586.

61. Beckler, G. S., Thompson, D., and Oosbree, T. V. (1995) In vitro translation using rabbit reticulocyte lysate. *Meth. Mol. Biol.* **37,** 215–232.

62. Goldenberg, D. M., Horowitz, J. A., Sharkey, R. M., Hall, T. C., Murthy, S., Goldenberg, H., Lee, R. E., Stein, R., Siegel, J. A., and Izon, D. O. (1991) Targeting, dosimetry, and radioimmunotherapy of B-cell lymphomas with iodine-131-labeled LL2 monoclonal antibody. *J. Clin. Oncol.* **9,** 548–564.

63. Ghetie, M. A., Richardson, J., Tucker, T., Jones, D., Uhr, J. W., and Vitetta, E. S. (1991) Antitumor activity of Fab' and IgG-anti-CD22 immunotoxins indisseminated human B lymphoma grown in mice with severe combined immunodeficiency disease effect on tumor cells in extranodal sites. *Cancer Res.* **51,** 5876–5880.

64. Kreitman, R. J., Hansen, H. J., Jones, A. L., FitzGerald, D. J. P., Goldenberg, D. M., and Pastan, I. (1993) Pseudomonas Exotoxin-based immunotoxins containing the antibody LL2 or LL2-Fab' induce regression of subcutaneous human B-cell lymphoma in mice. *Cancer Res.* **53,** 819–825.

# V

## ASSAYS USING NUCLEASES

# 27

## Restriction Endonucleases and Their Uses

**Raymond J. Williams**

## 1. Introduction

Restriction endonucleases, which cleave DNA in a site-specific manner, are a fundamental tool of molecular biology. The discovery of endonucleases began in the 1960s and led to commercial availability in the early 1970s. The number of characterized enzymes continues to grow, as does the number of vendors and the size of their product lines. Although many similarities exist among endonucleases in terms of their structures, mechanisms, and uses, important differences remain. Now a staple of molecular biology, restriction endonucleases are an area of active research as models of site-specific DNA recognition, cleavage mechanism, in vivo function, and evolutionary origins. New enzymes continue to be discovered or developed by using protein engineering to modify the specificity of existing enzymes.

### 1.1. Diversity and In Vivo Function

It is estimated that one in four bacteria examined contain one or more restriction endonucleases *(1)*. Approximately 3000 restriction enzymes are now known, which recognize 235 different DNA sequences *(2)*. Of these, fewer than 2900 enzymes are classified as Type II or one of its subclasses. Of the total, 37 homing endonucleases, so named because they are encoded by genes that are mobile, self-splicing introns or inteins, each with a unique recognition site, have been discovered.

Restriction endonucleases were originally named for their ability to restrict the growth of phage in a host bacterial cell by cleavage of the invading DNA. In this manner, they act as bacterial protection systems. They have now been located in bacterial genomes, plasmids, and phages. The DNA of the host is protected from restriction by the activity of a methylase(s), which recognizes the same sequence as the restriction enzyme and methylates a specific nucleotide (4-methylcytosine, 5-methylcytosine, 5-hydroxymethylcytosine, or 6-methyladenine) on each strand within this sequence. Once methylated, the host DNA is no longer a substrate for the endonuclease. Since both strands of the DNA are methylated and hemi-methylated DNA is protected, even freshly replicated host DNA is not digested by the endonuclease. *Dpn*I is the only commercially available restriction endonuclease that requires methylated DNA for cleavage.

From: *Methods in Molecular Biology, vol. 160: Nuclease Methods and Protocols*
Edited by: C. H. Schein © Humana Press Inc., Totowa, NJ

## *1.2. Genomic Organization*

Approximately 220 restriction/modification systems have been cloned and sequenced. Significantly, little if any sequence homology exists between the endonuclease and methyltransferase recognizing the same DNA sequence. Even restriction endonucleases recognizing the same sequence may show little or no homology, including the amino acids involved in recognition, and as such are excellent candidates for a comparative study of protein–DNA interaction. For example, the enzymes *Hha*II and *Hinf*I are both isolated from strains of *Haemophilus*, recognize GANTC, and cleave between the G and A. However, they share only 19% identity in their amino-acid sequence *(3)*. Endonuclease/methylase systems recognizing the same sequence may also exhibit different methylation patterns and restriction sensitivity. Only a limited common amino-acid motif, PD...D/EXK, has been proven by mutational or structural analysis to participate in catalysis for 10 endonucleases. However, the 10 enzymes include members that are classified as Type II, IIb, IIe, IIs, or intron-encoded endonucleases *(4)*. In contrast, general motifs have been found for 30 6-methyladenine, 4-methylcytosine, and 5-methylcytosine methylases *(5)*.

Frequently referred to as an R/M system, the restriction endonuclease and modification methylase genes lie adjacent to each other on the host DNA and may be oriented transcriptionally in a convergent, divergent, or sequential manner. The proximity of these genes appears to be universal and is utilized in a common cloning method sometimes referred to as the "Hungarian Trick" *(6)*. Basically, a library of clones is made by digestion of the genomic DNA with an endonuclease(s) from the bacteria containing the R/M system of interest. The expression vector used must contain the recognition site of the R/M system of interest. Purified plasmids from the clones are then subjected to the restriction enzyme in vitro. If a plasmid contains an expressed methylase gene, it will be resistant to cleavage. Often, the endonuclease is also expressed without a need for subcloning.

It is assumed that methylation must occur before restriction activity to protect the host DNA. One way bacteria limit the possibility of self-restriction is to significantly reduce the number of recognition sites in their genomes. Alternatively, methylase expression may precede that of the endonuclease. One manner in which this may be accomplished is through an open reading frame located upstream of the endonuclease gene and coding for a "C" or control protein. This *C* protein positively regulates the endonuclease gene and allows for the methylase activity which is not under such control to precede expression of the endonuclease *(7)*. Such *C* genes are frequently found in situations where the methylase and endonuclease genes are in divergent or convergent transcriptional orientations. Using the cloned restriction, methylase, and *C* genes from the *Bam*HI, *Sma*I, *Pvu*II, and *Eco*RV R/M systems, C genes were provided on a separate plasmid after the *C* gene within the cloned R/M system had been disrupted. *Bam*HI restiction activity was equally stimulated by the *Sma*I *C* and the *Bam*HI *C* gene, and only one order of magnitude less by the *Pvu*II *C* gene. The *Eco*RV *C* gene provided no stimulation. The *Bam*HI *C* gene stimulated *Pvu*II restriction activity as well as the *Pvu*II *C* gene *(8)*. Why some *C* genes stimulate expression of dissimilar endonucleases is not fully understood, but the phenomenon may have evolutionary implications.

## 1.3. Nomenclature and Commercial Availability

Individual enzymes are named in accordance with the proposal of Smith and Nathans *(9)*. Briefly, three letters in italics are derived from the first letter of the genus and the first two letters of the microbial species from which the enzyme was derived. An additional letter without italics or number may be used to designate a particular strain. This is followed by a roman numeral to signify the first, second, etc., enzyme discovered from the organism. As may be deduced from the large number of enzymes and the limited number of different DNA sequences they recognize, many enzymes from different biological sources recognize the same DNA sequence and are called isoschizomers. A special subset wherein two enzymes recognize the same DNA sequence but cleave in a different manner is referred to as neoschizomers.

For the common Type II and Type II subclasses, 468 endonucleases are commercially available, comprising 212 specificities. In addition, 7 of the homing endonucleases are commercially available. A database maintained by Dr. Richard J. Roberts at http://www.neb.com/rebase contains a complete listing of restriction endonucleases, including prototype enzyme, isoschizomers, recognition sequence, commercial sources, and references in a number of formats. A list of commercially available endonucleases is also published annually in *Nucleic Acids Research*.

## 1.4. Structure, Specificity, and Mechanism

### 1.4.1. Classification and General Mechanism

Restriction endonucleases are classified according to their structure, recognition site, cleavage site, cofactor(s), and activator(s). Sets of these criteria are used to define the different types (I, II, III, and IV) and subclasses (IIb, IIe, and IIs) which are explained in detail in **Table 1**. Multiple subunit and holoenzyme assemblies are possible to achieve the needed restriction, methylase, and specificity domains. However, common features are the requirement of $Mg^{2+}$ for endonuclease activity and AdoMet (also referred to as S-adenosyl methionine) for methylase activity. The majority of recognition sites are 4, 6, or 8 bases long and are palindromic. Some enzymes recognize sites with a limited degree of ambiguity or those consisting of interrupted palindromes. When the specificity domain allows ambiguities, the possible nucleotides at a particular position are limited, and others are strictly excluded. This results in palindromic and partially palindromic sites that are recognized and cleaved by Type II endonucleases. For example, the recognition site for *Sty*I is listed as CCWWGG. Therefore, the substrate sequences for *Sty*I can be palindromic (CCTAGG or CCATGG) or partially palindromic (CCTTGG or CCAAGG) *(21)*. This flexibility of recognition is not currently understood. Of particualr interest are the situations where allowed nucleotides can be either purine or pyrimidine, or when only a single nucleotide is excluded. The single-letter code for these ambiguities follows:

R = A or G    Y = C or T    M = A or C
K = G or T    S = G or C    W = A or T
B = not A (C or G or T)    D = not C (A or G or T)
H = not G (A or C or T)    V = not T (A or C or G)
N = A or C or G or T

**Table 1**
**Restriction Endonuclease Types and Their Properties**[a]

| Type | Example(s) | Subunit structure[b] | Cofactors[c] and activators | Recognition site | Cleavage site | Methylase properties |
|---|---|---|---|---|---|---|
| I (EC 3.1.21.3) | EcoKI, EcoAI, EcoBI, CfrAI, StySPI, and so forth | Usually a pentameric complex (2 R, 2 M, and 1 S) | $Mg^{2+}$, AdoMet, ATP (hydrolyzed) | Bipartite, interrupted | Distant and variable from recognition site, for example, EcoKI: $AAC(N_6)GTGC(N_{>400})\downarrow$ $TTG(N_6)CACG(N_{>400})\uparrow$ | May be heterodimer 1 M, 1 S or heterotrimer 2 M, 1 S |
| II (EC 3.1.21.4) | EcoRI, BamHI, HindIII, KpnI, NotI, PstI, SmaI, XhoI, and others | Homodimer[d] (2 R–S) | $Mg^{2+}$ | Palindromic or interrupted palindrome, ambiguity may be allowed[e] | Defined, within recognition site, may result in a 3' overhang, 5' overhang, or blunt end, for example, EcoRI: $G\downarrow AATTC$ $CTTAA\uparrow G$ | Separate, single, monomeric (M-S) methyl-transferase[f] |
| IIb | BcgI, Bsp24I, BaeI, CjeI, and CjePI | Heterotrimer[g] (2 R-M, 1 S) | $Mg^{2+}$, AdoMet | Bipartite, interrupted | Cuts both strands on both sides of recognition site a defined, symmetric, short distance | Same |

away and leaves 3′ overhangs, for example, $Bcg$I:

$$\downarrow_{10}(N)CGA(N)_6TCG(N)_{12}\downarrow$$
$$\uparrow_{12}(N)GCT(N)_6ACG(N)_{10}\uparrow$$

| Type | Enzymes | Subunit structure | Cofactor/activation | Recognition site | Cleavage | Methyltransferase |
|---|---|---|---|---|---|---|
| IIe[h] | NaeI, NarI, BspMI, HpaII, SacII, EcoRII, AtuBI, Cfr9I, SauBMKI, and Ksp632I | Homodimer (2 R-S) or monomer (R-S), similar to Type II or Type IIs | $Mg^{2+}$, A second recognition site, acting in cis or trans, binds to the endonuclease as an allosteric affector | Palindromic, palindromic with ambiguities, or nonpalindromic | Cuts in defined manner within the recognition site or a short distance, may need activator DNA for complete cleavage, for example, NaeI: GCC↓GGC CGG↑CCG | Separate, single, monomeric (M-S) methyltransferase |
| IIs | FokI, Alw26I, BbvI, BsrI, EarI, HphI, MboII, PleI, SfaNI, Tth111I, and so on | Monomeric (R-S) | $Mg^{2+}$ | Nonpalindromic, nearly always contiguous and without ambiguities | Cuts in defined manner with at least one cleavage site outside of the recognition site, rarely leaves blunt ends, for example, FokI: GGATG(N)$_9$↓ CCTAC(N)$_{13}$↑ | May be 1 monomeric (M-S) which methylates one or both strands, or two separate monomeric (M-S) methyltransferases, one for each strand, may also methyllate different nucleotides[i] |

(continued)

**Table 1** (*continued*)

| Type | Example(s) | Subunit structure[b] | Cofactors[c] and activators | Recognition site | Cleavage site | Methylase properties |
|---|---|---|---|---|---|---|
| III (EC 3.1.21.5) | *Eco*P15I, *Eco*PI, *Hinf*III, and *Sfy*LTI | Both R and M-S required | $Mg^{2+}$ (AdoMet)[a], ATP (not hydrolyzed), May require a second unmodified site in opposite orientation, variable distance away[k] | Nonpalindromic | Cuts approx 25 bases away from recognition site, may not cut to completion, for example, *Eco*P15I:<br>CAGCAG(N)$_{25-26}$↓<br>GTCGTC(N)$_{25-26}$↑ | Methylates adenines, only on one strand, in an independent manner |
| IV | *Eco*57I | R-M-S monomer | $Mg^{2+}$, (AdoMet)[a] | Nonpalindromic | Cuts in a defined manner a short distance away from recognition site, may not cut to completion, for example, *Eco*57I:<br>CTGAAG(N)$_{16}$↓<br>GACTTC(N)$_{14}$↑ | Separate, single, monomeric (M-S) methyl-transferase (methylase activity of restriction monomer only methylates one strand) |
| Intron or Intein encoded | I-*Ppo*I, I-*Ceu*I, I-*Hmu*I, I-*Sce*I, I-*Tev*I, PI-*Psp*I, F-*Sce*II, and others | Monomer, homodimer, other protein or RNA may be required | $Mg^{2+}$, may also bind $Zn^{2+}$ | 12–40 bp, tolerance for basepair substitutions exists | 3' and 5' overhangs from 1–10 bases, a few not yet determined, may cleave one strand preferentially or in the absence of $Mg^{2+}$, two enzymes cleave only one strand, for example, I-*Ppo*I:<br>CTCTCTTAA↓GGTAGC<br>GAGAG↑AATTCCATCG | None |

[a]The first five columns list examples and properties of the restriction endonuclease. The recognition and cleavage site of the first example is given under the column "Cleavage Site." The sequence of the top strand is given from 5' to 3'. Cleavage is indicated by the arrows. The last column refers to the methyltransferase activity. AdoMet, also referred to as S-adenosyl methionine, is always required for methylation. Information presented represents the knowledge to date, future discoveries may provide exceptions. For reviews of each type, see the following references: Type I (10), Type II (1,11), Type IIb (12,13), Type IIe (14,15), Type IIs (16), Type III (10), Type IV (17), and homing endonucleases (intron or intein encoded) (18).

[b]R, M, and S refer to restriction, methyltransferase, and substrate specificity domains which may exist as separate subunits (R, M, S) or be combined (R-S, M-S, R-M) in a single polypeptide. In the case of Type II systems, the primary sequence of the restriction endonuclease and methyltransferase specificity domains demonstrate little, if any, homology. Components in parentheses stimulate activity but are not required.

[c]Although they show a strong preference for $Mg^{2+}$, other divalent metals may substitute, usually $Mn^{2+}$ but also $Ca^{2+}$, $Co^{2+}$, $Fe^{2+}$, $Ni^{2+}$, and $Zn^{2+}$. However, specificity may be relaxed and cleavage rates significantly decreased.

[d]AatII and SfiI reported to exist as homotetramers.

[e]DpnI is the only Type I, II, or III enzyme known that requires 6-methyladenine in its recognition site of GATC for activity.

[f]Some systems contain two methyltransferases (1).

[g]This tertiary structure has only been shown for BcgI, whereas the structures of the other four systems of this type (Bsp24I, BaeI, CjeI, and CjePI) are unknown.

[h]Many isoschizomers exist which are common Type II. There is evidence to suggest that Eco57I could also be classified as Type IIe (14).

[i]This information is based on very limited information (methyltransferase activity characterized from 9 R/M systems).

[j]ATPase activity has been previously reported as <1% compared to Type I restriction activity and therefore ATP was regarded as a cofactor rather than a substrate. However, more recent evidence with EcoP15I suggests a need to investigate more closely possible ATPase activity of Type III restriction activities (19).

[k]In the host protection mechanism for EcoP15I, DNA is hemimethylated in the fully protected state and freshly replicated DNA is protected by the fact that a second, convergently orientated, and also totally unmodified site is required for cleavage. This host protection mechanism may be true for the other Type III systems as well (EcoPI, HinfIII, and StyLTI) (19,20).

Restriction enzymes cleave a specific phosphodiester bond in each strand. Briefly, a nucleophilic attack by an activated water occurs on the scissile phosphorous followed by a pentacovalent transition state, and finally inversion of the phosphorous retained on the 5' end as a proton is donated to the resultant 3' hydroxyl *(11,22)*. The cleavage position may generate a blunt end or a single-stranded 3' or 5' overhang of 1–4 bases. It should be noted that enzymes with ultimately different recognition sites may still produce overhangs that are complementary and therefore suitable for ligation, although the recognition site for one or both enzymes may be lost in the ligation product. For example, *Nar*I, *Msp*I, *Acy*I, *Taq*I, *Cla*I, *Csp*45I, *Hpa*II, and *Acc*I all produce a 5'-CG overhang, although each has a different recognition sequence.

### 1.4.2. Type II Endonucleases

The structure, mechanism, and kinetics of Type II endonucleases are described in detail in Chapter 19. Briefly, the common Type II endonucleases are homodimers (most between 25 and 35 kDa for the monomeric subunit), require only Mg$^{2+}$, and cleave within palindromes or partial palindromes. Despite dissimilar primary sequence, Type II endonucleases have a similar three-dimentional structure: a "U" shaped dimeric holoenzyme with each of the identical subunits contributing recognition and catalyic domains on the sides and bridging domains at the bottom. Only two are known to exist as homotetramers, *Sfi*I *(23)* and *Aat*II *(24)*, and these do not cleave a few residual possible sites even when enzyme is in excess relative to substrate. For *Sfi*I it has been shown that the homotetramer must interact with two intact recognition sites *(25)*. These sites must contain cleavable phosphodiester bonds as opposed to the nonhydrolyzed thiodiester bond in the activator sequences explained later for Type IIe enzymes *(26)*.

### 1.4.3. Type IIb Endonucleases

Little is known about the structure and mechanism of Type IIb endonucleases, which require both Mg$^{2+}$ and AdoMet for restriction activity. The only holoenzyme model proposed thus far is for *Bcg*I and it is not yet based on crystallographic data. In solution, the molecular weight determined by gel filtration suggests a heterohexamer consisting of two working units *(12)*. The working unit of this model, derived from sequence motifs, mutational and truncation analysis, and subunit stoichiometry, is a heterotrimer consisting of one specificity subunit with two restriction–methylation subunits. The restriction–methylation subunits are bound one on each side of the specificity subunit, positioning them both upstream and downstream of the recognition site. Double-stranded cleavage by each restriction–methylation subunit thereby excises the recognition site. Substrates containing a single site are cleaved at a much lower rate than those with two, suggesting that both recognition domains must be occupied *(27)*. A host recognition site which is hemimethylated, such as after recent replication, is preferentially methylated on the other strand. Conversely, a recognition site unmethylated on both strands such as foreign DNA is cleaved *(28)*.

### 1.4.4. Type IIs Endonucleases

Type IIs endonucleases are monomeric, 45–110 kDa, require only Mg$^{2+}$, recognize non-palindromic sequences, and cleave at least one of the two strands outside the recognition site. More structural information is available for these endonucleases, as the crystal structure of one member, *Fok*I, bound to DNA, has been determined. The amino terminal portion of *Fok*I contains the DNA recognition domain, and the carboxy terminal portion contains the cleavage domain. The crystal structure of *Fok*I bound to a 20-bp fragment containing the recognition site in the absence of $Mg^{2+}$ revealed two apparent anomalies. First, the cleavage domain was not in contact with the cleavage site. This has also been substantiated with footprinting studies. The cleavage domain is positioned away from the DNA, while the enzyme searches for its recognition site. When bound to its site, and in the presence of $Mg^{2+}$, the *Fok*I cleavage domain swings into an active position through a series of intramolecular shifts *(29)*. Second, there is only a single cleavage domain per monomer. It has recently been reported that in order to cleave both strands, the catalytic domain of a second monomer transiently dimerizes at the cleavage site. Structural similarity to the catalytic and bridging domains of the homodimeric Type II enzyme *Bam*HI may further substantiate this model *(30)*.

### 1.4.5. Homing Endonucleases

The homing endonucleases, sometimes referred to as intron and intein (protein intron) encoded endonucleases, are different from the standard restriction enzymes in several respects. They may be monomers or dimers, and may require other proteins or RNA for activity. They may tolerate base substitutions in their large recognition sequences, especially the outside regions, with only small changes in cleavage rates. They have been found in archaea and bacteria and, unlike typical endonucleases, even occur in eukaryotes. Their genomic location can be mitochondrial, chloroplast, chromosomal, or extrachromosomal. They can be subdivided into four groups based on sequence motifs. To date, 35 have been identified and characterized to varying degrees *(18)*.

### 1.4.6. Type IIe Endonucleases

The Type IIe endonucleases are similar to the common Type II or Type IIs in their structure, recognition patterns, and mechanisms. However, they are distinctive because they are activated to cleave slow or resistant sites by the binding of a second recognition sequence to a distal, noncatalytic site on the enzyme. Typically, these enzymes cleave incompletely at a subset of recognition sites. Isoschizomers of Type IIe endonucleases cleave completely. For *Eco*RII and pBR322 (six recognition sites per molecule) the ratio of enzyme to recognition sites in a reaction mix for optimal activity is 0.25–0.5, or 2–4 recognition sites per enzyme dimer. This suggests that each enzyme dimer binds a recognition sequence at its catalytic site and a second at the allosteric site *(31)*. Based on observed cleavage at particular sites, an original classification of Type IIe endonuclease activity in a 1-h digest is as follows: cleavable sites, >90% cleavage with one- to fivefold excess enzyme; slow sites, 5–90% cleavage with fivefold excess enzyme, and additional cleavage with 10- to 30-fold excess; resistant sites, <5% cleavage with fivefold excess enzyme and no additional cleavage with a 10- to 30-fold excess. The same enzyme may cleave one DNA site slowly while another, in the same or on different DNAs, is resistant to cleavage *(15)*.

The Type IIe enzymes can be separated into two classes in a more descriptive manner based on the change in cleavage kinetics upon binding of an affector sequence (which may be an oligonucleotide, linear phage, or supercoiled DNA). In the *K* class of enzymes (*Nar*I, *Hpa*II, *Sac*II), activator DNA binding decreases the $K_m$ without altering the $V_{max}$ of cleavage, indicating that cooperative binding induces a conformational shift, thereby increasing the affinity of the enzyme for substrate. In the V class (NaeI, BspMI), binding of activator DNA increases $V_{max}$ without changing $K_m$, indicating that the increased catalytic activity is not related to the affinity of the enzyme for substrate (15). It is assumed that the cleavage kinetics of different recognition sites is influenced by the flanking sequences for Type IIe enzymes. The flanking sequence preferences are not presently understood. However, sequences including a readily cleaved site and its flanking regions are a starting point to determine good activator sequences.

Incomplete digestion by Type IIe enzymes can make interpretation of banding patterns and subsequent applications difficult. Adding activators may improve cleavage. For example, oligos containing the recognition site of EcoRII that are uncleavable because of specific methylation or presence of nucleotide analogs can bind to the allosteric site and stimulate cleavage of refractory sites in pBR322 (32). A similar approach, developed by Topal and colleagues, used an oligonucleotide containing the recognition site for NaeI with a sulfur replacing phosphorus at the scissile bond (33). Complete substrate cleavage is achieved without consuming the activator oligonucleotide as the sulfur prevents hydrolysis by NaeI. The same strategy has also been used successfully for NarI, demonstrating the utility of this approach for both V and K class enzymes. Some of these enzymes are available commercially with the activating oligo premixed in the provided reaction buffer (e.g., Promega's Turbo NaeI and Turbo NarI). The presence of the oligo does not interfere with ligation or random primer labeling. A one-step purification yields a cleaved DNA suitable for end-labeling (34).

NaeI contains a 10 amino-acid region similar to a motif in human DNA ligase I. A leucine at position 43 in NaeI is a lysine in the ligase motif that, in the adenylated intermediate, is essential for ligation catalysis. A mutant, NaeI L43K, exhibits type I topoisomerase activity (cleavage, strand passage, and reunion). This suggests a possible origin for the activator DNA-binding site and a potential link with topoisomerases and recombinases (35,36).

### 1.4.7. Cleavage of ssDNA

All restriction endonucleases cleave double-stranded DNA, but a few enzymes hydrolyze ssDNA at significantly reduced rates. Two theories exist regarding the mechanism of apparent ssDNA cleavage. Although cleavage of actual ssDNA has been reported (37), in other cases, the enzyme may really act on transiently formed double-stranded DNA (38). One method for cleaving ssDNA uses an oligonucleotide adaptor and a Type IIs enzyme where the recognition site and cleavage site are significantly separated, such as FokI. The oligonucleotide contains a hairpin loop and a double stranded region with the recognition site of the enzyme. A single-stranded region of the oligonucleotide protrudes past the recognition site and is complementary to the ssDNA substrate. The endonuclease is then able to recognize the double-stranded region of the oligonucleotide and cleave in the region formed by the oligo:ssDNA hybrid as illustrated in **Fig. 1A**. After cleavage, the oligonucleotide fragment bound to the ssDNA can be heat-denatured. However, this method cannot be used to cleaved ssRNA (16). A similar approach for cutting ssDNA uses the enzyme XcmI, which recognizes the longest (9 nt) degenerative sequence known (5'-CCANNNNN/NNNNTGG-3'). An oligo is designed with two hairpin loops and a region of dsDNA containing the recognition

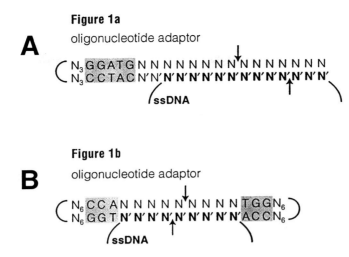

Fig. 1. Use of oligonucleotide adaptors for cleaving ssDNA with the endonucleases *Fok*I (**A**) and *Xcm*I (**B**). The ssDNA appears in bold, the recognition site of the endonuclease contained within the oligonucleotide is highlighted, and the cleavage positions are indicated by the arrows.

nucleotides but leaving the center, nonspecific nucleotides single-stranded. A complementary ssDNA that hybridizes to these nonspecific nucleotides will be cleaved as illustrated in **Fig. 1B** *(39)*.

### 1.4.8. Star Activity

Although endonucleases bind DNA nonspecifically, they exhibit a very high preference catalytically for their recognition site over sites with even a one-base-pair difference. A partial relaxation of specificity under suboptimal digest conditions is an inherent property of some enzymes that is commonly referred to as "star activity." Depending on the enzyme, star activity is most influenced by volume excluders (glycerol, ethylene glycol), pH, and/or substitution of $Mg^{2+}$ with another metal *(11)*. The number of water molecules normally present at the protein-DNA interface for *Eco*RI at noncognate sites is reduced at high osmotic pressure because of volume exclusion, and because of the tighter binding of the enzyme, the active conformation is more easily achieved at star sites *(40)*. For example, *Eco*RI cleaves its recognition site (5'-GAATTC-3') at a rate $10^5$ faster than the next best sequence (5'-TAATTC-3') under optimal conditions *(41)*. With increasing ethylene glycol concentrations, cleavage rates decrease at the cognate site, but increase at the next best site until the rates approach equivalence at 4 *M* ethylene glycol *(40)*. At higher pH, the high [OH⁻] may reduce the need for activated water formed at the catalytic site as the attacking nucleophile. All restriction endonucleases prefer for $Mg^{2+}$ for activity. A few can use a different divalent metal, usually $Mn^{2+}$ but occasionally $Ca^{2+}$ $Fe^{2+}$, $Co^{2+}$, $Ni^{2+}$, and $Zn^{2+}$, but cleavage with these ions is usually less specific and slower *(16)*. $Mn^{2+}$ bound $H_2O$ may be better than $Mg^{2+}$ bound $H_2O$ at providing the proton necessary for the leaving group 3' OH. For *Eco*RV, activity at the cognate site is $10^6$ higher than at the star site with $Mg^{2+}$, but only sixfold higher with $Mn^{2+}$ *(42)*. Alterations in ionic strength, the type of ions, trace organic solvents, and high enzyme-to-DNA ratios may also result in star activity. Read the information sent with commercial preparations to avoid star activity for those enzymes that are susceptible.

## 1.5. Strategies for Custom Design of Sequence Specific Cleavage

### 1.5.1. Protein Engineering

Despite the comprehensive sequence and 3D structural data available, attempts to mutate restriction enzymes to alter the DNA sequence they recognize have generally resulted only in relaxed specificity and/or decreased cleavage rates. However, moderate success has been achieved in engineering mutants of *Eco*RV that recognize and cleave different sequences with equal or greater specificity. Mutants have been made which respectively cleave recognition sites with a uracil instead of thymine by more than two orders of magnitude over wild-type (N188Q) *(43)*, or with a methylphosphonate in one of the phosphate backbone positions by three orders of magnitude over wild-type (T94V) *(44)*. In addition, A181K and A181E mutants preferentially cleave sites with a purine or thymine respectively 5' to the recognition site *(45)*. A directed evolution approach has produced a N97T/S183A/T222S mutant with a 20-fold preference for an oligonucleotide with a GC-rich flanking region and a K104N/A181T mutant with a sevenfold preference for an AT-rich flanking region *(46)*. Heterodimers of *Eco*RV containing a catalytically inactive mutant subunit act as a site-specific nickase *(47)*. More structural information is needed regarding the large number of enzyme-DNA contacts and intra-/intermolecular protein shifts to facilitate additional enzyme engineering. However, several other approaches to achieve custom-designed, sequence-specific cleavage have also been investigated.

### 1.5.2. Fusion Proteins

Hybrid enzymes can be constructed by fusing recognition and cleavage domains from different proteins. For example, using various spacers and constructs, the Type IIs *Fok*I catalytic domain has been combined with DNA-binding domains from the *Drosophila Ubx* homeodomain *(48)*, the zinc-finger region of the eukaryotic transcription factor Sp1, the designed zinc-finger consensus sequence protein QQR *(48)*, and the zinc-finger region of the yeast transcription factor Gal4 *(50)*. Interestingly, the fusions with Sp1 and QQR will also bind and cleave the DNA strand of DNA–RNA hybrids. Since the recognition sequence of Gal4 is palindromic and the protein dimerizes at the site, the strands are cleaved on opposite sides of the recognition site. This results in a ≥24-base, 5' overhang, which includes the recognition nucleotides. Sites for the *Ubx*, Sp1, and QQR hybrids are nonpalindromic, the fusion proteins act as monomers, and both strands are cleaved to one side of the recognition site.

### 1.5.3. Metal Catalyzed Cleavage

Another method uses either an oligonucleotide capable of forming a triple helix *(51)* or a DNA-binding protein to provide specificity. Covalently attached to them is a metal complex, usually EDTA-iron or *o*-phenanthroline-copper, which catalyzes phosphodiester cleavage in the presence of a reducing agent. Proteins successfully used to supply specificity include Cro *(52)*, the catabolite activator protein "CAP" *(53)*, and the Msx-1 homeodomain *(54)*.

## 1.5.4. Achilles Heel Cleavage

Both of the above strategies suffer from two main drawbacks: Cleavage at more than one phosphodiester bond results in a mixed population of overhangs, and cleavage is generally incomplete. A technique designed to eliminate these problems has been referred to as "Achilles heel cleavage." First, a target sequence is protected by a bound *Rec*A/oligo complex *(55,56)* or triple helix formation *(57)*. A methylase modifies all sites except the protected target. The methylase is removed by purification, followed by the protecting group. The target sequence is then specifically cleaved by a methyl-sensitive restriction enzyme.

## 1.6. Considerations for Use of Commercially Prepared Restriction Endonucleases

### 1.6.1. Unit Definition and Commercial Quality Control Assays

Commercial vendors of restriction endonucleases use relatively standard assays for unit definition and the core quality tests. The products of the activity, overdigest, cut-ligate-recut, and nickase reactions explained below are generally electrophoresed in agarose gels and detected by staining with ethidium bromide. An activity unit is defined as the amount of enzyme necessary to completely digest 1 μg of the defined substrate, usually λ DNA, in a 50-μL reaction volume in 1 h. If the number of sites is small (≤3), λ predigested with another enzyme (e.g., *Eco*RI) may be used to improve gel resolution. A different DNA such as Adenovirus 2 is used if there is a single site or no sites in λ. An example of a titration assay used to define a unit is provided in **Subheading 2.**

An overdigest or nonspecific exo- and endonuclease assay is performed in the same fashion, except a large excess of enzyme units is used (5–100 U) and incubation times are long (generally 16 h). Often the result is reported as (X-) fold-overdigestion (U × h), although the activity of the enzyme typically diminishes significantly over this length of time at the reaction temperature. Although star activity is an inherent property of the enzyme itself and not a separate, distinct contaminant, it does result in cleavage at noncognate sites that will interfere with downstream applications and therefore must be considered in determining the endpoint of this assay. Some suppliers only consider the absence of detectable contaminants in their specifications, which can be misleading.

The cut-ligate-recut assay is more sensitive with regard to contaminating exonuclease, and can additionally detect phosphatase activity. DNA is slightly overdigested, ligated with T4 ligase, and then recut. Dephosphorylated ends will not be ligated, and staggered ends which have been blunted by single-stranded exonuclease activity will exhibit less efficient ligation. Loss of any of the original terminal nucleotides after cleavage almost always results in the loss of the recognition site for recutting, even if the substrate religated. Although ligation of a one-base overhang is still high enough to be useful in T-vector cloning of PCR fragments, for unknown reasons the efficiency of ligation, as indicated by transformation efficiencies of plasmids, can be as much as two orders of magnitude lower than blunt-end ligation. In contrast, a four-base G-C overhang is stable enough to transform well, even without in vitro ligation.

Some vendors also test for the presence of nickases by incubating the enzyme and a supercoiled substrate (RF I form) that does not contain a recognition site to determine the amount of substrate converted to a relaxed, open circle (RF II form). However, the impact of nicking activity on the major applications of cloning and mapping is minimal. One application influenced by nicking is the generation of nested deletions with exonuclease III. Improperly purified DNA is a far more likely source of nicks than a contaminant activity of the endonuclease. Current purification methods and the sensitivity of other assays are such that rarely, if ever, is a nickase test warranted.

Another test used for the detection of exonucleases is digestion of $^3$H-ds and -ssDNA, TCA precipitate the remaining DNA, and detect released nucleotides through scintillation counting of the supernatant. If the enzyme cleaves the substrate within 30 bp of the DNA end, inefficient precipitation of the resultant small fragments may lead to an incorrect interpretation of exonuclease activity. Labeling the 5' or 3' ends of DNA with $^{32}$P is a sensitive way to detect contaminating phosphatase or exonuclease. However, this method requires frequent preparation of substrate.

The Blue/White assay combines excellent sensitivity and a verification of performance. A cloning plasmid is used that contains a multiple cloning site flanked by RNA polymerase promoter(s) within a coding sequence for the *lacZ* gene α-peptide and a separate selectable marker, such as ampicillin resistance. The plasmid is several-fold overdigested with an enzyme that has a single site within the multiple cloning region. The DNA is ligated (without insert) and then transformed into cells lacking the α-peptide region of *lacZ*. An agarose gel of cut, ligated, and recut DNA is also examined. If the integrity of the cut ends is perfectly maintained, ligation will produce mostly higher molecular-weight concatamers and a lesser amount of circularized monomer. After transformation and α-peptide expression, the functional *lacZ* gene product β-galactosidase is produced through α-complementation. When plated in the presence of IPTG and X-Gal, blue colonies will result. Expected transformation efficiency will be 1–2 orders of magnitude lower than with control supercoiled plasmid. Both phosphatase and exonuclease contamination will lower ligation efficiency and thereby decrease transformation efficiency. More importantly, loss of nucleotides at the cut site, even if ligatable, yields a mixed population of clones containing frame shifts and codon deletions in the *lacZ* gene α-peptide. These cause white colonies, i.e., false positives in a cloning experiment with an insert *(58)*. A special case is the loss of a single nucleotide. The resultant colonies are then able to use an alternative start codon that shifts to become in-frame. However, it produces weak translation initiation and/or improper complementation for fully active β-galactosidase, and the colonies develop a faint blue color that is easily mistaken for white. A contaminant that exists even in cloned preparations removes the 3' nucleotide from cut DNA and causes the faint or pale blue colonies. This is especially problematic with blunt-end cutting enzymes. Not all commercial vendors specifically assay for and remove this contaminant, which has not yet been positively identified *(59)*.

### 1.6.2. Reaction Conditions: Buffers, Volume, Units Needed for a Substrate, and Double Digests

The reaction conditions and enzyme units needed to digest a given substrate must be chosen carefully to ensure expected performance. The buffer systems provided with commercially obtained enzymes are designed to balance optimal individual enzyme performance and using a limited number of various buffers. Very few provided reaction buffers are actually optimized for a single enzyme, nor is there a true "universal" buffer. As with any group of similar enzymes, endonucleases are all unique at some level in their preferences for buffer components and concentration of cation ($Na^+$ or $K^+$), anion ($Cl^-$ or acetate), pH (7.2–8.5), stabilizer, such as BSA, detergent, or spermidine (*see* **Note 10**), and reducing agent. The storage buffer of the enzyme may also adversely affect use when it comprises an unusually large amount of the total reaction volume (volume exclusion of glycerol causing star activity, chelators for stability binding Mg in reaction, and so forth).

Because it may constitute a large percentage of the reaction volume, substrate preparation is critical for enzyme performance. Enzymes vary in their resistance to proteases, interference by DNA binding proteins, competitive inhibition from RNA, and tolerance for EDTA, PEG, SDS, CsCl, phenol, chloroform, and alcohols. Unit definitions are generally given for high-quality linear phage or viral DNA. It is important to keep in mind the number of pmoles of cut sites used to define a unit vs other substrates. **Table 2** suggests the enzyme units needed for complete cutting with *Bam*HI based strictly on the number of cut sites. Although no other parameters are taken into account, this approach can be a useful approximation. For PCR fragments, oligonucleotides, or double digests of sites in close proximity, *see* **Note 9** regarding cleavage near DNA ends.

Although it is impractical for commercial vendors to list the relative activity of each restriction enzyme in every possible reaction buffer, tables are usually listed in their catalogs for the most frequently used buffers to assist in choosing a buffer for multiple enzyme digests. Some vendors supply a multiple-digest buffer, which may be the best alternative when using more than one enzyme. In either case, avoid digestion conditions <50% of unit definition activity if at all possible. Also, certain digest conditions for a particular enzyme may concurrently give good activity but produce a loss of specificity or induce "star activity."

## 2. Materials

1. Test DNA (phage, plasmid, PCR fragment, or oligo).
2. Unit definition DNA (usually λ).
3. Commercially provided 10X reaction buffer.
4. BSA (acetylated to ensure no contaminating activities).
5. Commercially provided restriction endonuclease.
6. Nuclease-free water.
7. 6X stop solution/loading dye (6X concentration): 25% glycerol, 50 m$M$ EDTA, 0.5% SDS, 0.2% Orange G.
8. DNA markers (covering range appropriate for restriction fragment size).
9. Agarose.
10. Ethidium bromide (5 mg/mL).
11. Gel electrophoresis buffer (final concentration): 40 m$M$ Tris acetate, pH 7.9 at 25°C, 5 m$M$ sodium acetate, 1 m$M$ EDTA.

## 3. Methods

### 3.1. Determination of RE Units Necessary to Cleave a Specific DNA Substrate

1. Prepare appropriate enzyme dilutions (*see* **Subheading 1.6.2.**) using 1X reaction buffer containing 0.5 mg/mL BSA, including one to 1 U/μL. Mix before use by briefly "flicking" the eppendorf tube or repeated pipetting, do not vortex. Maintain dilutions and reaction mixes on ice until ready to begin incubation.

**Table 2**
**Approximate Units Nedded to Cut 1 μg of Various DNAs**

| DNA | Base pairs | Picomoles in 1 μg[a] | Cut sites (*Bam*HI) | Picomoles cut sites | Enzyme units needed |
|---|---|---|---|---|---|
| Unit definition (ex. λ) | 48,502 | 0.0317 | 5 | 0.159 | 1 |
| Plasmid | 3000 | 0.5 | 1 | 0.5 | 3–4[b] |
| PCR fragment | 700 | 2.2 | 1 | 2.2 | 12–15 |
| Oligonucleotide | 25 | 62.5 | 1 | 62.5 | 400 |

[a]Based on 650 daltons per base pair of DNA.
[b]Enzymes differ in their ability to cut supercoiled vs linearized plasmids.

2. Prepare reaction mix minus DNA such that the final reaction contains 1X reaction buffer and 0.1 mg/mL BSA, bring to volume with nuclease-free water and vortex briefly. For "*n*" reactions in the dilution series, make reaction mix for *n* + 3. The final volume after adding DNA and enzyme is ideally 50 μL (*see* **Note 6**).
3. Add 1 μg control DNA (unit definition), then 1 μL of 1 U/μL RE, to one aliquot of reaction mix (positive control).
4. Add test DNA to remaining reaction mix (usually 1 μg per reaction) and vortex for a few seconds.
5. Aliquot reaction mix + DNA, then add 1 mL of each enzyme dilution, saving one aliquot as a "no enzyme" (negative) control. Mix all reaction tubes briefly by "flicking" or pipeting, do not vortex.
6. Microfuge briefly, incubate at appropriate temperature for 1 h.
7. Stop with 10 μL 6X stop solution/loading dye.
8. Heat tubes 15 min, 65°C, to assist SDS denaturation of protein, and dissociate any complementary DNA overhangs.
9. Prepare agarose gel in electrophoresis buffer with a final concentration of 0.5–1.0 μg ethidium bromide/mL of gel. Alternatively, the gel can be stained after running samples by soaking in a solution of 2 μg/mL ethidium bromide in gel buffer. Use enough solution to cover the gel and shake gently for 10–30 min. Refer to **Note 12** for suggested agarose percentages.
10. Load 20–30 μL (0.33–0.5 μg DNA) onto gel (*see* **Note 13**), allow samples to settle for a minute(s), begin electrophoresis at 1–2X the normal voltage for several minutes. A lane(s) with DNA molecular-weight markers may be included if desired. Electrophoresis voltage and time will vary considerably, but 125 V for 150 min are ususally required for a full-size gel (approx 80-mm separation) or 80 V for 100 min for a mini-gel (approx 40-mm separation).
11. Use a transilluminator, set for midrange (300 nm) UV light, to see the gel bands stained with ethidium bromide. Photograph the gel on the box with a Polaroid camera or a video imaging system.

### 3.2. Digesting Multiple Substrates with the Same RE(s)

1. Prepare DNA substrates. Prepare enzyme(s) dilution if necessary as above. If using more than one enzyme, *see* **Subheading 1.6.2.** to determine which reaction buffer to use.
2. Prepare two reaction mixes so that the final reactions contain 1X reaction buffer and 0.1 mg/mL BSA. One mix consists of buffer, BSA, and nuclease-free water, and the other(s) of buffer, BSA, enzyme(s), and water. The final reaction volume after adding DNA is ideally 50 μL (*see* **Note 6**). For "*n*" DNA substrates to be tested, make *n* + 2 of each reaction mix.

3. Aliquot two tubes of reaction mix (with and without enzyme) per DNA sample to be tested.
4. Add 1 μg of each DNA to be digested to a reaction mix with and without enzyme.
5. Add 1 μg control DNA (unit definition) to a reaction mix with and without enzyme. There must be a minimum of 1 U of enzyme(s) per μg of control DNA. Mix all reaction tubes briefly by "flicking" or pipeting, do not vortex.
6. Follow **steps 6–11** in **Subheading 3.1.**

## 4. Notes

1. Achieving complete cleavage with *Nae*I, *Nar*I, *Sac*II, *Eco*RII, *Hpa*II, *Bsp*MI, *Cfr*9I, and *Ksp*632I is difficult because of the mechanism of Type IIe enzymes. *See* **Subheading 1.4.6.** for further explanation.
2. *Sac*II exhibits good activity at very low ionic strengths and lower pH (7.5) on linear substrates, but needs 50 m*M* salt and higher pH (7.9) for efficient cleavage of supercoiled plasmids containing a single site (Promega, unpublished observations).
3. An overlay of mineral oil, such as Sigma #M5904, is suggested for digestions >50°C if incubation times exceed 1 h. A single drop added just before incubation is sufficient.
4. A potassium dodecyl sulfate precipitate will form if the reaction buffer contains potassium and the stop buffer contains SDS. Brief centrifugation before loading is advisable.
5. Digestion of amplified DNA for many endonucleases may be carried out directly in PCR or RT-PCR buffer, thus eliminating DNA purification steps. However, the DNA will not retain the compatible end of the restriction cleavage caused by the fill-in, exonuclease, and/or 3' nucleotide addition properties of the polymerase *(60)*.
6. To reduce the concentration of unwanted components contributed by the enzyme storage buffer (e.g., glycerol) and DNA purification (e.g., alcohol, salt), use 50-μL reaction volumes. Also, commercially supplied buffers at 1X concentration may not supply sufficient amounts of $Mg^{2+}$ if the volume is significantly reduced. However, increasing $[Mg^{2+}]$ by adding $Mg^{2+}$ in solution can sometimes further inhibit the reaction (Promega, unpublished observations). Greatly reduced volumes significantly alter the ratio-to-surface area, increasing evaporation and possible denaturation of the endonuclease.
7. The unit definition parameters of complete digestion with 1 U in 1 h are not necessarily linear. In general, 4 U of enzyme, after a 15 min digest (except for some high-temperature enzymes),will usually give complete cleavage, but 0.25 U after a 4 h digest may not. Examples of enzymes that show such nonlinear cleavage rates are *Acc*I, *Acc*65I, *Bam*HI, *Csp*I, *Eco*52I, *Not*I, *Sal*I, *Spe*I, *Sph*I, and *Sty*I (Promega, unpublished observations).
8. Problems may arise in the digestion of genomic DNA in solution because of substrate purity ($A_{260}/A_{280}$ ratios should be ≥1.8), methylation, and high background caused by shearing. To reduce shearing, whole cells may be embedded in agarose, lysed, treated with protease, and then digested with endonuclease. However, in addition to substrate purity and methylation, problems may arise because of low diffusion into the agarose plug. Proper digestion conditions may need to be experimentally determined, but 5–20 U of enzyme per μg of DNA is a typical range. In agarose plugs, digest for 2–4 h. Avoid enzymes having overdigestion specifications of <20 U in 16 h. An abbreviated protocol for agarose-plug preparation and digestion is given in the 1998/1999 New England BioLabs catalog, pp. 276–278 *(61)*.
9. Endonucleases require contact with the DNA backbone for several bases adjacent to the recognition sequence. This must be considered when cutting oligos or near the end of a fragment. In general, the recognition site must lie three base pairs from the end to give good cleavage. Tables have been developed for a limited list of enzymes based on cutting

a short oligo *(60)*, cutting a PCR fragment near one end *(62)*, and double digests of adjacent sites in a polylinker *(63)*.

10. Although spermidine and detergents may stabilize the enzyme or prevent its binding to the reaction vessel, BSA is the general additive of choice and is used at least to some degree by most vendors (0.1 mg/mL in reactions, 0.5 mg/mL for enzyme dilutions suggested by Promega). BSA is not known to be inhibitory for any enzyme, and usually gives 1.5- to threefold higher activity, occasionally >sixfold, with various miniprep plasmid substrates (Promega, unpublished observations.). Acetylation of the BSA is sometimes used to inactivate possible enzymatic contaminants.

11. If the intended use of the digested DNA is cloning into a Blue/White plasmid selection system, *see* **Subheading 1.6.1.** for an explanation of potential false positives seen as pale or light blue colonies.

12. Suggested agarose percentages for resolving DNA of various lengths follows *(64)*:

| % Agarose (w/v) | DNA molecules separated (kb) |
| --- | --- |
| 0.3 | 5–60 |
| 0.6 | 1–20 |
| 0.7 | 0.8–10 |
| 0.9 | 0.5–7 |
| 1.2 | 0.4–6 |
| 1.5 | 0.2–3 |
| 2.0 | 0.1–2 |

13. Considerably smaller amounts of DNA loaded onto gels is sufficient to detect the major DNA species. However, it is recommended to load 0.33–0.5 µg of DNA when needing to detect small percentages of recognition sites that were potentially not cleaved, such as in determining the endpoint of a unit activity assay.

## Acknowledgment

The author would like to thank Michael Slater, Mark Klekamp, Doug Barron, and Elizabeth Murray for their review of the manuscript and many helpful suggestions.

## References

1. Roberts, R. J. and Halford, S. E. (1993) Type II restriction endonucleases, in *Nucleases,* 2nd ed. (Linn, S. M., Lloyd, S. R., and Roberts, R. J., eds.), Cold Spring Harbor Laboratory, Cold Spring Harbor, NY, pp. 35–88.
2. Roberts, R. J. and Macelis, D. (1998) REBASE—restriction enzymes and methylases. *Nucleic Acids Res.* **26,** 338–350.
3. Wilson, G. G. and Murray, N. E. (1991) Restriction and modification systems. *Annu. Rev. Genet.* **25,** 585–627.
4. Stahl, F., Wende, W., Jeltsch, A., and Pingoud, A. (1998) The mechanism of DNA cleavage by the Type II restriction enzyme *Eco*RV: Asp36 is not directly involved in DNA cleavage but serves to couple indirect readout to catalysis. *Biol. Chem.* **379,** 467–473.
5. Smith, H. O., Annau, T. M., and Chandrasegaran, S. (1990) Finding sequence motifs in groups of functionally related proteins. *Proc. Natl. Acad. Sci. USA* **87,** 826–830.
6. Szomolanyi, E., Kiss, A., and Venetianer, P. (1980) Cloning the modification methylase gene of *Bacillus sphaericus* R in *Escherichia coli. Gene* **10,** 219–225.
7. Tao, T., Bourne, J. C., and Blumenthal, R. M. (1991) A family of regulatory genes associated with Type II restriction-modification systems. *J. Bact.* **173,** 1367–1375.
8. Ives, C. L., Sohail, A., and Brooks, J. E. (1995) The regulatory C proteins from different restriction-modification systems can cross-complement. *J. Bact.* **177,** 6313–6315.

9. Smith, H. O. and Nathans, D. (1973) A suggested nomenclature for bacterial host modification and restriction systems and their enzymes. *J. Mol. Biol.* **81,** 419–423.

10. Bickle, T. A. (1993) The ATP-dependent restriction enzymes, in *Nucleases,* 2nd ed. (Linn, S. M., Lloyd, S. R., and Roberts, R. J., eds.), Cold Spring Harbor Laboratory, Cold Spring Harbor, NY, pp. 35–88.

11. Pingoud, A. and Jeltsch, A. (1997) Recognition and cleavage of DNA by type-II restriction endonucleases. *Eur. J. Biochem.* **246,** 1–22.

12. Kong, H. (1998) Analyzing the functional organization of a novel restriction modification system, the *Bcg*I system. *J. Mol. Biol.* **279,** 823–832.

13. Sears, L. E., Zhou, B., Aliotta, J. M., Morgan, R. D., and Kong, H. (1996) *Bae*I, another unusual *Bcg*I-like restriction endonuclease. *Nucleic Acids Res.* **24,** 3590–3592.

14. Reuter, M., Kupper, D., Pein, C. D., Petrusyte, M., Siksnys, V., Frey, B., and Kruger, D. H. (1993) Use of specific oligonucleotide duplexes to stimulate cleavage of refractory DNA sites by restriction endonucleases. *Anal. Biochem.* **209,** 232–237.

15. Oller, A. R., Broek, W. V., Conrad, M., and Topal, M. (1991) Ability of DNA and spermidine to affect the activity of restriction endonucleases from several bacterial species. *Biochem.* **30,** 2543–2549.

16. Szybalski, W., Kim, S. C., Hasan, N., and Podhajska, A. J. (1991) Class-IIS restriction enzymes—a review. *Gene* **100,** 13–26.

17. Janulaitis, A., Petrusyte, M., Maneliene, Z., Klimasauskas, S., and Butkus, V. (1992) Purification and properties of the *Eco*57I restriction endonuclease and methylase — prototypes of a new class (type IV). *Nucleic Acids Res.* **20,** 6043–6049.

18. Belfort, M. and Roberts, R. J. (1997) Homing endonucleases: keeping the house in order. *Nucleic Acids Res.* **25,** 3379–3388.

19. Meisel, A., Mackeldanz, P., Bickle, T. A., Kruger, D. H., and Schroeder, C. (1995) Type III restriction endonucleases translocate DNA in a reaction driven by recognition site-specific ATP hydrolysis. *EMBO J.* **14,** 2958–2966.

20. Kruger, D. H., Kupper, D., Meisel, A., Reuter, M., and Schroeder, C. (1995) The significance of distance and orientation of restriction endonuclease recognition sites in viral DNA genomes. *FEMS Microbiol. Rev.* **17,** 177–184.

21. Mise, K. and Nakajima, K. (1985) Purification of a new restriction endonuclease, StyI, from Escherichia coli carrying the hsd+ minplasmid. *Gene* **33,** 357–361.

22. Sam, M. D. and Perona, J. J. (1999) Catalytic roles of divalent metal ions in phosphoryl transfer by *Eco*RV endonuclease. *Biochem.* **38,** 6576–6586.

23. Wentzell, L. M., Nobbs, T. J., and Halford, S. E. (1995) The *Sfi*I restriction endonuclease makes a 4-strand DNA break at two copies of its recognition sequence. *J. Mol. Biol.* **248,** 581–595.

24. Sato, H., Suzuki, T., and Yamada, Y. (1990) Purification of restriction endonuclease from *Acetobacter aceti* IFO 3281 (*Aat*II) and its properties. *Agric. Biol. Chem.* **54,** 3319–3325.

25. Nobbs, T. J. and Halford, S. E. (1995) DNA cleavage at two recognition sites by the *Sfi*I restriction endonuclease: salt dependence of *cis* and *trans* interactions between distand DNA sites. *J. Mol. Biol.* **252,** 399–411.

26. Nobbs, T. J., Williams, S. A., Connolly, B. A., and Halford, S. E. (1998) Phosphorothioate substrates for the *Sfi*I restriction endonuclease. *Biol. Chem.* **379,** 599–604.

27. Kong, H. and Smith, C. L. (1998) Does *Bcg*I, a uniquerestriction endonuclease, require two recognition sites for cleavage? *Biol. Chem.* **379,** 605–609.

28. Kong, H. and Smith, C. L. (1997) Substrate DNA and cofactor regulate the activities of a multi-functional restriction-modification enzyme, *Bcg* I. *Nucl. Acids Res.* **25,** 3687–3692.

29. Wah, D. A., Hirsch, J. A., Dorner, L. F., Schildkraut, I., and Aggarwal, A. K. (1997) Structure of the multimodular endonuclease *Fok*I bound to DNA. *Nature* **388,** 97–100.

30. Bitinaite, J., Wah, D. A., Aggarwal, A. K., and Schildkraut, I. (1998) *Fok*I dimerization is required for DNA cleavage. *Proc. Natl. Acad. Sci.* **95,** 10,570–10,575.

31. Reuter, M., Kupper, D., Meisel, A., Schroeder, C., and Kruger, D. H. (1998) Cooperative binding properties of restriction endonuclease *Eco*RII with DNA recognition sites. *J. Biol. Chem.* **273,** 8294–8300.

32. Pein, C. D., Reuter, M., Meisel, A., Cech, D., and Kruger, D. H. (1991) Activation of restriction endonuclease *Eco*RII does not depend on the cleavage of stimulator DNA. *Nucleic Acids Res.* **19,** 5139–5142.

33. Conrad, M. and Topal, M. (1992) Modified DNA fragments activate *Nae*I cleavage of refractory DNA sites. *Nucleic Acids Res.* **20,** 5127–5130.

34. Senesac, J. H. and Allen, J. R. (1995) Oligonucleaotide activation of the Type IIe restriction enzyme *Nae*I for digestion of refractory sites. *BioTechniques* **19,** 990–993.

35. Jo, K. and Topal, M. D. (1995) DNA topoisomerase and recombinase activities in *Nae*I restriciton endonuclease. *Science* **267,** 1817–1820.

36. Jo, K. and Topal, M. D. (1998) Step-wise DNA relaxation and decatenation by *Nae*I-43K. *Nucleic Acids Res.* **26,** 2380–2384.

37. Yoo, O. J. and Agarwal, K. L. (1980) Cleavage of single strand oligonucleotides and bacteriophage phiX174 DNA by Msp I endonuclease. *J. Biol. Chem.* **255,** 10,559–10,562.

38. Blakesley, R. W., Dodgson, J. B., Nes, I. F., and Wells, R. D. (1977) Duplex regions in "single-stranded" phiX174 DNA are cleaved by a restriction endonuclease from Haemophilus aegypius. *J. Biol. Chem.* **252,** 7300–7306.

39. Shaw, P. C. and Mok, Y. K. (1993) *Xcm*I as a universal restriction enzyme for single-stranded DNA. *Gene* **133,** 85–89.

40. Robinson, C. R. and Sligar, S. G. (1998) Changes in solvation during DNA binding and cleavage are critical to altered specificity of the *Eco*RI endonuclease. *Proc. Natl. Acad. Sci. USA* **95,** 2186–2191.

41. Lesser, D. R., Kurpiewski, M. R., and Jen-Jacobson, L. (1990) The energetic basis of specificity in the *Eco*RI endonuclease-DNA interaction. *Science* **250,** 776–786.

42. Vermote, C. L. M. and Halford, S. E. (1992) *Eco*RV restriction endonuclease: communication between catalytic metal ions and DNA recognition. *Biochemistry* **31,** 6082–6089.

43. Wenz, C., Selent, U., Wende, W., Jeltsch, A., Wolfes, H., and Pingoud, A. (1994) Protein engineering of the restriction endonuclease *Eco*RV: replacement of an amino acid residue in the DNA binding site leads to an altered selectivity towards unmodified and modified substrates. *Biochim. Biophys. Acta.* **1219,** 73–80.

44. Lanio, T., Selent, U., Wnez, C., Wende, W., Schulz, A., Adiraj, M., Katti, S. B., and Pingoud, A. (1996) *Eco*RV-T94V: a mutant restriction endonuclease with an altered substrate specificity towards modified oligodeoxynucleotides. *Protein Eng.* **9,** 1005–1010.

45. Schottler, S., Wenz, C., Lanio, A., Jeltsch, A., and Pingoud, A. (1998) Protein engineering of the restriction endonuclease *Eco*RV: structure-guided design of enzyme variants that recognize the base pairs flanking the recognition site. *Eur. J. Biochem.* **258,** 184–191.

46. Lanio, T., Jeltsch, A., and Pingoud, A. (1998) Towards the design of rare cutting restriction endonucleases: using directed evolution to generate variants of *Eco*RV differing in their substrate specificity by two orders of magnitude. *J. Mol. Biol.* **283,** 59–69.

47. Stahl, F., Wende, W., Jeltsch, A., and Pingoud, A. (1996) Introduction of asymmetry in the naturally symmetric restriction endonuclease *Eco*RV to investigate intersubunit communication in the homodimeric protein. *Proc. Natl. Acad. Sci. USA* **93,** 6175–6180.

48. Kim, Y. G. and Chandrasegaran, S. (1994) Chimeric restriction endonuclease. *Proc. Natl. Acad. Sci USA* **91,** 883–887.

49. Kim, Y. G., Shi, Y., Berg, J. M., and Chandrasegaran, S. (1997) Site-specific cleavage of DNA-RNA hybrids by zinc finger/*Fok*I cleavage domain fusions. *Gene* **203,** 43–49.

50. Kim, Y. G., Smith, J., Durgesha, M., and Chandrasegaran, S. (1998) Chimeric restriction enzyme: Gal4 fusion to *Fok*I cleavage domain. *Biol. Chem.* **379,** 489–495.

51. Dervan, P. B. (1992) Reagents for the site-specific cleavage of megabase DNA. *Nature* **359,** 87,88.

52. Ebright, Y. W., Chen, Y., Pendergrast, P. S., and Ebright, R. H. (1992) Incorporation of an EDTA-metal complex at a rationally selected site within a protein: application to EDTA-iron DNA affinity cleaving with catabolite gene activator protein (CAP) and Cro. *Biochem.* **31,** 10,664–10,670.

53. Pendergrast, P. S., Ebright, Y. W., and Ebright, R. H. (1994) High-specificity DNA cleavage agent: design and application to kilobase and megabase DNA substrates. *Science* **265,** 959–962.

54. Shang, Z., Ebright, Y. W., Iler, N., Pendergrast, P. S., Echelard, Y., McMahon, A. P., Ebright, R. H., and Abate, C. (1994) DNA affinity cleaving analysis of homeodomain-DNA interaction: identification of homeodomain consensus sites in genomic DNA. *Proc. Natl. Acad. Sci. USA* **91,** 118–122.

55. Koob, M., Burkiewicz, A., Kur, J., and Szybalski, W. (1992) RecA-AC: single-site cleavage of plasmids and chromosones at any predetermined restriction site. *Nucleic Acids. Res.* **20,** 5831–5836.

56. Schoenfeld, T., Harper, T., and Slater, M. (1995) RecA cleavage and protection for genomic mapping and subcloning. *Promega Notes* **50,** 9–14.

57. Strobel, S. A. and Dervan, P. B. (1991) Single-site enzymatic cleavage of yeast genomic DNA mediated by triple helix formation. *Nature* **350,** 172–174.

58. Hung, L., Murray, E., Murray, W., Bandziulis, R., Lowery, R., Williams, R., and Noble, R. (1991) A blue/white cloning assay for quality control of DNA restriction and modifying enzymes. *Promega Notes* **33,** 12,13.

59. Murray, E., Singer, K., Cash, K., and Williams, R. (1993) Cloning-qualified blunt end restriction enzymes: causes and cures for light blue colonies. *Promega Notes* **41,** 1–5.

60. Turbett, G. R. and Sellner, L. N. (1996) Digestion of PCR and RT-PCR products with restriction endonucleases without prior purification or precipitation. *Promega Notes* **60,** 23–27.

61. *New England BioLabs 1998/99 Catalog* (1998) New England BioLabs, 258,259.

62. Dallas-Yang, Q., Jiang, G., and Sladek, F. M., (1998) Digestion of terminal restriction endonuclease recognition sites on PCR products. *BioTechniques* **24,** 582–584.

63. Moreira, R. F. and Noren, C. J. (1995) Minimum duplex requirements for restriction enzyme cleavage near the termini of linear DNA fragments. *BioTechniques* **19,** 57–59.

64. Sambrook, J., Fritsch, E. F., and Maniatis, T. (1989) Gel electrophoresis of DNA, in *Molecular Cloning, A Laboratory Manual,* 2nd ed., Cold Spring Harbor Laboratory, Cold Spring Harbor, NY, p. 65.

# 28

## Isolation and Characterization of an Unknown Restriction Endonuclease

### Raymond J. Williams

## 1. Introduction

Currently, there are approx 3000 restriction endonucleases known, recognizing 235 different sequences *(1)*. Although primarily found in bacteria, they also exist in archaea, viruses, and eukaryotes. An estimated 25% of bacteria examined contain at least one restriction endonuclease *(2)*, and therefore the probability of encountering new ones is relatively high. Indeed, many new enzymes have been "discovered" in contaminated bacterial cultures. The presence of three restriction activities in a single organism is not unusual. *Neisseria* strains appear to be particularly rich in restriction endonucleases and their corresponding methyltransferases. As many as seven different endonucleases from a single strain have been idenitified through cloning *(3)*. The first enzyme discovered which recognizes a unique sequence, although it may not be commercially available or commonly known, is designated the prototype. Although few new prototypes have been discovered recently, two potential four-base palindromes and seven potential six-base palindromes are not cleaved by any known restriction endonucleases. Most databases are arranged alphabetically by prototype, with isoschizomers listed under the prototype heading. A database of all known endonucleases, maintained by Dr. Richard J. Roberts, is available at http://www.neb.com/rebase. A number of formats are available, and references are provided. Detailed information on restriction endonuclease biology, classification, structure, specificity, and catalytic mechanism is provided elsewhere in this book (*see* Chapters 19, 27, and 29).

This chapter focuses on general guidelines for partially purifying and characterizing an unknown restriction activity. **Subheading 1.1.** outlines a plan to obtain sufficient quantity and purity of endonuclease for characterization. The degree of purity required for each characterization step is also explained, because in some cases, completing the purification plan may be unnecessary. **Subheadings 1.2.** and **1.3.** detail the assays for screening, partially optimizing reaction conditions, determining unit activity, mapping a recognition site, and determining the cleavage site of the unknown endonuclease. If a screening program is being designed, careful consideration should be given to the number of strains to be screened, the thoroughness of screening for each, the degree of characterization (and therefore the degree of purification) desired, and the time avail-

From: *Methods in Molecular Biology, vol. 160: Nuclease Methods and Protocols*
Edited by: C. H. Schein © Humana Press Inc., Totowa, NJ

able. The full purification and characterization of an unknown endonuclease described below will typically require 2–3 wk.

## 1.1. Purification

Initial screening for endonuclease activity may be done on crude lysates, and therefore requires only cell breakage. However, an endonuclease activity may occasionally be masked by high exonuclease activity in lysates. Once an activity has been demonstrated, the degree of purification needed is based on the degree of characterization desired and the ratio of the activity of the specific endonuclease of interest to contaminating activities (exonucleases, phosphatases, and other endonucleases). Careful analysis at multiple points in the purification may be needed to discriminate between star activity and more than one restriction activity. Clear and unambiguous DNA digest patterns, which can usually be obtained after one chromatography step, are needed to partially optimize reaction conditions, determine a unit activity, and map the recognition site. To determine the specific cleavage site, a second chromatography step is probably needed to sufficiently purify the enzyme.

### 1.1.1. Cell Growth, Cell Breakage, and Nucleic Acid Precipitation

For procedures, media, and references for the isolation, identification, and growth of strains, refer to **refs. 4–7**. Cell growth beyond late log phase should be avoided, as it may result in decreased endonuclease, increased exonuclease, and/or increased protease activity. Although 2–3 g of cell paste is enough to screen for activity from crude lysates, 25 g or more is recommended to complete the full purification and characterization. This allows for possible low endonuclease expression, low in-process stability, additional steps if there are stubborn contaminants, and/or the discrimination of possible star activity. A single buffer with varying salt concentrations is usually suitable for the entire purification. Although protease inhibitors, such as PMSF, generally do not inhibit restriction activity, their affect on increasing yield is usually minimal as well.

A number of methods may be employed to lyse the cells including sonication, mechanical (glass beads), pressure (French press, microfluidizer), and/or enzymatic-detergent lysis *(8,9)*. Mechanical and pressure methods are preferred, if a sufficient volume of cell paste is available, because of their ability to lyse resilient cell types and reproducibility over wide ranges of scale. Cell breakage must be monitored to assure maximum release of the enzyme and negligible shearing or denaturation caused by excessive processing.

Although not required, precipitating nucleic acids improves downstream chromatography steps. Nonspecific endonuclease binding to chromosomal DNA lowers absorption to cation exchange resins, and causes nucleic acids to compete with enzyme for binding to anion exchangers. The precipitant can then be removed, and protein can be concentrated by ammonium sulfate precipitation. Although some endonuclease activity will likely be lost in the process, significant amounts of protein and exonuclease activity will also be removed.

### 1.1.2. Chromatography

Cation exchangers are usually the first chromatography step, as they generally provide good capture, purification, and recovery for all endonucleases. Most of the protein

(>80% of the load) and exonuclease activity will be found in the column flowthrough and wash. Because of their broad applicability and high recovery, standard weak or strong anion exchangers are generally the best option for a second column. About one-third of the protein loaded and significant exonuclease activity will again be found in the column flowthrough and wash.

### 1.2. Screening, Determining Activity, and Mapping the Recognition Site

Screening for activities, defining the pattern of a complete digest, optimizing reaction conditions, and determining the unit activity for a particular substrate all employ the basic method described in **Subheading 1.6.1.** of Chapter 27. Modifications are detailed in **Subheading 3.**

#### 1.2.1. Screening

Restriction activity is detected as distinct DNA bands which cannot be attributed to uncut (or nicked if plasmid) substrate or genomic DNA of the strain. In screening for infrequent cutters, use multiple DNA substrates to increase the probability that at least one will contain a recognition site(s). For frequent cutters, use mulitple enzyme dilutions and one DNA substrate to increase the chances of obtaining a level of enzyme allowing distinction between bands and general smearing.

#### 1.2.2. Titrational Assays for Optimizing Reaction Conditions and Unit Activity

Once a restriction activity is detected, continue to purify the activity until clear band patterns are achieved. Different enzyme amounts are then tested to determine a range resulting in a consistent pattern indicating complete digestion. The lowest enzyme amount resulting in complete digestion is the assay endpoint. A partial optimization of reaction buffer is done by comparing the endpoints of four or five standard buffers. As much as a 10-fold increase in activity may be seen with buffers of different pH, ionic strength, and/or type of salt. Using the best reaction buffer, the unit activity is defined as the lowest amount of enzyme providing complete cleavage of 1 μg of substrate digested for 1 h in a 50 μL total volume and reported as U/μL.

#### 1.2.3. Mapping the Recognition Site with Multiple Substrates, Double Digests, and Partial Digests

First, establish the complete digest patterns of fully sequenced, commercially available substrates, such as λ, adenovirus 2, SV40, T7, φX174, pBR322, M13mp18, and pUC19. Estimate the number of cleavage sites. Increased intensity of a band may indicate that it actually consists of two distinct fragments of similar size. Tiny fragments may only be seen with the use of acrylamide gels. For frequent cutters, it may be virtually impossible to actually count bands. Methylation by the host used to prepare the substrate may protect recognition site(s) from cleavage and lead to underestimated site frequency. Therefore, a cleavable substrate, such as λ-prepared from wild-type and methylase deficient *Escherichia coli* strains, should be digested and compared.

The estimated site frequency of as many substrates as possible is compared to tables listing the number of recognition sites of known restriction endonucleases. Close matches can be compared directly in single and double digests to determine whether the unknown enzyme is an isoschizomer. If the unknown enzyme cannot be experimentally demonstrated as a isoschizomer, a similar approach of single and double digests is continued for mapping a recognition site(s) *(10)*. However, instead of looking for single and double digests giving exactly the same patterns, now the unknown enzyme is used with a series of known endonucleases in single and double digests to progressively narrow the region(s) containing the recognition site(s) of the unknown enzyme. Another technique, especially useful for frequently cleaving enzymes, is to end-label one strand of a DNA containing multiple sites. The labeled DNA is then partially digested, electrophoresed, and compared with markers to define specific regions containing recognition sites *(10)*. Alternatively, use a substrate generated by PCR with an end-labeled primer and a of mix standard and methylated dNTPs. The random incorporation of methylated nucleotides will yield the same "partial" digestion pattern even with excess endonuclease *(11)*. The more sequences containing a site and the more narrowly mapped those sites are, the easier it is to search for homologies. Computer-aided searches using sequence analysis software such as that provided by GCG and DNASTAR are helpful to check proposed recognition site(s). It must be remembered that endonuclease recognition sites may be 3, 4, 5, 6, 7, or 8 base pairs in length and consist of palindromes, partial palindromes, interrupted palindromes, and nonpalindromes, even within the common Type II and Type II subclasses.

### 1.2.4. Impact of Star Activity on Mapping

The inherent property of some endonucleases to cleave with relaxed specificity under suboptimal conditions or excessive amounts of enzyme, also known as star activity (*see* Subheading 1.4.8. of Chapter 27), may complicate efforts to map the recognition site, and is easily confused with the possibility of a second endonuclease with lower activity. As purification progresses, the relative level of star activity may increase slightly, but it will always perfectly copurify with the peak of the enzyme being followed. Also, nearly all endonucleases susceptible to star activity will exhibit a significant increase of cleavage at noncognate sites with increasing glycerol concentration. Stimulating the suspect cleavage(s) by titrating glycerol into the reactions provides strong evidence for star activity over the possibility of a second endonuclease, as the activity of a second enzyme is unlikely to be increased by glycerol.

### 1.2.5. Impact of a Second Endonuclease on Mapping

Confusing and misleading digest patterns often result when a strain contains more than one restriction activity. As activity levels of the enzymes rarely differ by orders of magnitude, the apparent endpoint at early stages of purification is probably the result of complete digestion by more than one enzyme. The key factor distinguishing another endonuclease(s) is that digest patterns will change as purification proceeds and the activities begin to separate. Although separate activity peaks may be identified by titration assays of the fractions, typically there will be some degree of overlap for a given purification step. For near-complete double digestion at one peak, one endonuclease is in excess and the other is limiting. For the other peak, the excess and limiting enzymes are reversed, and therefore the near complete double-digest patterns are different. Although another endonuclease(s) activity may present initial problems in mapping the recognition site, it is usually not a concern in sequencing the cleavage site. Even partial purification is likely to greatly enrich the endonuclease of interest relative to others. Also, a sequence for determining the cleavage site may be chosen that does not contain the recognition site of the other activity.

### 1.3. Determining the Cleavage Site

Endonucleases recognizing 4, 6, or 8 base continuous palindromes always produce blunt ends or overhangs of two or four bases. Enyzmes recognizing degenerate or interrupted palindromes wherein the total number of base pairs (both specific and nonspecific) within the sequence is an odd number produce overhangs of one or three bases. Therefore, for these situations, determining the point of cleavage in one strand defines the cleavage point on the other. As confirmation, however, it is recommended to sequence the cleavage position in both strands. For Type IIs endonucleases, the recognition sites are nonpalindromic and may contain an even or odd number of base pairs. These enzymes may cleave between defined or degenerate nucleotides, with at least one strand cleaved outside of the recognition sequence. The products can be blunt ends or overhangs of an even or odd number of bases. Therefore, after determination of the cleavage position in one strand, the cut site of the complementary strand cannot be inferred, and must also be experimentally defined.

#### 1.3.1. Cloning a Site and Preparation of the DNA

The easiest approach to sequencing the actual location of cleavage for an unknown endonuclease is to clone a fragment containing a single such site. For infrequent or moderately frequent occurring recognition sites, the results of the mapping studies may be applied to produce a fragment suitable for cloning. In the case of frequent cutters, PCR or synthesis may be more useful alternatives. If there are no recognition sites for the endonuclease of interest naturally present in the vector, the single site in the insert provides an easy conformation of successful cloning. However, even multiple sites within the vector backbone will not interfere with sequencing the cut site as long as only one appears between the primer sites.

Plasmid and fragment are ligated, cells transformed, clones selected, and DNA prepared according to established methods *(12,13)*. The DNA is split into two aliquots— one for generating the standard sequencing ladders and one to be digested by the endonuclease to sequence the cleavage site. Since the DNA digested with the unknown endonuclease will be linear, it is helpful to make the DNA used for the standard ladder also linear, with a cut in the vector backbone to equalize the sequencing reactions. After restriction digestions have taken place, the DNA is purified before sequencing.

#### 1.3.2. Sequencing the Cleavage Site

The sequencing itself is best done by thermal-cycle sequencing with end-labeled primers utilizing the dideoxy chain-termination method. Commercial kits, such as *fmol*® DNA Sequencing System (Promega, Madison, WI) or Thermo Sequenase Cycle Sequencing kit (Amersham Pharmacia, Arlington Heights, IL), contain extensive protocols and all necessary reagents except isotope and primers. Thermal cycle sequencing requires little DNA, eliminates alkali denaturation of dsDNA templates, increases the stringency of primer binding, and decreases possible secondary structure of the template. For defining the cleavage site on one strand, four sequencing reactions generate the standard ladder using template cut in the vector backbone and the fifth is a runoff primer extension with only dNTPs (no dideoxy NTPs) of the plasmid restricted with the endonuclease of interest. The cleavage position will be indicated by where the runoff band matches the standard ladder. The actual cleavage site is then defined as between the template bases represented at this position and the next largest termination product. If using Taq DNA polymerase, such as in the *fmol*® DNA Sequencing System, *see* **Note 5** regarding the nontemplate directed addition of a nucleotide by this polymerase.

## 2. Materials

1. >25 g cell paste.
2. Purification buffer: 50 m$M$ Tris-HCl, pH 7.4 at 25°C, 10 m$M$ EDTA, 1 m$M$ dithiothritol (DTT), 10% glycerol (*see* **Note 1**).
3. 5 $M$ NaCl (or 3 $M$ KCl).
4. 50% solution polyethylenimine (Sigma, St. Louis, MO), diluted with distilled water to 5% and the pH brought to 7.3–7.6 with HCl.
5. Ammonium sulfate.
6. Dialysis tubing or ultrafiltration membrane (should be <15,000 molecular weight cutoff).
7. Cation exchanger Macro Prep High S (Bio-Rad, Hercules, CA).
8. Anion exchanger Q sepharose FF (Amersham Pharmacia).
9. Storage buffer: 10 m$M$ Tris-HCl, pH 7.4 at 25°C, 0.1 m$M$ EDTA, 1 m$M$ DTT, 50 m$M$ NaCl, 50% glycerol.
10. BSA (acetylated to ensure no contaminating activities).
11. Restriction endonuclease 4-CORE® and MULTI-CORE™ buffer packs (Promega) or similar reaction buffers (*see* **Note 2**). Composition at 1X concentration as follows:
    a. RBA: 6 m$M$ Tris-HCl, pH 7.5 at 25°C, 6 m$M$ NaCl, 6 m$M$ MgCl$_2$,1 m$M$ DTT.
    b. RBC: 10 m$M$ Tris-HCl, pH 7.9 at 25°C, 50 m$M$ NaCl, 6 m$M$ MgCl$_2$,1 m$M$ DTT.
    c. RBD: 6 m$M$ Tris-HCl, pH 7.9 at 25°C, 150 m$M$ NaCl, 6 m$M$ MgCl$_2$, 1 m$M$ DTT.
    d. MULTI-CORE: 25 m$M$ Tris-acetate, pH 7.8, at 25°C, 100 m$M$ K-acetate, 10 m$M$ Mg(acetate)$_2$, 1 m$M$ DTT.
12. DNA substrates (lambda, unmethylated lambda, Ad-2, T7, pBR322, SV40, φX174, M13mp18, and pUC19).
13. Nuclease-free water (for assays).
14. DNA markers (various size ranges).
15. 6X stop solution/loading dye (6X concentration): 25% glycerol, 50 m$M$ EDTA, 0.5% SDS, 0.2% Orange G.
16. Agarose.
17. Ethidium bromide (5 mg/mL).
18. Gel electrophoresis buffer (final concentration): 40 m$M$ Tris acetate, pH 7.9, at 25°C, 5 m$M$ sodium acetate, 1 m$M$ EDTA.
19. Agar$ACE$® enzyme for agarose digestion of gel-purified fragment (Promega).
20. pGEM cloning vector (Promega).
21. T4 DNA ligase w/10X buffer.
22. Gel-purified, amplified, or synthesized fragment containing a recognition site of the endonuclease of interest.
23. Competent cells.
24. Wizard® *Plus* Miniprep DNA Purification System (Promega).
25. SP6 and T7 Primers (Promega).
26. Gamma $^{32}$P ATP.

27. fmol DNA Cycle Sequencing system (Promega) or Thermo Sequenase Cycle Sequencing kit (Amersham Pharmacia). Both kits include T4 PNK for labeling primers, d/ddNTP mixes, buffers, and sequencing grade Taq or modified T7 DNA polymerase, respectively.
28. Four dNTPs for runoff primer extension.

## 3. Methods

### 3.1. Purification

1. Maintain temperature as close to 4°C as possible. Avoid foaming as it contributes to denaturation.
2. Resuspend cells on ice in 2–3 mL of purification buffer per gram of cells. The NaCl concentration (or KCl) should be appropriate for the step after cell breakage: 250 m$M$ if the optional nucleic-acid precipitation step is to be performed or 50 m$M$ if to be directly loaded to a column. Stir until a smooth suspension is obtained.
3. Break cells according to the volume of the resuspension and equipment available *(8,9)*. Monitor the process by comparing time points taken during breakage with the unbroken cell resuspension. Cell breakage should be stopped when a qualitative microscopic inspection shows an estimated 70–90% reduction in whole cells or when an absorbance at 660 nm of a 100-fold dilution with distilled water has been reduced 50–70%. Alternatively, nucleic-acid release can be monitored by absorbance at 260 nm of the lysate supernatant. However monitored, the rate of change will decrease as breakage proceeds. Additional processing may actually result in slight decreases in endonuclease activity. If no nucleic-acid precipitation is performed, centrifuge for ≥15 min at ≥12,000$g$ to remove cell debris and proceed to **step 6**.
4. Nucleic-acid precipitation is optional. If it is to be performed, a small-scale titration should be done each time cells are broken *(14)*. To seven tubes with 250 μL of cell lysate each, add 10, 15, 20, 25, 30, 35, and 40 μL of 5% polyethylenimine. Vortex twice for 5–10 s each, let sit on ice for 2–10 min, microfuge for 5 min, and measure the absorbance at 260 nm of 100-fold dilutions of the supernatants with distilled water. Based on the sample with the minimum A$_{260}$ value (the titration curve should be roughly parabolic), calculate the amount of polyethylenimine needed for the bulk of the cell lysate and add slowly with stirring over 10 min. Stir an additional 20 min and centrifuge for 30 min at ≥12,000$g$. The pellet should be firm, and the supernatant easily poured off.
5. Ammonium sulfate precipitation is primarily used to remove the polyethylenimine before chromatography. Add dry ammonium sulfate to 70% of saturation or 472 mg/mL of polyethylenimine supernatant. Stir until completely in solution and continue stirring for at least an additional 30 min. Centrifuge for 30 min at ≥12,000$g$, pour off the supernatant, and resuspend the pellet in a minimal amount of purification buffer. Dialyze or diafilter against purification buffer with 50 m$M$ NaCl until the conductivity is equivalent to that of fresh 50 m$M$ NaCl buffer.
6. Equilibrate a Macro Prep High S column with purification buffer at 50 m$M$ NaCl. Use a minimum of 1–3 mL of resin (double if no nucleic-acid precipitation) per 10 g of cells broken. Load the crude or clarified lysate and wash with 3–6 column volumes of equilibration buffer. Elute with a step gradient of buffer with 500 m$M$ NaCl. If more than one endonuclease is suspected, a linear gradient of 8–16 column volumes total and 50–500 m$M$ NaCl is recommended. Pool active fractions so as to recover >50% of the units eluting from the column and attempting to minimize exonuclease (*see* **step 5** in **Subheading 3.3.**). If another column is needed, dialyze until the conductivity is equivalent to that of the column equilibration buffer. If no second column is needed, proceed to **step 8**.
7. Equilibrate a Q sepharose FF column with purification buffer at 50 m$M$ NaCl. Use a minimum of 0.1–0.3 mL of resin per 10 g (with or without nucleic acid precipitation) of cells

broken. Load the dialyzed pool from the previous column and wash with 3–6 column volumes of equilibration buffer. Elute with a step gradient of buffer with 400 m*M* NaCl. If more than one endonuclease is suspected, a linear gradient of 8–16 column volumes total and 50–400 m*M* NaCl is recommended. Pool active fractions so as to recover >50% of the units eluting from the column and attempting to minimize exonuclease (*see* **step 5** in **Subheading 3.3.**). If additional purification is needed, *see* **Note 3**.

8. Dialyze into storage buffer. Add BSA directly to the enzyme pool before dialysis to result in a final concentration of 0.5 mg/mL based on the 60% volume reduction when the sample is dialyzed from the 10% glycerol purification buffer to the 50% glycerol storage buffer.

## 3.2. Screening for Activity

1. Keep enzyme containing fraction and reaction mixes on ice until ready to begin incubation. If using an aliquot of crude lysate as the source of enzyme, microfuge for 5 min to pellet cellular debris. Use Promega reaction buffer RBC at 1X concentration or similar buffer (*see* **Note 2**) for reaction mixes. If screening for rare cutters, use a minimum of two amounts of enzyme (1 and 5 μL) with each of two substrates (λ and Adenovirus-2). If screening for frequent cutters, use a minimum of five amounts of enzyme (0.5, 1, 2, 4, and 8 μL with a single substrate (λ).

2. Prepare reaction mixes for each DNA such that the mix contains reaction buffer at 1X concentration, 1 μg substrate, 0.1 mg/mL BSA, and nuclease free water to 45 μL for each reaction. Vortex briefly. For "*n*" reactions, make reaction mix for *n* + 2.

3. Aliquot 45 μL reaction mix to reaction tubes and add enzyme, saving one aliquot as a "no enzyme" (negative) control. A "no substrate" control of enzyme and water may be beneficial to identify genomic DNA when using crude lysates. Mix all reaction tubes briefly by "flicking" or pipeting, do not vortex.

4. Microfuge briefly, incubate for 1 h at the same temperature used for growing cells (*see* **Note 4**).

5. Stop with 10-μL 6X stop solution/loading dye, prepare a sample with DNA molecular weight markers if desired.

6. Heat tubes 15 min, 65°C, to assist SDS denaturation of protein and dissociate any complementary DNA overhangs.

7. Prepare agarose gel in electrophoresis buffer with a final concentration of 0.5–1.0 μg ethidium bromide per mL of gel. Alternatively, the gel can be poststained using 2 μg/mL of ethidium bromide in a minimal volume of electrophoresis buffer with gentle shaking for up to 30 min. Refer to **Note 5** for suggested agarose percentages.

8. Load 20–30 μL (0.33–0.5 μg DNA) onto gel, allow samples to settle for a minute, begin electrophoresis at 1.5–2X the normal run voltage for several minutes and then reduce. Electrophoresis voltage and time will vary considerably, but 125 V for 150 min are usually required for a full-size gel (approx 80 mm dye migration) or 80 V for 100 min for a mini-gel (approx 40 mm dye migration).

9. Visualize with a transilluminator, usually employing midrange (300 nm) UV light. Discrete DNA bands representing substrate cleavage that cannot be attributed to genomic DNA indicates an endonuclease activity (*see* **Note 6**). For a permanent record, photograph the gel with a Polarid camera or use a video imaging system.

## 3.3. Titration Assays for Partial Reaction Buffer Optimization, Unit Activity Determination, and Functional Purity Assessment

1. Purity must be sufficient (usually requiring one chromatography step) to see clear and unambiguous band patterns in the range of a complete digest to proceed. In addition to Promega reaction buffer RBC, prepare reaction mixes with at least three additional buff-

ers, such as Promega reaction buffers RBA, RBD, and Multi-Core or similar buffers (*see* **Note 2**) with a single substrate. Prepare serial dilutions of enzyme using 1X RBC with 0.5 mg/mL BSA. An initial dilution range of approx 16-fold is recommended and should be based on a preliminary assessment of digestion completeness in the screening assays. Repeating with different dilutional ranges may be necessary.

2. Prepare reaction mixes for each buffer such that the mix contains reaction buffer at 1X concentration, 1 μg substrate, 0.1 mg/mL BSA, and nuclease-free water to 45 μL for each reaction. Vortex briefly. For "*n*" reactions, make reaction mix for *n* + 2. Aliquot 45 μL of reaction mix into each reaction tube.

3. Add 5 μL of diluted enzyme per reaction, mix as above, and follow **steps 4–9** of **Subheading 3.2.** The assay may be repeated using a single buffer to optimize the incubation temperature if desired.

4. Using the assay conditions (buffer and temperature) giving the highest activity, repeat **steps 1–3** using smaller dilutional intervals to quantitate the unit activity concentration to ±20%. (*See* **Subheading 1.2.2.** for help in determining the endpoint of the assay.) Further optimization of buffers may be done if desired.

5. A variation of this assay based on excess enzyme and time is useful as a general assessment of functional purity and helpful in identifying star activity. Once a unit activity is established, various excess enzyme units are incubated for 16 h instead of 1 h. When ≥5 U in this "overdigest" assay gives a clear, complete digest pattern and no smearing or extra bands, the purity should be sufficient for determining the cut site.

## *3.4. Mapping the Recognition Site*

1. Refer to Subheading 3. of Chapter 27 for suggestions regarding units necessary to completely cut different substrates and buffer choices for double digests.

2. Follow the method detailed in Subheading 3.2. of Chapter 27 using as many of lambda, unmethylated lambda, adenovirus-2, T7, pBR322, φX174, SV40, M13mp18, and pUC19 as are available. Use an excess amount of endonuclease, such as 10 U/μg of DNA, to expect complete cleavage of all substrates.

3. Compare the experimentally estimated number of cleavage sites for each substrate with tables listing the same DNAs and known endonucleases. The New England BioLabs catalog contains a relevant table in the Reference Appendix section.

4. Purchase close matches. Digest a substrate(s) with the known endonuclease(s), the unknown endonuclease, and a double digest with both enzymes. If the enzymes are isoschizomers, all three digests will appear exactly the same. If not isoschizomers, the three digests should all be at least slightly different. Repeat with a different substrate if necessary.

5. If no isoschizomers can be identified, perform a series of digests with known enzymes, the unknown enzyme, and double digests of both to define regions containing the recognition sites of the unknown enzyme. As the recognition sites become more narrowly localized to regions of DNA, use of polyacrylamide gels may be helpful to separate smaller DNA fragments. Refer to **Note 5** for suggested acrylamide percentages. Partial digests of an end-labeled substrate may be preferable for frequently cleaving enzymes. Appropriately sized DNA markers are helpful. Look for homologies in the regions mapped to contain the recognition site of the unknown enzyme.

6. Test proposed recognition sites using software packages, such as GCG and DNASTAR, by comparing the site frequency and computer-generated fragment sizes with those determined experimentally for a particular substrate. Also test sequences which did not experimentally appear to contain sites.

### *3.5. Determining the Cut Site*

1. Clone a fragment containing one recognition site. The site must be between 30 and 200 base pairs from each end of the fragment. A fragment produced by digestion with known endonucleases and gel-purified is most easily obtained from the agarose by use of the Agar*ACE* enzyme *(15)*. Alternatively, a suitable fragment may be generated by PCR or oligonucleotide synthesis. Ligate the isolated fragment with any of the pGEM vectors followed by transformation, cell growth, and plasmid preparation according to established protocols *(12,13)*.

2. Divide the plasmid into two aliquots. Digest one with a known endonuclease having a single recognition site in the vector backbone to linearize the plasmid, and the other with the endonuclease of interest. Both aliquots are then purified for sequencing with the Wizard Miniprep Purification system following the manufacturer's recommendations. A minimum of 8 and 2 fmol respectively with a concentration of at least 0.5 nanomolar is necessary. For a 3-kb plasmid, this translates to 16 and 4 ng at 1 ng/μL. Considerably higher amounts and concentrations are easily achieved and dilutions may be necessary.

3. End-label the SP6 and T7 primers with $\gamma$-$^{32}$P ATP according to the manufacturer's protocol with the reagents provided in the *fmol* DNA Cycle Sequencing system or Thermo Sequenase Cycle Sequencing kit.

4. Perform 10 primer extensions, five with each primer (*see* **Note 7**). For each primer, four primer extensions are for the standard sequencing ladders using the four dd/dNTP mixes and template linearized by cleavage in the vector backbone. A fifth primer extension is a runoff using only dNTPs and the plasmid digested by the endonuclease of interest as a template.

5. Run a sequencing gel and record on film according to the manufacturer's suggestions. Run the five reactions for each primer in adjacent lanes. The runoff of DNA cleaved with the enzyme of interest will likely need to be diluted before loading, as all the label will appear in a single band instead of distributed throughout the ladder. Where the runoff extension digested with the endonuclease of interest lines up with the standard ladder indicates the base on the template that comprises the 5' end of the cleavage site. The band of the standard ladder corresponding to a termination product with one additional nucleotide indicates the base on the template that comprises the 3' end of the cleavage site (*see* **Note 8**).

## 4. Notes

1. Stability during purification and long-term storage is always a consideration in buffer design *(16–18)*. Although usually not an issue for the two-column purification described here, in-process stability for some endonucleases may be increased by use of additives, such as 5 m*M* MgCl$_2$ and/or detergent (for example, 0.01% Triton). Another stabilization technique is to add BSA to the fraction tubes of a column before elution such that the final BSA concentration will be 0.1 mg/mL for the full fraction. Some endonucleases may be stable for many months at 4°C in the purification buffer. However, for long-term stability, dialyze or dilute into storage buffer. Most endonucleases will retain full activity for more than 1 yr at –20°C in storage buffer.

2. Similar premade reaction buffers are available from most vendors of restriction endonucleases. If purchased from a vendor other than Promega, substitute the buffers that most closely resemble those described in **Subheading 2.**

3. If additional purification is needed or as an option for the second column, there are several choices, depending on the situation. Affinity resins with ligands, such as heparin or dyes often provide excellent purification, especially for separating multiple restriction activi-

ties, but each also has potential drawbacks. For heparin, ligand isoforms and concentration may be variable. Recovery may be low for heparin resins if the column is oversized. For dyes, some endonucleases exhibit activity losses resulting from the high-salt concentrations (>500 m$M$) required for elution and very broad, diluting elution profiles. Although expensive, DNA agarose may also be effective in removing exonucleases and phosphatases.

4. The pH of Tris buffers decreases with increasing temperature. Therefore, Tris-based reaction buffers with higher pHs at 37°C, such as RBC and RBD, are more likely to be in the optimal range for endonucleases requiring 50–65°C digestion temperatures.

5. Suggested agarose percentages for resolving DNA of various lengths is given in Note 12 of Chapters 22 and 27. Suggested nondenaturing polyacrylamide gel percentages for resolving dsDNA of various lengths is as follows *(13)*: 3.5% for 1000–2000 bp, 5.0% for 80–500 bp, 8.0% for 60–400 bp, 12.0% for 40–200 bp, and 20.0% for 6–100 bp.

6. Gel artifacts caused by DNA-binding proteins may be seen, especially when assaying crude lysates. They may sterically interfere with restriction activity and/or reduce the mobility and resolution of electrophoresis.

7. Several DNA polymerases will work satisfactorily for primer extension sequencing, including T7, T4, Tli, Pfu, and Klenow in addition to the sequencing-grade Taq and modified T7 used in the kits referred to above.

8. Be aware that *Taq* DNA polymerase (and the sequencing grade *Taq* polymerase used in the *fmol* DNA Cycle Sequencing system) can add one nontemplate directed adenine to the 3' end of the runoff extension depending on reaction conditions and the terminal nucleotide *(19)*. This can be avoided in a number of ways. Cycle sequencing may done by substituting a proofreading thermostable polymerase (and its buffer), such as Tli or Pfu polymerase, or adding the proofreading enzyme to the existing reaction in addition to sequencing grade Taq. Dye terminator (Perkin Elmer, Norwalk, CT) or [33]P terminator (Amersham Pharmacia Biotech) dideoxynucleotides will also prevent the addition of a nontemplate directed addition.

## Acknowledgments

The author would like to thank Michael Slater, Mark Klekamp, Doug Barron, and Erik Gulliksen for their review of the manuscript and many helpful suggestions.

## References

1. Roberts, R. J. and Macelis, D. (1998) REBASE—restriction enzymes and methylases. *Nucleic Acids Res.* **26,** 338–350.
2. Roberts, R. J. and Halford, S. E. (1993) Type II restriction endonucleases, in *Nucleases,* 2nd ed. (Linn S. M., Lloyd, S. R., and Roberts, R. J., eds.), Cold Spring Harbor Laboratory, Cold Spring Harbor, NY, pp. 35–88.
3. Stein, D. C., Gunn, J. S., Radlinska, M., and Piekarowicz, A. (1995) Restriction and modification systems of *Neisseria gonorrhoeae. Gene* **157,** 19–22.
4. *Bergey's Manual of Systematic Bacteriology,* vol. 1–4 (1984, 1986, 1989) (Holt, J. G., ed.), Williams and Wilkins, Baltimore, MD.
5. Murray, R. G. E., Wood, W. A., and Krieg, N. R. (1993) *Methods for General and Molecular Bacteriology* (Gerhardt, P., ed.), ASM Press, Washington, DC.
6. Atlas, R. M. (1996) *Handbook of Microbiological Media,* 2nd ed. (Parks, L. C., ed.), CRC Press, Boca Raton, FL.
7. Balows, A. (1991) *The Prokaryotes,* 2nd ed. Springer Verlag, New York.
8. Scopes, R. K. (1994) *Protein Purification, Principles and Practice,* 3rd ed. Springer-Verlag, New York, pp. 26–34.

9. Engler, C. R. (1994) Cell breakage, in *Protein Purification Process Engineering* (Harrison, R. G., ed.), Marcel Dekker, New York, pp. 37–55.

10. Danna, K. J. (1980) Determination of fragment order through partial digests and multiple enzyme digests. *Methods Enzymol.* **65,** 449–467.

11. Wong, K. K., Markillie, L. M., and Saffer, J. D. (1997) A novel method for producing a partial restriction digestion of DNA fragments by PCR with 5-methyl-CTP. *Nucleic Acids Res.* **25,** 4169–4171.

12. Ausubel, F. M., Brent, R., Kingston, R. E., Moore, D. D., Seidman, J. G., Smith, J. A., and Struhl, K. (eds.) *Current Protocols in Molecular Biology.* Wiley, New York.

13. Sambrook, J., Fritsch, E. F., and Maniatis, T. (1989) *Molecular Cloning, A Laboratory Manual,* 2nd ed., Cold Spring Harbor Laboratory, Cold Spring Harbor, NY.

14. Jendrisak, J. J. and Burgess, R. R. (1975) A new method for large scale purification of wheat germ DNA-dependent RNA polymerase II. *Biochemistry* **14,** 4639–4645.

15. Knoche, K., Selman, S., Kobs, G., Brady, M., and Knuth, M. (1995) Agar*ACE* agarose-digesting enzyme part two: applications testing for cloning, sequencing, and PCR. *Promega Notes* **54,** 14–19.

16. Schein, C. H. (1990) Solubility as a function of protein structure and solvent components. *Bio/Technology* **8,** 308–317.

17. Santoro, M. M., Liu, Y., Khan, S. M. A., Hou, L., and Bolen, D. W. (1992) Increased thermal stability of proteins in the presence of naturally occurring osmolytes. *Biochemistry* **31,** 5278–5283.

18. Volkin, D. B. and Klibanov, A. M. (1989) Minimizing protein inactivation, in *Protein Function: A Practical Approach* (Creighton, T. E., ed.), IRL Press, Oxford, pp. 1–24.

19. Magnuson, V. L., Ally, D. S., Nylund, S. J., Karanjawala, Z. E., Rayman, J. B., Knapp, J. I., et al. (1996) Substrate nucleotide-determined non-templated addition of adenine by *Taq* DNA polymerase: implications for PCR-based genotyping and cloning. *BioTechniques* **21,** 700–709.

# 29

## Gene Modification with Hapaxoterministic Restriction Enzymes

### *Easing the Way*

**Shelby L. Berger**

## 1. Introduction

Subcloning can be extraordinarily easy or extremely difficult. The simplest case requires cleavage of the desired fragment from the DNA in which it is located with one or two restriction enzymes, cleavage of the plasmid that will serve as the ultimate recipient with the same enzymes, and ligation of the fragment to the plasmid. In a somewhat more complicated version of simple cloning, the investigator uses the polymerase chain reaction to amplify the DNA to be subcloned. In this case, restriction sites can be incorporated into the primers used for PCR, generating products with the desired restriction sites located near the ends of the newly synthesized DNA. After cleavage of those sites, the fragment is ready for insertion into an appropriately cut vector. The major disadvantage of the method is the need for sequencing the amplified material, because even the best of the thermostable polymerases makes mistakes that might render the DNA useless for the anticipated application.

Difficulties arise when the DNA of interest is not bounded by unique restriction sites. Regardless of whether the DNA is excised as a fragment or amplified from uncut DNA with defined primers, subcloning requires a single piece of DNA bearing the desired sequence, and ends compatible with those of the vector. In the worst-case scenario, the desired fragment is extremely large, unmapped, and unsequenced. Then, cleavage with most enzymes will result in multiple fragments with ends in common. When the pieces are mixed with the cleaved vector and ligated, they can reassemble in an almost random manner. As every novice cloner knows, insertion of several fragments in the correct order and orientation into a vector, assuming the correct order is known, can be a formidable job. What can an experienced investigator do?

The simple answer is to use hapaxoterministic restriction enzymes, known as hapaxomers. Although the definition of a hapaxomer will be provided in detail, the reader should understand that hapaxomers are the enzymes that people avoid, those that produce fragments with ends that have overhangs of a defined size but with

From: *Methods in Molecular Biology, vol. 160: Nuclease Methods and Protocols*
Edited by: C. H. Schein © Humana Press Inc., Totowa, NJ

unspecified sequences. Once the nature of hapaxomers is understood, the novice cloner will be able to use these enzymes to mutate, modify, shuffle, rearrange, and reassemble genes and gene parts, without concern for the fact that the DNA segment to be manipulated may consist of many fragments.

## 1.1. Symmetry

With nearly 3000 Type II restriction endonucleases described, almost 200 of which are commercially available, it is not difficult to cleave DNA at or near any location of choice. The problems occur when the many fragments obtained must be sorted and reassembled. Obviously, if a plasmid is cleaved with *Bam*HI into many pieces, those fragments will become linked to each other indiscriminately during the subsequent ligation reaction; the possibility that the original structure will be recovered is minimal. The reason lies in the nature of the recognition and cleavage sites of most Type II restriction endonucleases. Type II enzymes identify a defined palindrome-like sequence in double-stranded DNA and cleave both strands at precise locations within the recognition site *(1)*. Palindromes such as the word "redder," or the sentence, "Able was I ere I saw Elba," or the phrase, "A man, a plan, a canal—Panama," are identical when read in the forward and the reverse directions, ignoring punctuation, capital letters, and the spaces between words. A restriction site like that for *Bam*HI, GGATCC does not conform to the definition, because it reads CCTAGG in the reverse direction. In the jargon of symmetry, a palindrome has a mirror plane at the center. With the first half of the palindrome given, e.g., r-e-d, and a mirror placed at the position of the *vertical dotted line* in red:der the entire word is generated. However, because of the rules for base pairing, the first three letters of the *Bam*HI site also specify the entire site, but with a difference; in the "mirror" one "sees" not the image in the reverse orientation, but a backward version of its complement.

The *Bam*HI recognition site has a classical element of symmetry, a dyad or twofold rotation axis. If one takes one-half of the site, namely GGA, and rotates it 180° about an imaginary axis positioned after the "A" between the strands of DNA, and inserts the missing nucleotides using the rules of base complementarity, the double-stranded recognition site in **Fig. 1** (*left*) is generated. Here, the bases are symmetrically distributed about the axis of rotation (*diamond*), as are the cleavage sites (*arrows*). Furthermore, the staggered ends generated by cleaving a symmetrical site symmetrically almost always retain a part of the symmetry of the parent sequence. The sequence of the recognition site determines the sequence of the overhangs.

With a slight extension of the definition of a dyad axis of symmetry, one can include the so-called interrupted palindromes, such as the restriction site for *Bst*XI. This restriction endonuclease recognizes the sequence CCANNNN↓NTGG and cuts at the position of the *arrow*. When the N's represent any nucleotide and are therefore unidentified, the *Bst*XI site and the *Bam*HI site share the same symmetry of sequence and cleavage sites noted here (**Fig. 1**, *right*) However, on reentry into the real world, when the N's are replaced with bases and the DNA is cut, the symmetry is flawed in a most important way: the sequence of the overhangs is unrelated to that of the recognition site; there need not be an element of symmetry. You will not find *Bst*XI sites in multiple cloning sites (also called polylinkers) because a fragment generated with the enzyme from one DNA sequence cannot in all likelihood be inserted into another

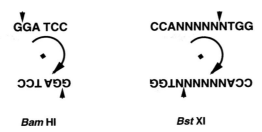

Fig. 1. The symmetry of recognition sites for estriction endonucleases using *Bam*HI and *Bst*XI as examples. (**Left**) An asymmetric unit designated GGA, in the example with *Bam*HI, is rotated 180° around an axis located between the strands of DNA. The same three bases now appear on the opposite strand. It should be understood that from left to right the bases will be read as AGG in their new location but, when the order is determined by the polarity of DNA, they remain GGA, read from 5' to 3', regardless of which strand is considered. Note that if the three bases occupy positions 1–3 on one strand, they will occupy positions 4–6 on the opposite strand. (**Right**) The symmetry of the site recognized by *Bst*XI, a restriction endonuclease that cleaves an "interrupted palindrome" within the recognition site. If the bases denoted N are ignored, the site is symmetrically equivalent to the *Bam*HI site. Arrowheads indicate the cleavage sites on both strands; diamonds indicate the dyad axes; and curved arrows show rotation about those axes. Note that a recognition and cleavage site on only one strand must be stipulated because of the existence of a twofold axis of symmetry.

sequence, cut with the same enzyme. That is, a *Bst*XI fragment can be ligated into the *Bst*XI-cut DNA from which it was derived, but not necessarily into any other *Bst*XI-cleaved DNA. Unlike the *Bam*HI fragments, which form random linear arrays when ligated, *Bst*XI fragments will rearrange themselves into the original order with high probability—they "remember" where they came from. The destruction of symmetry has created specificity and specificity, simplifies subcloning.

## 1.2. Hapaxoterministic Enzymes

Restriction endonucleases that generate unique ends are known as hapaxoterministomers or hapaxomers (2). The term "hapaxoterministomer" owes its origin primarily to the Greek as follows: "Hapaxo" is the expression used by classicists to designate a phrase or idea that occurs once in a subset of text; terministic, in both Latin and Greek means ended as in possessing ends; and *mer* means share. Thus, hapaxoterministomer connotes the *share* (of restriction enzymes) that produces *unique ends* in a *subset* (of DNA). To obtain unique termini, molecular biologists seek to minimize the sequence complexity of the DNA, namely, to limit the subset, while maximizing the size of the staggered ends.

### 1.2.1. Classification of Hapaxomers

Two classes of restriction enzymes contribute members to the hapaxomer group, the interrupted palindromes and the Type IIs enzymes, but all enzymes in these categories are not hapaxomers. Among the interrupted palindromes, only those with overhangs of three or more bases have been included. Enzymes that produce smaller overhangs are arbitrarily excluded, because, with only four bases from which to choose, the subset of

DNA would have to be drastically curtailed to obtain unique termini. The Type IIs enzymes need similar arbitrary constraints. These peculiar enzymes, *Fok*I and *Bsa*I, are examples, recognize sequences without symmetry, and precisely cleave the DNA asymmetrically outside of the recognition site. Again, the sequence recognized by the enzyme and that of the overhangs produced upon cleavage are unrelated. Those Type II enzymes that produce protrusions of three or more bases are designated as hapaxomers. **Table 1** lists commercially available hapaxomers, their recognition sequences, and the locations of the cut sites.

One enzyme listed in **Table 1** is literally in a class by itself—*Tsp*RI. This enzyme does not recognize an interrupted "palindrome," but an interrupted "palindrome" turned inside out. The five-base symmetric recognition site CA (C/G) TG is located in the center, with the C/G ambiguitity ignored by the dictates of symmetry, whereas single-strand cleavage sites are located on the opposite sides of the recognition site. The result is a 9-base 3' overhang with $4^1/_2$ unspecified bases that are not contiguous, when the ambiguity in the recognition site is taken into consideration. A bizarre arrangement, but no less a hapaxomer for that!

There is a new class of enzymes that contains one commercially available hapaxomer, *Bpl*I. These are the enzymes that excise their recognition sites from DNA by cleaving each strand of DNA twice: once on the 5' side of the recognition site and once on the 3' side. As displayed in **Table 1**, *Bpl*I excises a double-stranded DNA fragment of 32 bases with 3' unspecified overhangs of five bases.

### 1.2.2. Survey of the Termini Generated by Hapaxomers

Hapaxoterministic enzymes, unlike the classical restriction endonucleases, are unified not by the nature of their recognition sites, but by the rarity of the ends they create. That the recognition site does not determine the sequence at the ends is profoundly valuable. This means that a fragment can be ligated to the fragments that were its original neighbors, one on each side, but not to other fragments generated by cutting DNA with the same or a different enzyme. It also means that in synthetic DNA or in PCR DNA generated with synthetic primers, one can design a terminus capable of ligating to only one other fragment in a mixture. The idea of using the Type II enzymes, particularly *Hga*I with a five-base overhang, to regenerate cleaved structures from many fragments was recognized long ago, but was virtually ignored *(3,4)* because the yield of correctly ordered fragments was not quite 3% *(5)*. Twenty years later, the availability of many high-purity enzymes have converted hapaxomers into sophisticated tools for efficiently rearranging and reconstructing DNA.

At this point it should be clear that if the regions cleaved by hapaxomers are unspecified, the enzymes have the potential of generating ends with symmetry. For example, of the 256 possible staggered ends generated by *Bsa*I or *Bst*XI, any of the 16 symmetrical four-base sequences together with the 240 nonsymmetrical possibilities may be represented. The resurgence of symmetry means a return to some of the problems illustrated by the *Bam*HI fragments. However, the ability to generate symmetrical ends with hapaxomers can be also useful. DNA can be crafted and cleaved by hapaxomers to create three-, four-, or five-base 5' overhangs that are compatible with other overhangs of the same size or with such termini produced by any other enzyme. You could, for example, design primers for PCR with a *Bsa*I sites, which when cleaved,

**Table 1**
**Commerically Available Hapaxomers**

| | | | |
|---|---|---|---|
| Alw NI | CAGNNN▼CTG<br>GTC▲NNNGAC | Bst XI | CCANNNNN▼NTGG<br>GGTN▲NNNNNACC |
| Bbs I | GAAGACNN▼<br>CTTCTGNNNNNN▲ | Dra III | CACNNN▼GTG<br>GTG▲NNNCAC |
| Bbv I | GCAGCNNNNNNNN▼<br>CGTCGNNNNNNNNNNNN▲ | Ear I | CTCTTCN▼<br>GAGAAGNNNN▲ |
| Bgl I | GCCNNNN▼NGGC<br>CGGN▲NNNNCCG | Fok I | GGATGNNNNNNNNN▼<br>CCTACNNNNNNNNNNNNN▲ |
| Bsa I | GGTCTCN▼<br>CCAGAGNNNNN▲ | Hga I | GACGCNNNNN▼<br>CTGCGNNNNNNNNNN▲ |
| Bsl I | CCNNNNN▼NNGG<br>GGN▲NNNNNCC | Mwo I | GCNNNNN▼NNGC<br>CGN▲NNNNNCG |
| Bsm AI | GTCTCN▼<br>CAGAGNNNNN▲ | Pfl MI | CCANNNN▼NTGG<br>GGTN▲NNNNACC |
| Bsm BI | CGTCTCN▼<br>GCAGAGNNNNN▲ | Sap I | GCTCTTCN▼<br>CGAGAAGNNNN▲ |
| Bsm FI | GGGACNNNNNNNNNN▼<br>CCCTGNNNNNNNNNNNNNN▲ | Sfa NI | GCATCNNNNN▼<br>CGTAGNNNNNNNNN▲ |
| Bsp MI | ACCTGCNNNN▼<br>TGGACGNNNNNNNN▲ | Sfi I | GGCCNNNN▼NGGCC<br>CCGGN▲NNNNCCGG |
| Bst API | GCANNNN▼NTGC<br>CGTN▲NNNNACG | Tsp R I | NNCACTGNN▼<br> G<br>▲NNGTGACNN<br>   C |

| | |
|---|---|
| Bpl I | ▼NNNNNNNNGAGNNNNNCTCNNNNNNNNNNNNN▼<br>▲NNNNNNNNNNNNNCTCNNNNNGAGNNNNNNNN▲ |

---

would produce an overhang of AATT. Since *Eco*RI also generates AATT overhangs, the fragment could be ligated into an *Eco*RI cut-vector. Clever examples with a *Fok*I site have appeared *(6,7)*.

There are no blunt-end cutters of any type among the hapaxomers. The reason is that, with respect to reassembling fragments in the correct order, all blunt ends are alike despite the differing sequences at their termini. Put bluntly, blunt ends behave promiscuously in ligation reactions.

## 2. Materials

To use hapaxomers, one must procure the enzyme together with the correct buffer and use it at the proper temperature. An excellent compendium of restriction endonucleases and their recognition and cleavage sites can be found in The Restriction Endonuclease Database (REBASE) maintained by New England Biolabs, Inc. at http://www.neb.com/rebase.

Since protocols with hapaxomers make use of standard methods of cleaving, purifying, ligating, and cloning DNA, no specific techniques for these procedures will be presented. However, there are a few kitchen tricks that some might find useful:

1. To reassemble an array of designed single-stranded, phosphorylated oligomers into double stranded DNA, dissolve them at a concentration of ~0.1 μM in 100 mM Tris-HCl at pH 7.4 and 20 mM $MgCl_2$. Heat briefly at 90°C, and cool slowly over a period of hours to allow hybridization to occur. Ligate in 50 mM Tris-HCl at pH 7.4, 10 mM $MgCl_2$, 10 mM dithiothreitol (DTT), 1 mM ATP, and 10 U of T4 DNA ligase for ~16 h at ~15°C.
2. Use PCR to amplify DNA. Because the technique is notoriously fickle, specific conditions cannot be recommended. However, if the template has never been used for PCR and the fragments to be synthesized are ~ 500 bp, the suggested starting conditions are 1–2 μM of each primer, 1 ng of template, and 1 cycle consisting of 94°, 55°, and 72°C for 30 s, 15 s, and 1 min, respectively, followed by the same temperatures for 15 s, 15 s, and 1 min, respectively, for 49 cycles. At completion, incubation for 10 min at 72°C is suggested. A DNA thermocycler and the accessories included in any kit for PCR are required.
3. Modify conditions for PCR if one strand of a double-stranded fragment is to be used as a primer for PCR in the presence of its complementary strand. A reduction in the concentration of the double-stranded DNA to 1/20 the concentration of a single-stranded primer is recommended. Since single-stranded primers are usually used at a concentration of ~0.2 μM, the double-stranded DNA "primer" in the reaction should be ~0.05 μM.

## 3. Use of Hapaxomers to Solve Practical Problems in Molecular Biology

Hapaxomers and the hapaxoterministic concept can be applied to fabricating, manipulating, and rearranging DNA. Whereas PCR, together with synthetic oligomers bearing the desired mutations such as substitutions, additions, or subtractions, has the capability of generating any sequence of interest, difficulties remain in achieving the proper termini for subsequent ligation. If the new piece must replace an existing stretch of DNA in a very large segment of a gene, and if there are few, widely spaced, unique sites for insertion, the investigator will amplify long stretches of unchanged sequence adjacent to the sequence scheduled for alteration in order to bridge the gaps between user-friendly restriction sites. Then, the old, duplicated material as well as all of the newly minted matter must be sequenced to verify the fidelity of the polymerases. If you enjoy sequencing DNA and are set up to do 20 kb before breakfast, you will not need simpler methods. For everyone else, the aim is to create almost anything with fragments that are small enough to sequence by the most old-fashioned, nonautomated methods. On a good day, you may not need to sequence at all!

### 3.1. Preparation of Modified DNA

Two methods for obtaining modified DNA for use in subcloning projects will be described, both of which require enzymatic ligation either to assemble an array of fragments or to integrate the new material into existing constructs. In the first method (*see* **Subheading 3.1.1.**), oligonucleotide modules that are prepared synthetically are combined to build larger structures, whereas in the second method (*see* **Subheading 3.1.2.**), the modified DNA is a product of the polymerase chain reaction.

### 3.1.1. An Approach Without a Hapaxomer

This techniques utilizes oligomers that are hybridized and ligated to form double-stranded DNA in a method similar to laying bricks in a wall. The concept is illustrated in **Fig. 2** for eight fragments, and was used almost 30 yr ago to synthesize a gene for tRNA *(8)*. DNA fragments of hundreds of bases can be prepared by first assembling subfragments from synthetic oligomers, and then ligating the subfragments into larger fragments *(9)*. Because no enzyme, with its accompanying limitations, is involved in the formation of the oligomers, the internal "overhangs" can be generous, one-half the length of the oligomer, if necessary. The ultimate termini of the fully assembled fragment can be constructed to be compatible with those of any piece of DNA.

1. Obtain the complete sequence of the fragment to be synthetically generated.
2. Divide the sequence on paper into single-stranded fragments that are 30–40 bases long. By consulting **Fig. 2**, you will see that on one strand, the 3' end of one oligomer and the 5' end of the next should abut near the center of an oligomer on the complementary strand.
3. Synthesize the oligomers or obtain them commercially.
4. Phosphorylate the 5'-ends with T4 polynucleotide kinase and any convenient protocol.
5. Mix the fragments, hybridize, and ligate as specified in **step 1** of **Subheading 2.**

When a totally synthetic approach is employed to construct fragments, it may be unnecessary to sequence the final products. In general, the machines that synthesize DNA are at least as accurate as those that sequence it. Clearly, sequencing adds further certainty and should not be discouraged.

### 3.1.2. Use of Two Consecutive Polymerase Chain Reactions to Generate Mutant Fragments

When the sequence of a required piece of DNA is partly known, PCR may be the best method for manipulating it. **Figure 3** presents the use of two consecutive reactions to make a mutant fragment whose ultimate ends can be made compatible with those of any other fragment of DNA *(10)*.

1. Obtain an oligomer, called *mutant reverse* (*see* **Fig. 3**), which bears the desired mutation. Also obtain *primer forward*, which will define the upstream terminus of the finished product. Note that *primer forward* has a site for a hapaxomer denoted "H" located 5' to the region, which will hybridize with the template.
2. Obtain the template. The template sequence need not be known, with the exception of the regions that must hybridize with the primers and the oligomer.
3. Generate the subfragment, using PCR to amplify the correct DNA sequence. Instructions appear in **Subheading 2., step 2**. The subfragment contains the mutation near the downstream end and the recognition site for the chosen hapaxomer on both strands at the upstream end, hence, the enclosed H.
4. Gel-purify the subfragment.
5. Obtain fresh template (the purified subfragment from **step 4**) and *primer reverse*, which defines the downstream boundary of the fragment to be produced. Note that the

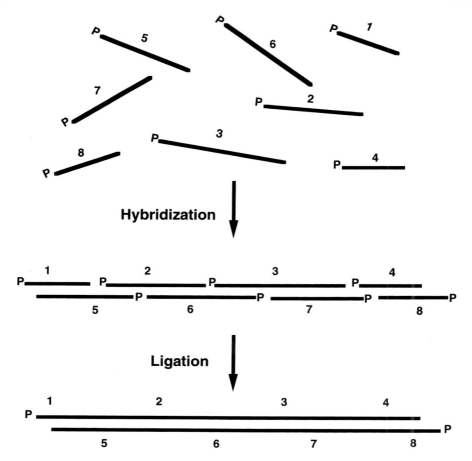

Fig. 2. Eight oligomers with 5' phosphates are assembled into a large fragment of DNA. With the exception of the 5' end fragments (oligomers 1 and 8), each oligomer hybridizes to two oligomers that constitute the opposite strand. After ligation, the nicks are sealed and an intact double-stranded fragment of DNA is obtained.

subfragment, upon denaturation, will contribute the forward primer for PCR in **step 6**. The complementary strand of the subfragment, although present, does not participate owing to the absence of a suitable primer in the forward direction. However, because the strands of the subfragment can hybridize to each other, the instructions in **step 3** of **Subheading 2.** should be consulted. *Primer reverse* in this example, bears a recognition site for the same hapaxomer used in *primer forward*.

6. Perform a second round of PCR. The result of polymerization is a blunt-ended fragment with an internal mutation and recognition sites for restriction enzymes at either end.
7. Cleave with the hapaxomer that recognizes the sites in the primers and clone into the desired vector (*see* **Notes 1** and **2**).
8. Sequence all PCR products to ensure fidelity.

### 3.2. Reassembly of Plasmids and the Substitution of Fragments

With powerful tools available for creating new or modified DNA, the problem of integrating that DNA into preexisting structures must be confronted. In general, the

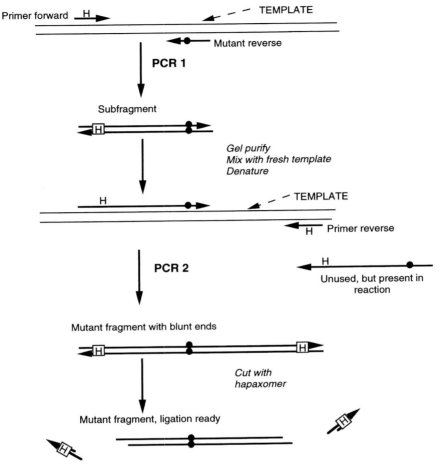

Fig. 3. A mutant fragment is synthesized using two rounds of the polymerase chain reaction. In the first round, a primer bearing the mutation, symbolized by ●, together with a second primer targeted upstream on a given template, is used to synthesize a mutant subfragment. In the second round, one strand of the mutated subfragment, together with a primer targeted downstream on the same template is used to synthesize a double-stranded, blunt-ended mutant fragment with hapaxoterministic restriction sites at both ends, denoted H. It is not necessary to remove the extraneous strand of the mutated subfragment from the reaction. Primer forward and primer reverse include single-stranded recognition sites for hapaxomers, denoted *H* (without the box). It should be clear that the mutant primer could have been oriented in the forward direction and used with primer reverse in the first PCR. Then, primer forward would be introduced into the second PCR.

goal here is to replace an existing fragment in a plasmid with an engineered one. The engineered fragment may have undergone insertions, deletions, or mutations. However, its termini must be identical to those of the fragment it has been fashioned to replace. The design plan that follows is an example of how one might proceed. It includes cleavage of the plasmid with a hapaxomer, a determination of whether the unmodified fragments will reassemble properly in high yield, and suggestions for reas-

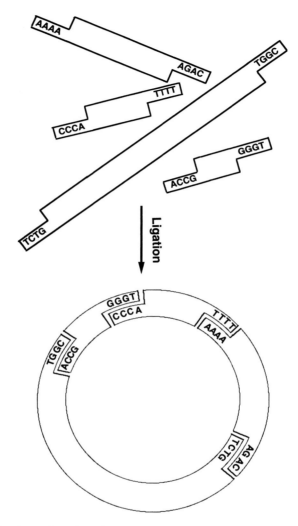

Fig. 4. Reconstitution of a plasmid by ligation after cleavage into four pieces with a hapaxomer, such as *Bst*XI, which generates four-base 3' protruding ends. Note that the pattern of assembly is unique.

sembly with altered fragments. An example in which a plasmid is cleaved with a hapaxomer and reconstituted is presented in **Fig. 4**. Reactions like this embody the essence of the hapaxoterministic concept.

1. Cleave the parent plasmid with a hapaxomer that isolates the targeted sequence in a fragment of <400 bp. Obviously, the smaller the fragment to be created using the techniques illustrated in **Figs. 2** and **3**, the easier subsequent manipulations will be. Cleavage with two hapaxomers to achieve the desired size is also an option.

2. Divide the digest into equal portions. Ligate one-half of the sample and label it #1; reserve the remainder as sample #2. Using a spin column or an ethanol precipitation, recover the DNA from both samples and dissolve them in separate tubes in a transfection-compatible buffer.

3. For comparison, cleave a fresh aliquot of the parent plasmid, with any enzyme that cuts only once. If the hapaxomer generates 5' overhangs of four bases, for example, a good choice would be a restriction enzyme that also generates similar overhangs of four bases. Ligate, exchange the ligation buffer for the transfection buffer (*see* **step 2**), and label it sample #3. Some investigtors are inclined to reserve an equal portion as an unligated control in the same buffer—sample #4.

4. Transfect bacteria with equal amounts of DNA from each of the three or four samples. In an experiment of this type with an 8-kb plasmid containing four sites for the hapaxomer, *Pfl*MI, we obtained 75 colonies from sample #2 (background) and 26,000 colonies from sample #1, using 1 ng DNA. The comparable results with *Hin*dIII, as the single-base cutter, were ~6000 colonies as background (sample #4) and 19,000 colonies from sample #3 *(10)*. (*Pfl*MI generates three base overhangs, but in the actual plasmid, there were no convenient unique sites that also generate three base overhangs. Instead, *Hin*dIII with a four-base overhang was used.) These results are typical; with the plasmid multi-cut with *Pfl*MI, the background should be much lower than that using the unique *Hin*dIII site, and the yield of reassembled plasmids should be higher, because each fragment produced by a hapaxomer can ligate only to fragments with which it was once contiguous. In contrast, the *Hin*dIII-cut DNA will never be cleaved to completion, and the linearized plasmid can form tandem arrays in the head-to-tail, head-to-head, and tail-to-tail orientations.

If the system reassembles properly with native DNA as a control, namely sample #1 generates many, many, more clones than sample #2, it should reassemble with modified DNA. Proceed as follows:

1. Cleave the parent DNA as above.
2. Add an engineered fragment of DNA made as in **Subheading 3.1.** This DNA will compete with a native fragment for a place in the finished product.
3. Ligate and clone.
4. Select for the mutant. For ease in selection, it is wise to generate a mutant whose pattern of restriction fragments differs from that of the parent construct. Then, DNA from separate clones can be analyzed to find the desired plasmid. If a new restriction site cannot be inserted or if an old one cannot be deleted, one might have to sequence in order to select the correct construct.

Although this strategy should be effective, it has not behaved as expected in practice (*see* **Note 3**).

An alternative strategy, more time-consuming but more likely to succeed, follows:

1. Cleave the parent plasmid as above.
2. Separate the fragments on a gel, recover the pieces that are to be retained, and discard the fragment to be replaced. Use low melting-temperature agarose for the matrix, if desired, and recover the fragments by any convenient means. In some cases, a single piece of gel can be retrieved with several fragments.
3. Mix the recovered fragments in equimolar amounts with an equimolar amount of the engineered fragment.
4. Ligate and clone. If the volume is small and the DNA concentration is large, ligations can be performed efficiently in low melting-temperature agarose without first isolating the DNA (*see* **Note 4**).
5. Select for the mutant.

If the desired construct if not obtained, *see* **Note 5**, which provides a strategy for trouble shooting, and **Note 6**, which modifies the strategy.

## 3.3. The Addition of Sequence Information to Genes

The use of hapaxomers can greatly facilitate the addition of a new sequence to an existing gene. The system requires genomic DNA, preferably cloned or mapped, a double-stranded blunt-ended fragment containing the new sequence for insertion, and the use of two different Type II hapaxoterministic enzymes that produce overhangs of the same type and length. Since these two hapaxomers will fashion the ends of the insert, the choice of one of them depends on the sequence of the gene near the point at which new DNA will be introduced. The investigator is free to choose the other enzyme. The example in **Fig. 5** displays a scheme for inserting a sequence coding for an epitope tag immediately upstream of the stop codon.

1. Find a recognition site for a hapaxomer upstream of the TAG stop codon. The closest site should be selected if it has a six-base recognition site that produces a four-base overhang. If there is no such site, look for a five-base recognition site with a four-base overhang. *Hga*I with a five-base overhang is generally not useful because there are no other hapaxomers that generate five-base 5' protrusions. At first glance, *Sap*I, with a seven-base recognition site and a three-base overhang, looks attractive, but the only other enzyme producing three-base protrusions is *Ear*I, which has as its recognition site a subset of the *Sap*I site. Hence, *Ear*I cleaves all *Sap*I sites, making it impossible to use the enzymes independently (*see* **Note 7**).

2. Cleave the DNA with the chosen enzyme (**Fig. 5, Part I**). Purify the fragment immediately upstream of the stop codon. It will serve as the attachment site for the engineered DNA. Reserve all remaining fragments separately or as a mixture. In the illustration in **Fig. 5, Part I**, the gene is cleaved with *Bsp*MI (underlined sequence) into several fragments, two of which are shown in the vicinity of the cut site. Set the genomic DNA aside.

3. Prepare the double-stranded, blunt-ended fragment for insertion by any convenient means. The fragment can be generated from oligomers or with the aid of PCR, but must have the following properties:
   a. A restriction endonuclease site (*Bsp*MI in the illustration) capable of generating an overhang which will be compatible with the overhang (boldface) of the isolated upstream fragment (here 5'-ATCA on the insert which will hybridize with genomic 3'-TAGT);
   b. The codons or parts of codons downstream of the genomic cleavage site with the stop codon omitted;
   c. The codons for the epitope in frame;
   d. A new stop codon, here TAA; and
   e. A site for a different restriction endonuclease that, when cleaved, will generate a downstream terminus compatible with the downstream genomic fragment shown in the illustration, namely 3'-TAGT which will hybridize with genomic 5'-ATCA.

   It is essential that the two hapaxoterministic enzymes used to cut the insert be different and preferably independent so that neither has the capabiity of cleaving the site of the other. An acceptable example is shown in **Fig. 5, Part II** with recognition sites for *Bsp*MI and *Bbs*I, both of which are underlined.

4. Cleave with the hapaxomer that recognizes the site upstream of the rebuilt codons and epitope tag, and eliminate the snipped end. In the illustration, the enzyme is *Bsp*MI and the ousted fragment contains its recognition site. There is no need to choose the same enzyme as that used to cleave the gene, but there is no reason to avoid it. Do not cleave with the downstream hapaxomer; this would create a fragment capable of forming multimers.

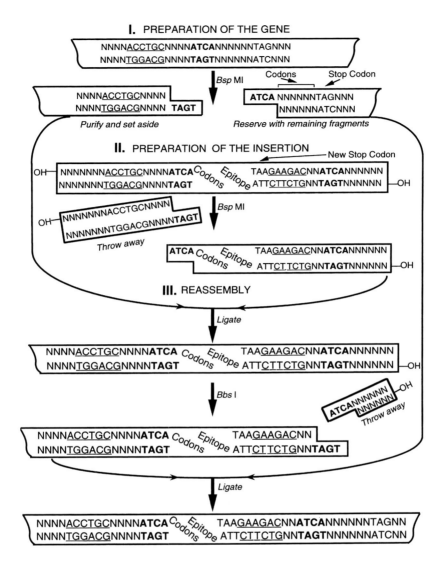

Fig. 5. Construction of a gene coding for a protein attached to an epitope tag at the carboxyl terminus. (**I**) The gene is cleaved by a hapaxomer, *Bsp*MI in the example, at a conveniently located site. The fragment immediately upstream of the cleavage site is purified; it will serve as the attachment site for the engineered DNA in (**II**). All other fragments are retained and reserved either separately or as a mixture. (**II**) A blunt-ended synthetic DNA containing the mutation, here an epitope tag, is cleaved to generate ends compatible with the upstream fragment. The snipped end is discarded. Note that the requirement is for compatible ends; the nature of the enzyme, here *Bsp*MI, is not important. (**III**) In the first step of reassembly, the upstream genomic fragment and the fragment bearing the epitope tag are ligated. Only one type of ligation event can occur, because the upstream end of the genomic fragment (not shown) and the downstream end of the mutant fragment cannot ligate to any DNA in the reaction mix. In the second step, the downstream end of the mutant fragment is cleaved, here with *Bbs*I, to be compatible with the genomic fragment located immediately downstream of the cleavage site shown in (I). The snipped end is discarded. It is essential that the two recognition sites in the mutant fragment are different. In the final step, all remaining fragments are added and ligated to reconstitute the gene. All fragments have 5'-phosphates, except where noted otherwise.

5. Reassemble the gene (**Fig. 5, Part III**). The process must be performed in three steps, the first of which is to ligate the reserved fragment of the gene, designated as the attachment site for the insertion, with the upstream protrusion of the prepared insert. Note that the downstream end of the insert is blunt and unphosphorylated to prevent self-ligation.
6. Cleave the newly-fused fragment with the hapaxomer that recognizes the downstream site, *Bbs*I in **Part III**). Discard the clipped end.
7. Add all remaining fragments of the gene in equimolar amounts and ligate.

With the successful assembly of the fragments into the correct order, one obtains a reconstituted gene containing the original sequence, added codons for the epitope, a new stop codon, a duplication of codons removed from the upstream genomic fragment by the cleavage in **step 1** of **Subheading 3.3.**, and the original stop codon. Since the duplicated material is downstream of the new, upstream stop codon, it will be irrelevant to the final protein. Note that in this scheme, a *Bbs*I site has been introduced into the engineered gene, thereby creating a cassette between the *Bsp*MI and *Bbs*I sites. This region can easily be replaced with a sequence coding for a different epitope, for example. Furthermore, only the insert (at most) must be sequenced, because all other fragments remained unmanipulated and unchanged.

## 4. Notes

1. The key to successful incorporation of the mutant fragment into large DNA lies in the design of *primer forward* and *primer reverse*. The choice of hapaxoterministic recognition sites residing at the ends of the blunt-ended fragment reflects two possibilities: the sites may appear in the template and, of necessity, on *primer forward* and *primer reverse* as well, or they may be selected by the investigator from among those in **Table 1**. Choose the first option if a fragment of the template bounded by hapaxomers is to be replaced by the PCR DNA product as discussed in **Subheading 3.2.** Choose the second option to make the ends compatible with any other staggered end.

   It is important to be aware that the Type II recognition sites have two orientations. For example, an *Ear*I site written as in **Table 1** will direct cleavage downstream of the recognition site but, if rotated by 180°, the *top* strand would read ↓NNNNGAAGAG-3', the bottom strand would read ↓NCTTCTC-5', and the cleavage would occur upstream of the site. Here then, is yet another advantage conferred by avoiding symmetry: It is possible to create seamless fusions of mutant and partner DNA without introducing new material at the junctions. However, because the recognition sites for the cleaving enzyme(s) in **Fig. 3** are oriented in a way that departs with the snipped debris, there is a disadvantage: ligation could be forever.

   Enzymes selected sites from those in **Table 1** can be used for insertion of the PCR DNA into a polylinker site. One might have incorporated sites for an enzyme such as *Bsp*MI into both *primer forward* and *primer reverse* to generate *Eco*RI ends, or *Hin*dIII ends or, for directional insertion into a polylinker, one of each. In the last case, one enzyme, here *Bsp*MI, could generate both staggered ends simultaneously. Unsuspected *Eco*RI or *Hin*dIII sites in the fragment, which could derail conventional subcloning by generating several fragments, would not be a concern because the enzymes that cleave them have been bypassed. In contrast, if there are unsuspected *Bsp*MI sites, those fragments will only be capable of ligating to each other to form contiguous DNA as it was in the template. There is no need to modify the cloning strategy to accommodate putative unsuspected *Bsp*MI sites.

2. There are limitations when using hapaxomers. All of the commercially available Type II hapaxomers leave 5' protruding ends. All of the commercially available hapaxoterministic, interrupted palindromes produce 3' protruding ends. Thus, sequences with 5'-overhangs can be joined perfectly, whereas sequences with the 3' protrusions generated by an interrupted palindrome retain a telltale half-recognition site among the newly fashioned fragments. With the discovery of additional enzymes, it may be possible to escape from these limitations because there is no *a priori* reason for them. Exceptions include *Bst*F5I, which cleaves 5'-GGATGNN↓ and 5'-NN↓CATCC to make a two-base 3' overhang, and *Tth*III, which cleaves the interrupted palindrome GACN↓NNGTC to generate a one-base 5'-overhang. Neither enzyme is a hapaxomer, because the overhangs are too small.

3. Reassembly of native DNA is not identical to assembly of native DNA to which an engineered fragment has been added. When engineered DNA is introduced into a mixture of cleaved fragments of the parent plasmid, there are two fragments which must compete for a place in the finished construct—the original DNA and the engineered piece. One might think that by adding an excess of the engineered DNA, a yield of the new construct in excess of that of the reassembled parent would be assured. I have not found this to be true; the new DNA seems to be at a competitive disadvantage. Although the cause is unclear, the proposed method has obvious ramifications: the fragments are no longer equimolar. Specifically, if fragments *A*, *B*, *C*, and *D* are to be reassembled into a circle, and if *M* represents excess mutant fragment designed to replace *C*, the products will include *M* ligated to both of its allowed partners to form *ABM* and *MD*. *ABM* and *MD* cannot ligate to one another to form *ABMD*. Because C is still present, *ABC* and *CD* will also be in the reaction. In the absence of *M*, the original fragments will be strictly equimolar so that there will be a *D* for every *ABC* product in the reaction and an *AB* partner for every *CD*. When an excess of any fragment has been introduced, recalling that *C* + *M* represents an excess of DNA slated for the position between *B* and *D*, the correct amount of the required partners will not be produced. One would expect abortive incomplete structures. If this were the only problem, the yield of correct constructs, with either *C* or *M* would be reduced, but those with *M* should be as likely as those with *C*, assuming *M* was equivalent to *C* (it is not). We obtained about 2% of the mutant construct in such an experiment.

4. There are three steps in ligation reactions, all of which depend on ATP. In the first step, the ligase reacts with ATP to form an adenylated enzyme intermediate. An ATP concentration of 0.1–0.4 m$M$ is chosen because it is sufficiently higher than the value of $K_m$ (10–50 µM) to drive the reaction. In the second step, an adenylate residue is transferred from the enzyme to each of the fragments. In this reaction, the *amount* rather than the *concentration* of ATP in the mixture is important because one adenylate residue is required for each 5' end to be ligated. In the last step, free enzyme performs the ligation. If the investigator has used low melting-temperature agarose, the volume probably exceeds 250 µL, the *amount* of ATP at the preferred concentration far exceeds the number of fragments, and there will be no free enzyme; ATP remaining after all 5' ends have been activated is inhibitory. The problem is resolved by using a huge excess of *ligase*, so a negligible percentage of free enzyme will drive the ligation reaction regardless of the concentration of ATP.

5. If reassembly with engineered DNA fails, seek evidence for missing fragments by attempting to reconstitute the reserved purified parent fragments, including the DNA which one planned to discard. A very poor yield of colonies after transfection is a strong indication of missing material.

6. A more laborious approach, which involves fewer fragments but more manipulation, requires cleavage with a hapaxomer, isolation of the targeted fragment, further cleavage of the isolated fragment with the second hapaxomer to generate a smaller segment of DNA, and elimination of the piece designated for replacement. Again, in order to recon-

stitute the plasmid, all of the fragments of the original plasmid with the exception of the discarded DNA must be recombined with the replacement fragment and ligated.

7. To use *Ear*I and *Sap*I as the hapaxomers in projects designed to add a fragment to a gene, the investigator would have to use *Ear*I, the less specific of the two enzymes, to cleave genomic DNA and *Sap*I to create the downstream protruding end of the insert, namely, the penultimate step of the scheme in **Fig. 5**. Thus, the more restrictive enzyme, *Sap*I, must be used under conditions of lesser sequence complexity, whereas the less restrictive enzyme is required to cleave highly complex genomic DNA.

## References

1. Roberts, R. J. and Halford, S. E. (1993) Type II restriction endonucleases, in *Nucleases* (Linn, S. M., Lloyd, R. S., and Roberts, R. J., eds.), Cold Spring Harbor Laboratory Press, Plainview, NY, pp. 35–88.
2. Berger, S. L. (1994) Expanding the potential of restriction endonucleases: use of hapaxoterministic enzymes. *Anal. Biochem.* **222,** 1–8.
3. Brown, N. L. and Smith, S. (1977) Cleavage specificity of the restriction endonuclease isolated from Haemophilus gallinarum (Hga I). *Proc. Natl. Acad. Sci. USA* **74,** 3213–3216.
4. Roberts, R. J. (1978) Restriction endonucleases, in *Microbiology* (Schlessinger, D., ed.), American Society for Microbiology, Washington, DC, pp. 5–9.
5. Moses, P. B. and Horiuchi, K. (1979) Specific recombination *in vitro* promoted by the restriction endonuclease *Hga*I. *J. Mol. Biol.* **135,** 517–524.
6. Szybalski, W. (1985) Universal restriction endonucleases: designing novel cleavage specificities by combining adapter oligodeoxynucleotide and enzyme moieties. *Gene* **40,** 169–173.
7. Mandecki, W. and Bolling, T. J. (1988) FokI method of gene synthesis. *Gene* **68,** 101–107.
8. Agarwal, K. L., Buchi, H., Caruthers, M. H., Gupta, N., Khorana, H. G., Kleppe, K., et al. (1970) Total synthesis of the gene for an alanine transfer ribonucleic acid from yeast. *Nature* **227,** 27–34.
9. Ivanov, I. G. (1990) Shotgun concatenation of synthetic genes: construction of concatemeric human calcitonin genes. *Anal. Biochem.* **189,** 213–216.
10. Berger, S. L., Manrow, R. E., and Lee, H. Y. (1993) Phoenix mutagenesis: one-step reassembly of multiply cleaved plasmids with mixtures of mutant and wild-type fragments. *Anal. Biochem.* **214,** 571–579.

# 30

# Techniques to Measure Nucleic Acid-Protein Binding and Specificity

*Nuclear Extract Preparations, DNase I Footprinting, and Mobility Shift Assays*

**Richard A. Rippe, David A. Brenner, and Antonio Tugores**

## 1. Introduction

DNA–protein interactions are the basis for the molecular mechanisms responsible for nucleic-acid replication, gene transcription, recombination, viral integration, and gene regulation in both normal and pathophysiological conditions. Initially, deletional analysis coupled with reporter gene assays are used to roughly map the locations of regulatory regions involved in gene expression. Relevant cis-regulatory elements may also be detected as areas of DNase I hypersensitivity in native chromatin. DNase I footprinting analysis is subsequently used to precisely locate sites of DNA–protein interactions within the identified regulatory regions. The proteins that interact with these regulatory regions can be detected and isolated using various modifications of the mobility shift assay. The protocols described below have been successfully used in our laboratory to analyze the regulatory regions of several genes *(1–6)*.

The first step in the analysis of DNA–protein interactions is to isolate nuclear extracts; however, it is sometimes more convenient to obtain whole-cell extracts (*see* **Note 1**). To isolate nuclear extracts, cultured cells or tissues are incubated in a hypotonic buffer, which weakens the integrity of the outer cellular membrane. The outer cellular membrane is then mechanically disrupted, using a dounce homogenizer, or treated with a detergent, such as NP-40, leaving the nucleus intact. The nuclei are then isolated by centrifugation and the nuclear membrane is disrupted using high-salt conditions. Although desalting is generally not required for most applications, the salt may be removed by dialysis or using one of several commercially available desalting columns according to the manufacture's recommended protocol. We have described protocols for preparing nuclear extracts from cultured cells (**Subheading 3.1.1.**) and tissue (**Subheading 3.1.2.**). Our protocol was developed for liver, but can be adapted to other tissues with minor modifications.

From: *Methods in Molecular Biology, vol. 160: Nuclease Methods and Protocols*
Edited by: C. H. Schein © Humana Press Inc., Totowa, NJ

The DNase I footprinting assay is used to determine where specific protein interactions occur on DNA, and is particularly useful for identifying specific DNA sites involved in regulating gene transcription. DNase I footprinting is based on the premise that protein binding protects the phosphodiester backbone of the DNA from DNase I hydrolysis. A protein extract is incubated with an end-labeled DNA fragment followed by partial digestion with DNase I. The reaction products are separated on a denaturing polyacrylamide gel of the type used for DNA sequencing. Since protein binding protects the labeled probe from DNase I digestion, the absence of DNA cleavage bands indicates the presence of protein binding activity. Radiolabeled probes, described in **Subheading 3.2.1.**, are stable for several weeks, but as probes age, the specific activity decreases and detection of weak DNA–protein interactions may become more difficult. A Maxam-Gilbert G + A chemical sequencing reaction of the labeled probe (**Subheading 3.2.2.**). Is run on the same resolving gel in order to determine the precise locations of DNA–protein interactions. This marker is stable for several weeks. When using a previously prepared marker, be sure to account for radioactive decay before loading the sample on the gel. We have used this footprinting technique to locate sites of DNA–protein interactions in several regions of the $\alpha1(I)$ collagen gene using extracts obtained from collagen producing NIH 3T3 fibroblast cells and from hepatic stellate cells (HSCs) *(4)* (**Fig. 1**). The HSC is the predominant cell in the liver responsible for excess collagen production during fibrosis.

PCR-based DNase I footprinting provides an alternative method for the detection of protein binding to DNA. This procedure is more experimentally laborious and time-consuming, and should not be attempted unless absolutely necessary. Situations suitable for PCR footprinting include working with promoters that contain a very high G + C content, which may cause problems when the conventional approach is taken for footprinting analysis, or when protein binding to a nonlinear DNA substrate is being studied. In the PCR-based footprinting assay, nuclear proteins are bound to a supercoiled plasmid template, which is then treated with DNase I. It is important to use clean preparations of the plasmid template for optimal results (*see* **Note 2**). Specific cleavage products are detected by performing primer extension on the denatured DNase I digested DNA and using a specific $^{32}$P-labeled oligonucleotide as a primer (**Subheading 3.3.1.**). Several different primers may be needed, depending on the size of the region to be examined (*see* **Note 3**). Using a thermostable DNA polymerase for the primer extension products allows several cycles of elongation to better detect the produces, and also ensures efficient progression through a G + C-rich region. A standard chain termination sequencing reaction is performed using the same primer on the plasmid template to provide a sequence marker of the region being examined. This allows for the precise identification of the protected region(s). Alternatively, a radiolabeled DNA ladder (i.e., *Msp*I digested pBR322) may be used to determine footprint locations. The PCR-based footprinting technique has been used in our laboratory to analyze protein binding to the human ferrochelatase gene promoter (**Fig. 2**).

The mobility shift assay is used to assess factors affecting the ability of proteins to bind to specific sites on DNA. This assay is based on the ability of a protein to retard the migration of a labeled DNA fragment through a nondenaturing polyacrylamide gel. A small radiolabeled DNA fragment is incubated with a nuclear extract (or other protein preparation), and the reaction products are separated in a nondenaturing gel where

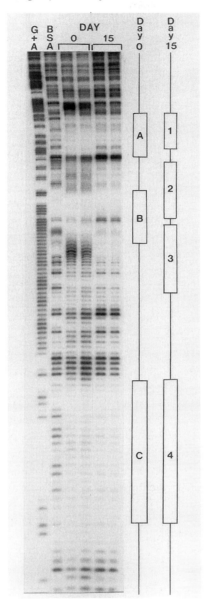

Fig. 1. DNase I footprinting analysis of proteins binding to the α1(I) collagen gene promoter in nuclear extracts (**Subheading 2.1.**) obtained from quiescent hepatic stellate cells (HSCs) (d 0) or culture-activated HSCs (d 15). The noncoding strand of the α1(I) collagen promoter region (–220 to +116) was radiolabeled at the 5' end (as per **Subheading 3.2.1.**) and incubated with 20 μg of BSA (lane 2) as a control for the DNase I digestion pattern or with 25 μg of nuclear extract from either freshly isolated HSCs (lanes 3 and 4) or from HSC cultured for 15 d and on plastic (lanes 5 and 6) to detect specific regions of DNA–protein interactions. The Maxam-Gilbert G + A sequencing reaction (**Subheading 3.2.2.**) is shown in lane 1, which is used to determine the specific sites of DNA–protein interactions on the DNA. Protected regions are indicated to the right of the footprint reactions. This figure has been previously published *(4)*, and is reproduced with permission from the publisher (WB Saunders Company, Orlando, FL).

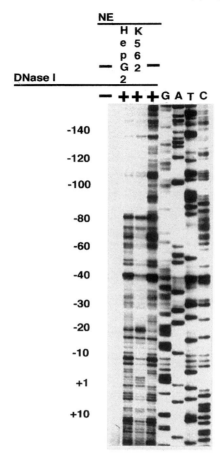

Fig. 2. PCR-based DNase I footprinting of the human ferrochelatase gene promoter. Plasmid DNA containing the HFC promoter fragment (–367 to +55) was incubated with 20 µg of nuclear extracts (**Subheading 3.1.**) from either K562 or Hep G2 cells and subjected to DNase I digestion as described in the text (**Subheading 3.3.**). The resulting digestion products were detected by primer extension from the reverse primer using *Taq* DNA polymerase. Sequencing reactions (using the Sequenase Method, USB, Cleveland, OH) were performed, using the same plasmid and primer, and simultaneously run in the gel to determine the precise locations of DNA–protein interactions. This figure has been previously published (*6*), and is reproduced with permission from the publisher (The American Society for Biochemistry and Molecular Biology, Inc.).

DNA–protein complexes migrate slower through the gel than the probe alone. The mobility shift assay can be used to optimize the binding conditions of the protein to DNA by systematically altering KCl, NaCl, or $Mg^{2+}$ concentrations and pH. The specificity of protein binding can be assessed using competition assays, or chemically modified probes, as in the methylation interference assay. Alternatively, mutated DNA fragments as either labeled probes or as cold competitors in the binding assays can identify specific nucleotides required for protein binding to a particular DNA sequence. The identity of the DNA-binding protein can be determined using specific antibodies in a modification of the mobility shift assay—the "supershift assay" (*see* **Subheading**

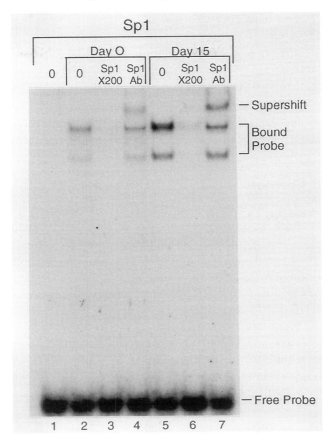

Fig. 3. Mobility shift assay showing increased binding activity of Sp1 following HSC activation. A radiolabeled double-stranded oligonucleotide (20-mer) (**Subheading 3.4.**) containing a consensus Sp1 binding site was incubated with 5 µg of nuclear extract obtained from either freshly isolated, quiescent HSCs (lanes 2–4) or with nuclear extracts obtained from culture-activated HSCs (lanes 5–7). Lane 1: probe incubated without nuclear extract. Lane 3 and 6, a 200-fold molar excess of unlabeled Sp1 oligonucleotide was used as a competitor in the binding reaction. A Sp1 specific MAB (PEP-2; Santa Cruz Biotechnology, Inc., Santa Cruz, CA) was included in the binding reactions in lanes 4 and 7, which resulted in a supershifted band as indicated. This figure has been previously published *(4)*, and is reproduced with permission from the publisher (WB Saunders Company, Orlando, FL).

**3.4.3.** and **Note 9**). Here, the antibody may block complex formation or slow the migration of the DNA–protein complex in the nondenaturing gel. The size of the protein can also be estimated by crosslinking the protein to the labeled probe and separating the complexes in standard SDS-polyacrylamide gels (described in **ref. 10**). Finally, the mobility shift assay can quantitatively determine the abundance of a particular protein in an extract, determine affinity binding of a protein to a particular binding sequence, and establish association- and dissociation-rate constants for the DNA-binding protein. We have used the mobility shift assay to demonstrate that HSC activation is accompanied by an increase in Sp1 binding activity *(4)* (**Fig. 3**).

The assay we describe uses a constant amount of probe CPM in the binding reaction, which facilitates the determination of exposure times. However, to determine protein-binding kinetics, it is better to use known amounts of probe DNA in the binding reaction (typically 5 pmol of probe is used). Radiolabeled probes can be prepared by two methods, kinasing (**Subheading 3.4.1.1.**) or by filling-in overhangs (**Subheading 3.4.1.2.**) and are stable for several weeks. When using previously prepared probes, be sure to account for radioactive decay, especially when using competing oligonucleotides, to assure proper molar ratios in the binding reactions.

## 2. Materials
### 2.1. Specialized Materials

The materials used for these assays are found in most laboratories; however, some special supplies used are worth mentioning. For preparation of nuclear extracts from tissue, a 55-mL Potter homogenizer, a Teflon pestle, and a Wheaton overhead stirrer to drive the Teflon pestle will be needed. Centrifuge tubes to fit SW28 rotor buckets (Beckman, polyallomer, Reference #326823) as well as Oak Ridge tubes (Nalgene, Reference #3431-2526) will be needed for the ultracentrifuge rotors SW28 and Ti55.2, respectively. A platform rocker will also be used. For the DNase I footprinting and mobility shift assay techniques, standard gel electrophoresis equipment is used. For DNase I footprinting, gels measuring $20 \times 42$ cm with 0.4-mm spacers are used, while mobility shift assays require gels with dimensions of $165–200 \times 200$ mm with 1.0–1.5-mm spacers. All electrophoresis equipment, including the 2000 V power supply, can be obtained from CBS Scientific Company, Inc., Del Mar, CA.

### 2.2. Reagents
#### 2.2.1. Preparation of Nuclear Extracts

1. Stock solutions:

| Solution | Storage | Sterilization |
|---|---|---|
| 1 $M$ HEPES, pH 7.6 | RT | Autoclave |
| 2 $M$ KCl | RT | Autoclave |
| 0.2 $M$ EDTA | RT | Autoclave |
| 1 $M$ spermidine | –20°C | |
| 0.5 $M$ spermine | –20°C | |
| 0.1 $M$ EGTA | RT | Autoclave |
| 1 $M$ dithiothreitol (DTT) | –20°C | |
| 100 m$M$ PMSF (in isopropanol) | –20°C | |
| 1 $M$ Na$_2$MoO$_4$ | RT | |
| 50% glycerol | RT | Autoclave |
| 1 $M$ MgCl$_2$ | RT | Autoclave |
| 4 mg/mL aprotinin | –20°C | |
| 4 mg/mL leupeptin | –20°C | |
| 4 mg/mL bestatin | –20°C | |

2. Add stock solutions of DTT, sodium molybdate, and the protease inhibitors PMSF, aprotinin, leupeptin, and bestatin, just before using the buffer.
3. Dignam A buffer: 10 m$M$ HEPES, pH 7.6, 1.5 m$M$ MgCl$_2$, 10 m$M$ KCl, 0.5 m$M$ DTT, 0.5 m$M$ PMSF, 2 µg/mL aprotinin, 0.5 µg/mL leupeptin, 40 µg/mL bestatin, and 10 m$M$ Na$_2$MoO$_4$. Make Dignam A buffer fresh before each use and keep on ice.

4. Dignam C buffer: 10 m$M$ HEPES, pH 7.6, 25% glycerol, 420 m$M$ NaCl, 1.5 m$M$ MgCl$_2$, 0.5 m$M$ DTT, 0.5 m$M$ PMSF, 2 µg/mL aprotinin, 0.5 µg/mL leupeptin, 40 µg/mL bestatin, and 10 m$M$ Na$_2$MoO$_4$. Make Dignam C buffer fresh before each use and keep on ice.

5. Final concentrations of phosphatase inhibitors for all buffers: 10 m$M$ PNPP (para-nitrophenylphosphate) (make a 2-$M$ stock solution), 20 m$M$ β-glycerolphosphate (make a 1-$M$ stock solution), 10 m$M$ sodium orthovanadate (for preparation, boil a 0.2-$M$ stock solution until in becomes translucent, pH to 10.0). Phosphatase inhibitors can be obtained from Boehringer Mannheim (Indianapolis, IN). Store all stock solutions at –20°C.

6. 10% NP-40 (Sigma, St. Louis, MO) in water.

7. Homogenization buffer: 0.3 $M$ sucrose, 10 m$M$ HEPES, pH 7.6, 10 m$M$ KCl, 0.75 m$M$ spermidine, 0.15 m$M$ spermine, 0.1 m$M$ EDTA, 0.1 m$M$ EGTA. This solution may be stored at –20°C. Immediately before use, adjust to 1 m$M$ DTT, 0.5 m$M$ PMSF, 2 µg/mL aprotinin, 0.5 µg/mL leupeptin, 40 µg/mL bestatin, 10 m$M$ Na$_2$MoO$_4$.

8. Cushion buffer: 2.2 $M$ sucrose, 10 m$M$ HEPES, pH 7.6, 10 m$M$ KCl, 0.75 m$M$ spermidine, 0.15 m$M$ spermine, 0.1 m$M$ EDTA, 0.1 m$M$ EGTA. This solution may be stored at –20°C. Immediately before use, adjust to 1 m$M$ DTT, 10 m$M$ Na$_2$MoO$_4$, 0.5 m$M$ PMSF, 2 µg/mL aprotinin, 0.5 µg/mL leupeptin, 40 µg/mL bestatin.

9. Nuclear storage buffer (NSB): 25% glycerol, 50 m$M$ HEPES, pH 7.6, 3 m$M$ MgCl$_2$, 0.1 m$M$ EDTA. Immediately before use, adjust to 1 m$M$ DTT, 0.1 m$M$ PMSF.

10. Nuclear lysis buffer (NLB): 10% glycerol, 10 m$M$ HEPES, pH 7.6, 100 m$M$ KCl, 3 m$M$ MgCl$_2$, 0.1 m$M$ EDTA.

11. Storage reconstitution buffer (SRB): 7.5% glycerol, 2 m$M$ HEPES, pH 7.6, 116 m$M$ KCl, 3 m$M$ MgCl$_2$, 0.1 m$M$ EDTA. Immediately before use, adjust to 1 m$M$ DTT, 0.1 m$M$ PMSF, 10 m$M$ Na$_2$MoO$_4$, 2 µg/mL aprotinin, 0.5 µg/mL leupeptin, 40 µg/mL bestatin.

12. Nuclear dialysis buffer (NDB): 20% glycerol, 20 m$M$ HEPES, pH 7.6, 0.1 $M$ KCl, 0.2 m$M$ EDTA. Immediately before use, adjust to 2 m$M$ DTT, 0.1 m$M$ PMSF, 10 m$M$ Na$_2$MoO$_4$, 2 µg/mL aprotinin, 0.5 µg/mL leupeptin, 40 µg/mL bestatin.

## 2.2.2. Standard DNase I Footprinting Assay

1. Plasmid DNA, which contains the region to be assessed for DNA-protein interaction.
2. Restriction endonucleases to excise the region of interest.
3. 10X restriction endonuclease buffer.
4. 3 $M$ sodium acetate, pH 5.2.
5. 10X reverse transcriptase buffer: 0.5 $M$ Tris-HCl, pH 8.3, 0.5 $M$ NaCl, 50 m$M$ MgCl$_2$, 20 m$M$ DTT.
6. Reverse transcriptase (20 U M-MLV; 10–20 U AMV)
7. [$^{32}$P]α-dCTP, 3000 Ci/mmol (or other labeled base, depending on the sequence of the overhang created by the restriction endonuclease).
8. 100 m$M$ EDTA.
9. 5 $M$ ammonium acetate.
10. 75 and 100% ethanol.
11. Agarose.
12. Gel loading dye: 0.25% bromophenol blue, 0.25% xylene cyanol FF, 30% glycerol.
13. 10X TAE buffer: 0.4 $M$ Tris-acetate, 0.01 $M$ EDTA.
14. Radiolabeled double-stranded DNA probe (**Subheading 3.2.**).
15. 2% formic acid diluted in water (store at room temperature; stable for 6 mo).
16. 10 mg/mL salmon sperm DNA (Boehringer Mannheim, cat. no. 1 467 140).
17. Formamide dye: 95% formamide, 20 m$M$ EDTA, 0.05% bromophenol blue, 0.05% xylene cyanol FF.

18. 5 mg/mL poly[d(I-C)] dissolved in water (Boehringer Mannheim; cat. no. #219 847).
19. 10% polyvinyl alcohol.
20. 1 $M$ MgCl$_2$.
21. Reaction buffer: 20 m$M$ HEPES, pH 7.6, 100 m$M$ KCl, 0.2 m$M$ EDTA, 2 m$M$ DTT, 20% glycerol.
22. Nuclear extract (*see* **Subheadings 2.1.** and **2.2.**)
23. 10 mg/mL BSA.
24. 1 mg/mL stock solution of DNase I (Worthington Biochemical Corporation) prepared in 0.15 $M$ NaCl, 50% glycerol. Store at –20°C. This solution must always be kept cold!
25. HEPES/CaCl$_2$ solution: 10 m$M$ HEPES, pH 7.6, 5 m$M$ CaCl$_2$.
26. Stop solution: 200 m$M$ NaCl, 20 m$M$ EDTA, 1% SDS, 50 µg/mL yeast tRNA.
27. Tris-HCl saturated phenol (Boehringer Mannheim).
28. Chloroform: isoamyl alcohol (24:1).
29. Formamide dye: 95% formamide, 20 m$M$ EDTA, 0.05% bromophenol blue, 0.05% xylene cyanol FF.
30. Standard 8% acrylamide/8 $M$ urea sequencing gel.
31. 10X TBE: 89 m$M$ Tris-base, 89 m$M$ boric acid, 0.02 $M$ EDTA.
32. Gel fix solution: 10% acetic acid, 10% methanol.

## 2.2.3. PCR-Based DNase I Footprinting Assay

1. 25 pmol of oligonucleotide.
2. 10X T4 Polynucleotide kinase buffer: 700 m$M$ Tris-HCl, pH 7.5, 100 m$M$ MgCl$_2$.
3. 1 $M$ DTT.
4. T4 polynucleotide kinase.
5. [$^{32}$P]γ-ATP (6000 Ci/mmol).
6. 100 m$M$ EDTA, pH 8.0.
7. 5 $M$ ammonium acetate (filter sterilize).
8. 10 mg/mL yeast tRNA.
9. 75 and 100% ethanol.
10. Nuclear extracts (**Subheading 3.1.**).
11. Nuclear dialysis buffer (NDB): 20% glycerol, 20 m$M$ HEPES, pH 7.6, 0.1 $M$ KCl, 0.2 m$M$ EDTA.
12. 5 mg/mL poly [d(I-C)] in 10 m$M$ Tris-HCl, pH 7.5 (Boehringer Mannheim; cat. no. 219 847).
13. HEPES/CaCl$_2$ buffer: 10 m$M$ HEPES, pH 7.6, 5 m$M$ CaCl$_2$.
14. 10 mg/mL DNase I (dissolved in 0.15 $M$ NaCl, 50% glycerol, and stored at –20°C).
15. Stop buffer: 200 m$M$ NaCl, 20 m$M$ EDTA, 1% SDS, 50 µg/mL yeast tRNA.
16. Phenol:chloroform (24:1).
17. Recombinant AmpliTaq DNA polymerase (Perkin Elmer Cetus).
18. PCR mixture: 50 m$M$ KCl, 10 m$M$ Tris-HCl, pH 8.3, 2 m$M$ MgCl$_2$, 100 µ$M$ each of the four deoxyribonucleoside triphosphates.
19. 10X TBE buffer: 89 m$M$ Tris base, 89 m$M$ boric acid, 0.02 $M$ EDTA.
20. 20% acrylamide solution (29:1) in 1X TBE, 50% urea.
21. 1X TBE, 50% urea solution.
22. Formamide-loading dye: 95% formamide, 20 m$M$ EDTA, 0.05% bromophenol blue, 0.05% xylene cyanol FF.

## 2.2.4. Mobility Shift Assay

1. Single-stranded or double-stranded oligonucleotide.
2. 10X kinase buffer: 0.5 $M$ Tris-HCl, pH 7.5, 0.1 $M$ MgCl$_2$, 0.05 $M$ DTT.

3. T4 polynucleotide kinase.
4. 10 mg/mL yeast tRNA.
5. 5 *M* ammonium acetate.
6. 75 and 100% ethanol.
7. [$^{32}$P]γ-dATP; 6000 Ci/mmol (or 3000 Ci/mmol).
8. 10X reverse transcriptase buffer: 0.5 *M* Tris-HCl, pH 8.3, 0.5 *M* NaCl, 50 m*M* MgCl$_2$, 20 m*M* DTT.
9. 5 m*M* dNTP mix: Dilute 100 m*M* stock dNTP solution in water and prepare fresh each time. Use each of the bases in the nucleotide mix, except the radiolabeled base.
10. [$^{32}$P]α-dCTP, 3000 Ci/mmol.
11. Reverse transcriptase (20 U M-MLV; 10 U AMV) (Gibco-BRL).
12. 10 mg/mL yeast tRNA.
13. 100 m*M* EDTA.
14. 5 *M* ammonium acetate.
15. 75 and 100% cold ethanol.
16. 30% acrylamide solution (29:1).
17. 0.4X TBE (1X TBE: 89 m*M* Tris-base, 89 m*M* boric acid, 2 m*M* EDTA).
18. 10% ammonium persulfate.
19. TEMED.
20. Radiolabeled double-stranded oligonucleotide probe (or short DNA fragment).
21. 5 mg/mL poly[d(I-C)] (Boehringer Mannheim, cat. no. 219 847) dissolved in water.
22. Cold competitor DNA, either 100 ng/μL or 10 ng/μL (depending on the specific activity of the probe and the amount of competitor desired in the reaction).
23. Nuclear extract (**Subheading 3.1.**).
24. Dye mixture (10 m*M* HEPES, pH 7.6, 10% glycerol, 0.01% bromophenol blue).

## 3. Methods
### 3.1. Nuclear Extracts
#### 3.1.1. Nuclear Extracts from Cultured Cells

Adherent cells can be grown in tissue-culture dishes ranging from 24-well size to 150-mm-diameter dishes. Cells grown in suspension can be cultured in volumes >50 mL. For most applications, the use of protease inhibitors (0.5 mm PMSF, 2 μg/mL aprotinin, 40 μg/mL bestatin, 0.5 μg/mL leupeptin) in the buffers is adequate. If the extracts are to be used for in vitro kinase assays, or if phosphorylation of the protein of interest is important for binding activity, the addition of phosphatase inhibitors (*see* **Subheading 2.2.1.5.**) to all buffers is required.

1. Remove all but approx 0.5 mL of the growth medium from the cells and scrape the cells from the tissue-culture dish using a cell lifter (cat. no. 3008; Costar, Cambridge, MA). Transfer the cell suspension to a microfuge tube. Centrifuge the cells at 1500*g* at 4°C for 5 min. If using nonadherent cells, centrifuge the cell suspension at 1500*g* at 4°C for 5 min. Wash the cells once with PBS, and pellet the cells by centrifugation at 1500*g* at 4°C for 5 min. Do not trypsinize the cells, as residual trypsin may damage the integrity of the proteins in the nuclear extract.
2. Measure the packed cells volume (PCV) by comparing the volume to a precalibrated microfuge tube, then decant the supernatant.
3. If the PCV is less than or equal to 100 μL, add 400 μL of ice-cold Dignam A buffer to the pellet and suspend the cells by pipeting. If the PCV is greater than 100 μL, suspend the cells using a proportional amount of ice-cold Dignam A buffer.

4. Allow the cells to swell on ice for 15 min.
5. Add 25 µL 10% NP-40 per 400 µL of swelled cells and vortex the tube for 10 s.
6. Immediately centrifuge the cellular homogenate in a microfuge at 16,000*g* at room temperature for 30 s.
7. Discard the supernatant (this contains cytoplasm and RNA; this fraction may be saved for other analysis).
8. The nuclear pellet is lysed by adding 50 µL (or approximately an equal volume of the pelleted nuclei) ice-cold Dignam C buffer and gently rocking the tube at 4°C for 15 min. To obtain a nuclear extract with higher protein concentration, use less Dignam C buffer. To ensure complete lysis, flick the tub rather hard to dislodge the nuclei form the bottom of the tube. Do not vortex, as this may damage protein integrity.
9. Centrifuge the nuclear extract in a microfuge at 16,000*g* at 4°C for 5 min.
10. Collect the supernatant and aliquot into several tubes to prevent repeated freezing and thawing, as this will damage some proteins. Store aliquots at –80°C.
11. Determine the protein concentration of 1–4 µL of the extract using the Bradford assay (Bio-Rad Laboratories, Hercules, CA). Typical protein concentrations will range from 1 to 10 µg/µL.

### 3.1.2. Nuclear Extracts from Tissue

The following is a detailed protocol to isolate nuclear proteins from liver tissue. A brief overview of the procedure is initially outlined, followed by a list of the preparation steps to be done before starting the isolation procedure. Once the nuclei have been isolated, they may either be frozen for later use (**Subheadings 3.1.2.4.** and **3.1.2.5.**), or the protocol can be continued to isolate nuclear extracts from the freshly isolated nuclei (**Subheading 3.1.2.6.**).

#### 3.1.2.1. ABBREVIATED PROTOCOL

For one rat liver (or six mice livers):

1. Mince and homogenize liver in 25 mL of homogenization buffer.
2. Transfer to 100-mL graduated cylinder.
3. Add 50 mL cushion buffer to bring the total volume to ~75–80 mL. Carefully layer the homogenate onto three SW28 centrifuge tubes (Beckman, polyallomer, ref. no. 326823) containing 10 mL cushion buffer.
4. Centrifuge at 76,200*g* at 1°C for 1 h.

#### 3.1.2.2. DETAILED PROTOCOL—PRELIMINARY PREPARATIONS

1. Prepare three cushions for each liver (10 mL each) with cushion buffer (2.2 *M* sucrose without DTT and PMSF) in SW28 tubes. Set them on ice.
2. Precool, in cold room:
   a. ~100 mL liver of homogenization buffer (0.3 *M* sucrose).
   b. ~50 mL liver of cushion buffer (2.2 *M* sucrose) in a graduated cylinder.
   c. 55 mL Potter homogenizer and Teflon pestle.
   d. 100 mL graduated cylinder.
   e. 250 mL glass beaker, on ice.
   f. Ti55.2 rotor and SW28 rotor with buckets.
3. Have a balance in the cold room to equilibrate the weight of the centrifuge tubes.
4. Just before starting, prepare 50 mL of homogenization buffer (*see* **Subheading 2.2.1.**).

### 3.1.2.3. ISOLATION OF NUCLEI

1. Anesthetize the rat, cut the jugular vein, and bleed the animal until it stops (running cold water over the incision helps). The goal is to rid the liver of as much blood as possible.
2. Free the liver from as much nondesirable tissue as you can. Place it into a glass beaker containing ~10 mL of homogenization buffer. The beaker must always be on ice!
3. Mince the liver into small pieces with scissors, and wash away the remaining blood with homogenization buffer. Wash out as much blood as possible,
4. Pour the minced tissue into the 55-mL Potter homogenizer containing 25–30 mL homogenization buffer.
5. Homogenize the tissue with 2–3 slow strokes using a motor-driven Teflon pestle.
6. Transfer the homogenate to the precooled 100-mL graduated cylinder by filtering through cheesecloth to remove cellular debris.
7. Add cushion buffer to the liver homogenate to a total volume of 50 mL.
8. Cover the cylinder with Parafilm and mix well by gentle inversion.
9. Add 10 µL 1 $M$ DTT and 50 µL 100 m$M$ PMSF to the centrifuge tubes containing the sucrose cushion and mix well.
10. Carefully layer the homogenate of the cushions, place the tubes in the SW28 buckets, and equilibrate the weight of the tubes using the balance. Place the buckets on the SW28 rotor.
11. Centrifuge at 76,200$g$ at 1°C for 1 h.
12. After centrifugation, check for the presence of a nuclear pellet. Typically, the pellet is white and uniform in shape. If you do not see a pellet, spin for another 60 min.
13. Remove the top layer (fat and unbroken cells) with a spatula.
14. Vacuum aspirate, leaving a bit of the liquid covering the nuclear pellet (about one-half of the sucrose cushion). Decant the remaining liquid covering the nuclear pellet and place the tubes "under" ice, covering the nuclear pellet with ice. Be sure that ice does not get into the tube or touch the nuclear pellet. (Push a hole in the ice with the centrifuge buckets, then carefully place the inverted tubes into the holes. Cover the bottom of the tubes with ice where the nuclear pellet is located.)
15. After leaving the tubes in an inverted position for 5–10 min, wipe the inner walls of the tubes with a Kimwipe, using forceps. Keep the tubes, especially the nuclear pellet, always on ice.
16. After **step 15**, you can continue with the protocol for preparation of nuclear extracts, or you can freeze the nuclei for preparation of nuclear extracts at a later date. Alternatively, these nuclei may be used for nuclear run-on assays or for DNase I hypersensitivity analysis *(9)*.

### 3.1.2.4. FREEZING THE NUCLEI

1. Suspend the nuclear pellets carefully, avoiding bubbles, in a total volume of 5 mL NSB. Transfer to 15-mL polypropylene tubes, always keeping these tubes on ice.
2. Flash freeze the tubes in liquid N$_2$ and store at –80°C. The isolated nuclei are very stable, and can be stored for several months under these conditions.

### 3.1.2.5. PROCESSING OF FROZEN NUCLEI

1. Precool a 40-mL Wheaton dounce homogenizer on ice.
2. Thaw the frozen nuclei on ice.
3. When thawed, briefly vortex the samples. Transfer 5 mL of nuclei + 25 mL SRB to a precooled 40-mL Wheaton dounce homogenizer on ice. After adding SRB, the final KCl, HEPES, and glycerol concentrations should be the same as in NLB.
4. Homogenize the nuclei with one stroke using a precooled B type pestle. Transfer the suspended nuclei to a 30-mL Oak Ridge tube (Nalgene, ref. no. 3431-2526). Fill each tube volume to 20 mL, if necessary, with NLB. Ideally, lyse nuclei obtained from one rat liver in each tube.

### 3.1.2.6. Processing Isolated Nuclei Without Freezing

1. Prepare NLB with stock solutions of DTT, PMSF, aprotinin, leupepin, bestatin, and Na$_2$MoO$_4$ (*see* **Subheading 2.2.1.**).
2. Suspend each nuclear pellet in 3 mL NLB. Transfer the nuclei to a 40-mL Wheaton dounce homogenizer. Rinse the tube with 2 mL NLB and add to the homogenizer. Adjust the total volume to 20 mL with NLB (one rat liver = 20 mL).
3. Dounce to suspend the nuclei (2–3 strokes).
4. Transfer this nuclear suspension to 30 mL Oak Ridge tubes (20 mL per tube).
4. Add 2 mL of 4 *M* (NH$_4$)$_2$SO$_4$ to each tube. Immediately cap the tube and mix well by inversion. The nuclei are lysing if chromatin flakes appear and the solution becomes viscous.
6. Place the tube horizontal on and under ice, and place the ice bucket on a rocker platform in the cold room. Continue lysis for 30–60 min with gentle agitation (60–80 rpm).
7. Centrifuge the tubes at 206,000*g* at 1°C for 60 min in a Ti55.2 rotor.
8. Carefully remove the supernatant, which should be clear and nonviscous, and transfer to a clean Oak Ridge tube (~21 mL/tube), on ice, using a 10-mL pipet to measure the exact volume. Be very careful to avoid DNA contamination from the pellet.
9. Add 0.33 g finely powdered (NH$_4$)$_2$SO$_4$ (crush the granular (NH$_4$)$_2$SO$_4$ supplied by the manufacturer using a mortar and pestle) per mL of supernatant. Close the tubes, lay them down on and under ice, and shake gently on a rocker platform in the cold room (~60–80 rpm) until all of the (NH$_4$)$_2$SO$_4$ is completely dissolved (about 60 min). Then place the tube vertically in the ice and let it sit for about 15 min, until all the liquid goes to the bottom of the tube.
10. While the (NH$_4$)$_2$SO$_4$ is being dissolved, prepare 500 mL and 10 mL of NDB (without DTT, PMSF, aprotinin, leupeptin, bestatin, and Na$_2$MoO$_4$) and cool the buffers on ice.
11. Centrifuge the tubes at 101,000*g* at 1°C for 20 min in a Beckman Ti55.2 rotor. Mark the tubes on the outer side to locate the protein pellet following centrifugation.
12. After centrifugation, discard the supernatant by inverting the tubes for 5–10 min on ice. Make sure that the protein pellet is always in contact with the ice. Afterwards, wipe the tube mouth with Kimwipes using forceps.
13. Dissolve the nuclear pellet in about 300 µL filter-sterilized NDB. Be sure to dissolve the protein on the sides of the tube by carefully pipeting. Avoid bubble formation, as this may damage proteins. Let the tube sit on ice for 5 min to allow the protein pellet to dissolve. You may accelerate dissolving the protein by placing the tubes horizontally on ice on a platform shaker and gently rocking for about 10 min.
14. Wash a sterile clean dialysis tube with 4–5 mL sterile NDB.
15. Transfer the nuclear extract to the dialysis tube, wash the tube with ~100 µL of NDB, and transfer this to the dialysis tube.
16. Prepare 250 mL of NDB by adding 500 µL 1 *M* DTT, 250 µL 100 m*M* PMSF, and 250 µL 1 *M* Na$_2$MoO$_4$ to the buffer and mix well.
17. Place the dialysis tube in a beaker containing 250 mL NDB and a magnetic stirrer. The beaker should be placed in an ice-water bath. Slowly stir in a cold room for 1.5 h.
18. Transfer the dialysis tube to fresh 250 mL NDB and continue dialysis as in **step 17**.
19. After 1.5 h, recover the extract from the dialysis tube and centrifuge at 16,000*g* in a microfuge at 4°C for 10 min to remove insoluble material.
20. Aliquot the extract into small volumes (25–50 µL) into screw-capped Eppendorf tubes. Reserve a small aliquot to determine the protein concentration. Protein concentrations typically range from 4–10 µg/µL.
21. Flash freeze the extract in liquid N$_2$. Store the sample in liquid N$_2$ or at –80°C (liquid N$_2$ is best).

## 3.2. Standard DNase I Footprinting

### 3.2.1. Preparation of Probe

Probes used for DNase I footprinting are typically obtained from regulatory regions of genes that have been cloned. Convenient restriction endonuclease cleavage sites are usually present, which allows for easy radiolabeling, release, and purification of the region of interest to be assessed for protein binding. To generate the probe, the plasmid is initially digested with a restriction endonuclease, preferable one that linearizes the plasmid. The linearized plasmid is then radiolabeled and subsequently digested with a second restriction endonuclease to generate a small, 200–500 bp radiolabeled fragment that contains the segment of DNA to be assayed for protein binding and a larger labeled fragment. The desired band is purified from the gel and used as the probe in the footprinting reactions. Caution: The radiolabeled base must be complementary to one of the bases in the overhang left by restriction digestion!

1. Digest approx 25 µg plasmid DNA with the first restriction enzyme (the site to be radiolabeled) in a screw-capped Eppendorf tube (to minimize chances of radioactive contamination).
2. Precipitate DNA after digestion with the addition of 0.1 vol 3 $M$ sodium acetate, pH 5.2, and 2.5 vol 100% ethanol.
3. Set the tube at –20°C for 10 min.
4. Centrifuge the tube at 16,000$g$ in a microfuge at room temperature for 10 min, decant the supernatant, and wash the pellet with 1.0 mL 75% ethanol.
5. Briefly dry the DNA pellet.
6. Dissolve the DNA in 40 µL water and add:
   a. 6 µL 10X reverse transcriptase buffer.
   b. 6 µL 5 m$M$ dNTP mix (if label = dCTP, then use dATP, dGTP, dTTP).
   c. 7 µL [$^{32}$P]α-dCTP, 3000 Ci/mmol (or other labeled base, depending on the sequence of the overhang).
   d. 1.0 µL reverse transcriptase (20 U M-MLV; 10–20 U AMV).
7. Incubate the labeling reaction at 37°C for 45 min.
8. Stop the labeling reaction with the addition of 6 µL 100 m$M$ EDTA.
9. Precipitate the radiolabeled DNA by adding 34 µL water, 100 µL 5 $M$ ammonium acetate, and 500 µL 100% ethanol.
10. Set the tube at –20°C for 10 min.
11. Centrifuge the tube at 16,000$g$ in a microfuge at room temperature for 10 min, decant the supernatant, and reprecipitate the labeled DNA by dissolving in 100 µL water, and adding 100 µL 5 $M$ ammonium acetate and 500 µL 100% ethanol.
12. Set the tube at –20°C for 10 min.
13. Centrifuge the tube at 16,000$g$ in a microfuge at room temperature for 10 min, decant the supernatant, and wash the DNA pellet with two 1.0-mL washes using 75% ethanol, then briefly dry the DNA pellet.
14. Digest the radiolabeled DNA with the second restriction endonuclease by dissolving the DNA in 24 µL water and adding 3 µL 10X restriction endonuclease buffer and 3 µL of the appropriate restriction enzyme.
15. Incubate the restriction digest at 37°C for 1.5 h.
16. Add 5 µL of gel-loading dye to the restriction digest and separate the radiolabeled fragments in an agarose gel using 1X TAE buffer.
17. Excise the appropriate band and elute the DNA fragment using a commercially available kit (we use AUIquick Gel Extraction Kit, Quigen, Inc., Valencia, CA).

### 3.2.2. Maxam-Gilbert G + A Chemical Sequencing Reaction

1. To 200,000 CPM of probe (from **Subheading 3.2.1.**) add 0.5 µL salmon sperm DNA and water to a total volume of 27 µL.
2. Add 2 µL of 2% formic-acid solution.
3. Incubate the reaction mixture at 37°C for 10 min.
4. Set the sample at –80°C for 10 min.
5. Dry the sample in a Speed-Vac.
6. Dissolve the pellet in 40 µL of water.
7. Dry the sample in a Speed-Vac.
8. Dissolve the DNA pellet in 40 µL of formamide dye and store at –20°C.

### 3.2.3. Binding Reaction

1. In an Eppendorf tube, prepare a probe mix for each binding reaction that contains 1 µg poly[d(I-C)], 2% polyvinyl alcohol, 20 m*M* MgCl$_2$, 20,000–40,000 CPM probe, and water to 25 µL. Prepare for *n* + 1 samples, so for nine binding reactions make enough reaction mix for 10. Poly[d(I-C)] is used to bind nonspecific DNA binding proteins, abundant in the cell nucleus, which could interfere with determining specific DNA–protein interactions.
2. Add reaction buffer and nuclear extract (or an empirically determined amount of BSA as a control reaction), 25 µL combined final volume, to the tube containing the probe mix. Binding reactions typically contain between 10 and 50 µg protein, but may contain more or less protein.
3. Incubate the binding reaction at room temperature for 20 min. (The binding reaction may be incubated on ice if necessary).

### 3.2.4. DNase I Digestion

The dilution of the DNase I must be experimentally determined, and will often vary with different preparations of nuclear extracts (primarily due to varying quality of nuclear extract preparations) and probes being used (*see* **Notes 4** and **5**). A good starting point is to use two extract concentrations (i.e., 15 and 30 µg total protein for each reaction) and 1:400 to 1:1000 dilutions of the stock DNase I solution.

1. Prepare dilutions of the stock DNase I solution in water and store on ice.
2. To each 50 µL binding reaction add 50 µL of the HEPES/CaCl$_2$ solution.
3. Add freshly diluted DNase I solutions to each of the tubes and incubate at room temperature exactly 1.5 min. The amount of DNase I added to the BSA containing binding reactions are typically 0.1X the concentration of that added to the binding reactions containing nuclear extract.
4. After DNase I digestion, add 100 µL of stop solution.
5. Phenol extract each digestion reaction twice with 200 µL of phenol saturated with Tris-HCl and once with 200 µL chloroform:isoamyl alcohol (24:1).
6. Transfer the aqueous phase to a new microfuge tube and precipitate the reaction products with the addition of 500 µL 100% ethanol.
7. Incubate the tubes at room temperature for 15 min, then centrifuge at 16,000*g* in a microfuge at room temperature for 15 min.
8. Decant the supernatant, wash the pellets with 75% ethanol, and dry in a Speed-Vac.
9. Dissolve the samples in 4 µL of formamide dye.

### 3.2.5. Gel Electrophoresis

1. Heat the samples at 100°C for 5 min.
2. Place the tubes in an ice bath to cool, then apply the samples to the 8% sequencing gel. Prerun the gel for approx 30 min before loading the samples.

3. Electrophorese the samples at 1500–2000 V for approx 1.5 h (until the leading bromophenol blue dye reaches the bottom of the gel).
4. After electrophoresis, fix the gel in gel fix solution for 15 min.
5. Dry the gel and expose to X-ray film overnight at –80°C using an intensifying screen. Once the results are acceptable, expose the gel to X-ray film at room temperature without an intensifying screen. This will require approximately double the exposure time, but the bands will be sharper.

### 3.3. PCR-Based DNases I Footprinting

### 3.3.1. Labeling Primer

1. In a screw-capped microfuge tube (to prevent radioactive contamination), combine:
   a. 25 pmol oligonucleotide.
   b. 3 μL 10X T4 polynucleotide kinase buffer.
   c. 1.5 μL 0.1 *M* DTT.
   d. 7.5 μL [$^{32}$P]γ-ATP (6000 Ci/mmol).
   e. 1 μL T4 polynucleotide kinase (10 U).
   f. Water to 29 μL.
2. Incubate the reaction mixture at 37°C for 1 h.
3. Stop the reaction by adding 3 μL 100 m*M* EDTA, 32.5 μL 5 *M* ammonium acetate, and 195 μL 100% ethanol.
4. Set the tube at room temperature for at least 30 min.
5. Centrifuge the tube at 16,000*g* in a microfuge at room temperature for 10 min.
6. Dissolve the pellet in 50 μL water and precipitate the oligonucleotide as in **step 3**.
7. Wash the pellet with 1.0 mL 75% ethanol and dry the sample in a Speed-Vac.
8. Dissolve the pellet in 50 μL of water (0.5 pmol/μL) and count 1 μL in a scintillation counter. Store the labeled oligonucleotide at –20°C.

### 3.3.2. Sequence Marker

To locate the precise positions of DNA–protein interactions on the template DNA being analyzed, the radiolabeled oligonucleotide used for the PCR-based footprinting reaction can be used as a primer to perform standard dideoxy sequencing reactions with the template used for footprinting analysis. Secondary structure of the region being analyzed may prevent sequencing. If sequencing fails, a radiolabeled DNA size marker can be used to determine the precise locations of DNA–protein interactions. One of many commercially available DNA ladder can be radiolabeled and used in the assay. Alternatively, one can digest plasmid, pBR322 with *Msp*I to generate the following size ladder (in base pairs): 622, 527, 404, 307, 242, 238, 217, 201, 190, 180, 160 (×2), 147 (×2), 123, 110, 90, 76, and 67. The fragments can be labeled by filling the 5' overhang ends with [α$^{32}$P]-dCTP as described in **Subheading 3.4.1.2.** for double-stranded oligonucleotides. A labeling reaction containing 100 ng of DNA should yield enough radiolabeled marker for many assays.

### 3.3.3. Binding Reaction, PCR Amplification, and Gel Electrophoresis

1. Prepare a 50-μL binding reaction that contains:
   a. 20 fmol of plasmid DNA (*see* **Note 2**).
   b. 1 μg poly[d(I-C)].
   c. 5 m*M* MgCl$_2$.
   d. 20 μg of nuclear proteins in total volume of 25 μL in NDB (*see* **Note 6**).

2. Incubate the binding reaction at room temperature for 20 min.
3. Add and equal volume of HEPES/CaCl$_2$ buffer to the samples.
4. Add diluted DNase I to the samples and digest at room temperature for 1.5–2 min. Use amounts of DNase I similar to those required in standard footprinting assay (10–100 ng/ reaction).
5. Stop the digestion with the addition of an equal volume of stop buffer, and extract the samples with an equal volume (200 μL) of phenol:chloroform (24:1).
6. Transfer the aqueous layer to a new tube, and precipitate the DNA by adding 2.5 vol of 100% ethanol; set at room temperature for 30 min.
7. Centrifuge the samples at 16,000$g$ in a microfuge at room temperature for 15 min. Decant the supernatant, and wash the pellet with 70% ethanol. Briefly air-dry the sample.
8. Suspend each sample in 48.5 μL of PCR mixture, transfer to a thin-wall PCR tube, and add 1 μL of [$^{32}$P]-labeled primer (0.5 pmol), and 0.5 μL (2.5 U) of Taq DNA polymerase (*see* **Note 6**).
9. Perform six cycles of amplification (1 min at 95°C, 1 min at the calculated T$_m$ of the oligonucleotide [*see* formula in **ref. *10***], and 2 min at 72°C). Sufficient product should result so that it will be visible on a film after an overnight exposure of the wet gel at –80°C using an intensifying screen. Fewer cycles (but longer film exposure times) should be tried if background levels are unacceptable (i.e., there are too many bands in the non-DNase-PCR-treated sample or resolution of the restriction enzyme-digested control is poor).
10. After OCR amplification, add 50 μL of DNase I stop buffer to the samples and extract them with an equal volume of phenol:chloroform (24:1). Precipitate the samples with the addition of 2.5 vol of 100% ethanol.
11. Incubate the samples at room temperature for 30 min.
12. Centrifuge at high speed (16,000$g$) in a microfuge at room temperature for 15 min.
13. Carefully decant the supernatant and suspend the pellet in 4 μL of formamide dye.
14. Heat the samples at 100°C for 5 min, briefly chill the samples on ice, and load on a 4–6% denaturing gel (29:1 crosslinking ratio of acrylamide to bis-acrylamide). Load 3000–5000 CPM of radiolabeled size marker in another lane on the gel.
15. Electrophorese the gel at 55–60 W constant power until the xylene cyanol band (the slowest migrating dye) is approx 5 cm from the bottom. Be aware that the lower buffer chamber and the lower part of the gel will be very radioactive (the lower buffer chamber will contain most of the free radiolabeled oligonucleotide). Discard the radioactive buffer following general safety guidelines for $^{32}$P.
16. Carefully separate the glass plates and gently press a piece of Whatman 3MM paper, slightly larger than the gel, onto its surface. Make a mark on the filter paper beforehand to keep the orientation of the gel from being lost. Carefully lift the gel from the gel plate and dry the gel on a gel drier without fixing. After drying, a small amount of baby powder can be applied to the surface of the gel (by gently sprinkling the powder evenly on the gel), or it can be covered with a layer of plastic wrap to prevent sticking to the film during autoradiography. Alternatively, the gel can be transferred from the glass onto an old film, then covered with plastic wrap and exposed directly at –80°C without any further treatments. This is the most convenient way to deal with these gels, and does not necessarily result in a loss of resolution unless you freeze-thaw them multiple times to obtain different exposures.
17. Expose the gel to X-ray film at –80°C with an intensifying screen. An overnight exposure should be sufficient to see the footprint pattern.

## 3.4. Mobility Shift Assay

### 3.4.1. Probe Preparation

Probes used for the mobility shift assay should be relatively small, to give better separation of DNA–protein complexes from one another, and from the free DNA probe in the resolving gel. Short double-stranded oligonucleotides (15–50 bp) or DNA fragments (<200 bp) are used. Double-stranded oligonucleotides can be labeled using several methods. If the ends of the oligonucleotides are blunt or contain 3'-overhangs, they may be kinased with [$^{32}$P]γ-dATP using T4 polynucleotide kinase. Double-stranded oligonucleotide with 5'-overhangs may be labeled with [$^{32}$P]γ-dATP in the same way or by filling in using Klenow fragment of M-MLV or AMV reverse transcriptase (*see* **Subheading 3.2.1.**). Either single-stranded or double-stranded oligonucleotides may be labeled with T4 polynucleotide kinase. Once labeled, single-stranded oligonucleotides can be annealed to the complementary strand prior to use in the mobility shift assay.

#### 3.4.1.1. LABELING PROBE USING T4 POLYNUCLEOTIDE KINASE

1. In an Eppendorf tube add:
   a. 200 ng of double-stranded or single-stranded oligonucleotide.
   b. 5 μL [$^{32}$P]γ-dATP (6000 Ci/mmol [or 3000 Ci/mmol]).
   c. 4 μL 10X kinase buffer.
   d. 1 μL T4 polynucleotide kinase (10 U).
   e. Water to 39 μL.
2. Incubate the reaction at 37°C for 45 min.
3. After incubation, precipitate the reaction mixture with the addition of 1 μL 10 mg/mL yeast tRNA (as earlier), 40 μL 5 *M* ammonium acetate, 200 L 100% ethanol.
4. Set at –20°C for 10 min.
5. Centrifuge at 16,000*g* in a microfuge at room temperature for 10 min, then decant the supernatant.
6. Dissolve the pellet in 50 μL water and precipitate again by adding 50 μL 5 *M* ammonium acetate, and 250 μL 100% ethanol.
7. Incubate at –20°C for 10 min, then centrifuge the tube at 16,000*g* in a microfuge at room temperature for 10 min. Decant the supernatant, and wash the pellet twice with 1.0 mL 75% ethanol.
8. Dry the pellet in a Speed-Vac, then dissolve it in 100 μL water.
9. Count 1 μL in a scintillation counter.

#### 3.4.1.2. LABELING PROBE USING THE FILL-IN REACTION

This procedure can be used to radiolabel double-stranded oligonucleotides with either reverse transcriptase (M-MLV or AMV) or Klenow fragment. By precipitating the radiolabeled oligonucleotide with ammonium acetate, free isotope is selectively eliminated from the radiolabeled product.

1. In an Eppendorf tube add:
   a. 200 ng of double-stranded oligonucleotide.
   b. 6 μL [α$^{32}$P]-dCTP, 3000 Ci/mmol (or other labeled base, depending on the nucleotide sequence of the overhang).
   c. 3 μL 10X reverse transcriptase buffer.
   d. 1 μL reverse transcriptase (20 U M-MLV; 10–20 U AMV).

    e. 3 μL 5 mM dNTP mix (if label = dCTP, then use dATP, dGTP, dTTP).

    f. Water to 30 μL.

2. Incubate the reaction at 37°C for 45 min.
3. After incubation, precipitate the reaction mixture with the addition of 2 μL 10 mg/mL yeast tRNA (as earlier), 3 μL 100 mM EDTA, 25 μL water, 60 μL 5 M ammonium acetate, and 300 μL 100% ethanol.
4. Set at –20°C for 10 min.
5. Centrifuge at 16,000g in a microfuge at room temperature for 10 min, then decant the supernatant.
6. Dissolve the pellet in 60 μL water and precipitate again by adding 60 μL 5 M ammonium acetate, and 300 μL 100% ethanol to the tube and repeating **steps 4** and **5**.
7. Wash the pellet twice with 1.0 mL 75% ethanol, then briefly dry the pellet.
8. Dissolve in 100 μL TE (10 mM Tris-HCl, pH 8.0, 1 mM EDTA) or water.
9. Count 1 μL in a scintillation counter.

### 3.4.2. Gel Preparation

1. Clean and assemble glass plates (approx w = 165–200 mm; l = 200 mm) using 1.0–1.5-mm-thick spacers.
2. Use a 4% native polyacrylamide gel (probes 150–300 bp in size) or a 5% native gel (probes 18–150 bp in size).
3. Use the following table to prepare a gel mixture

|  | 5% gel | 4% gel |
|---|---|---|
| 30% acrylamide (29:1) | 10 mL | 8 mL |
| 10X TBE | 2.4 mL | 2.4 mL |
| Water | 47.2 mL | 49.2 mL |
| 10% ammonium persulfate | 400 μL | 400 μL |
| TEMED | 60 μL | 60 μL |

4. Quickly pour the gel mixture between the glass plates and let the gel polymerize for 1–2 h.
5. Prerun the gel using 0.4X TBE as the running buffer for approx 1 h at 200 V, constant voltage. The amperage usually starts out at approx 35 mA, when it drops about in half, to approx 15–20 mA, the gel is ready to run.
6. After prerunning the gel, discard the running buffer and replace with fresh 0.4X TBE.

### 3.4.3. Binding Reaction

The presence and concentration of $MgCl_2$ and other salts is dependent on the DNA-binding protein being assessed. Proteins usually require an optimal salt concentration between 50 and 200 mM KCl/NaCl that must be empirically determined. Poly[d(I-c)] in the binding reaction is intended to capture nonspecific DNA binding proteins from the extract; if using purified protein, omit this.

A 10- to 200-fold molar excess of cold competing oligonucleotide is used to confirm the specificity of complex formation. For supershift assays, which are used to confirm the identity of a protein forming a specific complex, add the antibody after adding the probe (*see* **Note 9**).

1. Prepare an 18-μL binding reaction that contains 2–20 μg nuclear extract, 1–5 μg poly[d(I-C)], 10 mM HEPES, pH 7.6, 50–150 mM KCl, 0.1 mM EDTA, 1 mM DTT, 10% glycerol, 0–5 mM $MgCl_2$. Combine the reaction mixture after adding each component using a pipet tip; avoid bubbles, as this may damage proteins.
2. Incubate the binding reaction on ice for 10 min.

3. Add 2 μL (20,000 CPM) of the radiolabeled probe diluted in water.
4. Incubate the reaction mixture on ice for 20 min.
5. Add 1.5 μL of dye mixture to each tube.
6. Load 20 μL of the reaction mixture on the nondenaturing prerun acrylamide gel.

### 3.4.4. Gel Electrophoresis

The low ionic strength of the running buffer (0.4X TBE) increases the affinity and thus the stability of most DNA–protein interactions, resulting in longer associations between the two. Preelectrophorese the gel at 200 V until the initial current in milliamps drops by on-half, then change the running buffer before loading the binding reaction samples on the gel. The gel should be run at a constant voltage of 200 V, until the gel dye migrates approximately three-quarters down the gel (this is dependent on the size of the probe—larger probes can be run longer to allow separation of the complexes). The buffers from the upper and lower compartments should be mixed every 30 min during electrophoresis to keep the pH constant; DNA–protein complexes are very pH-dependent. After electrophoresis, dry the gel and expose gel to X-ray film, using an intensifying screen at –80°C overnight.

## 4. Notes

1. It is often more convenient to isolate whole-cell extracts rather than nuclear extracts, especially when few cells are available. Whole-cell extracts work well in most applications; however, the results in mobility shift assays are not as "clean" (producing more nonspecific bands), and in footprinting protocols, generally more protein is needed. To prepare whole-cell extracts from cultured cells, follow the protocol outlined in **Subheading 3.1.1.**, but omit the swelling step, using Dignam A buffer and equal volume of Dignam C buffer to the PCV after washing and pelleting the cells.
2. High-quality plasmid DNA is needed to reduce background in the PCR-based footprinting method. Plasmid DNA can be prepared by the alkaline lysis method, as described by Sambrook et al. *(10)* with an additional protein removal step (phenol:chloroform extraction) prior to isopropanol precipitation. RNA should be removed from the preparation by RNase A treatment followed by PEG precipitation, as described by Hattori and Sakaki *(11)*. The resulting plasmid DNA should be further purified through a CsCl gradient, and desalted by size-exclusion chromatography.
3. Since PCR-based footprinting can assess 150–200 nucleotides for each oligonucleotide used, several synthetic oligonucleotides may be required to completely analyze DNA–protein interactions in a DNA fragment of interest. The quality of the primer is important and purification of the primer is strongly recommended *(10)*. The choice of the primer should follow the general guidelines for designing a sequencing or PCR primer. Primers 20–25 nucleotides in length are sufficient. The distance from the primer hybridization site to the region of interest should be approx 50 nucleotides.
4. Using this technique, it is important to use as little probe DNA and as much protein as possible in the footprinting reaction to increase the possibility of detecting weak DNA–protein interactions. Therefore, the labeled DNA (probe) used in the footprinting reactions should be of as high specific activity as possible.
5. The results obtained in the footprinting assay are affected by the concentration of the probe, protein, and DNase I enzyme used. Therefore, this assay requires titration of these components. To minimize the use of the nuclear extract, it is best to test different concentrations of the probe incubated with bovine serum albumin (BSA) and various DNase I

concentrations. In a good digestion, there is a significant amount of undigested probe at the top of the gel, and an even ladder of bands extending to the bottom of the gel. Once a suitable ladder of DNA cleavage products is obtained, tirations using the extract can be done. A control reaction is performed where the DNA is incubated with BSA and digested in an identical fashion with DNase I. The banding pattern generated with BSA is then compared to that of the extract to locate regions of protections. Bands that are missing in the experimental extract, compared to the bands in the BSA lane, indicate positions of DNA–protein interactions.

6. The amount of protein may need to be determined by titration. This template is more complex than an isolated DNA fragment, and cryptic binding sites in the plasmid vector DNA may "compete" for protein binding to the cloned region of interest. Thus, higher concentrations of protein/mol of templates must be used in this procedure.

7. Other thermostable DNA polymerases can be used, but be aware that they may possess proofreading-associated exonuclease activity, which could degrade your labeled oligonucleotide, resulting in a significant loss of signal.

8. When assessing complex specificity, add cold competing oligonucleotide between 10- and 500-fold molar excess.

9. For supershift assays, add 1–2 µL of antibody (more if needed, depending on the concentration of antibody used and the binding affinity of the antibody to the protein being assayed) to the binding reaction, 10 min after adding the probe. Incubate the binding reaction containing the antibody for an additional 30–60 min on ice. Then add the dye mixture and load the samples on the nondenaturing acrylamide gel. Some antibodies will disrupt complex formation, but others will alter the mobility of the DNA–protein complex in the gel, resulting in a "supershifted" complex. Additionally, some antibodies may require preincubation with the nuclear extract prior to the addition of the radiolabeled probe.

## References

1. Rippe, R. A., Lorenzen, S.-I., Brenner, D. A., and Breindle, M. (1989) Regulatory elements in the 5' flanking region and the first intron contribute to transcriptional control of the mouse alpha 1 type I collagen gene. *Mol. Cell. Biol.* **9**, 2224–2227.

2. Brenner, D. A., Rippe, R. A., and Veloz, L. (1989) Analysis of the collagen $\alpha 1(I)$ promoter. *Nucleic Acids Res.* **17**, 6055–6064.

3. Nehls, M. C., Rippe, R. A., Veloz, L., and Brenner, D. A. (1991) Transcription factors NF-I and Sp1 interact with the murine collagen alpha l(I) promoter. *Mol. Cell. Biol.* **11**, 4065–4073.

4. Rippe, R. A., Almounajed, G., and Brenner, D. A. (1995) Sp1 binding activity increases in activated Ito cells. *Hepatology* **22**, 241–251.

5. Rippe, R. A., Kimball, J., Breindl, M., and Brenner, D. A. (1997) Binding of USF-1 to an E-box in the 3' flanking region stimulates $\alpha 1(I)$ collagen gene expression. *J. Biol. Chem.* **272**, 1753–1760.

6. Tugores, A., Magness, S. T., and Brenner, D. A. (1994) A single promoter directs both housekeeping and erythroid preferential expression of the human ferrochelatase gene. *J. Biol. Chem.* **269**, 30,789–30,797.

7. Schreiber, E., Matthias, P., Muller, M. M., and Schaffner, W. (1989) Rapid detection of octamer binding proteins with "mini-extracts," prepared from a small number of cells. *Nucleic Acids Res.* **17**, 6419.

8. Hattori, M., Tugores, A., Veloz, L., Karin, M., and Brenner, D. A. (1990) A simplified method for the preparation of transcriptionally active liver nuclear extracts. *DNA Cell Biol.* **9**, 777–781.

9. Magness, S. T., Tugores, A., Diala, E. S., and Brenner, D. A. (1998) Analysis of the human ferrochelatase promoter in transgenic mice. *Blood* **91,** 320–328.
10. Sambrook, J., Fritsch, E. F., and Maniatis, T., eds. (1989) *Molecular Cloning: A Laboratory Manual*, 2nd ed. Cold Spring Harbor Laboratory, Cold Spring Harbor, NY.
11. Hattori, M. and Sakaki, Y. (1986) Dideoxy sequencing method using denatured plasmid templates. *Anal. Biochem.* **152,** 232–238.

# 31

# Analyzing the Developmental Expression of Sigma Factors with S1-Nuclease Mapping

## Jan Kormanec

## 1. Introduction

### 1.1. Identification of the Genes Encoding Sigma Factors of RNA Polymerase in Streptomyces aureofaciens

Streptomycetes are mycelial, Gram-positive soil bacteria which produce a variety of biologically active secondary metabolites, including the majority of known antibiotics. In addition to this physiological differentiation, they undergo a complex cycle of morphological differentiation. Spores after germination on solid-growth media form a network of vegetative "substrate mycelium" that grows into the medium. After unknown signal(s), the substrate mycelium differentiates, and white aerial mycelium is formed that grows into the air. Later it undergoes septation and, in the last step of the process, forms chains of spores. The differentiation of *Streptomyces* is controlled on several levels, of which heterogeneity of σ factors of RNA polymerase seems to play an important role *(1)*. Therefore, we examined the *Streptomyces* differentiation through the study of these key proteins in this process.

We used the high homology among different σ factors to isolate new σ factor genes participating in the differentiation of *S. aureofaciens*. Most sigma factors can be divided into four conserved domains *(2)*. We chose the second domain, which is the most highly conserved for preparation of several oligonucleotide probes. After colony blot hybridization of two independent *S. aureofaciens* genomic libraries, we identified more then 100 positive signals. Sequence analysis identified six genes, *hrdA, hrdB, hrdD, hrdE, rpoZ,* and *sigF*, encoding sigma factors in *S. aureofaciens* *(3–5)*. Disruption of the genes in the chromosome of *S. aureofaciens* has revealed that at least two of these are critical for the early *(rpoZ)* and late *(sigF)* stages of differentiation *(4,6)*.

### 1.2. Analysis of Expression of Sigma Factors by S1-Nuclease Mapping

We next analyzed the expression of all sigma factors in the course of differentiation. There are several techniques to study expression of genes in bacteria, including fusion of promoter regions into reporter genes, dot blot, and Northern blot hybridization analysis, primer extension analysis, and S1-nuclease mapping (**Fig. 1**). The last two in vitro

From: *Methods in Molecular Biology, vol. 160: Nuclease Methods and Protocols*
Edited by: C. H. Schein © Humana Press Inc., Totowa, NJ

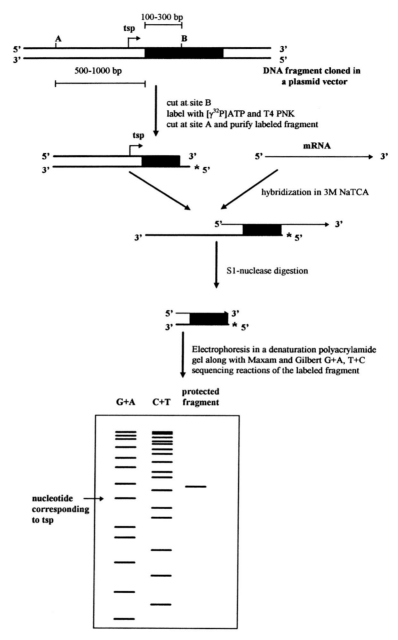

Fig. 1. Flow chart of S1-nuclease mapping procedure. Thick lines represent DNA fragment, and thin lines represent mRNA. Filled box indicates a coding region. Bent arrow represents position and direction of an apparent TSP from promoter. The asterisk indicates 5' protruding [$^{32}$P]-labeled end after digestion with a restriction endonuclease B.

techniques are the most sensitive for the estimation of expression level and to determine transcription start point (TSP) in the nucleotide sequence, thereby localizing promoter regions. We have found S1-nuclease mapping of TSP as the most reliable technique for *S. aureofaciens* and other *Streptomyces* strains. Primer extension analysis in many cases failed to localize the TSP, possibly because the high percentage of GC pairs in the *Streptomyces* genome (about 73%) may induce formation of secondary structures at the lower temperature that hinder reverse transcriptase (primer extension using reverse transcriptase is usually done at 42°C). As in S1-nuclease mapping, since the hybridization of the heteroduplexes of RNA and DNA is performed at the higher temperature, secondary structures are less likely to form.

We isolated RNA from the cells of *S. aureofaciens* grown on solid Bennet medium to several developmental phases: vegetative substrate mycelium, the beginning of aerial mycelium formation, and the late aerial mycelium at approximately the time of septation into spores. The S1-mapping technique described here was then used to identify the TSP upstream of each sigma factor coding region in the different developmental stages. Whereas *hrdB* was expressed through all the differentiation phases from two tandem promoters (one is internal to coding region), *hrdD* is expressed from two tandem promoters that are active only in the vegetative stage, and transcription of two tandem *hrdA* promoters temporally correlates with the aerial mycelium formation. Two tandem promoters identified upstream of the *rpoZ* gene were active in distinct developmental stages, the weaker was active only in substrate mycelium, and the stronger was induced at the beginning of aerial mycelium formation. S1-nuclease mapping of *sigF* revealed two proximal apparent TSP detectable in the late stage of aerial mycelium formation *(7,8)*, while *hrdE* was not expressed at any stage. The results are summarized in **Fig. 2**.

## 1.3. Methodology of S1-Nuclease Mapping

S1-nuclease mapping is a widely used technique for detection of 5'- and 3'-ends of mRNA on DNA templates, and for locating 5'- and 3'-junctions of introns and exons in eukaryotic genomes. The technique is very sensitive, and can be used for quantitation of the amount of a given mRNA. The procedure was originally described by Berk and Sharp *(9)*, and several modifications and improvements of the technique has been published since the appearance of the paper *(10–13)*. As outlined in **Fig. 1**, for detection of TSP of mRNA on the template DNA fragment, the RNA to be analyzed is hybridized to a [$^{32}$P] 5'-labeled DNA fragment from the template overlapping proposed 5'-end of the coding region. The length of the proposed RNA:DNA hybrid should be larger then 100 bp. After hybridization, the RNA:DNA heteroduplexes are digested with S1-nuclease under conditions that cleave only single-stranded nucleic acids. The nuclease resistant fragments are analyzed by high-resolution denaturing polyacrylamide gel electrophoresis, along with Maxam and Gilbert *(14)* sequencing ladders prepared from labeled DNA fragments. The 5'-terminus of mRNA (TSP) corresponds to the sequencing-ladder fragment with a mobility 1.5 nucleotide faster, as the sequencing ladder fragments contain an additional 3'-end phosphate. Originally, hybridization was performed in high concentration of formamide (80%), which favors the formation of RNA:DNA over DNA:DNA duplexes. However, the difference between the melting temperatures of the duplexes is only about 5°C at these conditions; therefore, considerable care had to be taken to optimize hybridization temperature. Replacing formamide with the chaotropic salt sodium trichloracetate (NaTCA) *(11)* increases the thermal stability of RNA:DNA heteroduplexes relative to DNA:DNA duplexes by >20°C.

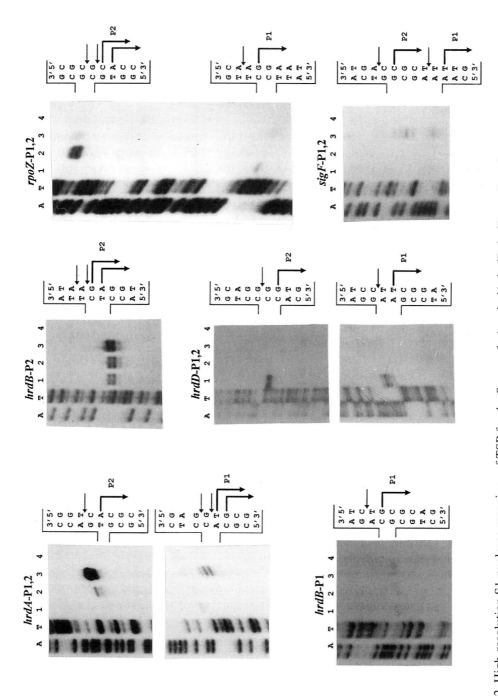

Fig. 2. High-resolution S1-nuclease mappings of TSP for the *S. aureofaciens hrdA*, *hrdB*, *hrdD*, *rpoZ*, and *sigF* genes with RNA isolated from surface cultures of *S. aureofaciens*. Tracks A and T are G+A and T+C Maxam and Gilbert sequencing ladders. 5′-labeled DNA fragments were hybridized with RNA from surface culture of *S. aureofaciens* from 13 h (lane 1), 19 h (lane 2) and 36 h (lane 3) from surface culture of *S. aureofaciens*, and with tRNA as a control (lane 4). Thin horizontal arrows indicate the position of RNA-protected fragments and thick bent vertical arrows indicate the nucleotide corresponding to apparent TSP after subtraction of 1.5 nt.

Partial digest by S1-nuclease at the ends of RNA:DNA duplexes ("end-nibbling") may influence the interpretation of the results. Therefore the amount of S1-nuclease must be optimized for each new probe used. We give suggestions for optimizing the amount of S1-nuclease, detecting end-nibbling, or underdigestion in S1-nuclease mappings. The enzyme concentration used in our protocols can serve as a starting point. Some end-nibbling and underdigestion are usually seen as weaker bands around the stronger RNA-protected fragment (**Fig. 2**, *hrdD*-P1 promoter).

PCR technology can simplify S1-nuclease mapping in several ways. First, if no suitable restriction site is found to 5'-end-label the probe, the labeled fragment can be prepared by PCR amplification using the template DNA and reverse oligonucleotide (complementary to the region about 100 bp downstream of the 5'-end of the proposed coding region) labeled at the 5' end by $[\gamma^{32}P]ATP$ and T4 PNK, and unlabeled direct oligonucleotide upstream of the 5'-end of the proposed coding region. The length of the latter depends on the size of the region to be screened; usually ~500 bp upstream is sufficient. Second, the dideoxynucleotide sequencing ladders prepared from the template DNA and $[^{32}P]$-labeled reverse oligonucleotide can be used for sizing the protected fragment. This is easier than preparing Maxam and Gilbert sequencing ladders, and the length of the fragments can be directly compared, as they share a common 3'-end.

**Subheading 3.** describes our modification of this NaTCA-based method for detection of 5'-ends (TSP) of mRNA, which we have used to locate potential promoters directing expression of several sigma factor genes in the course of differentiation of *S. aureofaciens*. Improved procedures for purification of $[^{32}P]$ 5'-labeled DNA fragments and for total RNA preparation *(14)* are also described. Our procedure increases the quality and yield of RNA from *Streptomyces* mycelia and spores, and can also be used for other bacteria (*Escherichia coli, Bacillus subtilis*), and yeast.

## 2. Materials
### 2.1. Bacterial Strains

*S. aureofaciens* CCM3239 wild-type (ATCC 10762) used in this study was from Czechoslovak Collection of Microorganisms, Brno, Czech Republic. For RNA isolation from surface culture, $10^8$ spores of *S. aureofaciens* wild-type strain were spread on sterile cellophane membranes placed on Bennet medium (0.1% Difco Yeast extract, 0.1% Difco meat extract, 1% maltose, 0.2% Difco Trypton, 1.5% Difco agar, pH 7.2), and grown for: 13 h (substrate mycelium), 19 h (beginning of aerial mycelium), and 36 h (septation of aerial mycelium into spores). *E. coli* SURE™ (Stratagene) was used as host for all *E. coli* cloning experiments.

### 2.2. Reagents and Solutions
#### 2.2.1. Isolation of RNA

1. Diethyl pyrocarbonate (DEPC), (Sigma, cat. no. D5758), store at 4°C.
2. Guanidine isothiocyanate (Gibco-BRL, ultra pure, enzyme grade, cat. no. 5535UA), store at 4°C.
3. Guanidine solution: 4 $M$ guanidine isothiocyanate, 25 m$M$ sodium citrate pH 7.0, 0.5% *N*-laurylsarcosine, 0.1 $M$ 2-mercaptoethanol. Mix in DEPC-treated beaker with magnetic stirrer (*see* **Note 1**): 40 g guanidine isothiocyanate, 48.5 mL DEPC-treated water, 2.8 mL DEPC-treated 0.75 $M$ sodium citrate pH 7.0 (Sigma, cat. no. S4641), 0.5 g *N*-lauryl-

sarcosine (Sigma, cat. no. L5777). Filtrate through Nalgene disposable filter (0.2 mm), and store at 4°C. Stable for 6 mo. Add 2-mercaptoethanol (Sigma, cat. no. M6250) before use (14.4 mL/2 mL of solution).

4. Na-acetate solution: 3 $M$ sodium acetate (Sigma, cat. no. S2889), pH 4.0. Treat with DEPC (*see* **Note 1**). Store at room temperature.
5. Chloroform. Store in the dark at room temperature in DEPC-treated bottle.
6. Isopropanol. Store at 4°C in DEPC-treated bottle.
7. Ethanol 70% (v/v). Mix with DEPC-treated water, and store in DEPC-treated bottle at –20°C.
8. Phenol (Sigma, cat. no. P5566). Mix in DEPC-treated bottle: 55 g phenol, 50 mg 8-hydroxyquinoline (Sigma, cat. no. H6878) and 15 mL DEPC-treated water. After dissolving at 65°C, store at –20°C. Just before use, equilibrate to 65°C. Take lower phase.
9. Phenol/chloroform: Mix in DEPC-treated bottle: 55 g phenol, 50 mg 8-hydroxyquinoline, 15 mL DEPC-treated water, and 55 mL chloroform. After dissolving at room temperature, store at 4°C. Take lower phase.
10. Formamide (Sigma, cat. no. F5786). Deionize 4 h by stirring with AG 501-X8(D) resin (Bio-Rad, cat. no. 143-6425), filter through 0.2-μm nylon filter, and store in aliquots at –20°C.

## 2.2.2. Preparation of the $^{32}$P-Labeled DNA Probes

Unless otherwise stated, all solutions were prepared in bidest water, filtered through 0.2-μm filter, and autoclaved.

1. Suitable restriction endonuclease enzymes, and optimal buffers supplied by manufacturer (New England Biolabs).
2. Calf intestinal alkaline phosphatase (CIP) (New England Biolabs, cat. no. 290S).
3. 10X CIP buffer: 0.5 $M$ Tris-HCl (Trizma base, Sigma, cat. no. T6791), 10 m$M$ MgCl$_2$, 1 m$M$ ZnCl$_2$, 10 m$M$ spermidine (Sigma, cat. no. S2626), pH 9.4. Store at –20°C.
4. Alkaline phenol/chloroform: phenol:chloroform:isoamyl alcohol: 8-hydroxyquinoline, 50:50:1:0.1, equilibrated with 50 m$M$ Tris-HCl, pH 8.3. Store at 4°C. Stable for 2 mo.
5. TE buffer: 10 m$M$ Tris-HCl, 1 m$M$ ethylenediamine tetraacetic acid (EDTA) (Sigma, cat. no. E5391), pH 8.0. Store at 4°C.
6. 3 $M$ sodium acetate, pH 5.2. Store at 4°C.
7. 10X buffer 1: 100 m$M$ Tris-HCl, 100 m$M$ MgCl$_2$, 10 m$M$ DTT, pH 7.0. Store at –20°C.
8. 10X STOP buffer: 0.1% xylene cyanol (XC, Sigma, cat. no. X4126), 0.1% bromphenol blue (BPB, Sigma, cat. no. B8026), 0.2 M EDTA, 1% SDS, 60% glycerol, pH 8.0. Store at room temperature, warm for 5 min at 65°C before use.
9. Agarose (Gibco-BRL, ultra-pure, electrophoresis grade, cat. no. 15510-027), store at 4°C.
10. 50X TAE buffer: 242 g Tris, 57.1 mL glacial acetic acid, 100 mL 0.5 $M$ EDTA pH 8.0, adjust with bidest water to 1000 mL. Store at 4°C.
11. 10X TBE buffer: 108 g Tris, 55 g boric acid (Sigma, cat. no. B7901), 40 mL 0.5 $M$ EDTA, pH 8.0, adjust with bidest water to 1000 mL. Store at 4°C.
12. [γ$^{32}$P]ATP, 3000 Ci/mmol (Amersham). Store at –70°C, half-life 14.3 d.
13. T4 polynucleotide kinase (T4-PNK) (10,000 U/mL; New England Biolabs, cat. no. 201S). Store at –20°C.
14. 10X KIN buffer: 0.5 $M$ Tris-HCl, 0.1 $M$ MgCl$_2$, 1 m$M$ spermidine, 1 m$M$ EDTA, 50 m$M$ DTT (dithiothreitol, Sigma, cat. no. D9779), pH 7.5. Store at –20°C.
15. 1 $M$ NaCl. Store at 4°C.
16. DEAE paper (ion exchange chromatography paper DE81, Whatman, cat. no. 3658 915).
17. Sonicated carrier DNA, solution of 10 mg/mL (Sigma, cat. no. D7290). Store at –20°C.

## 2.2.3. S1-Nuclease Mapping

1. 3 *M* sodium acetate pH 7.0. Treat with DEPC (*see* **Note 1**). Store at 4°C.
2. NaTCA solution: 3 *M* NaTCA, 50 m*M* Pipes-NaOH pH 8.0, 5 m*M* EDTA, pH 8.0. To DEPC-treated microcentrifuge tube add 574 mg NaTCA, 100 mL 0.5 *M* Pipes-NaOH pH 7 (Pipes, Sigma, cat. no. P9219, prepared in DEPC-treated water, adjust with NaOH to pH 7.0, and sterilize by filtration through 0.2-μm filter), 10 mL 0.5 *M* EDTA, pH 8.0 (DEPC-treated), and DEPC-treated water. After dissolving adjust volume to 1 mL with DEPC-treated water. Extract with chloroform, and store at –20°C. The solution is stable for at least 1 yr. Preparation of sodium trichloracetate (NaTCA): Titrate 50 mL of 1 *M* Na$_2$CO$_3$ solution with 2 *M* trichloracetic acid (TCA) (Sigma, cat. no. T9159) (~47 mL) to pH 7.0 under nitrogen (solution must be bubbled with nitrogen). Stir the solution for 4 h (in the dark) with activated acid-washed charcoal (Sigma, cat. no. C4386), then overnight with Chelex 100 Resin (Bio-Rad, cat. no. 142-2832), and filter through 0.2-μm filter. Concentrate the solution by rotary evaporation at 50°C, until crystals began to form and then cool to 4°C. Filter crystals through Whatman no. 1 paper and dry under vacuum. Store NaTCA at room temperature in the dark.
3. 5X S1 solution: 1.25 *M* NaCl, 0.2 *M* sodium acetate, pH 4.5, 5 m*M* ZnCl$_2$. Store at –20°C.
4. S1-nuclease, 250,000 U/mL (Sigma, cat. no. N7385). Store at –20°C.
5. S1 termination buffer: 2.5 *M* ammonium acetate, 50 m*M* EDTA, 20 mg/mL tRNA. Store at –20°C.
6. 10 mg/mL tRNA (Sigma, cat. no. R1753). Store at –20°C.
7. Maxam loading buffer: 80% formamide, 10 m*M* NaOH, 1 m*M* EDTA, 0.05% XC, 0.05% BPB. Store at –20°C.
8. Isopropanol, ethanol, and 70% ethanol (*see* **Subheading 2.2.1.**).

## 2.2.4. Analysis of Nuclease-Protected Fragments

1. Pyridinium formiate solution: 1 *M* formic acid (Sigma, cat. no. F4636) adjust to pH 2.0 with pyridine (Sigma, cat. no. P4036). Store at 4°C.
2. Hydrazine anhydrous (Sigma, cat. no. H2761). **Toxic and explosive.** Store at 4°C.
3. Hydrazine stop solution: 0.3 *M* sodium acetate, pH 7.0, 0.1 m*M* EDTA, 25 mg/mL tRNA. Store at 4°C, and add tRNA just before use.
4. 0.4 *M* NaCl. Store at 4°C.
5. 1 *M* piperidine: 10X dilute piperidine (Sigma, cat. no. P5881) with water. Prepare fresh before use.
6. 0.3 *M* sodium acetate. Store at 4°C.
7. Urea (Sigma, cat. no. U5378).
8. 40% acrylamide (Sigma, cat. no. A9099)/1.3% *N,N'* methylenebisacrylamide (Sigma, cat. no. M2022). Deionize the solution 4 h by stirring with AG 501-X8(D) resin, filter through 0.2-μm nylon filter, and store at –4°C. A potent neurotoxin.
9. *N,N,N',N'*-tetramethylethylenediamine (TEMED) (Sigma, cat. no. T7024). Store at 4°C.
10. 10% (w/v) ammonium persulfate (Sigma, cat. no. A9164). Prepare fresh before use, or store up to 1 mo at –20°C.
11. Dimethyldichlorsilane (DDS) (Sigma, cat. no. D3879).

## 2.4. Equipment

1. Whatman no. 1 and 3MM paper (VWR Scientific Products).
2. 12-mL Greiner disposable tube (Greiner, cat. no. 187261), or Falcon 15-mL tubes (cat. no. 2096).
3. Eppendorf 1.5-mL microcentrifuge tubes.

   4. Screw-cap 1.5-mL microcentrifuge tubes.
   5. DNase/RNase-free MultiFit Guard Filter tips 10, 20, 200, 1000 µL (Sorenson BioScience, Inc).
   6. Glass balls, 3.5–4.5-mm diameter (Kavalier-Glassexport a.s., Liberec, Czech Republic). Boil 10 min in 1 *M* HCl, then 3 × 10 min in water. Treat with DEPC (*see* **Note 1**), and dry 4 h at 200°C.
   7. Autoradiographic film (XAR-5, Kodak).
   8. Horizontal electrophoresis apparatus (for agarose-gel electrophoresis, available from several suppliers).
   9. Vertical electrophoresis apparatus (for acrylamide-gel electrophoresis, available from several suppliers).
  10. Glass plates (15 × 15 cm, 2-mm spacers and comb) for preparative polyacrylamide electrophoresis.
  11. Sequencing electrophoresis apparatus (available from several suppliers).
  12. Glass plates (18 × 50 cm, 0.4-mm spacers and comb) for sequencing electrophoresis.
  13. Power supply, Feather Volt 3000 (Stratagene).
  14. Gel dryer (Model 583, Bio-Rad).
  15. Water bath with water-cooling system (available from several suppliers).

## 3. Methods

### 3.1. Isolation of RNA from Different Developmental Stages of Streptomyces aureofaciens

   This procedure for RNA isolation is a modification of our previously published method *(15)*. Isolating RNA from all developmental stages of *S. aureofaciens* in the presence of strong inhibitors of ribonucleases, guanidine isothiocyanate and hot phenol, substantially increased the yield and decreased the degradation of isolated RNA. The RNA is also almost free of contaminated DNA. DNase I treatment, used in several other methods as a final step in RNA purification, is unnecessary, and the RNA can be used directly for S1-nuclease mapping.

   The protocol is the following:

   1. Grow cells of *S. aureofaciens* on cellophane discs placed on solid medium, scrape off the cells with a spatula, and filter through Whatman paper No.1 in Buchner funnel (200–400 mg of wet wt; usually from 3 to 8 plates of 82-mm-diameter, depending on time of growth). Wash cells four times with ice-cold DEPC-treated water (with about 4 × 20 mL) (*see* **Note 2**).
   2. Immediately scrape the cell paste into a 12–15 mL disposable tube (on ice) containing ice-cold mixture of 2 mL guanidine solution, 150 mL Na-acetate solution, and 1 mL chloroform, and 1 mL of glass balls.
   3. Immediately add 2 mL hot phenol (equilibrated to 65°C) and vortex the mixture shortly. Equilibrate to 65°C in water bath for 2 min (tube must be partially open). Thoroughly vortex the mixture 5 times for 1 min with 1-min pause for heating at 65°C in water bath (*see* **Note 3**).
   4. Place the tube on ice for 15 min, centrifuge for 10 min at 10,000*g* at 2°C.
   5. Transfer the aqueous phase to new 12–15-mL tube and extract with 2 mL phenol/chloroform. Centrifuge the mixture as in **step 4**, transfer the aqueous phase to new tube, and precipitate with an equal volume of isopropanol for 1 h at –20°C.
   6. Pellet the RNA by centrifugation as in **step 4**, remove the supernatant completely, dissolve the pellet in 0.6 mL of guanidine solution.
   7. Transfer the mixture to a microcentrifuge tube, extract with an equal volume of phenol/chloroform, and centrifuge as in **step 4**. Repeat the extraction until no precipitate is detected in the interphase.

8. Divide the mixture into 3–5 equal aliquots in new microcentrifuge tubes and precipitate with an equal volume of isopropanol for at 1 h at –20°C (could be also overnight).

9. Pellet the RNA by centrifugation as in **step 4**, wash pellet two times with 70% ethanol, and store at –70°C under the 70% ethanol.

10. To characterize the isolated RNA, pellet one aliquot by centrifugation, remove supernatant completely, dry pellet leaving 10 min open tube in clean bench. Dissolve the RNA in 50 μL formamide and determine concentration spectrophotometrically by measuring $A_{260}$ of 5 μL of RNA sample in 1 mL water against water (1 $A_{260}$ unit corresponds to 40 μg/mL) (*see* **Note 4**).

## 3.2. Preparation of the $^{32}$P-Labeled DNA Probes for S1-Nuclease Mapping

DNA fragments used as probes for S1-nuclease mapping were labeled with [$\gamma^{32}$P]ATP and T4-PNK at 5' protruding end 100–300 bp downstream of the 5'-end of a proposed coding region, after dephosphorylation with CIP. After labeling, the fragment was digested by restriction endonuclease from the upstream part (approx 500 bp– 1 kb upstream of the 5'-end of a proposed coding region), and the labeled fragment was purified by polyacrylamide electrophoresis, transferred on DEAE-paper, and eluted with 1 *M* NaCl.

The protocol for fragment labeling and purification is the following:

1. Digest from 6 to 10 μg of plasmid DNA (depending on the length of the plasmid, usually 3 pmol of DNA) in 50 μL of appropriate restriction buffer with 20 U of restriction endonuclease for 4 h in optimal temperature for restriction endonuclease used (*see* **Note 5**).

2. Add 5 μL of 10X CIP buffer and 2 U of CIP and incubate for 30 min at 37°C.

3. Add another aliquot of CIP and incubate for further 30 min.

4. Extract the mixture with alkaline phenol/chloroform. Reextract organic phase with 50 μL TE buffer, and extract the pooled aqueous phases with chloroform. Precipitate the fragment by adding 10 μL of 3 *M* sodium acetate, pH 5.2, and 2.5 vol of ethanol and leaving 1 h at –20°C (*see* **Note 6**).

5. Pellet the fragment by centrifugation, wash with 70% ethanol, and dry for 5 min in Speed-Vac.

6. Dissolve the fragment in 26 μL water. Add 3 μL of 10X restriction buffer and 10 U of particular restriction endonuclease, and incubate for 1 h in optimal temperature for restriction endonuclease used.

7. Add 3 μL of 10X CIP buffer and 2 U of CIP and incubate for further 30 min at 37°C. Stop the reaction by adding 3 μL 10X STOP buffer, and load sample on 0.8% agarose gel in TAE with 0.5 μg/mL ethidium bromide.

8. After electrophoresis, isolate the corresponding fragment from agarose gel by any of commercial kit for DNA-fragment isolation (for instance, GeneClean technique, BIO 101, La Jolla, CA) (*see* **Note 7**).

9. Estimate the concentration by electrophoresis of 1 μL of the DNA fragment prepared along with the known amount (100 ng) of DNA standard.

10. Label the 5' protruding ends of the fragment by mixing 1 pmol of the fragment (corresponds to 660 ng of 1 kb fragment, or 2.64 μg of 4 kb fragment) with at least 20 pmol [$\gamma^{32}$P]ATP (3000 Ci/mmol, 60 μCi), and 20 U of T4-PNK in 30 μL of KIN solution. After 30 min at 37°C, inactivate the activity of T4 PNK by 15 min incubation at 70°C.

11. Add 1 μL of 10X buffer 1, adjust salt (NaCl, KCl) for the optimal concentration for second restriction enzyme, adjust volume to 40 μL with water, and incubate the sample for 2 h with 20 U of restriction endonuclease (*see* **Note 8**).

12. Add 4 μL of 10X STOP buffer, heat the sample for 5 min at 65°C, and electrophorese through a preparative 5% polyacrylamide gel (15 × 15 cm, 2-mm-thick) in TBE until XC is about 4 cm before end of the gel (*see* **Note 9**).

13. Cover the gel, laying on the glass plate, by Saran wrap, label by hot ink (0.5 μL on a stick label) (*see* **Note 10**), and expose 10 min to autoradiography film.

14. Cut the region in the gel corresponding to the signal with scalpel, remove the piece of the Saran wrap, stick the piece of the DEAE paper having approximately the same size as the signal on the gel, and remove the piece of the gel from the gel. Put the same-size Whatman 3MM paper on the other side of the gel piece, put such a sandwich into the wide slot in a small 2% agarose gel, and electrophorese in TBE buffer for 30 min at 9 V/cm (*see* **Note 11**).

15. Remove the sandwich by forceps, remove the DEAE paper, wash it briefly in DEPC-treated water, dry briefly on DEPC-treated Whatman paper, and put into microcentrifuge tube with a small hole in the bottom containing 300 μL 1 *M* NaCl, and standing on the second microcentrifuge tube. Disperse the DEAE paper by a tip and incubate the tubes for 15 min at 37°C in a cabinet. Insert the tubes into 12–15 mL tube and shortly centrifuge (about 20 s at 10,000*g*) to elute the eluted fragment through the hole into the second microcentrifuge tube. The DEAE paper remains in the upper tube (*see* **Note 12**).

16. Extract the eluted fragment with chloroform, centrifuge (2 min at 10,000*g*), remove the aqueous phase to a new microcentrifuge tube. Store the labeled fragment at –20°C until the use (*see* **Note 13**).

17. For Maxam and Gilbert G+A, and T+C chemical reactions, pipet 40 μL of the labeled fragment into a new microcentrifuge tube containing 2 μL of sonicated carrier DNA. Precipitate the fragment with 2.5 vol of ethanol for at least 1 h at –20°C.

### *3.3. S1-Nuclease Mapping*

The following method of S1-nuclease mapping uses the NaTCA hybridization method. The [$^{32}$P] 5'-labeled DNA fragment prepared as described above is coprecipitated with the optimized amount of total RNA of the interest. The mixture is further hybridized in the NaTCA hybridization solution, digested with S1-nuclease, and further purified before loading on denaturation polyacrylamide-sequencing gel. The temperature of hybridization and amount of S1-nuclease is optimal for low background and minimizing "end-nibbling." Therefore, it is recommended as a starting experiment with any fragment and RNA. However, if there is high background, or significant "end-nibbling," it is recommended to optimize the amount of S1-nuclease in the reaction.

The mappings were done by the protocol described in **steps 3–5**.

1. Pellet a known amount of RNA in a prepared aliquot (*see* **Subheading 3.1.**) by centrifugation, remove supernatant completely, and dry pellet, leaving the tube open for 10 min in a clean bench. Dissolve the RNA in ice-cold DEPC-treated water to final concentration 1 mg/mL (*see* **Note 14**). Coprecipitate 40 μg of RNA with 0.02 pmol of labeled fragment (*see* **Subheading 3.2.**) (with the activity in the range 30,000–200,000 cpm), 4 μL 3 *M* sodium acetate, pH 7.0, and 3 vol of ethanol for at least 1 h at –20°C (*see* **Note 15**).

2. Centrifuge the samples for 10 min at 10,000*g* and 2°C, washed two times with 70% ethanol, and dry as in **step 1**.

3. Dissolve the samples in 20 μL NaTCA solution (*see* **Note 16**). Seal the microcentrifuge tubes with parafilm and incubate in water bath at 65°C for 5 min. Decrease the temperature of the water bath quickly (by water cooling) to about 57°C, and then slowly to 45°C (for about 30 min, without water cooling). Hybridize the samples for at least 4 h at 45°C.

4. Open the tubes in water bath, immediately add 300 µL of ice-cold S1 solution with 100 U of S1-nuclease and 20 µg/mL carrier DNA (*see* **Note 17**), transfer tubes to ice, and incubate for 30 s. Finally incubate samples 30 min at 37°C.
5. Stop the reaction with 75 µL of S1 termination buffer, 1 µL 10 mg/mL tRNA, and 0.4 mL alkaline phenol/chloroform. Vortex the mixture for 3 min, and centrifuge for 10 min at 10,000*g* at room temperature. Transfer an aqueous phase to a new tube, and precipitate by equal volume of isopropanol 30 min at –20°C. Pellet the RNA-protected DNA fragment by centrifugation as in **step 2**, wash with 70% ethanol, dry for 2 min in a Speed-Vac, and dissolve in 5-µL Maxam loading buffer.

## 3.4. Analysis of Nuclease-Protected Fragments

To detect TSP of each promoter, the RNA-protected DNA fragments were analyzed on sequencing urea-polyacrylamide gel, along with Maxam and Gilbert sequencing ladders. For exact location of TSP it was sufficient to load only G+A and T+C sequencing reactions. The reactions were prepared by modification of Maxam and Gilbert sequencing reactions *(14)*.

1. Coprecipitate the labeled fragment with the carrier DNA (*see* **Subheading 3.2.**), centrifuge for 10 min at 10,000*g* and 2°C, wash pellet two times with 70% ethanol, and dry 2 min in a Speed-Vac.
2. Dissolve the fragment in 19 µL water and divide to two siliconized microcentrifuge tubes (8 µL for G+A reaction; 11 µL for T+C reaction).
3. Add 10 µL water and 3 µL of pyridinium formiate solution to the G+A reaction, incubate the sample 13 min at 37°C, freeze in liquid nitrogen, and lyophilize at least 2 h.
4. During the lyophilization, add 15 µL water and 30 µL hydrazine (*see* **Note 18**) to the C+T, and incubate the sample 5 min at 18°C. Stop the modification reaction with 750 µL ice-cold hydrazine stop solution, 2 µL 10 mg/mL tRNA, and 750 µL cold ethanol. After vortex mixing, incubate the sample in ethanol/dry ice bath (–80°C) for 10 min, centrifuge as in **step 1**, and dissolve pellet briefly in 250 µL ice-cold hydrazine stop solution. Precipitate the sample with 750 µL cold ethanol 10 min in –80°C as above, centrifuge as in **step 1**, and dissolve the pellet in 50 µL 0.4 *M* NaCl. Precipitate the sample with 1 mL cold ethanol 10 min in –80°C, and centrifuge as in **step 1**. Wash the pellet with 1 mL 95% ethanol, and dry 2 min in a Speed-Vac.
5. Dissolve the both reactions (A+G, C+T) in 50 µL 1 *M* piperidine, transfer to siliconized 1.5-mL screw microcentrifuge tubes, and incubate 30 min at 90°C.
6. Precipitate the samples with 50 µL of 0.3 *M* sodium acetate and 0.4 mL ethanol for 10 min at –80°C as in **step 4**. Wash the pellet with 95% ethanol, dry for 2 min in a Speed-Vac, dissolve in 20 µL water, lyophilize 4 h, and dissolve in 5 µL Maxam loading buffer.
7. Heat the RNA-protected DNA fragments (*see* **Subheading 3.3.**), G+A, and T+C sequencing reactions for 2.5 min at 95°C, cool immediately on ice, and load 1.5-µL aliquots on 6% polyacrylamide/urea sequencing gel (*see* **Subheading 3.4.1.**) after 30 min prerun (for heating the glass plates). Run the gel in TBE buffer at ~50 V/cm (55 W, temperature of glass plates must be about 65°C) until BPB is out of gel (*see* **Note 19**).
8. Transfer the gel on Whatman 3MM paper, overlay with Saran wrap, and dry in gel dryer for 3 h at 80°C. Expose the dried gel on Kodak XAR-5 autoradiographic film.

**Figure 2** shows the results of S1-nuclease mapping of TSP of all identified sigma factor genes in *S. aureofaciens*, using RNA prepared from all developmental stages. Using the 5'-labeled probes for *hrdA, hrdB, hrdD, rpoZ* and *sigF*, RNA-protected fragments were identified with RNA isolated from different developmental stages. The fragments corresponded to the TSP of tandem promoters termed *hrdA*-P1,2; *hrdB*-P1,2; *hrdD*-P1,2; *rpoZ*-P1,2; *sigF*-P1,2. No RNA-protected fragment was identified with tRNA as a control (lane 4).

### 3.4.1. Preparation of 6% Polyacrylamide/7 M Urea Sequencing Gel

Dissolve 42 g of urea (by stirring and heating) with 10 mL of 10X TBE buffer and 15 mL 40% acrylamide/1.3% *N,N*′ methylenebisacrylamide in final volume 100 mL, and filter through 0.2-μm filter. After degassing, add 0.5 mL of freshly prepared 10% ammonium persulfate solution, and 27 μL TEMED. After mixing, immediately pour the mixture into assembled glass-plates sandwich (18 × 50 cm, 0.4-mm thick), and insert a teflon comb. Allow the gel to polymerize for 1 h at room temperature (*see* **Note 20**).

## 4. Notes

1. To DEPC-treat water and solutions, add DEPC to 0.1%, mix, treat overnight, and autoclave twice. To DEPC-treated equipment (microcentrifuge tubes, beakers, bottles, magnetic stirrers), soak in 0.1% DEPC, treat overnight, autoclave twice, and dry at 60°C.
2. To use liquid-cultured *Streptomyces* cells grown as mycelium, cool the cells on ice for 10 min, filter through Whatman paper, and wash with ice-cold DEPC-treated water. For other liquid-cultured cells (*E. coli,* yeast), cool the cell on ice for 10 min, centrifuge for 10 min at 2°C, and 5000*g*, and wash four times with ice-cold DEPC-treated water.
3. Wear latex gloves during all purification steps. Vortex with caution, as hot phenol is a strong irritant and is also toxic. Mix the glass balls through the whole tube, not only at the bottom, and hold the tube aslant on the Vortex.
4. RNA stored under 70% ethanol at –70°C is stable, no degradation was detected after 1 yr. To test the RNA, dissolve it in formamide, which inactivates RNases. RNA dissolved in formamide can be stored at –20°C for at least 1 yr without degradation. Before S1 mapping, check the integrity of RNA by standard electrophoresis for RNA in 1.2% agarose gel containing 2.2 *M* formaldehyde (by loading of 5 μL of test sample directly on gel after 5 min heating at 60°C), and staining with ethidium bromide. Sharp bands of rRNA indicate that the preparation is intact.
5. Select a restriction site where cleavage will yield a 5′ protruding end for effective dephosphorylation and end labelling. Also restriction sites producing blunt or 3′ protruding ends can be dephosphorylated by modifying the condition. Add CIP, incubate the mixture for 15 min at 54°C, and then 15 min at 37°C. Add another aliquot of CIP and incubate further 15 min at 54°C and 15 min at 37°C.
6. This purification step and the following digestion and dephosphorylation increase the specific activity of the labeled fragment, since more 5′ ends are dephosphorylated.
7. Agarose-gel purification of the fragment before labeling is extremely important, since most plasmid preparations (CsCl gradient purification, QIAGEN kit purification) contain the rest of small RNA fragments after RNase treatment (invisible on agarose electrophoresis), and these can also be labeled by T4 PNK.
8. See the conditions for restriction-enzyme digestion in Chapters 27–29.
9. Preparative polyacrylamide gel: mix 10 mL of 10X TBE buffer, 12.5 mL 40% acrylamide/ 1.3% *N,N*′ methylenebisacrylamide, and 76.4 mL water. Filter the solution through 0.2-μm filter. After degassing, add 1 mL of freshly prepared 10% ammonium persulfate solution and 60 μL TEMED. After mixing, pour the mixture immediately into assembled glass-plates sandwich (15 × 15 cm, 2-mm thick), and insert teflon comb. Allow the gel to polymerize for 20 min at room temperature. The run (in 1X TBE) of polyacrylamide

electrophoresis depends on the length of the isolated fragment. The mobility of XC and BPB in this gel are 260 and 65 bp, respectively.

10. Prepare the hot ink by pipetting about 10 µL of ink into the microcentrifuge tube used for labeling.

11. The usual size of the gel piece is 12 × 5 mm, the slot in the agarose gel has the same size, and the thickness of the slot is about 3 mm. The agarose gel is 6-mm-thick, and electrophoresis is run with the buffer not overloading the gel.

12. All solutions, microcentrifuge tubes, and tips used in this and following steps are DEPC-treated. Burn forceps in a flame to remove potential contamination by RNases. The microcentrifuge tubes used for all further steps are, before DEPC-treatment, siliconized by filling them with 5% DDS in chloroform for 5 min, and washing with water. Make a hole in the bottom of microcentrifuge tube by inserting a 20-gage × 1$^1/_2$ in. needle until the point of the needle is seen inside the tube.

13. To determine the activity of the labeled fragment, spot 1 µL of the eluted fragment on a small piece of Whatman 3MM paper, dry and measure radioactivity in scintilation counter. Calculate the chemical concentration from the starting amount of the fragment (1 pmol) with the estimated yield of about 80%.

14. After adding the ice-cold DEPC-treated water to the RNA, incubate microcentrifuge tube in water bath at 65°C, repeatedly vortex the sample for about 5 min (with 20 s pause for incubation at 65°C), and after dissolving put back on ice.

15. 40 µg of *E. coli* tRNA is used as a control sample.

16. Dissolve the pellet at room temperature by pipeting (about 100 times) through tip. Briefly warm the sample (2 min at 60°C), thoroughly vortex, and briefly centrifuge.

17. Prepare the S1 solution before use from 5X S1 solution and DEPC-treated water. After cooling on ice, add fresh denatured sonicated carrier DNA (boiled for 10 min and cooled on ice) to a final concentration 20 µg/mL. Finally add the calculated amount of S1-nuclease, mix solution and use immediately.

18. **Anhydrous hydrazine is volatile, toxic, and explosive.** We have found that the hydrazine hydrate (Sigma, cat. no. H0883) can also be used without any negative effect. In this case 45 µL is directly added to the 11 µL of the labeled fragment.

19. For best resolution, the run of denaturing polyacrylamide electrophoresis depends on the proposed mobility of the protected fragment. The mobility of XC and BPB in this gel are 106 bp and 26 bp, respectively. At the beginning, it is recommended to run the gel until BPB is out of gel, and to run the second gel based on the results after exposition.

20. Siliconize glass plates by soaking in 2% DDS in chloroform for 10 min. Clean the plates with ethanol and assemble.

## Acknowledgments

This work was supported by a Grant 2/4007/98 from Slovak Academy of Sciences.

## References

1. Chater, K. F. (1993) Genetics of differentiation in *Streptomyces. Annu. Rev. Microbiol.* **47,** 685–713.

2. Lonetto, M., Gribskov, M., and Gross, C. A. (1992) The $\sigma^{70}$ family: sequence conservation and evolutionary relationships. *J. Bacteriol.* **174,** 3843–3849.

3. Kormanec, J., Farkasovsky, M., and Potuckova, L. (1992) Four genes in *Streptomyces aureofaciens* containing a domain characteristic of principal sigma factors. *Gene* **122,** 63–70.

4. Kormanec, J., Potuckova, L., and Rezuchova, B. (1994) The *Streptomyces aureofaciens* homologue of the *whiG* encoding a putative sigma factor essential for sporulation. *Gene* **143,** 101–103.

5. Potuckova, L., Kelemen, G. H., Findlay, K. C., Lonetto, M. A., Buttner, M. J., and Kormanec, J. (1995) A new RNA polymerase sigma factor, $\sigma^F$, is required for the late stages of morphological differentiation in *Streptomyces* spp. *Mol. Microbiol.* **17,** 37–48.

6. Rezuchova, B., Barak, I., and Kormanec, J. (1997) Disruption of a sigma factor gene, *sigF*, affects an intermediate stage of spore pigment production in *Streptomyces aureofaciens*. *FEMS Microbiol. Let.* **153,** 371–377.

7. Kormanec, J. and Farkasovsky, M. (1993) Differential expression of principal sigma factor homologues of *Streptomyces aureofaciens* correlates with developmental stage. *Nucleic Acids Res.* **21,** 3647–3652.

8. Kormanec, J., Homerová, D., Potuckova, L., Novakova, R., and Rezuchova, B. (1996) Differential expression of two sporulation specific σ factors of *Streptomyces aureofaciens* correlates with the developmental stage. *Gene* **181,** 19–27.

9. Berk, A. J. and Sharp, P. A. (1977) Sizing and mapping of early adenovirus mRNA by gel electrophoresis of S1 endonuclease hybrids. *Cell* **12,** 721–732.

10. Burke, J. F. (1984) High-sensitivity S1 mapping with single-stranded [$^{32}$P]DNA probes synthesized from bacteriophage M13mp templates. *Gene* **30,** 63–68.

11. Murray, M. G. (1986) Use of sodium trichloracetate and mung bean nuclease to increase sensistivity and precision during transcript mapping. *Anal. Biochem.* **158,** 165–170.

12. Aldea, M., Claverie-Martin, F., Diaz-Torres, M. R., and Kushner, S. R. (1988) Transcript mapping using [$^{35}$S]DNA probes, trichloracetate solvent and dideoxy sequencing ladders: a rapid method for identification of transcriptional start points. *Gene* **65,** 101–110.

13. Weaver, R. F. and Weissmann, C. (1979) Mapping of RNA by a modification of the Berk-Sharp procedure: the 5' termini of 15 S beta-globin mRNA precursor and mature 10 S beta-globin mRNA have identical map coordinates. *Nucleic Acids Res* **7,** 1175–1193.

14. Maxam, A. M. and Gilbert, W. (1980) Sequencing end-labelled DNA with base specific chemical cleavages. *Methods Enzymol* **65,** 449–560.

15. Kormanec, J. and Farkasovsky, M. (1994) Isolation of total RNA from yeast and bacteria and detection of rRNA in Northern blots. *BioTechniques* **17,** 838–842.

# 32

# Detection and Quantitation of mRNAs Using Ribonuclease Protection Assays

## Ellen A. Prediger

## 1. Introduction

The ribonuclease protection assay (RPA) is a sensitive and straightforward method for detecting, quantitating, and mapping specific mRNA transcripts (5' and 3' ends, intron:exon boundaries; *1–6*). The method is an adaptation of the S1 nuclease assay (*7,8*; *see also* Chapter 31) where RNA, instead of DNA probes are used, and single-stranded specific ribonucleases replace S1 nuclease. Treatment of RNA:RNA duplexes with ribonuclease is more reproducible than treatment of RNA:DNA duplexes with S1 nuclease *(9)*. However, both S1 assays and RPAs are robust techniques, and can be used fairly interchangeably to analyze mRNA transcript abundance and structure. Solution hybridization assays such as the RPA and S1 nuclease assay provide greater sensitivity than hybridization protocols that rely on RNA bound to a solid support (e.g., Northern blots, dot blots) *(10–12)*. Multiple probes (as many as 12; *13–15*; **Fig. 1**) can be used in a single reaction as long as the protected fragment sizes are significantly different. Because of the high resolution of the denaturing polyacrylamide gels used to analyze the protected fragments, RPAs are well-suited for mapping transcript ends and intron:exon junctions *(7,16)*.

The RPA procedure begins by hybridization of a molar excess of single-stranded antisense RNA probe to a sample containing up to 100 µg of total or poly(A) RNA. After hybridization, excess unhybridized probe, portions of the probe not homologous to target, and nontarget sample RNA are degraded by single-strand specific ribonuclease, usually RNase A, RNase T1, and RNase 1, either alone or in combination *(6,17)*. The ribonuclease is then inactivated and/or removed, and the probe:target hybrids are separated by denaturing polyacrylamide urea-gel electrophoresis (*see* **Note 1**). The gel is exposed to film (or, if nonisotopic probes are used, the RNA is transferred to membrane, processed through secondary detection, and exposed to film) for detection and quantitation. **Figure 1** shows the detection of five oncogenes and two internal controls (β-actin and cyclophilin) simultaneously within a single sample using this procedure.

From: *Methods in Molecular Biology, vol. 160: Nuclease Methods and Protocols*
Edited by: C. H. Schein © Humana Press Inc., Totowa, NJ

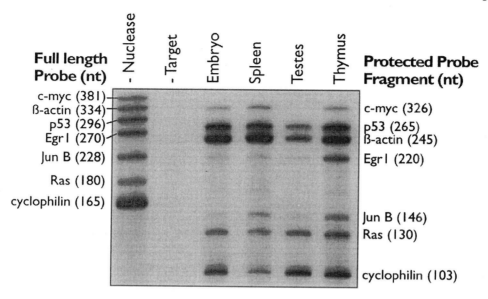

Fig. 1. Simultaneous quantitation of multiple mRNAs using a ribonuclease protection assay. Ten micrograms of various mouse tissue total RNAs were hybridized overnight with 50,000 cpm each of seven distinct probe transcripts, and then digested with RNase A/T1 following the Ribonuclease Protection Assay III™ protocol (Ambion). Protected products were assessed on a denaturing 6% polyacrylamide gel exposed to film for 4 h at –80°C.

## 1.1. RNA Sample

Total or poly(A) RNA (up to 50 μg) can be used for RPA experiments, although use of poly(A) RNA is typically unnecessary. Unlike RT-PCR, small amounts of contaminating DNA will typically not influence RPA data. Thus, RNA samples do not have to be DNase-treated. Samples from cells transfected with a plasmid are an exception. Use of a sense-strand probe, which would only protect DNA, can be used to detect contaminating DNA in the sample.

Most total RNA isolation methods start with sample/cell disruption in a solution containing a chaotrope (e.g., guanidinium, LiCl), that simultaneously inactivates RNases. The RNA is then separated from the other cellular components by RNA-specific binding to silica, differential extraction with phenol, or centrifugation through a density gradient. The protocol provided here uses acidified phenol:chloroform extraction of a guanidinium cell lysate to separate RNA from cellular components and DNA. The reagents are standard in most laboratories, and the procedure can be scaled up or down to accommodate different sample sizes. It is important to note that the method of RNA isolation used for RPAs is not critical and there are a variety of other protocols that will work equally well *(9,18,19)*.

## 1.2. Probe Preparation

The cloning of phage RNA T7, T3, and SP6 polymerases and definition of their respective promoter sequences made possible in vitro transcription of defined, high specific activity RNA transcripts *(20–24)*. RPA probes should each be equal length, single-stranded RNA of antisense strand sequence approx 50–600 nt long. The RNA probe is synthesized from a DNA template which is later destroyed with DNase I. The polymerase promoter must be adjacent to the downstream end of the insert or PCR fragment for transcription of antisense sequence. Promoters can be added to cDNA or genomic inserts, or PCR fragments by appending the 19-base minimal promoter sequence to the 5' end of a downstream PCR primer, and performing PCR (*see* **Note 2**). Probe transcripts should contain some additional sequence not homologous to the target mRNA, so that the shorter, homologous, and thus protected, portion of the probe can be distinguished from a full-length probe. The specific activity of the probe determines the sensitivity of the assay. To detect rare transcripts, high-specific-activity probes are used. When included in the assay, internal control probes (e.g., β-actin, GAPDH, cyclophilin), which are typically of high abundance, are synthesized to 200- to 500-fold lower specific activity (*see* **Note 3**).

## 1.3. Hybridization Conditions

Hybridization is typically done in a formamide-based hybridization buffer that serves to lower the hybridization temperature, allowing stringent RNA:RNA hybridization at the relatively low temperatures of 42–55°C (*25–28*). 42°C is sufficient for most experiments; higher temperatures are necessary only when probe or target sequences are GC-rich, or multiple probes show interaction (*29*; Ambion, personal communication). The hybridization reaction is incubated overnight for rare messages; shorter times can be used for more abundant target transcripts.

For quantitative results, it is important for the probe to be in molar excess over target RNA. If the amount of target is known, one can calculate the amount of probe necessary; however, this is rarely the case. Most target mRNAs will be less abundant than β-actin. By using sufficient probe to be in molar excess of β-actin, one can assume that they will be in molar excess for their mRNA of interest. β-actin makes up approx 0.1% of all mRNA. Therefore, 10 μg of sample RNA would contain 0.4 fmol of β-actin, assuming that the target is 2.1 kb long and that the mRNA represents about 3% of total RNA. A fivefold excess of a 300-base probe with a specific activity of $3 \times 10^8$ cpm/μg would thus require 2.0 fmol, or $6 \times 10^4$ cpm, corresponding to about 200 pg of transcript.

## 1.4. RNase Choice

RNase A, RNase T1, RNase T2, and RNase I have all been used for RPAs. RNase A cleaves after C and U residues (*6*) and RNase T1 cleaves after G residues (*17*). RNase A/T1 cocktails have shown more complete digestion of unhybridized probe than either of these nucleases alone. Although both RNase T2 and RNase I cleave after all four ribonucleotides, RNase T2 preferentially cuts after As (*30*), and RNase I has a narrower concentration range for unhybridized probe degradation than RNase A/T1 mixtures (Ambion, personal communication; *see* **Note 4**). In addition, RNase A is able to cleave single-stranded looped-out regions, and can also cleave single-base mismatches resulting from point mutations (*31*). An RNase A/T1 cocktail will be used in this protocol.

## 1.5. Quantitation and Analysis of Results

RPAs can be used for relative or absolute quantitation. Relative quantitation compares transcript expression level across the samples under study. An internal control transcript which has invariant expression across the samples is included in the study to normalize the signal from the experimental probes (**Fig. 1**). For absolute quantitation, either calculate yield of RNA in the protected bands by comparing probe specific activity to the decays per minute (dpm) measured in the protected fragment bands cut out of the final gel, or create a dilution series of a known sense-strand RNA within the assay (sense-strand dilutions should fall between 10 fg and 50 pg), hybridize probe to both the sense-strand dilutions and the experimental sample RNAs, and compare dpm measured in the protected fragment bands to that from the sense-strand dilution series bands.

When mapping the 5' end of a transcript, the probe template must contain the genomic DNA sequence, extending from upstream of the putative transcript start site to 50–400 bp downstream of the putative transcript start site. Only the length of probe complementary to the mRNA target transcript (5' end and downstream sequence) will be left after hybridization and digestion. Since the position of the downstream end of the probe is known, the length of the protected probe fragment will determine how many bases upstream from this position the mRNA transcript starts.

RNA markers are used to estimate protected fragment size for RPAs, as denatured DNA markers migrate 10–20% faster than RNA under normal gel running conditions *(9)*. A single-base RNA "ladder," used for exact sizing in mapping studies, is generated by treating the RNA probe with dilute NaOH. A second "G" ladder is generated by digesting the probe with RNase T1. The two ladders are run side-by-side on the final gel next to the protected fragments. Alternatively, a DNA sequencing reaction can be used as the single-base ladder, although one cannot infer size information from the sequencing reaction.

The protected fragment bands are quantitated by direct phosphoimaging of the gel, cutting out the gel bands and counting them in a scintillation counter, or densitometry of the exposed film. When included in the experiment, the signal from the internal control probe is used to normalize the signal from each of the experimental samples.

## 2. Materials

### 2.1. Buffers and Solutions

1. Lysis solution: 4 $M$ guanidinium isothiocyanate, 25 m$M$ sodium citrate, 0.1 $M$ β-mercaptoethanol. Make this from stock solutions directly before use.
2. Phenol/CHCl$_3$/isoamylalcohol in the ratio of 25:24:1 (v:v:v).
3. 3.0 $M$ sodium acetate (NaOAc), pH 4.5; pH with acetic acid.
4. Acid-phenol:chloroform, 5:1, pH 4.5 *(32)*. It is also available commercially; commercial versions contain stabilizers that increase the lifespan of the solution.
5. Isopropanol, molecular biology grade.
6. 0.1 m$M$ EDTA, made up in RNase-free H$_2$O.
7. DEPC-treated H$_2$O: Add diethylpyrocarbonate (DEPC) to double-distilled, deionized dH$_2$O to a concentration of 0.1% (1 mL DEPC/L of dH$_2$O). Mix well, incubate at least 2 h, and then autoclave for 45 min, or until DEPC odor is gone.
8. 10X transcription buffer: Use the transcription buffer provided with the polymerase. Some sources of RNA phage polymerase provide salt within the 10X transcription buffer, while others add it to the polymerase.
9. Separate solutions of 10 m$M$ rATP, 10 m$M$ rCTP, 10 m$M$ rGTP, 10 m$M$ rUTP
10. [α-$^{32}$P]UTP (800 Ci/mmol, 10 mCi/mL). [α-$^{32}$P]UTP (cat. no. BL507X) or [α-$^{32}$P]CTP (cat. no. BL508X) from New England Nuclear is recommended.

11. Template DNA: linearized plasmid DNA (1 μg; 0.5 μg/μL), or PCR fragment (0.2 μg; 100 ng/μL) in RNase-free $H_2O$ or TE. Extract the DNA with phenol to remove inhibitory proteins and trace RNases.

12. 5 *M* $NH_4OAc$: Dissolve 38.55 g ammonium acetate in RNase-free $H_2O$ to a final volume of 100 mL.

13. 100% EtOH; molecular-biology grade.

14. 70% EtOH: Prepare 70% ethanol from 100% EtOH and DEPC-treated water, and store chilled at –20°C.

15. 5% polyacrylamide urea gel: For 15 mL, enough for a 13 × 15 cm × 0.75 mm gel, dissolve 7.2 g high quality molecular biology grade urea in 1.5 mL 10X TBE, add 2.5 mL 30% acrylamide (acrylamide:*bis*-acrylamide = 19:1) and $dH_2O$ to 15 mL. Add 120 μL 10% ammonium persulfate, made up fresh in $dH_2O$. Finally, add 16 μL TEMED, mix briefly, and pour gel immediately. Adding the last reagent will start the polymerization reaction.

16. Denaturing gel-loading buffer: 95% formamide, 18 m*M* EDTA, 0.025% SDS, 0.025% xylene cyanol, 0.025% bromophenol blue. Can be used as either a 1X or a 2X solution.

17. 1X TBE: 89 m*M* Tris-HCl, 89 m*M* boric acid, 2 m*M* EDTA.

18. Elution buffer: 0.1 m*M* EDTA made up in RNase-free $H_2O$.

19. Scintillation fluid.

20. Sheared yeast RNA (5 mg/mL).

21. 50 m*M* ammonium carbonate, pH 9.5.

22. Gel-purified, radiolabeled antisense RNA probe; prepared in **Subheading 3.2.**

23. Hybridization solution: 80% deionized formamide, 40 m*M* PIPES, pH 6.4, 400 m*M* NaCl, 1 m*M* EDTA.

24. RNase digestion buffer: 10 m*M* Tris-HCl, pH 7.5, 300 m*M* NaCl, 5 m*M* EDTA.

25. RNase digestion solution: dilute RNase Cocktail 1:100 in RNase digestion buffer

26. 20% SDS in $dH_2O$.

## 2.2. Enzymes and Proteins

1. T7, T3, or SP6 phage RNA polymerase (20 U/μL); use the transcription buffer provided with the enzyme to ensure correct salt concentration.

2. DNase I, RNase-free (2 U/μL).

3. T1 RNase (1000 U/μL).

4. RNase stock solution: 0.5 mg/mL RNase A (Ambion), 10,000 U/mL RNase T1 (Ambion), 10 m*M* Tris-HCl, pH 7.5, 20 m*M* NaCl, 50% glycerol.

5. Proteinase K Solution: 5 mg/mL Proteinase K, 50 m*M* Tris-HCl, pH 8.0, 3 m*M* $CaCl_2$, 50% glycerol.

## 2.3. Equipment and Supplies

1. Drawn-out Pasteur pipet: Flame-pull the thin neck of the pipet by heating it over a flame and drawing out with hand and forcep. Break at the thinnest spot.

2. 1.5 mL RNase-free microfuge tubes.

3. RNase-free pipeter tips for 20-, 200-, and 1000-μL pipeters.

4. 20-, 200-, and 1000-μL pipeters.

5. Scintillation counter.

# 3. Methods

## 3.1. RNA Isolation

1. To the fresh tissue or cell sample, or to pulverized frozen tissue (*see* **Note 5**), add 10 mL lysis solution per 1.0 g tissue, or 10 mL per $1 \times 10^8$ cells, and thoroughly disrupt the cells.

2. Measure the volume of the lysate ("starting volume").
3. Add one "starting volume" of phenol/CHCl₃/isoamyl alcohol to the lysate and vortex for 2 min.
4. Incubate the mixture on ice (4°C) for 5 min. for microfuge tubes, or 10 min for 10–15-mL tubes.
5. Centrifuge at 10,000–12,000$g$ (e.g., full speed in a microfuge; 10,000 rpm in SS34 [Sorvall] rotor or its equivalent) for 5 min for microfuge tubes or for 15 min for larger tubes, to separate the aqueous and organic phases.
6. Transfer the top aqueous phase to a new tube. Discard the organic phase.
7. Add 1/10 aqueous phase volume of 3.0 $M$ NaOAc to the aqueous phase and mix by inverting several times.
8. Add one "starting volume" of acid phenol to the aqueous phase (even if aqueous phase is greater in volume) and vortex for 2 min.
9. Incubate the solution on ice (4°C) for 5 min for microfuge tubes, or 10 min for 10–15-mL tubes.
10. Centrifuge to separate the aqueous and organic phases as before in **step 5**.
11. Transfer the top aqueous phase to a new RNase-free tube and measure its volume. Avoid including any of the interface.
12. Add an equal volume of isopropanol and mix by inverting several times.
13. Incubate the preparation at –20°C for at least 30 min for microfuge tubes, or for 1 h or more for larger tubes.
14. Centrifuge at 10,000–12,000$g$ (e.g., full speed in a microfuge; 10,000 rpm in SS34 [Sorvall] rotor or its equivalent) for 15 min for microfuge tubes or for 30 min for larger tubes, to precipitate the RNA.
15. Carefully remove the supernatant using a glass Pasteur pipet with a drawn-out tip. Allow the pellet to air dry for 5–10 min at room temperature.
16. Resuspend the RNA pellet in 50–100 µL of 0.1 m$M$ EDTA/100 mg tissue ($1 \times 10^7$ cells).
17. Determine the concentration of the sample by reading absorbence at 260 nm (1 $A_{260}$ = 40 µg RNA). The RNA sample integrity can also be check by running an aliquot on an agarose gel and staining with EtBr. The 28S rRNA band should be about twice the intensity of the smaller 18S rRNA band. Aliquot the RNA and store at –20°C short-term, or –80°C long-term.

## 3.2. RNA Probe Synthesis

This protocol is designed for the efficient synthesis of probes 400 nt or less. Longer probes will require a higher concentration of the limiting rNTP (the one that is being used for labeling) (*see* **Note 3**).

1. Assemble the following components at room temperature in the order shown below:
   To 20 µL    RNase-free H₂O.
   2 µL        10X transcription buffer (use buffer provided with the polymerase).
   1 µL        rATP (500 µ$M$).
   1 µL        rCTP (500 µ$M$). These NTPs can be stored as a mixture.
   1 µL        rGTP (500 µ$M$).
   X µL        linear DNA template (1-µg plasmid template or 0.2-µg PCR template; *see* **Note 6**).
   5 mL        [α-³²P]rUTP (800 Ci/mmol, 10 mCi/mL) (*see* **Note 7**).
   2 µL        RNA polymerase (SP6 or T7 or T3).
2. Flick tube to mix, and spin down briefly to remove droplets from the tube wall.
3. Incubate 1 h at 37°C.
4. Remove an aliquot from the reaction for quantitation (total input counts).

5.  Precipitate the transcription reaction with 0.5 *M* NH$_4$OAc and 2.5 vol of 100% EtOH. Incubate at –20°C for 30 min, then collect the pellet by microfuging at 4°C at maximum speed for 30 min. Aspirate off the supernatant and wash the pellet with 70% EtOH.
6.  Remove template DNA by dissolving the pellet in 20 µL of 1X transcription buffer. Add 2 µL DNase I (5 U/µL) to the reaction, flick to mix, and incubate at 37°C for 1 h.
7.  Add an equal volume of denaturing gel-loading buffer to the transcription reaction and gel-purify the RNA probe in a polyacrylamide urea gel (*see* **Note 8**). Use 1X TBE as the gel running buffer.
8.  Leave the gel on the bottom glass plate; cover with plastic wrap, and expose to film (30 s–2 min). Align the film under the gel and cut out the gel-purified transcript. Cut out the smallest gel fragment possible; the gel can always be reexposed to determine whether or not the transcript has been completely removed.
9.  Chop the gel fragment into several small pieces and transfer to a microfuge tube. Cover with approx 350 µL of elution buffer and incubate the tube overnight at 37°C, or for several hours at 65°C (*see* **Note 9**).
10. Transfer the solution containing the eluted RNA away from the gel fragment into a new microfuge tube.
11. Remove an aliquot from the eluted RNA for quantitation (recovered incorporated counts).
12. Count cpm of the two aliquots in scintillation fluid. The percentage recovered, full-length cpm is determined by total #cpm recovered from gel/total #cpm in reaction. This number will typically be in the range of 10–50% (only a proportion of incorporated counts are incorporated into full-length product). Counts falling within this range suggest that the transcription reaction was efficient. Use cpm recovered from gel/µL to determine how many cpm to use in the RPA-mapping experiment. Typical counts should be in the range of 10,000–50,000 cpm/µL elution buffer. There should be a total of 3–17.5 × 10$^6$ cpm, sufficient counts to perform several hundred RPA mapping reactions.
13. Store the probe at –20°C.

### *3.3. Generating RNA Ladders*

#### *3.3.1. G Ladder*

1.  Add 20,000 cpm (100–800 pg) probe to 2.5 µg unlabeled yeast RNA.
2.  Add 1 U T1 RNase to the RNA and incubate at 55°C for 15 min.
3.  Add denaturing gel-loading buffer and store at –20°C until ready to use on gel.

#### *3.3.2. Single Base Ladder*

1.  Incubate the RNA probe in 50 m*M* ammonium carbonate (pH 9.5) at 95°C for 5 min.
2.  Add denaturing gel-loading buffer and store at –20°C until ready to use on gel.

### *3.4. RPA Protocol for Detection and Quantitation*

#### *3.4.1. Hybridization of Probe and RNA Sample*

1.  For each experimental sample, add 2–8 × 10$^4$ cpm (i.e., 100–800 pg) antisense RNA probe and sample RNA (usually 1–20 µg total RNA) to an RNase-free 1.5-mL microfuge tube (*see* **Note 10**). Also set up two no-template control tubes for each probe, containing probe and yeast RNA (or RNA from a species where no hybridization would be expected) instead of sample RNA.
2.  Ethanol-precipitate the samples as in **Subheading 3.2., step 5**.
3.  Remove the supernatant, then respin for 3 s and remove any remaining liquid with a drawn-out Pasteur pipet. Air-dry the pellets for 5–10 min at room temperature.

4. Resuspend the RNA pellets in 20 µL hybridization solution using heat and vortexing as necessary.
5. Heat all tubes at 94°C for 3–4 min, then incubate in a 42–45°C heat block or water bath overnight (*see* **Note 11**).

### 3.4.2. RNase Digestion of Probe: Target Hybrids

1. Prepare working dilutions of RNase by diluting RNase stock 1:100 in RNase digestion buffer.
2. Remove the hybridizing samples from the 45°C heat block or water bath and microfuge briefly to bring reaction and any condensation to bottom of tube.
3. Add 200 µL of the working dilutions of RNase to all experimental tubes and one of each pair of no-template control tubes.
4. Add 200 µL of RNase digestion buffer *without* RNase to the remaining no-template control tube(s), one for each different probe used.
5. Briefly vortex and microfuge all tubes. Incubate at 37°C for 30 min.
6. Mix equal volumes of Proteinase K Solution and 20% SDS so that 20 µL of the final solution can be added to each tube.
7. Add 20 µL of the Proteinase K:SDS mixture to each tube, vortex, and spin briefly. Incubate at 37°C for 15 min.
8. Add 250 µL phenol/CHCl$_3$/isoamyl alcohol to each tube; vortex for 2 min and microfuge for at least 1 min to completely separate phases.
9. Remove the aqueous phase(s) to fresh tube(s).
10. Add 5 µg of carrier nucleic acid or glycogen to each reaction.
11. Add 625 µL 100% EtOH to each tube. Vortex and incubate at –20°C for 30 min.

### 3.4.3. Separation and Detection of Protected Fragments

1. Prepare a 5–8% denaturing polyacrylamide urea gel (*see* **Note 8**).
2. Remove the tubes from the freezer and microfuge for 15 min at maximum speed at 4°C.
3. Carefully remove the EtOH supernatant from each tube using a flame-pulled Pasteur pipet and air-dry for 5–10 min at room temperature.
4. Resuspend each pellet in denaturing gel-loading buffer by vigorous vortexing and heating. Microfuge briefly.
5. Heat all tubes, including RNA markers or ladders, at 95°C for 3–4 min and then place on ice.
6. Load samples and markers on the denaturing polyacrylamide urea gel and run using 1X TBE as the gel-running buffer.
7. Transfer gel to filter paper, cover with plastic wrap, and expose to X-ray film with one intensifying screen at –80°C. Gels can also be dried on a gel dryer if desired; this is a good idea if more than one exposure will be required.
8. Develop the film. Size the experimental bands using the RNA markers. Full-length probe should be present in the "no target, no RNase" control, and the "no target, plus RNase" control lane should be blank. For mapping studies, the length of the protected fragment will determine the transcription start site(s) of the RNA transcript.
   If it is unnecessary to have a visual readout of the RPA results, gel separation can be omitted, and samples can be directly TCA-precipitated after RNase digestion (*33,34*).
9. Quantitate the protected fragment bands by direct phosphoimaging of the gel, densitometry of the film, or by cutting out the gel bands and counting them in a scintillation counter. If an internal control was used, divide the counts from the experimental band by those of the control for each sample.

## 4. Notes

1. While non-denaturing polyacrylamide gels can be used for protected fragment analysis, the probes and control lanes (probe + yeast RNA +/– RNase) cannot be run on these gels. Therefore, it is recommended that denaturing gels be used until the system is well-defined.

2. The minimal promoter sequence for T7 polymerase is 5'-TAATACGACTCACTATAGG-3'.

3. Synthesize high specific-activity probes by including little to no unlabeled NTP matching the NTP used for labeling (e.g., no UTP; 3 µM [α-$^{32}$P]UTP). Synthesize internal control probes to lower specific activity by including a 50:1 to 500:1 ratio of unlabeled NTP:labeled NTP (e.g. 500 µM UTP and 1 µM [α-$^{32}$P]UTP).

4. Supporting data can be found in Ambion's Technical Bulletin #152, *Choice of ribonucleases for ribonuclease protection assays.*

5. Cellular disruption of the sample is enhanced by mechanical shearing (Vortex, sonicator, dounce, Polytron®, Bead Beater®, mortar and pestle, and others). Cultured cells other than plant cells, yeast, fungi, and bacteria typically require only vortexing or sonicating in lysis solution. Animal and plant tissues often require the stronger shearing forces provided by a dounce and pestle, a Polytron, or a Bead Beater. For extremely tough samples (e.g., bone, hard tumors, plant roots, and woody tissues), tissues are frozen and then ground to a fine powder with a mortar and pestle or bag and hammer. The frozen powder is processed like fresh tissue or cells. Cells with tough exteriors (Gram-positive bacteria, yeast, and some fungi) may require enzymatic treatment to remove outer protective cell walls and layers prior to disruption. To do this, resuspend pelleted cells in 100 µL TE containing 1 mg/mL lysozyme. Incubate for 15 min at 37°C. Centrifuge for 5 min and discard the supernatant. Resuspend the cells in lysis buffer and vortex vigorously to lyse the cells.

6. Phenol extract plasmid templates after linearization with restriction enzymes and prior to transcription to remove inhibitory proteins and RNases. PCR templates can typically be added directly from the PCR reaction to the transcription reaction.

7. Alternatively, labeled rCTP and unlabeled rUTP could be substituted for labeled rUTP and unlabeled rCTP.

8. A 5–6% gel will effectively separate transcripts of 100–600 nt in length. A mini-gel is sufficient to separate protected fragments that differ by at least 20 bases in length. Use larger gels (e.g., sequencing gels) for finer resolution.

9. Sufficient probe is eluted in two hours to proceed to the hybridization step (**Subheading 3.4.1.**). Transfer the gel elution buffer to a new tube and count an aliquot. Additional gel elution buffer can then be added to the gel fragment, and elution continued overnight to recover addition probe.

10. To perform relative or absolute quantitation, probe must be in excess of the target molecules. Typically, $2–8 \times 10^4$ cpm (or 200–800 pg probe) per 10 µg total RNA sample (or 0.3 µg poly(A) RNA) is sufficient probe to be in excess of target for targets of less than or equal abundance to β-actin mRNA.

11. The time required for complete hybridization will vary. Less abundant transcripts require longer hybridization times, and should be hybridized overnight in initial experiments.

## Acknowledgments

I thank Marianna Goldrick (Ambion) for protocol development and technical advice, and Lori Martin for critical editing of this manuscript.

## References

1. Zinn, K., Dimaio, D., and Maniatis, T. (1983) Identification of two distinct regulatory regions adjacent to the human β-interferon gene. *Cell* **34**, 865–879.

2. Fischer, J. A. and Maniatis, T. (1985) Structure and transcription of the *Drosophila mulleri* alcohol dehydrogenase genes. *Nucleic Acids Res.* **13**, 6899–6917.
3. Dolci, S., Grimaldi, P., Geremia, R., Pesce, M., and Rossi, P. (1997) Identification of a promoter region generating Sry circular transcripts both in germ cells from the male adult mice and in male mouse embryonal gonads. *Biol. Reprod.* **57(5)**, 1128–1135.
4. Pagenstecher, A., Stalder, A. K., and Campbell, I. L. (1997) RNase protection assays for the simultaneous and semiquantitative analysis of multiple murine matrix metallo-proteinase (MMP) and MMP inhibitor mRNAs. *J. Immunol. Meth.* **206**, 1–9.
5. Blais, A., Labrie, Y., Pouliot, F., Lachance, Y., and Labrie, C. (1998) Structure of the gene encoding the human cyclin-dependent kinase inhibitor p18 and mutational analysis in breast cancer. *Biochem Biophys. Res. Comm.* **247(1)**, 146–153.
6. Melton, D. A., Kreig, P. A., Rebagliati, M. R., Maniatis, T., Zinn, K., and Green, M. R. (1984) Efficient *in vitro* synthesis of biologically active RNA and RNA hybridization probes from plasmids containing a bacteriophage SP6 promoter. *Nucleic Acids Res.* **12**, 7035–7056.
7. Berk, A. J. and Sharp, P. A. (1977) Sizing and mapping of early adenovirus mRNAs by gel electrophoresis of S1 endonuclease-digested hybrids. *Cell* **12**, 721–732.
8. Weaver, R. F. and Weissmann, C. (1979) Mapping of RNA by modification of the Berk-Sharp procedure: The 5′ termini of 15S β-globin mRNA and mature 10S β-globin mRNA have identical map coordinates. *Nucleic Acids Res.* **7(5)**, 1175–1193.
9. Maniatis, T., Fritsch, E. F., and Sambrook, J. (1989) *Molecular Cloning. A Laboratory Manual,* 2nd ed. Cold Spring Harbor Laboratory, Cold Spring Harbor, NY.
10. Lee, J. J. and Costlow, N. A. (1987) A molecular titration assay to measure transcript prevalence levels. *Meth. Enzymol.* **152**, 611–632.
11. Frayan, K. N., Langin, D., Holm, C., and Belfrage, P. (1993) Hormone-sensitive lipase: quantitation of enzyme activity and mRNA level in small biopsies of human adipose tissue. *Clin. Chim. Acta* **216**, 183–189.
12. Lee, P. J., Washer, L. L., Law, D. J., Boland, C. R., Horon, I. L., and Feinberg, A. P. (1996) Limited up-regulation of DNA methyltransferase in human colon cancer reflect-ing increased cell proliferation. *Proc. Natl. Acad. Sci. USA* **93**, 10,366–10,370.
13. Ngai, J., Dowling, M. M., Buck, L., Axel, R., and Chess, A. (1993) The family of genes encoding odorant receptors in the channel catfish. *Cell* **72**, 657–666.
14. Stalder, A. K. and Campbell, I. L. (1994) Simultaneous analysis of multiple cytokine receptor mRNAs by RNase protection assay in LPS-induced endotoxemia. *Lymphokine Cytokine Res.* **13(2)**, 107–112.
15. Duncan, S. R., Elias, D. J., Roglic, M., Pekny, K. W., and Theofilopoulos, A. N. (1997) T-cell receptor biases and clonal proliferations in blood and pleural effusions of patients with lung cancer. *Hum. Immunol.* **53(1)**, 39–48.
16. Calzone, F. J., Britten, R. S., and Davidson, E. H. (1987) Mapping of gene transcript by nuclease protection assays and cDNA primer extension. *Meth. Enzymol.* **152**, 611–632.
17. Winter, E., Yamamoto, F., Almognesa, C., and Perucho, M. (1985) A method to detect and characterize point mutations in transcribed genes: amplification and over expression of the mutant c-Ki-ras allele in human tumor cells. *Proc. Natl. Acad. Sci. USA* **82**, 7575–7579.
18. Ausubel, F. M., Brent, R., Kingston, R. F., Moore, D. D., Seidman, J. G., Smith, J. A., and Struhl, K. (eds.) (1987) *Current Protocols in Molecular Biology.* Greene Publishing Associates and Wiley-Interscience, New York.
19. Rapley, R. and Manning, D. L. (eds.) (1998) *Methods in Molecular Biology, vol. 86: RNA Isolation and Characterization Protocols,* Humana Press, Totowa, NJ.
20. Butler, E. T. and Chamberlin, M. J. (1982) Bacteriophage SP6-specific RNA polymerase. I. Isolation and characterization of the enzyme. *J. Biol. Chem.* **257**, 5772–5778.

21. Davanloo, P., Rosenberg, A. H., Dunn, J. J., and Studier, F. W. (1984) Cloning and expression of the gene for bacteriophage T7 RNA polymerase. *Proc. Natl. Acad. Sci. USA* **81(7),** 2035–2039.

22. Grachev, M. A. and Pletnev, A. G. (1984) Phage T7 RNA polymerase: gene cloning and its structure. *Bioorg. Khim.* **10(6),** 824–843.

23. Morris, C. E., Klement, J. F., and McAllister, W. T. (1986) Cloning and expression of the bacteriophage T3 RNA polymerase gene. *Gene* **41,** 193–200.

24. Kasseveis, G. A., Butler, E. T., Roulland, D., and Chamberlin, M. J. (1982) Bacteriophage SP6-specific polymerase. II. Mapping of SP6 DNA and selective *in vitro* transcription. *J. Biol. Chem.* **257,** 5779–5787.

25. Belin, D. (1998) The use of RNA probes for the analysis of gene expression: Northern blot hybridization and ribonuclease protection assay in *Methods in Molecular Biology, vol. 86: RNA Isolation and Characterization Protocols* (Rapley, R. and Manning, D. L., eds.), Humana Press, Totowa, NJ, pp. 87–102.

26. Farrell, R. E., Jr. (1998) Quantification of specific messenger RNAs by nuclease protection in *RNA Methodologies,* 2nd ed, Academic Press, San Diego, CA, pp. 385–405

27. Goldrick, M., Kessler, D., and Winkler, W. (1996) in *A Laboratory Guide to RNA: Isolation, Analysis and Synthesis* (Krieg, P., ed.), Wiley-Liss, New York, pp. 105–132.

28. Jones, P., Qiu, J., and Rickwood, D. (1994) Determination of RNA sequences, in *RNA, Isolation and Analysis,* Bios Scientific Publishers, Oxford, pp. 95–108.

29. Saccomanno, C. F., Bordonaro, M., Chen, J. S., and Nordstrom, J. L. (1992) A faster ribonuclease protection assay. *Biotechniques* **13(6),** 847–849.

30. Uchider, T. and Egami, F. (1967) The specificity of ribonuclease T2. *J. Biochem.* **61,** 44–49.

31. Myers, R. M., Larin, Z., and Maniatis, T. (1985) Detection of single-base substitutions by ribonuclease cleavage at mismatches in RNA:DNA duplexes. *Science* **230,** 1242–1246.

32. Chomczynski, P. and Sacchi, N. (1987) Single-step method of RNA isolation by acid guanidinium thiocyanate-phenol-chloroform extraction. *Analyt. Biochem.* **162,** 156–159.

33. Pham, K., LaForge, S., and Kreek, M. J. (1998) Comparison of methods for quantitation of radioactivity in protected hybrids in rnase protection assay. *BioTechniques* **25,** 198–206.

34. Personal communication from M. Ganjeizadeh; data printed in Ambion's *TechNotes* **4(1),** 7 (1997).

# Index